CRM Series in Mathematical Physics

Springer
New York
Berlin
Heidelberg
Barcelona
Hong Kong
London
Milan
Paris
Singapore
Tokyo

CRM Series in Mathematical Physics

Robert Conte
Editor

The Painlevé Property

One Century Later

With 11 Illustrations

 Springer

Robert Conte
Service de Physique de l'etat Condensé
CEA Saclay
F91191 Gif–sur–Yvette Cedex
France

Library of Congress Cataloging-in-Publication Data
The Painlevé property : one century later / editor, Robert Conte.
    p.   cm. — (CRM series in mathematical physics)
    Includes bibliographical references and index.
    ISBN 978-0-387-98888-7 (alk. paper)
    1. Painlevé equations. 2. Mathematical physics. I. Conte,
  Robert, 1943–   . II. Series: CRM series in mathematical physics.
  QC20.7.D5P34 1999
  530.1'535—dc21   99-16039

Printed on acid-free paper.

Production managed by MaryAnn Cottone; manufacturing supervised by Jeffrey Taub.
Photocomposed copy prepared from CRM's LaTeX files.

9 8 7 6 5 4 3 2 1

ISBN 978-0-387-98888-7 Springer-Verlag New York Berlin Heidelberg   SPIN 10731750

To Florent Bureau (1906–1999)

1 Alexandr Orlov
2 Monica Ugaglia
3 Tamara Grava
4 Andrei K. Svinin
5 Marta Mazzocco
6 Serguei A. Zykov
7 Vladimir V. Tsegel'nik
8 Isidore Ndayirinde
9 Annie Touchant

10 Nikolai A. Kudryashov
11 Milena A. Khlabystova
12 Brenton LeMesurier
13 Marina Cooks
14 Philippe Di Francesco
15 Gilbert Mahoux
16 Zora Thomova
17 Pavel Winternitz
18 Robert Conte

19 Hiroyuki Kawamuko
20 Kazuo Okamoto
21 Thierry Lehner
22 Martin D. Kruskal
23 Vladimir V. Sokolov
24 Adrian-S Cârstea
25 Luciano Seta
26 Valerii I. Gromak
27 Metin Ünal

28 Micheline Musette
29 Simonetta Abenda
30 Vangelis Marinakis
31 Christian Scheen
32 Maciej Dunajski
33 Jarmo Hietarinta
34 Xing-biao Hu
35 Valeria Ricci

Participants. The e-mail is at press date. The address is the affiliation during the school (name is in *italics* if changed).

Name	Address		e-mail
Simonetta Abenda	U. di Bologna, Dipart. di Matematica Piazza San Donato 5–40127 Bologna	I	Abenda@vsgte6.bo.infn.it
Mark J. Ablowitz	U. of Colorado, Program in Appl. Math. Campus box 526–Boulder, Co 80309	USA	MarkJAb@newton.colorado.edu
Iadh Ayari	U. de Montréal, Centre de rech. math. CP 6128, Succ. Centre ville–Montréal H3C 3J7	CDN	Ayaria@crm.umontreal.ca
Serge Bouquet	CEA Bruyères, Service DPTA/PPE BP 12–91680 Bruyeres	F	Bouquet@Bruyeres.cea.fr
Annalisa *Calini*	Case Western Reserve U., Dept. of Math 10900 Euclid Avenue–Cleveland, OH 44106	USA	CaliniA@math.cofc.edu
Adrian-S Cârstea	Inst. de phys. atomique, Phys. théor. BP MG–6, Magurele–Bucarest	ROU	ACarst@theor1.ifa.ro
Robert Conte	CEA Saclay, Service de phys. de l'état condensé 91191 Gif/Yvette Cedex	F	Conte@spec.saclay.cea.fr
Marina Cooks	U. of New South Wales, Dept. Appl. Math Sydney, NSW 2052	AUS	MarinaK@maths.unsw.edu.au
Silvana De Lillo	U. di Perugia, Dipart. di Fisica Via A. Pascoli–06100 Perugia	I	DeLillo@perugia.infn.it
Philippe Di Francesco	CEA Saclay, Service de physique théorique 91191 Gif/Yvette Cedex	F	Philippe@spht.saclay.cea.fr
Valerii S. Dryuma	Ac. Sci. Moldavia, Inst. of Math. Akademitchesky ul. 5–277028 Kichinev	MOL	15Valery@mathem.moldova.su
Boris Dubrovin	Scuola intern. sup. di studi avanzati Via Beirut 4–34014 Trieste	I	Dubrovin@tsmi19.sissa.it
Maciej Dunajski	U. of Oxford, Mathematical Institute 24–29 St Giles–Oxford OX1 3LB	GB	Dunajski@maths.ox.ac.uk

Name	Address		e-mail
Tamara *Grava*	Scuola intern. sup. di studi avanzati, Via Beirut 4–34014 Trieste	I	Grava@math.umd.edu
Valerii I. Gromak	Bielorussian state U., Mechanics and Math. Prospekt F. Skarina 4–220050 Minsk	BY	Grom@mmf.bsu.unibel.by
Amina Helmi	Rijks U. Leiden, Sterewacht Leiden, Niels Bohrweg 2, Postbus 9513–2300 RA Leiden	NL	AHelmi@strw.LeidenUniv.nl
Jarmo Hietarinta	U. of Turku, Dept. of Physics, 20014 Turku	FIN	Hietarin@utu.fi
Xing-biao Hu	Ac. Sinica, Institute of Comput. Math, PO Box 2719–100080 Beijing	CHI	HXb@indigo31.cc.ac.cn
Nalini *Joshi*	U. of New South Wales, Dept. Appl. Math. Sydney, NSW 2052	AUS	NJoshi@maths.adelaide.edu.au
Hiroyuki *Kawamuko*	U. of Tokyo, Dept. of Math. Sciences, 3-8-1 Komaba, Meguroku–Tokyo 153–8914	JPN	Kawam@Poisson.ms.u-tokyo.ac.jp
Milena A. *Khlabystova*	St. Petersburg U., Math. & Comput. Phys. 198904 St. Petersburg	RU	Milena@cuper.niif.spb.ru
Martin D. Kruskal	Rutgers U., Dept. of Mathematics, Hill Center, Busch Campus–New Brunswick, NJ 08903	USA	Kruskal@math.rutgers.edu
Nikolai A. Kudryashov	Moscow Eng. Phys. Inst., Applied Math. 31 Kashirskoe Avenue–115409 Moscow	RU	NKud@dpt31.mephi.msk.su
Stéphane Lafortune	U. de Montréal, Centre de rech. math. CP 6128, succ. Centre-ville–Montréal H3C 3J7	CDN	LafortuS@crm.umontreal.ca
Thierry Lehner	École polytechnique, Phys. matière ionisée 91128 Palaiseau Cedex	F	Lehner@lpmi.polytechnique.fr
Shanna *Lutzewitsch*	U. Paderborn, Fachbereich 17–Math. Warburger Str. 100, Postfach 1621–33098 Paderborn	D	
Wen-xiu *Ma*	U. Paderborn, Fachbereich 17–Math. Warburger Str. 100, Postfach 1621–33098 Paderborn	D	MaWx@cityu.edu.hk

Name	Address		e-mail
Gilbert Mahoux	CEA Saclay, Service de physique théorique 91191 Gif/Yvette Cedex	F	Mahoux@spht.saclay.cea.fr
Vangelis Marinakis	U. of Patras, Dept. of Mathematics 26110 Patras	GR	Vagelis@math.upatras.gr
Marta *Mazzocco*	Scuola intern. sup. di studi avanzati Via Beirut 4–34014 Trieste	I	Mazzocco@neumann.sissa.it
Viktor K. Mel'nikov	JINR, Bogoliubov Lab of Theor. Phys. 141980 Dubna	RU	Melnikov@thsun1.jinr.ru
Micheline Musette	Vrije U. Brussel, Theoretische Natuurkunde Pleinlaan 2–1050 Brussel	B	MMusette@vub.ac.be
Isidore *Ndayirinde*	U. Antwerpen, Departement Natuurkunde Universiteitsplein 1–B–2610 Wilrijk	B	
Frank W. Nijhoff	U. of Leeds, Dept. of Applied Math. Leeds LS2 9JT	GB	Frank@amsta.leeds.ac.uk
Kazuo Okamoto	U. of Tokyo, Dept. of Math. Sciences 3-8-1 Komaba, Meguroku–Tokyo 153–8914	JPN	Okamoto@Poisson.ms.u-tokyo.ac.jp
Alexandr *Orlov*	U. de Montréal, Centre de rech. math. CP 6128, succ. Centre-ville–Montréal H3C 3J7	CDN	OrlovS@wave.sio.rssi.ru
Andrew *Pickering*	U. of Kent, Inst. of Math. and Stat. Cornwallis bldg–Canterbury CT2 7NF	GB	APickeri@maths.adelaide.edu.au
K. Porsezian	Anna U., Dept. of Physics Chennai–600005	IND	Annalib@sirnetm.ernet.in
Alfred Ramani	École polytechnique, Centre de phys. théor. 91128 Palaiseau Cedex	F	Ramani@cpht.polytechnique.fr
Mark Ratter	U. of Glasgow, Dept. of Mathematics University gardens–Glasgow G12 8QW	GB	MR@maths.gla.ac.uk
Jean Reignier	Vrije U. Brussel, Theoretische Natuurkunde Pleinlaan 2–1050 Brussel	B	JReignie@vub.ac.be

Name	Address		e-mail
Valeria Ricci	U. di Roma I, Dipart. di Fisica P.le Aldo Moro 2–00185 Roma	I	Valeria.Ricci@roma1.infn.it
Colin Rogers	U. of New South Wales, Dept. Appl. Math. Sydney, NSW 2052	AUS	C.Rogers@unsw.edu.au
Christian Scheen	U. de Liège, Institut d'astrophysique Avenue de Cointe, 5–4000 Liège 1	B	Scheen@astro.ulg.ac.be
Constance Schober	U. of Colorado, Program in Appl. Math. Campus box 526–Boulder, Co 80309	USA	Schober@math.odu.edu
Luciano Seta	U. di Palermo, Dipart. di Matematica Via Archirafi 34–90123 Palermo	I	LSeta@gremat.math.unipa.it
Vladimir V. Sokolov	Russian Ac. of Sciences, Inst. of Math. Chernyshevsky ul. 112–Ufa	RU	Sokolov@nkc.bashkiria.su
Andrei K. Svinin	Inst. of System Dynamics & Control Theory Lermontov ul. 134. PO Box 1233–664033 Irkutsk	RU	Svinin@k517.icc.ru
Zora Thomova	U. de Montréal, Centre de rech. math. CP 6128, Succ. Centre-ville–Montréal H3C 3J7	CDN	ThomovZ@sunyit.edu
Vladimir V. Tsegel'nik	Bielorussian State U., Mathematics P. Brovka ul. 6–220027 Minsk	BY	math@micro.rei.minsk.by
Monica Ugaglia	Scuola intern. sup. di studi avanzati Via Beirut 4–34014 Trieste	I	Ugaglia@neumann.sissa.it
Metin Ünal	U. of Glasgow, Dept. of Mathematics University gardens–Glasgow G12 8QW	GB	Metin@maths.gla.ac.uk
Luc Vinet	U. de Montréal, Centre de rech. math. CP 6128, succ. Centre-ville–Montréal H3C 3J7	CDN	Vinet@crm.umontreal.ca
Pavel Winternitz	U. de Montréal, Centre de rech. math. CP 6128, succ. Centre-ville–Montréal H3C 3J7	CDN	Wintern@crm.umontreal.ca
Serguei A. Zykov	Inst. of Metal Physics Kovalevskaya ul. 18, GSP–170–620219 Ekaterinburg	RU	Serg@imp.uran.ru

Series Preface

The Centre de recherches mathématiques (CRM) was created in 1968 by the Université de Montréal to promote research in the mathematical sciences. It is now a national institute that hosts several groups, holds special theme years, summer schools, workshops, postdoctoral program. The focus of its scientific activities ranges from pure to applied mathematics, and includes satistics, theoretical computer science, mathematical methods in biology and life sciences, and mathematical and theoretical physics. The CRM also promotes collaboration between mathematicians and industry. It is subsidized by the Natural Sciences and Engineering Research Council of Canada, the Fonds FCAR od the Province of Québec, the Canadian Institute for Advanced Research and has private endowments. Current activities, fellowships, and annual reports can be found on the CRM web page at http://www.CRM.UMontreal.CA/.

The CRM Series in Mathematical Physics will publish monographs, lecture notes, and proceedings base on research pursued and events held at the Centre de recherches mathématiques.

Yvan Saint-Aubin
Montréal

Preface

The subject of this three-week school was the explicit integration, that is, analytical as opposed to numerical, of all kinds of *nonlinear* differential equations (ordinary differential, partial differential, finite difference). The result of such integration is ideally the "general solution," but there are numerous physical systems for which only a particular solution is accessible, for instance the solitary wave of the equation of Kuramoto and Sivashinsky in turbulence.

Nonlinear differential equations describe many physical phenomena whose behavior can be either integrable (such as solitonic equations), partially integrable (the complex Ginzburg–Landau equation of nonlinear optics), or chaotic (dynamical systems such as the Lorenz model). In statistical physics, the models that are defined on a lattice are described by finite difference equations, in short, discrete equations.

This subject is quite important for applications, because any analytic result (particular solution, first integral, etc.) is preferable to a numerical computation, which by definition is long, costly, and, worse, intrinsically subject to numerical error. But the distinction is even stronger. Indeed, the analytic approach, which was the subject of this school, often provides *global* knowledge of the solution, while the numerical approach is always *local* and hence insufficient: This is not because some initial state generates a "regular" régime that there do not exist other initial conditions with a "chaotic" régime. This question can then be settled only by a global study.

This explicit integration is not based on recipes, as we learned, but on quite powerful methods, most often not taught in universities, that were developed by Henri Poincaré and mainly Paul Prudent Painlevé in his famous *Leçons de Stockholm* of 1895. The guideline is the in-depth study of singularities, because "les fonctions, comme les êtres vivants, sont caractérisées par leurs singularités" (Montel). The landmark is the discovery, by Painlevé and his student Bertrand Gambier, of six new functions, each defined by a second-order nonlinear differential equation, the first of which, (P1), is as simple as $\mathrm{d}^2u/\mathrm{d}x^2 = 6u^2 + x$.

Initially qualified by Poincaré as "île originale et splendide dans l'océan voisin [du continent des mathématiques]," i.e., without any link with the exterior world, this discovery for a long time excited the curiosity only of mathematicians (mainly from Belgium, Japan, and Russia). Its second youth dates from about thirty years ago, and this renewal of interest has never decreased from then on. It arises from three, apparently disjoint, fields: statistical physics and field theory (Ising model, Wu and McCoy, Brézin and Kazakov), solitonic evolution equations (Kruskal, Ablowitz, Ramani, and Segur), and dynamical systems (Lorenz model, etc.). In all these fields were encountered the six Painlevé functions, whether in integrable or partially integrable cases; for instance, the self-dual Yang–Mills equations, so fundamental in field theory, admit reductions to the *six* functions (P1)–(P6). The physicists then became strongly interested in understanding the reason for that, and they took over from the mathematicians to make progress in the theory: They are responsible for extension of the theory to partial differential equations, in which the integrability manifests itself as *solitons*, and more recently to discrete equations.

The different courses presented at Cargèse intentionally alternated between mathematics and physics. These courses were aimed at gradually bringing the audience (thirty participants were near the doctorate) to the level of current research. In this book as well as during the school, the sequence of contributions appears in the following logical order (linear before nonlinear, elementary before advanced, etc.) In the following we abbreviate ordinary differential equation as ODE, partial differential equation as PDE.

- A reminder of more or less well known results on singularities (Fuchsian, non-Fuchsian) of linear ODEs (Reignier)

- A long course on the theory and practice of explicit integration of nonlinear ODEs by the study of only their singularities in the complex plane ("Painlevé analysis," properly said) (Conte)

- Two successive long courses on the questions of monodromy and the connection of linear, then nonlinear, ODEs, in the modern formalism of the Japanese school (Mahoux, Joshi).

- A long course on statistical physics, on the approach to bidimensional quantum gravity with the random matrices formalism, and how one naturally encounters the Painlevé functions (Di Francesco)

- Another long course of "physics" on bidimensional topological field theory and its link with the Painlevé functions (Dubrovin)

- A long course on the discretization of the equations of Painlevé, a blooming subject under the impulse of lattice models of statistical physics (Nijhoff and Ramani)

- A medium length course on the extension of Painlevé methods to PDEs, whether integrable or not (Musette)

- An introduction to the method of the inverse spectral transform, which extends to the nonlinear case the well-known Fourier transform, followed by a generalization of the famous system of Darboux and Halphen (Ablowitz)

- A long course on the method of symmetry and reduction, very useful to derive particular analytic solutions to PDEs by reducing them to ODEs (Winternitz, Clarkson)

- Three more specialized courses dealing with the six equations of Painlevé: one on their Hamiltonian formalism (Okamoto), one on their groups of transformations (transformations of Schlesinger) (Gromak), one on their Hirota bilinear form (Hietarinta)

- Finally, a brief contribution aimed at revising some conventional ideas and proposing new ones (Kruskal)

We deeply regret the cancellation of the course to be given by Florent Bureau, one of the last true disciples of the Painlevé school, due to health problems. This was also the reason for the absence of Peter Clarkson, whose joint course with Pavel Winternitz was given by the latter.

This school would not have been possible without the generous support of the following organizations, to which we express our deepest gratitude: Centre national de la recherche scientifique; Commission européenne; Collectivité territoriale de Corse; Commissariat à l'énergie atomique; Direction des recherches, études et techniques (DRET); ministère de la Recherche (programme ACCES); ministère des Affaires étrangères (bureau des congrès); Logovaz foundation (International Science Foundation).

Robert Conte
Gif-sur-Yvette

Contents

14 "Completeness" of the Painlevé Test—General Considerations—Open Problems   789
Martin D. Kruskal

Contributors

Mark J. Ablowitz Program in Applied Mathematics, University of Colorado, Campus box 526, Boulder, CO 80309, USA
MarkjAb@newton.colorado.edu

Sarby Chakravarty Program in Applied Mathematics, University of Colorado, Campus box 526, Boulder, Co 80309, USA
Chuck@newton.colorado.edu

Peter A. Clarkson Institute of Mathematics and Statistics, University of Kent at Canterbury, Cornwallis bldg, Canterbury CT2 7NF, UK
P.A.Clarkson@ukc.ac.uk

Robert Conte Service de physique de l'état condensé, CEA Saclay, F-91191 Gif-sur-Yvette Cedex, France
Conte@spec.saclay.cea.fr

Philippe Di Francesco Service de physique théorique, CEA Saclay, F-91191 Gif-sur-Yvette Cedex, France
Philippe@spht.saclay.cea.fr

Boris Dubrovin SISSA, Via Beirut 2, I-34013 Trieste, Italy
Dubrovin@tsmi19.sissa.it

Basile Grammaticos GMPIB, Université Paris VII Denis Diderot, 2, place Jussieu, F-75251 Paris Cedex 05, France
Grammati@paris7.jussieu.fr

Valerii I. Gromak Department of mechanics and mathematics, Prospekt F. Skarina 4, Bielorussian State University of Informatics and Radioelectronics, 220050 Minsk, Bielorussia
Grom@mmf.bsu.unibel.by

Rod Halburd Program in Applied Mathematics, University of Colorado, Campus box 526, Boulder, Co 80309, USA
Rod@newton.colorado.edu

Jarmo Hietarinta Department of physics, University of Turku, FIN-20014 Turku, Finland
Hietarin@utu.fi

Nalini Joshi Department of Pure Mathematics, University of Adelaide, Adelaide 5005, Australia
NJoshi@maths.adelaide.edu.au

Martin D. Kruskal Department of Mathematics, Hill Center, Busch Campus, Rutgers University, New Brunswick, NJ 08903, USA
Kruskal@math.rutgers.edu

Gilbert Mahoux Service de physique théorique, CEA Saclay, F-91191 Gif-sur-Yvette Cedex, France
Mahoux@spht.saclay.cea.fr

Micheline Musette Dienst Theoretische Natuurkunde, Vrije Universiteit Brussel, Pleinlaan 2, B-1050 Brussel, Belgium
MMusette@vub.ac.be

Frank W. Nijhoff Department of Applied Mathematics, University of Leeds, Leeds LS2 9JT, UK
Frank@amsta.leeds.ac.uk

Kazuo Okamoto Department of Mathematical Sciences, The University of Tokyo, 3-8-1 Komaba, Meguroku, Tokyo 153–8914, Japan
Okamoto@Poisson.ms.u-tokyo.ac.jp

Alfred Ramani Centre de physique théorique , École polytechnique, F-91128 Palaiseau Cedex, France
Ramani@orphee.polytechnique.fr

Jean Reignier Dienst Theoretische Natuurkunde, Vrije Universiteit Brussel, Pleinlaan 2, B-1050 Brussel, Belgium
JReignie@vub.ac.be

Pavel Winternitz Centre de recherches mathématiques, Université de Montréal, C.P. 6128. Succ. Centre ville, Montréal, Québec H3C 3J7, Canada
Wintern@CRM.UMontreal.CA

1

Singularities of Ordinary Linear Differential Equations and Integrability

Jean Reignier

ABSTRACT The Cauchy existence theorem for solutions of ordinary analytic differential equations concerns only the *local* existence. Integrability requires global solutions, i.e., analytic continuation on larger domains. It is then useful to have a priori some information on the positions of the singularities of the solutions. The simplest case corresponds to linear equations: The singularities of the solutions can arise only where the coefficients of the equation are themselves singular. For an isolated singular point, one can derive the (local) general structure of the solutions from a simple algebraic study. This leads to a natural distinction between two important cases: weakly singular and strongly singular equations, according to the absence or presence of essential singularities in the local general solution (essential singularities make more difficult the calculation of the solution by efficient constructive algorithms). Fuchs's theorem gives a simple criterion to differentiate the two cases. It also provides an efficient constructive algorithm for the weakly singular case. If the equation is completely "Fuchsian," it is then easily integrated. The alternative case of strongly singular equations will also be discussed for second-order differential equations. The Thomé method classifies these strongly singular equations, and it gives an efficient algorithm to construct formal local solutions. These turn out to be asymptotic of true solutions. Summation methods (like the Borel method) can then be used to construct these true solutions.

1 Generalities

The problem is to construct solutions of ordinary analytical differential equations in a domain as large as possible. It is generally presented in the canonical Cauchy form:

$$\begin{array}{ll} \text{(i)} & \text{differential equation:} \quad Y' = F(x; Y); \\ \text{(ii)} & \text{auxiliary condition:} \quad Y(x_0) = Y_0. \end{array} \qquad (1.1)$$

Capital letters represent sets of n-functions:

$$Y = (y_1, y_2, \ldots, y_n), \quad F = (f_1, f_2, \ldots, f_n). \qquad (1.2)$$

The f_k functions are assumed to be holomorphic in a common domain of the x and y_k complex variables:

$$f_k(x; y_1, y_2, \ldots, y_n) : \text{hol. in } D = D_x \otimes D_1 \otimes \cdots \otimes D_n. \qquad (1.3)$$

The y_k solutions are locally holomorphic in the complex variable x; they also depend on the auxiliary constants

$$y_k(x; x_0, y_{10}, y_{20}, \ldots, y_{n0}) \qquad (1.4)$$

and on the parametric constants that can be present in the functions f_k. It is, of course, assumed that the initial point $(x_0, y_{10}, y_{20}, \ldots, y_{n0})$ belongs to the domain D. In principle, the requirement of constructing the solution in a domain *as large as possible* corresponds ideally to a construction in the full domain D_x, without exceeding in value the limits of the domains D_k. This system of n first-order equations is the most general case, since an equation of order r can always be replaced by r first-order equations.

Cauchy's theorem guarantees the local existence and uniqueness of the solution of problem (1.1) in some neighborhood of the initial point x_0:

Let us consider a set of associated circles of analyticity common to all the functions f_k, with centers at the initial point

$$\{(x_0, a), (y_{10}, b_1), \ldots, (y_{n0}, b_n)\}$$

(the notation for a circle being (center, radius)); let b be the smallest of the radii b_i, and let M be the largest of the values of the $|f_k|$ on the boundaries of the circles, i.e., when $|x - x_0| = a$, $|y_k - y_{k0}| = b$ $(k = 1, \ldots, n)$; then the solution Y exists and is unique in the circle

$$|x - x_0| < r_0 = a(1 - \exp[-b/(n+1)aM]). \qquad (1.5)$$

This solution is not only holomorphic in x inside the indicated circle, it is furthermore (locally) holomorphic in the initial data and in the possible parametric constants on which the f_k analytically depend (Poincaré's theorem).

We see from (1.5) that the domain of existence of the solution is generally *smaller* than the domain where the problem is posed: $r_0 < a$. This means that the challenge to construct the solution in a domain "as large as possible" requires, in general, the analytic continuation of this "germ" of a solution. If one really succeeds in constructing the solution $Y(x)$ in the circle $|x - x_0| < r_0$, one can proceed as follows:

One considers some point x_1 belonging to the circle $|x - x_0| < r_0$, and one computes the numerical value of the solution Y at that point; let $Y_1 = Y(x_1)$ be this value.

One poses and solves as before a second Cauchy's problem with the same differential equation $(1 - i)$ and the *new* auxiliary condition $Y(x_1) = Y_1$.

The solution of this new Cauchy's problem exists, mutatis mutandis, in a circle $|x - x_1| < r_1$; if this new circle contains points of D_x that lie outside the previously found circle $(|x - x_0| < r_0)$, the solution of this second problem is an analytical continuation of the solution of the first one.

Can we then reasonably hope to succeed in constructing the solution of the global problem in this way, i.e., to construct the solution of Cauchy's problem in the *full* domain D_x? It is not that certain! The analyticity of the functions f_k does, of course, guarantee the feasibility of a step-by-step analytical continuation of these functions f_k, from the initial circles $\{(x_0, a), (y_{10}, b), \ldots, (y_{n0}, b)\}$ through the full domain D. But as was already noted, at each step of the calculation of the solution we get for this solution a circle of existence *smaller* than the circle where the problem was posed. This means that in the process of analytic continuation of the solution we can meet singularities of the solution that were not expected when we were considering the equations. This difficulty is illustrated by the following examples:

1. $y' = y^2$, $y(x_0) = y_0$; the domain D is clearly the double complex plane $C \otimes C$; the solution can be built by elementary integration: $y = (y_0^{-1} - x + x_0)^{-1}$. It has one first-order pole at $x = x_0 + y_0^{-1}$. This singularity could not have been foreseen from the equation and the initial data.

2. Let us now slightly modify the problem by introducing a parameter α in a way that doesn't change the holomorphy of the problem: $y' = y^2 + \alpha^2$. Again, the solution is easily built by elementary integration: $y = a \tan[\alpha(x - x_0) + \arctan(y_0/\alpha)]$. It has an infinity of first-order poles located at $x = x_0 + [(k + \frac{1}{2})\pi + \arctan(y_0/\alpha)]/\alpha$.

When comparing these two examples, we see that the *global* singularity structure of the solution is strongly changed by the apparently harmless introduction of a parameter. One can just as easily introduce essential singularities, or cuts, whose positions depend on the initial data and on the parameters in a way that cannot be foreseen a priori.

The problem of the *position* of the singularities of the solution is much simpler in the case of *linear* differential equations:

$$
\begin{array}{lll}
\text{(i)} & \text{differential equation:} & Y' = AY + B; \\
\text{(ii)} & \text{auxiliary condition:} & Y(x_0) = Y_0.
\end{array}
\qquad (1.6)
$$

In (1.6) we use the following convention (as in linear algebra): Y is the column vector of the solutions y_k, A is the $n \times n$ matrix of the coefficients $a_{kl}(x)$ of the homogeneous part of the equation, and B is the column vector of the coefficients $b_k(x)$ of the inhomogeneous part of the equation. A and B are holomorphic in the domain D_x. In this case, the singularities of the solution can occur only where the coefficients A and B of the linear equation are themselves singular. The *positions* of the *possible* singularities of the solution are known a priori. In other words, the analytic continuation of the solution in the full domain D_x is always possible. This remarkable theorem is proved as follows: One poses Cauchy's problem in the circle (x_0, a), where "a" is the distance from x_0 to the closest singularity of the coefficients (a_{kl}, b_k), i.e., the shortest distance to the boundary of D_x; by using essentially the same methods as in the general case, one can show that the radius r_0 of existence and uniqueness of the solution is now *equal* to a. Therefore, the analytic continuation of the solution through the full domain D_x becomes (in principle) possible: It is just like the analytic continuation of the coefficients (a_{kl}, b_k).

I now recall briefly some elementary properties of linear differential equations (see any text book on the subject). The solutions of the homogeneous equation (i.e., $B = 0$) belong to an n-dimensional vector space. Therefore, it is necessary and sufficient to construct n linearly independent solutions $(Y_1, Y_2, \ldots, Y_n)$ in order to find all the solutions of the equation. Furthermore, once this fundamental set of solutions is constructed, the solutions of the inhomogeneous equation can be obtained by the method of variation of constants, which requires only the computation of ordinary integrals. I also recall that in the case of one single nth-order linear homogeneous equation, a necessary and sufficient condition for the linear independence of n solutions is that the Wronskian of these solutions be nonzero. (The Wronskian $W(x)$ is formed by the n solutions and their derivatives up to order $n - 1$.) This determinant is itself a solution of a first-order linear homogeneous equation whose unique coefficient is the coefficient of $y^{(n-1)}$ in the original equation:

$$y^{(n)} + a_1(x)y^{(n-1)} + a_2(x)y^{(n-2)} + \cdots + a_n(x)y = 0, \qquad (1.7)$$
$$W' + a_1(x)W = 0. \qquad (1.8)$$

The algorithmic construction of a fundamental set of solutions around some regular point x_0 proceeds as follows. It can easily be worked out with symbolic calculation programs.

One expands each of the coefficients $a_i(x)$ in its Taylor series around the point x_0:

$$a_i(x) = \sum_{k=0}^{\infty} \frac{a_i^{(k)}(x_0)}{k!}(x - x_0)^k. \qquad (1.9)$$

One substitutes in the equation the (unknown) Taylor series of the solution around x_0:

$$y(x) = \sum_{k=0}^{\infty} \frac{c_k}{k!}(x - x_0)^k. \qquad (1.10)$$

One rewrites the equation as an ascending power series, and one sets all its coefficients to zero. It turns out that the first n coefficients of the solution y (i.e., $c_0, c_1, \ldots, c_{n-1}$) can be arbitrarily chosen, while the next ones (i.e., $c_n, c_{n+1}, \ldots$) are recurrently defined as linear combinations of the first ones. Therefore, the construction of a *fundamental set* of solutions amounts to choosing n linearly independent vectors $(c_0, c_1, \ldots, c_{n-1})$.

It can happen that one recognizes in the Taylor series (1.10) thus constructed some already known analytic function (or some combination of known functions). The solution then becomes explicit, and its analytic continuation is known. If this is not the case, one meets a "new" analytic function, which is defined by its germ inside $|x - x_0| < a$. This germ provides a "sufficient solution" inside this circle, in the sense that the solution is there numerically known with an arbitrary high precision, the only limitation coming from the power of our computers. However, this germ of a solution must now be analytically continued outside the circle.

It is well known that the analytic continuation of a germ of an analytic function is in general not a well-posed problem, and so it must be regularized. Fortunately, the linear differential equation provides its own regularizing method! One constructs a second fundamental set of solutions, centered at some point x_1 not too far from x_0, such that the two circles intersect. Inside the first circle the solution is represented by some *known* linear combination of the functions of the first fundamental set of solutions. Inside the second circle it is represented by some *unknown* combination of the functions of the second fundamental set. Then the problem of analytic continuation from the first circle to the second amounts to determining the unknown coefficients of the second representation! In the intersection the solution is equally well represented by linear combinations of solutions belonging to the first or alternatively to the second of the two fundamental sets. Choosing freely n points in this intersection, one identifies the numerical values of the solution as given by the two representations, and one easily finds the unknown coefficients by solving a linear Cramér's algebraic system (i.e., by solving a well-posed problem, which can be treated with arbitrarily high numerical precision!).

Example 1.1. Let us consider a second-order linear differential equation and let us suppose that the singularities are such that we can construct two fundamental sets of solutions inside intersecting circles: (u_1, u_2) inside the circle (x_0, r_0), and (v_1, v_2) inside the circle (x_1, r_1). The general problem amounts to continuing (u_1, u_2) in the second circle, i.e., to finding the coefficients (α_i, β_i) of the linear combinations that continue these functions in the second circle:

$$u_i(x) = \alpha_i v_1(x) + \beta_i v_2(x). \tag{1.11}$$

We consider slightly smaller circles ($r_0' < r_0$ and $r_1' < r_1$), and we assume that these smaller circles still intersect. At any point of a "smaller" circle, the Taylor series representation of an analytic function allows one to compute the numerical value of this function to arbitrarily high precision. Choosing (freely) two points "a" and "b" of the intersection, one computes the numerical values of the four functions (u_1, u_2) and (v_1, v_2) at these two points. Then the coefficients (α_i, β_i) are found by solving the Cramér's system

$$u_i(a) = \alpha_i v_1(a) + \beta_i v_2(a),$$
$$u_i(b) = \alpha_i v_1(b) + \beta_i v_2(b) \quad (i = 1, 2).$$

One gets in that way the analytic continuation of the fundamental set (u_1, u_2) inside the circle where one constructed the fundamental set (v_1, v_2). Any solution defined by Cauchy's initial conditions at x_0 is now known in the union of the two circles.

I apologize for insisting so much on well-known properties of linear equations. My purpose is to make clear that the essential point in the construction of global solutions is to obtain firstly *fundamental sets of solutions in intersecting circles*. Up to now we constructed these fundamental sets around regular points. We shall now consider the same problem around isolated singular points of the equations.

2   Structure of the Solutions of the Homogeneous Equation Around an Isolated Singular Point

Let us consider the case of an isolated singularity located at the origin, and let us assume that the coefficients of the homogeneous nth-order equation (1.7) are single-valued functions around this singularity.[1] In this case there exists some circle with center at the origin and radius r_0 where the coefficients are regular, except for the possible singularity located at the origin, which is either a pole or an essential singularity (Laurent's representation theorem). Let x_0 be some point of this circle, and let us assume that we have built a fundamental set of solutions around this point: $U = (u_1, u_2, \ldots, u_n)$.

---

[1]An isolated singularity located at some point a of the plane is brought back to the origin by the transformation: $x \mapsto x' = x - a$; if the singularity is located at infinity, one uses the transformation: $x \mapsto x' = 1/x$. A p-valued function around the origin becomes single-valued by the transformation $x \mapsto x' = x^{1/p}$. For the simultaneous uniformization of several multivalued functions one takes for p the smallest common multiple of the different multiplicities. In case of an infinitely valued function, one uses the transformation $x \mapsto x' = 1/\log x$.

Using the method described at the end of the previous paragraph, we can continue analytically this fundamental set of solutions in the full open circle $0 < |x| < r_0$. We know that whenever some function f defined in such a domain is single-valued, Laurent's representation theorem gives its general structure:

$$f(x) \text{ single-valued in } 0 < |x| < r_0 \iff f(x) = \sum_{n=-\infty}^{+\infty} c_n x^n, \qquad (2.1)$$

with either zero or some finite number or infinitely many negative powers; these three possibilities correspond, respectively, to the cases of a function that is either regular or has a pole or has an essential singularity at the origin. Unfortunately, the general theorem that indicates the possible positions of the singularities of the solutions of a linear differential equation says nothing about the *nature* of these singularities. Therefore, we have to consider the most general possibility, i.e., that the isolated singularity of the u_i functions located at the origin is a branch point. This means that when we perform the analytic continuation of some solution u_i around the origin and reach again after this rotation the original small Cauchy's circle around x_0, we do not necessarily recover the original value $u_i(x)$, but possibly some other value $v_i(x)$. The new functions $V = (v_1, v_2, \ldots, v_n)$ are solutions of the original equation, since the coefficients of this equation are themselves single-valued. Furthermore, these v-solutions are linearly independent because the Wronskian, which evolves according to (1.8), never vanishes during the process of analytic continuation:

$$W(x) = W(x_0) \exp\left[-\int_{x_0}^{x} a_1(x')\,dx' \right]. \qquad (2.2)$$

Therefore, the new set of solutions is again a fundamental set, and the transformation $U \mapsto V$ is simply a change of basis in the n-dimensional vector space of the solutions of the homogeneous equation. This means that there exists an $n \times n$ regular matrix of complex numbers $G = (g_{kl})$ that describes this transformation:

$$V = GU \iff U = G^{-1}V. \qquad (2.3)$$

Let us now consider some solution $y(x)$ of the homogeneous equation. It can be written as a linear combination of the fundamental solutions $u_i(x)$, with appropriate coefficients c_i. If we perform its analytic continuation around the origin, we get after this rotation a new solution $z(x)$ that can be written as the *same* linear combination of the fundamental solutions $v_i(x)$:

$$y(x) = \sum_{i=1}^{n} c_i u_i(x) \mapsto z(x) = \sum_{i=1}^{n} c_i v_i(x). \qquad (2.4)$$

Of course, this solution $z(x)$ can also be written as a linear combination of the solutions $u_i(x)$; thanks to the transformation matrix G, we get

$$z(x) = \sum_{i=1}^{n} c_i \left[\sum_{i=1}^{n} g_{ij} u_j(x) \right]. \tag{2.5}$$

Comparing with the original solution $y(x)$, we see that $y(x)$ will be single-valued if and only if

$$\forall j = 1, \ldots, n, \quad \sum_{j=1}^{n} g_{ij} c_i = c_j, \tag{2.6}$$

i.e., if and only if $C = (c_1, \ldots, c_n)$ is an eigenvector of the transposed matrix G^T, with eigenvalue equal to *one*. There is a priori no reason why the matrix G^T would have this eigenvalue one. In full generality, we know only that the regular matrix G^T has at least one eigenvector and that the possible eigenvalues λ are certainly nonzero:

$$\exists \lambda \neq 0 \quad \text{and} \quad C : G^T C = \lambda C. \tag{2.7}$$

Let $y(x)$ be the solution that corresponds to this linear combination C; it transforms like

$$y(x) \mapsto z(x) = \lambda y(x). \tag{2.8}$$

The general structure of a function that undergoes such a transformation when one performs a 2π rotation around the origin is easily found. One has at one's disposal the following model:

$$x^s \mapsto e^{2i\pi s} x^s, \tag{2.9}$$

and one sees that by choosing the exponent s as

$$s = \frac{1}{2\pi i} \log \lambda, \tag{2.10}$$

the ratio $y(x)/x^s$ is then single-valued. In this way we prove that there exists in the open circle $0 < |x| < r_0$ at least one solution $y(x)$ with general structure

$$y(x) = x^s f(x) = x^s \sum_{n=-\infty}^{n=+\infty} c_n x^n. \tag{2.11}$$

Let us remark that this assertion of existence is not at all equivalent to a construction: The parameter s and the coefficients c_i of formula (2.11) are unknown![2] Furthermore, the construction of the fundamental matrix G^T can prove to be a rather tedious task; in practice, one never proceeds to this explicit construction. Nevertheless, we can assume that it was done (since it is mathematically feasible) and proceed to the research of other solutions of the same type. Let us call these solutions: "eigensolutions." It is clear that there are as many linearly independent eigensolutions as the number of linearly independent eigenvectors of the matrix G^T. The discussion of the number of linearly independent eigenvectors of a matrix is an easy matter when this matrix is transformed into its "canonical Jordan form." Linear algebra tells us that this transformation is always possible within the general linear group. One finds then the eigenvalues along the principal diagonal, the numbers 0 and 1 along one of the closest secondary diagonals, and only 0 for the rest of the matrix. Once this form is achieved, the matrix has been decomposed into a set of "Jordan submatrices" placed along the principal diagonal:

$$\left\{ \begin{matrix} (J_1) & & & \\ & (J_2) & & \\ & & \ddots & \\ & & & (J_k) \end{matrix} \right\}, \quad J_i = \left\{ \begin{matrix} \lambda_i & 0 & 0 & 0 \\ 1 & \lambda_i & 0 & 0 \\ 0 & 1 & \lambda_i & 0 \\ 0 & 0 & 1 & \lambda_i \end{matrix} \right\}. \tag{2.12}$$

Each of these submatrices contains only one eigenvalue λ, but we might find this same eigenvalue λ in several submatrices. It is important to notice that each submatrix has only one eigenvector! Therefore, there are as many eigenvectors corresponding to a given eigenvalue λ as the number of submatrices with this eigenvalue λ. And altogether, there are as many eigenvectors as the number of submatrices. One can also consider the dimensions of the submatrices, but this discussion is not much useful for our purpose. Again, we do not have to perform explicitly the Jordan transformation. It is enough to know that this transformation exists and that the Jordan canonical form is unique, modulo a permutation of the places of the different submatrices along the principal diagonal.

Let us therefore assume that we have found a linear transformation R that transforms G into its Jordan canonical form G':

$$G' = RGR^{-1}. \tag{2.13}$$

If we apply this transformation to equation (2.3), we obtain the equivalent equation (2.14), which clearly concerns a new fundamental set of solutions: $U' = RU$, where some of the solutions u' are now eigensolutions,

$$V' = RV = R(GU) = (RGR^{-1})RU = G'U'. \tag{2.14}$$

---

[2]Notice that the exponent s is defined only modulo 1; this corresponds to the arbitrary choice of the determination of $\log \lambda$. This indeterminacy can always be compensated by an ad hoc translation of the indices i of the coefficients c_i.

We have indeed, when considering a Jordan submatrix of G' of dimension k, $k \geq 1$, the following transformation scheme:

one "eigensolution": $\quad u_1' \mapsto v_1' = \lambda u_1',$

$k - 1$ solutions such that $\quad u_i' \mapsto v_i' = \lambda u_i' + u_{i-1}' \ (i = 2, \ldots, k).$ $\qquad$ (2.15)

The first relation indicates that u_1' is an eigensolution. We already know the structure of this kind of solution (cf. (2.11)):

$$u_1'(x) = x^s f_1(x), \qquad (2.16)$$

where $f_1(x)$ is some single-valued function (Laurent series). Let us show that the other relations of (2.15) also allow us to find the structure of the other solutions. We start with the first of these relations, which describes the analytic continuation of u_2' around the origin and which concerns only u_2' and the eigensolution u_1'. We use this relation together with relation (2.16) to define the analytic continuation of the ratio $u_2'(x)/x^s$ around the origin, and we obtain

$$\frac{u_2'}{x^s} \mapsto \frac{u_2'}{x^s} + \frac{1}{\lambda} f_1. \qquad (2.17)$$

We see that this ratio $u_2'(x)/x^s$ is single-valued except for the addition of the single-valued function f_1/λ. It is now an easy matter to get the structure of a function that transforms this way. We have at our disposal the following model:

$$\log x \mapsto \log x + 2\pi i, \qquad (2.18)$$

and it is immediately seen that the difference $u_2'/x^s - (1/2\pi i\lambda)f_1 \log x$ is some single-valued function f_2, i.e., some new Laurent series. Therefore, we get the general structure of the first of the "noneigensolutions":

$$u_2'(x) = x^s \left(\frac{1}{2\pi i \lambda} f_1(x) \log x + f_2(x) \right). \qquad (2.19)$$

We proceed in the same way with the next relation of (2.15), which concerns u_3' and the function u_2', whose structure is now known. Let us pose for simplicity $L = (2\pi i)^{-1} \log x$. We obtain

$$\frac{u_3'}{x^s} \mapsto \frac{u_3'}{x^s} + \frac{1}{\lambda} \frac{u_2'}{x^s} = \frac{u_3'}{x^s} + \frac{1}{\lambda^2} f_1 L + \frac{1}{\lambda} f_2. \qquad (2.20)$$

We see that the ratio $u_3'(x)/x^s$ is single-valued, except for the addition of the sum of a single-valued function and a single-valued function times a logarithm. For such a function, we have at our disposal the following model:

$$aL^2 + bL \mapsto a(L+1)^2 + b(L+1) = aL^2 + bL + 2aL + (a+b), \qquad (2.21)$$

where a and b are arbitrary single-valued functions. Comparing (2.20) and (2.21), we see that the difference

$$\frac{u_3'}{x^s} - \left[\frac{1}{\lambda^2} \frac{1}{2} L(L-1)f_1 + \frac{1}{\lambda} Lf_2 \right] \tag{2.22}$$

is some single-valued function f_3, i.e., some new Laurent series. Therefore, we get the general structure of the second of the "noneigensolutions":

$$u_3' = x^s \left[\frac{1}{\lambda^2} \frac{1}{2} L(L-1)f_1 + \frac{1}{\lambda} Lf_2 + f_3 \right]. \tag{2.23}$$

One can proceed in the same way for the other members of the set of solutions related to the same Jordan matrix. It is, however, easier to check directly the following recurrence relation:[3]

$$u_k' = x^s \left[\frac{1}{\lambda^{k-1}} P_{k-1}(L)f_1 + \frac{1}{\lambda^{k-2}} P_{k-2}(L)f_2 + \cdots \right.$$
$$\left. + \frac{1}{\lambda} P_1(L)f_{k-1} + f_k \right], \tag{2.24}$$

with

$$P_k(L) = \frac{1}{k!} L(L-1) \cdots (L-k+1) \quad (P_0(L) = 1). \tag{2.25}$$

Let us now summarize this discussion. We have shown that there exists, in the open circle $0 < |x| < r_0$, a fundamental set of solutions whose general analytical structure is as follows:

There exists at least one eigensolution of the type (2.16); in fact, there exists one (and only one) such solution for each Jordan submatrix.

If this first set of solutions is too small, it can be completed by new solutions that contain the previous ones times $\log x$; this happens when the Jordan canonical reduction introduces matrices of dimension higher than one, i.e., whenever the original transformation matrix G cannot be diagonalized; there exists exactly one such solution for each Jordan submatrix of dimension higher than one.

If necessary, the fundamental set of solutions is completed, step by step, by new solutions containing higher powers of $\log x$; this happens whenever higher-dimensional Jordan submatrices appear in the canonical reduction of G; if the largest dimension is k $(1 \le k \le n)$, one finds solutions containing $\log x$ up to the power $k - 1$.

---

[3] Start with checking that $P_k(L+1) = P_k(L) + P_{k-1}(L)$.

Each of these solutions contains some new unknown single-valued function, i.e., some unknown Laurent series. One could imagine substituting formally the Laurent expansions of the coefficients and the a priori structures of the solution in the differential equation, and to proceed as in the regular case. Unfortunately, unlike the regular case, this procedure does not necessarily lead to the definition of a recurrence, simply because unlike Taylor series, Laurent series do not necessarily have a first term! We meet in this way a possible natural classification of the equations, i.e., a classification based on the following alternative: either the total absence of essential singularities for the coefficients and the solutions or the presence of at least one such essential singularity. In fact, because it is always possible to recompute the coefficients from ad hoc algebraic combinations of the solutions and their derivatives, this classification is made on the basis of the solutions alone. The isolated singular point is called a:

Weak singularity if all the Laurent series found in the fundamental set of solutions have a first term (i.e., they represent functions that either are regular at the origin or have a pole at that point). Notice that an ad hoc translation of the characteristic exponents (i.e., $s_i \mapsto s_i - n_i$) transforms the Laurent series into Taylor series that do not vanish at the origin. These weakly singular equations are also called "locally Fuchsian." We shall see that their treatment can be conducted in much the same way as for regular equations (see Section 3).

Strong singularity if at least one of the single-valued functions of the fundamental set of solutions has an essential singularity at the origin. The issue of solutions of these equations then becomes much more complicated. We shall discuss only the case of second-order equations whose coefficients have non-Fuchsian polar singularities (Thomé's equations, see Section 4).

3 Weakly Singular Equations (Fuchs)

The previous definition of a weakly singular equation is rather indirect: It is based on the alternative of the *exclusive* presence of Taylor series (times powers x^s and $\log x$) in the fundamental set of solutions, while this fundamental set of solutions is still to be built. The purpose of Fuchs's theorem is precisely to transform this rather implicit condition into an equivalent explicit condition on the coefficients of the equation

Theorem 3.1 (Fuchs's Theorem). *A necessary and sufficient condition for the differential equation*

$$y^{(n)} + a_1(x)y^{(n-1)} + a_2(x)y^{(n-2)} + \cdots + a_n(x)y = 0 \qquad (3.1)$$

to be weakly singular at the origin is that, the function $x^k a_k(x)$ be regular at the origin for each of the coefficients $a_k(x)$ (in other words, the coefficients $a_k(x)$ may well be singular at the origin, but this singularity cannot be "worse" than a pole of order k).

One proves this theorem as follows. One first shows that the backwards computation of the coefficients from the solutions gives, in the case of a weakly singular equation, polar singularities as indicated (i.e., the condition is necessary). Conversely, if the coefficients are such, one can build explicitly a fundamental set of solutions that characterizes a weakly singular equation (i.e., the condition is sufficient). In order to make the proof as simple as possible, we first consider the case of a second-order equation and generalize later on the proof to higher-order equations.

Necessary condition

Let us consider the differential equation

$$y'' + p(x)y' + q(x)y = 0, \tag{3.2}$$

where $p(x)$ and $q(x)$ are single-valued around the origin; let us assume that $u(x)$ and $v(x)$ are linearly independent solutions that correspond to a weakly singular equation, i.e., either

$$\begin{aligned} u(x) &= x^s f(x), \\ v(x) &= x^{s'} g(x), \end{aligned} \tag{3.3}$$

or, alternatively

$$\begin{aligned} u(x) &= x^s f(x), \\ v(x) &= x^{s'} g(x) + u(x) \log x, \end{aligned} \tag{3.4}$$

where f and g are holomorphic and do not vanish at the origin. Please remember that the alternative case (3.4) corresponds to a doubly degenerate eigenvalue λ and that the difference $s' - s$ is then an integer. In the next lemma I shall use the expression "y is of type u ... " in order to characterize any function y that, like the function u of (3.3), is the product of some power x^s times a holomorphic function f that does not vanish at the origin.

Lemma 3.1.

(i) *If u and v are "of type u," then their product uv, their quotient u/v, and the derivatives u' and v' are also "of type u;"*

(ii) *If y is "of type u," then, xy'/y and $x^2 y''/y$ are regular (i.e., holomorphic) at the origin.*

The first part of this lemma is evident. The second part is also evident, since

$$xy'/y = s + xf'/f,$$
$$x^2 y''/y = s(s-1) + 2sxf'/f + x^2 f''/f, \tag{3.5}$$

with $f(0) \neq 0$.

Returning to our problem, we compute the coefficient $p(x)$ with the help of the equation of the Wronskian (cf. (1.8)),

$$p(x) = -W'/W, \tag{3.6}$$

and we write the Wronskian in the form

$$W = uv' - u'v = u^2 (v/u)'. \tag{3.7}$$

It now becomes clear that in the case of a weakly singular equation, the Wronskian is "of type u":

This is, of course, evident from the first part of the lemma if u and v are "of type u" (case (3.3)).

If alternatively v contains a logarithm (case (3.4)), the ratio v/u is the sum of $\log x$ and a polar Laurent series (remember that $s' - s$ is an integer), and therefore, the derivative of this sum is again "of type u."

According to (3.6) and the second part of the lemma, the function $xp(x)$ is then holomorphic at the origin.

Next, we compute the coefficient $q(x)$ directly from the differential equation, where we substitute the solution $u(x)$:

$$q(x) = -(u'' + pu')/u. \tag{3.8}$$

Applying again the second part of the lemma and our knowledge of $p(x)$, we conclude that $x^2 q(x)$ is holomorphic at the origin.

Sufficient condition

Let us consider again the differential equation

$$y'' + p(x)y' + q(x)y = 0, \tag{3.9}$$

where the coefficients $p(x)$ and $q(x)$ now are represented in the open circle $0 < |x| < r_0$ by the following series:

$$p(x) = \frac{1}{x} \sum_{n=0}^{\infty} a_n x^n, \quad q(x) = \frac{1}{x^2} \sum_{n=0}^{\infty} b_n x^n. \tag{3.10}$$

We have to show that it is possible to build in this circle a weakly singular fundamental set of solutions. Let us pose formally

$$y(x) = x^s \sum_{n=0}^{\infty} c_n x^n, \quad c_0 \neq 0, \tag{3.11}$$

and let us (formally) substitute in the equation. We rewrite the equation as an ascending power series,

$$x^s \sum_{n=0}^{\infty} x^n \left[c_n(n+s)(n+s-1) \right.$$
$$\left. + \sum_{n=0}^{n} c_k[(k+s)a_{n-k} + b_{n-k}] \right] = 0, \tag{3.12}$$

and we set all its coefficients equal to zero: If $n = 0$, then

$$c_0[s(s-1) + a_0 s + b_0] = 0, \tag{3.13}$$

while if $n \geq 1$, then

$$c_n[(s+n)(s+n-1) + a_0(s+n) + b_0]$$
$$= -\sum_{k=0}^{n-1} c_k[(k+s)a_{n-k} + b_{n-k}]. \tag{3.14}$$

Because $c_0 \neq 0$, equation (3.13) is to be considered as a "characteristic second-degree equation," which determines two possible values of the parameter s, say s_1 and s_2. For each of these two values, the relations (3.14) define a recurrence that gives each new coefficient c_n ($n \geq 1$) times the value of the characteristic polynomial in $s + n$, as a linear combination of the previous coefficients. Therefore, we get the coefficients c_n in a recurrent way, provided that we do not have to divide by the possible value zero of this characteristic polynomial. This difficulty can happen only if the difference of the roots of the characteristic equation is an integer: If $s_1 + r = s_2$, $r \in \mathbb{N}_+$, then the computation of the coefficient c_r of the series corresponding to the root s_1 is not possible.

We obtain the following results:

If $r = s_1 - s_2$ is not an integer, then there exist, at least formally,[4] two linearly independent eigensolutions.

If $r = s_1 - s_2$ is an integer, then there exists, at least formally, one eigensolution that corresponds to the root s with the larger real value.

---

[4]The formal series defines a "true" solution if (and only if) it converges in the circle $|x| < r_0$.

It is easy to prove the convergence of the formal series in some neighborhood of the origin by using the usual majorization techniques:

Let $M = \max\{|xp(x)|, |x^2 q(x)|; |x| < r_0\}$; then the Taylor coefficients $|a_n|$ and $|b_n|$ are bounded by M/r_0^n (Cauchy's inequalities).

We make a first majorization of the recurrent coefficients,

$$c_n = \frac{-1}{n(n \pm r)} \sum_{k=0}^{n-1} c_k [(s+k)a_{n-k} + b_{n-k}], \qquad (3.15)$$

by making use of absolute values, and we make a second majorization by using the Cauchy bounds

$$|c_n| \leq \frac{M}{n^2 e} \sum_{k=0}^{n-1} |c_k| \big[\, |s| + k + 1 \big] \frac{1}{r_0^{n-k}}, \qquad (3.16)$$

where $e = \min\{|1 \pm r/n|; n \in \mathbb{N}_+\}$ (notice that $e > 0$, since we assume that we can indeed build the formal series). We majorize once more:

$$|c_n| \leq \frac{K}{r_0^n} \frac{1}{n} \sum_{k=0}^{n-1} |c_k| r_0^k, \qquad (3.17)$$

with $K = M(|s|+1)/e$. It is then easily shown by recurrence that if the coefficients $|c_n|$ are bounded as indicated by (3.17) and if $K \geq 1$, then

$$|c_n| \leq \frac{K^n}{r_0^n} |c_0| \qquad (3.18)$$

(if $K < 1$, one makes a last majorization by setting $K = 1$).

Formula (3.18) proves the convergence of the formal series in the circle $|x| < r_0/K$. Therefore, the solution exists as a true solution inside this open circle. But since we know that existing solutions can be singular only where the coefficients of the equations are singular, this "smaller" circle cannot be smaller than the original circle, and the solution exists in the full open circle $0 < |x| < r_0$.

To achieve the proof, we have to construct a second (linearly independent) solution in the pathological case where the difference of the characteristic values is an integer, and we have to show that this second solution gives, together with the first one, a weakly singular fundamental set of solutions. Let $u(x)$ be the eigensolution that we have just constructed; it corresponds to the characteristic root s_+ with largest real value: $r = s_+ - s_- \in \mathbb{N} = \{0, 1, 2, \dots\}$. We build a second solution by using the method of variation of constants. We pose $v = uz$, and get a first-order linear equation for the derivative z' of the unknown z:

$$(z')' + (p + 2u'/u)(z') = 0. \qquad (3.19)$$

This equation is then trivially integrated:

$$
\begin{aligned}
z' &= \frac{1}{u^2} \exp\left[-\int_0^x p(t)\,dt\right] \\
&= \frac{x^{-2s_+}}{\left(\sum_0^\infty c_n x^n\right)^2} \exp\left[-a_0 \log x - C - \sum_{k=1}^\infty \frac{a_k}{k} x^k\right] \\
&= C x^{-2s_+ - a_0} \frac{\exp\left[-\sum_{k=1}^\infty a_k x^k / k\right]}{\left(\sum_0^\infty c_n x^n\right)^2},
\end{aligned}
\tag{3.20}
$$

where C is some constant. The combination of Taylor series that appears in the last expression is nonzero at the origin, and it is regular in some neighborhood of the origin. Therefore, it can be replaced by a single Taylor series whose first coefficient d_0 is different from zero. We also notice that $2s_+ + a_0 = r + 1 \in \mathbb{N}_+ = \{1, 2, \dots\}$. Therefore, z' is a Laurent series with a pole of order $r + 1$:

$$
z' = x^{-r-1} \sum_{k=0}^\infty d_k x^k \quad (d_0 \neq 0),
\tag{3.21}
$$

and its primitive function z is the sum of a Laurent series with a pole of order r, and a $\log x$, which corresponds to the integration of the power x^{-1}, i.e., to the coefficient d_r.

This construction clearly achieves the demonstration of Fuchs's theorem for second-order equations. Notice that in the pathological case of characteristic roots differing by an integer, the logarithm is present in the second solution only if the coefficient d_r is nonzero. Therefore, it is certainly present in the case of equal characteristic roots ($d_0 \neq 0$), but it is not necessarily present when $r \geq 1$. If d_r turns out to be zero, then the second solution is a new eigensolution.

We can now summarize the practical rules of construction of a fundamental set of solutions as follows:

(i) Compute the characteristic roots.

(ii) If the difference of the roots is *not* an integer, construct both eigensolutions u and v by substitution in the equation and construction of the recurrences.

(iii) If the difference of the characteristic roots is a positive integer, i.e., $r = s_+ - s_- \in \mathbb{N}_+ = \{1, 2, \dots\}$, construct first the solution u corresponding to s_+ and then *try* to construct in the same way the solution v corresponding to s_-, computing recurrently the coefficients up to the catastrophic coefficient c_r. If you obtain $c_r = (\neq 0)/0$, the problem is impossible, and you know that a logarithm will be necessary; if alternatively you obtain $c_r = 0/0$, the problem is possible but simply undetermined, and no logarithm has to be introduced. In that case you put, e.g., $c_r = 0$,[5] and you simply go on with the construction of the recurrence.

(iv) If a logarithm is necessary (i.e., either $r = 0$ or $r = 1, 2, \ldots$ and the previous computation of c_r turned out to be impossible), you introduce this logarithm a priori in the general form of the solution v,

$$v(x) = \alpha u(x) \log x + x^{s_-} \sum_{k=0}^{\infty} d_k x^k, \qquad (3.22)$$

with some unknown coefficient α that you choose later on when you reach the recurrent construction of the coefficient d_r of the Taylor series of (3.22): α is to be chosen in such a way that d_r becomes undetermined, i.e., $d_r = 0/0$; you put, e.g., $d_r = 0$, and you continue the construction of the recurrence.

Remark. It happens frequently that one succeeds in constructing the solution v containing a logarithm as an application of the Poincaré theorem that asserts the analyticity of the solutions in the parameters that are analytically present in the equation. One first eliminates the degeneracy $s_+ - s_- \in \mathbb{N}$ by some ad hoc perturbation of the coefficients of the characteristic equation, and one computes both eigensolutions of the perturbed equation:

$$(a_0, b_0) \to (a_0', b_0'), \quad (s_1, s_2) \to (s_1', s_2'), \quad r \to r' \notin \mathbb{Z}. \qquad (3.23)$$

One studies carefully the unperturbed limit of the perturbed solution v' (i.e., the eigensolution corresponding to the root s_-'). It is easily seen that the critical coefficient $c_r'^-$ becomes singular like $1/(r - r')$:

$$c_r'^- = \frac{1}{r - r'} \left(\frac{1}{r} \sum_{k=0}^{r-1} c_k'^- \left[(k + s'^-) a_{r-k}' + b_{r-k}' \right] \right) = \frac{\alpha'}{r - r'}, \qquad (3.24)$$

the limit $\alpha' \to \alpha$ being regular. The same kind of singular behavior happens with all the upper coefficients $c_{r+k}'^-$, with this particularity that the regular part of their limit is proportional to the coefficient c_k^+ of the eigensolution u:

$$c_{r+k}'^- = d_{r+k}^- + \frac{\alpha}{r - r'} c_k^+ + O(r - r') \quad (c_0'^+ = 1). \qquad (3.25)$$

One eliminates the singular part by subtraction of the eigensolution u'

---

[5] Any other choice of this coefficient is, of course, allowed. It amounts to adding to our solution v some contribution proportional to the solution u (exercise).

times $\alpha/(r - r')$, and one takes the limit. One then obtains the logarithm:

$$v' - \frac{\alpha}{r - r'}u' = x^s \sum_{n=0}^{\infty} d_n^- x^n + \frac{\alpha}{r - r'}[x^{s'_-+r} - x^{s'_+}] \sum_{n=0}^{\infty} c_n^+ x^n + O(r - r')$$

$$\rightarrow x^{s-} \sum_{n=0}^{\infty} d_n^- x^n + \alpha u \log x. \quad (3.26)$$

In summary, one first computes perturbed eigensolutions u' and v' and the coefficient α'; once this is done, one computes the limit for $r' \rightarrow r$ of the ad hoc linear combination of the perturbed eigensolutions, i.e., $v' - \alpha'u'/(r - r')$, where u' is normalized by $c_0'^+ = 1$.

Higher-order equations

The generalization of the previous discussion to higher-order equations is rather straightforward. The complications are indeed only technical. Let us assume that the coefficients $a_k(x)$ of equation (3.1) do satisfy the Fuchs condition. We substitute in this equation the corresponding Taylor series representations of these coefficients and also the formal structure of an eigensolution:

$$x^k a_k(x) = \sum_{i=0}^{\infty} a_{k,i} x^i, \quad (3.27)$$

$$u(x) = x^s \sum_{i=0}^{\infty} c_i x^i \quad (c_0 \neq 0). \quad (3.28)$$

We rewrite the equation as an ascending power series, and we set all its coefficients equal to zero. We obtain in that way

(i) for the lowest power, the characteristic equation $P_n(s) = 0$, where $P_n(s)$ is the polynomial

$$P_n(s) = s(s - 1) \cdots (s - n + 1) + a_{1,0} s(s - 1) \cdots (s - n + 2)$$
$$+ \cdots + a_{n-1,0} s + a_{n,0}, \quad (3.29)$$

(ii) for the other powers, the recurrence relation

$$c_k P_n(k + s) = \text{linear combination of } \{c_0, c_1, \ldots, c_{k-1}\}, \quad (3.30)$$

which gives each new coefficient c_k times the value of the characteristic polynomial in $s + k$ as a linear combination of the previous coefficients. Therefore, we get the coefficients c_k in a recurrent way, provided that we do not have to divide by the possible value zero of this characteristic polynomial. This difficulty happens only if among the differences between the roots of the characteristic equation, some of them are integers. Therefore, we have to partition the set of roots into subsets of roots that are equal modulo an integer. Let us consider, e.g., the subset corresponding to some root s' and let us suppose that we find there $k + 1$ roots that we write according to a nonincreasing order of their real values, i.e., $\{s', s' - r_1, s' - r_2, \ldots, s' - r_k\}$, where the r_i are natural integers in nondecreasing order. We shall see that this subset corresponds to the $k + 1$ solutions that in the general discussion of Section 2 correspond to the same eigenvalue λ (remember that according to formula (2.10), the exponent s is defined by the eigenvalue λ, modulo an integer; cf. footnote 2). They were there, in general, related to several Jordan submatrices; and correspondingly, this leads here to the different possibilities that we shall now discuss. Let us first consider the nondegenerate case where all the roots of the subset s' are unequal (i.e., let us consider that the differences r_i are positive and unequal) and let us consider the feasibility of the construction of the recurrences:

For the first element of the subset, i.e., s', we of course never meet a zero of the characteristic polynomial when we construct the recurrence (3.30); therefore, this recurrence defines an eigensolution u:

$$u = x^{s'} \sum_{i=0}^{\infty} c_i x^i; \tag{3.31}$$

it corresponds to the first element of some Jordan submatrix.

For the next element, i.e., $s' - r_1$, the recurrence poses a problem when we try to compute the coefficient with index r_1, because we have to divide by zero; we have then the following alternative:

Either the linear combination of the right-hand side is also zero; in this case the coefficient with index r_1 is arbitrary (choose, e.g., $c_{r1} = 0$), and we obtain then a second eigensolution:

$$v = x^{s'-r_1} \sum_{i=0}^{\infty} c_i x^i; \tag{3.32}$$

it corresponds to the first element of a second Jordan submatrix with the same eigenvalue λ;

or the linear combination of the right-hand side is not zero; in this case we have to introduce a priori some contribution $\alpha u \log x$, with a parameter α that is to be determined by requiring indeterminacy when we compute the d-coefficient with indices r_1 of the new input:

$$v = x^{s'-r_1} \sum_{i=0}^{\infty} d_i x^i + \alpha u \log x. \qquad (3.33)$$

This solution v corresponds to the second element of the first Jordan submatrix.

The same kind of alternative is met *two times* when we consider the recurrence corresponding to the root $s - r_2$: We have to face the possibility of a division by zero, a first time when we compute the coefficient with index $r_2 - r_1$, and a second time when we compute the coefficient with index r_2. We obtain then the following alternatives, depending on the value (0 or $\neq 0$) of the linear combination of the right-hand side:

either an eigensolution if both linear combinations are equal to zero; this eigensolution corresponds to the first element of a new Jordan submatrix;

or a solution v containing a term $\alpha u \log x$ if one (and only one) of the linear combinations is equal to zero; this solution v corresponds to the second element of a Jordan submatrix;

or a solution w containing terms $\alpha u (\log x)^2$ and $\beta v \log x$, with two parameters α and β that are to be determined by requiring indeterminacy when we compute the d'-coefficients with indices $r_2 - r_1$ and r_2 of the new input:

$$w = x^{s'-r_2} \sum_{i=0}^{\infty} d'_i x^i + \alpha u (\log x)^2 + \beta v \log x; \qquad (3.34)$$

this solution w corresponds to the third element of the first Jordan submatrix.

The construction proceeds as for the other roots of the subset, introducing whenever necessary higher powers of $\log x$. The convergence of the formal series that appear in all these formulae can be proved just in the same way as for the second-order equation. The crucial point in this convergence proof is that the degree of the characteristic polynomial P_n is one unit higher than the degree of the polynomials that appear in the linear combination of the right-hand side. This remark, together with the usual majorizations, suffices to prove the convergence of the series in some neighborhood of the origin, i.e., to prove that the constructed formal solutions are true solutions.

The case of degenerate roots of the characteristic polynomial can be considered as a limiting case where some of the differences between the roots vanish. It is clear that one has then to introduce a priori the ad hoc logarithmic contributions (i.e., $\log x$ and possibly also higher powers of $\log x$) in order to recover a full set of solutions. The computation of the complementary series proceeds then as above[6].

To conclude, we see that the construction of a fundamental set of solutions of the homogeneous equation around a Fuchsian singularity is algorithmically nearly as simple as the construction of a fundamental set of solutions around a regular point. In particular, it can be programmed and made automatic by means of symbolic computation software. The solutions are analytically constructed with known functions and convergent series. It allows us to reach a numerical precision in the open circle $0 < |x| < r_0$, which is limited only by the capacity of the computers at our disposal. The analytic continuation from one such neighborhood to another intersecting neighborhood can be made just as in the regular case. The problem of the *complete* integration of equations with only Fuchsian singularities will be considered in Section 5.

4   Thomé's Equations

Let us consider again the second-order differential equation

$$y'' + p(x)y' + q(x)y = 0, \tag{4.1}$$

where $p(x)$ and $q(x)$ are single-valued around the origin. Let us suppose that this equation is strongly singular (i.e., non-Fuchsian) at the origin. Let us eliminate the coefficient of the first derivative by the transformation

$$y(x) = z(x) \exp\left[-\frac{1}{2}\int_0^x p(t)\,dt\right], \tag{4.2}$$

where the new unknown $z(x)$ is a solution of the differential equation

$$z'' + \left(q - \frac{1}{4}p^2 - \frac{1}{2}p'\right)z = 0. \tag{4.3}$$

It can happen that this new differential equation is regular or Fuchsian at the origin; in that case, the construction of a fundamental set of solutions around the origin can be achieved by the methods described in the previous paragraphs. We dismiss these cases, and we assume that equation (4.3) remains strongly singular at the origin. If the new coefficient has a *polar* singularity

$$\tilde{q} = q - \frac{1}{4}p^2 - \frac{1}{2}p' = \frac{1}{x^n}\sum_{k=0}^{\infty} a_k x^k \quad (a_0 \neq 0, n > 2), \tag{4.4}$$

---

[6]One can also recover these logarithmic contributions through a perturbation approach from the nondegenerate case (see the remark following the second-order case).

then one can construct a fundamental set of solutions around the origin by the *method of Thomé*. In order to make more precise the non-Fuchsian character of this equation, one considers the difference $r = n - 2$; it is called the Thomé rank of the equation. This equation can be integrated as follows:

If the Thomé rank is even, one puts a priori

$$z(x) = \exp\left[Q_{r/2}\left(\frac{1}{x}\right)\right] x^s \sum_{k=0}^{\infty} c_k x^k, \qquad (4.5)$$

where $Q_{r/2}$ is some (unknown) polynomial of degree $r/2$ and where $c_0 \neq 0$. Substitution of this form in the equation allows us to construct two linearly independent formal solutions, i.e., to determine two polynomials Q and correspondingly two exponents s and two recurrences defining the power series of (4.5).

If the Thomé rank is odd, the change of variable $x \mapsto t = x^{1/2}$ gives a new differential equation whose Thomé rank is just doubled, so that one comes back to the previous case.

Some explanation is in order. One knows from Fuchs's theorem that at least one of any pair of linearly independent solutions around the origin contains a Laurent series with an essential singularity. The Thomé method is based on a modeling of the essential singularity. It seems a bit extraordinary that one can in this way replace an infinite series of negative powers by the exponentiation of a polynomial of negative powers, i.e., by some recurrence based on a finite number of coefficients. In fact, the method is not as good. In general, the Thomé solutions are not exact solutions in the ordinary sense of complex analysis, but they are asymptotic forms of solutions. The formal series that appears in the representation (4.5) of the solutions has, in general, a vanishing radius of convergence. Fortunately, it is always possible to extract from this formal series enough information to construct a true solution through summation methods like the Borel summation method.

In order to make the proof as simple as possible, we first consider the case of an equation of rank 2, and we shall later on generalize this proof to higher-rank equations. Let us consider the equation

$$y'' + \frac{1}{x^4} \sum_{k=0}^{\infty} a_k x^k y = 0 \quad (a_0 \neq 0), \qquad (4.6)$$

and let us substitute the Thomé a priori form

$$y = \exp\left[\frac{\omega}{x}\right] x^s \sum_{k=0}^{\infty} c_k x^k \quad (c_0 \neq 0). \qquad (4.7)$$

We factorize $\exp[\omega/x]x^s$, rewrite the equation as an ascending power series, and set all its coefficients equal to zero. We obtain in that way, successively,

(i)

$$(\omega^2 + a_0)c_0 = 0, \tag{4.8}$$

which always defines two distinct roots,

$$\omega = \pm(-a_0)^{1/2}; \tag{4.9}$$

(ii) for *each* of these two roots, the next relation,

$$[-2\omega(s-1) + a_1]c_0 = 0, \tag{4.10}$$

defines univocally the corresponding exponent s,

$$s = 1 + a_1/2\omega; \tag{4.11}$$

(iii) the next relations define then a recurrence $(n \geq 1)$,

$$2\omega n c_n = [(n+s-1)(n+s-2) + a_2]c_{n-1} + \sum_{k=0}^{n-2} a_{n+1-k}c_k, \tag{4.12}$$

which univocally gives all coefficients c_n of the formal series in terms of the parameters a_k of the equation, and proportionally to the first coefficient c_0. Unfortunately, this recurrence has the asymptotic form

$$c_n \propto n c_{n-1} \iff c_n \propto n!, \tag{4.13}$$

which, generally speaking, characterizes a vanishing radius of convergence. Although it is not at all excluded that the series converges (in particular, the recurrence can well terminate after some finite number of terms: $c_n = 0$ when $n > n_0$), one concludes that in general, the Thomé method leads only to *formal* solutions, which still must be manipulated in order to obtain exact solutions. The Borel summation method is particularly convenient for this purpose. Let us now see how it works:

(i) One divides each coefficient c_n by $\Gamma(\alpha+n)$, where Γ is the gamma function and where α is some parameter that one conveniently chooses later on;[7]

---

[7]The Borel method is generally presented with $\alpha = 1$. However, the parameter α can be usefully used to obtain the analytic continuation of the series $B(x; \alpha)$ or to identify this function among already known functions.

(ii) One considers the analytic function $B(x; \alpha)$, which is completely defined by its germ, i.e., its Taylor series:

$$B(x; \alpha) = \sum_{n=0}^{\infty} \frac{c_n}{\Gamma(\alpha + n)} x^n \quad (|x| < R), \qquad (4.14)$$

which converges in some finite circle $|x| < R$.

(iii) If one succeeds in performing the analytic continuation of this germ along the positive real axis (or equivalently, along some topological equivalent path), one gets the "Borel associated function $B(x; \alpha)$," which is to be introduced in the following integral:

$$I(x; \alpha) = \int_0^{\infty} e^{-t} t^{\alpha-1} B(tx; \alpha) \, dt. \qquad (4.15)$$

This integral defines an analytic function $I(x; \alpha)$. A formal term-by-term integration gives back the original series; therefore, according to ordinary complex analysis, this term-by-term integration is simply meaningless. However, it *is meaningful* in the framework of asymptotic analysis! In this latter framework, the formal integration proves that the original series is a local, sectorial, asymptotic representation of the analytic function $I(x, \alpha)$:

$$I(x; \alpha) \propto \sum_{n=0}^{\infty} c_n x^n. \qquad (4.16)$$

It is clear that the analytic function

$$u(x) = \exp\left[\frac{\omega}{x}\right] x^s I(x; \alpha), \qquad (4.17)$$

which is singular at the origin, possesses there the original Thomé solution as an asymptotic representation. One can show that it is effectively a true solution of the differential equation.[8]

If one is interested in obtaining numerical values of the solutions, one can alternatively consider numerical summation methods of the original Thomé series. One can, for example, consider Padé's approximants, which can give in some sectorial neighborhood of the origin a sufficiently high numerical precision to proceed then to an analytic continuation via other fundamental sets of solutions (as explained here above).

Let us now consider the general case of a Thomé's equation of even rank $r = 2r'$, $r' \in \mathbb{N}_+$. We write the equation in the form

$$x^{2r'+2} y'' + \sum_{k=0}^{\infty} a_k x^k y = 0 \quad (a_0 \neq 0), \qquad (4.18)$$

---

[8]Cf. E. Borel, *Leçons sur les séries divergentes*, Paris, 1901.

and we substitute the Thomé a priori form

$$y = \exp\left[\frac{\omega_{r'}}{x^{r'}} + \frac{\omega_{r'-1}}{x^{r'-1}} + \cdots + \frac{\omega_1}{x}\right] x^s \sum_{k=0}^{\infty} c_k x^k \quad (c_0 \neq 0). \qquad (4.19)$$

We factorize $\exp[P_{r'}(1/x)]x^s$, rewrite the equation as an ascending power series, and set all its coefficients equal to zero. We obtain in this way successively:

(i) a first relation

$$(r'^2 \omega_{r'}^2 + a_0)c_0 = 0, \qquad (4.20)$$

which defines two *distinct* roots $\omega_{r'} = \pm(1/r')(-a_0)^{1/2}$;

(ii) for *each* of these two roots a set of $r' - 1$ relations $(r' > 1)$ that define successively and univocally the other coefficients ω_i of the polynomial:

$$\omega_{r'-1} = -a_1/[2r'(r'-1)\omega_{r'}], \quad \text{(etc.)}; \qquad (4.21)$$

(iii) a relation that defines the exponent s:

$$s = \frac{1}{2r'\omega_{r'}}\left[a_{r'} + r'(r'+1)\omega_{r'} + \sum_{i=1}^{r'-1} \omega_i \omega_{r'-1} i(r'-1)\right]; \qquad (4.22)$$

(iv) a recurrence $(n \geq 1)$:

$$2r'\omega_{r'} n c_n = \text{linear combination of } \{c_0, c_1, \ldots, c_{n-1}\}, \qquad (4.23)$$

which defines univocally all coefficients c_n of the formal series in terms of the parameters a_k of the equation, proportionally to the first coefficient c_0.

When we look for the origin of the different terms of this recurrence relation, we notice that the left-hand side comes from the product of the first derivative of the exponential times the first derivative of the formal series. That is why it linearly increases with n. We find on the right-hand side some terms that come from the second derivative of the formal series. Consequently, the contribution of these terms to the coefficient of c_{n-k} increases like $(n-k)^2$. Therefore, we find, in this general case, essentially the same asymptotic behavior as in the simpler rank-2 case,

$$c_n \propto n c_{n-1} \iff c_n \propto n!, \qquad (4.24)$$

which, generally speaking, characterizes a vanishing radius of convergence. Although it is not at all excluded that the series converges, we can conclude that in general the Thomé method leads only to formal solutions that still must be manipulated in order to obtain exact solutions (cf. the discussion above).

5 Global Considerations

In the previous sections we always considered the problem of constructing a fundamental set of solutions around some regular or singular point. In practice, one often has to face the rather different problem of constructing some well-defined solution on a domain that is a priori given. For example, in the quantum-mechanical problem of the hydrogen atom, one looks for solutions of the Legendre equation that are regular on the interval $[-1, +1]$, and for solutions of the radial equation that are regular on the interval $[0, \infty)$ and decrease asymptotically to zero at infinity.[9] When one imposes such a regularity condition at a singular point of a differential equation, one generally selects a particular solution. If furthermore, one imposes the simultaneous regularity of the solution at another singular point, this puts restrictions on the parameters that are present in the equation. This is the essence of quantization in Schrödinger's wave mechanics. In order to carry out such a program, it is necessary to dispose of global solutions that cover simultaneously several singular points. This study requires at first a classification of equations with respect to the *global* structure of their singularities.

Completely Fuchsian Equations

The assumption that an equation has only Fuchsian singularities on the Riemann sphere (i.e., the complex plane and the point at infinity), reduces considerably the arbitrariness in the choice of the coefficients. Let us consider a second-order equation and let us put $n - 1$ Fuchsian singularities at definite points a_i of the complex plane and also an nth singularity at infinity. It is easily seen that the coefficients $p(x)$ and $q(x)$ have then necessarily the form

$$p(x) = \sum_{i=1}^{n-1} \frac{A_i}{x - a_i}, \tag{5.1}$$

$$q(x) = \sum_{i=1}^{n-1} \left[\frac{B_i}{x - a_i} + \frac{C_i}{(x - a_i)^2} \right], \tag{5.2}$$

with the constraint $\sum_{i=1}^{n-1} B_i = 0$, in order that the equation be Fuchsian at infinity. This equation contains only $3n - 4$ parameters, aside, of course, from the places of the singularities. For example, one checks easily that except for trivial transformations of the Riemann sphere (i.e., plane translations and inversions), the only second-order completely Fuchsian equation with one singularity is $y'' = 0$. Similarly, the only completely Fuchsian

---

[9]These conditions come from the general requirement that the wave function be quadratically integrable.

equations with two singularities are Euler's equations with two arbitrary parameters A and C:

$$y'' + (A/x)y' + (C/x^2)y = 0. \tag{5.3}$$

Next, we meet the second-order equations with three Fuchsian singularities. If we put the singularities at places a_i in the plane and choose as free parameters the roots s_i and s_i' of the characteristic equations in these places, we get the Riemann equation

$$y'' + y' \sum_{i=1}^{3} \frac{1 - s_i - s_i'}{x - a_i}$$
$$+ y \sum_{i=1}^{3} \frac{s_i s_i'(a_i - a_j)(a_i - a_k)}{x - a_i} \prod_{m=1}^{3} \frac{1}{x - a_m} = 0, \tag{5.4}$$

where the constraint $\sum_{i=1}^{3}(s_i + s_i') = 1$ has to be imposed in order that the equation be regular at infinity. This equation can be written in a much more transparent way as a table that summarizes the complete information at our disposal: variable x, places of the singularities a_i, and roots s_i and s_i' at these points. This is Papperitz's symbolic writing, which is very useful to study these equations:

$$y = P \left\{ \begin{matrix} a_1 & a_2 & a_3 & \\ s_1 & s_2 & s_3 & x \\ s_1' & s_2' & s_3' & \end{matrix} \right\}. \tag{5.5}$$

This symbolic writing is clearly invariant for a permutation of the columns corresponding to the singularities, and also for a permutation of the two roots s and s' of a definite column. The Papperitz symbol is not only a convenient representation of equation (5.4), it is also, an implicit convenient representation of the set of its solutions. It is in this double sense that one should consider the transformations that follow.

One can freely move the three singularities on the Riemann sphere by homographic transformations. These transformations are most easily written in implicit form:

For an arbitrary displacement in the plane (i.e., $a_i \mapsto a_i'$, $i = 1, 2, 3$),

$$\frac{x' - a_1'}{x' - a_2'} \cdot \frac{a_3' - a_2'}{a_3' - a_1'} = \frac{x - a_1}{x - a_2} \cdot \frac{a_3 - a_2}{a_3 - a_1}. \tag{5.6}$$

If one of the three singularities is sent to infinity (i.e., $a_i \mapsto a_i'$ for $i = 1, 2$; $a_3 \mapsto \infty$),

$$\frac{x' - a_1'}{x' - a_2'} = \frac{x - a_1}{x - a_2} \cdot \frac{a_3 - a_2}{a_3 - a_1}. \tag{5.7}$$

The Riemann–Papperitz equation is invariant for these transformations:

$$y = P \left\{ \begin{matrix} a_1 & a_2 & a_3 \\ s_1 & s_2 & s_3 & x \\ s_1' & s_2' & s_3' \end{matrix} \right\} = P \left\{ \begin{matrix} a_1' & a_2' & a_3' \\ s_1 & s_2 & s_3 & x' \\ s_1' & s_2' & s_3' \end{matrix} \right\}. \tag{5.8}$$

One can also shift by a constant amount both roots corresponding to one singular point, but this shift has to be compensated by a similar shift of the other roots in order to maintain the constraint $\sum_{i=1}^{3}(s_i + s_i') = 1$. For example, we can shift the roots corresponding to a_1 by λ and the roots corresponding to a_2 by μ, but we then have to shift the roots of the third column by $-(\lambda + \mu)$. We get the following formulae:

$$y = P \left\{ \begin{matrix} a_1 & a_2 & a_3 \\ s_1 & s_2 & s_3 & x \\ s_1' & s_2' & s_3' \end{matrix} \right\}$$

$$= \frac{(x - a_3)^{\lambda + \mu}}{(x - a_1)^{\lambda}(x - a_2)^{\mu}} P \left\{ \begin{matrix} a_1 & a_2 & a_3 \\ s_1 + \lambda & s_2 + \mu & s_3 - \lambda - \mu & x \\ s_1' + \lambda & s_2' + \mu & s_3' - \lambda - \mu \end{matrix} \right\}, \tag{5.9}$$

$$y = P \left\{ \begin{matrix} a_1 & a_2 & \infty \\ s_1 & s_2 & s_3 & x \\ s_1' & s_2' & s_3' \end{matrix} \right\}$$

$$= \frac{1}{(x - a_1)^{\lambda}(x - a_2)^{\mu}} P \left\{ \begin{matrix} a_1 & a_2 & \infty \\ s_1 + \lambda & s_2 + \mu & s_3 - \lambda - \mu & x \\ s_1' + \lambda & s_2' + \mu & s_3' - \lambda - \mu \end{matrix} \right\}, \tag{5.10}$$

which connect the solutions of two different Riemann–Papperitz equations with the same singular points and shifted roots.

The transformations (5.8)–(5.10) allow us to reduce the general Riemann–Papperitz equation to a standard form, with singularities located at some conventional places (generally, 0, 1, ∞) and with roots that depend on three complex parameters only. This equation is Gauss's hypergeometric equation

$$y = P \left\{ \begin{matrix} 0 & 1 & \infty \\ 0 & 0 & a & x \\ 1 - c & c - a - b & b \end{matrix} \right\}, \tag{5.11}$$

$$x(1 - x)y'' + [c - (a + b + c + 1)x]y' - aby = 0, \tag{5.12}$$

with three complex parameters, a, b, and c. One solution of this equation is regular at the origin; it is defined by the hypergeometric series

$$_2F_1(a, b \mid c \mid x) = 1 + \frac{ab}{c} \frac{x}{1!} + \frac{a(a+1)b(b+1)}{c(c+1)} \frac{x^2}{2!} + \cdots$$

$$= \frac{\Gamma(c)}{\Gamma(a)\Gamma(b)} \sum_{k=0}^{\infty} \frac{\Gamma(a+k)\Gamma(b+k)}{\Gamma(c+k)} \frac{x^k}{k!} \quad (|x| < 1). \tag{5.13}$$

This local solution can be extended by analytic continuation in the full complex plane, except for a cut from 1 to infinity; one has the following formulae:

$$_2F_1(a,b \mid c \mid x) = \frac{\Gamma(c)\Gamma(c-a-b)}{\Gamma(c-a)\Gamma(c-b)} {}_2F_1(a,b \mid a+b-c+1 \mid 1-x)$$

$$+ \frac{\Gamma(c)\Gamma(c-a-b)}{\Gamma(c-a)\Gamma(c-b)} {}_2F_1(c-a,c-b \mid c-a-b+1 \mid 1-x), \quad (5.14)$$

$$_2F_1(a,b \mid c \mid x) = \frac{\Gamma(c)\Gamma(b-a)}{\Gamma(b)\Gamma(c-a)} (-x)^{-a} {}_2F_1(a,1+a-c \mid 1+a-b \mid 1/x)$$

$$+ \frac{\Gamma(c)\Gamma(a-b)}{\Gamma(a)\Gamma(c-b)} (-x)^{-b} {}_2F_1(b,1+b-c \mid 1+b-a \mid 1/x), \quad (5.15)$$

which define the analytic continuations in the circle $|1-x| < 1$ and outside the unit circle $|x| = 1$, respectively.

A second solution, linearly independent of the first,[10] is easily found, either by direct Fuchsian integration around the origin or, alternatively, by a Papperitz permutation of the corresponding roots; it is

$$x^{1-c} {}_2F_1(a+1-c, b+1-c \mid 2-c \mid x).$$

This second solution is easily extended in the cut plane by using again formulae (5.14) and (5.15).

Equations with One Thomé Singularity

If we introduce *one* non-Fuchsian singularity, we have, of course, to take its Thomé rank into consideration. It is convenient to place a priori this Thomé singularity at infinity, and to locate the possible Fuchsian singularities in the plane. We write the equation in the standard Thomé form (i.e., no first-order derivative); the coefficient $q(x)$ is then the product of x^{r-2} (where r is the Thomé rank) times a function holomorphic at infinity. Assuming that all other singularities are Fuchsian singularities reduces the arbitrariness of the choice of this function just as in the pure Fuchsian case. For a Thomé singularity of rank r at infinity and $n-1$ Fuchsian singularities located at points a_i in the plane, we get

$$q(x) = P_{r-2}(x) + \sum_{i=1}^{n-1} \left[\frac{B_i}{x-a_i} + \frac{C_i}{(x-a_i)^2} \right], \quad (5.16)$$

where $P_{r-2}(x)$ is some polynomial of degree $r-2$. This equation contains therefore $2n+r-3$ arbitrary parameters, aside, of course, from the places of the Fuchsian singularities.

---

[10]We consider here general values of the parameters a, b, and c; exceptional cases are discussed in books cited in the References.

For example, one meets in quantum mechanics the case of a rank-2 singularity at infinity (corresponding to the energy of the particle), and one Fuchsian singularity at the origin (corresponding to the angular momentum of the particle). If no other singularity exists, the corresponding general equation is

$$z'' + z[A_0 + A_1/x + A_2/x^2] = 0. \qquad (5.17)$$

One can eliminate one of the three parameters A by an ad hoc dilatation: $x \mapsto x' = \alpha x$. The equation then becomes equivalent to the confluent hypergeometric equation[11] with two parameters a and c, which is the standard equation of the case

$$xy'' + (c - x)y' - ay = 0 \qquad (5.18)$$

[the corresponding parameters of (5.17) are $A_0 = -\frac{1}{4}$ (standardized by the dilatation), $A_1 = -a + c/2$, $A_2 = (1 - c/2)c/2$].

One solution of this equation is regular at the origin; it is defined by the confluent hypergeometric series

$$\begin{aligned}
{}_1F_1(a \mid c \mid x) &= 1 + \frac{a}{c}\frac{x}{1!} + \frac{a(a+1)}{c(c+1)}\frac{x^2}{2!} + \cdots \\
&= \frac{\Gamma(c)}{\Gamma(a)} \sum_{k=0}^{\infty} \frac{\Gamma(a+k)}{\Gamma(c+k)} \frac{x^k}{k!},
\end{aligned} \qquad (5.19)$$

and because this series converges in the full complex plane, there is no problem of analytic continuation ("entire" function). Nevertheless, it is interesting to obtain its asymptotic form, which is, of course, some definite linear combination of the Thomé solutions around infinity:

$$\begin{aligned}
{}_1F_1(a \mid c \mid x) &\propto \frac{\Gamma(c)}{\Gamma(c-a)}(-x)^{-a}{}_2F_0(a, a+1-c \parallel -1/x) \\
&+ \frac{\Gamma(c)}{\Gamma(a)}e^x x^{a-c}{}_2F_0(1-a, c-a \parallel 1/x). \qquad (5.20)
\end{aligned}$$

The notation ${}_pF_q$ represents a *formal* generalized hypergeometric series:

$$\begin{aligned}
{}_pF_q(a_1, \ldots, a_p \mid c_1, \ldots, c_q \mid x) \\
= \frac{\Gamma(c_1)\ldots\Gamma(c_q)}{\Gamma(a_1)\ldots\Gamma(a_p)} \sum_{k=0}^{\infty} \frac{\Gamma(a_1+k)\ldots\Gamma(a_p+k)}{\Gamma(c_1+k)\ldots\Gamma(c_q+k)} \frac{x^k}{k!}. \qquad (5.21)
\end{aligned}$$

---

[11]This equation results in the "confluence" of the Fuchsian singularities located at 1 and infinity of the hypergeometric equation (5.11)–(5.12). Technically, this confluence corresponds to the change of variables $x \mapsto x' = bx$, followed by the limit $b \to \infty$.

This series defines an analytic function if $q \geq p - 1$ (entire function if $q > p - 1$, holomorphic function in the circle $|x| < 1$ if $q = p - 1$). The series is only formal, and not numerical (vanishing circle of convergence), if $q < p - 1$. Nevertheless, one can still use this formal series to define new functions if one uses ad hoc summation methods.

For example, one can define the analytic function $_2F_0$, by a Borel transformation of the formal series $_2F_0$. Let us consider the "Borel associated function $B(x; a, b; \alpha)$," defined in the circle $|x| < 1$ by the convergent series

$$B(x; a, b; \alpha) = \frac{1}{\Gamma(a)\Gamma(b)} \sum_{n=0}^{\infty} \frac{\Gamma(a+n)\Gamma(b+n)}{\Gamma(\alpha+n)\Gamma(1+n)} x^n. \qquad (5.22)$$

Let us choose $\alpha = a$ (alternatively, $\alpha = b$), so that the series takes the familiar form of a geometric series; the Borel associated function is then known in the plane, cut from 1 to infinity:

$$B(x; a, b; a) = \frac{1}{\Gamma(a)\Gamma(b)} \sum_{n=0}^{\infty} \frac{\Gamma(b+n)}{\Gamma(1+n)} x^n = \frac{1}{\Gamma(a)}(1-x)^{-b}. \qquad (5.23)$$

We obtain the analytic function $_2F_0(a, b \mid\mid x)$ as the Borel's transform

$$_2F_0(a, b \mid\mid x) = \frac{1}{\Gamma(a)} \int_0^{\infty} e^{-t} t^{a-1} (1 - tx)^{-b} \, dt. \qquad (5.24)$$

One checks easily (either by direct calculation or alternatively by using the Laplace method of integration of linear differential equations) that the analytic functions so defined,

$$x^{-a} {}_2F_0(a, a + 1 - c \mid\mid -1/x) \qquad (5.25)$$

and

$$e^x x^{a-c} {}_2F_0(1 - a, c - a \mid\mid 1/x), \qquad (5.26)$$

are indeed solutions of the confluent hypergeometric equation. This clarifies the meaning of the linear combination of the right-hand side of (5.20).

Using Fuchs's method around the origin, one finds a second solution of the confluent hypergeometric equation, linearly independent of the first solution $_1F_1(a \mid c \mid x)$ above ((5.19)). It is $x^{1-c} {}_1F_1(a + 1 - c \mid 2 - c \mid x)$, and it is defined in the cut plane. Its asymptotic form can be derived from formula (5.20).

6   References

[1] A. Erdélyi, W. Magnus, F. Oberhettinger, F. Tricomi, *Higher Transcendental Functions*, three volumes (McGraw-Hill, New York, 1953).

[2] A.R. Forsyth, *Theory of Differential Equations*, six volumes bound as three (Dover, New York, 1959).

[3] E.L. Ince, *Ordinary Differential Equations* (Longmans, Green, and Co., London and New York, 1926). Reprinted (Dover, New York, 1956).

[4] E.D. Rainville, *Intermediate Differential Equations* (Chelsea, New York, 1972).

[5] B. Spain, M.G. Smith, *Functions of Mathematical Physics* (Van Nostrand Reinhold, London, 1970).

2

Introduction to the Theory of Isomonodromic Deformations of Linear Ordinary Differential Equations with Rational Coefficients

Gilbert Mahoux

ABSTRACT The general solution of a linear ordinary differential equation (ODE) with rational coefficients is generically multivalued. This property is described by a representation of the fundamental homotopy group of the complex plane deprived of the singular points, the *monodromy representation*. The idea of isomonodromic deformations, which traces back to Riemann in the middle of the nineteenth century, is to construct a family of linear ODEs that share a given monodromy representation. This leads to systems of linear partial differential equations, the integrability conditions of which are nonlinear differential equations that enjoy the Painlevé property. It is thus a powerful tool to associate linear with integrable nonlinear equations.

The aim of these lectures is to get a first insight into the problem and to provide explicit algorithms to solve it, starting with linear ODEs that possess regular as well as irregular singularities. The first part is devoted to Schlesinger's theorem, which solves the regular case. As an application, the Lax pair of P_{VI} is constructed. The second part is devoted to the theorems of M. Jimbo, T. Miwa, and K. Ueno, dealing with irregular singularities. The application chosen in that case is the construction of a Lax pair for P_I. Finally, the direct isomonodromy method that exploits these results to solve connection problems is outlined.

1 Introduction

The idea of deforming a linear ordinary differential equation (ODE), while preserving its associated *monodromy representation*, traces back to Riemann in the middle of the nineteenth century. From this idea has now emerged a powerful mathematical tool, used extensively to tackle efficiently various problems, as for example the difficult *connection problems* posed by nonlinear differential equations.

The starting point is a homogeneous linear ODE, say of order N, that is most conveniently written as a set of N coupled linear ODEs of first order:

$$\frac{\mathrm{d}}{\mathrm{d}x}y(x) = A(x)y(x). \tag{1.1}$$

Here $y(x)$ is an N-component vector, and $A(x)$ is an $N \times N$ matrix, *rational* in x. It is convenient to deal with *fundamental matrix solutions* $Y(x)$, which are $N \times N$ matrices built with a given set of N linearly independent vector solutions, which constitute the columns of $Y(x)$. Obviously, $Y(x)$ satisfies the same equation

$$\frac{\mathrm{d}}{\mathrm{d}x}Y(x) = A(x)Y(x). \tag{1.2}$$

The N vector solutions are linearly independent if their *Wronskian*

$$W(Y;x) = \det Y(x)$$

does not vanish identically. Choosing another set of linearly independent vector solutions amounts to multiplying $Y(x)$ on the right by a nonsingular constant matrix.

If $A(x)$ is analytic at $x = x_0$, so is $Y(x)$. The converse is not true![1] Let the poles of $A(x)$ be localized at the points a_ν ($\nu = 1, \ldots, n$) and at $a_\infty = \infty$. All the singularities of $Y(x)$ belong to this set of points. A sufficient but not necessary condition for a singularity to be Fuchsian is that it be a simple pole of $A(x)$. An ODE, or a differential system, is called Fuchsian when all its singularities are Fuchsian. A differential system (1.2) with simple poles (DSSP) is Fuchsian. But a Fuchsian system cannot always be written with only simple poles.

The singularities of $Y(x)$ are generically critical (branch points). To describe the multivaluedness of $Y(x)$, we define the following objects [1]. We call $\mathbb{P}_a^1$ the projective plane deprived of the singular points, $\overline{\mathbb{P}_a^1}$ its universal covering, and $\pi \colon \overline{\mathbb{P}_a^1} \to \mathbb{P}_a^1$ the covering map. $Y(x)$ is single-valued on $\overline{\mathbb{P}_a^1}$.

Let γ be a path in $\overline{\mathbb{P}_a^1}$, starting at the point x and ending at x_γ such that $\pi(x_\gamma) = \pi(x)$ (that is to say, $\pi(\gamma)$ is a closed path in $\mathbb{P}_a^1$). $Y(x_\gamma)$ still satisfies (1.2), thus there exists a nonsingular constant matrix M_γ such that

$$Y(x_\gamma) = Y(x)M_\gamma.$$

M_γ is a function of the homotopy class $[\gamma]$ of the path γ. The mapping $[\gamma] \mapsto M_\gamma$ defines a *representation of the fundamental homotopy group of* $\mathbb{P}_a^1$, the so-called *monodromy representation* associated with the differential system (1.2).

It is known that any representation of the fundamental homotopy group of $\mathbb{P}_a^1$ is the monodromy representation of a *Fuchsian* system. The problem

---

[1] See Appendix A (Section 7).

that is often referred to as the Riemann–Hilbert problem is the following: Is it possible to realize any given representation of the fundamental homotopy group of $\mathbb{P}_a^1$ as the monodromy representation of a system *with only simple poles*? This problem has been definitively settled only recently [2], and the answer is no.[2]

A related problem, the one that will be our concern throughout these notes, is the following: Given a differential system (1.2), is it possible to deform it while preserving its monodromy representation? The answer is that to ensure the isomonodromy of the deformation, $Y(x)$, as a function of *deformation parameters*, has to satisfy a set of linear partial differential equations, or equivalently, $A(x)$, as a function of the same deformation parameters, has to satisfy a set of *completely integrable nonlinear* differential equations. This establishes a deep connection between linear and nonlinear but completely integrable differential equations.

Section 2 is devoted to the Fuchsian case of this *isomonodromic deformation problem*, where the differential system has only simple poles. It was solved at the beginning of the century by Schlesinger [4]. As an application, we treat in Section 3 the deformation problem that leads to P_{VI} [5]. The extension to non-Fuchsian systems is more recent [6,7]. Section 4 is devoted to this extension, and in Section 5, we apply it to a differential system that leads to P_I. We conclude by explaining how one can take advantage of these results to solve connection problems by the so-called *direct isomonodromy method*.

Coming back to the monodromy representation, let us choose for each singular point a_ν a path γ_ν as described above with endpoints x and x_ν, which encircles a_ν once counterclockwise. The generators of the fundamental homotopy group are represented by the *monodromy matrix* M_ν defined by

$$Y(x_\nu) = Y(x)M_\nu \quad (\nu = 1, \ldots, n, \infty).$$

A deformation is isomonodromic if and only if it leaves invariant all the M_ν's.

It is always possible to choose the γ_ν's in such a way that the product $\gamma_1 \cdots \gamma_n \gamma_\infty$ is homotopic to a point (Figure 2.1). Then the following *monodromy constraint* holds:

$$M_1 \cdots M_n M_\infty = 1. \tag{1.3}$$

A general property of $Y(x)$ is that its Wronskian has no zeros in $\overline{\mathbb{P}_a^1}$. Indeed, if ∂ is a first-order differential operator that acts on a matrix M, then one has the useful formula

$$\partial \log \det M = \operatorname{tr}(M^{-1}\partial M). \tag{1.4}$$

---

[2] A clear and complete review of the question can be found in [3]. See Appendix A.

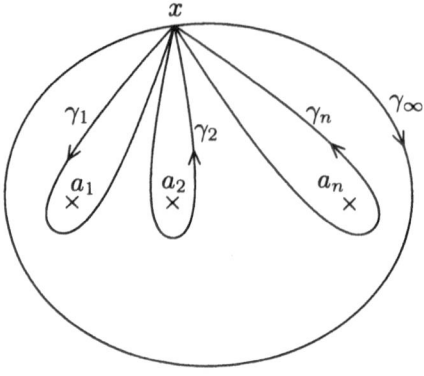

FIGURE 2.1.

Thus

$$\frac{\mathrm{d}}{\mathrm{d}x}\log W(Y;x) = \operatorname{tr} A(x).$$

Let x_0 be a point where the Wronskian does not vanish. Then

$$W(Y;x) = W(Y;x_0)\exp\left(\int_{x_0}^{x}\operatorname{tr} A(z)\,\mathrm{d}z\right).$$

The integral on the right-hand side depends on the integration path from x_0 to x. This formula defines the Wronskian as single-valued and without zeros on $\overline{\mathbb{P}_a^1}$. As a consequence, $Y(x)$ is *invertible* on $\overline{\mathbb{P}_a^1}$.

Note that if $\operatorname{tr} A(x)$ vanishes identically, the Wronskian does not depend on x. In that case $\det M_\nu = 1$ for all ν's, that is to say, the monodromy representation is unimodular.

2   Isomonodromic Deformations of Linear ODEs with Fuchsian Singularities

Let us first define some notation. We write

$$A(x) = \sum_{\nu=1}^{n}\frac{A_\nu}{x - a_\nu},$$

where the A_ν's are given $N \times N$ matrices. If the matrix

$$A_\infty = -\sum_{\nu=1}^{n} A_\nu \tag{2.1}$$

does not vanish, the point $a_\infty = \infty$ is a (Fuchsian) singularity of (1.2). We assume that *all matrices A_ν are diagonalizable* (**Assumption A.2.1**),[3], and choose a set of invertible matrices G_ν such that

$$A_0^{(\nu)} = G_\nu^{-1} A_\nu G_\nu$$

is diagonal for $\nu = 1, \ldots, n, \infty$. Without loss of generality, we assume that A_∞ is diagonal, so that $G_\infty = 1$.

The behavior of $Y(x)$ about its singularities plays a central role in the following. All these singularities being Fuchsian, the singular part of any solution of (1.2) around a_ν is contained in a factor $(x - a_\nu)^{A_\nu}$ or $(1/x)^{A_\infty}$. Indeed, we can define particular solutions $Y^{(\nu)}(x)$, which around a_ν expand as follows:

$$Y^{(\nu)}(x) = G_\nu \left[1 + \sum_{j=1}^{\infty} Y_j^{(\nu)}(x_\nu)^j \right] (x_\nu)^{A_0^{(\nu)}}, \qquad (2.2)$$

where $(x_\nu) = (x - a_\nu)$ for $\nu = 1, \ldots, n$, and $(x_\infty) = 1/x$.

To determine the coefficients $Y_j^{(\nu)}$, we need the expansion of $A(x)$ about a_ν. It is convenient to write

$$A(x) = \begin{cases} G_\nu \sum_{j=0}^{\infty} A_j^{(\nu)}(x - a_\nu)^{j-1} G_\nu^{-1}, & \nu = 1, \ldots, n, \\ -\sum_{j=0}^{\infty} A_j^{(\infty)}(1/x)^{j+1}. \end{cases} \qquad (2.3)$$

Plugging the expansions (2.2) and (2.3) into (1.2) provides us, for $\nu = 1$, $\ldots, n, \infty$, with the equations

$$Y_1^{(\nu)} + [Y_1^{(\nu)}, A_0^{(\nu)}] = A_1^{(\nu)},$$

$$2Y_2^{(\nu)} + [Y_2^{(\nu)}, A_0^{(\nu)}] = A_2^{(\nu)} + A_1^{(\nu)} Y_1^{(\nu)},$$

$$\vdots$$

$$pY_p^{(\nu)} + [Y_p^{(\nu)}, A_0^{(\nu)}] = A_p^{(\nu)} + A_{p-1}^{(\nu)} Y_1^{(\nu)} + \cdots + A_1^{(\nu)} Y_{p-1}^{(\nu)},$$

$$\vdots$$

If solved sequentially, all these equations have the form

$$pY + [Y, A] = B,$$

where A and B are known matrices, A is diagonal, and p is a positive integer. The solution is unique:

$$Y_{ij} = \frac{B_{ij}}{p - A_{ii} + A_{jj}}.$$

---

[3] We shall treat in Section 5 a problem with a nondiagonalizable matrix.

It exists if and only if no denominator $p - A_{ii} + A_{jj}$ vanishes. We are thus led to assume that the diagonal elements of the $A_0^{(\nu)}$'s, that is *the eigenvalues of the A_ν's, for $\nu = 1, \ldots, n, \infty$, are distinct modulo the nonzero integers* (**Assumption A.2.2**). This assumption is a *no logarithm condition*. If it is not satisfied, the expansion (2.2) has to be modified to account for logarithms. Note that degenerate eigenvalues are not excluded by the assumption, which just says that the difference of two eigenvalues is different from $\pm 1, \pm 2, \ldots$.

The series in (2.2) converges for $|x - a_\nu|$ or $|1/x|$ smaller than some *nonzero* radius of convergence ρ_ν or ρ_∞^{-1} (which is determined by the nearest singularity). And inside the circle of convergence, the right-hand side of (2.2) is multivalued.

Any solution of (1.2) can be written $Y^{(\nu)}(x)C_\nu$, with C_ν some invertible constant matrix. We now choose $Y^{(\infty)}(x)$ as our fundamental matrix solution:

$$Y(x) = \begin{cases} Y^{(\infty)}(x) \\ Y^{(\nu)}(x)C_\nu & \nu = 1, \ldots, n. \end{cases} \tag{2.4}$$

This unambiguously defines the *connection matrices* C_ν, as soon as in these equations the determinations of $Y^{(\infty)}(x)$ and $Y^{(\nu)}(x)$ have been chosen.

According to the expansion (2.2), in a transformation $x \mapsto x_\nu$ along a path γ_ν around the singularity a_ν as defined in Section 1, $Y^{(\nu)}(x)$ transforms into

$$Y^{(\nu)}(x_\nu) = Y^{(\nu)}(x)e^{2\pi i A_0^{(\nu)}}. \tag{2.5}$$

This implies that under the same conditions $Y(x)$ becomes

$$Y(x_\nu) = Y(x)M_\nu, \tag{2.6}$$

where the *monodromy matrix* M_ν is given by

$$M_\nu = C_\nu^{-1}e^{2\pi i A_0^{(\nu)}}C_\nu. \tag{2.7}$$

Before stating the problem of isomonodromic deformations of the system (1.2), we introduce further notation. We call *singularity data* (SD), the set of data that specifies the system, that is,

$$SD = \{a_\nu, A_0^{(\nu)}, G_\nu; \nu = 1, \ldots, n\},$$

with the constraint that $\sum_{\nu=1,\ldots,n} G_\nu A_0^{(\nu)} G_\nu^{-1}$ is diagonal. We call *monodromy data* (MD), the set of data that characterize the monodromy properties of the fundamental matrix solution $Y(x)$, that is,

$$MD = \{a_\nu, A_0^{(\nu)}, C_\nu; \nu = 1, \ldots, n, \infty\},$$

where $a_\infty = \infty$, $C_\infty = 1$, and the relation (1.3) with (2.7) holds. These two sets have indeed the same dimension, namely $(n-1)N^2 + (n+1)N + n$.

We shall say that $Y(x)$ *has the monodromy properties MD if*

1. *it is holomorphic and invertible in* $\overline{\mathbb{P}^1_a}$;

2. $Y(x)C_\nu^{-1}\left(\frac{x-a_\nu}{1/x}\right)^{-A_0^{(\nu)}}$ *is holomorphic at* $x = a_\nu$ *for any* ν.

Such a matrix $Y(x)$ satisfies (2.6) with M_ν given by (2.7).

Note that there is some arbitrariness in the definition of SD and MD. First, the diagonal matrices $A_0^{(\nu)}$ are defined up to a permutation of their eigenvalues. Once a choice of the $A_0^{(\nu)}$'s is made, the matrices G_ν and C_ν are defined up to a nonsingular matrix D_ν that commutes with $A_0^{(\nu)}$: $G_\nu \mapsto G_\nu D_\nu$ and $C_\nu \mapsto D_\nu^{-1}C_\nu$.

We have shown that the MD are determined by the SD. The converse is true. More precisely, one has the following result.

Proposition 2.1. *If there exists a matrix* $Y(x)$ *that has the monodromy properties MD, then it is unique. Furthermore, it satisfies a differential equation of the form* (1.2), *with a matrix* $A(x)$ *that is rational and has only simple poles.*

Proof. Assume that $Y_1(x)$ and $Y_2(x)$ have the same MD. Then, since $Y_2(x)$ is invertible in $\overline{\mathbb{P}^1_a}$, $P(x) = Y_1(x)Y_2(x)^{-1}$ is holomorphic in $\overline{\mathbb{P}^1_a}$. Indeed, since $Y_1(x)$ and $Y_2(x)$ have the same singular parts at the $n+1$ points a_ν, $P(x)$ is holomorphic in $\mathbb{P}^1$. It is thus a constant. Furthermore, with the normalization (2.4) at infinity, this constant is 1. Quite similarly, one finds that $(\partial/\partial x)Y(x)Y(x)^{-1}$ is uniform in $\mathbb{P}^1_a$ and that it has simple poles at the a_ν's. $\square$

Corollary 2.1. *The SD are uniquely defined by the MD.*

Proof. It suffices to remark that G_ν is known as soon as $Y(x)$ and the MD are known:

$$G_\nu = \left[Y(x)C_\nu^{-1}(x-a_\nu)^{-A_0^{(\nu)}}\right]_{x=a_\nu} \quad (\nu = 1,\ldots,n).$$ $\square$

We are now in a position to state the problem of the isomonodromic deformations of the (Fuchsian) differential system (1.2): Can we continuously deform the matrix $A(x)$ without modifying the monodromy matrices M_ν? Or equivalently, can we continuously deform the SD while preserving the *partial monodromy data*

$$\text{PMD} = \{A_0^{(\nu)}, C_\nu; \nu = 1,\ldots,n,\infty; C_\infty = 1\}?$$

As a first remark, we immediately infer from the above corollary that the only parameters we are allowed to deform independently are the a_ν's.

Therefore, from now on, $Y(x)$ and the G_ν's become functions of the a_ν's (which we denote collectively by a): $Y(x,a)$ and $G_\nu(a)$. The matrices A_ν through the G_ν's also become functions of a.

We now prove the following result.

Theorem 2.1 (Schlesinger [4]). *The deformations of the system of linear differential equations*

$$\frac{\partial}{\partial x}Y(x,a) = \sum_{\nu=1}^{n}\frac{A_\nu(a)}{x-a_\nu}Y(x,a) \tag{2.8}$$

are isomonodromic if and only if $Y(x,a)$ satisfies the following set of linear PDEs:

$$\frac{\partial}{\partial a_\nu}Y(x,a) = -\frac{A_\nu(a)}{x-a_\nu}Y(x,a) \quad (\nu = 1,\dots,n). \tag{2.9}$$

An equivalent condition is that the $A_\nu(a)$'s satisfy the integrability conditions of (2.8) and (2.9), namely the completely integrable set of nonlinear PDEs

$$\frac{\partial}{\partial a_\mu}A_\nu = \frac{[A_\mu, A_\nu]}{a_\mu - a_\nu} \quad (\mu \neq \nu), \qquad \frac{\partial}{\partial a_\nu}A_\nu = -\sum_{\substack{\mu=1\\\mu\neq\nu}}^{n}\frac{[A_\mu, A_\nu]}{a_\mu - a_\nu}. \tag{2.10}$$

Proof.

1. *Condition (2.9) is necessary:* Let $\eth$ be the exterior differentiation with respect to the a_ν's:

$$\eth = \sum_{\nu=1}^{n}\mathrm{d}a_\nu\frac{\partial}{\partial a_\nu}.$$

Consider the matrix of 1-forms

$$\Omega(x,a) = \big(\eth Y(x,a)\big)Y(x,a)^{-1}.$$

It is holomorphic in the variable x in $\overline{\mathbb{P}^1_a}$. Assume that the PMD are independent of a. Then (2.6) implies

$$\eth Y(x_\nu, a) = \eth Y(x,a)M_\nu,$$

so that $\Omega(x,a)$ is single-valued in $\mathbb{P}^1_a$. To determine the nature of its singularities, we calculate its expansion about a_ν, by using (2.2). Keeping only the divergent terms, we get

$$\Omega(x,a) = \begin{cases} -\dfrac{A_\nu}{x-a_\nu}\mathrm{d}a_\nu + O(1), & \nu = 1,\dots,n, \\ O(1/x), & \nu = \infty. \end{cases} \tag{2.11}$$

$\Omega(x, a)$ is thus meromorphic in x, with simple poles of residue $-A_\nu da_\nu$ at a_ν. Since furthermore it goes to zero at infinity,

$$\Omega(x, a) = -\sum_{\nu=1}^{n} \frac{A_\nu(a)}{x - a_\nu} da_\nu. \tag{2.12}$$

This is (2.9).

2. *Condition* (2.9) *is sufficient:* A straightforward calculation of the integrability conditions of (2.8) and (2.9) leads to the equations (2.10), which can be rewritten in a compact way as

$$\partial A_\nu = \sum_{\substack{\mu=1 \\ \mu \neq \nu}}^{n} [A_\mu, A_\nu] \frac{da_\mu - da_\nu}{a_\mu - a_\nu} \quad (\nu = 1, \ldots, n). \tag{2.13}$$

As for the compatibility conditions of the set of PDE (2.10), they are identically satisfied (check it), that is, the 1-forms ∂A_ν are closed:

$$\partial\partial A_\nu \equiv 0 \quad (\nu = 1, \ldots, n).$$

Equations (2.10) are thus integrable. Let a set of A_ν's be a solution, the a_ν's ranging in some domain $\mathcal{D}$ (we shall assume that conditions A.2.1 and A.2.2 are fulfilled in $\mathcal{D}$, and that there is no collapse of singularities). Then $Y(x, a)$, a solution of (2.8) and (2.9), exists.

So, let $Y(x, a)$ and a set of $A_\nu(a)$'s be a solution of (2.8), (2.9) and (2.10). Let s be a complex number not belonging to the spectrum of A_ν. According to the identity (1.4), one has for $\nu = 1, \ldots, n$,

$$\partial \log \det(A_\nu - s) = \sum_{\substack{\mu=1 \\ \mu \neq \nu}}^{n} \frac{da_\mu - da_\nu}{a_\mu - a_\nu} \, \mathrm{tr}\big((A_\nu - s)^{-1}[A_\mu, A_\nu]\big).$$

The traces on the right-hand side all vanish. Thus $\partial \det(A_\nu - s) = 0$, which means that the spectrum of A_ν does not depend on a, that is,

$$\partial A_0^{(\nu)} = 0 \quad (\nu = 1, \ldots, n). \tag{2.14}$$

Next, from (2.1) and (2.13) we get

$$\partial A_\infty = \sum_{\substack{\mu,\nu=1 \\ \mu \neq \nu}}^{n} [A_\nu, A_\mu] \frac{da_\nu - da_\mu}{a_\nu - a_\mu}.$$

Note that the quantity that is summed up on the right-hand side changes sign when μ and ν are exchanged. Thus $\partial A_\infty = 0$. Now, we can always rewrite (2.8) in such a way that $A_\infty(a)$ is diagonal for any a in $\mathcal{D}$. In that case $G_\infty(a) \equiv 1$, and

$$\partial A_0^{(\infty)} = 0. \tag{2.15}$$

It remains to prove that the C_ν's also do not depend on a. From the relation $A_\nu = G_\nu A_0^{(\nu)} G_\nu^{-1}$, we first get

$$\partial A_\nu = G_\nu [G_\nu^{-1} \partial G_\nu, A_0^{(\nu)}] G_\nu^{-1} \quad (\nu = 1, \ldots, n).$$

Comparing with (2.13), which we rewrite as

$$\partial A_\nu = G_\nu \left[-G_\nu^{-1} \sum_{\substack{\mu=1 \\ \mu \neq \nu}}^{n} A_\mu \frac{da_\nu - da_\mu}{a_\nu - a_\mu} G_\nu, A_0^{(\nu)} \right] G_\nu^{-1},$$

we see that $G_\nu^{-1} \partial G_\nu$ can be written

$$G_\nu^{-1} \partial G_\nu = -G_\nu^{-1} \sum_{\substack{\mu=1 \\ \mu \neq \nu}}^{n} A_\mu \frac{da_\nu - da_\mu}{a_\nu - a_\mu} G_\nu + d_\nu \quad (\nu = 1, \ldots, n),$$

where d_ν is some matrix that commutes with $A_0^{(\nu)}$. But we know that G_ν is defined up to such a matrix. So, we could have expected the presence of this d_ν, which is indeed arbitrary and which we can choose at our convenience. Let it be zero. Then

$$\partial G_\nu G_\nu^{-1} = -\sum_{\substack{\mu=1 \\ \mu \neq \nu}}^{n} A_\mu \frac{da_\nu - da_\mu}{a_\nu - a_\mu} \quad (\nu = 1, \ldots, n). \tag{2.16}$$

Let us now turn back to the expansion of $\Omega(x, a)$ about a_ν, as in (2.11). But this time, we go one order further, and keep all the terms that do not vanish when $x \to a_\nu$. Keeping in mind that C_ν might still depend on a, we get from (2.2), (2.4), and (2.5):

$$\Omega(x, a) = \begin{cases} -G_\nu \left[\dfrac{A_0^{(\nu)}}{x - a_\nu} + A_1^{(\nu)} \right] G_\nu^{-1} da_\nu + Y C_\nu^{-1} \partial C_\nu Y^{-1} \\ \quad + \partial G_\nu G_\nu^{-1} + O(x - a_\nu), \quad \nu = 1, \ldots, n, \\ Y C_\infty^{-1} \partial C_\infty Y^{-1} + O(1/x). \end{cases} \tag{2.17}$$

The matrix $A_1^{(\nu)}$, which appears on the right-hand side, is given by

$$G_\nu A_1^{(\nu)} G_\nu^{-1} = \sum_{\substack{\mu=1 \\ \mu \neq \nu}}^{n} \frac{A_\mu}{a_\nu - a_\mu} da_\mu \quad (\nu = 1, \ldots, n). \tag{2.18}$$

With (2.16) and (2.18), equation (2.17) becomes

$$
\Omega(x,a) = \begin{cases}
-\dfrac{A_\nu}{x - a_\nu}\mathrm{d}a_\nu + \displaystyle\sum_{\substack{\mu=1 \\ \mu\neq\nu}}^{n} \dfrac{A_\mu}{a_\mu - a_\nu}\mathrm{d}a_\mu + YC_\nu^{-1}\eth C_\nu Y^{-1} \\[4mm]
\qquad + O(x - a_\nu), \quad \nu = 1, \ldots, n, \\[3mm]
YC_\infty^{-1}\eth C_\infty Y^{-1} + O(1/x).
\end{cases}
$$

This result is to be compared with (2.12), which provides us with the same expansions of $\Omega(x,a)$ about a_ν and a_∞, but without terms containing $\eth C_\nu$ or $\eth C_\infty$. As a consequence,

$$
\eth C_\nu = 0 \quad (\nu = 1, \ldots, n, \infty).
$$

With (2.14) and (2.15), this last equation expresses the independence in a of the PMD. It ends the proof of Schlesinger's theorem (2.1).   □

Remarks.

1. By a homographic transformation on x, we can fix the position of three singularities. One is already at infinity, so two at finite distance can be fixed. If there are no more singularities at finite distance, no deformation parameter is left. In that case, Schlesinger's theorem becomes trivial (check it!). The simplest nontrivial case corresponds to $n = 3$, and then there remains only one deformation parameter, which is the position of the singularity that cannot be fixed.

2. If for a given ν we change A_ν into $A_\nu + \alpha_\nu \mathbf{1}$, where α_ν is a complex number, the solution $Y(x,a)$ of (2.8) is multiplied by a factor $(x - a_\nu)^{\alpha_\nu}$, $A_0^{(\nu)}$ is changed into $A_0^{(\nu)} + \alpha_\nu \mathbf{1}$, and G_ν and C_ν stay unchanged. This simple transform gives us the freedom to choose at our convenience the trace of A_ν, or one of its eigenvalues. We can even do that for all values of a with a constant α_ν, since the eigenvalues of A_ν do not depend on a. Note that Schlesinger's equations (2.10) are invariant under that transform.

τ *Function*

The 1-form

$$
\omega = \frac{1}{2}\sum_{\substack{\mu,\nu=1 \\ \mu\neq\nu}}^{n} \mathrm{tr}(A_\mu A_\nu)\frac{\mathrm{d}a_\mu - \mathrm{d}a_\nu}{a_\mu - a_\nu}
$$

has the remarkable property of being closed. Let us prove it. Using (2.13), we first get

$$\partial\omega = \sum_{\mu\neq\nu,\lambda\neq\nu} \mathrm{tr}(A_\mu[A_\nu, A_\lambda]) \frac{da_\lambda - da_\nu}{a_\lambda - a_\nu} \wedge \frac{da_\mu - da_\nu}{a_\mu - a_\nu}$$

$$= \sum_{\lambda\neq\mu\neq\nu} \mathrm{tr}(A_\mu A_\nu A_\lambda - A_\lambda A_\nu A_\mu) \frac{(da_\mu \wedge da_\nu + da_\nu \wedge da_\lambda + da_\lambda \wedge da_\mu)}{(a_\mu - a_\nu)(a_\lambda - a_\nu)}.$$

The trace and the numerator of the fraction in this last formula are invariant under a circular permutation of the three indices. So we can replace $1/[(a_\mu - a_\nu)(a_\lambda - a_\nu)]$ by

$$\frac{1}{3}\left(\frac{1}{(a_\mu - a_\nu)(a_\lambda - a_\nu)} + \frac{1}{(a_\nu - a_\lambda)(a_\mu - a_\lambda)} + \frac{1}{(a_\lambda - a_\mu)(a_\nu - a_\mu)} \right).$$

But this quantity vanishes identically. Thus

$$\partial\omega = 0.$$

As a consequence, there exists a function τ of the deformation parameters satisfying

$$\partial \log \tau = \omega.$$

This function plays an important role in the development of the theory. It has been generalized [7] to the case of ODEs with irregular singularities.

3   The Isomonodromic Deformation Problem for Painlevé VI

In this section we elaborate the content of Schlesinger's equations (2.10) in the simplest nontrivial case, namely (1) the A_ν's are 2×2 matrices, and (2) there are three singularities at finite distance [5]. We fix two of them at $x = 0$ and 1, and the last one at $x = t$; t is the unique deformation parameter of the problem.

$$\frac{\partial}{\partial x} Y(x, t) = \left[\frac{A_0(t)}{x} + \frac{A_1(t)}{x-1} + \frac{A_t(t)}{x-t} \right] Y(x, t), \qquad (3.1)$$

$$\frac{\partial}{\partial t} Y(x, t) = -\frac{A_t(t)}{x-t} Y(x, t). \qquad (3.2)$$

To simplify the writing, we denote by $\dot{f}(t)$ and $\ddot{f}(t)$ the first and second derivatives of a function $f(t)$ with respect to t. Equations. (2.10) become

$$\dot{A}_0 = -\frac{1}{t}[A_0, A_t],$$

$$\dot{A}_1 = \frac{1}{1-t}[A_1, A_t],$$

$$\dot{A}_t = \frac{1}{t}[A_0, A_t] - \frac{1}{1-t}[A_1, A_t].$$

We take advantage of (2.1), which can be written as

$$A_0 + A_1 + A_t + A_\infty = 0,$$

to remove A_t from the above equations:

$$\dot{A}_0 = \frac{1}{t}[A_0, A_1 + A_\infty], \tag{3.3}$$

$$\dot{A}_1 = -\frac{1}{1-t}[A_1, A_0 + A_\infty]. \tag{3.4}$$

At this stage we have to parameterize the matrices A_0, A_1, A_t, and A_∞. We choose them (arbitrarily, but without loss of generality) traceless, and call $\pm\theta_0/2$, $\pm\theta_1/2$, $\pm\theta_t/2$, and $\pm\theta_\infty/2$, their eigenvalues. From (3.3) and identity (1.4), we get

$$\frac{\partial}{\partial t}\log\det A_0 = \operatorname{tr} A_0^{-1}\dot{A}_0 = \frac{1}{t}\operatorname{tr} A_0^{-1}[A_0, A_1 + A_\infty] = 0.$$

Thus $\det A_0$, and similarly $\det A_1$ and $\det A_t$, are t-independent. As for the matrix A_∞, it is itself t-independent: $\dot{A}_\infty = 0$. Consequently, the four numbers θ_0, θ_1, θ_t, and θ_∞, are constant parameters. We have thus recovered, in the present particular case of Schlesinger's theorem, the fact that the isomonodromy equations (2.10) entail the t-independence of the matrices $A_0^{(\nu)}$.

Since A_t disappears from the game, we have to keep track of its eigenvalues:

$$\det(A_0 + A_1 + A_\infty) = -\frac{1}{4}\theta_t^2.$$

Note first that if A and B are 2×2 matrices, and one of them at least is traceless, then $\det(A + B) = \det A + \det B - \operatorname{tr} AB$. So we can advantageously replace the above equation by

$$\operatorname{tr}(A_0 A_1 + A_1 A_\infty + A_\infty A_0) = \frac{1}{4}(\theta_t^2 - \theta_0^2 - \theta_1^2 - \theta_\infty^2). \tag{3.5}$$

Second, (3.3) and (3.4) entail

$$\frac{d}{dt}\operatorname{tr}(A_0 A_1 + A_1 A_\infty + A_\infty A_0) = \operatorname{tr}\left[\dot{A}_0(A_1 + A_\infty) + \dot{A}_1(A_0 + A_\infty)\right] = 0.$$

Therefore, (3.5) is nothing but a first integral of the differential system (3.3) and (3.4), and the constant θ_t, which does not appear in this system, plays the role of integration constant.

The diagonal matrix A_∞ takes the form $\begin{pmatrix} \theta_\infty/2 & 0 \\ 0 & -\theta_\infty/2 \end{pmatrix}$. Let us parameterize A_0 and A_1 as follows:

$$A_\nu = \frac{1}{2}\begin{pmatrix} z_\nu & u_\nu(\theta_\nu - z_\nu) \\ \dfrac{\theta_\nu + z_\nu}{u_\nu} & -z_\nu \end{pmatrix} \quad (\nu = 0, 1), \tag{3.6}$$

where z_0, z_1, u_0, and u_1, are functions of t. Our problem now amounts to writing equations (3.3), (3.4), and (3.5) in terms of these four functions.

However, it is convenient to replace u_0 and u_1 by two equivalent linear combinations k and y, defined as follows from the component 12 of $A(x,t)$:

$$A(x,t)_{12} = \frac{k(x-y)}{2x(x-1)(x-t)}, \tag{3.7}$$

where

$$k = tu_0(z_0 - \theta_0) - (1-t)u_1(z_1 - \theta_1), \tag{3.8}$$
$$ky = tu_0(z_0 - \theta_0). \tag{3.9}$$

The function $y(t)$ is indeed the function on which we will finally focus our attention.

Similarly, two linear combinations of z_0 and z_1 appear naturally in the calculations, namely

$$\xi = z_0 + z_1, \tag{3.10}$$
$$\zeta = t(1-y)z_0 + (1-t)yz_1. \tag{3.11}$$

Each matrix equation (3.3), (3.4) provides us with only two scalar equations, because first, both sides are traceless, and second, we have already exploited them to show that θ_0 and θ_1 do not depend on t. Choosing the components 11 and 12 of both equations, we get a linear system for $\dot{z}_0$, $\dot{z}_1$, $\dot{y}$, $\dot{k}$, whose unique solution is

$$\dot{z}_0 = -\frac{Z}{2t}, \tag{3.12}$$

$$\dot{z}_1 = -\frac{Z}{2(1-t)}, \tag{3.13}$$

$$\dot{y} = \frac{(1-\theta_\infty)y(1-y) - \zeta}{t(1-t)}, \tag{3.14}$$

$$\frac{\dot{k}}{k} = (1-\theta_\infty)\frac{(y-t)}{t(1-t)}, \tag{3.15}$$

where

$$Z = -\left[\frac{1}{t(1-y)} + \frac{1}{(1-t)y}\right]\frac{\zeta^2}{y-t} + \frac{2\zeta\xi}{y-t} - \frac{t(1-y)}{(1-t)y}\theta_0^2$$
$$+ \frac{(1-t)y}{t(1-y)}\theta_1^2. \tag{3.16}$$

As for the first integral (3.5), it can be written

$$2\theta_\infty\xi = \theta_t^2 - \theta_\infty^2 - \left[\frac{\theta_0^2}{(1-t)y} - \frac{\theta_1^2}{t(1-y)}\right](y-t)$$
$$- \frac{\zeta^2}{t(1-t)y(1-y)}. \tag{3.17}$$

The elimination of any two of the three functions z_0, z_1, y between (3.12), (3.13), (3.14) leads to a second-order differential equation for the third one. It turns out that among these three equations, the only one that is linear in the second derivative is that for y. Let us build it. As a necessary intermediate result, we differentiate both sides of (3.14) with respect to t:

$$\ddot{y} = \frac{1}{t(1-t)(y-t)} \left[\frac{Z}{2}(y-t) - \zeta\xi - \theta_\infty y(1-y)\xi + (1-\dot{y})\zeta \right.$$

$$\left. + (y-t)\left[(1-\theta_\infty)(1-2y) - (1-2t)\right]\dot{y} \right]. \quad (3.18)$$

By eliminating Z, ζ, and ξ between equations (3.14), (3.16), (3.17), and (3.18), we get the differential equation satisfied by $y(t)$:

$$\ddot{y} = \frac{1}{2}\left(\frac{1}{y} + \frac{1}{y-1} + \frac{1}{y-t} \right)\dot{y}^2 - \left(\frac{1}{t} + \frac{1}{t-1} + \frac{1}{y-t} \right)\dot{y}$$

$$+ \frac{y(y-1)(y-t)}{t^2(t-1)^2}\left[\alpha + \beta\frac{t}{y^2} + \gamma\frac{t-1}{(y-1)^2} + \delta\frac{t(t-1)}{(y-t)^2} \right], \quad (3.19)$$

with

$$\alpha = \frac{(1-\theta_\infty)^2}{2}, \quad \beta = -\frac{\theta_0^2}{2}, \quad \gamma = \frac{\theta_1^2}{2}, \quad \delta = \frac{1-\theta_t^2}{2}.$$

It is the *Painlevé VI equation*.

Thus, solving Riemann's problem in the simplest case has led us to the sixth Painlevé equation [5], *and to an associated Lax pair* (3.1) *and* (3.2), *with an appreciable bonus, the knowledge of the existence of monodromy invariants (the connection matrices C_0, C_1, and C_t).*

Remarks.

1. Let us complete the solution of Schlesinger's equations. Once y is known by integration of (3.19), ζ is calculated by using (3.14), and k by integrating (3.15). Next, if θ_∞ *is not equal to zero*, ξ is given by the first integral (3.17). Once ξ and ζ are known, z_0 and z_1 are easily calculated, and finally, u_0 and u_1 are obtained from (3.8) and (3.9).

   Alternatively, we can write from (3.12), (3.13), and (3.16) the differential equation satisfied by ξ

$$t(1-t)\dot{\xi} + \frac{\zeta\xi}{y-t} = \left[\frac{1}{t(1-y)} + \frac{1}{(1-t)y} \right]\frac{\zeta^2}{2(y-t)} + \frac{t(1-y)}{(1-t)y}\frac{\theta_0^2}{2}$$

$$- \frac{(1-t)y}{t(1-y)}\frac{\theta_1^2}{2}, \quad (3.20)$$

   which can be integrated *regardless of the value of θ_∞*.

2. When θ_∞ vanishes, ξ disappears from (3.17). Then, eliminating ζ between (3.11) and (3.17), we find that y satisfies the following nonlinear *first-order* ODE:

$$[t(1-t)\dot{y} - y(1-y)]^2 + [\theta_0^2 t(1-y) - \theta_1^2(1-t)y](y-t)$$
$$- \theta_t^2 t(1-t)y(1-y) = 0. \quad (3.21)$$

Thus, when $\alpha = \frac{1}{2}$, P$_{VI}$ admits a one-parameter family of solutions [8] that satisfy[4] (3.21).

3. Schlesinger's equations (3.1) and (3.2) have been derived under the assumption that the A_ν's are diagonalizable.[5] Let us now solve them in the general case where this assumption is relaxed. We first note that as long as the eigenvalues of A_ν are different, namely $\theta_\nu \neq 0$, A_ν is diagonalizable. We note also that the parameterization (3.6) of A_0 and A_1 is valid even if these matrices are not diagonalizable. This is not true for A_∞, the assumed diagonal form of which should be replaced by

$$A_\infty = \frac{1}{2} \begin{pmatrix} \theta_\infty & 0 \\ \tau & -\theta_\infty \end{pmatrix},$$

where the parameter τ can be fixed to 1 if $\theta_\infty = 0$, and 0 otherwise (Jordan canonical forms). With this new parameterization of A_∞, a few changes appear in the previous equations, namely, (3.12), (3.13), (3.17), and (3.20) become

$$\dot{z}_0 = -\frac{Z}{2t} - \tau \frac{ky}{2t^2}, \quad (3.22)$$

$$\dot{z}_1 = -\frac{Z}{2(1-t)} - \tau \frac{k(1-y)}{2(1-t)^2}, \quad (3.23)$$

---

[4]The differential equation (3.21) can be solved explicitly. Indeed, when θ_∞ vanishes, (3.1) no longer has a singularity at infinity, and it becomes a hypergeometric system with three singularities at 0, 1, and t. Its matrix solution $Y(x,t)$, as well as the matrices $A_\nu(t)$ and thus also $y(t)$, can be expressed in terms of hypergeometric functions. One finds that the general solution of (3.21) takes the form

$$y(\theta_0, \theta_1, \theta_t; t) = \frac{u(abu + ct\dot{u})}{abu^2 + [c - (a+b)(1-t)]u\dot{u} - t(1-t)\dot{u}^2},$$

where $u(t) = u(a, b; c; t)$ is a solution of the hypergeometric equation with parameters a, b, and c (usual notations) related to the θ's by

$$\theta_0 = c, \quad \theta_1 = a + b - c, \quad \theta_t = a - b.$$

Obviously, (3.21) and the hypergeometric equation are related by a birational transform [9].

[5]The fearless reader is encouraged to investigate their validity when Assumption A.2.1 is not fulfilled.

$$2\theta_\infty \xi = \theta_t^2 - \theta_\infty^2 - \left[\frac{\theta_0^2}{(1-t)y} - \frac{\theta_1^2}{t(1-y)} \right](y-t)$$

$$- \frac{\zeta^2}{t(1-t)y(1-y)} + \tau \frac{k(y-t)}{t(1-t)}, \tag{3.24}$$

$$t(1-t)\dot{\xi} + \frac{\zeta\xi}{y-t} = \left[\frac{1}{t(1-y)} + \frac{1}{(1-t)y} \right] \frac{\zeta^2}{2(y-t)} + \frac{t(1-y)}{(1-t)y} \frac{\theta_0^2}{2}$$

$$- \frac{(1-t)y}{t(1-y)} \frac{\theta_1^2}{2} - \frac{\tau k}{2} \left[\frac{(1-t)y}{t} + \frac{t(1-y)}{1-t} \right]. \tag{3.25}$$

Of course, $y(t)$ still satisfies the P$_{VI}$ equation (3.19).

Now we can complete the solution of (3.1) and (3.2) in the case where $\theta_\infty = 0$ and $\tau \neq 0$. Once y is known by integration of (3.19), and ζ by using (3.14), k is given by eq. (3.24). Next, ξ is obtained by integrating (3.25). Finally, z_0, z_1, u_0, and u_1 are easily calculated.

Lax Pairs for Painlevé Equations I to V

It is well known that the first five Painlevé equations can be constructed from the sixth one by successive confluences of singularities. This procedure enables us to extend the above results to equations P$_I$ to P$_V$, and in particular to construct their Lax pairs from that of P$_{VI}$. Let us do this for P$_V$. The substitution [10]

$$(t, y, \alpha, \beta, \gamma, \delta) \mapsto (1 + \varepsilon t, y, \alpha, \beta, \varepsilon^{-1}\gamma - \varepsilon^{-2}\delta, \varepsilon^{-2}\delta), \tag{3.26}$$

followed by the limit $\varepsilon \to 0$, transforms (3.19) into P$_V$:

$$\ddot{y} = \left(\frac{1}{2y} + \frac{1}{y-1} \right)\dot{y}^2 - \frac{\dot{y}}{t} + \frac{(y-1)^2}{t^2} \left(\alpha y + \frac{\beta}{y} \right) + \gamma \frac{y}{t} + \delta \frac{y(y+1)}{y-1}.$$

To obtain its Lax pair from (3.1) and (3.2), we need the extension to the A_ν's of the above substitution rule. A careful analysis of equations. (3.12) to (3.20), which solve Schlesinger's equations, leads to the following extension of the substitution rule (3.26):

$$(z_0, u_0, \theta_0, z_1, u_1, \theta_1, \theta_\infty) \to (z_0, u_0, \theta_0, \varepsilon^{-1}z_1, \varepsilon^{-1}u_1, \varepsilon^{-1}\theta_1, \theta_\infty).$$

It implies for the A_ν's

$$(A_0, A_1, A_t, A_\infty) \mapsto (A_0, \varepsilon^{-1}A_1, -A_0 - A_\infty - \varepsilon^{-1}A_1, A_\infty).$$

This last substitution, with the variable x kept unchanged, transforms (3.1) and (3.2) into

$$\frac{\partial Y}{\partial x} = \left[\frac{A_0}{x} - \frac{A_0 + A_\infty}{x-1} - \frac{tA_1}{(x-1)^2} \right]Y,$$

$$\frac{\partial Y}{\partial t} = \frac{A_1}{x-1}Y. \tag{3.27}$$

This is a Lax pair for P_V. The interesting point is the appearance of a *double pole* in the first equation. Multiple poles, that is irregular singularities, are present also in the Lax pairs of the other Painlevé equations, P_I to P_{IV}.

In the process of singularity confluence that just led us to the Lax pair of P_V, a question still remains: What are the monodromy invariants associated with P_V? To answer this, we could look at the limits in the above process of the monodromy invariants associated with P_{VI}, namely the connection matrices C_0, C_1, and C_t. This is indeed a difficult task. A better way to answer the question is to solve directly the isomonodromic deformations problem of the linear ODE (3.27), the first equation of the Lax pair for P_V. But to be able to do this, we first have to learn how to handle *irregular singularities of finite rank* (Thomé singularities). That is the object of the next section.

4   Isomonodromic Deformations of Linear ODEs with Thomé Singularities

The $N \times N$ matrix $A(x)$, rational in x, has now multiple poles:

$$A(x) = \sum_{\nu=1}^{n} \sum_{k=0}^{r_\nu} \frac{A_{\nu,k}}{(x - a_\nu)^{k+1}} + \sum_{k=1}^{r_\infty} A_{\infty,k} x^{k-1}.$$

The nonnegative integer r_ν is the *Poincaré rank* of the singularity at $x = a_\nu$ ($\nu = 1, \ldots, n, \infty$).

The coefficient of the leading term at $x = a_\nu$ is the matrix A_{ν,r_ν}. We assume that *all matrices A_{ν,r_ν} are diagonalizable* (Assumption A.4.1), and choose a set of invertible matrices G_ν such that

$$A_{\nu,r_\nu} = G_\nu A_{-r_\nu}^{(\nu)} G_\nu^{-1} \quad (\nu = 1, \ldots, n, \infty)$$

with $A_{-r_\nu}^{(\nu)}$ diagonal. Without loss of generality, we assume that A_{∞,r_∞} is itself diagonal, so that $G_\infty = \mathbf{1}$.

With Assumption A.4.1, if $r_\nu > 0$, the singularity is irregular.

To simplify the writing, we introduce the following notation:

$$\xi_\nu = \begin{cases} x - a_\nu & \nu = 1, \ldots, n \\ 1/x, & \nu = \infty. \end{cases}$$

4.1   Formal Expansion Around an Irregular Singularity

The generalization of (2.2) to irregular singularities of finite rank is the *Thomé asymptotic expansion*:

$$\widetilde{Y}^{(\nu)}(x) = G_\nu \left(1 + \sum_{j=1}^{\infty} Y_j^{(\nu)} \xi_\nu^j \right) \exp \left(\sum_{j=-r_\nu}^{-1} \frac{1}{j} T_j^{(\nu)} \xi_\nu^j + T_0^{(\nu)} \log(\xi_\nu) \right), \qquad (4.1)$$

where the matrices $T_j^{(\nu)}$ are diagonal.

$\widetilde{Y}^{(\nu)}(x)$ is a *formal* solution of (1.2), formal in the sense that the series is asymptotic and generically does not converge. Therefore, $\widetilde{Y}^{(\nu)}(x)$ *does not define a solution of* (1.2), as in the Fuchsian case. We shall see in a moment the use to which it can be put.

We first give an alternative form for $\widetilde{Y}^{(\nu)}(x)$, more convenient for the calculation of its coefficients:

$$\widetilde{Y}^{(\nu)}(x) = G_\nu \left(1 + \sum_{j=1}^{\infty} U_j^{(\nu)} \xi_\nu^j \right)$$

$$\times \exp \left(\sum_{j=-r_\nu}^{-1} \frac{1}{j} T_j^{(\nu)} \xi_\nu^j + T_0^{(\nu)} \log(\xi_\nu) + \sum_{j=1}^{\infty} \frac{1}{j} T_j^{(\nu)} \xi_\nu^j \right). \qquad (4.2)$$

It differs from (4.1) by an infinite series in the exponential. With such a form, the matrices $T_j^{(\nu)}$, now infinite in number, are diagonal, and the matrices $U_j^{(\nu)}$ are diagonal-free.[6]

The expansion of $A(x)$ about a_ν is conveniently written as follows (compare with (2.3)):

$$A(x) = G_\nu \frac{\mathrm{d}\xi_\nu}{\mathrm{d}x} \sum_{j=-r_\nu}^{\infty} A_j^{(\nu)} \xi_\nu^{j-1} G_\nu^{-1} \quad (\nu = 1, \ldots, n, \infty). \qquad (4.3)$$

Expressing that (4.2) satisfies equation (1.2), we get (to simplify the writing, we have removed everywhere the index ν)

$$\left(1 + \sum_{j=1}^{\infty} U_j \xi^j \right) \left(\sum_{i=-r}^{\infty} T_i \xi^{i-1} \right) - \left(\sum_{i=-r}^{\infty} A_i \xi^{i-1} \right) \left(1 + \sum_{j=1}^{\infty} U_j \xi^j \right)$$

$$+ \sum_{j=1}^{\infty} j U_j \xi^{j-1} = 0. \qquad (4.4)$$

This gives the set of equations

$$T_{-r} = A_{-r},$$
$$T_{-r+1} + [U_1, A_{-r}] = A_{-r+1},$$
$$T_{-r+2} + [U_2, A_{-r}] = A_{-r+2} + (A_{-r+1}U_1 - U_1 T_{-r+1}),$$

---

[6]We call a matrix diagonal-free if its the diagonal elements all vanish.

$$\cdots$$

$$T_{-r+k} + [U_k, A_{-r}] = A_{-r+k} + \sum_{j=1}^{k-1} (A_{-r+k-j}U_j - U_j T_{-r+k-j})$$

$$- \begin{cases} 0, & k = 0, 1, \ldots, r, \\ (k-r)U_{k-r} & k = r+1, \ldots, \infty, \end{cases}$$

$$\cdots$$

If r is strictly positive, all these equations but the first can be written

$$T + [U, A] = B,$$

where A and B are known matrices, and A is diagonal. We look for a solution with T diagonal and U diagonal free. The solution is then unique:

$$T_{ij} = \begin{cases} B_{ii}, & i = j \\ 0, & i \neq j, \end{cases} \quad U_{ij} = \begin{cases} 0, & i = j, \\ \dfrac{B_{ij}}{A_{jj} - A_{ii}} & i \neq j. \end{cases}$$

It exists if and only if no denominator $A_{jj} - A_{ii}$ vanishes, which leads us to generalize Assumption A.2.2 as follows: *If $r_\nu > 0$, the eigenvalues of A_{ν,r_ν} are distinct, and if $r_\nu = 0$, they are distinct modulo the nonzero integers* (**Assumption A.4.2**).

Once formula (4.2) is obtained, it is easily transformed into (4.1) by rearranging the following product of series

$$\left(1 + \sum_{j=1}^{\infty} U_j \xi^j \right) \exp\left(\sum_{j=1}^{\infty} \frac{1}{j} T_j \xi^j \right) = \left(1 + \sum_{j=1}^{\infty} Y_j \xi^j \right).$$

An interesting relation can be obtained from equation (4.4). Multiplying both sides by $(1 + \sum_{j=1}^{\infty} U_j \xi^j)^{-1}$ and taking the trace, we get

$$\operatorname{tr} \sum_{j=-r}^{\infty} (T_j - A_j)\xi^j = -\operatorname{tr}\left(1 + \sum_{j=1}^{\infty} U_j \xi^j \right)^{-1} \left(\sum_{j=1}^{\infty} j U_j \xi^{j-1} \right).$$

The right-hand side is equal to $\operatorname{tr}(-U_1 + O(\xi))$. It is thus of order of ξ. As a consequence,

$$\operatorname{tr} T_{-k}^{(\nu)} = \operatorname{tr} A_{-k}^{(\nu)} \quad (\nu = 1, \ldots, n, \infty; k = 0, 1, \ldots, r_\nu).$$

Note that $A_0^{(\infty)}$, the coefficient of $-1/x$ in the expansion of $A(x)$ at infinity, is given by

$$A_0^{(\infty)} = -\sum_{\nu=1}^{n} G_\nu A_0^{(\nu)} G_\nu^{-1}. \tag{4.5}$$

From the last two equations we get the Fuchs relation

$$\sum_{\nu=1,\dots,n,\infty} \operatorname{tr} T_0^{(\nu)} y = 0. \tag{4.6}$$

Remarks.

1. In equation (4.1), the singular part of $\widetilde{Y}^{(\nu)}(x)$ is entirely contained in the argument of the exponential. The fact that this argument is a diagonal matrix is essential: It allows an easy calculation of derivatives. Its significance is indeed clear. Thanks to it, all the elements of a particular column of the matrix $\widetilde{Y}^{(\nu)}(x)$ contain the same singular exponential factor. Thus, $\widetilde{Y}^{(\nu)}(x)$ corresponds to a choice of particular independent formal vector solutions, such that each one is characterized by one and only one singular exponential factor.

2. The logarithm in the argument of the exponential is responsible for the multivaluedness of $\widetilde{Y}^{(\nu)}(x)$. It was already present in (2.2), although written differently. Its coefficient $T_0^{(\nu)}$, which controls the branching of $\widetilde{Y}^{(\nu)}(x)$, is called the *exponent of formal monodromy* in [7]. The rest of the argument, a polynomial in the variable $1/\xi_\nu$, is new. It is responsible for the exponential growth of $\widetilde{Y}^{(\nu)}(x)$ near the singularity, and for the so-called *Stokes phenomenon* [11,12].

3. We had no problem in Section 2 differentiating the convergent series (2.2). Here, we have to deal with asymptotic series that are generically divergent. It has to be known that it is permitted to differentiate termwise an asymptotic series with respect to its expansion variable. To do this with respect to a parameter (by using a generalization of the operator ∂), the asymptotic series has to be *uniform* in this parameter.[7]

4.2 The Stokes Phenomenon and Stokes Multipliers

In order to simplify the discussion of the Stokes phenomenon, we restrict ourselves to the case of a second-order ODE, so that we deal with 2×2 matrices. We call ω_1 and ω_2 the eigenvalues of A_{ν,r_ν}. We also remove everywhere the index ν.

Let us call $\tilde{y}_1(x)$ and $\tilde{y}_2(x)$ the columns of $\widetilde{Y}(x)$: $\widetilde{Y} = \{\tilde{y}_1, \tilde{y}_2\}$. They are formal vector solutions of (1.1), and their dominant singular terms are $\exp(\omega_1 \xi^{-r})$ and $\exp(\omega_2 \xi^{-r})$, respectively. When $|\xi| \to 0$, one solution dominates the other according to the sign of the real part of $(\omega_1 - \omega_2)\xi^{-r}$. This leads us to the following geometrical construction. Let $\mathcal{L}_\ell$ (ℓ integer), be the rays in $\overline{\mathbb{P}^1_a}$ around the singular point, the so-called *Stokes lines*,

---

[7]See Appendix B (Section 8).

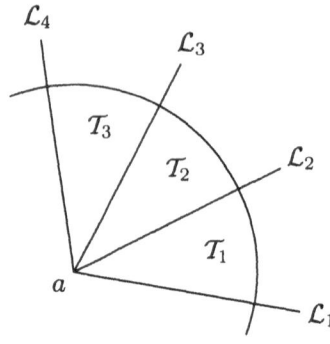

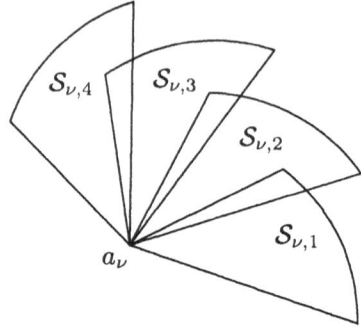

FIGURE 2.2.   FIGURE 2.3.

where the real part of $(\omega_1 - \omega_2)\xi^{-r}$ vanishes

$$\mathcal{L}_\ell = \left\{ x \in \overline{\mathbb{P}^1_a}; \operatorname{phase}(\xi) = \varphi + \frac{\pi}{2r} + \frac{\ell\pi}{r} \right\} \quad (\ell \text{ integer})$$

($\varphi = \operatorname{phase}(\omega_1 - \omega_2)/r$). Let $\mathcal{V}$ be a neighborhood of the singular point, minus the singular point itself. Its inverse image by the covering map, $\overline{\mathcal{V}} = \pi^{-1}(\mathcal{V})$, is divided by the Stokes lines into sectors $\mathcal{T}_\ell$ (Figure 2.2):

$$\mathcal{T}_\ell = \left\{ x \in \overline{\mathcal{V}}; \varphi - \frac{\pi}{2r} + \frac{\ell\pi}{r} < \operatorname{phase}(\xi) < \varphi + \frac{\pi}{2r} + \frac{\ell\pi}{r} \right\}.$$

Inside each sector, when $|\xi| \to 0$, one of the above formal vector solutions dominates the other exponentially. We say that it is dominant, and that the other is recessive. When one crosses a Stokes line, dominant and recessive solutions exchange.

Let $Y(x) = \{y_1(x), y_2(x)\}$ be a matrix solution (not formal this time), constructed with two vector solutions $y_1(x)$ and $y_2(x)$. Imposing that it behaves like $\widetilde{Y}(x)$ in a particular sector $\mathcal{T}_\ell$ does not determine it completely: One can always add to the dominant vector solution a recessive component, this does not change its asymptotic behavior. To be more specific, assume that y_1 is dominant, and thus y_2 recessive, in the sector $\mathcal{T}_0$. If $Y(x)$ behaves like $\widetilde{Y}(x)$ in $\mathcal{T}_0$, so does $Y(x) \left(\begin{smallmatrix} 1 & 0 \\ \alpha & 1 \end{smallmatrix} \right)$, for any complex α. Similarly, if $Y(x)$ behaves like $\widetilde{Y}(x)$ in the next sector $\mathcal{T}_1$, so does $Y(x) \left(\begin{smallmatrix} 1 & \beta \\ 0 & 1 \end{smallmatrix} \right)$, for any complex β. One proves [11] that there exists a solution $Y(x)$ that behaves like $\widetilde{Y}(x)$ in both sectors $\mathcal{T}_0$ and $\mathcal{T}_1$. Obviously, it is *unique*.

This construction of completely defined solutions is general. Given two contiguous sectors $\mathcal{T}_{\ell-1}$ and $\mathcal{T}_\ell$, let us call $\mathcal{S}_\ell$ their union. More precisely, $\mathcal{S}_\ell = \mathcal{T}_{\ell-1} \cup \mathcal{T}_\ell \cup (\mathcal{L}_\ell \cap \overline{\mathcal{V}})$. Then there exists a unique $Y^{(\ell)}(x)$ such that[8]

$$Y^{(\ell)}(x) \sim \widetilde{Y}(x) \quad \text{in } \mathcal{S}_\ell.$$

---

[8]The symbol $\sim$ means here "behaves asymptotically like."

Since $Y^{(\ell)}$ and $Y^{(\ell+1)}$ have the same asymptotic expansion in $\mathcal{T}_\ell$, the constant matrix S_ℓ that connects them and is defined by

$$Y^{(\ell+1)} = Y^{(\ell)} S_\ell \tag{4.7}$$

has the form $\left(\begin{smallmatrix} 1 & 0 \\ \alpha_\ell & 1 \end{smallmatrix}\right)$ or $\left(\begin{smallmatrix} 1 & \beta_\ell \\ 0 & 1 \end{smallmatrix}\right)$, and this alternatively from one sector to the next one. The sectors $\mathcal{S}_\ell$ are called *Stokes sectors*, and the matrices S_ℓ *Stokes multipliers*.

A striking consequence of these results is that *the asymptotic expansion of a given solution in a given Stokes sector varies from sector to sector.* For example,

$$Y^{(0)}(x) \sim \widetilde{Y}(x) S_\ell^{-1} S_{\ell-1}^{-1} \cdots S_1^{-1} \quad \text{in } \mathcal{S}_\ell.$$

This is the *Stokes phenomenon*.

We note that $\pi(\mathcal{S}_\ell) = \pi(\mathcal{S}_{\ell+2r})$. Thus, to any point x in $\mathcal{S}_1$ corresponds one point x_+ in $\mathcal{S}_{2r+1}$ with the same projection $\pi(x) = \pi(x_+)$. Obviously,

$$\widetilde{Y}(x_+) = \widetilde{Y}(x) e^{2\pi i T_0}.$$

Since $Y^{(1)}(x) \sim \widetilde{Y}(x)$ and $Y^{(2r+1)}(x_+) \sim \widetilde{Y}(x_+)$, one has

$$Y^{(2r+1)}(x_+) = Y^{(1)}(x) e^{2\pi i T_0}.$$

From this formula and (4.7) we get

$$Y^{(1)}(x_+) = Y^{(1)}(x) e^{2\pi i T_0} S_{2r}^{-1} \cdots S_1^{-1}. \tag{4.8}$$

This is to be compared with (2.5).

These results can be extended to the general case where the dimension N of the matrices is not restricted to 2. Around each singularity $x = a_\nu$ ($\nu = 1, \ldots, n, \infty$), one constructs on $\overline{\mathbb{P}_a^1}$ a set of Stokes sectors $\mathcal{S}_{\nu,\ell}$ with the following properties (Figure 2.3)

1. The intersection $\mathcal{S}_{\nu,\ell} \cap \mathcal{S}_{\nu,\ell'}$ is nonempty if and only if $|\ell - \ell'| = 1$.

2. $\pi(\mathcal{S}_{\nu,\ell}) = \pi(\mathcal{S}_{\nu,\ell+2r_\nu})$.

3. $\pi(\cup_{\ell=1,\ldots,2r_\nu} \mathcal{S}_{\nu,\ell})$ is a neighborhood of the singular point $x = a_\nu$, minus the point itself.

A difference with the case $N = 2$ is that the opening angle θ_ν of the Stokes sectors $\mathcal{S}_{\nu,\ell}$ has to be chosen in general smaller than $2\pi/r_\nu$ (indeed, $\pi/r_\nu < \theta_\nu \leq 2\pi/r_\nu$) in order to have the following result.

Proposition 4.1. *There exists a unique solution $Y^{(\nu,\ell)}(x)$ of (1.2), holomorphic and invertible in $\overline{\mathbb{P}_a^1}$, such that*

$$Y^{(\nu,\ell)}(x) \sim \widetilde{Y}^{(\nu)}(x) \text{ in } \mathcal{S}_{\nu,\ell} \quad (\nu = 1, \ldots, n, \infty; \ \ell \text{ integer}).$$

The Stokes multipliers $S_{\nu,\ell}$ are defined by

$$Y^{(\nu,\ell+1)} = Y^{(\nu,\ell)} S_{\nu,\ell}.$$

They can be made triangular matrices with all diagonal elements equal to 1 by a permutation of the lines and the same permutation of the columns. This permutation, which depends only of ν, is the one that sorts the real parts of $\omega_{\nu,i}\xi^{-r}$ for any $\xi \in S_\ell \cap S_{\ell+1}$, where $\omega_{\nu,i}$ is the ith diagonal element of $A_{-r_\nu}^{(\nu)}$. As a consequence, the Stokes multipliers are unimodular matrices:

$$\det(S_{\nu,\ell}) = 1.$$

Equation (4.8) can now be written

$$Y^{(\nu,1)}(x_\nu) = Y^{(\nu,1)}(x) e^{2\pi i T_0^{(\nu)}} S_{2r_\nu}^{-1} \cdots S_1^{-1},$$

where x and x_ν are as explained in the Section 1.

To define connection matrices as in (2.4), we have to distinguish, for each ν, one particular solution $Y^{(\nu,\ell)}(x)$. We choose $Y^{(\infty,1)}(x)$ as our fundamental matrix solution, and write

$$Y(x) = Y^{(\infty,1)}(x) \tag{4.9}$$
$$= Y^{(\nu,1)}(x) C_\nu, \quad (\nu = 1, \ldots, n). \tag{4.10}$$

Defining the monodromy matrices as in (2.6), we get from the last three equations

$$M_\nu = C_\nu^{-1} e^{2\pi i T_0^{(\nu)}} S_{2r_\nu}^{-1} \cdots S_1^{-1} C_\nu. \tag{4.11}$$

This is to be compared with (2.7). Note that because of the unimodularity of the Stokes multipliers,

$$\det(M_\nu) = e^{2\pi i \operatorname{tr} T_0^{(\nu)}}.$$

Then, the Fuchs relation (4.6) entails

$$\det(M_1 \cdots M_n M_\infty) = 1.$$

As a consequence, the matrix monodromy constraint (1.3) is equivalent to $N^2 - 1$ scalar constraints.

4.3 Singularity Data, Monodromy Data, and Deformation Parameters

The singularity data SD, which specify the differential system, are

$$SD = \{a_\nu, A_{-k_\nu}^{(\nu)}, G_\nu; \nu = 1, \ldots, n, \infty; k_\nu = 0, 1, \ldots, r_\nu\},$$

$$a_\infty = \infty, G_\infty = 1\},$$

where the matrices $A_{-r_\nu}^{(\nu)}$ are diagonal, and the relation (4.5) holds.

The monodromy data MD, which characterize the monodromy properties of the fundamental matrix solution $Y(x)$, are

$$\mathrm{MD} = \{a_\nu,\; T_{-k_\nu}^{(\nu)},\; S_{\nu,\ell_\nu},\; C_\nu;\; \nu = 1,\ldots,n,\infty;\; k_\nu = 0,\ldots,r_\nu;$$

$$\ell_\nu = 1,\ldots,2r_\nu;\; a_\infty = \infty,\; C_\infty = 1\},\quad (4.12)$$

where (1) the matrices $T_{-k_\nu}^{(\nu)}$ are diagonal, (2) modulo a convenient re-ordering of their lines and columns, the matrices S_{ν,ℓ_ν} are triangular with diagonal elements equal to 1, and (3) the Fuchs relation (4.6) and the monodromy constraint (1.3) hold.

A careful counting shows that these two sets have the same dimension, namely $(n+r-1)N^2 + (n+1)N + n$, where $r = \sum_{\nu=1,\ldots,n,\infty} r_\nu$.

We shall say that $Y(x)$ *has the monodromy properties MD* if

1. *it is holomorphic and invertible in* $\overline{\mathbb{P}_a^1}$,

2. *there are invertible matrices* G_ν *($\nu = 1,\ldots,n$) and sectors* $S_{\nu,\ell}$ *as described previously such that*

$$Y(x)C_\nu^{-1}S_{\nu,1}\cdots S_{\nu,\ell-1}$$

$$\sim G_\nu\left(1 + O(\xi)\right)\exp\left(\sum_{j=-r_\nu}^{-1}\frac{1}{j}T_j^{(\nu)}\xi_\nu^j + T_0^{(\nu)}\log(\xi_\nu)\right)\quad (4.13)$$

in $S_{\nu,\ell}$, *for* $\nu = 1,\ldots,n,\infty$, *and* $\ell = 1,\ldots,2r_\nu+1$, *with* $G_\infty = 1$.

A matrix $Y(x)$ that enjoys these properties satisfies (2.6), with a monodromy matrix M_ν given by (4.11).

As in the Fuchsian case, there is here some arbitrariness in the definition of singularity and monodromy data. In particular, G_ν, C_ν, and $S_{\nu,\ell}$ can be changed into $G_\nu D_\nu$, $D_\nu^{-1}C_\nu$ and $D_\nu^{-1}S_{\nu,\ell}D_\nu$, where D_ν is a nonsingular matrix that commutes with $T_0^{(\nu)}$.

Proposition 2.1 and its corollary 2.1 extend without difficulty to the present situation, with the only difference that $A(x)$ has multiple poles. Thus, as long as Assumptions A.4.1 and A.4.2 are satisfied, *the two sets SD and MD are homeomorphic*. In the course of the proof, one finds that $Y_1(x)Y_2(x)^{-1}$ has the same asymptotic expansion in all sectors $T_{\nu,\ell}$ around the singularity. As a consequence, the expansion *does converge*.

From the expression (4.11) of the monodromy matrices we define the set of partial monodromy data that will be preserved in isomonodromic deformations as containing (1) the exponents of formal monodromy $T_0^{(\nu)}$, (2) the Stokes multipliers $S_{\nu,\ell}$, and (3) the connection matrices C_ν:

$$\mathrm{PMD} = \{T_0^{(\nu)}, S_{\nu,\ell_\nu}, C_\nu : \nu = 1,\ldots,n,\infty; \ell_\nu = 1,\ldots,2r_\nu; C_\infty = 1\}.$$

The remaining partial monodromy data constitute the set of *deformation parameters* (DP):

$$\mathrm{DP} = \{a_\nu, T^{(\nu)}_{-k_\nu} : \nu = 1, \ldots, n, \infty; k_\nu = 1, \ldots, r_\nu; a_\infty = \infty\}.$$

The dimension of this last set is $n + rN$.

Considering the SD as independent variables and the MD as dependent variables, the isomonodromy problem is formulated as follows: Can we continuously deform the SD while preserving the PMD? Alternatively, let us consider the MD as independent variables, and the SD as dependent variables. The formulation of the problem becomes: What are the deformations of the SD under any continuous variation of the DP, the PMD being kept fixed?

From now on we shall write an explicit dependence in t, like $Y(x,t)$, $G_\nu(t) \ldots$, to indicate that Y, G_ν, ... depend on the DP, which we denote collectively by t. As in Section 2, we introduce the exterior differentiation with respect to the $n + rN$ deformation parameters:[9]

$$\eth = \sum_{\nu=1}^{n} da_\nu \frac{\partial}{\partial a_\nu} + \sum_{\nu=1,\ldots,n,\infty} \sum_{k_\nu=1}^{r_\nu} dT^{(\nu)}_{-k_\nu} \frac{\partial}{\partial T^{(\nu)}_{-k_\nu}}.$$

We then have the following theorem.

Theorem 4.1 (M. Jimbo et al. [7]).   *Under Assumptions A.4.1 and A.4.2, the deformations of the linear differential equation*

$$\frac{\partial}{\partial x} Y(x,t) = \left(\sum_{\nu=1}^{n} \sum_{k=0}^{r_\nu} \frac{A_{\nu,k}(t)}{(x - a_\nu)^{k+1}} + \sum_{k=1}^{r_\infty} A_{\infty,k}(t) x^{k-1} \right) Y(x,t) \quad (4.14)$$

are isomonodromic if and only if $Y(x,t)$ and $G_\nu(t)$ satisfy the total differential equations

$$\eth Y(x,t) = \Omega(x,t) Y(x,t) \tag{4.15}$$

and

$$\eth G_\nu(t) = \Theta_\nu(t) G_\nu(t), \tag{4.16}$$

where the 1-form $\Omega(x,t)$, rational in x, is given by (4.20), (4.18), and (4.21), and the 1-form $\Theta_\nu(t)$ is given by (4.22).

Proof. We shall be content here to construct $\Omega(x,t)$ and $\Theta(t)$, and to prove that (4.15) and (4.16) are necessary conditions. We refer the reader to [7] for the proof that they are sufficient.

---

[9]If M and N are two matrices of same dimension, we use the compact notation $M\partial/\partial N$ for the differential operator $\sum_{i,j} M_{ij} \partial/\partial N_{ij}$.

Let $Y(x,t)$ be the matrix solution of (4.14), as defined in (4.9) and (4.10), for t ranging in some open set $\mathcal{D}$ of the space of the DP. Assume that for t in $\mathcal{D}$, conditions A.4.1 and A.4.2 are satisfied, and the PMD are constant. Then in the Stokes sector $\mathcal{S}_{\nu,\ell}$,

$$Y(x,t) \sim G_\nu(t)\left(1 + \sum_{j=1}^{\infty} Y_j^{(\nu)}(t)\xi_\nu^j\right)\exp\left(\sum_{j=-r_\nu}^{-1}\frac{1}{j}T_j^{(\nu)}\xi_\nu^j + T_0^{(\nu)}\log(\xi_\nu)\right)$$
$$\times S_{\nu,\ell-1}^{-1}\cdots S_{\nu,1}^{-1}C_\nu. \quad (4.17)$$

The proof follows closely that of Schlesinger's theorem. Consider the matrix of 1-forms $\eth Y(x,t)Y(x,t)^{-1}$, which we call $\Omega(x,t)$. Since (1) Y is invertible in $\overline{\mathbb{P}_a^1}$, and (2) M_ν does not depend on t, Ω is holomorphic in the variable x in $\mathbb{P}_a^1$. To determine the nature of its singularities, we calculate its asymptotic expansion by termwise differentiation of (4.17) with respect to the parameters t. To be able to do this, we need one further assumption,[10] namely that the asymptotic expansion (4.17) is uniform in $t \in \mathcal{D}$. Once again, to simplify the writing, we suppress the index ν. We get

$$\Omega(x,t) \sim G\left(1+\sum_1^\infty Y_j\xi^j\right)\left[\sum_{-r}^0 T_j\xi^{j-1}\mathrm{d}\xi + \sum_{-r}^{-1}\frac{1}{j}\xi^j\mathrm{d}T_j\right]\left(1+\sum_1^\infty Y_j\xi^j\right)^{-1}G^{-1}$$
$$+ G\left[\sum_1^\infty(jY_j\xi^{j-1}\mathrm{d}\xi + \xi^j\eth Y_j)\right]\left(1+\sum_1^\infty Y_j\xi^j\right)^{-1}G^{-1} + \eth GG^{-1}.$$

On the right-hand side, the coefficient of $\mathrm{d}\xi$ is nothing but $\mathrm{d}\widetilde{Y}/\mathrm{d}\xi$. It is easily calculated by expressing that $\widetilde{Y}$ satisfies (1.2) (compare with (4.4)). Using (4.3), we get

$$\frac{\mathrm{d}\widetilde{Y}^{(\nu)}}{\mathrm{d}\xi_\nu}\widetilde{Y}^{(\nu)-1} = G_\nu\sum_{j=-r_\nu}^{\infty}A_j^{(\nu)}\xi_\nu^{j-1}G_\nu^{-1} \quad (\nu = 1,\ldots,n,\infty).$$

Note that, from the definition of ξ_ν, $\mathrm{d}\xi_\nu = \begin{cases} -\mathrm{d}a_\nu, & \nu = 1,\ldots,n \\ 0, & \nu = \infty \end{cases}$.   □

As for the terms proportional to the differential elements $\mathrm{d}T_j^{(\nu)}$, we define new 1-forms $\Psi_j^{(\nu)}$ by[11]

---

[10]See Appendix B (Section 8).

[11]Note that we have excluded from the definition of $\Psi_j^{(\nu)}$ all $\mathrm{d}a_\nu$ terms, the explicit expression of which is simple, and which for this reason we write separately. Our $\Psi_j^{(\nu)}$ differ from the $\Phi_j^{(\nu)}$ of M. Jimbo et al. [7] in this respect. Compare also the expression (4.20) of $\Omega(x,t)$ with their formula (3.14).

$$\left(1+\sum_1^\infty Y_j^{(\nu)}\xi_\nu^j\right)\left[\sum_{-r}^{-1}\frac{1}{j}\xi_\nu^j dT_j^{(\nu)}\right]\left(1+\sum_1^\infty Y_j^{(\nu)}\xi_\nu^j\right)^{-1}=\sum_{-r}^\infty \xi_\nu^j \Psi_j^{(\nu)}. \quad (4.18)$$

Finally, keeping only the terms that do not vanish when $\xi \to 0$, we get

$$\Omega(x,t)\sim G_\nu\left(\sum_{-r}^1 A_j^{(\nu)}\xi^{j-1}d\xi+\sum_{-r}^0\Psi_j^{(\nu)}\xi^j\right)G_\nu^{-1}+\eth G_\nu G_\nu^{-1}+O(\xi). \quad (4.19)$$

All these singularities are poles: $\Omega(x,t)$ is rational in x, and to reconstruct it, we sum up all the contributions of poles at finite distance and infinity, which gives

$$\Omega(x,t) = -\sum_{\nu=1}^n\sum_{k=0}^{r_\nu}\frac{A_{\nu,k}(t)}{(x-a_\nu)^{k+1}}da_\nu +\sum_{\nu=1}^n\sum_{k=1}^{r_\nu}\frac{\Psi_{\nu,k}(t)}{(x-a_\nu)^k}$$
$$+\sum_{k=0}^{r_\infty}\Psi_{\infty,k}(t)x^k, \quad (4.20)$$

where the 1-forms $\Psi_{\nu,k}$ are defined by

$$\Psi_{\nu,k} = G_\nu\Psi_{-k}^{(\nu)}G_\nu^{-1}. \quad (4.21)$$

Furthermore, the constant term in the above expansion (4.19) is equal to

$$G_\nu(A_1^{(\nu)} + \Psi_0^{(\nu)})G_\nu^{-1} + \eth G_\nu G_\nu^{-1}.$$

But this constant term can now be calculated by expanding (4.20) about a_ν, thus providing the expression of the 1-form $\Theta_\nu = \eth G_\nu G_\nu^{-1}$. One obtains

$$\Theta_\nu(t) = \sum_{\substack{\mu=1\\\mu\neq\nu}}^n\sum_{k=0}^{r_\mu}A_{\mu,k}(t)\frac{da_\nu - da_\mu}{(a_\nu-a_\mu)^{k+1}} - \Psi_{\nu,0}(t)$$
$$+\sum_{\substack{\mu=1\\\mu\neq\nu}}^n\sum_{k=1}^{r_\mu}\frac{\Psi_{\mu,k}(t)}{(a_\nu-a_\mu)^k} +\sum_{k=0}^{r_\infty}\Psi_{\infty,k}(t)a_\nu^k. \quad (4.22)$$

Formula (4.20) is an important result. It enables us to construct the second element of a Lax pair, knowing the first one. In the next section we apply this technique to an ODE that leads to the first Painlevé equation.

The integrability conditions of the system of PDEs (4.14) and (4.15) can be written

$$\eth A(x,t) = \frac{\partial}{\partial x}\Omega(x,t) + [\Omega(x,t), A(x,t)], \quad (4.23)$$

$$\eth\Omega(x,t) = \Omega(x,t) \wedge \Omega(x,t). \quad (4.24)$$

Both sides of these equations are rational in the variable x. Writing that they are identically satisfied provides a set of equations involving only the deformation parameters t, which are the deformation equations of the singularity data. We state without proof the following theorems of the same authors, which complete the extension of Schlesinger's theorem to ODEs with irregular singularities:

Theorem 4.2 ([7]). *The deformations of the linear differential system (4.14) are isomonodromic if and only if the singularity data satisfy the total differential equations (4.23) and (4.16).*

Theorem 4.3 (] [7][). *The system of PDEs (4.23) and (4.16) is completely integrable.*

This system of differential equations is nonlinear. Its general solution has singularities when either two poles of (4.14) coalesce, or Assumption A.4.1 or A.4.2 is not satisfied. One has the following important result:

Theorem 4.4 (T. Miwa [13]). *The general solution of the deformation equations has no movable critical points.*

This means that the construction of nonlinear differential equations by the technique of isomonodromic deformations leads to equations that enjoy the Painlevé property.

5   An Isomonodromic Deformation Problem for Painlevé I

As an exercise, we apply the ideas of the previous section to the following 2×2 matrix:

$$A(x) = A_0 + xA_1 + x^2 A_2, \qquad (5.1)$$

where

$$A_0 = \begin{pmatrix} -v & y^2 + t/2 \\ -4y & v \end{pmatrix}, \quad A_1 = \begin{pmatrix} 0 & y \\ 4 & 0 \end{pmatrix}, \quad A_2 = \begin{pmatrix} 0 & 1 \\ 0 & 0 \end{pmatrix}. \qquad (5.2)$$

Here, t, y, and v are three parameters, which we are free to deform. $A(x)$ has only one singular point, at infinity, which is irregular. This matrix is the first one of a classical Lax pair for P_I [14].

In this section we shall first complete the pair by solving the isomonodromic deformation problem associated with $A(x)$, and second, we shall analyze the contents of the monodromy invariants, the PMD, encountered in the problem.

We first remark that Assumption A.4.1 is not satisfied, since A_2, the most singular part of $A(x)$ at infinity, is not diagonalizable. Moreover, it is nilpotent, and consequently does not control alone the behavior at infinity of the solution $Y(x)$. If it did, $Y(x)$ would behave like $\exp(A_2 x^3/3)$, but the nilpotency of A_2 kills the exponential dependence in x: $\exp(A_2 x^3/3) = 1 + A_2 x^3/3$. Now, it is natural to expect that the behavior at infinity would be controlled by the sum of the two most singular terms in $A(x)$. This leads us to diagonalize $A_1 + x A_2 = \left(\begin{smallmatrix} 0 & x+y \\ 4 & 0 \end{smallmatrix}\right)$. Since we are interested in the limit of large x, we neglect y in the sum $x + y$ and diagonalize the matrix

$$\begin{pmatrix} 0 & x \\ 4 & 0 \end{pmatrix} = T \begin{pmatrix} 2\sqrt{x} & 0 \\ 0 & -2\sqrt{x} \end{pmatrix} T^{-1}.$$

A first consequence is that there appears a singularity at $x = 0$, although $Y(x)$ is not singular there. From now on, we switch from x to the new variable[12] $\zeta = \sqrt{x}$.

It is convenient to choose a unimodular matrix T to perform the above diagonalization

$$T(\zeta) = \begin{pmatrix} \sqrt{\zeta}/2 & \sqrt{\zeta}/2 \\ 1/\sqrt{\zeta} & -1/\sqrt{\zeta} \end{pmatrix}.$$

A difference with the case where Assumption A.4.1 is satisfied is that T is no longer a constant matrix. Its square root singularity in the variable ζ will turn out to be harmless.

We now define

$$Z(\zeta) = T^{-1}(\zeta) Y(\zeta^2).$$

It satisfies the system of equations

$$\frac{\partial}{\partial \zeta} Z(\zeta) = B(\zeta) Z(\zeta), \tag{5.3}$$

where[13]

$$B(\zeta) = -(2v\zeta + \frac{1}{2\zeta})\tau_1 - (4y\zeta^2 + 2y^2 + t)\tau_2 + (4\zeta^4 + 2y^2 + t)\tau_3,$$

with

$$\tau_1 = \begin{pmatrix} 0 & 1 \\ 1 & 0 \end{pmatrix}, \quad \tau_2 = \begin{pmatrix} 0 & 1 \\ -1 & 0 \end{pmatrix}, \quad \tau_3 = \begin{pmatrix} 1 & 0 \\ 0 & -1 \end{pmatrix}.$$

We have thus replaced the initial problem by a new one, where now Assumption A.4.1 is satisfied: The leading term at infinity, $4\zeta^4 \tau_3$, in $B(\zeta)$ is diagonal. But we also now have two singularities, one irregular at infinity, with rank $r_\infty = 5$, and one regular at the origin.

---

[12]Compare with the situation where the *Thomé rank* [16] is odd.
[13]Thanks to the choice of a unimodular matrix $T(\zeta)$, $B(\zeta)$ is traceless.

Note the following symmetry property of $B(\zeta)$:

$$B(-\zeta) = -\tau_1 B(\zeta)\tau_1. \tag{5.4}$$

Let us write the Thomé asymptotic expansion at infinity in the form of (4.2), with diagonal-free terms in front of the exponential;

$$\widetilde{Z}(\zeta) = (1 + f_1(\zeta)\tau_1 + f_2(\zeta)\tau_2)\exp(f_0(\zeta)1 + f_3(\zeta)\tau_3).$$

As consequences of (5.4), $f_0(\zeta)$ and $f_1(\zeta)$ are even functions of ζ, $f_2(\zeta)$ and $f_3(\zeta)$ are odd, and no logarithmic term is allowed in the argument of the exponential (vanishing of the exponent of formal monodromy). One readily obtains

$$f_1(\zeta) = \frac{y}{2\zeta^2} + O\left(\frac{1}{\zeta^4}\right), \tag{5.5}$$

$$f_2(\zeta) = \frac{v}{4\zeta^3} + O\left(\frac{1}{\zeta^5}\right), \tag{5.6}$$

$$f_3(\zeta) = \frac{4}{5}\zeta^5 + t\zeta + \left(2y^3 + ty - \frac{v^2}{2}\right)\frac{1}{\zeta} + O\left(\frac{1}{\zeta^3}\right), \tag{5.7}$$

$$f_0(\zeta) = \frac{y^2}{8\zeta^4} + O\left(\frac{1}{\zeta^6}\right). \tag{5.8}$$

Note that it is not necessary to calculate the expansion of $Z(\zeta)$ around its singularity at $\zeta = 0$, since we know that $Y(x)$ is analytic[14] at $x = 0$.

Following the method of the previous section, we now have to calculate the 1-form $\Omega(x; y, v, t) = (\eth Y)Y^{-1} = T(\eth Z)Z^{-1}T^{-1}$, where

$$\eth = dy\frac{\partial}{\partial y}\bigg|_{\zeta,v,t} + dv\frac{\partial}{\partial v}\bigg|_{\zeta,y,t} + dt\frac{\partial}{\partial t}\bigg|_{\zeta,y,v}.$$

The Thomé expansion of Z and the explicit expression of T provide us with the asymptotic expansion of Ω. Keeping all terms that do not vanish at infinity, we get

$$\Omega \underset{x\to\infty}{\sim} T(\eth\widetilde{Z})\widetilde{Z}^{-1}T^{-1} = \mathcal{M}(\zeta)\eth f_3(\zeta) + O\left(\frac{1}{\zeta^2}\right),$$

where the matrix $\mathcal{M}(\zeta)$ is given by

$$\mathcal{M}(\zeta) = T(1 + f_1\tau_1 + f_2\tau_2)\tau_3(1 + f_1\tau_1 + f_2\tau_2)^{-1}T^{-1}$$

$$= \begin{pmatrix} 0 & \zeta/2 + y/2\zeta \\ 2/\zeta & 0 \end{pmatrix} + O\left(\frac{1}{\zeta^3}\right)$$

---

[14] As an exercise, calculate a few terms of this expansion, and check that $T(\zeta)Z(\zeta)$ is analytic in x at $x = 0$.

and

$$\eth f_3(\zeta) = \zeta dt + ((6y^2 + t)dy + ydt - vdv)\frac{1}{\zeta} + O\left(\frac{1}{\zeta^3}\right).$$

Finally, since Ω has no singularity in x at finite distance (it is a polynomial), we get its exact expression by dropping from its asymptotic expansion all terms that vanish at infinity:

$$\Omega = \begin{pmatrix} 0 & (y + x/2)dt + (3y^2 + t/2)dy - vdv/2 \\ 2dt & 0 \end{pmatrix}.$$

It remains to express the integrability conditions (4.23) and (4.24). First, the vanishing of

$$\eth\Omega - \Omega \wedge \Omega = \begin{pmatrix} -(6y^2 + t)dy \wedge dt + vdv \wedge dt & dy \wedge dt \\ 0 & (6y^2 + t)dy \wedge dt - vdv \wedge dt \end{pmatrix}$$

entails

$$dy \wedge dt = dv \wedge dt = 0.$$

These two relations mean that among the three deformation parameters y, v, and t, only one is independent. Let it be t. Then y and v are functions of t, $y(t)$, and $v(t)$, and the 1-form Ω now becomes

$$\Omega(x, t) = \begin{pmatrix} 0 & (y + x/2) + ((6y^2 + t)\dot{y} - v\dot{v})/2 \\ 2 & 0 \end{pmatrix} dt,$$

where the dot means derivative with respect to t.

Second, the vanishing for any x of

$$\eth A - \frac{\partial\Omega}{\partial x} + [A, \Omega]$$
$$= \begin{pmatrix} 6y^2 + t - \dot{v} - 2(x - y)\alpha & (x + 2y)(\dot{y} - v) - v\alpha \\ 4(v - \dot{y}) & -(6y^2 + t - \dot{v}) + 2(x - y)\alpha \end{pmatrix} dt,$$

where $\alpha = (6y^2 + t)\dot{y} - v\dot{v}$, entails

$$v = \dot{y},$$
$$\dot{v} = 6y^2 + t.$$

Among the two equations for $y(t)$ and $v(t)$,

$$\ddot{y} = 6y^2 + t,$$
$$(\dot{v} - 1)^2 = 24v^2(\dot{v} - t),$$

the one for $y(t)$ is *Painlevé* I.

Furthermore, after simplification of the expression for Ω, this gives us the second equation of the Lax pair associated with P_I:

$$\frac{\partial Y}{\partial t} = \begin{pmatrix} 0 & y + x/2 \\ 2 & 0 \end{pmatrix} Y.$$

This result is in agreement with [14, (C.2)].

At this point we have solved the problem of the isomonodromic deformations of the ODE (1.2), with $A(x)$ given by (5.1) and (5.2). In order to understand how such a result can be exploited to solve connection problems for P_I [15], we now investigate the contents of the set of PMD encountered in the problem.

With only one singularity, irregular, the only PMD of the problem are the Stokes matrices at this singularity. It is more convenient, and anyway equivalent, to look at the Stokes matrices of the ODE (5.3), written in the variable ζ.

We first remark that because of the symmetry property (5.4), if $Z(\zeta)$ is a solution of (5.3), then so is $\tau_1 Z(e^{i\pi}\zeta)$. Changing ζ into $e^{i\pi}\zeta$ does not change x, and since $Y(x)$ is an entire function in x,

$$T(e^{i\pi}\zeta)Z(e^{i\pi}\zeta) = T(\zeta)Z(\zeta).$$

From the expression for $T(\zeta)$, one finds that $T(e^{i\pi}\zeta) = iT(\zeta)\tau_1$. Thus

$$\tau_1 Z(e^{i\pi}\zeta) = -iZ(\zeta). \tag{5.9}$$

As for the Thomé asymptotic expansion $\widetilde{Z}(\zeta)$, it satisfies the equation

$$\tau_1 \widetilde{Z}(e^{i\pi}\zeta) = \widetilde{Z}(\zeta)\tau_1. \tag{5.10}$$

According to Section 4, we define, for any integer ℓ, sectors $\mathcal{T}_\ell$ and $\mathcal{S}_\ell$ around the point at infinity ($r_\infty = 5$):

$$\mathcal{T}_\ell = \left\{ |\zeta| > \Delta; -\frac{\pi}{10} + \frac{\ell\pi}{5} < \text{phase}(\zeta) < \frac{\pi}{10} + \frac{\ell\pi}{5} \right\},$$

$$\mathcal{S}_\ell = \left\{ |\zeta| > \Delta; -\frac{3\pi}{10} + \frac{\ell\pi}{5} < \text{phase}(\zeta) < \frac{\pi}{10} + \frac{\ell\pi}{5} \right\}.$$

Then there exists a unique matrix solution $Z^{(\ell)}(\zeta)$ of (5.3) that behaves asymptotically like $\widetilde{Z}(\zeta)$ in $\mathcal{S}_\ell$. The Stokes matrices S_ℓ are defined by the relation $Z^{(\ell+1)} = Z^{(\ell)}S_\ell$.

Let us call $\tilde{z}_1(\zeta)$ and $\tilde{z}_2(\zeta)$ the columns of $\widetilde{Z}(\zeta)$, formal vector solutions of (5.3): $\widetilde{Z} = \{\tilde{z}_1, \tilde{z}_2\}$. The dominant solution in the sector $\mathcal{T}_\ell$ is $\tilde{z}_1$ if ℓ is even, and $\tilde{z}_2$ if ℓ is odd. Consequently,

$$S_\ell = \begin{pmatrix} 1 & 0 \\ s_\ell & 1 \end{pmatrix} \quad \text{if } \ell \text{ is even}, \qquad S_\ell = \begin{pmatrix} 1 & s_\ell \\ 0 & 1 \end{pmatrix} \quad \text{if } \ell \text{ is odd}.$$

We note that if $\zeta \in \mathcal{S}_\ell$, then $e^{i\pi}\zeta \in \mathcal{S}_{\ell+5}$. Thus, for $\zeta \in \mathcal{S}_\ell$

$$Z^{(\ell+5)}(e^{i\pi}\zeta) \sim \widetilde{Z}(e^{i\pi}\zeta) = \tau_1 \widetilde{Z}(\zeta)\tau_1,$$

and since $Z^{(\ell)}(\zeta) \sim \widetilde{Z}(\zeta)$,

$$Z^{(\ell+5)}(e^{i\pi}\zeta) = \tau_1 Z^{(l)}(\zeta)\tau_1.$$

With (5.9), this becomes

$$Z^{(\ell+5)}(\zeta) = iZ^{(\ell)}(\zeta)\tau_1,$$

and with the relation $Z^{(\ell+5)} = Z^{(\ell)}S_\ell S_{\ell+1}S_{\ell+2}S_{\ell+3}S_{\ell+4}$, it entails

$$S_\ell S_{\ell+1}S_{\ell+2}S_{\ell+3}S_{\ell+4} = i\tau_1.$$

Comparing this last equation with the same written with ℓ changed into $\ell + 1$, we get

$$S_{\ell+5} = \tau_1 S_\ell \tau_1.$$

It remains to plug the explicit expressions of the Stokes matrices into the last two equations, to obtain the following formulae, valid for any ℓ:

$$s_{\ell+5} = s_\ell,$$
$$s_{\ell+3} = i(1 + s_\ell s_{\ell+1}).$$

The reader will easily convince himself that as a consequence of these two formulae, it suffices to know two of the numbers s_ℓ to know all of them. Thus, the set of quantities that are preserved in the monodromy deformations reduce to *two and only two independent invariants*, which are any two of the five quantities s_1, s_2, s_3, s_4, and s_5.

This situation is general for all Painlevé equations, which are *second-order differential equations, the general solution of which depends on *two* arbitrary constants.

6   Conclusion

In this chapter we have learned how to deform isomonodromically a linear ODE with rational coefficients, whether its singularities are Fuchsian or not. To be able to undergo these deformations, the ODE has to be flexible enough, that is, its coefficients have to be functions of sufficiently many parameters. We have found that necessary and sufficient condition for the deformations to preserve the monodromy is that the coefficients satisfy non-linear differential equations that possess the Painlevé property. In this way, a deep link has been established between Painlevé equations and linear differential equations, with their associated monodromy invariants. In the last section we have verified in the case of P_I that the number of independent monodromy invariants is equal to the order of the Painlevé equation.

We conclude this chapter by showing how the knowledge of monodromy invariants associated with a Painlevé equation enables one to solve *connection problems* for this equation. To be specific, assume that we know a parameterization of the asymptotic behavior of $y(t)$, the general solution of the equation, near one of its singular points, in terms of two parameters, say α and β. The matrix $A(x)$ and the solution $Y(x)$ of the associated ODE become asymptotically functions of these parameters, $A(x; \alpha, \beta)$ and $Y(x; \alpha, \beta)$, and in principle, we are able to calculate the two monodromy invariants, say s_1 and s_2, as functions[15] of α and β:

$$s_i = f_i(\alpha, \beta), \quad i = 1, 2.$$

Assume furthermore that we know a parameterization of the asymptotic behavior of $y(t)$ near another singular point, in terms of parameters α' and β', and that we have calculated the monodromy invariants as functions of these new parameters:

$$s_i = g_i(\alpha', \beta'), \quad i = 1, 2.$$

We immediately see that the knowledge of the two invariants s_1 and s_2 enables us to connect the two sets of parameters, α and β on the one hand, and α' and β' on the other hand:

$$f_i(\alpha, \beta) = g_i(\alpha', \beta'), \quad i = 1, 2.$$

These equations solve the connection problem between the two considered singular points of $y(t)$. Indeed, with the two invariants s_1 and s_2, we have at our disposal a *complete set of independent first integrals* of the Painlevé equation.

The method just outlined here is known as the *direct isomonodromy method*. It was used for the first time by B.M. McCoy, C.A. Tracy, and T.T. Wu [17], to derive connection formulae for P_{III}, and by M.J. Ablowitz and H. Segur [18], for P_{II}. Since then, many other connection formulae have been obtained. See 19, 20 for references.

The case of P_I was completely solved recently, by A.A. Kapaev and A.V. Kitaev [15]. They derived formulae connecting the asymptotic behavior of the general solution in different sectors around its essential singularity at $t = \infty$.

---

[15]Obviously, we could not have found more than two independent invariants.

Acknowledgments: These notes have greatly benefited from discussions with the participants of the Cargèse school. The author is particularly grateful to R. Conte, B.A. Dubrovin, and K. Okamoto for valuable comments and suggestions. He also thanks M.L. Mehta for a careful reading of the manuscript.

7   Appendix A: Matrix Versus Scalar Formalisms and Fuchs's Theorem

Consider the homogeneous linear ODE of order N:

$$z^{(N)} + a_1(x)z^{(N-1)} + a_2(x)z^{(N-2)} + \cdots + a_N(x)z = 0. \qquad (7.1)$$

It can be written as a set of N coupled homogeneous linear ODEs of first order:

$$\frac{\mathrm{d}}{\mathrm{d}x}y(x) = A(x)y(x), \qquad (7.2)$$

where $A(x)$ is an $N \times N$ matrix, and $y(x)$ an N-component vector. Reciprocally, given a system (7.2), each component of $y(x)$ satisfies an ODE of the form (7.1). For example, if $N = 2$, the first component $y_1(x)$ satisfies the second-order ODE

$$y_1'' - \left(\operatorname{tr} A + \frac{A_{12}'}{A_{12}}\right)y_1' + \left(\det A - A_{11}' + A_{11}\frac{A_{12}'}{A_{12}}\right)y_1 = 0. \qquad (7.3)$$

The second component satisfies a similar but different equation.

The object of this appendix is to compare the *scalar* equation (7.1) with the equivalent *matrix* equation (7.2).

The coefficients of (7.1) and (7.2) are assumed to be analytic in x in some domain. At a point of analyticity of these coefficients, the general solutions of both equations are also analytic. The converse is not true. For example, let $A(x)$ be the matrix $\left(\begin{smallmatrix} n/x & 0 \\ 1/x^n & 0 \end{smallmatrix}\right)$. It has a pole of order n arbitrarily large at the origin. A fundamental matrix solution, as defined in Section 1, of the corresponding equation (7.2) is $Y(x) = \left(\begin{smallmatrix} x^n & 0 \\ x & 1 \end{smallmatrix}\right)$, and it is analytic at the origin.

7.1   Criterion for a Fuchsian Singularity

A singularity of an ODE is called *weak*, or *regular*, or *Fuchsian*, if the general solution has polynomial growth[16] at this point. For the scalar equation

---

[16] A function $f(x)$ with a norm $|f(x)|$ has polynomial growth at $x = x_0$ if there exists an integer p such that $|x - x_0|^p|f(x)| \to 0$ when $x \to x_0$.

(7.1), the important Fuchs theorem [16, 21] carries this definition, given in terms of the general solution, into a criterion on the coefficients of the equation itself: It gives simple necessary and sufficient conditions for a singularity of a scalar ODE (7.1) to be Fuchsian.

Such a simple characterization of a Fuchsian scalar equation does not exist for a Fuchsian matrix system. A *sufficient* condition for a system (7.2) to have a Fuchsian singularity at some point is that this point be a simple pole of $A(x)$. But the condition is *not necessary*, as is shown by the following example, where the matrix $A(x) = \begin{pmatrix} n/x & 0 \\ 1/x^{n+1} & 0 \end{pmatrix}$ has a pole of order arbitrarily large at the origin, whereas the corresponding fundamental matrix solution $Y(x) = \begin{pmatrix} x^n & 0 \\ \log x & 1 \end{pmatrix}$ exhibits a Fuchsian singularity.

It has been proven by J. Horn [22] that if the point x_0 is a Fuchsian singularity of (7.2), one can find a matrix $T(x)$, meromorphic at $x = x_0$, with a determinant not identically zero, such that the new vector

$$\hat{y}(x) = T(x)y(x) \qquad (7.4)$$

satisfies a new system of equations

$$\frac{\mathrm{d}}{\mathrm{d}x}\hat{y}(x) = \hat{A}(x)\hat{y}(x),$$

where the matrix

$$\hat{A} = T^{-1}AT - T^{-1}\frac{\mathrm{d}}{\mathrm{d}x}T$$

has a *simple* pole at x_0.

Horn's theorem is an existence theorem, and unfortunately it provides no method for constructing a matrix $T(x)$, and consequently no practical criterion to decide whether a singularity is Fuchsian or not. Since then, different criteria have been devised (J. Moser [23], W.B. Jurkat and D.A. Lutz [24], N. Katz [25], B. Malgrange [26], R. Gérard and A.H.M. Levelt [27]). None of them has the simplicity of Fuchs's theorem, but at least they allow the determination of the nature of a singularity of a matrix system (7.2) in a finite number of steps.

Note that Horn's theorem is a *local* theorem. It does not say that if (7.2) is Fuchsian, there exists a *global* matrix $T(x)$ such that $\hat{A}(x)$ has simple poles at *all* singular points and is analytic elsewhere. If it were true, then any Fuchsian system could be written as a DSSP. And this is wrong, as we now see.[17]

---

[17]Note that because $T(x)$ is meromorphic at x_0, the *shearing* transformation (7.4) leaves invariant the generator M_0 of the monodromy representation associated with the singular point x_0. A global $T(x)$ would preserve the whole monodromy representation.

7.2   The Riemann–Hilbert Problem

The answer to the celebrated Riemann–Hilbert problem has long been controversial. It is only recently that it has been definitively settled. A. Bolibruch [2] has explicitly built Fuchsian differential systems with multiple poles (DSMP), the monodromy representations of which are not isomorphic to the monodromy representation of any differential system with simple poles (DSSP). Thus, there exist representations of the fundamental homotopy group of $\mathbb{P}_a^1$ that cannot be realized as monodromy representations of DSSPs.

7.3   Apparent Singularities

A glance at equation (7.3) shows that its coefficients are singular at any zero of the matrix element A_{12}. More generally, zeros of nondiagonal elements of $A(x)$ show up as singularities of the associated scalar equations. If these zeros do not coincide with poles of $A(x)$, they are analyticity points of the general solution. The poles they induce in the scalar equations are thus *apparent singularities*. To check this, consider a zero of order r in A_{12}. It induces a simple pole in (7.3), and the calculation of the two indices at this pole gives the values 0 and $r + 1$, which shows that $y_1(x)$ is indeed analytic there.

Poincaré has given a general definition of apparent singularities (*points à apparence singulière*) [28]. A singular point of a differential equation (7.1) is said to be apparent if the ratio of any two particular solutions is single-valued in a neighborhood of this point. The reader will easily convince himself that with this definition, a singular point of a differential system (7.2) is apparent if its associated monodromy matrix is a multiple of the unit matrix. If the singularity is Fuchsian, it means that the eigenvalues of its exponent of formal monodromy differ by integers. Poincaré has shown the necessity of such singularities if one wants to be able to deform isomonodromically a differential equation. If they are absent, the equation has not enough parameters. It is too rigid.

Note that the system (3.1), which led us to P_{VI}, has an apparent singularity[18] at $x = y(t)$. This is not the case with Heun's equation [29], which has four Fuchsian singular points as (3.1), but no apparent singularity, and which cannot be deformed isomonodromically. The reader will easily check that the singularity at $\zeta = 0$ of the system (5.3), which led us to P_I, is apparent:[19] According to Poincaré this singularity is unavoidable.

---

[18] See (3.7).

[19] The associated monodromy matrix is equal to -1.

8   Appendix B: Asymptotic Power Series

In this appendix we collect a few definitions and theorems that are relevant to Section 4. They concern representations of analytic functions of a complex variable by divergent power series.[20]

To simplify the writing, we deal with expansions around the origin.

Let S be a set of points having the origin as accumulation point. Typically, S is a sector $\theta_1 \leq \arg x \leq \theta_2$, $0 < |x| \leq \delta$. If $\theta_2 - \theta_1 > 2\pi$, S is understood as a sector on a multisheeted Riemann surface. Then, one says that *the possibly divergent power series $\sum_{r=0}^{\infty} f_r x^r$ is the asymptotic expansion (AE) of the function $f(x)$ in S if*

$$x^{-m}\left[f(x) - \sum_{r=0}^{m} f_r x^r\right] \to 0,$$

when $x \to 0$ in S, for all $m \geq 0$. This is expressed by the shorthand notation

$$f(x) \sim \sum_{r=0}^{\infty} f_r x^r.$$

We note first that the AE of $f(x)$ in S is unique, and second that there is an infinite number of functions that possess the same AE in S. Thus, the knowledge of the AE in one sector S is not sufficient to determine the function $f(x)$, *even if the series does converge in S.*

If $f(x)$ is holomorphic in a whole annular neighborhood $0 < |x| < \delta$ of the origin, and if its AE is valid for all values of $\arg x$, then the series is convergent, converges to $f(x)$ in that annular neighborhood, and $f(x)$ is indeed holomorphic at the origin. This is why the validity of a divergent AE is limited to a sector.

Formal algebraic operations on AEs, like linear combinations, multiplication, composition, inverse when the constant term f_0 does not vanish, are allowed without restriction. The same is true for termwise integration. If the opening angle $\theta_2 - \theta_1$ of S does not vanish, termwise differentiation is also allowed, and is valid inside any proper subsector of S.

If $f(x)$ is analytic in S, and if all its derivatives $f^{(r)}(x)$ have a limit $f^{(r)}(0)$ when $x \to 0$ in S, then the formal Taylor series

$$\sum_{r=0}^{\infty} \frac{f^{(r)}(0)}{r!} x^r$$

is the AE of $f(x)$ in S.

Rather unexpectedly, given any formal power series and any sector S, there exist analytic functions that admit this power series as AE in S.

---

[20]See the classical textbook of W. Wasow [11] for details and proofs.

Consider now a function $f(x,t)$ of two variables x and t, analytic in x. Let T be a closed bounded region in the complex t-plane. Assume that, for any t fixed in T, $f(x,t)$ admits the AE

$$f(x,t) \sim \sum_{r=0}^{\infty} f_r(t)x^r. \tag{8.1}$$

This means that

$$x^{-m}\left[f(x,t) - \sum_{r=0}^{m} f_r(t)x^r\right] \to 0 \tag{8.2}$$

when $x \to 0$ in S, for all $m \geq 0$ and all $t \in T$. If the convergence in (8.2) is uniform in t, for $t \in T$, one says that *the asymptotic expansion (8.1) is uniform in t, for $t \in T$*. This notion of uniform AE is relevant when one needs to differentiate with respect to the variable t. The following theorem holds:

Let $f(x,t)$ be holomorphic in both variables x and t, in the direct product $S \times T$. Assume furthermore that it admits the asymptotic expansion (8.1), uniformly in T. Then, all the coefficients $f_r(t)$ are analytic in T, and

$$\frac{\partial f(x,t)}{\partial t} \sim \sum_{r=0}^{\infty} \frac{\mathrm{d}f_r(t)}{\mathrm{d}t}x^r,$$

uniformly in every compact proper subset of T.

9   References

[1] L.S. Pontryagin, *Topological Groups* (Gordon and Breach, New York, 1966). See Chapter 9.

[2] A. Bolibruch, *Fuchsian systems with reducible monodromy and the Riemann–Hilbert problem*, Global Analysis—Studies and Applications. V, Lecture Notes in Math. **1520** (Springer Verlag, Berlin, 1992), pp. 139–155.

[3] A. Beauville, *Monodromie des systèmes différentiels linéaires à pôles simples sur la sphère de Riemann*, Séminaires Bourbaki, 45e année, Astérisque **216** (1993), 103–119.

[4] L. Schlesinger, *Über eine Klasse von Differentialsystemen beliebiger Ordnung mit festen kritischen Punkten*, J. für R. und Angew. Math. **141** (1912), 96–145.

[5] R. Fuchs, *Sur quelques équations différentielles linéaires*, C. R. Acad. Sc. Paris **141** (1905), 555–558.

[6] H. Flaschka and A.C. Newell, *Monodromy and spectrum-preserving deformations I*, Comm. Math. Phys. **76** (1980), 65–116.

[7] M. Jimbo, T. Miwa, and K. Ueno, *Monodromy preserving deformation of linear ordinary differential equations with rational coefficients*, Physica **2D** (1981), 306–352.

[8] R. Garnier, *Étude de l'intégrale générale de l'équation VI de M. Painlevé*, Ann. Éc. Norm. Paris **34** (1917), 239–353. See also V.I. Gromak, Chapter 12, this volume.

[9] B. Gambier, *Sur les équations différentielles du second ordre et du premier degré dont l'intégrale générale est à points critiques fixes*, Thesis, Paris (1909); Acta Math. **33** (1910), 1–55.

[10] P. Painlevé, *Sur les équations différentielles du second ordre à points critiques fixes*, C. R. Acad. Sc. Paris **143** (1906), 1111–1117.

[11] W. Wasow, *Asymptotic Expansions for Ordinary Differential Equations* (Wiley Interscience, New York, 1965).

[12] A. Erdélyi, *Asymptotic Expansions* (Dover, New York, 1956).

[13] T. Miwa, *Painlevé property of monodromy preserving deformation equation and analyticity of τ functions*, Publ. RIMS, Kyoto Univ., **17** (1981), 703–721.

[14] M. Jimbo and T. Miwa, *Monodromy preserving deformation of linear ordinary differential equations with rational coefficients. II*, Physica **2D** (1981), 407–448.

[15] A.A. Kapaev and A.V. Kitaev, *Connection formulae for the first Painlevé transcendent in the complex domain*, Letters in Mathematical Physics **27** (1993), 243–252.

[16] J. Reignier, Chapter 1, this volume.

[17] B.M. McCoy, C.A. Tracy, and T.T. Wu, *Painlevé functions of the third kind*, J. Math. Phys. **18** (1977), 1058–1092.

[18] M.J. Ablowitz and H. Segur, *Exact linearization of a Painlevé transcendent*, Phys. Rev. Lett. **38** (1977), 1103–1106.

[19] A.R. Its and V.Y. Novokshenov, *The Isomonodromic Deformation Method in the Theory of Painlevé Equations* (Springer-Verlag, Berlin, 1986).

[20] A.R. Its, *Connection formulae for the Painlevé transcendants*, The Stokes Phenomenon and Hilbert's 16th Problem, eds. B.L.J. Braaksma, G.K. Immink, and M. van der Put (World Scientific, Singapore, 1996), pp. 139–166.

[21] E.L. Ince, *Ordinary Differential Equations* (Longmans, Green, and Co., London and New York, 1926). Reprinted (Dover, New York, 1956).

[22] J. Horn, *Zur Theorie der Systeme linearer Differentialgleichungen mit einer unabhängigen Veränderlichen, II*. Math. Annalen **40** (1892), 5527–5550.

[23] J. Moser, *The order of a singularity in Fuchs' theory*, Math. Zeitschrift **72** (1960), 379–398.

[24] W.B. Jurkat and D.A. Lutz, *On the order of solutions of analytic linear differential equations*, Proc. London Math. Soc. **3** (1971), 465–482.

[25] P. Deligne, *Équations différentielles à points singuliers réguliers*, Lecture Notes in Mathematics **163** (Springer-Verlag, Berlin–New York, 1970).

[26] B. Malgrange, *Remarques sur les points singuliers des équations différentielles*, C. R. Acad. Sc. Paris **273** Série A (1971), 1136–1137.

[27] R. Gérard and A.H.M. Levelt, *Invariants mesurant l'irrégularité en un point singulier des systèmes d'équations différentielles linéaires*, Ann. Inst. Fourier, Grenoble **23** (1973), 157–195.

[28] H. Poincaré, *Sur les groupes des équations linéaires*, Acta Math. **4** (1884), 201–311.

[29] Y. Brihaye, S. Giller, and P. Kosinski, *Heun equations and quasi exact solvability*, J. Phys. A **28** (1995), 421–431.

3

The Painlevé Approach to Nonlinear Ordinary Differential Equations

Robert Conte

ABSTRACT The "Painlevé analysis" is quite often perceived as a collection of tricks reserved to experts. The aim of this chapter is to demonstrate the contrary and to unveil the simplicity and the beauty of a subject that is, in fact, *the* theory of the (explicit) integration of nonlinear differential equations.
To achieve our goal, we will *not* start the exposition with a more or less precise "Painlevé test." On the contrary, we will finish with it, after a gradual introduction to the rich world of singularities of nonlinear differential equations, so as to remove any cooking recipe.
The emphasis is put on embedding each method of the test into the well-known theorem of perturbations of Poincaré. A summary can be found at the beginning of each section.

1 Introduction

This chapter is a major revision of a series of lectures delivered in Chamonix [9]. Let us start with a few realistic applications.

1.1 A Few Elementary Examples

1.1.1 Linearization of the Riccati Equation

The Riccati equation

$$u' = a_2(x)u^2 + a_1(x)u + a_0(x), \quad a_2 \neq 0, \quad \text{with } ' = \frac{\mathrm{d}}{\mathrm{d}x}, \qquad (1.1)$$

is known to be linearizable, but how is one to retrieve the explicit formula that maps it onto a linear equation? One just looks for the *first* coefficient u_0 of an expansion for u that describes the dependence on the integration constant (a simple pole), i.e., a Laurent series in some expansion function $\varphi(x)$: $u = \varphi^{-1}(u_0 + u_1\varphi + \cdots)$. It is given by balancing the two lowest-degree terms $-u_0\varphi'\varphi^{-2} = a_2(u_0\varphi^{-1})^2$, and the linearizing transformation is then simply the change of function $u \to \varphi$ defined by the *singular part transformation* $u = u_0 \varphi^{-1}$:

$$u = -\frac{\varphi'}{a_2\varphi}, \quad \varphi'' - \left[\frac{a_2'}{a_2} + a_1\right]\varphi' + a_0 a_2 \varphi = 0. \tag{1.2}$$

This will be justified in Section 7.3.

1.1.2   A First Integral of the Lorenz Model

The Lorenz model of atmospheric circulation

$$\frac{dx}{dt} = \sigma(y - x), \quad \frac{dy}{dt} = rx - y - xz, \quad \frac{dz}{dt} = xy - bz(x - y) \tag{1.3}$$

admits for $(b, \sigma, r) = (0, \frac{1}{3}, \text{arbitrary})$ the first integral

$$[-9x^4 + 16x(y - x) + 12(z - r + 1)x^2]e^{4t/3}/144 = K, \tag{1.4}$$

but can one go further, i.e., can one obtain more first integrals or even explicitly integrate? The answer is yes [107]. By elimination of (y, z), one first builds the second-order equation for $x(t)$

$$\frac{d^2x}{dt^2} = \frac{1}{x}\left(\frac{dx}{dt}\right)^2 - \frac{x^3}{4} - \frac{4K}{x}e^{-4t/3}. \tag{1.5}$$

For $K = 0$ this equation admits the first integral

$$\frac{1}{x^2}\left(\frac{dx}{dt}\right)^2 + \frac{x^2}{4} = A^2, \tag{1.6}$$

and the general solution $x = (1/(2A))\cosh(t - t_0)$. For $K \neq 0$ this equation for $x(t)$ is equivalent, as shown below, to the following equation for $X(T)$:

$$X'' = \frac{X'^2}{X} - \frac{X'}{T} + \frac{\alpha X^2 + \beta}{T} + \gamma X^3 + \frac{\delta}{X}, \quad (\alpha, \beta, \gamma, \delta) \text{ constant}, \tag{1.7}$$

which is the third of six irreducible equations discovered between 1900 and 1906 by Paul Prudent Painlevé and his student Bertrand Gambier, and according to the theory of Painlevé developed in this chapter, the integration is then achieved ("parfaite," says Painlevé). Two words may not be familiar to the reader: "equivalent" and "irreducible." "Equivalent" means that the transformation law from the physical variables (x, t) to the variables (X, T) that satisfy (P3) should not result from a good guess, but should be looked for within a precise set of transformations (mathematically the homographic transformations (3.5) defined Section 3.3) designed so as not to alter the structure of singularities (poles, branch points, ...). In this case, one finds that the transformation

$$x = a(t)X, \quad T = \tau(t), \tag{1.8}$$

with

$$a = \frac{2ic}{3}e^{-t/3}, \quad \tau = ce^{-t/3}, \quad c^4 = (3K)^4, \tag{1.9}$$

maps the equation for $x(t)$ to the equation (P3) for $X(T)$ with the parameter values for (P3) $\alpha = \beta = 0$, $\gamma = \delta = 1$.

As to "irreducible," it means that there exists no transformation, again within a precise class (Drach, Umemura, see Section 3.5), reducing any of the six (Pn) equations either to a linear equation or to a first-order equation. Consequently, the general solution of (Pn) has no "explicit expression"; it is just *defined by the equation itself.* There is absolutely no difference between defining the "exponential function" from the general solution of $u' = u$ and defining the "P3 function" from the general solution of the equation (P3).

1.1.3   A Reduction of the Boussinesq Equation

The Boussinesq equation of fluid mechanics

$$u_{tt} + \left[u^2 + \frac{1}{3}u_{xx}\right]_{xx} = 0, \tag{1.10}$$

in the stationary case where u does not depend on time, reduces to an ordinary differential equation (ODE) that admits two first integrals

$$u'' + 3u^2 + K_1 x + K_2 = 0, \tag{1.11}$$

and depending on K_1, this ODE is equivalent either to the (P1) equation

$$u'' = 6u^2 + x \tag{1.12}$$

or, after one more integration, to an equation introduced by Weierstrass, the *elliptic equation*

$$u'^2 = 4u^3 - g_2 u - g_3, \quad (g_2, g_3) \text{ complex constants.} \tag{1.13}$$

Both equations have a general solution single valued in the whole complex plane.

1.2   "Solvable" Models, "Integrable" Equations, and So On

Two main fields contributed to the recent interest in the Painlevé theory. The first one is statistical physics. When Ising solved his one-dimensional model and found the partition function $F = -(1/\beta)\log(2\cosh(\beta J))$, $\beta^{-1} = k_B T$, the result was a posteriori not surprising. But when Barouch, McCoy, and Wu [68] expressed the correlation function of the two-dimensional Ising model with a (P3) function, this strongly contributed to a revival of interest in these six functions, which now appear in any "solvable model" of statistical physics (see Di Francesco, Chapter 5, this volume). Retrospectively, the cosh function of Ising is a quite elementary output of the Painlevé theory.

The second field is that of partial differential equations (PDE), as shown by the above Boussinesq example. After the extension of the Fourier transform to nonlinear PDEs [57], called *inverse spectral transform* (IST), Ablowitz and Segur [2] noticed a link between those "IST-integrable PDEs" and the theory of Painlevé, a link expressed by Ablowitz, Ramani, and Segur [4] as the conjecture, "Every ODE obtained by an exact reduction of a nonlinear PDE solvable by the IST method has the Painlevé property." For more details, see the book by Ablowitz and Clarkson [1] and [85].

1.3   Insufficiency of Quadratures: The Need for a Theory

Let us return to our main subject, the explicit, analytic integration of ODEs. An exceptionally clear introduction *ad usum Delphini* is the *Leçon d'ouverture* [96, vol. I, p. 199] given by Painlevé in 1895 before starting the *Leçons de Stockholm*. For centuries, the question of integration has been formulated as follows: Find enough first integrals in order to reduce the problem to a sequence of *quadratures*. But even the simple example of the pendulum shows the insufficiency of this point of view. Its motion is reducible to a quadrature defined by the integral

$$t - t_0 = \int_{u_0}^{u} \frac{du}{\sqrt{(1 - u^2)(1 - k^2 u^2)}}, \quad k \text{ constant}, \qquad (1.14)$$

giving the time t as a "function" of the position u. However, this *elliptic integral* does not provide the desired result, i.e., the position as a function of time, and, worse, nothing ensures the existence of such an expression. This classical problem (the inversion of the elliptic integral) could be solved by Abel and Jacobi only by going to the complex domain, leading to a unique value $u(t, t_0, u_0) = u_0 + \mathrm{sn}(t - t_0, k)$. The symbol sn does deserve the name of *function* (this is one of the twelve Jacobi elliptic functions, equivalent to the unique Weierstrass function) because for any complex k, the map $t \mapsto \mathrm{sn}(t, k)$ is single-valued.

Following an idea of Briot and Bouquet, this led Painlevé to remark (*Leçon* no. 1): But the importance of this class of equations [the general solution of which is single valued] becomes even clearer if one observes that most auxiliary transcendents, whose role is so considerable (exponential function, elliptic functions, Fuchsian functions, etc.), are the integrals of algebraic differential equations which are quite simple. Differential equations therefore appear as the source for the most remarkable single valued transcendents, potentially able to contribute to the integration of other differential equations whose general solution is no more single valued.

This is the famous "double interest" of differential equations: One may consider them either as the source for defining new functions or as a class of equations to be integrated with the existing functions available.

1.4 What Can "To Integrate" Mean? The Painlevé Property

Any converging Taylor series defined on some part of the real line, representing for instance a solution of an ODE on some interval $-R < x < R$, defines in fact an analytic function inside the disk $|x| < R$. Therefore, even when their variables are real, differential equations and their solutions are naturally defined in the complex plane.

To integrate an ODE is to acquire a global knowledge of its general solution, not only the local knowledge ensured by the existence theorem of Cauchy. So, the most demanding possible definition for the "integrability" of an ODE is the single-valuedness of its general solution, so as to adapt this solution to any kind of initial conditions. Since even linear equations may fail to have this property, e.g., $2xu' + u = 0$, $u = cx^{-1/2}$, a more reasonable definition is the following one.

Definition 1.1. The *Painlevé property* (PP) of an ODE is the uniformizability of its general solution.

Following Bureau [13], we will call *stable* an equation with the PP. In the above example, uniformization is achieved, for instance, by removing from the complex plane any line joining the two branch points 0 and ∞.

1.5 Singularities of Ordinary Differential Equations

"Les fonctions, comme les êtres vivants, sont caractérisées par leurs singularités" (Paul Montel). Singularities are responsible for the limitation of the domain of validity of Taylor or Laurent expansions, so their study is mandatory.

There exists a deep difference among the singularities of solutions of differential equations according as whether these equations are linear or nonlinear. In the linear case, the general solution (GS) has no other singularities than those of the coefficients of the equation once solved for the highest derivative. These singularities have a location independent of the arbitrary coefficients of integration, and they are called *fixed*.

On the contrary, solutions of nonlinear ODEs may have other singularities, then called *movable*, at locations depending on the arbitrary coefficients. Thus, the equations [90]

$$\frac{du}{dx} + \frac{u}{x^2} = 0, \quad u = ce^{1/x}, \tag{1.15}$$

$$\frac{du}{dx} + \frac{u^2}{x} = 0, \quad u = \frac{1}{c + \log x}, \tag{1.16}$$

$$\frac{du}{dx} - \frac{\sqrt{1 - u^2}}{x} = 0, \quad u = \sin(c + \log x), \tag{1.17}$$

where c is the arbitrary constant of integration, all have a fixed singularity in their general solution at $x = 0$ (isolated essential singular point for the first one, logarithmic branch point for the two others). In addition, among the last two, which are nonlinear, the second has movable simple poles, and the third has no movable singularity. All three have the PP.

The possible singularities of differential equations have been classified by Mittag-Leffler: In addition to the familiar ones (poles, branch points, essential singular points), there can exist essential singular lines, analytic or not, or perfect sets of singular points, as illustrated by the Fuchsian and Kleinean functions of Poincaré. One example is Chazy's equation of class III

$$u''' - 2uu'' + 3u'^2 = 0, \tag{1.18}$$

whose general solution is defined only inside or outside a circle characterized by the three initial conditions (two for the center, one for the radius); this solution is holomorphic in its domain of definition and cannot be analytically continued beyond it. This equation therefore has the PP, and the only singularity is a movable analytic essential singular line, which is a natural boundary.

1.6   Outline and Basic References

For the outline, we refer the reader to the detailed table of contents; the choice made is to develop the construction of necessary conditions, at the expense of the explicit integration methods, only briefly introduced in Section 7.

The basic texts are due to Painlevé and his students, and we abbreviate their references as *Leçons* ([90] *Leçons de Stockholm*, 1895, a high-level course on nonlinear ODEs), BSMF ([91] first memoir, 1900, on first- and second-order ODEs), Acta ([93] second memoir, 1902, on second- and higher-order ODEs), CRAS ([95], 1906, an addendum after Gambier discovered the functions (P4), (P5), (P6)), Gambier ([56] thesis, 1909, on second-order ODEs), Chazy ([22] thesis, 1910, on third- and higher-order ODEs), Garnier ([58] thesis, 1911, on higher-order ODEs). Most Painlevé works are reprinted in Œuvres ([96] three volumes 1973, 74, 76, again available from CNRS-Éditions). For a global overview of these results, see the book of Hille [64], preferably to the one of Ince [67]. For a detailed exposition (indeed, in the classical period, it rather fashionable to avoid details) and additional results, see the three memoirs of Bureau M. I [14], M. II [15], M. III [16]. Peter Clarkson maintains an extensive bibliography [26] covering both the classical and the recent period, reproduced in [1].

2   The Meromorphy Assumption

2.1   Specificity of the Elliptic Function

A very deep result of L. Fuchs, Poincaré ([99], cf. [96, vol. III, p. 189])
and Painlevé (*Leçon* no. 7 p. 107) is that the class of first-order ODEs
$F(u', u, x) = 0$, with F polynomial in u' and u, analytic in x, defines one
and only one function, from the general solution of (1.13). This function is
not historically new, since this is precisely the *elliptic function* $\wp$ introduced
earlier by Weierstrass, i.e., the particular solution of

$$\wp'^2 = 4\wp^3 - g_2\wp - g_3 = 4(\wp - e_1)(\wp - e_2)(\wp - e_3), \quad (g_2, g_3, e_\alpha) \in \mathbb{C}, \qquad (2.1)$$

which admits a pole at the origin:

$$\wp(x, g_2, g_3) = x^{-2} + \frac{g_2}{20}x^2 + \frac{g_3}{28}x^4 + O(x^6). \qquad (2.2)$$

The novelty of $\wp$ is elsewhere: This is the impossibility of reducing the
elliptic equation to a linear equation. According to the addition formula of
$\wp$, $\wp(x - x_0)$ has a rational dependence on $\wp(x_0)$ with coefficients rational
in $\wp(x)$, $\wp'(x)$, but if one does not allow $\wp(x)$, $\wp'(x)$ in the expression for
$\wp(x - x_0)$, the dependence on x_0 is transcendental. Among the many nice
properties of elliptic functions (see, e.g., [5]), the most interesting to us
is their structure of singularities. These are doubly periodic meromorphic
functions (which is their usual definition), and there exists an *entire* function σ, i.e., without any singularity at a finite distance, whose $-\wp$ is the
second logarithmic derivative:

$$\wp = -\frac{\mathrm{d}}{\mathrm{d}x}\zeta, \quad \zeta = \frac{\mathrm{d}}{\mathrm{d}x}\log\sigma, \quad \zeta''^2 + 4\zeta'^3 - g_2\zeta' + g_3 = 0. \qquad (2.3)$$

Therefore, the only singularities of the general solution $\wp(x - x_0, g_2, g_3)$
of (1.13) come from the zeros of σ and are a lattice of movable double
poles located at $x_0 + 2m\omega + 2n\omega'$, with m and n integers, ω, ω' the two
half-periods.

2.2   The Meromorphy Assumption

The Laurent expansion (2.2) certainly motivated two students of Weierstrass, Paul Hoyer and Sophie Kowalevski, to investigate further the possibility for the GS of an ODE to be represented by a Laurent series with
a finite *principal part*, so as to exclude essential singularities. This meromorphy assumption, briefly said, consists in checking the existence of the
Laurent series and its ability to represent the GS, i.e., to contain enough
arbitrary parameters. But, since a Laurent series is defined only inside its
annulus of convergence, this study is only local, and one cannot dispense

with a further study in order to explicitly integrate, using completely different means.

The first attempt was by Hoyer in 1879 [65] with the system

$$
\frac{d}{dt} \begin{pmatrix} x_1 \\ x_2 \\ x_3 \end{pmatrix} = \begin{pmatrix} a_1 & a_2 & a_3 \\ b_1 & b_2 & b_3 \\ c_1 & c_2 & c_3 \end{pmatrix} \begin{pmatrix} x_2 x_3 \\ x_3 x_1 \\ x_1 x_2 \end{pmatrix} ,
\tag{2.4}
$$

under the restriction that neither the determinant nor any of its first- or second-order diagonal minors vanishes; he even generalized the assumption to the Puiseux series

$$
x_i = \sum_{j=0}^{+\infty} A_{ij} \{ (t - t_0)^{1/r} \}^{-n+j}, \quad i = 1, 2, 3,
$$

with n and r positive integers and $(A_{10}, A_{20}, A_{30}) \neq (0, 0, 0)$, but in fact the numerous cases of integrability by elliptic functions that he discovered were found by a direct ansatz, and not as necessary conditions for the Laurent series to exist. Continued by Kowalevski with a quite similar system except that it is six-dimensional, which will be seen Section 2.5, the method will get its final shape only with Gambier in 1910 (see pages 9 and 49 of his thesis).

2.3   A Flavor of the Meromorphy Test

We must warn the reader that this section is *not* the algorithm to apply, but just a flavor of it; the final algorithm will be given only in Section 6.6.

Let us start with a single equation; the case of a system is not different, apart from technical complications.

Consider the equation

$$
E(x, u) \equiv -\frac{d^2 u}{dx^2} + 6u^2 + g(x) = 0,
\tag{2.5}
$$

with g analytic.

Assume that u has polar behavior at some location x_0 distinct from any of the possible singularities of the coefficients of the equation, here $g(x)$; such a pole is therefore movable. One has to check the existence of *all* possible Laurent series with a finite principal part

$$
u = \sum_{j=0}^{+\infty} u_j \chi^{j+p}, \quad \chi = x - x_0, \quad u_0 \neq 0,
\tag{2.6}
$$

in which $-p$ is the order of the pole, which must be an integer, and the coefficients u_j are independent of x.

After insertion of this series in (2.5), which is polynomial in u and its derivatives, and replacement of $g(x)$ by its Taylor series in the neighborhood of x_0, the left-hand side, as a sum of Laurent series, is itself a Laurent series with a finite principal part

$$E = [p(p-1)u_0\chi^{p-2} + (p+1)pu_1\chi^{p-1} + \cdots] + 6[u_0^2\chi^{2p} + 2u_0u_1\chi^{2p+1} + \cdots]$$
$$+ [g(x_0) + g'(x_0)\chi + \cdots], \quad (2.7)$$

which we denote more generally by

$$E = \sum_{j=0}^{+\infty} E_j\chi^{j+q}, \quad (2.8)$$

q being the smallest integer of the list $(p-2, 2p, 0)$. The method consists in expressing the conditions for this series to identically vanish: $\forall j \in \mathbb{N}$: $E_j = 0$.

First Step. Determine all possible *families of movable singularities* (u_0, p). This is expressed with three conditions:

condition $u_0 \neq 0$ equality of at least two elements of the list $(p-2, 2p, 0)$ (q denotes their common value), the involved terms of E being called *dominant* and denoted by $\widehat{E}$;

dominance condition inferiority of q to the other elements of the list;

vanishing Laurent series condition vanishing of the coefficient E_0 of the lowest power χ^q, which involves only the dominant terms

$$E_0 \equiv \lim_{\chi\to 0} \chi^{-q}\widehat{E}(x, u_0\chi^p) = 0, \quad u_0 \neq 0, \quad (2.9)$$

i.e., respectively, one linear equation for p by the pair of terms considered, several linear inequalities for p, one algebraic equation for (u_0, p).

A necessary condition to prevent multivaluedness is then the following:

C0. All possible values for p are integral.

If there exists no family that is truly singular (p negative), the method stops without concluding.

Here, the various possibilities for these linear equations and inequalities are

$$q = p - 2 = 2p \quad \text{and} \quad q \leq 0, \quad (2.10)$$
$$q = 2p = 0 \quad \text{and} \quad q \leq p - 2, \quad (2.11)$$
$$q = p - 2 = 0 \quad \text{and} \quad q \leq 2p. \quad (2.12)$$

Their geometric representation is known as the *Puiseux diagram* or *Newton's polygon* ([64, Section 3.3], [67, Section 12.61]). The two solutions $(p, q) = (-2, -4)$, $(2, 0)$ satisfy the condition **C0**, and the second one must be rejected, as being nonsingular. So, the dominant part is here $\widehat{E} = -u'' + 6u^2$. The algebraic equation $E_0 = 0$,

$$E_0 \equiv -6u_0 + 6u_0^2 = 0, \quad u_0 \neq 0, \tag{2.13}$$

has only one root, $u_0 = 1$.

For $j = 1, 2, \ldots$, each successive equation $E_j = 0$ has then the form

$$\forall j \geq 1 : E_j \equiv P(u_0, j)u_j + Q_j(\{u_l \mid l < j\}) = 0, \tag{2.14}$$

where

$$P(u_0, j) = -(j - 2)(j - 3) + 12u_0 = -(j + 1)(j - 6), \tag{2.15}$$

$$Q_1 = 0, \quad Q_2 = 6u_1^2, \quad Q_3 = 12u_1u_2, \tag{2.16}$$

$$Q_j = \frac{g^{(j-4)}(x_0)}{(j - 4)!} + 6\sum_{k=1}^{j-1} u_k u_{j-k}, \quad \forall j \geq 4. \tag{2.17}$$

So the sequence $E_j = 0$, $j \geq 1$, is just *one* linear equation with different right-hand sides, and it can be solved recursively for u_j. Whenever the positive integer j is a zero of P, two subcases occur: Either Q_j does not vanish and the Laurent series does not exist, or Q_j vanishes and u_j is arbitrary. Since x_0 is already arbitrary, in order to represent the GS, one wants $N - 1$ additional arbitrary coefficients to enter the expansion, where N is the order of the ODE. Let us admit for a moment that the value $j = -1$ is always a zero of P, a result whose general proof (given Section 5.5) needs prerequisite notions of perturbation theory; this value $j = -1$ will be seen to represent the arbitrary location of x_0. Hence the following steps.

Second Step. For each family, determine the polynomial P (do not compute Q_j yet) and require the necessary conditions [102]:

C1. For each family, all the zeros of P with a positive real part are integral.

C2. There exists a family with $N - 1$ positive zeros of P, counting their multiplicity.

C3. There exists a family with $N - 1$ distinct positive zeros of P.

If **C1** is violated, the method stops, and one concludes failure, for the general solution cannot be meromorphic. If either **C2** or **C3** is violated, the method stops without concluding, because it cannot provide a meromorphic representation of the general solution.

The zeros of P are called *indices*, and $P = 0$ itself is the *indicial equation*. Indeed, anticipating the exposition of the general theory in Sections 5.6

and 5.7, they are the Fuchs indices i near $\chi = 0$ of a linear equation introduced by Darboux [45] under the name "équation auxiliaire," so the indicial equation is computed as follows [52]. Take the derivative of $\widehat{E}(x, u)$ with respect to u,

$$\forall v : \widehat{E}'(x, u)v \equiv \lim_{\lambda \to 0} \frac{\widehat{E}(x, u + \lambda v) - \widehat{E}(x, u)}{\lambda}, \qquad (2.18)$$

which leads to

$$\widehat{E}'(x, u) \equiv -\partial_x^2 + 12u; \qquad (2.19)$$

evaluate this linear operator at the point $u = u_0\chi^p$, which defines the "auxiliary equation" (i.e., the linearized equation at the leading term)

$$\forall v : \widehat{E}'(x, u_0\chi^p)v = 0, \qquad (2.20)$$

which leads to

$$\forall v : \widehat{E}'(x, u_0\chi^{-2})v \equiv (-\partial_x^2 + 12u_0\chi^{-2})v = 0; \qquad (2.21)$$

establish the indicial equation of this linear ODE near its Fuchsian singularity $\chi = 0$,

$$P(i) = \lim_{\chi \to 0} \chi^{-i-q} \widehat{E}'(x, u_0\chi^p)\chi^{i+p} = 0, \qquad (2.22)$$

which leads to

$$\begin{aligned} P(i) &= \lim_{\chi \to 0} \chi^{-i+4}(-\partial_x^2 + 12u_0\chi^{-2})\chi^{i-2} \\ &= -(i-2)(i-3) + 12u_0 \\ &= -(i+1)(i-6). \end{aligned} \qquad (2.23)$$

The shift $i \mapsto i + p$ in the above equation is just a convention aimed at not producing an unfortunate difference between the Fuchs index i and the index j of the recursion relation $E_j = 0$.

Now, one just has to check the existence of the Laurent series.

Third Step. For each family, for each positive integer zero i of P (a Fuchs index), require the condition.

C4. $\forall i \in \mathbb{N}, P(i) = 0$: $Q_i = 0$.

This is done by successively solving the recursion relation up to the greatest positive integer Fuchs index. As soon as a **C4** condition is violated, one stops and concludes failure: The ODE does not have the PP. After the greatest positive integer Fuchs index has been checked, the method is finished.

Here, one finds

$$u_0 = 1, \quad u_1 = u_2 = u_3 = 0, \quad u_4 = -\frac{g_0}{10}, \quad u_5 = -\frac{g_0'}{6}, \tag{2.24}$$

and the condition **C4** at index $i = 6$ is

$$Q_6 \equiv g_0''/2 = 0, \tag{2.25}$$

i.e., since x_0 is arbitrary, $g'' = 0$. The ODE (2.5) is restricted to be (1.11), which has been seen to have a meromorphic GS, so in this case the generated necessary conditions are sufficient.

Remarks.

1. We have retained the classical vocabulary ("famille" is used by Gambier, p. 38 of his thesis [56]; "indices" is used by Gambier, Chazy [22], and Bureau [13]), rather than the one more recently introduced [3,4] ("branch," "resonances"). Indeed, "branch" has another meaning in classical analysis, where it denotes a determination of a multivalued map, which may create some confusion. As to "resonance," its identification with a basic notion of a linear theory, the Fuchs indices, makes useless the introduction of such a term.

2. Conditions $Q_i = 0$ at Fuchs indices i are often referred to as "no-logarithm conditions" because if some of them are not satisfied, there exists a generalization of the Laurent series, called a ψ-series [64, Chap. 7], which is a double expansion in χ and $\log \chi$. This series contains no logarithms (i.e., reduces to the Laurent series) iff all Q_i vanish.

3. One must prove that the radius R of the punctured disk $|x - x_0| < R$ in which the series converges is nonzero.

4. As indicated by Gambier [56, p. 50], there is no need to expand the coefficients $g(x)$ of the equation around x_0. This is achieved [30] by taking for the expansion variable not $x-x_0$, but a dummy variable χ with the only property $\chi_x = 1$. Coefficients u_j in (2.24) become dependent on x instead of x_0:

$$u_0 = 1, \quad u_1 = u_2 = u_3 = 0, \quad u_4 = -\frac{g(x)}{10}, \quad u_5 = -\frac{g'(x)}{15},$$
$$Q_6 \equiv \frac{g''(x)}{2} = 0. \tag{2.26}$$

Let us insist again on the danger of using the present test as it is. An example of Chazy makes evident the necessity for a more reliable test: The equation with a single-valued general solution [22, p. 360]

$$(u''' - 2u'u'')^2 + 4u''^2(u'' - u'^2 - 1) = 0,$$

$$u = e^{c_1 x + c_2}/c_1 + \frac{c_1^2 - 4}{4c_1} x + c_3$$

possesses a logarithmic family $u \sim -\log(x - x_0)$.

Exercise 2.1. Handle the equation

$$2uu'' - 3u'^2 = 0, \quad u = c_1/(x - c_2)^2. \tag{2.27}$$

Solution.

$$2p - 2 = 2p - 2, \quad E_0 \equiv 2u_0^2 p(p - 1) - 3u_0^2 p^2 = 0. \tag{2.28}$$

Hence $p = -2$, u_0 arbitrary.   □

2.4 Extension to a System

If the differential equation is defined by a system

$$\mathbf{E}(x, \mathbf{u}) = 0 \tag{2.29}$$

(boldface characters represent multicomponent quantities), the scalar equations of Section 2.3 become systems: a linear system for the components of $\mathbf{p}$, an algebraic system for the components of $\mathbf{u}_0$, a linear system with a rhs for $\mathbf{u}_j$, a determinant for the indicial equation, etc. Take the example of the Euler system (diagonal Hoyer system)

$$E_1 \equiv \frac{dx_1}{dt} - \alpha x_2 x_3 = 0, \quad E_2 \equiv \frac{dx_2}{dt} - \beta x_3 x_1 = 0,$$

$$E_3 \equiv \frac{dx_3}{dt} - \gamma x_1 x_2 = 0. \tag{2.30}$$

First Step. The necessary condition on $(\mathbf{p}, \mathbf{u}_0)$ is the following.

C0. All components of $\mathbf{p}$ are integral, no component of $\mathbf{u}_0$ vanishes.

Of course, if a component of $\mathbf{u}_0$ is zero, one must increase by one the associated component of $\mathbf{p}$ until the new component of $\mathbf{u}_0$ becomes nonzero. If there exists no truly singular family (at least one component of $\mathbf{p}$ negative), the method stops without concluding.

Here, the unique solution $\mathbf{p}$ of the linear system $q_1 = p_1 - 1 = p_2 + p_3$ and cyclically is thus $p_1 = p_2 = p_3 = -1$, $q_1 = q_2 = q_3 = -2$. The algebraic system for $\mathbf{u}_0$,

$$\mathbf{E}_0 \equiv \lim_{\chi \to 0} \chi^{-\mathbf{q}} \widehat{\mathbf{E}}(x, \mathbf{u}_0 \chi^{\mathbf{p}}) = 0, \quad \mathbf{u}_0 \neq 0, \tag{2.31}$$

is written

$$E_{1,0} \equiv -x_{1,0} - \alpha x_{2,0} x_{3,0} = 0 \text{ and cyclically,} \qquad (2.32)$$

and it defines four families,

$$x_{1,0}^2 = \frac{1}{\beta\gamma}, \quad x_{2,0}^2 = \frac{1}{\gamma\alpha}, \quad x_{3,0}^2 = \frac{1}{\alpha\beta}, \quad x_{1,0}x_{2,0}x_{3,0} = -\frac{1}{\alpha\beta\gamma}, \qquad (2.33)$$

which we gather in the unique algebraic notation $x_{1,0} = a$, $x_{2,0} = b$, $x_{3,0} = c$.

Second Step. The linear system

$$\forall j \geq 1 : \mathbf{E}_j \equiv \mathbf{P}(j)\mathbf{u}_j + \mathbf{Q}_j(\{\mathbf{u}_l \mid l < j\}) = 0 \qquad (2.34)$$

generates the indicial equation

$$\det \mathbf{P}(i) = 0, \quad \mathbf{P}(i) = \lim_{\chi \to 0} \chi^{-i-q} \hat{\mathbf{E}}'(x, \mathbf{u}_0 \chi^{\mathbf{p}}) \chi^{i+\mathbf{p}}, \qquad (2.35)$$

where

$$\mathbf{P}(i) = \begin{pmatrix} i-1 & -\alpha c & -\alpha b \\ -\beta c & i-1 & -\beta a \\ -\gamma b & -\gamma a & i-1 \end{pmatrix}, \quad \det \mathbf{P}(i) = (i+1)(i-2)^2 = 0. \quad (2.36)$$

Classical results from linear algebra on the resolution of the matrix equation $AX = B$ give the following necessary conditions:

C1. For each family, all the zeros of $\det \mathbf{P}$ with a positive real part are integral.

C2. There exists a family with $N-1$ positive zeros of $\det \mathbf{P}$, counting their multiplicity.

C3. There exists a family with, apart from -1 counted once, only positive zeros of $\det \mathbf{P}$, each with a multiplicity equal to the dimension of the kernel of $\mathbf{P}(i)$.

Here, each of the four families has the same indices $(-1, 2, 2)$, and for the double index $i = 2$, the three rows of matrix $\mathbf{P}(2)$ are proportional, so its kernel has dimension two.

Third Step. For each family, for each positive integer zero i of $\det \mathbf{P}$ (a Fuchs index), require the following condition:

C4. $\forall i \in \mathbb{N}$, $\det \mathbf{P}(i) = 0$: the vector $\mathbf{Q}_i$ is orthogonal to the kernel of the adjoint of the operator $\mathbf{P}(i)$.

Here, the condition **C4** is satisfied at index two, and the Laurent series are finally

$$x_1 = a\chi^{-1} + a_2\chi + O(\chi^2), \quad \chi = t - t_0, \tag{2.37}$$

$$x_2 = b\chi^{-1} + b_2\chi + O(\chi^2), \tag{2.38}$$

$$x_3 = c\chi^{-1} + c_2\chi + O(\chi^2), \quad a_2 + b_2 + c_2 = 0, \tag{2.39}$$

with (t_0, b_2, c_2) arbitrary.

2.5 Motion of a Rigid Body Around a Fixed Point

The motion of a rigid body around a fixed point is ruled by the system

$$A\frac{d\omega_1}{dt} + (C - B)\omega_2\omega_3 + (x_3k_2 - x_2k_3) = 0,$$

$$B\frac{d\omega_2}{dt} + (A - C)\omega_3\omega_1 + (x_1k_3 - x_3k_1) = 0,$$

$$C\frac{d\omega_3}{dt} + (B - A)\omega_1\omega_2 + (x_2k_1 - x_1k_2) = 0,$$

$$\frac{dk_1}{dt} - \omega_3k_2 + \omega_2k_3 = 0, \tag{2.40}$$

$$\frac{dk_2}{dt} - \omega_1k_3 + \omega_3k_1 = 0,$$

$$\frac{dk_3}{dt} - \omega_2k_1 + \omega_1k_2 = 0,$$

depending on six parameters: the components (A, B, C) of the diagonal inertia momentum I which must positive, and the components (x_1, x_2, x_3) of the vector $\overrightarrow{OG}$ linking the fixed point O to the center of mass G which must be real. Because it admits the three first integrals

$$K_1 = (I\overrightarrow{\Omega}).\overrightarrow{\Omega} - 2\overrightarrow{OG}.\overrightarrow{k} = A\omega_1^2 + B\omega_2^2 + C\omega_3^2 - 2(x_1k_1 + x_2k_2 + x_3k_3),$$

$$K_2 = (I\overrightarrow{\Omega}).\overrightarrow{k} = A\omega_1k_1 + B\omega_2k_2 + C\omega_3k_3,$$

$$K_3 = \overrightarrow{k}.\overrightarrow{k} = k_1^2 + k_2^2 + k_3^2,$$

and a last Jacobi multiplier equal to 1,

$$\sum_{j=1}^{3} \partial_{\omega_j}\left(\frac{d\omega_j}{dt}\right) + \sum_{j=1}^{3} \partial_{k_j}\left(\frac{dk_j}{dt}\right) = 0, \tag{2.41}$$

a sufficient condition of reducibility to quadratures (i.e., to separation of variables, which implies neither meromorphy nor single-valuedness) is the existence of a single additional first integral independent of time.

Before Kowalevski, the only such known cases were

- the isotropy case, with

$$A = B = C : K_4 = \overrightarrow{OG}.\overrightarrow{\Omega} = x_1\omega_1 + x_2\omega_2 + x_3\omega_3, \qquad (2.42)$$

- the case of Euler (1750) and Poinsot (1851), G at the fixed point O ($x_1 = x_2 = x_3 = 0$) with

$$G = O : K_4 = |I\overrightarrow{\Omega}|^2 = A^2\omega_1^2 + B^2\omega_2^2 + C^2\omega_3^2, \qquad (2.43)$$

- the case of Lagrange (1788) and Poisson (1813),

$$A = B, x_1 = x_2 = 0 : K_4 = \omega_3, \qquad (2.44)$$

and for these three cases the general solution is elliptic, hence meromorphic [60].

Let us denote a family as ($t - t_0$ is abbreviated as t)

$$\omega_l = \sum_{j=0}^{+\infty} \omega_{l,j} t^{n_l+j}, \quad k_l = \sum_{j=0}^{+\infty} k_{l,j} t^{m_l+j}, \quad \omega_{1,0}\omega_{2,0}\omega_{3,0}k_{1,0}k_{2,0}k_{3,0} \neq 0,$$

with $l = 1, 2, 3$ and $(\omega_{l,j}, k_{l,j})$ complex. There exist numerous families, some of them with $n_l - n_k, m_l - m_k$ not integral. For the sake of brevity, let us restrict to the case where all the differences $n_l - n_k, m_l - m_k$ are integral, and redefine a family as

$$\omega_l = t^n \sum_{j=0}^{+\infty} \omega_{l,j} t^j, \quad (\omega_{1,0}, \omega_{2,0}, \omega_{3,0}) \neq (0,0,0),$$

$$k_l = t^m \sum_{j=0}^{+\infty} k_{l,j} t^j, \quad (k_{1,0}, k_{2,0}, k_{3,0}) \neq (0,0,0). \qquad (2.45)$$

One such family is defined [73] by the exponents $n_l = -1, m_l = -2$, and the sextuplets $(\omega_{l,0}, k_{l,0})$, solutions of the algebraic system

$$A\omega_{1,0} + (B - C)\omega_{2,0}\omega_{3,0} + x_2 k_{3,0} - x_3 k_{2,0} = 0,$$
$$B\omega_{2,0} + (C - A)\omega_{3,0}\omega_{1,0} + x_3 k_{1,0} - x_1 k_{3,0} = 0,$$
$$C\omega_{3,0} + (A - B)\omega_{1,0}\omega_{2,0} + x_1 k_{2,0} - x_2 k_{1,0} = 0,$$
$$2k_{1,0} + \omega_{3,0}k_{2,0} - \omega_{2,0}k_{3,0} = 0,$$
$$2k_{2,0} + \omega_{1,0}k_{3,0} - \omega_{3,0}k_{1,0} = 0,$$
$$2k_{3,0} + \omega_{2,0}k_{1,0} - \omega_{1,0}k_{2,0} = 0,$$

and the linear system for $j \geq 1$ is

$$\begin{pmatrix} (j-1)A & (C-B)\omega_{3,0} & (C-B)\omega_{2,0} & 0 & x_3 & -x_2 \\ (A-C)\omega_{3,0} & (j-1)B & (A-C)\omega_{1,0} & -x_3 & 0 & x_1 \\ (B-A)\omega_{2,0} & (B-A)\omega_{1,0} & (j-1)C & x_2 & -x_1 & 0 \\ 0 & k_{3,0} & -k_{2,0} & j-2 & -\omega_{3,0} & \omega_{2,0} \\ -k_{3,0} & 0 & k_{1,0} & \omega_{3,0} & j-2 & -\omega_{1,0} \\ k_{2,0} & -k_{1,0} & 0 & -\omega_{2,0} & \omega_{1,0} & j-2 \end{pmatrix} \begin{pmatrix} \omega_{1,j} \\ \omega_{2,j} \\ \omega_{3,j} \\ k_{1,j} \\ k_{2,j} \\ k_{3,j} \end{pmatrix}$$
$$+ \mathbf{Q}_j = 0.$$

The determinant $\det \mathbf{P}$ must have five positive zeros.

In the generic case (A, B, C) all different and $G \neq O$, there exists a unique solution to the algebraic system, depending on one arbitrary parameter and the root of an eight-degree equation [74], but the determinant

$$\det \mathbf{P} = ABC(j+1)j(j-2)(j-4)(j^2 - j - \mu), \qquad (2.46)$$

where μ is an algebraic expression of (A, B, C, x_1, x_2, x_3), has five positive integer zeros iff $\mu = 0$, which corresponds to inadmissible values for the six parameters (A, B, C must be real positive, x_1, x_2, x_3 real).

A thorough discussion of the nongeneric cases of this family $n_l = -1$, $m_l = -2$ led Kowalevski to retrieve the three known cases, as expected, and finally to find the subcase

$$A = B, \quad (x_1, x_2) \neq (0, 0), \quad \omega_{1,0}^2 + \omega_{2,0}^2 = 0, \qquad (2.47)$$

for which the unique solution is

$$\omega_{1,0} = -\frac{iC}{2(x_1 + ix_2)\lambda}, \quad \omega_{2,0} = i\omega_{1,0}, \quad \omega_{3,0} = 2i, \quad i^2 = -1,$$

$$k_{1,0} = -\frac{2C}{x_1 + ix_2}, \quad k_{2,0} = ik_{1,0}, \quad k_{3,0} = 0,$$

$$\det \mathbf{P} = ABC(j+1)(j-2)(j-3)(j-4)(j+1-2C/A)(j-2+2C/A),$$

in which λ is defined by the relation

$$2C - A - 4\lambda x_3 = 0, \quad \lambda \neq 0. \qquad (2.48)$$

There exist five positive integer indices iff $A = 2C, x_3 = 0$, and the first integral

$$A = B = 2C, \quad x_3 = 0 : K_4 = |C(\omega_1 + i\omega_2)^2 + (x_1 + ix_2)(k_1 + ik_2)|^2 \qquad (2.49)$$

terminates the proof of reducibility to quadratures. Sophie Kowalevski then managed to explicitly integrate with hyperelliptic integrals and to prove the meromorphy of the general solution, a feat that won her instantaneous fame.

Remark. Neither Hoyer nor Kowalevski enforced conditions **C3** and **C4**. This was done for the first time by Appelrot [6], who found another family $n_l = -1, m_l = 0$ with the indices $(-1, -1, 0, 1, 2, 2)$ and, despite the absence of five positive integer indices, computed the next terms and found at the double index 2 the no-log condition

$$x_1\sqrt{A(C-B)} + x_2\sqrt{B(A-C)} + x_3\sqrt{C(B-A)} = 0, \qquad (2.50)$$

whose real and imaginary parts yield, with the convention $A > B > C$,

$$x_2 = 0, \quad x_1\sqrt{A(B-C)} + x_3\sqrt{C(A-B)} = 0. \qquad (2.51)$$

Nekrasov [87] and Lyapunov then proved the multivaluedness of this case by exhibiting yet another family with complex exponents.

2.6 Insufficiency of the Meromorphy

Here is the opinion of Painlevé ([93, pp. 10 and 83], Œuvres III, pp. 196 and 269): Mrs. Kowalevski addresses the problem of finding all the cases where the rigid body motion is defined by *meromorphic functions of t which indeed admit poles*. Her procedure misses the cases when these functions would be single valued without possessing poles, either because they would be *holomorphic*, or because all their singularities would be transcendental.

Moreover, after having built all the conditions for the existence of movable poles, Mrs. Kowalevski notices that these conditions imply the integrability of the equations of motion, a fact which allows her to fully solve the question. But this remark misses a case for which there exist poles and which is not an integrability case. However, the Russian geometers have later proven that, in this case, the general solution of the equations of motion is not single valued.

The results of Mrs. Kowalevski therefore remain. But, however interesting be the road followed by Mrs. Kowalevski, it was desirable to tackle the question again in a more rational way. This is what is allowed by the procedures which I have used for the second order equations : they provide in the most natural and simple way the necessary conditions *to represent this motion by single valued functions of t*, without having to make any assumption on these functions. The conditions one arrives at in this way are not essentially different from those of Mrs. Kowalevski. For this particular problem, one therefore does not obtain any new cases.

Painlevé thus insists that poles are not privileged: They are just one kind of singularity among many possible others.

"Une discussion qui écarterait d'avance certaines singularités comme invraisemblables serait *inexistante*." (Painlevé, [93, p. 6], Œuvres III, p. 192).

"In the statement of the problem, *poles are not mentioned*; if in the final result the particular integrals prove to be meromorphic, it is a *result* of the research. Likewise, no mention is made of one or another type of critical or singular point." (Bureau [19, p. 105]).

Thus, definitely, the meromorphy assumption has to be waived as a *global* property, although it may be, and indeed is, quite useful at the *local* level. The only relevant property based on singularities is the Painlevé property as defined in Section 1.4, and the goal is to build a rigorous theory without any a priori assumptions on the movable singularities. Despite the pessimistic opinion of Picard, who thought the task impossible, Painlevé built that theory by a clever application of the theorem of perturbations of Poincaré and Lyapunov.

2.7 A Few Examples to Be Settled

The theory to come and the resulting "Painlevé test" should be able to handle the following differential systems, for which the meromorphy test of Section 2.3 is inconclusive or even erroneous. We give the location where the solution can be found.

1. ("Either they would be *holomorphic*") Extend the test to handle the equation $2uu'' - u'^2 = 0$, with general solution $u = (c_1 x + c_2)^2$. Solution: Section 5.4.

2. ("Either all their singularities would be transcendental"). Extend the test to handle the equations $uu'' - u'^2 = 0$ and $2u^2 u' u''' - 3u^2 u''^2 + u'^4 = 0$, whose general solutions are, respectively, $u = e^{c_1 x + c_2}$ and $u = c_1 e^{1/(c_2 x + c_3)}$. One may notice that the second equation is the Schwarzian derivative of $\log u$. Solution in [91, Section 37].

3. The "uncoupled" system with a meromorphic general solution
$$\frac{du}{dx} + u^2 = 0, \quad \frac{dv}{dx} + v^2 = 0 \tag{2.52}$$
admits two families (modulo the exchange of u and v)

(F1) $u \sim \chi^{-1}, \quad v \sim \chi^{-1}, \quad$ indices $(-1, -1),$ (2.53)

(F2) $u \sim \chi^{-1}, \quad v \sim v_0, \ v_0$ arbitrary,   indices $(-1, 0),$ (2.54)

of which the first one fails the condition **C2**. Solution: Section 5.4.

4. ("A case for which there exist poles and which is not an integrability case") The Bianchi IX cosmological model
$$(\log A)'' = A^2 - (B - C)^2 \text{ and cyclically}, \quad ' = d/d\tau, \tag{2.55}$$
admits for $B = C$ a particular four-parameter meromorphic solution [109]
$$A = \frac{k_1}{\sinh k_1(\tau - \tau_1)}, \quad B = C = \frac{k_2^2 \sinh k_1(\tau - \tau_1)}{k_1 \sinh^2 k_2(\tau - \tau_2)}. \tag{2.56}$$

Prove the absence of the PP by studying the family $\mathbf{p} = (0, -2, -2)$, which has only four Fuchs indices $(-1, 0, 1, 2)$. Solution: Section 5.8.3.

5. The Bianchi IX model (2.55) admits a family $\mathbf{p} = (-1, -1, -1)$ with the indices $(-1, -1, -1, 2, 2, 2)$. Prove the absence of the PP by studying this family. Solution: Section 5.7.6.

6. The Chazy equation of class III (1.18) admits a Laurent series that terminates $u = u_0/(x - x_0)^2 - 6/(x - x_0)$, with (x_0, u_0) arbitrary. From the study of this family, decide about the meromorphy of the general solution. Solution: Section 5.8.1.

7. In a problem in the geometry of surfaces, Darboux [44] encountered the system

$$\mathrm{d}x_1/\mathrm{d}t = x_2 x_3 - x_1(x_2 + x_3) \text{ and cyclically,} \qquad (2.57)$$

excluded by Hoyer, cf. Section 2.2, and found the two-parameter meromorphic solution

$$x_1 = c/(t - t_0)^2 + 1/(t - t_0), \quad x_2 = x_3 = 1/(t - t_0),$$
$$(t_0, c) \text{ arbitrary.} \quad (2.58)$$

For $c = 0$, the Fuchs indices are $(-1, -1, -1)$. Extend the test to build no-log conditions at this triple -1 index. Solution: Section 5.7.

8. The equations

$$-2uu'' + 3u'^2 + d_3 u^3 = 0, \quad d_3 \neq 0, \qquad (2.59)$$
$$u''' + uu'' - 2u'^2 = 0, \qquad (2.60)$$
$$u'''' + 2uu'' - 3u'^2 = 0, \qquad (2.61)$$

have no dominant behavior. Prove the absence of the PP for each of them [24]. Solution: Section 5.9.

3   The True Problems

In this section we state the true problems and manage logically to obtain the only correct definition for the Painlevé property (PP): "absence of movable critical points in the general solution," equivalent to that already given in Section 1.4. This includes:

the two classifications of singularities of differential equations (fixed or movable, critical or noncritical), the two differences between linear and nonlinear (movable singularity, singular solution),

the statement of the ambitious program proposed by Painlevé, a first, quick look at the method of resolution (the "double method" and the "double interest") and the results for first-order (equation of Riccati, function of Weierstrass) and second-order (classification of Gambier, the six Painlevé functions).

All the ODEs considered are defined on $\mathbb{C}$ or on the Riemann sphere (i.e., the complex plane compactified by addition of the point at infinity). Firstly, a more precise definition of the term "to integrate" is required.

Definition 3.1. *To integrate* an ODE, in the "modern sense" of Painlevé, is to find for the general solution a finite expression, possibly multivalued, in a finite number of functions, valid in the whole domain of definition.

The important terms in this definition are "finite" and "function."

Example 3.1 (nonintegrated ODE). The general solution of the ODE $u' + u^2 = 0$ is represented by the Taylor series $u = u_0 \sum_{j=0}^{+\infty} [-(x - x_0)u_0]^j$ for the Cauchy solution, but this representation is local and the integration is not achieved until one has found the radius of convergence, performed the summation, analytically continued the sum everywhere this is possible, and identified the analytic continuation to the meromorphic function $(x - x_1)^{-1}$.

Example 3.2 (integrated ODE). The ODE $2uu' - 1 = 0$ has for general solution $u = (x - x_0)^{1/2}$ a multivalued finite expression built from the "multivalued function" (see below) $z \mapsto z^{1/2}$.

Representations by an integral, a series, or an infinite product are acceptable iff they amount to a *global*, as opposed to *local*, knowledge of the solution.

A prerequisite to the integration in the sense of the above definition is therefore to extend the set of available functions, to serve as a reservoir from which to build finite expressions. At this stage, one must go back to the term "function."

Definition 3.2 ([12]). A *function* is a map of a set of objects into a set of images that maps a given object onto one *and only one* image.

In other words, a function is characterized by its single-valuedness, and terms such as "multivalued function" should be carefully avoided. In our context, a function is a single-valued map of the Riemann sphere onto itself.

Definition 3.3 (Painlevé [91, p. 206]). A *critical point* of a map of the Riemann sphere onto itself is any singular point, isolated or not, around which at least two determinations are permuted. Common synonyms are: for critical point, branch point, point of ramification; for determination, branch. Such a point is an obstacle for a map to be a function.

Examples. The maps $x \mapsto \sqrt{x-a}$ and $x \mapsto \log(x-a)$ both have exactly two critical points, a and ∞. Around each of them are permuted respectively two determinations and a countable infinity of determinations.

Remark. An *essential singular point* is not necessarily a critical point, since essential singularities, isolated or not, can be critical or not. Examples of critical essential singularities are $x = 0$ for $\tan(\log x)$ (nonisolated) or $\sin(C + \log x)$ (isolated and transcendental, *Leçons de Stockholm*, pp. 5–6, [93], [67, Section 14.1, p. 317]). Examples of noncritical essential singularities are $x = \infty$ for e^x or equivalently $x = x_0$ for $e^{1/(x-x_0)}$ (isolated), $x = \infty$ for $\tan x$ (nonisolated). Although, according to a classical theorem of Picard, an analytic function assumes all values except for at most two (∞ and another one) in a neighborhood of an isolated essential singularity, a noncritical essential singularity is *not* an obstacle to single-valuedness.

3.1 First Classification of Singularities, Uniformization

Definition 3.4. The *first classification of singularities* is the distinction critical or noncritical between singular points of maps. Note that it does not involve differential equations.

Consider a multivalued map of the Riemann sphere onto itself. There exist two classical methods, called *uniformizations*, to define from it a single-valued map, i.e., a function.

The first one is to restrict the object space by subtracting some lines, called *cuts*, so as to forbid local turns around critical points; for the above two examples, one removes any line joining the two points a and ∞.

The second method is to extend the object space to a *Riemann surface*, made of several copies, called *sheets*, of the Riemann sphere, cut and pasted. A point of the image space may then have several antecedents on the Riemann surface defining the object space. Example: two sheets for $x \mapsto \sqrt{x}$, a countable infinity of sheets for $x \mapsto \log x$.

As a consequence, to fill the reservoir of functions, one accepts all uniformizable maps, at the price of either restricting the object space by cuts or defining a Riemann surface in the object space.

Theorem 3.1. *The general solution of a linear ODE is uniformizable.*

Proof. Let

$$E \equiv \sum_{k=0}^{N} a_k(x) \frac{d^{(k)}u}{dx^k} = 0, \quad a_N(x) = 1 \tag{3.1}$$

be such an Nth-order ODE. Its general solution

$$u = \sum_{j=1}^{N} c_j u_j, \quad c_j \text{ an arbitrary constant,} \tag{3.2}$$

has for its only singularities those of the N independent particular solutions u_j, a subset of the singularities of the coefficients a_k [67, Section 15.1]. One knows where to make cuts or to paste the sheets of a Riemann surface in order to uniformize the general solution.   □

This has important consequences. Firstly, any linear ODE defines a function (Airy, Bessel, Gauss, Legendre, ...). Secondly, in the needed reservoir of functions one can put all solutions of all linear ODEs. Thirdly, a nonlinear ODE is considered as integrated if it is linearizable (of course, via a finite linearizing expression). Fourthly, in order to extend the list of known functions by means of ODEs, it is necessary to consider nonlinear ODEs.

Hence the problem stated by L. Fuchs and Poincaré.

Problem 1. Define new functions by means of ODEs, necessarily nonlinear.

3.2 Second Classification of Singularities, Different Kinds of Solutions

There exist two features of nonlinear ODEs without counterpart in the linear case. They concern the location of singularities of solutions and the possible existence of solutions in addition to the general solution.

The singularities of the solutions of nonlinear ODEs may be located at a priori unknown locations, which depend on the constants of integration.

Definition (already given in Section 1). A singular point of a solution of an ODE is called *movable* (resp. *fixed*) if its location in the complex plane depends (resp. does not depend) on the integration constants.

The point at ∞ is to be considered as fixed. A linear ODE has no movable singularities, the zeros of its general solution depend on the integration constants and are sometimes called for this reason movable zeros.

Definition 3.5. The *second classification of singularities* is the distinction movable or fixed between singularities of solutions of ODEs.

Among the four structures (critical or noncritical and fixed or movable) of singularities of solutions of ODEs, only one is an obstacle for this solution to be uniformizable and hence to define a function. This is the presence of singularities at the same time critical and movable. Indeed, in such a case, one knows neither where to make cuts nor where to paste the Riemann sheets, and uniformization is impossible.

But let us come to the second distinction between linear and nonlinear ODEs: Unlike linear ODEs, nonlinear ODEs may have several kinds of solutions.

Definition 3.6.

1. The *general solution* (GS) of an ODE of order N is the set of all solutions mentioned in the existence theorem of Cauchy (Section 5.3), i.e., determined by the initial value. It depends on N arbitrary independent constants.

2. A *particular solution* is any solution obtained from the general solution by giving values to the arbitrary constants. A synonym in English is *special solution*.

3. A *singular solution* is any solution that is not particular. Linear ODEs have no singular solution.

Example 3.3. The Clairaut-type equation $2u'^2 - xu' + u = 0$ has the general solution $cx - 2c^2$, a particular solution $x - 2$, and the singular solution $x^2/8$.

A singular solution can exist only when the ODE

$$E(u^{(N)}, u^{(N-1)}, \ldots, u', u, x) = 0, \tag{3.3}$$

considered as an equation for the highest derivative $u^{(N)}$, possesses at least two determinations (branches), whose coincidence may define a singular solution. This is a generalization of the notion of envelope of a one-parameter family of curves. A practical criterion to detect the singular solutions will be given in Section 5.1.

Painlevé stated the following program ([91, p. 201], Œuvres vol. III, p. 123; [93, p. 2], Œuvres vol. III, p. 188): "To determine all algebraic differential equations of first order, then second order, then third order, etc., the general solution of which is single valued."

One notices that singular solutions are excluded from this statement. Indeed, they present no interest at all for the theory of integration, for according to the above theorem, they satisfy an ODE of a strictly lower order than the ODE under consideration and have therefore been encountered at a lower order in the systematic program stated by Painlevé.

This problem (single-valuedness of the general solution) splits into two successive problems whose methods of solution are completely different: absence of movable critical points, then absence of fixed critical points. Hence the final statement.

Problem 2. Determine all the algebraic differential equations of first order, then second order, then third order, etc., whose general solution has no movable critical points.

This class of equations is often called "with fixed critical points."

We have reached the usual definition, equivalent to that of Section 1.4.

Definition 3.7. One calls the absence of movable critical singularities in its general solution the *Painlevé property* of an ODE.

3.3 Groups of Invariance of the PP

In the fulfillment of the program of Painlevé, it is sufficient to take one representative equation by class of equivalence of the PP. There exist two relations of equivalence for the PP, defined in Sections 3.3.1 and 3.3.2. Other relations of equivalence are defined in Section 3.3.3, but they violate the PP.

3.3.1 The Homographic Group

Theorem 3.2. *The only bijections (one-to-one mappings) of the Riemann sphere are the homographic transformations*

$$z \mapsto \frac{\alpha z + \beta}{\gamma z + \delta}, \quad \alpha\delta - \beta\gamma \neq 0, \ (\alpha, \beta, \gamma, \delta) \ \text{arbitrary complex constants.}$$

$$(3.4)$$

Proof. See any textbook. These transformations define a six-parameter group $\mathcal{H}$ called the Möbius group, also named $PSL(2, \mathbb{C})$. This group plays a fundamental role in the present theory. Given two triplets of points, there exists a unique homographic transformation mapping one triplet onto the other one. $\qquad\square$

Theorem 3.3. *The PP of an ODE $E(u, x) = 0$ is invariant under an arbitrary homographic transformation of the dependent variable u and an arbitrary holomorphic change of the independent variable x:*

$$(u, x) \mapsto (U, X): u = \frac{\alpha(x)U + \beta(x)}{\gamma(x)U + \delta(x)}, \quad X = \xi(x), \quad \alpha\delta - \beta\gamma \neq 0, \quad (3.5)$$

where α, β, γ, δ, ξ are arbitrary analytic (synonym: holomorphic) functions.

Proof. Let x_0 be a regular point of $(\alpha, \beta, \gamma, \delta, \xi)$, and $X_0 = \xi(x_0)$ its transform. In some neighborhood of x_0, the transformation between u and U is close to a homographic transformation with constant coefficients, and according to the previous theorem, the first classification (critical, noncritical) is invariant: If x_0 is critical (resp. noncritical) for u, then X_0 is critical (resp. noncritical) for U, and vice versa. Since $(\alpha, \beta, \gamma, \delta, \xi)$ do not depend on x_0, the second classification (fixed or movable) is also invariant. Thus the PP, which depends only on these two classifications, is invariant. $\qquad\square$

We will denote by $T(\alpha, \beta, \gamma, \delta; \xi)$ an element of this *homographic group* (3.5) or simply $T(\alpha, \beta; \xi)$ in the case $(\gamma = 0, \delta = 1)$. The representative equation is chosen so as to "simplify" some expression, e.g., a three-pole rational fraction the poles of which can be set at predefined locations like $(\infty, 0, 1)$.

Exercise 3.1. Choose a representative for the Riccati equation (1.1) in its equivalence class under the homographic group.

Solution. Two coefficients can be made numeric and the equation reduced to

$$\frac{dU}{dX} + U^2 + \frac{S(X)}{2} = 0, \tag{3.6}$$

under the linear transformation $T(\alpha, \beta; \xi)$

$$u = \alpha U + \beta, \quad X = x, \quad \alpha = -\frac{1}{a_2}, \quad \beta = -\frac{a_2' + a_1 a_2}{2a_2^2}. \tag{3.7}$$

□

This canonical form, in which S is called the *Schwarzian*, will be encountered again in Section 6.3.

3.3.2   The Birational Group

The PP is also invariant under a larger group [56,90], namely the group of birational transformations, in short, the *birational group* $(u, x) \leftrightarrow (U, X)$:

$$\begin{aligned}
u &= r(x, U, dU/dX, \ldots, d^{N-1}U/dX^{N-1}) = 0, \quad x = \Xi(X), \\
U &= R(X, u, du/dx, \ldots, d^{N-1}u/dx^{N-1}) = 0, \quad X = \xi(x),
\end{aligned} \tag{3.8}$$

(N is the order of the equation; r and R are rational in U, u and their derivatives, analytic in x, X).

For instance, given the ODE $u'' - 2u^3 = 0$ and the new dependent variable $U = u' + u^2$, the algebraic elimination of (u', u'') among these two equations and the derivative of the second one yields the inverse transformation $u = U'/(2U)$, which, once inserted in the direct transformation, yields the transformed equation $UU'' - U'^2/2 - 2U^3 = 0$.

3.3.3   Groups of Point Transformations (Cartan Equivalence Classes)

The definition of *to integrate* (Definition 3) allows transformations $(u, x) \leftrightarrow (U, X)$ outside the above two groups, which therefore may alter the PP. For instance, the unstable ODE $2uu' - 1 = 0$ is made stable by $u^2 \mapsto U$. One such group of point transformations, studied by Roger Liouville [83], Tresse [110], and Cartan [21], is defined as (it includes hodograph transformations)

$$u = f(X, U), \quad x = g(X, U), \quad U = F(x, u), \quad X = G(x, u), \tag{3.9}$$

and the variables u and x are equivalent geometric coordinates. This approach provides an additional insight to that of Painlevé, which forbids the exchange of the dependent and the independent variables; see Section 1.3.

The subgroup of "fiber-preserving" transformations

$$u = f(X, U), \quad x = g(X), \quad U = F(x, u), \quad X = G(x), \qquad (3.10)$$

whose equivalence classes are called *Cartan equivalence classes*, has been extensively studied by Kamran et al.; see, e.g., [66].

3.4   The Double Interest of Differential Equations

Let us return to the above problem. At each differential order of the program, the results are twofold (this is the "double interest" of differential equations):

1. some *new functions* (defined from the general solution of a stable ODE that is reducible neither to a lower-order nor to a linear equation),

2. an exhaustive list (i.e., a *classification*) of stable ODEs, which includes the ones defining new functions.

Of course, each equation is characterized by one representative in its equivalence class. Thus, as seen in the introduction, the ODE for $x(t)$ in the case $b = 0, \sigma = \frac{1}{3}$ of the Lorenz model is not distinct, under the homographic group, from the (P3) equation with $\alpha = \beta = 0, \gamma = \delta = 1$.

For instance, the result at order one and degree one (the *degree* of an algebraic ODE is the polynomial degree in the highest derivative) is no new function, one and only one stable equation, the Riccati equation (1.1).

3.5   The Question of Irreducibility

The classical definition of irreducibility as given by the "groupe de rationalité" of Jules Drach (Drach, in Œuvres vol. III, p. 14, [91, p. 246], [101]) relied on the infinite-dimensional differential Galois theory, to whose development Picard and Vessiot contributed, so difficult that it has not yet achieved. Painlevé believed that the theory could be achieved soon, and so used unproven results of Drach in his argumentation. This was pointed out by Roger (not Joseph) Liouville in a passionate discussion with Painlevé in the *Comptes Rendus* (see Œuvres vol. III). A precise, purely algebraic, definition of irreducibility has been given by Umemura [111]; see Okamoto, Chapter 13 of this volume. This definition shares many features with the algorithm of Risch and Norman in computer algebra (which decides whether the primitive of a class of expressions, e.g., rational fractions, is inside or outside the class).

3.6   The Double Method of Painlevé

To solve his problem, Painlevé split it into two parts (this is the "double méthode" [93] p. 11):

1. (a local study) construction of *necessary* conditions for stability,

2. (a global study) proof of their *sufficiency*, by finding the general solution as a finite expression of a finite number of elementary functions (solutions of linear equations, ...) or by proving the irreducibility of the general solution and its freedom from movable critical points.

The methods pertaining to each part are different. If some necessary condition is violated in the first part, one stops and proceeds to the next equations. If one has exhausted the construction of necessary conditions, or if one believes so (indeed, this process, although probably finite, is sometimes not bounded), one turns to the second part, i.e., practically, one tries to integrate (no irreducible equation has been discovered since 1906).

For a good presentation of ideas, see the book of E. Hille [64].

Remark. The reason why movable essential singularities create difficulties lies in the nonexistence of methods to express conditions that they be noncritical (*Leçons*, pp. 519 ff.).

3.7 The Physicist's Point of View

The physicist is not interested in establishing a classification or in finding new functions. Usually, some differential system, whether ordinary or partial, is imposed by physics, and the problem is to "integrate" it in some loose sense. By the way, this loose objective is certainly the main responsible for the numerous, of course divergent, interpretations of "integrable" to be found in the physicists' world.

The best applicability of the present theory arises when one knows nothing or very little about the possible analytical results: first integral, conservation laws of PDEs, particular solutions, ... Then, the first part of the double method of Section 3.6 happens to be a precious *integrability detector*. We have already seen a rough version of it: the meromorphy test of Section 2.3, the final version of which will be the Painlevé test of Section 6.6.

The loose objective of the physicist implies performing the test to its end, even if at some point it fails and should be stopped. One thus gathers a lot of information in the form of necessary conditions for a piece of local single-valuedness to exist. This *partial integrability detector* can be called the "partial Painlevé test" and will be exemplified in Section 6.7.

Examining each condition separately, i.e., independently of the others, or simultaneously, one *may* then find pieces of global information like a first integral or a particular closed-form solution.

4   The Classical Results (L. Fuchs, Poincaré, Painlevé)

The problem of determining all stable equations has been completely studied for several classes of ODEs, while others are still unfinished. We review here the main results achieved to date.

4.1   ODEs of Order One

The completely studied class is (L. Fuchs, Poincaré, Painlevé)

$$E \equiv P(u', u, x) = 0, \quad P \text{ a polynomial in } (u', u), \text{ analytic in } x. \quad (4.1)$$

When the degree is one, i.e., for the class $u' = R(u, x)$ with R rational in u and analytic in x, one finds one and only one stable equation, the Riccati equation (1.1). Since it is linearizable, this case defines no new function.

When the degree is greater than one, one finds one and only one new function, the elliptic function $\wp$ of Weierstrass, defined from (1.13). The stable equations are all the ODEs whose general solution has an algebraic dependence on the arbitrary constant, plus five binomial equations with constant coefficients (i.e., $u'^n = P_m(u)$, $(m, n) \in \mathbb{N}$, P_m a polynomial of degree m). Historically found by Briot and Bouquet, these binomial equations have the following solution (see, e.g., [84, Table 1, p. 73]):

$$u'^n = (u-a)^{n+1}(u-b)^{n-1}, \; n \geq 2, \qquad \frac{u-b}{u-a} = \left[\frac{b-a}{n}(x-x_0)\right]^n, \qquad (4.2)$$

$$u'^2 = (u-a)^2(u-b)(u-c), \qquad \frac{1}{u-a} = A\cosh[B(x-x_0)] + C, \quad (4.3)$$

$$u'^2 = (u-a)(u-b)(u-c)(u-d), \qquad \frac{1}{u-a} = A\wp(x-x_0, g_2, g_3) + B, \quad (4.4)$$

$$u'^3 = [(u-a)(u-b)(u-c]^2, \qquad \frac{1}{u-a} = A\wp'(x-x_0, 0, g_3) + B, \quad (4.5)$$

$$u'^4 = (u-a)^3(u-b)^3(u-c)^2, \qquad \frac{1}{u-c} - \frac{1}{a-c} = A\wp^2(x-x_0, g_2, 0), \quad (4.6)$$

$$u'^6 = (u-a)^5(u-b)^4(u-c)^3, \qquad \frac{1}{u-a} = A\wp^3(x-x_0, 0, g_3) + B, \quad (4.7)$$

in which a, b, c, d are complex constants and (A, B, C, g_2, g_3) algebraic expressions in (a, b, c, d).

Remarks.

- Equation (4.3) is a degeneracy of (4.4). The T transformation $u \rightarrow a_i + 1/u$, with a_i a zero of P_{2n}, generates eleven other equations with $m < 2n$, among them the Weierstrass equation (1.13).

- If the Weierstrass equation had not been known, it would have been discovered at this order as one of the systematic processes of Painlevé.

4.2 ODEs of Order Two, Degree One

The study of the class

$$u'' = R(u', u, x), \quad R \text{ rational in } u', \text{ algebraic in } u, \text{ analytic in } x, \quad (4.8)$$

was started by Painlevé [91,93,95] and finished by his student Gambier [56].

This class provides six new functions, the functions of Painlevé, defined by the ODEs

(P1) $u'' = 6u^2 + x,$

(P2) $u'' = 2u^3 + xu + \alpha,$

(P3) $u'' = \dfrac{u'^2}{u} - \dfrac{u'}{x} + \dfrac{\alpha u^2 + \beta}{x} + \gamma u^3 + \dfrac{\delta}{u},$

(P4) $u'' = \dfrac{u'^2}{2u} + \dfrac{3}{2}u^3 + 4xu^2 + 2(x^2 - \alpha)u + \dfrac{\beta}{u},$

(P5) $u'' = \left[\dfrac{1}{2u} + \dfrac{1}{u-1}\right]u'^2 - \dfrac{u'}{x} + \dfrac{(u-1)^2}{x^2}\left[\alpha u + \dfrac{\beta}{u}\right] + \gamma\dfrac{u}{x} + \delta\dfrac{u(u+1)}{u-1},$

(P6) $u'' = \dfrac{1}{2}\left[\dfrac{1}{u} + \dfrac{1}{u-1} + \dfrac{1}{u-x}\right]u'^2 - \left[\dfrac{1}{x} + \dfrac{1}{x-1} + \dfrac{1}{u-x}\right]u'$
$$+ \dfrac{u(u-1)(u-x)}{x^2(x-1)^2}\left[\alpha + \beta\dfrac{x}{u^2} + \gamma\dfrac{x-1}{(u-1)^2} + \delta\dfrac{x(x-1)}{(u-x)^2}\right],$$

depending on respectively 0, 1, 2, 2, 3, 4 complex parameters, since the homographic group allows one to restrict to $\gamma(\gamma - 1) = 0$, $\delta(\delta - 1) = 0$ for (P3), and to $\delta(\delta - 1) = 0$ for (P5).

The stable equations (4.8) define 53 equivalence classes under the homographic group, including, of course, the six above ones. They split into 50 with R rational in u and 3 with R algebraic in u, and their list can be found in the original articles of Gambier [54, 55], [95], Gambier 1910 thesis [56], Ince 1926 [67] (caution: The numbers 5, 6, 48, 49, 50 of Gambier are changed to 6, 5, 49, 50, 48 in Ince), Murphy 1960 [84], Davis 1961 [46], Bureau M. I 1964, Cosgrove 1993 [38]. In fact, the historical list of Gambier mixes two notions on purpose, namely the irreducibility and the homographic group, which makes this number 53 rather arbitrary; for instance, the classes numbered 1–4, 7–9 by Gambier have been united by Garnier [58] into the single class

$$u'' = \delta(2u^3 + xu) + \gamma(6u^2 + x) + \beta u + \alpha, \quad (4.9)$$

a stable equation admitting a second-order Lax pair.

The 50 stable equations (4.8) with R rational in u define 24 equivalence classes [56] under the birational group and fewer than 24 Cartan equivalence classes [66] under the group of point transformations (3.10).

This extremely important result (the discovery of six new functions, nowadays frequently encountered in physics, and the exhaustive list of 50+3 equations) deserves several comments depending on the field of interest.

Practical usage Given an algebraic second-order, first-degree ODE, either it is transformable by a T transformation (3.5) into one of the $50+3$ equations or not. If it is not, it has movable critical points. If it is, it is *explicitly integrated*, and by looking in the table of Gambier, one knows its general solution as a known finite single-valued expression made of the following functions: solutions of linear equations of order at most four, Weierstrass, (P1) to (P6).

Movable singularities The only movable (and of course noncritical) singularities of these ODEs are poles for $50 + 2$ of them, in addition a nonisolated essential singularity for 1 of them. This result of Painlevé and Gambier (poles are the only movable singularities of stable second-order, first-degree equations rational in (u', u) and analytic in x) is often believed to be more general, leading to wrong definitions for the PP; it is no more true for third order, degree one, or even second order, degree higher than one.

Fixed critical singularities (P1), (P2), (P4) have none, (P3) and (P5) have two transcendental critical points $(\infty, 0)$, both removable by the uniformizing transformation $x \mapsto e^x$. (P6) has three transcendental critical points $(\infty, 0, 1)$.

Dependence on the arbitrary constants What characterizes the 6 Painlevé equations among the 24 is the transcendental (i.e., not algebraic) dependence of their general solution on both constants of integration. The $24 - 6$ equations whose general solution does not involve (P1)–(P6) have either an algebraic dependence on both constants or a semitranscendental dependence (algebraic for one, transcendental for the other one).

Confluence By a confluence process (Painlevé [95], Gambier [54]; see Mahoux, Chapter 2 of this volume), (P6) generates the five others and (P1) generates the Weierstrass equation, so up to now algebraic equations have defined only one master function.

Monodromy (P6) was found independently by R. Fuchs [53] and Schlesinger [106] in the twenty-first problem of Riemann. Given the second-order linear ODE $y''(t) + a_1(t, x, u)y'(t) + a_2(t, x, u)y = 0$ with four Fuchsian singular points $t = (\infty, 0, 1, x)$ (see definitions Section 5.2.1) and an apparently singular point u, a necessary and sufficient condition for the group of monodromy (see Mahoux, Chapter 2 of this volume, for definitions) to be independent of x is that $u(x)$ satisfy equation (P6).

4.3   ODEs of Higher Order or Degree

Painlevé's opinion was that no new function should be expected at third order and that one should go to fourth order. In fact, despite huge efforts, no new function has yet been found.

ODEs of order two, degree higher then one

Only some subclasses have been studied, and their classification is nearly finished. See Chazy thesis [22], Bureau [16], Cosgrove (1993 [40, 41, 43]).

Those of degree two have the necessary form

$$[u'' + E_0 u'^2 + E_1 u' + E_2]^2 = F_0 u'^4 + F_1 u'^3 + F_2 u'^2 + F_3 u' + F_4,$$

with (E_k, F_k) rational in u and analytic in x. Its binomial subset $(E_0 = E_1 = E_2 = 0)$ is classified in [43].

The binomial subset $(u'')^n = F(u', u, x)$ of equations of degree $n \geq 3$ is classified in [40].

Finally, Garnier [59] has classified the set of two-dimensional systems $u' = F(u, v), v' = G(u, v)$, with F and G homogeneous polynomials.

ODEs of order three, degree one

For the general approach, see BSMF p. 252, Acta p. 67. The classification is nearly finished. See the theses of Garnier and Chazy, Bureau (M. II, [20]), Cosgrove [41].

ODEs of order four, degree one

The classification has just begun. See Chazy [22], Bureau [15].

For an account of similar work on PDEs, see [85].

5   Construction of Necessary Conditions. The Theory

The reader interested only in *using* the Painlevé test *may* skip this section, whose relevant parts will anyway be referred to in subsequent sections. By so doing, however, his/her confidence in the Painlevé test will falter at the first encounter of one of the innumerable so-called exceptions, counterexamples, and so on, which are published every year.

This section describes all the methods to build necessary conditions (NC) for the absence of movable critical points in the general solution. Most methods are analytic, and we unify their presentation by describing each of them as a perturbation in a small complex parameter ε, to which can then be applied the theorem of perturbations of Poincaré, itself a generalization of the existence theorem of Cauchy. One of them is arithmetic and leads to Diophantine conditions on the Fuchs indices of a linear differential equation.

What we try to emphasize is the quite small amount of nonlinear features in these methods. Indeed, most of the information is obtained by well-known theories concerning linear equations, whether differential or algebraic.

The detection of singular solutions is first explained in Section 5.1. The linear ODEs are then reviewed from the point of view of interest to us. Then we state the fundamental theorems at the origin of all the methods. For comparison purposes, two equations are defined that will be later processed by all the methods. Finally, we describe each method and apply it to the two examples.

Unless otherwise stated, the class of DEs considered is made of DEs (2.29) polynomial in $\mathbf{u}$ and its derivatives, analytic in x, with $(\mathbf{E}, \mathbf{u})$ multidimensional.

5.1 Removal of Singular Solutions

Since the PP excludes the consideration of singular solutions, one must discard them as early as possible.

Let us give a practical criterion to detect singular solutions.

Theorem 5.1. *A necessary condition for a solution of an ODE to be singular is the existence of a common finite root $u^{(N)}$ to $E = 0$ and its partial derivative with respect to $u^{(N)}$. If $E(u, x) = 0$ depends polynomially on the two highest derivatives $u^{(N)}, u^{(N-1)}$, after factorization of this polynomial existence condition in $u^{(N-1)}$ (called the discriminant), it is necessary that the vanishing factor have an odd multiplicity.*

Proof. See, e.g., Chazy [22]. The condition is not sufficient, and details and examples can be found in [105, Chapter 10].   □

Hence the method: Compute the discriminant, factorize it, discard the even factors, test each odd factor to check whether it defines a solution to the equation.

Chazy [22, p. 358] was the first to notice the absence of correlation between the structure of singularities of the GS and of the SS. Here are some examples of this phenomenon.

Single-valued GS, SS with a movable critical point (Chazy, [22, p. 360]):

$$(u''' - 2u'u'')^2 + 4u''^2(u'' - u'^2 - 1) = 0,$$
$$\text{discriminant} = -16u''^2(u'' - u'^2 - 1),\tag{5.1}$$
$$\text{GS}: u = \frac{e^{c_1 x + c_2}}{c_1} + \frac{c_1^2 - 4}{4c_1}x + c_3, \quad \text{SS}: u = C_2 - \log\cos(x - C_1).$$

GS with a movable critical point, single-valued SS [112, Section 148]:

$$27uu'^3 - 12xu' + 8u = 0,$$
$$\text{discriminant} = -12^3 \times 27^2 u^2 (27u^3 - 4x^3), \tag{5.2}$$
$$\text{GS} : u^3 = c(x - c)^2, \quad \text{SS} : u^3 = (4/27)x^3.$$

Single-valued GS and single-valued SS (Painlevé BSMF p. 239)

$$u''^2 + 4u'^3 + 2(xu' - u) = 0,$$
$$\text{discriminant} = -8(2u'^3 + xu' - u), \tag{5.3}$$
$$\text{GS} : u = \frac{1}{2}v'^2 - 2v^3 - xv, \quad v'' = 6v^2 + x, \quad \text{SS} : u = Cx + 2C^3.$$

5.2 Linear Equations Near a Singularity

Our only interest here is to decide about the local single-valuedness near a singularity $x = x_0$, put for convenience at the origin by a homographic transformation ($x \mapsto x - x_0$ or $x \mapsto 1/x$ according as x_0 is at a finite distance or not).

These results are detailed in Chapter 1.

Consider the most general linear system, put in a form solved for all first-order derivatives (the canonical form of Cauchy):

$$x\frac{d\mathbf{U}}{dx} = \mathbf{AU}, \tag{5.4}$$

with $\mathbf{U}$ a column vector of N components and $\mathbf{A}$ a square matrix rational in x. This can be the representation of the general scalar ODE (3.1), with $U_k = x^k u^{(k)}$, $k = 0, \ldots, N - 1$, and $b_k = x^{N-k} a_k$, e.g., for $N = 3$

$$x\frac{d}{dx}\begin{pmatrix} U_0 \\ U_1 \\ U_2 \end{pmatrix} = \begin{pmatrix} 0 & 1 & 0 \\ 0 & 1 & 1 \\ -b_0 & -b_1 & 2 - b_2 \end{pmatrix}\begin{pmatrix} U_0 \\ U_1 \\ U_2 \end{pmatrix}. \tag{5.5}$$

Definition 5.1. The point $x = 0$ is called *Fuchsian* iff all solutions of (5.4) have polynomial growth near it. It is called *non-Fuchsian* if at least one solution has nonpolynomial growth.

Example 5.1. For the ODE $u' + ax^n u = 0$, n integer, a nonzero, the Fuchsian case is $n \geq -1$, and the non-Fuchsian case is $n \leq -2$. The solution $u(n)$ for $n = -2, -1, 0$ is $u(-2) = e^{a/x}$, $u(-1) = x^{-a}$, $u(0) = e^{-ax}$, and its singularity at $x = 0$ is respectively an isolated noncritical essential point, a critical point or pole or zero (depending on a), a regular point.

Remarks.

1. This definition is the one of modern authors [7]. It involves a property of the solutions, not of the coefficients of the equation. *Fuchsian* denotes at the same time a case with the solutions $u = (x, x^2)$ (classically called *regular point*) and a case with $u = (x^{-1}, x^2)$ (classically called *singular regular point*). The motivation for such a definition is the difficulty to recognize it in matrix notation. While in the scalar case (3.1) the canonical form defined by setting $a_{N-1} = 0$ provides an easy criterion to decide about the nature of the singularity, in the matrix case the example

$$\mathbf{A}/x = \begin{pmatrix} n/x & 0 \\ x^{-n} & 0 \end{pmatrix}, \quad \mathbf{U}_1 = \begin{pmatrix} x^n \\ x \end{pmatrix}, \quad \mathbf{U}_2 = \begin{pmatrix} 0 \\ 1 \end{pmatrix} \qquad (5.6)$$

shows the difficulty in doing so.

2. We avoid the usual synonyms *regular singularity* and *irregular singularity* because of their built-in conflict.

Definition 5.2. Given a point x_0, a *fundamental set of solutions* of a linear ODE of order N is any set of N linearly independent solutions defined in a neighborhood of x_0.

5.2.1 Linear Equations Near a Fuchsian Singularity

Definition 5.3. Given a Fuchsian point $x = 0$, the eigenvalues i of $\mathbf{A}(0)$ are called *Fuchs indices*. The *indicial equation* is the characteristic equation of the linear operator $\mathbf{A}(0)$,

$$\lim_{x \to 0} \det(\mathbf{A}(x) - i) = 0. \qquad (5.7)$$

Near a Fuchsian point $x = 0$, there exists a fundamental set of solutions

$$x^{\lambda_i} \sum_{j=0}^{m_i} \varphi_{ij}(x)(\log x)^j, \quad i = 1, N, \qquad (5.8)$$

in which the λ_i's are complex numbers (the Fuchs indices), m_i positive integers (their multiplicity), φ_{ij} convergent Laurent series of x with finite principal parts.

Series (5.8) are the simplest examples of ψ-series.

A necessary and sufficient condition of local single-valuedness of the general solution of the linear equation is λ_i all integers, no log terms.

In the scalar case (3.1), the indicial equation is

$$0 = \lim_{x \to 0} x^{N-i} E(x, x^i) = \lim_{x \to 0} \det(\mathbf{A}(x) - i)$$
$$= b_0(0) + b_1(0)i + b_2(0)i(i - 1)$$
$$+ \cdots + b_N(0)i(i - 1) \cdots (i - N + 1). \qquad (5.9)$$

Theorem 5.2. *Given a Fuchsian point of the scalar ODE* (3.1), *necessary and sufficient conditions for the general solution to be locally single-valued near it are*

o *the N indices are distinct integers;*

o *$N(N-1)/2$ conditions for the absence of logarithms are satisfied.*

Proof. See Hille [64] or any other textbook.   □

In the matrix case (5.4), these conditions are replaced by

o the N indices are integers;

o the multiplicity of each index i is equal to the dimension of the kernel of $\mathbf{A}(0) - i$;

o all conditions for the absence of logarithms are satisfied.

The search for the no-log conditions can be achieved in one loop, by requiring the existence of a Laurent series extending from the lowest Fuchs index i_1 to $+\infty$,

$$x^{i_1} \sum_{j=0}^{+\infty} u_j x^j, \tag{5.10}$$

and containing N arbitrary independent coefficients; this is a finite process, which terminates when $j + i_1$ reaches the highest Fuchs index. Consider, for instance, the third-order ODE admitting the three solutions u_1, u_2, u_3,

$$x^{-2}, \quad x^{-3} + ax^{-2}\log x, \quad x^{-4} + bx^{-3}\log x + \frac{ab}{2}x^{-2}(\log x)^2,$$

namely

$$\begin{vmatrix} u & u_1 & u_2 & u_3 \\ u' & u_1' & u_2' & u_3' \\ u'' & u_1'' & u_2'' & u_3'' \\ u''' & u_1''' & u_2''' & u_3''' \end{vmatrix} = 0.$$

Its three Fuchs indices $-4, -3, -2$ are simple; it is sufficient that j runs from 0 to 2 with $i_1 = -4$; the condition $b = 0$ is found at $j = 1$ and the condition $a = 0$ at $j = 2$.

In some of the next sections we will encounter nonhomogeneous ODEs in which the rhs is itself a Laurent series with a finite principal part, so we will have to express the single-valuedness of a particular solution as well. This can be incorporated in the single loop described above, provided that it starts from the smallest of the two values i_1 and the singularity order of the particular solution, imposed by the rhs.

5.2.2   Linear Equations Near a Non-Fuchsian Singularity

Near a non-Fuchsian singular point $x = 0$, there exist N linearly independent solutions

$$e^{Q_i(1/z_i)} x^{s_i} \sum_{j=0}^{m_i} \varphi_{ij}(z_i)(\log x)^j, \quad z_i = x^{1/q_i}, \quad i = 1, N, \qquad (5.11)$$

in which the q_i's are positive integers, Q_i polynomials, s_i complex numbers called *Thomé indices*, φ_{ij} *formal* Laurent series with a finite principal part. The question of local single-valuedness of the general solution cannot be settled so easily, because formal series are generically divergent.

5.3   The Two Fundamental Theorems

Theorem I (Cauchy, Picard). *Consider an ODE of order N, of degree one in the highest derivative, defined in the canonical form*

$$\frac{\mathrm{d}\mathbf{u}}{\mathrm{d}x} = \mathbf{K}[x, \mathbf{u}], \quad x \in \mathbb{C}, \quad \mathbf{u} \in \mathbb{C}^N. \qquad (5.12)$$

Let $(x_0, \mathbf{u}_0)$ be a point in $\mathbb{C} \times \mathbb{C}^N$ and D be a domain containing $(x_0, \mathbf{u}_0)$. If $\mathbf{K}$ is holomorphic in D,

- *there exists a solution $\mathbf{u}$ satisfying the initial condition $\mathbf{u}(x_0) = \mathbf{u}_0$,*

- *it is* unique,

- *it is* holomorphic *in a domain containing $(x_0, \mathbf{u}_0)$.*

Proof. See any textbook. For delicate points on this classical theorem, see *Leçons*, p. 394. The contribution of Picard is to have moved the holomorphy property from the hypothesis to the conclusion.   □

There exists an important complement to the theorem of Cauchy, due to Poincaré: The Cauchy solution is also holomorphic in the Cauchy data.

Remark. More practically, the canonical form can also be defined as

$$\frac{\mathrm{d}^N u}{\mathrm{d}x^N} = K[x, u, u', \dots, u^{(N-1)}]. \qquad (5.13)$$

The theorem says nothing whenever the holomorphy of $\mathbf{K}$ is violated, as in the following two cases.

Case 1. $\mathrm{d}u/\mathrm{d}x = u/(u - 1)$, at $u_0 = 1$, a point of meromorphy for $\mathbf{K}$. The only way to possibly remove this singularity without altering the structure of singularities is to perform a T transformation (3.5). The homography $T : 1/(u - 1) = U$ yields a new $\mathbf{K}$, defined by $\mathrm{d}U/\mathrm{d}x = -U^2 - U^3$, which is indeed holomorphic in $\mathbb{C} \times \mathbb{C}$, now making the theorem applicable. In order to shorten the exposition, this step of a homographic transformation will be omitted in Sections 5 and 6, and only recalled for the synthesis of all the methods into the Painlevé test, Section 6.6.

Case 2. $du/dx = \sqrt{4(u - e_1)(u - e_2)(u - e_3)}$, at $u_0 = e_j$, $j = 1, 2, 3$, critical points for **K**.

Example 5.2. $du/dx + u^2 = 0$, with the datum $u = u_0$ at $x = x_0$. The Cauchy solution is represented by the (infinite) Taylor series $u = u_0 \times \sum_{j=0}^{+\infty} [-(x - x_0)u_0]^j$, a geometric series whose sum depends on one, not two, arbitrary constants, the arbitrary location $x_1 = x_0 - u_0^{-1}$ of the movable simple pole; it exists only locally, inside the disk of convergence centered at x_0 with radius $|u_0|^{-1}$. This sum is also represented by the Laurent series $(x - x_1)^{-1}$. One notices the enormous advantage of the Laurent series: It reduces to one term, and it has a much larger domain of definition (the whole complex plane but one point).

Lemma 5.1 (Poincaré, *Mécanique céleste* [100]). *Consider an ODE of order N, of degree one in the highest derivative, depending on a small complex parameter ε, defined in the canonical form*

$$\frac{d\mathbf{u}}{dx} = \mathbf{K}[x, \mathbf{u}, \varepsilon], \quad x \in \mathbb{C}, \quad \mathbf{u} \in \mathbb{C}^N, \quad \varepsilon \in \mathbb{C}. \qquad (5.14)$$

Let $(x_0, \mathbf{u}_0, 0)$ be a point in $\mathbb{C} \times \mathbb{C}^N \times \mathbb{C}$ and D be a domain containing $(x_0, \mathbf{u}_0, 0)$. If $\mathbf{K}$ is holomorphic in D, the Cauchy solution exists and is unique and holomorphic in a domain containing $(x_0, \mathbf{u}_0, 0)$.

Proof. See any textbook. Note that $\mathbf{K}$ may be independent of ε. ☐

Definition 5.4. Given $\mathbf{x}$, the map $\mathbf{u} \mapsto \mathbf{E}(\mathbf{x}, \mathbf{u})$, and some point $\mathbf{u}^{(0)}$, one calls the *differential* of $\mathbf{E}$ at point $\mathbf{u}^{(0)}$ the linear map, denoted by $\mathbf{E}'(\mathbf{x}, \mathbf{u}^{(0)})$, defined by

$$\forall \mathbf{v} : \mathbf{E}'(\mathbf{x}, \mathbf{u}^{(0)})\mathbf{v} = \lim_{\lambda \to 0} \frac{\mathbf{E}(\mathbf{x}, \mathbf{u}^{(0)} + \lambda \mathbf{v}) - \mathbf{E}(\mathbf{x}, \mathbf{u}^{(0)})}{\lambda}. \qquad (5.15)$$

This notion is known under various names: Gâteaux derivative, linearized map, tangent map, Jacobian matrix, and sometimes Fréchet derivative.

Definition 5.5. Given a DE $\mathbf{E}(\mathbf{x}, \mathbf{u}) = 0$ and a point $\mathbf{u}_0$, the linear DE

$$\mathbf{E}'(\mathbf{x}, \mathbf{u}^{(0)})\mathbf{v} = 0 \qquad (5.16)$$

in the unknown $\mathbf{v}$ is called the *linearized equation* in a neighborhood of $\mathbf{u}^{(0)}$ associated to the equation $\mathbf{E}(\mathbf{x}, \mathbf{u}) = 0$.

This is precisely the *équation auxiliaire* (2.20) of Darboux. The auxiliary equation of a linear equation is the linear equation itself.

Let us define the formal Taylor expansions

$$\mathbf{u} = \sum_{n=0}^{+\infty} \varepsilon^n \mathbf{u}^{(n)}, \quad \mathbf{K} = \sum_{n=0}^{+\infty} \varepsilon^n \mathbf{K}^{(n)}. \tag{5.17}$$

The single equation (5.14) is equivalent to the infinite sequence

$$n = 0: \quad \frac{d\mathbf{u}^{(0)}}{dx} = \mathbf{K}^{(0)} = \mathbf{K}[x, \mathbf{u}^{(0)}, 0], \tag{5.18}$$

$$n \geq 1: \quad \frac{d\mathbf{u}^{(n)}}{dx} = \mathbf{K}^{(n)} = \mathbf{K}'[x, \mathbf{u}^{(0)}, 0]\mathbf{u}^{(n)}$$
$$+ \mathbf{R}^{(n)}(x, \mathbf{u}^{(0)}, \dots, \mathbf{u}^{(n-1)}). \tag{5.19}$$

At order zero, the equation is nonlinear.

At order one, the equation, in the particular important case when $\mathbf{K}$ is independent of ε, is the linearized equation (without rhs, $\mathbf{R}^{(1)} = 0$) canonically associated to the nonlinear equation.

At higher orders, this is the same linearized equation with different rhs $\mathbf{R}^{(n)}$ arising from the previously computed terms, and only a particular solution is needed to integrate.

Theorem II (Poincaré 1890, Painlevé BSMF 1900, p. 208, Bureau 1939, M. I). *Take the assumptions of the previous lemma. If the general solution of* (5.14) *is single-valued in D except maybe at $\varepsilon = 0$, then*

- $\varepsilon = 0$ *is no exception, i.e., the general solution is also single-valued there;*

- *every $\mathbf{u}^{(n)}$ is single-valued.*

Proof. See BSMF, p. 208. The main difficulty is to prove the convergence of the series. This theorem remains valid if one replaces "single-valued" (Painlevé version) by "periodic" (Poincaré version) or "free from movable critical points" (Bureau version). □

Remarks.

- This feature (one nonlinear equation (5.18), one linear equation (5.19) with different rhs) is a direct consequence of perturbation theory, it is common to all methods aimed at building necessary stability conditions. The equations may be differential like (5.18)–(5.19), or simply algebraic. Moreover, all the methods that we are about to describe (except the one of Painlevé) will reduce the differential problems to algebraic problems keeping the same feature, and the overall difficulty will be to solve *one* nonlinear algebraic equation, then *one* linear algebraic equation with a countable number (practically, a finite number) of rhs.

- The two theorems and the lemma express a local property, not a global one; therefore, they cannot serve to prove integrability as defined in Definition 3. Conversely, they can be used to disprove the PP. In the same spirit, it is generally useless to try to sum the Taylor or Laurent series that will be defined. Indeed, these series only serve as generators of necessary stability conditions. The proof of sufficiency is achieved by completely different methods.

- The two theorems apply only to ODEs written in the canonical form of Cauchy.

As a summary, the equations successively involved are

- the original DE $\mathbf{E}(x, \mathbf{u}) = 0$, also called unperturbed DE because ε will be introduced into the equation from the outside;

- the perturbed DE $\mathbf{E}(X, \mathbf{U}, \varepsilon)$, obtained from the preceding one by some transformation $(x, \mathbf{u}, \mathbf{E}) \mapsto (X, \mathbf{U}, \mathbf{E}, \varepsilon)$ called perturbation;

- a canonical form (it is not unique) $d\mathbf{U}/dX = \mathbf{K}(X, \mathbf{U}, \varepsilon)$ of the perturbed equation, also called abusively perturbed equation;

- the infinite sequence (5.18)–(5.19) of equations independent of ε.

The methods described in the following sections for establishing necessary stability conditions consist in building one or two perturbed equations from the original unperturbed equation, then applying Theorem II at a point x_0 that is *movable*. This movable point can be either regular (method of Painlevé) or singular noncritical (all the others), which will require its previous transformation to a regular point (by a transformation close to $u \mapsto u^{-1}$) for Theorem II to apply. One is thus led to the equations (5.19), i.e., to *one linear* DE with a sequence of rhs. In order to avoid movable critical points in the original equation, one requires single-valuedness in a neighborhood of x_0 for the general solution of the linear homogeneous equation, and a particular solution of each of the successive linear nonhomogeneous equations.

One must therefore express that a very special class of linear nonhomogeneous DEs has a general solution single-valued in a neighborhood of x_0. Their lhs (homogeneous part) is the linearized equation (équation auxiliaire) of a nonlinear equation that has already passed the requirement $n = 0$ of Theorem II. The rhs (nonhomogeneous part) of equation n depends rationally on $\{\mathbf{u}^{(k)}, k = 0, \dots, n - 1\}$ and their derivatives, all single-valued near x_0 since the necessary conditions have been fulfilled until $n - 1$.

For the coefficients of the homogeneous linear DE, the point x_0 will appear to be either a point of holomorphy (method of Painlevé) or a singular noncritical point (the other methods). In the latter case, both situations (Fuchsian, non-Fuchsian; see Sections 5.2.1 and 5.2.2) will occur.

5.3.1 Two Examples: Complete (P1), Chazy's Class III

Example 5.3 ("Complete (P1)"; [91, p. 224], [67, Section 14.311, p. 329], [14, p. 267], [76]):

$$E \equiv -\frac{d^2u}{dx^2} + c\frac{du}{dx} + eu^2 + fu + g = 0, \qquad (5.20)$$

with (c, e, f, g) analytic in x, and e nonzero. This equation arises in the systematic study of class (4.8) and has led to the discovery of (P1).

Under a transformation $T(\alpha, \beta; \xi)$ (3.5), equation (5.20) is form-invariant [14, p. 267]:

$$-\frac{d^2U}{dX^2} + \left[c - \frac{\xi''}{\xi'} - 2\frac{\alpha'}{\alpha}\right]\xi'^{-1}\frac{dU}{dX} + e\frac{\alpha}{\xi'^2}U^2 + \left[f + c\frac{\alpha'}{\alpha} + 2e\beta - \frac{\alpha''}{\alpha}\right]\frac{U}{\xi'^2}$$

$$+ \left[g + f\beta + c\beta' + e\beta^2 - \beta''\right]\alpha^{-1}\xi'^{-2} = 0, \quad \alpha\xi' \neq 0.$$

This allows us to assign simple predefined values to as many coefficients as gauges in T, i.e., three. For any value of (c, e, f, g) it is possible to choose for the coefficients of dU/dX, U^2, U the values 0, 6, 0, and this requires solving two quadratures and one linear-algebraic equation for $(\alpha, \beta; \xi)$:

$$(\log \alpha)' = \frac{2}{5}\left[c - \frac{e'}{2e}\right], \qquad (5.21)$$

$$\xi'^2 = \frac{e\alpha}{6}, \quad \beta = \frac{1}{2e}[f + c(\log \alpha)' + (\log \alpha)'' + (\log \alpha)'^2]. \qquad (5.22)$$

Consequently, in what follows, one always assumes $c = 0, e = 6, f = 0$ in (5.20).

Example 5.4 (Chazy complete equation of class III [22]).

$$-u_{xxx} + \frac{a}{2}(2uu_{xx} - 3u_x^2) + a_1u_{xx} + c_1uu_x + c_0u_x + d_3u^3$$

$$+ d_2u^2 + d_1u + d_0 = 0, \quad (5.23)$$

where (a, a_i, c_i, d_i) are analytic in x, and a nonzero. This one led Chazy to the discovery of his equation (1.18).

Under a transformation $T(\alpha, \beta; \xi)$, equation (5.23) is form-invariant. For any value of (a, a_i, c_i, d_i) it is possible to choose the values 2, 0, 0 for the coefficients of UU_{XX}, U_{XX}, U^2, and this requires solving the coupled ODEs for $(\alpha, \beta; \xi)$ [notation $\Lambda = \alpha'/\alpha$]:

$$2\Lambda' - \Lambda^2 + 2a^{-3}[(a^2c_1 + 18ad_3)\Lambda + a^2d_2 + 9a'd_3 - 3a^2a_1d_3] = 0, \quad (5.24)$$

$$\xi' = \frac{a\alpha}{2}, \quad \beta = a^{-2}(6a\Lambda + 3a' - aa_1), \qquad (5.25)$$

i.e., one Riccati equation followed by two quadratures. Consequently, in what follows, one always assumes $a = 2$, $a_1 = 0$, $d_2 = 0$ in (5.23).

Neither of these two examples has singular solutions.

Exercise 5.1. Show the impossibility to cancel d_3 in (5.23) by choosing α, β, ξ.

5.4   The Method of Pole-Like Expansions

The method of pole-like expansions is a reliable version of the meromorphy test given in Sections 2.3 and 2.4.

Consider a movable singular point x_0 of either the general solution or a particular solution. Since $\mathbf{u}(x_0)$ is not finite, the theorems of Section 5.3 cannot be applied. It is nevertheless immediate to check that the perturbation

$$x = x_0 + \varepsilon X, \quad u = (\varepsilon X)^p \sum_{n=0}^{+\infty} (\varepsilon X)^n u^{(n)}(x),$$
$$E = (\varepsilon X)^q \sum_{n=0}^{+\infty} (\varepsilon X)^n E^{(n)}(x), \tag{5.26}$$

in which the key point is the dependence of $u^{(n)}$ on x, not X, generates equations $E^{(n)} = 0$, which differ from the algebraic equations $E_j = 0$ defined by (2.8) only by the replacement of x_0 by x. The identification is even complete if χ is defined by $\chi_x = 1$ instead of $\chi = x - x_0$; see Remark page 88.

Fortunately, the method we are about to describe has been made by Bureau (1939) [13], an application of Theorems I and II, as will be seen in Section 5.6. This *method of pole-like expansions* is the most widely used in Painlevé analysis. Initiated by Paul Hoyer [65] and Sophie Kowalevski [73,74], it has been formalized by Gambier [56]; revived by Ablowitz et al. [4], who applied it to wide classes of physical equations; extended to partial differential equations (PDEs) by Weiss et al. [113] (WTC), with technical simplifications by Kruskal [69] and Conte [28,29]. Painlevé himself never used "le procédé connu de Madame Kowalevski ... dont le caractère nécessaire n'était pas établi" (Acta, pp. 10, 83, Œuvres III pp. 196, 269); see Section 2.6.

We now rephrase the steps and generated conditions of Sections 2.3 and 2.4 so as to adapt them to the new objective: the PP. The expansion is written

$$\mathbf{u} = \sum_{j=0}^{+\infty} \mathbf{u}_j \chi^{j+\mathbf{P}}, \quad \mathbf{u}_0 \neq \mathbf{0}, \quad \chi' = 1. \tag{5.27}$$

First Step. Determine all possible families $(\mathbf{p}, \mathbf{u}_0)$. The necessary condition on $(\mathbf{p}, \mathbf{u}_0)$ is the following:

C0. For each family not describing a singular solution, all components of $\mathbf{p}$ are integral.

It there exists no truly singular family (at least one component of $\mathbf{p}$ negative), the method stops without concluding.

Remarks.

- Some components of $\mathbf{u}_0$ can be zero, or even some components of $\mathbf{u}$.

- To avoid missing some family, one should first put the ODE under a canonical form of Cauchy (5.12) or (5.13), so as to enumerate all the points $\mathbf{u}$ that make inapplicable the existence theorem of Cauchy; second, for each such point build a transformed ODE under a homography making the point regular for the Cauchy theorem; third, determine families of the transformed ODE as in Section 2.3.

- The derivative of order k of χ^p does not behave like χ^{p-k} if p is positive and $p - k$ negative.

Second Step. For each family, compute the indicial polynomial $\det \mathbf{P}$. None of the conditions **C1**, **C2**, **C3** of Section 2.4 is required for the existence of the Laurent series, since we also accept particular solutions and exclude only singular solutions.

Third Step. Unchanged as compared to Section 2.4. The condition **C4** is unchanged.

The resulting expansion (5.27) thus contains as many arbitrary coefficients $\mathbf{u}_i$ as the sum of the multiplicities of the distinct positive indices, in addition to the arbitrary location of x_0, associated to the index -1.

For indices that are not positive integers, the method says nothing, not even that they should be integers, and in such a case the expansion (5.27) represents only part of the general solution, without indication about some possible multivaluedness in the missing part.

Remarks.

- The semi-infinite Laurent expansion (5.27) for u about the singular point x_0 is equivalent to an expansion for u^{-1} about a regular point, an expansion, however, different from the Taylor one. This is used in Bureau's method of Section 5.6.

- We prefer the terms "pole-like singularity" to "pole singularity," for the actual singularity of the *general* solution may not be a pole, as shown by the example of Chazy's equation (1.18), for which it is a movable noncritical essential singularity.

- Index -1 also corresponds to an arbitrary coefficient, but since the general solution cannot depend on more than N such arbitrary coefficients, some renormalization occurs. In the example $du/dx + u^2 = 0$ already considered in Section 3.1, the Cauchy solution near the regular point x_0 can be reexpanded,

$$u = \sum_{j=-\infty}^{0} (-u_0)^j (x - x_0)^{j-1} \tag{5.28}$$

[if the example were not so simple, this would be a doubly infinite Laurent series], so as to exhibit an arbitrary coefficient at index $j = i = -1$, the only index of this too simple ODE. Note the "pole-like" singularity x_0, which is in fact an apparent, inessential singularity, in this case a regular point!

5.4.1   The Two Examples

Example 5.5 ("Complete (P1)" equation (5.20)). Already handled in Section 2.3.

Example 5.6 (Chazy's equation (1.18)).

First Step. The dominant terms are among $-u'''$, $2uu'' - 3u'^2$, d_3u^3; hence two possible families:

$$(p, q) = (-1, -4) \quad u_0 = -6 \quad \widehat{E} \equiv u''' - 2uu'' + 3u'^2,$$
$$(p, q) = (-2, -6) \quad d_3u_0^3 = 0 \quad \widehat{E} \equiv -2uu'' + 3u'^2 + d_3u^3.$$

The second family exists only if $d_3 = 0$. See Section 5.9 for a direct proof that $d_3 = 0$ is a necessary stability condition.

Second Step. The indicial polynomial of the first family is

$$\chi^{-(i-4)}[-\partial_x^3 + 2u_0\chi^{-1}\partial_x^2 - 6(u_0\chi^{-1})_x\partial_x + 2(u_0\chi^{-1})_{xx}]\chi^{i-1}$$
$$= -(i-1)(i-2)(i-3) - 12(i-1)(i-2) - 36(i-1) + 2(-1)(-2)$$
$$= -(i+3)(i+2)(i+1), \tag{5.29}$$

and the indices are -3, -2, -1; the algorithm stops here, due to the absence of positive integer indices.

Provided that $d_3 = 0$, the second family has the indices -1, 0.

Third Step (only for the second family, provided that $d_3 = 0$). At the index 0, the condition **C4** $Q_0 = 0$ is satisfied and the algorithm stops.

Exercise 5.2. Find the families and indices of the following equations

$$u'' - 2 = 0, \quad u = (x-a)(x-b), \tag{5.30}$$

$$uu'' - 2u'^2 = 0, \quad u = a(x-x_0)^{-1}, \tag{5.31}$$

$$u'' + 3uu' + u^3 = 0, \quad u = \frac{1}{x-a} + \frac{1}{x-b}. \tag{5.32}$$

5.4.2 Nongeneric Essential-Like Expansions

Just like (5.27), the expansion

$$\mathbf{u} = \sum_{-j=0}^{\infty} \mathbf{u}_j \chi^{j+\mathbf{p}}, \quad \mathbf{u}_0 \neq \mathbf{0}, \tag{5.33}$$

valid outside a disk centered at x_0, i.e., in a neighborhood of the point ∞, is locally single-valued. From this downward Laurent series one could conceive a "method of essential-like expansions" quite similar to the method of pole-like expansions, in order to generate necessary stability conditions, this time from the negative integer indices only.

However, for most equations this method is not applicable. For instance, with the example $-u''' + 2uu'' - 3u'^2 + d_3 u^3 = 0$ (a subset of (1.18)), none of the two expansions (5.33) with $p = -1$ or $p = -2$ exists, unless $d_3 = 0$, which is *not* a reason to conclude that d_3 must vanish.

It applies only to the very restricted class of equations with constant coefficients invariant under a scaling law $(x, \mathbf{u}) \mapsto (kx, k^{\mathbf{p}}\mathbf{u})$, having at least one pole-like family with a negative integer index other than -1. Even then, its failure to detect the movable logarithm in numerous equations that have one makes it of very little use. Such equations are (5.119), (5.134), or the equation $u''' - 7uu'' + 11u'^2 = 0$, whose single family $p = -1$, $u_0 = -2$ has only negative integer indices $(-6, -1, -1)$.

5.5 The α-Method of Painlevé

Consider an ordinary differential equation (2.29), a regular point x_0 (i.e., a point of holomorphy of the function $\mathbf{K}$ when (2.29) is written in the canonical form (5.12)); define a small nonzero complex parameter (which Painlevé denoted by α) and the perturbation

$$\alpha \neq 0 : x = x_0 + \alpha X, \quad \mathbf{u} = \alpha^{\mathbf{p}} \sum_{n=0}^{+\infty} \alpha^n \mathbf{u}^{(n)}, \quad \mathbf{E} = \alpha^{\mathbf{q}} \sum_{n=0}^{+\infty} \alpha^n \mathbf{E}^{(n)} = 0, \tag{5.34}$$

where $\mathbf{p}$ is a sequence of constant integers to be chosen optimally (see example below), and $\mathbf{q}$ another sequence of constant integers determined by $\mathbf{p}$. Then apply Theorem II to the equation for $\mathbf{u}(X, \alpha)$.

At perturbation order zero:

- All the explicit dependence of coefficients on X is removed, i.e., all coefficients of the equation are constant.

- For a suitable choice of **p**, only a few terms survive.

- The equation is invariant under the scaling transformation $(X, \mathbf{u}^{(0)}, \mathbf{E}^{(0)})$ $\mapsto (kX, k^{\mathbf{p}}\mathbf{u}^{(0)}, k^{\mathbf{q}}\mathbf{E}^{(0)})$ (such an equation is called scaled or weighted by physicists).

Definition 5.6. The *simplified equation* (équation simplifiée) associated to a given perturbation (5.34) is the equation of order zero $\mathbf{E}^{(0)}(x_0, \mathbf{u}^{(0)}) = 0$ in the unknown $\mathbf{u}^{(0)}(X)$.

The simplified equation admits the one-parameter solutions

$$\mathbf{u}_0^{(0)}(X - X_0)^{\mathbf{p}},$$

where $\mathbf{u}_0^{(0)}$ are constants. Its above properties usually make it easy to study.

Definition 5.7. The *complete equation* (équation complète), as opposed to the simplified equation, is the equation itself (2.29).

The value $\alpha = 0$ is forbidden in (5.34), but Theorem II takes care of that. The constants **p** and **q** must be integers, chosen so as to satisfy the holomorphy assumption in the small complex parameter of the above lemma. Moreover, since a linear ODE has no movable singularity and since all successive equations $\mathbf{E}^{(n)} = 0$, $n \geq 1$, are linear, the only way to have movable singularities, in order to test their single-valuedness, is to select simplified equations that are truly nonlinear.

The successive steps of the α-method and the generated necessary conditions for stability are (BSMF p. 209, Section 7 and footnote 1)

First Step. Find all sequences **p** of integers satisfying the holomorphy assumptions of Theorem II for the perturbation (5.34). Retain only those defining a truly nonlinear simplified equation. For each sequence **p** perform the next steps.

Remark. If the ODE (2.29) has degree one and order N, the holomorphy assumptions of Theorem II require that the highest derivative contribute to the simplified equation.

Second Step. Find the general solution $\mathbf{u}^{(0)}$ of the simplified equation.

C0. Require $\mathbf{u}^{(0)}$ to be free from movable critical points.

The general solution **v** of the auxiliary equation (5.16) of the simplified equation is then (BSMF p. 209, footnote 1)

$$\mathbf{v} = \sum_{k=1}^{N} d_k \frac{\partial \mathbf{u}^{(0)}}{\partial c_k}, \quad d_k \text{ arbitrary constants,} \tag{5.35}$$

and since $\mathbf{u}^{(0)}$ has no movable critical points, **v** has no movable critical points either (theorem *Leçons* p. 445).

Third Step. For each $n \geq 1$, define $\mathbf{u}^{(n)}$ as a particular solution of equation $\mathbf{E}^{(n)} = 0$ (linear with a rhs), by the classical method of the variation of constants.

C1. Require each $\mathbf{u}^{(n)}$ to be free from movable critical points.

These steps amount to a requirement of stability for the entire sequence of perturbed equations, exactly as formulated in Theorem II.

Remarks.

- In the second step one can take for $\mathbf{u}^{(0)}$ either the general solution or a particular one, but not a singular solution. The drawback of a particular solution will be a lesser number of generated necessary stability conditions. This may be useful when the quadratures of the third step are difficult with the general solution and easy with a particular solution.

- At order $n = 1$ equation $\mathbf{E}^{(1)} = 0$ may contain a rhs, making it different from the auxiliary equation.

- If in the second step only a particular solution has been chosen, it is better that $\mathbf{u}^{(1)}$ be taken as the general solution of the equation $\mathbf{E}^{(1)} = 0$.

Many people have intuitively used the α-method. Let us give two recent examples.

Example 5.7. In the Lorenz model (1.3), the simultaneous change of variables $(x, y, z) \to (\xi, \eta, \zeta)$ and parameters $(b, \sigma, r) \mapsto (b, \sigma, \varepsilon)$ defined by

$$\xi = \varepsilon x, \quad \eta = \varepsilon^2 \sigma y, \quad \zeta = \varepsilon^2 \sigma z, \quad \varepsilon^2 \sigma r = 1 \qquad (5.36)$$

led Robbins [103] to believe to have found a new integrable case, defined by

$$\sigma \neq 0, \quad \varepsilon = 0, \quad r = \infty : \text{first integrals } \xi^2 - 2\zeta, \quad -\xi^2 + \eta^2 + \zeta^2, \quad (5.37)$$

while in fact the new dynamical system is just a simplification of the original one, integrable by elliptic functions.

Example 5.8. The transformation $t \mapsto t^2 \log t$ with "$t \to 0$" from the Lorenz model to the system (24abc) of [82] is in fact the α-transformation $(x, y, z, t) = (\varepsilon^{-1} X, \varepsilon^{-2} Y, \varepsilon^{-2} Z, t_0 + \varepsilon T)$, resulting in the system

$$\frac{dX}{dT} = \sigma(Y - X), \quad \frac{dY}{dT} = -Y - XZ, \quad \frac{dZ}{dT} = XY, \qquad (5.38)$$

whose general solution is elliptic.

5.5.1 The Two Examples

Example 5.9 ("Complete (P1)" equation (5.20) (BSMF p. 224, Section 15)). The Cauchy form of the perturbed equation is

$$-\frac{\mathrm{d}^2(u^{(0)} + \cdots)}{\mathrm{d}X^2} + 6\alpha^{p+2}[u^{(0)} + \cdots]^2 + \alpha^{2-p}[g(x_0) + \alpha X g'(x_0) + \cdots] = 0.$$

First Step. The holomorphy requirement $-2 \le p \le 2$, $p \in \mathbb{Z}$, selects five values for p, and the requirement of a nonlinear simplified equation retains only $p = -2$, i.e., $q = p - 2 = 2p = -4$. The $g(x_0)$ term, which could vanish, does not contribute to the simplified equation

$$E^{(0)} \equiv -\frac{\mathrm{d}^2 u^{(0)}}{\mathrm{d}X^2} + 6u^{(0)^2} = 0. \tag{5.39}$$

Second Step. The general solution of this particular Weierstrass equation is

$$u^{(0)} = \wp(X - c_0, 0, g_3), \quad (c_0, g_3) \text{ arbitrary}. \tag{5.40}$$

The auxiliary equation of the simplified equation is a Lamé equation

$$E'(X, u^{(0)})v \equiv -\frac{\mathrm{d}^2 v}{\mathrm{d}X^2} + 12\wp(X - c_0, 0, g_3)v = 0, \tag{5.41}$$

whose general solution is a linear combination of $\partial\wp/\partial g_3$ and $\partial\wp/\partial c_0$ [5],

$$v = c_1(X\wp' + 2\wp) + c_2\wp', \tag{5.42}$$

without any movable critical singularity.

Third Step. The successive linear equations with their rhs are

$$E'(X, u^{(0)})u^{(1)} = 0, \tag{5.43}$$

$$E'(X, u^{(0)})u^{(2)} = -6u^{(1)^2}, \tag{5.44}$$

$$E'(X, u^{(0)})u^{(3)} = -12u^{(1)}u^{(2)}, \tag{5.45}$$

$$E'(X, u^{(0)})u^{(4)} = -12u^{(1)}u^{(3)} - 6u^{(2)^2}, -g_0 \tag{5.46}$$

$$E'(X, u^{(0)})u^{(5)} = -12(u^{(1)}u^{(4)} + u^{(2)}u^{(3)}) - g_0'X, \tag{5.47}$$

$$E'(X, u^{(0)})u^{(6)} = -12(u^{(1)}u^{(5)} + u^{(2)}u^{(4)}) - 6u^{(3)^2} - \tfrac{1}{2}g_0''X^2, \tag{5.48}$$

with the particular solutions

$$u^{(n)} = 0, \quad n = 1, 2, 3, \tag{5.49}$$

$$u^{(4)} = \frac{g_0}{24}\left[2X\wp\wp' + 2\wp^2 - \zeta\wp'\right], \tag{5.50}$$

$$u^{(5)} = \frac{g_0'}{24}\left[2X^2\wp\wp' + 2X\wp^2 + (X\wp' + 2\wp)\varsigma\right],\tag{5.51}$$

$$u^{(6)} = \frac{g_0''}{48}\left[(X\wp' + 2\wp)(X^2 + 2X\varsigma - 2\log\sigma)\right.$$
$$\left. + \left(X^3\wp + X^2\varsigma - 2X\log\sigma + 2\int\log\sigma\mathrm{d}X\right)\wp'\right],\tag{5.52}$$

where the functions ς and σ obey $\varsigma' = -\wp$, $\sigma' = \varsigma\sigma$. To prevent movable logarithms at $n = 6$ it is necessary that $g''(x_0) = 0$. Since x_0 is arbitrary, this condition is $\forall x : g''(x) = 0$, and Painlevé proved it to be sufficient, thus defining the (new in the sense of Section 3.2) function (P1) with the choice $g = x$.

Remarks.

- The reason why $u^{(1)}$, $u^{(2)}$, $u^{(3)}$ can be chosen zero is given by Theorem II. The reason given in [76, p. 120] is not correct: Even if the general solutions $u^{(1)}$, $u^{(2)}$, $u^{(3)}$ are meromorphic, they can in principle (this does not occur for the ODE under study) generate some multivaluedness further up in the computation. The theorem proven in *Leçons* p. 445 is quite profound: If a [second-order in *Leçons*] ODE is stable, its general solution has a single-valued dependence on the integration constants.

- Taking the particular solution $u^{(0)} = 1/(X - c_0)^2$ instead of the general one $\wp$ (see first remark, Section 5.5) makes all computations immediate ($\varsigma = 1/(X - c_0)$, $\sigma = X - c_0$). For this particular equation, one would not miss the generation of the only necessary stability condition.

Example 5.10 (Chazy complete equation of class III (5.23)).

First Step. For the Cauchy canonical form of the perturbed equation

$$-\frac{\mathrm{d}^2(u^{(0)} + \cdots)}{\mathrm{d}X^2} + 2\alpha^{p+1}u^{(0)}\frac{\mathrm{d}^2 u^{(0)}}{\mathrm{d}X^2} - 3\alpha^{p+1}\left[\frac{\mathrm{d}^2 u^{(0)}}{\mathrm{d}X^2}\right]^2$$
$$+ \cdots + \alpha^{3-p}d_0(x_0) + \cdots = 0,$$

the holomorphy condition is $-1 \le p \le 3$, $p \in \mathbb{Z}$, which the condition for a truly nonlinear simplified equation restricts to $p = -1$, $q = -4$. The value $p = -2$ [31,52] of the method of pole-like expansions is therefore forbidden.

Second Step. The simplified equation is that of Chazy (1.18), whose general solution $u^{(0)}$ is [17, 18] an algebraic transform (finite single-valued expression) of the Hermite modular function $y(X)$,

$$u^{(0)} = [\log(y_X^3 y^{-2}(y - 1)^{-2})]_X,\tag{5.53}$$

evaluated at the point $(c_1 X + c_2)/(c_3 X + c_4), c_1 c_4 - c_2 c_3 = 1$, and thus obviously depending on three arbitrary constants.

The auxiliary equation of Chazy's simplified equation

$$E^{(0)'} v \equiv [-\partial_X^3 + 2u^{(0)}\partial_X^2 - 6u_X^{(0)}\partial_X + 2u_{XX}^{(0)}]v = 0 \qquad (5.54)$$

has the three independent solutions $\partial u^{(0)}/\partial_{c_i}$, $i = 1, 2, 3$, all single-valued.

Instead of the general solution $u^{(0)}$, which would make the computations rather involved, let us restrict to the two-parameter particular solution

$$u^{(0)} = -6\chi^{-1} + c\chi^{-2}, \quad \chi = X - x_0, \quad (x_0, c) \text{ arbitrary constants}, \qquad (5.55)$$

for which the auxiliary equation admits the general solution

$$v = k_2\chi^{-2} + k_3\chi^{-3} + k_4 v_4, \quad v_4 = c^{-2}\chi^{-2}(e^{-2c/\chi} - 1 - 2c\chi^{-1}), \qquad (5.56)$$

with (k_2, k_3, k_4) arbitrary constants.

Third Step. The successive linear equations with their rhs are

$$-E^{(0)'} u^{(1)} = c_{1,0} u^{(0)} u^{(0)'} + d_{3,0} u^{(0)^3}, \qquad (5.57)$$

$$-E^{(0)'} u^{(2)} = 2u^{(1)} u^{(1)''} - 3u^{(1)'^2} + c_{1,1} X u^{(0)} u^{(0)'} + c_{1,0} u^{(0)'}(u^{(0)} + u^{(1)})$$
$$+ d_{3,1} X u^{(0)^3} + 3d_{3,0} u^{(0)^2} u^{(1)} + c_{0,0} u^{(0)'}. \qquad (5.58)$$

A particular solution of the first one is provided by the method of variation of the constants $u^{(1)} = K_2(X)\chi^{-2} + K_3(X)\chi^{-3} + K_4(X)v_4$:

$$K_2'\chi^{-2} + K_3'\chi^{-3} + K_4'v_4 = 0, \qquad (5.59)$$

$$-2K_2'\chi^{-3} - 3K_3'\chi^{-4} + K_4'v_4' = 0, \qquad (5.60)$$

$$6K_2'\chi^{-4} + 12K_3'\chi^{-5} + K_4'v_4'' = c_{1,0}(-6\chi^{-1} + c\chi^{-2})(6\chi^{-2} - 2c\chi^{-3})$$
$$+ d_{3,0}(-6\chi^{-1} + c\chi^{-2})^3. \qquad (5.61)$$

To prevent a movable logarithm in K_2 (resp. K_3), it is necessary that in the rhs of last equation the coefficients of χ^{-5} and χ^{-6} vanish:

$$\forall(x_0, c) : -2c^2 c_1(x_0) - 18c^3 d_3(x_0) = 0, \quad c^3 d_3(x_0) = 0; \qquad (5.62)$$

hence the two necessary stability conditions $\forall x$: $d_3(x) = c_1(x) = 0$, obtained at the perturbation order $n = 1$. We leave it as an exercise to check that after completion of $n = 4$, one has obtained all the conditions $(c_1 = c_0 = d_3 = d_1 = d_0 = 0)$ that Chazy proved to be necessary and sufficient.

Theorem 5.3. *For any family of the method of pole-like expansions, the value $i = -1$ is a Fuchs index.*

Proof. Let $\mathbf{u} \sim \mathbf{u}_0 \chi^{\mathbf{p}}$ be such a family and $\widehat{\mathbf{E}}(x, \mathbf{u})$ be the dominant terms. The equation $\widehat{\mathbf{E}}(x_0, \mathbf{u}) = 0$ admits as a particular solution the monomial $X \mapsto \mathbf{u} = \mathbf{u}_0(x_0)(X - X_0)^{\mathbf{p}}$; therefore the linearized equation at the leading term (2.20) admits as a particular solution its derivative with respect to X_0 : $X \mapsto \text{const} \times \partial_{X_0}(X - X_0)^{\mathbf{p}}$. Since at least one component of $\mathbf{p}$ is negative, the associated component of $\partial_{X_0}(X - X_0)^{\mathbf{p}}$ is proportional to $(X - X_0)^{\mathbf{p}-1}$; therefore $i = -1$ is a root of the indicial equation (2.35). $\quad\square$

5.5.2 General Stability Conditions (ODE of Order m and Degree 1)

Using his method, Painlevé could obtain quite general necessary stability conditions for algebraic ODEs of arbitrary order and degree; cf. BSMF [91, p. 258], Acta [93, p. 74], Chazy (thesis) [22]. Consider the class, defined in the Cauchy canonical form,

$$u^{(m)} = R(u^{(m-1)}, u^{(m-2)}, \dots, u', u, x), \tag{5.63}$$

with R rational in u and its derivatives, analytic in x [for R algebraic, and for arbitrary order and degree, cf. Acta pp. 73, 77]. *Necessary stability conditions* are:

C1. As a rational fraction of $u^{(m-1)}$, R is a polynomial of degree at most two:

$$u^{(m)} = Au^{(m-1)^2} + Bu^{(m-1)} + C. \tag{5.64}$$

C2. As a rational fraction of $u^{(m-2)}$, A has only simple poles a_i with residues r_i equal to $1 - 1/n_i$, n_i nonzero integers possibly infinite:

$$A = \sum_i \frac{1 - 1/n_i}{u^{(m-2)} - a_i}. \tag{5.65}$$

The above sum is finite.

For second-order $m = 2$ the fraction A has at most four simple poles, and the set of their residues can only take the five values of Table 3.1,

$$A = \sum_{i=1}^{4} \frac{r_i}{u - a_i}, \quad \sum_{i=1}^{4} r_i = 2, \quad r_i = 1 - \frac{1}{n_i}, \quad n_i \in \mathbb{Z} \text{ or } n_i = \infty. \tag{5.66}$$

Note the one-to-one correspondence between Table 3.1 and the list of powers of the five Briot–Bouquet equations (4.3)–(4.7).

Exercise 5.3. For the six equations (Pn), determine the set (a_i, r_i) of simple poles with their residues.

TABLE 3.1. Order two, degree one. Number of poles (nonzero r_i), list of their residues. The poles may be located at ∞ and may not be distinct. The type numbering convention is that of [84, Table I, p. 169)]. The least common multiplier (lcm) is shown for convenience.

Type	lcm(r_i)	r_1	r_2	r_3	r_4
I	$n \geq 1$	$1 + 1/n$	$1 - 1/n$	0	0
III	2	1/2	1/2	1/2	1/2
IV	3	2/3	2/3	2/3	0
V	4	3/4	3/4	1/2	0
VI	6	5/6	2/3	1/2	0

Solution.

(P6)	$(\infty, 1/2),\ (0, 1/2),\ (1, 1/2),\ (x, 1/2),$	(5.67)
(P5)	$(\infty, 1/2),\ (0, 1/2),\ (1, 1),$	(5.68)
(P4)	$(\infty, 3/2),\ (0, 1/2),$	(5.69)
(P3)	$(\infty, 1),\ (0, 1),$	(5.70)
(P2)	$(\infty, 2),$	(5.71)
(P1)	$(\infty, 2).$	(5.72)

For instance, (P4) belongs to type I of Table 3.1, and it is also a confluent case of types III, V, VI.   □

The similar finite lists of admissible values of A for any order m can be found in Painlevé ($m = 3$ Acta p. 68, Œuvres vol. III p. 254; $m \geq 4$ Acta p. 75, Œuvres vol. III p. 261).

C3. As rational fractions of $u^{(m-2)}$, B and C have no other poles than those of A, and these poles are all simple. Writing B, C as rational fractions of $u^{(m-2)}$ whose denominators are that of A, this implies the degree limitations

$$\text{(order 2, degree 1)} : \deg \text{num } B \leq 1, \ \deg \text{num } C \leq 3. \qquad (5.73)$$

C4. (Chazy, Thesis) Every ODE $u^{(m-2)} - a_i = 0$ (a denominator of A) is stable.

C5. ([24]). All polynomial degrees in $u^{(k)}$, $k = 0, \ldots, m - 2$ (of the numerator and denominator of A, B, C written as irreducible fractions of u and its derivatives) are limited, except in the "Fuchsian" case $n_i = -2$, $r_i = \frac{3}{2}$ (see [24] for details).

For additional conditions, see [92].

5.6 The Method of Bureau

Firstly, this method exhibits a linear differential equation with a Fuchsian singularity that allows us to interpret the indices i in the recursion relation of Kowalevski as Fuchs indices. Secondly, it brings rigor to the heuristic method of Kowalevski and Gambier. However, the generated no-log conditions are identical to those of the method of pole-like expansions.

Consider an Nth-order ODE $E(x, u) = 0$ (for simplicity, one assumes u and E unidimensional; the multidimensional case is handled in [19]) and a movable noncritical singular point x_0 where the general solution behaves like $u \sim u_0(x - x_0)^p$, with p a negative integer to be determined.

The integer p is computed by the method of pole-like expansions, and the highest derivative is required to contribute (M. II, p. 9) in order to be sure that one deals with the general, not a singular, solution.

One wants to apply the two fundamental theorems. Since the singularity x_0 violates the holomorphy assumption of Theorem I, one defines an equivalent differential system (in fact, two systems) for which x_0 is a point of holomorphy. These systems will depend on a perturbation parameter ε.

One first defines two new dependent variables (z, U) by the relations (Gambier, Thesis [56, p. 50], Bureau 1939 [13])

$$u = sz^p, \quad \frac{\mathrm{d}z}{\mathrm{d}x} = 1 + Uz, \quad s \neq 0. \tag{5.74}$$

Elimination of u and the derivatives of z (M. II, pp. 13, 77):

$$z^{-p}u = s, \tag{5.75}$$

$$z^{-p+1}\frac{\mathrm{d}u}{\mathrm{d}x} = ps + \left(\frac{\mathrm{d}s}{\mathrm{d}x} + psU\right)z, \tag{5.76}$$

$$z^{-p+2}\frac{\mathrm{d}^2u}{\mathrm{d}x^2} = p(p-1)s + \left(2p\frac{\mathrm{d}s}{\mathrm{d}x} + p(2p-1)sU\right)z$$
$$+ \left(\frac{\mathrm{d}^2s}{\mathrm{d}x^2} + 2p\frac{\mathrm{d}s}{\mathrm{d}x}U + p^2sU^2 + ps\frac{\mathrm{d}U}{\mathrm{d}x}\right)z^2, \tag{5.77}$$

etc., transforms E into

$$E \equiv E\left(x, U, \frac{\mathrm{d}U}{\mathrm{d}x}, \ldots, \frac{\mathrm{d}^{(N-1)}U}{\mathrm{d}x^{N-1}}, s, \frac{\mathrm{d}s}{\mathrm{d}x}, \ldots, \frac{\mathrm{d}^{(N)}s}{\mathrm{d}x^N}, z\right) = 0, \tag{5.78}$$

an equation for U of order $N-1$ polynomial in z. For the equivalent system (5.74), (5.78) made of two ODEs of orders one and $N-1$ in the unknowns (z, U), the point $z = 0$ is still a point of meromorphy; see examples below.

To remove it, one introduces a dependence in a small nonzero parameter ε to obtain a perturbed system to which Theorem II can be applied. Two such perturbations have been defined [13].

First Perturbation of Bureau.

$$x = x_0 + \varepsilon X, \quad z = \varepsilon Z, \quad U \text{ unchanged} : E \equiv (\varepsilon Z)^q \sum_{n=0}^{+\infty} (\varepsilon Z)^n E^{(n)} = 0, \qquad (5.79)$$

where the positive integer $-q$ is the singularity order of E. The coefficients must be expanded as Taylor series as in the α-method:

$$s(x) = s_0 + (\varepsilon X)s_0' + \cdots, \quad s_0^{(k)} = \frac{d^{(k)} s}{dx^k}(x_0),$$

$$a(x) = a_0 + (\varepsilon X)a_0' + \cdots. \qquad (5.80)$$

Expansions up to order one in ε for the above derivatives (5.75)–(5.77) are

$$z^{-p}u = s_0 + (s_0' X)\varepsilon + O(\varepsilon^2),$$

$$z^{-p+1}\frac{du}{dx} = ps_0 + ((ps_0')X + (s_0' + ps_0 U)Z)\varepsilon + O(\varepsilon^2),$$

$$z^{-p+2}\frac{d^2u}{dx^2} = p(p-1)s_0$$

$$+ p\left((p-1)s_0'X + \left(2s_0' + (2p-1)s_0 U + s_0 Z\frac{dU}{dX}\right)Z\right)\varepsilon + O(\varepsilon^2),$$

etc., together with $dZ/dX = 1 + \varepsilon ZU = 1 + O(\varepsilon)$.

Order zero is an algebraic equation $E^{(0)}(x_0, s_0) = 0$ for the nonzero coefficient s_0.

Order one is subtle: It filters out all terms nonlinear in U and its derivatives $d^{(k)}U/dX^k$, and it extracts the contribution of $d^{(k)}U/dX^k$ from the term z^{k+1} in the expansions (5.75)–(5.77). This results in

$$E^{(1)} \equiv A\frac{X}{Z} + B + \sum_{k=0}^{N-1} c_k Z^k \frac{d^{(k)}U}{dX^k} = 0, \quad (A, B, c_k) \text{ constant.} \qquad (5.81)$$

Since dZ/dX is unity at this order, this is a linear nonhomogeneous ODE of order at most $N-1$ for U with constant coefficients, whose homogeneous part is by construction of Fuchsian type (exactly one singular point $Z = 0$, of the singular regular type) and even Eulerian type.

In order to be sure of dealing with the general solution of the original nonlinear ODE, the linear ODE (5.81) must have exactly the order $N-1$; a necessary stability condition for the nonlinear ODE is the single-valuedness of the general solution of the linear ODE (5.81). Hence the necessary conditions, for each value of (p, s_0):

the order of the linear ODE at perturbation order one is exactly $N-1$;

its $N-1$ Fuchs indices are distinct integers;

if 0 is an index, the rhs vanishes ($A = B = 0$ condition for the particular solution to contain no logarithm).

Since (5.81) is Eulerian, these conditions are sufficient for the general solution of the linear ODE (5.81) to be single-valued, but only necessary for the stability of the nonlinear ODE.

Higher perturbation orders yield no information. The reasoning is then that any condition thus found at $x = x_0$, such as $s_0 = 1$, is valid at any x, since x_0 is arbitrary.

Second Perturbation of Bureau.

$$x \text{ unchanged, } z = \varepsilon Z, \ U = \sum_{n=1}^{+\infty} (\varepsilon Z)^{n-1} U^{(n)} :$$

$$E \equiv (\varepsilon Z)^q \sum_{n=0}^{+\infty} (\varepsilon Z)^n E^{(n)}. \quad (5.82)$$

Expansions for the above derivatives (5.75)–(5.77) are

$$z^{-p} u = s$$

$$z^{1-p} \frac{du}{dx} = ps + (psU^{(1)} + s')\varepsilon Z + psU^{(2)}(\varepsilon Z)^2 + psU^{(3)}(\varepsilon Z)^3 + O(\varepsilon^4),$$

$$z^{2-p} \frac{d^2 u}{dx^2} = p(p-1)s + p((2p-1)sU^{(1)} + 2s')\varepsilon Z,$$

$$+ \left(p^2 s \big(2U^{(2)} + U^{(1)^2} \big) + 2ps' U^{(1)} + s'' + ps \frac{dU^{(1)}}{dx} \right) (\varepsilon Z)^2 + O(\varepsilon^3),$$

etc., together with $\varepsilon dZ/dx = 1 + \varepsilon Z U^{(1)} + (\varepsilon Z)^2 U^{(2)} + O(\varepsilon^3)$.

Equation $E^{(0)}(x, s) = 0$, $s \neq 0$, is the same algebraic equation as above for the unknown $s(x)$, not $s(x_0)$. Each perturbation order $n \geq 1$ defines a linear algebraic equation

$$\forall n \geq 1 : P(n) U^{(n)} + Q_n\big(x, U^{(1)}, \ldots, U^{(n-1)}\big) = 0, \quad (5.83)$$

where $P(n)$ is the indicial polynomial of Fuchsian equation (5.81), and Q_n depends on the previously computed coefficients. Necessary stability conditions $Q_i = 0$ arise at every value of i that is also one of the $N - 1$ Fuchs indices. These conditions are identical to those of the method of pole-like expansions, as proven in Section 5.6.1.

The successive steps and generated necessary conditions of the method of Bureau are as follows:

Step a. Determine all possible p as in the method of pole-like expansions (details M. I p. 256, M. II p. 9). For all p satisfying (**C0, C1**), perform Step b.

C0. All p are integers.

C1. The linear ODE (order one of first perturbation) has exactly order $N - 1$ [this holomorphy condition excludes, for instance, $p = -2$ in Chazy]. This implies the necessity for the highest derivation order to contribute to the dominant part during the computation of p.

Step b. Solve the algebraic equation for s_0 at order zero of first perturbation. For all nonzero s_0 perform Steps c and d.

Step c. Solve the linear nonhomogeneous Euler equation for $U(Z)$ at order one of first perturbation.

C2. Its $N - 1$ Fuchs indices are distinct integers.

C3. If 0 is an index, the nonhomogeneous part vanishes.

Step d. Solve the linear algebraic equation (5.83) (order n of second perturbation) from $n = 1$ to the highest Fuchs index.

C4. Whenever the order n in Step d is a Fuchs index i, Q_i is zero.

As compared with the α-method, these stability conditions are directly taken at x, not at x_0. However, the method provides no conditions from the negative integer indices.

5.6.1 Bureau Expansion vs. Pole-Like Expansion

Let us first prove the existence of a one-to-one correspondence between the coefficients $U^{(n)}$ of Bureau (second perturbation) and those u_j of the method of pole-like expansions. The relations defining Bureau coefficients are

$$u = sz^p, \tag{5.84}$$

$$\frac{\mathrm{d}z}{\mathrm{d}x} = 1 + U^{(1)}z + U^{(2)}z^2 + O(z^3), \tag{5.85}$$

and those defining the pole-like expansion are

$$u = \chi^p(u_0 + u_1\chi + u_2\chi^2 + O(\chi^3)), \tag{5.86}$$

$$\frac{\mathrm{d}\chi}{\mathrm{d}x} = 1. \tag{5.87}$$

The property $\chi_x = 1$ of χ first ensures $s = u_0$ [taking $\chi = x - x_0$ would just create useless complications]. The elimination of u between (5.84) and (5.86) yields

$$z = \chi\left(1 + \frac{u_1}{u_0}\chi + \frac{u_2}{u_0}\chi^2 + O(\chi^3)\right)^{1/p}$$

$$= \chi\left(1 + \frac{u_1}{pu_0}\chi + \frac{2pu_2 + (1-p)u_1^2}{2p^2u_0^2}\chi^2 + O(\chi^3)\right). \tag{5.88}$$

Let us invert this Taylor series z of χ into a Taylor series χ of z:

$$\chi = z\left(1 - \frac{u_1}{pu_0}z + \frac{-2pu_2 + (3+p)u_1^2}{2p^2u_0^2}z^2 + O(z^3)\right). \qquad (5.89)$$

One finally substitutes this χ and dz/dx, both Taylor series in z, into (5.87) to obtain

$$\frac{d\chi}{dx} = 1$$
$$= \left(1 - \frac{2u_1}{pu_0}z + 3\frac{-2pu_2 + (3+p)u_1^2}{2p^2u_0^2}z^2 + O(z^3)\right)$$
$$\times \left(1 + U^{(1)}z + U^{(2)}z^2 + O(z^3)\right) - \frac{1}{p}\frac{d}{dx}\left(\frac{u_1}{u_0}\right)z^2 + O(z^3). \qquad (5.90)$$

The identification of the lhs and rhs as series in z provides the correspondence between the two sets of coefficients:

$$s = u_0, \qquad (5.91)$$

$$U^{(1)} = \frac{2u_1}{pu_0}, \qquad (5.92)$$

$$U^{(2)} = \frac{3u_2}{pu_0^2} + \left(\frac{2u_1}{pu_0}\right)^2 - (3p+1)\frac{u_1^2}{2p^2u_0^2} + \frac{1}{p}\frac{d}{dx}\left(\frac{u_1}{u_0}\right), \qquad (5.93)$$

or

$$u_0 = s, \qquad (5.94)$$

$$u_1 = \frac{p}{2}sU^{(1)}, \qquad (5.95)$$

$$u_2 = \frac{p}{3}s^2U^{(2)} + p\frac{3p+1}{24}s^2U^{(1)^2} - \frac{p}{6}s^2\frac{dU^{(1)}}{dx}. \qquad (5.96)$$

This bijection between the coefficients induces a bijection between the equations $E^{(n)} = 0$ of the expansion of Bureau (second perturbation) and the equations $E_j = 0$ of the method of pole-like expansions, hence a bijection between the no-log conditions.

This proves the equivalence between the method of pole-like expansions and the second perturbation of Bureau. As to the first perturbation of Bureau, it brings quite important information *not* obtainable by the method of pole-like expansions.

5.6.2   The Two Examples

Example 5.11 ("Complete (P1)" equation (5.20). See [14, equations (17.3), (21.2)].). The unperturbed equivalent meromorphic system

(5.74)–(5.78) in (z, U) is, in Cauchy form,

$$\frac{dz}{dx} = 1 + Uz,$$

$$\frac{dU}{dx} = 6s^2 z^p - (p-1)z^{-2} - \left[2\frac{s'}{s} + (2p-1)U\right]z^{-1} \tag{5.97}$$

$$- \frac{s'pU + s''}{ps} + \frac{g}{ps}z^{-p}.$$

Step a. The only value is $p = -2$, integer. The original ODE then reads, by increasing powers of z,

$$E \equiv 6s(1-s)z^{-4} + 2s\left[5U - \frac{2}{s}\frac{ds}{dx}\right]z^{-3}$$

$$- 2s\left[\frac{dU}{dx} + \frac{2}{s}\frac{ds}{dx}U - \frac{1}{2s}\frac{d^2s}{dx^2} - 2U^2\right]z^{-2} - g = 0. \tag{5.98}$$

Step b. Equation $E^{(0)} \equiv 6s_0(1 - s_0) = 0$ has for its only nonzero solution $s_0 = 1$.

Step c. At order one,

$$\frac{E^{(1)}}{2s_0} \equiv -Z\frac{dU}{dX} + 5U - 2\frac{s_0'}{s_0} + 3\frac{s_0'}{s_0}\frac{X}{Z} = 0, \quad \frac{dZ}{dX} = 1 + O(\varepsilon Z). \tag{5.99}$$

The only Fuchs index is $i = 5$. There is no condition on the rhs.

Step d. The computation presents no difficulty:

$$s = 1, \quad U^{(1)} = U^{(2)} = U^{(3)} = 0, \quad U^{(4)} = \frac{g}{4}, \quad U^{(5)} = \frac{g'(x)}{4}, \quad Q_6 \equiv -\frac{g''(x)}{2}.$$

Remark. On the Cauchy form (5.97) with $p = -2$, $s = 1$, one sees easily how perturbations I and II remove the meromorphy.

Example 5.12 (Chazy complete equation of class III (5.23)).

Step a. The two solutions are $p = -1, p = -2$. For $p = -2$ the computation of the linear equation (5.81) yields a zero coefficient for d^2U/dX^2, thus violating condition **C1**.

For $p = -1$ the original ODE then reads, by increasing powers of z,

$$E \equiv s(s+6)z^{-4} + s\left[6U - \frac{6}{s}\frac{ds}{dx} + 2\frac{ds}{dx} - c_1s + d_3s^2\right]z^{-3}$$

$$+ s\left[-2(s+2)\frac{dU}{dx} + \left(\left(2 - \frac{9}{s}\right)\frac{ds}{dx} - c_1s\right)U - c_0\right.$$

$$\left. + c_1\frac{ds}{dx} - \frac{3}{s}\left(\frac{ds}{dx}\right)^2 + \left(2 + \frac{3}{s}\right)\frac{d^2s}{dx^2} + (7-s)U^2\right]z^{-2}$$

$$+ s\left[\frac{\mathrm{d}^2 U}{\mathrm{d}x^2} + 3\frac{\mathrm{d}s}{\mathrm{d}x}\frac{\mathrm{d}U}{\mathrm{d}x} + \left(\frac{3}{s}\frac{\mathrm{d}^2 s}{\mathrm{d}x^2} - c_0\right)U + d_1\right.$$

$$\left.+ \frac{c_0}{s}\frac{\mathrm{d}s}{\mathrm{d}x} - \frac{1}{s}\frac{\mathrm{d}^3 s}{\mathrm{d}x^3} - \frac{3}{s}\frac{\mathrm{d}s}{\mathrm{d}x}U^2 - 3U\frac{\mathrm{d}U}{\mathrm{d}x} + U^3\right]z^{-1} + d_0. \quad (5.100)$$

Step b. Equation $E^{(0)} \equiv s_0(s_0 + 6) = 0$ has for its only nonzero solution $s_0 = -6$.

Step c. At order one,

$$\frac{E^{(1)}}{s_0} \equiv Z^2\frac{\mathrm{d}^2 U}{\mathrm{d}X^2} - 2(s_0 + 2)Z\frac{\mathrm{d}U}{\mathrm{d}X} + 12U - s_0 c_{1,0} - s_0^2 d_{3,0} + \left(2 - \frac{6}{s_0}\right)s_0'$$

$$+ \left(2 + \frac{6}{s_0}\right)s_0'\frac{X}{Z} = 0, \quad \frac{\mathrm{d}Z}{\mathrm{d}X} = 1 + O(\varepsilon Z).$$

The Fuchs indices are $i = -4, -3$, there is no condition on the rhs, and the algorithm stops here, due to the absence of positive integer indices.

5.7 The Fuchsian Perturbative Method

The Fuchsian perturbative method allows one to extract the information contained in the negative indices [52], thus building infinitely many necessary conditions for the absence of movable critical singularities of the logarithmic type [31].

The perturbation that describes it is close to the identity

$$x \text{ unchanged}, \quad \mathbf{u} = \sum_{n=0}^{+\infty}\varepsilon^n \mathbf{u}^{(n)}, \quad \mathbf{E} = \sum_{n=0}^{+\infty}\varepsilon^n \mathbf{E}^{(n)} = 0, \quad (5.101)$$

where, as for the α-method, the small parameter ε is not in the original equation.

Then, the single equation (2.29) is equivalent to the infinite sequence

$$n = 0 \quad \mathbf{E}^{(0)} \equiv \mathbf{E}(x, \mathbf{u}^{(0)}) = 0, \quad (5.102)$$

$$\forall n \geq 1 : \quad \mathbf{E}^{(n)} \equiv \mathbf{E}'(x, \mathbf{u}^{(0)})\mathbf{u}^{(n)} + \mathbf{R}^{(n)}(x, \mathbf{u}^{(0)}, \ldots, \mathbf{u}^{(n-1)}) = 0, \quad (5.103)$$

with $\mathbf{R}^{(1)}$ identically zero. From Theorem II, necessary stability conditions are:

the general solution $\mathbf{u}^{(0)}$ of (5.102) has no movable critical points;

the general solution $\mathbf{u}^{(1)}$ of (5.103) has no movable critical points;

for every $n \geq 2$ there exists a particular solution of (5.103) without movable critical points.

Order zero is just the complete equation for the unknown $\mathbf{u}^{(0)}$, so to get some information one must apply Theorem II for a perturbation different from (5.101). Since Bureau has proven that the method of pole-like expansions, with more rigorous assumptions, can be cast into an application of the two basic theorems, one uses it at order zero *only* to obtain the leading term $\mathbf{u}^{(0)} \sim \mathbf{u}_0^{(0)} \chi^{\mathbf{p}}$ of all the families of movable singularities.

First Step. Determine all possible families $(\mathbf{p}, \mathbf{u}_0^{(0)})$

$$\mathbf{u}^{(0)} \sim \mathbf{u}_0^{(0)} \chi^{\mathbf{p}}, \quad \mathbf{E}^{(0)} \sim \mathbf{E}_0^{(0)} \chi^{\mathbf{q}}, \quad \mathbf{u}_0^{(0)} \neq \mathbf{0} \qquad (5.104)$$

that do not describe a singular solution, by solving the algebraic equation

$$\mathbf{E}_0^{(0)} \equiv \lim_{\chi \to 0} \chi^{-\mathbf{q}} \hat{\mathbf{E}}(x, \mathbf{u}_0^{(0)} \chi^{\mathbf{p}}) = \mathbf{0}. \qquad (5.105)$$

C0. All components of $\mathbf{p}$ are integral.

If there exists no family that is truly singular (at least one component of $\mathbf{p}$ negative), the method stops without concluding.

Second Step. For each family, compute the indicial equation (2.35) and require the following necessary conditions:

C2. Every zero of $\det \mathbf{P}$ (a Fuchs index) is integral.

C3. Every zero i of $\det \mathbf{P}$ has a multiplicity equal to the dimension of the kernel of $\mathbf{P}(i)$:

$$\forall \text{ index } i : (\text{multiplicity of } i) = \dim \operatorname{Ker} \mathbf{P}(i). \qquad (5.106)$$

Remark. There is no such condition like **C1** on page 86, i.e., the indicial polynomial may have degree smaller than N. If the indicial equation has degree N, the conditions **C2** and **C3** (N distinct integers in the one-dimensional case) are slightly stronger than the conditions in Bureau ($N-1$ distinct integers).

The next step is easily computerizable [27,47] if one represents $\mathbf{u}^{(0)}$, $\mathbf{u}^{(1)}$, ... , as Laurent series bounded from below: $\mathbf{u}^{(0)}$ with powers in the range $(\mathbf{p} : +\infty)$, $\mathbf{u}^{(1)}$ with powers in the range $(\rho + \mathbf{p} : +\infty)$, where ρ denotes the smallest Fuchs index, an integer less than or equal to -1, etc..

Order $n = 0$ is identical to the method of pole-like expansions, and the Laurent series for $\mathbf{u}^{(0)}$,

$$\mathbf{u}^{(0)} = \sum_{j=0}^{+\infty} \mathbf{u}_j^{(0)} \chi^{j+\mathbf{p}}, \qquad (5.107)$$

represents a particular solution containing a number of arbitrary coefficients equal to one (index -1) plus the number of positive Fuchs indices, counting their multiplicity.

Order $n = 1$ is identical to the "équation auxiliaire" of Darboux,

$$\mathbf{E}^{(1)} \equiv \mathbf{E}'(x, \mathbf{u}^{(0)})\mathbf{u}^{(1)} = 0, \qquad (5.108)$$

and the Laurent series for $\mathbf{u}^{(1)}$,

$$\mathbf{u}^{(1)} = \sum_{j=\rho}^{+\infty} \mathbf{u}_j^{(1)} \chi^{j+\mathbf{P}}, \qquad (5.109)$$

represents a particular solution containing a number of arbitrary coefficients equal to the number of Fuchs indices, counting their multiplicity. If $\det \mathbf{P}(i)$ has degree N, it represents the general solution of (5.108).

Consequently, the sum $\mathbf{u}^{(0)} + \varepsilon\mathbf{u}^{(1)}$ is already, in a neighborhood of $(\chi, \varepsilon) = (0, 0)$, a *local representation of the greatest particular solution* of (2.29) available in this method (the general solution if $\det \mathbf{P}(i)$ has degree N), and this is a Laurent series with a strictly larger extension $(\rho + \mathbf{p} : +\infty)$ than that for the unperturbed expansion $(\mathbf{p} : +\infty)$.

At each order $n \geq 2$, the singularity order of the particular solution of the linear nonhomogeneous equation (5.103) $\mathbf{E}^{(n)} = 0$ is dictated by the contribution $\mathbf{R}^{(n)}$ of the previously computed coefficients: It is increased by ρ at each order n and is equal to $n\rho + \mathbf{p}$:

$$\forall n \geq 0 : \mathbf{u}^{(n)} = \sum_{j=n\rho}^{+\infty} \mathbf{u}_j^{(n)} \chi^{j+\mathbf{P}}. \qquad (5.110)$$

Third Step. Solve the recurrence relation for $\mathbf{u}_j^{(n)}$ for all values of $(n, j) \neq (0, 0)$:

$\forall n \geq 0 \ \forall j \geq n\rho$,

$$(n, j) \neq (0, 0) : \mathbf{E}_j^{(n)} \equiv \mathbf{P}(j)\mathbf{u}_j^{(n)} + \mathbf{Q}_j^{(n)}(x, \{\mathbf{u}_{j'}^{(n')}\}) = 0. \quad (5.111)$$

The generated necessary stability conditions are as follows:

C4. $\forall n \geq 0 \ \forall$ index i,

$$(n, i) \neq (0, 0) : \mathbf{Q}_i^{(n)} \text{ orthogonal to } \operatorname{Ker} \operatorname{adj} \mathbf{P}(i). \qquad (5.112)$$

These orthogonality conditions must be satisfied for any value of the previously introduced arbitrary coefficients. For a single equation, the condition **C4** is simply $Q_i^{(n)} = 0$.

$Q_j^{(n)}$ depends on all $\mathbf{u}_{j'}^{(n')}$ with $n' \leq n$, $j' - n'\rho \leq j - n\rho$, $(n', j') \neq (n, j)$, and this is the only ordering to be respected during the resolution. The costless ordering on (n, j) is the one that generates stability conditions the sooner, and it depends on the structure of indices of the DE under study.

In order to avoid introducing more than N arbitrary coefficients, the precise rule is:

- if $n = 0$ or $(n = 1$ and $i < 0)$, assign arbitrary values to $\text{mult}(i)$ components of $\mathbf{u}_i^{(n)}$ defining a basis of $\text{Ker}\,\mathbf{P}(i)$;

- if $(n = 1$ and $i \geq 0)$ or $n \geq 2$, assign the value 0 to $\text{mult}(i)$ components of $\mathbf{u}_i^{(n)}$ defining a basis of $\text{Ker}\,\mathbf{P}(i)$.

The resulting double expansion (Taylor in ε, Laurent in χ at each order in ε) can be rewritten as a *Laurent series in χ extending to both infinities*:

$$\forall n \geq 0 : \mathbf{u}^{(n)} = \sum_{j=n\rho}^{+\infty} \mathbf{u}_j^{(n)} \chi^{j+\mathbf{P}}, \quad \mathbf{E}^{(n)} = \sum_{j=n\rho}^{+\infty} \mathbf{E}_j^{(n)} \chi^{j+\mathbf{q}}, \tag{5.113}$$

$$\mathbf{u} = \sum_{n=0}^{+\infty} \varepsilon^n \left[\sum_{j=n\rho}^{+\infty} \mathbf{u}_j^{(n)} \chi^{j+\mathbf{P}} \right] = \sum_{j=-\infty}^{+\infty} \mathbf{u}_j \chi^{j+\mathbf{P}}. \tag{5.114}$$

Remarks.

1. The Fuchsian perturbative method (as well as the non-Fuchsian one, which will be seen in Section 5.8) is useful if and only if the zeroth-order $n = 0$ fails to describe the general solution. This may happen for two reasons. The most common one is a negative Fuchs index in addition to -1 counted once; the second, less common, one is a multiplicity higher than one for some family, as in the example of Section 5.7.3.

2. We do not know of an upper bound for n, but there exists a lower bound. Indeed, in the linear nonhomogeneous ODE (5.103), logarithms can arise only when some precise powers of χ, depending only on the homogeneous part, are present in the rhs Laurent series $R^{(n)}$. The lower bound n results from the condition that the lowest Fuchs index and the highest one, once forced to interfere by the nonlinear terms, start to contribute to such dangerous powers. An example of such a condition is given in Section 5.7.5.

   Even if all Fuchs indices are positive (except -1 counted once), the lower bound on n may be greater than 0, as in the example of Section 5.7.3.

3. *Remark on index -1.* In the case of a single equation, since indices must be distinct integers, the condition $Q_\rho^{(1)} = 0$ at the smallest Fuchs index $i = \rho$ is identically satisfied. Nevertheless, the frequently encountered statement "resonance -1 is always compatible" is erroneous, and numerous nonzero stability conditions $Q_{-1}^{(n)} = 0$ can be found in the examples of [31]. Indeed, even at first perturbation order, the stability condition

at index -1 may not be satisfied: Just as Fuchs index $i = \rho$ provides an identically satisfied stability condition, Painlevé "resonance" -1 has the same property if and only if $\rho = -1$, i.e., if -1 is the smallest integer index.

If ρ is different from -1, Painlevé resonance -1 *seems to* be satisfied, but it is only because a Laurent series ranging from power p to $+\infty$ cannot represent the *general* solution, thus preventing the building of Painlevé stability condition $Q_{-1}^{(0)} = 0$.

5.7.1 Fuchs Indices, Painlevé "Resonances," or Kowalevski Exponents?

Given a nonlinear algebraic DE of order N, one can define three sets of at most N numbers associated to it:

1. the Fuchs indices of the auxiliary equation of Darboux (Section 5.7),

2. the "resonances" of the nonlinear equation (Section 5.4),

3. the Kowalevski exponents, defined only if the nonlinear equation is invariant under a scaling transformation $(x, \mathbf{u}, \mathbf{E}) \mapsto (kX, k^{\mathbf{p}}\mathbf{u}, k^{\mathbf{q}}\mathbf{E})$.

We have already seen the identity of the first two sets, defined for each family of movable singularities. Let us prove that the third notion is not distinct. The third set is defined as follows (for an introduction, see [10]). The invariance implies the particular solution ("scaling solution") $\mathbf{u}^{(0)} = \text{const } (x - x_0)^{\mathbf{p}}$, which is identical to a family of movable singularities. The Kowalevski exponents ρ are defined as the characteristic exponents of the linearized system near this solution, which proves the identity of the three notions.

Said differently, all these numbers are Fuchs indices, and this link to the theory of *linear* DEs proves the uselessness of the notions of Kowalevski exponents and Painlevé resonances.

5.7.2 Understanding Negative Fuchs Indices

The ODE with a meromorphic general solution [31]

$$E \equiv u_{xx} + 3uu_x + u^3 = 0, \quad u = \frac{1}{x - a} + \frac{1}{x - b}, \tag{5.115}$$

a and b arbitrary, has two families:

(F1) $p = -1$, $u_0^{(0)} = 1$, indices $(-1, 1)$,

(F2) $p = -1$, $u_0^{(0)} = 2$, indices $(-2, -1)$,

and this provides a clear comprehension of negative Fuchs indices, since the index -2 must coexist with the meromorphy. Indeed, the representation of the general solution (5.115) as a Laurent series of $x - x_0$ is the sum of two copies of an expansion of $1/(x - c)$, and there exist two expansions of $1/(x - c)$:

$$(x - c)^{-1} = \sum_{j=-\infty}^{-1} (c - x_0)^{-1-j}(x - x_0)^j, \quad |c - x_0| < |x - x_0| \quad (5.116)$$

$$= \sum_{j=0}^{+\infty} -(c - x_0)^{1-j}(x - x_0)^j, \quad |x - x_0| < |c - x_0|. \quad (5.117)$$

The family (F1) corresponds to the sum (first expansion with $c = a = x_0$) plus (second expansion with $c = b$), while the family (F2) corresponds to the sum (first expansion with $c = a$) plus (second expansion with $c = b$). This can be checked by a direct application of the algorithm, which for family (F2) gives [31]

$$u = 2\chi^{-1} + \varepsilon(A_1\chi^{-3} + B_1\chi^{-2}) + \varepsilon^2\left(\frac{A_1^2}{2}\chi^{-5} + \frac{3A_1B_1}{2}\chi^{-4}\right)$$

$$+ \varepsilon^3\left(\frac{A_1^3}{4}\chi^{-7} + \frac{5A_1^2B_1}{4}\chi^{-6} + A_1B_1^2\chi^{-5} - \frac{1}{2}B_1^3\chi^{-4}\right) + O(\varepsilon^4)$$

$$= 2\chi^{-1} + \varepsilon B_1\chi^{-2} + \varepsilon A_1\chi^{-3} + \left(\frac{3}{2}\varepsilon^2 A_1 B_1 - \frac{1}{2}\varepsilon^3 B_1^3\right)\chi^{-4} + O(\chi^{-5})$$

$$= \frac{2\chi - \varepsilon B_1}{\chi^2 - \varepsilon B_1\chi + (-\varepsilon A_1 + \varepsilon^2 B_1^2)/2}, \quad (5.118)$$

where A_1 and B_1 are the arbitrary coefficients at order one. The simple pole $\chi = 0$ with residue 2 has been "unfolded" by the perturbation into two simple poles with residue 1, at the two arbitrary locations $\frac{1}{2}[\varepsilon B_1 \pm \sqrt{2\varepsilon A_1 - \varepsilon^2 B_1^2}]$, both close to 0.

For other examples, see [97] and conference proceedings referenced in [31].

5.7.3   The Simplest Constructive Example

The equation

$$u'' + 4uu' + 2u^3 = 0 \quad (5.119)$$

is the simplest constructive example, because

1. there exists a movable logarithm, as shown by the α-method (BSMF Section 13, p. 221);

2. the method of pole-like expansions fails to find it;

3. the assumption of a "descending" Laurent series (5.33) fails to find it;

4. the Fuchsian perturbative method finds it after a very short computation, as we now show.

There exists a single family

$$p = -1, \quad E_0^{(0)} = u_0^{(0)}(u_0^{(0)} - 1)^2 = 0, \quad \text{indices } (-1, 0), \qquad (5.120)$$

with the puzzling fact that $u_0^{(0)}$ should be at the same time equal to 1 according to the equation $E_0^{(0)} = 0$, and arbitrary according to the index 0. The application of the method provides

$$u^{(0)} = \chi^{-1} \text{ (the series terminates)}, \qquad (5.121)$$

$$E'(x, u^{(0)}) = \partial_x^2 + 4\chi^{-1}\partial_x + 2\chi^{-2}, \qquad (5.122)$$

$$u^{(1)} = u_0^{(1)}\chi^{-1}, \quad u_0^{(1)} \text{ arbitrary}, \qquad (5.123)$$

$$E^{(2)} = E'(x, u^{(0)})u^{(2)} + 6u^{(0)}u^{(1)^2} + 4u^{(1)}u^{(1)'}$$
$$= \chi^{-2}(\chi^2 u^{(2)})'' + 2u_0^{(1)^2}\chi^{-3} = 0, \qquad (5.124)$$

$$u^{(2)} = -2u_0^{(1)^2}\chi^{-1}(\log \chi - 1). \qquad (5.125)$$

The movable logarithmic branch point is therefore detected in a systematic way at order $n = 2$ and index $i = 0$.

The necessity to perform a perturbation arises from the multiple root of the equation for $u_0^{(0)}$, responsible for the insufficient number of arbitrary parameters in the zeroth-order series $u^{(0)}$.

5.7.4   The Two Examples

Example 5.13 ("Complete (P1)" equation (5.20)). The method is useless.

Example 5.14 (Chazy complete equation of class III (5.23)). For the *second family* (in the case $d_3 = 0$), the method is useless.

For the *first family*, since all indices are negative, one must start the perturbation process at $n = 1$.

To obtain the stability conditions up to a given order $n \geq 1$, we need to compute only the first $3n$ coefficients of each element:

$$u^{(r)} = \sum_{j=-3r}^{-3r+3n-1} u_j^{(r)}\chi^{j-1}, \quad r = 0, \ldots, n, \qquad (5.126)$$

i.e.,

$$u_{0:3n-1}^{(0)}, \quad u_{-3:3n-4}^{(1)}, \quad \ldots, \quad u_{-3n+3:2}^{(n-1)}, \quad u_{-3n:-2}^{(n)}, \qquad (5.127)$$

where $j_1 : j_2$ denotes a range of j values. The most efficient way to perform the double loop on (n, j) is to perform the outside loop in the variable $k = j - n\rho$, with $\rho = -3$, and the precise double loop is as follows: for

$k = 0$ to k_{max} do for $n =$(if $k = 0$ then 1 else 0) to n_{max} do solve the linear algebraic equation (5.111) for $u_j^{(n)}$, $j = k + n\rho$.

Let us compute all the stability conditions at first and second order. The computer printout reads (full details are given in [31]):

$$k = 0: \quad Q_{-3}^{(1)} \equiv 0, \quad u_{-3}^{(1)} \text{ arbitrary}, \tag{5.128}$$

$$k = 1: \quad Q_{-2}^{(1)} \equiv -6(5c_1 + 42d_3)u_{-3}^{(1)} = 0, \quad u_{-2}^{(1)} \text{ arbitrary}, \tag{5.129}$$

$$k = 2: \quad Q_{-1}^{(1)} \equiv -12(c_1 + 9d_3)u_{-2}^{(1)} + 18(c_1 - 8d_3)u_{-3}^{(1)'}$$
$$+ \frac{6}{5}(2c_0 - 3c_1^2 - 117c_1d_3 - 594d_3^2 + 18c_1' + 108d_3')u_{-3}^{(1)} = 0,$$
$$u_{-1}^{(1)} \text{ arbitrary}, \tag{5.130}$$

$$k = 3: \quad Q_{-3}^{(2)} \equiv -\frac{66}{5}d_1 u_{-3}^{(1)^2} = 0, \tag{5.131}$$

$$k = 4: \quad Q_{-2}^{(2)} \equiv \frac{1}{7}(8d_0 + 57d_1')u_{-3}^{(1)^2} - 12d_1 u_{-3}^{(1)} u_{-2}^{(1)} = 0, \tag{5.132}$$

$$k = 5: \quad Q_{-1}^{(2)} \equiv -\frac{1}{35}(18d_0' + 99d_1'')u_{-3}^{(1)^2} - \frac{24}{5}d_1 u_{-2}^{(1)^2}$$
$$+ \frac{3}{35}(16d_0 + 72d_1')u_{-3}^{(1)} u_{-2}^{(1)} = 0. \tag{5.133}$$

Five conditions are obtained, three at order one, equivalent to $d_3 = c_1 = c_0 = 0$, and two at order two, equivalent to $d_1 = d_0 = 0$ [in order to simplify expressions, we have put the first-order conditions in the above expressions for $k = j + 3n \geq 3$], after seventeen values of (n, j). These conditions were given without any detail by Chazy [22]. They restrict the complete ODE (5.23) to the simplified ODE (1.18), modulo (3.5).

Chazy proved the general solution of (1.18) to be single-valued inside or outside a circle whose center and radius depend on the choice of the three arbitrary constants; it is holomorphic in this domain, and the only singularity is a movable natural boundary ("coupure essentielle") defined by this circle. He also gave a parametric representation of the general solution $u(x)$ in terms of two solutions of the (linear) hypergeometric equation, but single-valuedness is not at all apparent in this representation.

The direct explicit solution of Bureau [17, 18] is given in Section 5.5.1.

5.7.5 An Example Needing Order Seven to Conclude

The following equation, isolated by Bureau [15, p. 79],

$$u'''' + 3uu'' - 4u'^2 = 0, \tag{5.134}$$

possesses the two families

$$p = -2, \quad u_0^{(0)} = -60, \quad \text{ind. } (-3, -2, -1, 20), \quad \widehat{K} = u'''' + 3uu'' - 4u'^2,$$

$$p = -3, \quad u_0^{(0)} \text{ arbitrary}, \quad \text{indices } (-1, 0), \quad \widehat{K} = 3uu'' - 4u'^2. \quad (5.135)$$

The *second family* has a Laurent series $(p : +\infty)$, which happens to terminate [31]:

$$u^{(0)} = c(x - x_0)^{-3} - 60(x - x_0)^{-2}, \quad (c, x_0) \text{ arbitrary}. \quad (5.136)$$

The Fuchsian perturbative method is useless, for the two arbitrary coefficients corresponding to the two Fuchs indices are already present at zeroth order.

The *first family* provides, at zeroth order, only a two-parameter expansion, and when one checks the existence of the perturbed solution

$$u = \sum_{n=0}^{+\infty} \varepsilon^n \left[\sum_{j=0}^{+\infty} u_j^{(n)} \chi^{j-2-3n} \right], \quad (5.137)$$

one finds that coefficients $u_{20}^{(0)}, u_{-3}^{(1)}, u_{-2}^{(1)}, u_{-1}^{(1)}$ can be chosen arbitrarily, and at order $n = 7$ one finds two violations [31]:

$$Q_{-1}^{(7)} \equiv u_{20}^{(0)} u_{-3}^{(1)^7} = 0, \quad Q_{20}^{(7)} \equiv u_{20}^{(0)^2} u_{-3}^{(1)^6} u_{-2}^{(1)} = 0, \quad (5.138)$$

implying the existence of a movable logarithmic branch point.

Remark ([86]). The value $n = 7$ is the root of the linear equation $n(i_{\min} - p) + (i_{\max} - p) = -1$, with $p = -2, i_{\min} = -5, i_{\max} = 18$, linking the pole of order p in the Fuchsian case $c = 0$, the smallest and the greatest Fuchs indices. It expresses the condition for the first occurrence of a power χ^{-1}, leading by integration to a logarithm, in the rhs $R^{(n)}$ of the linear nonhomogeneous equation (5.103), rhs created by the nonlinear terms $3uu'' - 4u'^2$.

5.7.6 Closed-Form Solutions of the Bianchi IX Model

In this example the no-log conditions are used in a constructive way, in order to isolate all possible single-valued solutions.

The Bianchi IX cosmological model [79] is a system of three second-order ODEs,

$$(\log A)'' = A^2 - (B - C)^2 \text{ and cyclically}, \quad ' = d/d\tau, \quad (5.139)$$

or equivalently,

$$(\log \omega_1)'' = \omega_2^2 + \omega_3^2 - \omega_2^2\omega_3^2/\omega_1^2, \quad A = \omega_2\omega_3/\omega_1,$$
$$\omega_1^2 = BC \text{ and cyclically}. \quad (5.140)$$

One of the families [36, 80]

$$A = \chi^{-1} + a_2\chi + O(\chi^3), \quad \chi = \tau - \tau_2,$$
$$B = \chi^{-1} + b_2\chi + O(\chi^3), \qquad \qquad (5.141)$$
$$C = \chi^{-1} + c_2\chi + O(\chi^3),$$

has the Fuchs indices -1, -1, -1, 2, 2, 2. The Fuchsian perturbative method

$$A = \chi^{-1} \sum_{n=0}^{N} \varepsilon^n \sum_{j=-n}^{2+N-n} a_j^{(n)} \chi^j, \quad \chi = \tau - \tau_2, \quad \text{and cyclically,} \qquad (5.142)$$

then gives a failure of condition **C4** at $(n, i) = (3, -1)$ and $(5, -1)$ [80], and the satisfaction of these no-log conditions generates the three solutions

$$(b_2^{(0)} = c_2^{(0)} \text{ and } b_{-1}^{(1)} = c_{-1}^{(1)}) \text{ or cyclically,} \qquad (5.143)$$
$$a_2^{(0)} = b_2^{(0)} = c_2^{(0)} = 0, \qquad (5.144)$$
$$a_{-1}^{(1)} = b_{-1}^{(1)} = c_{-1}^{(1)}. \qquad (5.145)$$

These are constraints that reduce the number of arbitrary coefficients to four, three, and four, respectively, thus defining particular solutions that may have no movable critical points. The question is, Do they define additional solutions to what is known?

The only three closed-form solutions that are known are single-valued. They are defined as the general solution of the following three subsystems:

1. The 4-dim axisymmetric case $B = C$ [109], whose general solution (2.56) is trigonometric.

2. The 3-dim *Darboux–Halphen system* [44, 62]

$$\omega_1' = \omega_2\omega_3 - \omega_1\omega_2 - \omega_1\omega_3, \quad \text{and cyclically.} \qquad (5.146)$$

3. The 3-dim *Euler system* (1750) [8], describing the motion of a rigid body around its center of mass,

$$\omega_1' = \omega_2\omega_3, \quad \text{and cyclically,} \qquad (5.147)$$

whose general solution is elliptic [8]; see (2.30) and (2.43).

The first constraint (5.143) implies the equality of two of the components (A, B, C) at every order and thus represents the four-parameter solution of Taub (2.56).

The second constraint (5.144) represents the three-parameter solution of the Darboux–Halphen system (5.146).

The third, and last, constraint (5.145) represents an extrapolation to four parameters of the three-parameter solution to the Euler system described by $a_2^{(0)} + b_2^{(0)} + c_2^{(0)} = 0$. This would-be four-parameter, global, closed-form, single-valued exact solution has not yet been found.

5.8   The Non-Fuchsian Perturbative Method

Whenever the family under study has a number of Fuchs indices smaller than the differential order N, the Fuchsian perturbation method fails to build a representation of the general solution, thus possibly missing some stability conditions. Examples are (5.135) and the second family of (5.23) in the case $d_3 = 0$. The missing solutions of the auxiliary equation (5.108) are then non-Fuchsian solutions; see Section 5.2.2.

There is no difficulty to algorithmically compute the non-Fuchsian expansions (5.11), but these are of no immediate help, due to their generic divergence.

There is one situation where some stability conditions can be generated *algorithmically* (indeed, we are not interested in computations adapted to a given equation, only in computerizable methods). It occurs when the two following conditions are met [86]:

1. There exists a particular solution $\mathbf{u} = \mathbf{u}^{(0)}$ that is known globally, is meromorphic, and has at least one movable pole at a finite distance denoted by x_0.

2. The only singular points of the linearized equation $\mathbf{E}^{(1)} = 0$ are $x = x_0$, non-Fuchsian, and $x = \infty$, Fuchsian.

Then the property that a fundamental set of solutions $\mathbf{u}^{(1)}$ is locally single-valued near $\chi = x - x_0 = 0$ is equivalent to the same property near $\chi = \infty$. This is the global nature of $\mathbf{u}^{(0)}$, which allows the study of the point $\chi = \infty$, which is easy to perform with the Fuchsian perturbation method.

An important technical bonus is the lowering of the differential order N of equation $\mathbf{E}^{(1)} = 0$ by the number M of arbitrary parameters c that appear in $\mathbf{u}^{(0)}$. Indeed, again since $\mathbf{u}^{(0)}$ is in closed form, its partial derivatives $\partial_c \mathbf{u}^{(0)}$ are in closed form and are particular solutions of $\mathbf{E}^{(1)} = 0$, which allows this lowering of the order.

At each higher perturbation order $n \geq 2$, one similarly builds particular solutions $\mathbf{u}^{(n)}$ as expansions near $\chi = \infty$, and one requires the same properties.

5.8.1   An Explanatory Example: Chazy's Class III ($N = 3$, $M = 2$)

The simplified equation (1.18), which possesses the PP [22] and for which therefore no $u^{(n)}$ is multivalued, is quite useful just to understand the method. This equation admits the global two-parameter solution (5.55) $u^{(0)} = c\chi^{-2} - 6\chi^{-1}$. The linearized equation

$$E^{(1)} \equiv E'(x, u^{(0)})u^{(1)} \equiv [\partial_x^3 - 2u^{(0)}\partial_x^2 + 6u_x^{(0)}\partial_x - 2u_{xx}^{(0)}]u^{(1)} = 0 \quad (5.148)$$

possesses the two single-valued global solutions $\partial_{x_0} u^{(0)}$, $\partial_c u^{(0)}$, i.e., $u^{(1)} = \chi^{-3}$, χ^{-2}, and it has only two singular points $\chi = 0$ (Fuchsian) and $\chi = \infty$

(non-Fuchsian with Thomé rank two). The lowering by $M = 2$ units of the order of the linearized equation results from the change of function

$$u^{(1)} = \chi^{-3}v: \ E^{(1)} \equiv \chi^3[\partial_x + 3\chi^{-1} - 2c\chi^{-2}]v'' = 0, \qquad (5.149)$$

and the study of the Fuchsian point $\chi = \infty$ yields an integer Fuchs index, which proves the *global* single-valuedness of the general solution $u^{(1)}$.

Remarks.

- The local study of $\chi = 0$ provides a formal expansion (5.11), which happens to terminate, a nongeneric situation, thus providing the fundamental set of *global* solutions at perturbation order $n = 1$:

$$\forall \chi \ \forall c: u^{(1)} = \chi^{-2}, \ \chi^{-3}, \ (e^{-2c/\chi} - 1 + 2c\chi^{-1})\chi^{-2}/(2c^2). \quad (5.150)$$

  This proves the existence of an essential singularity at $\chi = 0$ [67, Chapter XVII].

- Proceeding on with the formalism of Painlevé's lemma at higher orders constitutes the rigorous mathematical framework of the local representation of the general solution obtained by Joshi and Kruskal [71]:

$$u = -6\chi^{-1} + c\chi^{-2}\left(1 + z - \frac{z^2}{8} + \frac{z^3}{144} - \frac{7z^4}{13824} + O(\varepsilon^5)\right),$$

$$z = \frac{\varepsilon}{c}e^{-2c/\chi}.$$

  This representation reduces to the one given by Chazy (Taylor series in $1/\chi$) if one starts from the Fuchsian family $u \sim -6\chi^{-1}$.

5.8.2 The Fourth-Order Equation of Bureau ($N = 4, M = 2$)

In Section 5.7.5 the fourth-order equation (5.134) has been proven to be unstable after a computation practically intractable without a computer. Let us now prove this result without computation at all [86]. For the global two-parameter solution (5.136), the linearized equation

$$E^{(1)} = E'(x, u^{(0)})u^{(1)} \equiv [\partial_x^4 + 3u^{(0)}\partial_x^2 - 8u_x^{(0)}\partial_x + 3u_{xx}^{(0)}]u^{(1)} = 0 \quad (5.151)$$

has only two singular points $\chi = 0$ (non-Fuchsian) and $\chi = \infty$ (Fuchsian); it admits the two global single-valued solutions $\partial_{x_0}u^{(0)}$ and $\partial_c u^{(0)}$, i.e., $u^{(1)} = \chi^{-4}, \chi^{-3}$. The lowering by $M = 2$ units of the order of the linearized equation (5.151) is obtained with

$$u^{(1)} = \chi^{-4}v: [\partial_x^2 - 16\chi^{-1}\partial_x + 3c\chi^{-3} - 60\chi^{-2}]v'' = 0, \qquad (5.152)$$

and the local study of $\chi = \infty$ is unnecessary, since one recognizes the confluent hypergeometric equation. The two other solutions in global form

are

$$c \neq 0 : v_1'' = \chi^{-3}{}_0F_1(24; -3c/\chi) = \chi^{17/2}J_{23}(\sqrt{12c/\chi}), \qquad (5.153)$$

$$v_2'' = \chi^{17/2}N_{23}(\sqrt{12c/\chi}), \qquad (5.154)$$

where the hypergeometric function ${}_0F_1(24; -3c/\chi)$ is single-valued and possesses an isolated essential singularity at $\chi = 0$, while the function N_{23} of Neumann is multivalued because of a $\log\chi$ term.

Remark. The local study of (5.151) near $\chi = 0$ provides the formal expansions (5.11) for the two non-Fuchsian solutions

$$\chi \to 0, \ c \neq 0 : u^{(1)} = e^{\pm\sqrt{-12c/\chi}}\chi^{31/4}(1 + O(\sqrt{\chi})), \qquad (5.155)$$

detecting the presence in (5.151) of an essential singularity at $\chi = 0$, but the generically null radius of convergence of the formal series forbids one to conclude the multivaluedness of $u^{(1)}$. A nonobvious result is the existence, as seen above, of a linear combination of the two formal expansions (5.155) that is single valued.

5.8.3 An Example in Cosmology: Bianchi IX ($N = 6$, $M = 4$)

The Bianchi IX cosmological model in vacuum (5.139) does not possess the PP [37,80]. Let us prove this rapidly [80,86]. Taub [109] found the general solution of the axisymmetric case of two equal components, a meromorphic expression (2.56) depending on the four arbitrary parameters $(k_1, k_2, \tau_1, \tau_2)$. The linearized system generated by the perturbation

$$A = A^{(0)}(1 + \varepsilon A^{(1)} + O(\varepsilon^2)) \text{ and cyclically} \qquad (5.156)$$

has the differential order $N = 6$, which is then lowered by $M = 4$ units by the change of function dictated by the symmetry of the system: $P^{(1)} = B^{(1)} + C^{(1)}$, $M^{(1)} = B^{(1)} - C^{(1)}$,

$$A^{(1)''} - 2A^{(0)^2}A^{(1)} = 0, \qquad (5.157)$$

$$P^{(1)''} - 2A^{(0)}B^{(0)}P^{(1)} = 4(A^{(0)}B^{(0)} - A^{(0)^2})A^{(1)}, \qquad (5.158)$$

$$M^{(1)''} + 2(A^{(0)}B^{(0)} - 2B^{(0)^2})M^{(1)} = 0. \qquad (5.159)$$

Indeed, the four single-valued global solutions

$$(A^{(1)}, P^{(1)}) = \partial_c(\log A^{(0)}, \log(B^{(0)} + C^{(0)})), \quad c = k_1, k_2, \tau_1, \tau_2, \qquad (5.160)$$

are those of the equations (5.157)–(5.158),

$$M^{(1)} = 0, \ (A^{(1)}, P^{(1)} + 2A^{(1)}) = \begin{cases} ((\tau - \tau_1)\coth k_1(\tau - \tau_1) - 1/k_1, 0), \\ (0, (\tau - \tau_2)\coth k_2(\tau - \tau_2) - 1/k_2), \\ (\coth k_1(\tau - \tau_1), 0), \\ (0, \coth k_2(\tau - \tau_2)), \end{cases}$$

and there remains the equation (5.159) only to study. It has a countable infinity of singular points: $\tau - \tau_2 = im\pi/k_2$, $m \in \mathbb{Z}$ (non-Fuchsian, of Thomé rank two), accumulating at $\tau = \infty$. This uneasy situation can be overcome by taking the limit $k_1 = k_2 = 0$; it is indeed sufficient to exhibit a movable logarithm in this limit, for it will persist for $(k_1, k_2) \neq (0,0)$. In this limit

$$k_1 = k_2 = 0 : \frac{\mathrm{d}^2 M^{(1)}}{\mathrm{d}t^2} + \left(\frac{2}{t^2} - \frac{4(t-1)^2}{t^4} \right) M^{(1)} = 0, \quad t = \frac{\tau - \tau_2}{\tau_1 - \tau_2},$$

the only singular points are $t = 0$ (non-Fuchsian) and $t = \infty$ (Fuchsian), the optimal situation. The Fuchs indices being -2 and 1, the computation of three terms is sufficient to exhibit a logarithm, and this proves the absence of the Painlevé property for the Bianchi IX model in vacuum.

Remarks.

1. The two solutions are globally known [80]:

$$k_1 = k_2 = 0 : M^{(1)} = e^{-2/t} t^{-1}, \quad e^{-2/t} t^{-1} \int^{1/t} z^{-4} e^{4z} \mathrm{d}z, \quad (5.161)$$

   which shows the presence of a logarithmic branch point at $t = 0$, or at $t = \infty$ as well.

2. The two formal non-Fuchsian solutions are

$$\tau - \tau_2 \to 0 : M^{(1)} = e^{\alpha/(\tau - \tau_2)} \sum_{k=0}^{+\infty} \lambda_k (\tau - \tau_2)^{k+s}, \quad \lambda_0 \neq 0, \quad (5.162)$$

   with

$$\alpha = \pm 2k_1^{-1} \sinh k_1 (\tau_2 - \tau_1), \quad s = 1 \mp 2 \cosh k_1 (\tau_2 - \tau_1). \quad (5.163)$$

   The two generically irrational values for the Thomé exponents s allow one to conclude only if the divergent series $\lambda_k (\tau - \tau_2)^k$ can be summed.

5.9 Miscellaneous Perturbations

The differential complexity of the α-method explains why it usually succeeds in case of failure of all the other methods, which have only algebraic complexity. Consider the ODEs, none of which admits a power-law leading behavior,

$$-2uu'' + 3u'^2 + d_3 u^3 = 0, \quad d_3 \neq 0, \qquad (5.164)$$

$$u''' + uu'' - 2u'^2 = 0, \qquad (5.165)$$

$$u'''' + 2uu'' - 3u'^2 = 0, \qquad (5.166)$$

and let us prove that each of them has movable logarithms. The first one is extracted from Chazy's class III (5.23) by the perturbation $u = \varepsilon^{-1}U$, $x = x_0 + \varepsilon X$, and it represents its second family; see Section 5.4.1. The second and third ones were considered by Chazy [23, 24], who had to establish a special theorem, using divergent series, to exhibit the movable logarithms. Having degree one, none of these ODEs admits singular solutions.

The first equation (5.164) is classically processed by the α-method:

$$u = \varepsilon^{-1} \sum_{n=0}^{+\infty} \varepsilon^n u^{(n)}, \quad E = \varepsilon^{-4} \sum_{n=0}^{+\infty} \varepsilon^n E^{(n)}, \quad x = x_0 + \varepsilon X, \qquad (5.167)$$

resulting in

$$E^{(0)} \equiv -2u^{(0)}u^{(0)\prime\prime} + 3u^{(0)\prime 2} = 0, \tag{5.168}$$

$$u^{(0)} = c(X - X_0)^{-2}, \quad (X_0, c) \text{ arbitrary}, \tag{5.169}$$

$$E^{(1)} \equiv c(X - X_0)^{-5}[-2((X - X_0)^3 u^{(1)})\prime\prime + c^2 d_3/(X - X_0)] = 0, \tag{5.170}$$

$$u^{(1)} = c^2 d_3(X - X_0)^3[(X - X_0)\log(X - X_0) - (X - X_0)]/2, \tag{5.171}$$

and proving the instability at perturbation order one.

For equations (5.165) and (5.166) there exists no perturbation satisfying the assumptions of Theorem II, page 115, there exist only singular perturbations, i.e., those that discard the highest derivative. However, since they give the correct information, it would be desirable to extend Theorem II in that direction. Meanwhile, the results below cannot be considered as proven.

Equation (5.165) is handled by the singular perturbation

$$u = \varepsilon^{-1} \sum_{n=0}^{+\infty} \varepsilon^n u^{(n)}, \quad E = \varepsilon^{-2} \sum_{n=0}^{+\infty} \varepsilon^n E^{(n)}, \tag{5.172}$$

which excludes $u\prime\prime\prime$ from the simplified equation

$$E^{(0)} \equiv u^{(0)}u^{(0)\prime\prime} - 2u^{(0)\prime 2} = 0, \tag{5.173}$$

$$u^{(0)} = c\chi^{-1}, \quad \chi = x - x_0, \quad (x_0, c) \text{ arbitrary}, \tag{5.174}$$

$$E^{(1)} \equiv c(\chi^{-3}(\chi^2 u^{(1)})\prime\prime - 6\chi^{-4}) = 0, \tag{5.175}$$

$$u^{(1)} = \chi^{-1}(\log\chi - 1). \tag{5.176}$$

This same perturbation (5.172) solves the case of (5.166):

$$E^{(0)} \equiv 2u^{(0)}u^{(0)\prime\prime} - 3u^{(0)\prime 2} = 0, \tag{5.177}$$

$$u^{(0)} = c\chi^{-2}, \tag{5.178}$$

$$E^{(1)} \equiv c(2\chi^{-5}(\chi^3 u^{(1)})\prime\prime + 120\chi^{-6}) = 0, \tag{5.179}$$

$$u^{(1)} = -60\chi^{-2}(\log\chi - 1). \tag{5.180}$$

5.10   The Perturbation of the Continuum Limit of a Discrete Equation

Discrete equations can be considered as functional equations linking the values taken by some field variable u at a finite number $N+1$ of points, either arithmetically consecutive, $x+kh$, $k-k_0 = 0, 1, \ldots, N$, or geometrically consecutive, xq^k, $k - k_0 = 0, 1, \ldots, N$, where h or q is the lattice step size, assumed to lie in some neighborhood of, respectively, 0 or 1, and k_0 is just some convenient origin.

Definition 5.8 ([35]). A discrete equation is said to possess the *discrete Painlevé property* if there exists a neighborhood of $h=0$ at every point at which the general solution $x \mapsto u(x, h)$ has no movable critical singularities.

Consider an arbitrary discrete equation (5.181),

$$\forall x \; \forall h : E(x, h, \{u(x + kh), k - k_0 = 0, \ldots, N\}) = 0, \qquad (5.181)$$

algebraic in the values of the field variable, with coefficients analytic in x, the step size and some parameters a. Let $(x, h, u, a) \mapsto (X, H, U, A, \varepsilon)$ be an arbitrary perturbation admissible by the suitable extension of the theorem of Poincaré to discrete systems. Two such perturbations are well known, the *autonomous limit*

$$x = x_0 + \varepsilon X, \quad h = \varepsilon H, \quad u = U, \quad a = \text{ analytic } (A, \varepsilon), \qquad (5.182)$$

and the *continuum limit*

$$x \text{ unchanged}, \quad h = \varepsilon, \quad u = U, \quad a = \text{ analytic } (A, \varepsilon). \qquad (5.183)$$

The latter can be extended to a *perturbation of the continuum limit* [35]

$$x \text{ unchanged}, \quad h = \varepsilon, \quad u = \sum_{n=0}^{+\infty} \varepsilon^n u^{(n)}, \quad a = \text{ analytic } (A, \varepsilon), \qquad (5.184)$$

entirely analogous to the Fuchsian (Section 5.7) or non-Fuchsian (Section 5.8) perturbative method of the continuous case.

This perturbation generates an infinite sequence of (continuous) differential equations $E^{(n)} = 0$ whose first one $n = 0$ is the continuum limit. The next ones $n \geq 1$, which are linear nonhomogeneous, have the same homogeneous part $E^{(0)'} u^{(n)} = 0$ independent of n, defined by the derivative of the equation of the continuum limit, while their nonhomogeneous part $R^{(n)}$ ("right-hand side") comes at the same time from the nonlinearities and the discretization.

Let us handle just the Euler scheme for the Bernoulli equation

$$E \equiv \frac{\bar{u} - u}{h} + u^2 = 0, \qquad (5.185)$$

(notation is $u = u(x)$, $\bar{u} = u(x + h)$), i.e., the logistic map of Verhulst, a paradigm of chaotic behavior. Let us expand the terms of (5.185) according to the perturbation (5.184) up to an order in ε sufficient to build the first equation $E^{(1)} = 0$ beyond the continuum limit $E^{(0)} = 0$:

$$u = u^{(0)} + u^{(1)}\varepsilon + O(\varepsilon^2), \tag{5.186}$$

$$u^2 = u^{(0)^2} + 2u^{(0)}u^{(1)}\varepsilon + O(\varepsilon^2), \tag{5.187}$$

$$\bar{u} = u + u'h + \frac{1}{2}u''h^2 + O(h^3), \tag{5.188}$$

$$\frac{\bar{u} - u}{h} = u^{(0)'} + \left(u^{(1)'} + \frac{1}{2}u^{(0)''}\right)\varepsilon + O(\varepsilon^2). \tag{5.189}$$

The equations of orders $n = 0$ and $n = 1$ are written as

$$E^{(0)} = u^{(0)'} + u^{(0)^2} = 0, \tag{5.190}$$

$$E^{(1)} = E^{(0)'}u^{(1)} + \frac{1}{2}u^{(0)''} = 0, \quad E^{(0)'} = \partial_x + 2u^{(0)}. \tag{5.191}$$

Their general solution is

$$u^{(0)} = \chi^{-1}, \chi = x - x_0, \quad x_0 \text{ arbitrary}, \tag{5.192}$$

$$u^{(1)} = u^{(1)}_{-1}\chi^{-2} - \chi^{-2}\log\psi, \quad \psi = x - x_0, \quad u^{(1)}_{-1} \text{ arbitrary}, \tag{5.193}$$

and the movable logarithm proves the instability as soon as order n is equal to 1, at the Fuchs index $i = -1$.

5.11 The Diophantine Conditions

In fulfilling the systematic program of Painlevé, one encounters the following kind of Diophantine equation:

$$\sum_{k=1}^{p} \frac{1}{n_k} = \frac{1}{n}, \tag{5.194}$$

with n and p given integers, whose unknowns (n_k) are Fuchs indices, which must therefore be either integral or infinite. They admit a finite set of solutions, which allows all cases to be further examined. Details can be found in [14, 15].

Such a Diophantine condition always arises when there exists more than one family, as the constraint that *simultaneously* all Fuchs indices of all families be integral. Let us give just one example [31]. The Hamiltonian Hénon–Heiles system [63] in two coupled variables (q_1, q_2),

$$H \equiv (1/2)(q_{1,x}^2 + q_{2,x}^2 + c_1 q_1^2 + c_2 q_2^2) + \alpha q_1 q_2^2 - (1/3)\beta q_1^3 = E, \tag{5.195}$$

$$q_{1,xx} + c_1 q_1 - \beta q_1^2 + \alpha q_2^2 = 0, \tag{5.196}$$

$$q_{2,xx} + c_2 q_2 + 2\alpha q_1 q_2 = 0, \tag{5.197}$$

defines by elimination the fourth-order ODE in $v = q_1$ [50]

$$v_{xxxx} + (8\alpha - 2\beta)vv_{xx} - 2(\alpha + \beta)v_x^2 - (20/3)\alpha\beta v^3$$
$$+ (c_1 + 4c_2)v_{xx} + (6\alpha c_1 - 4\beta c_2)v^2 + 4c_1 c_2 v + 4\alpha E = 0, \quad (5.198)$$

with $(\alpha, \beta, c_1, c_2, E)$ constants.

There exist two families

$$p = -2, v_0 = \frac{3}{\alpha}, \quad \text{indices } (-1, 10, r_1, r_2), \quad (5.199)$$

$$p = -2, v_0 = -\frac{6}{\beta}, \quad \text{indices } (-1, 5, s_1, s_2), \quad (5.200)$$

in which r_i and s_i satisfy the equations

$$r^2 - 5r + 12 + 6\gamma = 0, \quad s^2 - 10s + 24 + 48\gamma^{-1} = 0, \quad \gamma = \beta/\alpha. \quad (5.201)$$

The Diophantine equations to be solved are

$$(r_1 - r_2)^2 = (2k - 1)^2, \quad (s_1 - s_2)^2 = (2l)^2,$$
$$(r_i + 1)(r_i - 10)(s_i + 1)(s_i - 5) \neq 0, \quad (5.202)$$

with k and l two strictly positive integers. Since $(r_1 - r_2)^2 = (2r - 5)^2$, $(s_1 - s_2)^2 = 4(s - 5)^2$, the elimination of γ between (5.201) yields

$$\gamma = \frac{48}{1 - l^2}, \quad l^2 = 1 + \frac{1152}{23 + (2k - 1)^2}, \quad (5.203)$$

and this provides sharp bounds for $l : 1 < l^2 \leq 49$. One thus obtains the following four solutions for β/α, (k, l), (r_1, r_2), (s_1, s_2):

$$-1: \quad (1, 7), \quad (2, 3), \quad (-2, 12) \quad \text{(SK)}, \quad (5.204)$$
$$-2: \quad (3, 5), \quad (0, 5), \quad (0, 10), \quad (5.205)$$
$$-6: \quad (6, 3), \quad (-3, 8), \quad (2, 8) \quad \text{(KdV5)}, \quad (5.206)$$
$$-16: \quad (10, 2), \quad (-7, 12), \quad (3, 7) \quad \text{(KK)}. \quad (5.207)$$

Let us assume for simplicity $c_1 = c_2 = 0$. Three of the four solutions restrict the ODE to the stationary reduction of well-known soliton equations, thus proving the PP: Sawada–Kotera (SK [104]), higher-order Korteweg–de Vries (KdV5, [81]) and Kaup–Kupershmidt (KK [51,72]) equations.

The case $\beta = -2\alpha$ is similar to that of the ODE in Section 5.7.3: v_0 is a double root of its algebraic equation and is not arbitrary, although 0 is an index. The results of the Fuchsian perturbative method are also similar; listed by increasing cost (number of needed values of (n, i)), the first stability conditions $Q_i^{(n)} = 0$ are

$$Q_0^{(1)} \equiv 0, \qquad\qquad \text{cost} = 2, \qquad (5.208)$$

$$Q_0^{(2)} \equiv -40\alpha u_0^{(1)^2} = 0, \qquad \text{cost} = 5, \qquad (5.209)$$

$$Q_{10}^{(0)} \equiv -30\alpha^3 u_5^{(0)^2} = 0, \qquad \text{cost} = 10, \qquad (5.210)$$

$$Q_5^{(1)} \equiv -120\alpha u_5^{(0)} u_0^{(1)} = 0, \quad \text{cost} = 12. \qquad (5.211)$$

To detect the instability, the method of pole-like expansions is here sufficient, but the Fuchsian perturbative method is much cheaper.

6 Construction of Necessary Conditions. The Painlevé Test

This section makes a synthesis of all the methods of Section 5 in order to define a usable end product that makes obsolete the meromorphy test of Section 2.3. This end product is widely known as the *Painlevé test*. Before detailing the steps of this algorithm in Section 6.6, for ODEs as well as for PDEs, some prerequisite technical developments are needed: implementation of physicists' desiderata (Section 6.1), technicalities to simplify the computations (Section 6.2), and the quite important feature of the invariant Painlevé analysis (Sections 6.3, 6.4, and 6.5).

6.1 *Physical Considerations*

Some DEs encountered in physics are unstable, although integrable or partially integrable in some obvious physical sense. It is then extremely important not to discard them; this is achieved by relaxing some of the mathematical requirements.

Firstly, nonpolynomial DEs can be made polynomial by transformations on u as in Section 3.3.3, necessarily outside the groups of invariance of the PP defined in Sections 3.3.1 and 3.3.2.

Example 6.1 (sine–Gordon).

(sine–Gordon) $u_{xt} = \sin u, \quad e^{iu} = v, \quad 2(vv_{xt} - v_x v_t) - v^3 + v = 0. \qquad (6.1)$

Example 6.2 (Benjamin–Ono). The nonlocal, nonpolynomial PDE

$$u_t + uu_x + \mathrm{H}(u_{xx}) = 0, \quad \mathrm{H}(v) = \frac{1}{\pi} \mathrm{pp} \int_{-\infty}^{+\infty} \frac{v(x', t)}{x' - x} \, \mathrm{d}x', \qquad (6.2)$$

in which H is the Hilbert transform and pp the Cauchy principal value distribution, is equivalent [61, 102] to the local and polynomial system

$$u_t + uu_x + u_{xy}, \quad u_{xx} + u_{yy} = 0, \qquad (6.3)$$

in one additional independent variable y.

Secondly, unstable polynomial DEs may be made stable and polynomial by transformations like (3.9).

Example 6.3 (parity invariance). The Ermakov–Pinney ODE [49,98]

$$u_{xx} - \alpha^2 u + \beta^2 u^{-3} = 0 \qquad (6.4)$$

is unstable (algebraic branch point $p = \frac{1}{2}$) and invariant by parity on u: the transformation $u \mapsto u^2$ or u^{-2} preserves its polynomial form and makes it stable.

6.2 Technicalities

A careful *choice of the dependent variables* can save many computations.

Example 6.4 (dynamical systems). These systems of first-order ODEs possess sometimes an *equivalent* scalar ODE. This is the case of the Lorenz model (1.3), equivalent to [108]

$$xx''' - x'x'' + x^3 x' + (b + \sigma + 1)xx'' + (\sigma + 1)(bxx' - x'^2) + \sigma x^4$$
$$+ b(1 - r)\sigma x^2 = 0 \quad (6.5)$$

and of the Hénon–Heiles Hamiltonian system in (q_1, q_2) (5.195), which implies the fourth-order ODE in q_1 only (5.198). This offers two advantages. The first one is to reduce the matrix recurrence relation to a scalar one. The second one is much more interesting: The scalar ODE has a number of families less than or equal to that of the DS, which saves a lot of useless cases to consider; thus, in the HH system, the leading powers for (q_1, q_2) are $(-2, -2)$, $(-2, -1)$, $(-2, 0)$, while the equivalent fourth-order ODE for q_1 has only one leading power, -2.

Choosing an *integrated dependent variable* for the computations saves a lot. The principle is that if a DE for u is to be stable, this allows the presence of *one* movable logarithm in its primitive $v = \int u \, dx$. If changing u to v_x allows the DE to be integrated once or more, expressions are shortened.

6.3 Equivalence of Three Fundamental ODEs

Let S be a given analytic function of a complex variable x, and let us consider the following three differential equations in φ, χ, ψ:

$$\frac{\varphi_{xxx}}{\varphi_x} - \frac{3}{2}\left(\frac{\varphi_{xx}}{\varphi_x}\right)^2 = S, \qquad (6.6)$$

$$\omega = \chi^{-1}, \quad -2\omega_x - 2\omega^2 = S, \qquad (6.7)$$

$$-2\frac{\psi_{xx}}{\psi} = S. \qquad (6.8)$$

The first one is the Schwarz equation; if read backwards, it defines S as the Schwarzian $\{\varphi; x\}$ of φ. The second one is the Riccati equation in its normalized form (equations for ω or χ are equivalent and both of Riccati type). The third one is the second-order linear Sturm–Liouville ODE in its normalized form.

Each of these three ODEs possesses a fundamental uniqueness property. As shown by S. Lie, the Schwarzian is the unique elementary homographic differential invariant of a function φ, i.e., the unique elementary function of the derivatives $D\varphi$ of φ, excluding φ itself, invariant under the 6-parameter group $\mathcal{H}$ (or Möbius group, or $PSL(2, \mathbb{C})$) of homographic transformations:

$$\mathcal{H} : \varphi \mapsto \frac{a\varphi + b}{c\varphi + d},$$

(a, b, c, d) arbitrary complex constants, $ad - bc = 1$. (6.9)

Among nonlinear first-order ODEs in the class

$$u' = R(u, x), \tag{6.10}$$

where R is rational in u and analytic in x, the Riccati equation is the unique one whose general integral has no movable critical points. As to equation (6.8), its uniqueness lies in its linear form.

It is a classical result due to Painlevé ([90, p. 230], [96, vol. I]) that the three ODEs (6.6), (6.7), (6.8) are equivalent: It is sufficient to integrate one in order to integrate the two others. Consequently, any ODE reducible to one of these three ODEs can be considered as *explicitly linearizable*. The six ODEs obtained by elimination of S between any two of the three ODEs have the general solution

$$\omega(\varphi) = \frac{c_1 \varphi_x}{c_1 \varphi + c_2} - \frac{\varphi_{xx}}{2\varphi_x}, \tag{6.11}$$

$$\psi(\varphi) = (c_1 \varphi + c_2)\varphi_x^{-1/2}, \tag{6.12}$$

$$\varphi(\omega) = \frac{c_1(\omega_2 - \omega_1) + c_2(\omega_3 - \omega_1)}{c_3(\omega_2 - \omega_1) + c_4(\omega_3 - \omega_1)}, \quad c_1 c_4 - c_2 c_3 = 1, \tag{6.13}$$

$$\psi(\omega) = c_1\psi_1 + c_2\psi_2, \quad \psi_1^2 = \frac{\omega_2 - \omega_3}{(\omega_2 - \omega_1)(\omega_3 - \omega_1)}, \quad \psi_2 = \psi_1\frac{\omega_3 - \omega_1}{\omega_3 - \omega_2}, \tag{6.14}$$

$$\varphi(\psi) = \frac{c_1\psi_1 + c_2\psi_2}{c_3\psi_1 + c_4\psi_2}, \quad c_1 c_4 - c_2 c_3 = 1, \tag{6.15}$$

$$\omega(\psi) = \frac{c_1\psi_{1,x} + c_2\psi_{2,x}}{c_1\psi_1 + c_2\psi_2}, \tag{6.16}$$

where the c_i's are arbitrary constants, ω_i and ψ_i particular solutions of (6.7) and (6.8).

Only two of these six solutions, namely $\chi(\varphi)$ and $\psi(\varphi)$, equations (6.11)–(6.12), are expressed with a single function; therefore, among the three equivalent functions, φ is the most elementary one, and we are going to see that the two others, $\chi(\varphi)$ and $\psi(\varphi)$, are the basic building blocks of the invariant Painlevé analysis of both PDEs and ODEs.

If the space of independent variables is multidimensional, for each additional independent variable t let us define a function $C(x, t, \dots)$ by

$$-\frac{\varphi_t}{\varphi_x} = C. \tag{6.17}$$

As seen from (6.11)–(6.12), the t-dependence of the three equivalent functions is then characterized by the three equivalent *linear* PDEs (two homogeneous, one nonhomogeneous)

$$\varphi_t + C\varphi_x = 0, \tag{6.18}$$

$$\omega = \chi^{-1}, \quad \omega_t + \left(C\omega - \frac{1}{2}C_x\right)_x = 0, \tag{6.19}$$

$$\psi_t + C\psi_x - \frac{1}{2}C_x\psi = 0. \tag{6.20}$$

The linearity of these PDEs, as well as the invariance of C under the change of function $\varphi \mapsto F(\varphi)$, F arbitrary, shows that all independent variables but one give rise to *linear* equations.

Systems (6.6)–(6.8) and (6.18)–(6.20) require the cross-derivative condition

$$
\begin{aligned}
\varphi_x^{-1}\left((\varphi_{xxx})_t - (\varphi_t)_{xxx}\right) &= 2((\chi^{-1})_t)_x - 2((\chi^{-1})_x)_t \\
&= 2\psi^{-1}((\psi_t)_{xx} - (\psi_{xx})_t) \\
&= S_t + C_{xxx} + 2C_x S + C S_x = 0.
\end{aligned}
\tag{6.21}
$$

6.4 Optimal Choice of the Expansion Variable

A PDE has movable singularities that are not isolated, as opposed to the case of an ODE, but which lie on a codimension-one manifold

$$\varphi(x, t, \dots) - \varphi_0 = 0, \tag{6.22}$$

in which φ is an arbitrary function of the independent variables and φ_0 an arbitrary movable constant. Even in the ODE case, the movable singularity can be defined as $\varphi(x) - \varphi_0 = 0$, since the implicit function theorem allows this to be inverted to $x - x_0 = 0$; this provides a gauge freedom to be used later on in Section 7.

The singular manifold and the expansion variable play two different roles, and there is no a priori reason to confuse them, so let us denote by φ the function that defines the movable singular manifold $\varphi - \varphi_0 = 0$, and χ the expansion variable. The only requirement on χ is that it must vanish as $\varphi - \varphi_0$ and be a single-valued function of $\varphi - \varphi_0$ and its derivatives.

The Laurent series for u and E are defined as

$$u = \sum_{j=0}^{+\infty} u_j \chi^{j+p}, \quad -p \in \mathbb{N}, \tag{6.23}$$

$$E = \sum_{j=0}^{+\infty} E_j \chi^{j+q}, \quad -q \in \mathbb{N}^*. \tag{6.24}$$

To illustrate our point, let us take as an example the Korteweg–de Vries equation

$$E \equiv -u_t + u_{xxx} + 6uu_x = 0 \tag{6.25}$$

(this is one of the very rare locations where this equation can be taken as an example; indeed, usually, things work so nicely for KdV that it is hazardous to draw general conclusions from its isolated study).

With the choice $\chi = \varphi - \varphi_0$ [113], the coefficients (u_j, E_j) are invariant under the two-parameter group of translations $\varphi \mapsto \varphi + b$, b an arbitrary complex constant, and therefore they depend only on the differential invariant φ_x of this group and its derivatives:

$$u = -2\varphi_x^2 \chi^{-2} + 2\varphi_{xx} \chi^{-1} + \frac{\varphi_t}{6\varphi_x} - \frac{2}{3}\frac{\varphi_{xxx}}{\varphi_x} + \frac{1}{2}\left[\frac{\varphi_{xx}}{\varphi_x}\right]^2 + O(\chi). \tag{6.26}$$

[The quantity $C = -\varphi_t/\varphi_x$ is invariant under $\varphi \mapsto F(\varphi)$, F an arbitrary function, and therefore is uninteresting for the moment.]

With the choice $\chi = (\varphi - \varphi_0)/\varphi_x$, always possible, since the gradient of φ has at least one nonzero component, the invariance is extended to the four-parameter group of affine transformations $\varphi \mapsto a\varphi + b$, (a, b) arbitrary complex constants, with accordingly a dependence on the differential invariant φ_{xx}/φ_x and its derivatives:

$$u = -2\chi^{-2} + 2\frac{\varphi_{xx}}{\varphi_x}\chi^{-1} + \frac{\varphi_t}{6\varphi_x} - \frac{2}{3}\left[\frac{\varphi_{xx}}{\varphi_x}\right]_x - \frac{1}{6}\left[\frac{\varphi_{xx}}{\varphi_x}\right]^2 + O(\chi). \tag{6.27}$$

Let us extend this invariance to the six-parameter homographic group.

Eliminating φ_0 between χ and χ_x for each of the two choices of χ, one obtains the following ODEs of order one for χ:

$$\chi_x - \varphi_x = 0, \quad \chi = \varphi - \varphi_0, \tag{6.28}$$

$$\frac{1 - \chi_x}{\chi} - \frac{\varphi_{xx}}{\varphi_x} = 0, \quad \chi = \frac{\varphi - \varphi_0}{\varphi_x}, \tag{6.29}$$

whose coefficients depend only on the respective differential invariants. Now, one knows since S. Lie the differential invariant of the homographic group

$$S = \{\varphi; x\} = \left[\frac{\varphi_{xx}}{\varphi_x}\right]_x - \frac{1}{2}\left[\frac{\varphi_{xx}}{\varphi_x}\right]^2 \tag{6.30}$$

and ipso facto the associated ODE of order one. Its general solution leads, by taking the homographic transform that vanishes as $\varphi - \varphi_0$, to *the good expansion variable* [29]

$$\chi = \frac{\varphi - \varphi_0}{\varphi_x - \varphi_{xx}(\varphi - \varphi_0)/(2\varphi_x)} = \left[\frac{\varphi_x}{\varphi - \varphi_0} - \frac{\varphi_{xx}}{2\varphi_x}\right]^{-1}. \tag{6.31}$$

Check: Due to the homographic dependence of χ on φ, grad χ is a polynomial of degree two in χ with coefficients homographic invariants. Denoting by t an arbitrary independent variable, possibly equal to x, one obtains

$$\chi_t = -C + C_x\chi - \frac{1}{2}(CS + C_{xx})\chi^2, \tag{6.32}$$

$$\chi_x = 1 + \frac{S}{2}\chi^2, \tag{6.33}$$

$$(\log\psi)_t = -C\chi^{-1} + \frac{1}{2}C_x = -C(\log\psi)_x + \frac{1}{2}C_x. \tag{6.34}$$

Again, (6.32)–(6.33) are not different from (6.7), (6.19).

The only price to pay for invariance is to privilege some coordinate x.

For our KdV example, the final Laurent series, to be compared with the initial one (6.26), is remarkably simple:

$$u = -2\chi^{-2} - \frac{C}{6} - \frac{2S}{3} + O(\chi). \tag{6.35}$$

The successive values of χ and the corresponding subgroup items are gathered in the Table 3.2.

TABLE 3.2.

Group	Invariant I	Riccati(χ)	Solution for χ
$\varphi + b$	φ_x	$\chi_x = I$	$\varphi - \varphi_0$
$a\varphi + b$	φ_{xx}/φ_x	$\chi_x = 1 - I\chi$	$(\varphi - \varphi_0)/\varphi_x$
$\dfrac{a\varphi + b}{c\varphi + d}$	$\{\varphi; x\}$	$\chi_x = 1 + (I/2)\chi^2$	$\dfrac{\varphi - \varphi_0}{\varphi_x - \varphi_{xx}(\varphi - \varphi_0)/(2\varphi_x)}$

Kruskal's Choice

Kruskal [69] indicated the very simple choice $\chi = x - f(t, \dots)$ of expansion variable to make the practical computations as short as possible. This choice is equivalent in our formalism to a choice of gauge, namely $S = 0$, $C_x = 0$, and φ is then an arbitrary homographic function of $x - f(t, \dots)$ with constant coefficients. The choice of Kruskal is really a choice of the expansion variable, *not* of the singular manifold, i.e., $\chi_x = 1$, not $\varphi_x = 1$.

Caution: This choice should be used only at the stage of building necessary conditions for the PP, and never at the stage of sufficiency because of the constraints put on (S, C).

6.5 Unified Invariant Painlevé Analysis (ODEs, PDEs)

This is a reference section containing all the items of the version of Painlevé analysis that is common to ODEs and PDEs and that generates the simplest possible expressions, due to its built-in invariance.

Consider a DE

$$\mathbf{E}(\mathbf{u}, \mathbf{x}) = \mathbf{0}, \tag{6.36}$$

polynomial in $\mathbf{u}$ and its derivatives, analytic in $\mathbf{x}$ ($\mathbf{E}, \mathbf{u}, \mathbf{x}$ multidimensional), and the Laurent series for $\mathbf{u}$ and $\mathbf{E}$ around the movable singular manifold $\varphi - \varphi_0 = 0$:

$$\mathbf{u} = \mathbf{u}_{-\mathbf{p},1} \log \psi + \sum_{j=0}^{+\infty} \mathbf{u}_j \chi^{j+\mathbf{p}}, \quad -\mathbf{p} \in \mathbb{Z}, \tag{6.37}$$

$$\mathbf{E} = \sum_{j=0}^{+\infty} \mathbf{E}_j \chi^{j+\mathbf{q}}, \quad -\mathbf{q} \in \mathbb{Z}. \tag{6.38}$$

The coefficient $\mathbf{u}_{-\mathbf{p},1}$ can be nonzero only if $\mathbf{E}$ does not explicitly depend on $\mathbf{u}$. Let us denote by x any independent variable such that $\varphi_x \neq 0$. In order to establish the most general formulae, we need two other independent variables, t and y. The gradient of expansion variables χ and ψ is (auxiliary notation is $\omega = \chi^{-1}$)

$$\chi_x = 1 + \frac{S}{2}\chi^2, \tag{6.39}$$

$$\chi_t = -C + C_x \chi - \frac{1}{2}(CS + C_{xx})\chi^2, \tag{6.40}$$

$$\chi_y = -K + K_x \chi - \frac{1}{2}(KS + K_{xx})\chi^2, \tag{6.41}$$

$$(\log \psi)_x = \chi^{-1}, \tag{6.42}$$

$$(\log \psi)_t = -C\chi^{-1} + \frac{1}{2}C_x = -C(\log \psi)_x + \frac{1}{2}C_x, \tag{6.43}$$

$$(\log \psi)_y = -K\chi^{-1} + \frac{1}{2}K_x = -K(\log \psi)_x + \frac{1}{2}K_x, \tag{6.44}$$

$$\omega_x = -\omega^2 - \frac{S}{2}, \tag{6.45}$$

$$\omega_t = C\omega^2 - C_x \omega + \frac{1}{2}(CS + C_{xx}) = \left(-C\omega + \frac{1}{2}C_x\right)_x, \tag{6.46}$$

$$\omega_y = K\omega^2 - K_x \omega + \frac{1}{2}(KS + K_{xx}) = \left(-K\omega + \frac{1}{2}K_x\right)_x \tag{6.47}$$

(note that (6.40) generates the eight others), where S, C, K are elementary homographic differential invariants linked by the cross-derivative conditions

$$\varphi_x^{-1}\big((\varphi_{xxx})_t - (\varphi_t)_{xxx}\big) = S_t + C_{xxx} + 2C_x S + CS_x = 0, \qquad (6.48)$$

$$\varphi_x^{-1}\big((\varphi_{xxx})_y - (\varphi_y)_{xxx}\big) = S_y + K_{xxx} + 2K_x S + KS_x = 0, \qquad (6.49)$$

$$\varphi_x^{-1}\big((\varphi_y)_t - (\varphi_t)_y\big) = C_y - K_t + C_x K - CK_x = 0. \qquad (6.50)$$

Kruskal's choice is implemented by putting $S = 0$, $C = f_t$, $K = f_y, \ldots$ in (6.39)–(6.47), thus reducing each rhs to one term and making (6.48)–(6.50) useless.

The function $\varphi - \varphi_0$ never appears in the above formulae. Similarly, the explicit expressions of χ, ψ, S, C, K as functions of $\varphi - \varphi_0$ are *not* needed during the computations. We recall them here only for reference:

$$\chi = \left(\frac{\varphi_x}{\varphi - \varphi_0} - \frac{\varphi_{xx}}{2\varphi_x}\right)^{-1}, \qquad (6.51)$$

$$\psi = (\varphi - \varphi_0)\varphi_x^{-1/2}, \qquad (6.52)$$

$$S = \{\varphi; x\} = \frac{\varphi_{xxx}}{\varphi_x} - \frac{3}{2}\left(\frac{\varphi_{xx}}{\varphi_x}\right)^2 = \left(\frac{\varphi_{xx}}{\varphi_x}\right)_x - \frac{1}{2}\left(\frac{\varphi_{xx}}{\varphi_x}\right)^2$$

$$= -2\left(\frac{\varphi_x}{\varphi - \varphi_0} - \frac{\varphi_{xx}}{2\varphi_x}\right)_x - 2\left(\frac{\varphi_x}{\varphi - \varphi_0} - \frac{\varphi_{xx}}{2\varphi_x}\right)^2, \qquad (6.53)$$

$$C = -\frac{\varphi_t}{\varphi_x}, \qquad (6.54)$$

$$K = -\frac{\varphi_y}{\varphi_x}. \qquad (6.55)$$

In some applications, it is necessary to choose for χ the most general homographic transform of (6.51) that vanishes as $\varphi - \varphi_0$,

$$\operatorname{grad} \chi = \mathbf{X}_0 + \mathbf{X}_1 \chi + \mathbf{X}_2 \chi^2, \qquad (6.56)$$

$$\operatorname{grad} \omega = -\mathbf{X}_2 - \mathbf{X}_1 \omega - \mathbf{X}_0 \omega^2, \qquad (6.57)$$

$$\operatorname{grad} \log \psi = \mathbf{X}_0 \chi^{-1} + \frac{1}{2}\mathbf{X}_1. \qquad (6.58)$$

The vectorial coefficients $\mathbf{X}_i$ depend on $(S, C, K, \ldots)$ and two additional arbitrary functions. The auxiliary expansion variable ψ is defined by its logarithmic gradient and by the condition that it should vanish as $\varphi - \varphi_0$.

6.6   The Painlevé Test

The synthesis of the different methods to generate necessary conditions for the PP produces the following algorithm, called the "Painlevé test."

Consider a DE (2.29) of order N, already transformed if necessary (see Sections 6.1 and 6.2), so as to be polynomial in $\mathbf{u}$ and its derivatives, analytic in x. The Painlevé test consists of the following steps.

Step 0. Perform a transformation (3.5) in order to reduce the number of terms in the equation (details in Section 5.3.1 and [14] and, for PDEs, [39]).

Example: Equation $u_x + u_t + u_{xxt} + u_x u_t = 0$, under the translation $u = U - x - t$, becomes $U_{xxt} + U_x U_t + 1 = 0$.

Step 1. Require the satisfaction of the very general necessary conditions obtained by Painlevé (details in Section 5.5.2).

Example [75]: $-3u^2 u' u''' + 5u^2 u''^2 - uu'^2 u'' - u'^4 = 0$. Unstable for $\frac{5}{3}$ has not the required value $1 - 1/n$.

Example [25]: $(1 + u^2)u_{xx} - 2uu_x^2 + u_t^2 = 0$. The ODE obtained by the reduction $(x, t) \mapsto x - ct$ has an A with two simple poles $u = \pm i$ and residues $1 \pm ic^2/2$. The ODE is unstable, and so is the PDE.

Step 2. If the degree is greater than one, establish the ODE satisfied by the singular solutions (details in Section 5.1).

Step 3. Put the DE in Cauchy canonical form; find all the exceptional points where the Cauchy theorem fails; for each such point, define a homographic transformation (3.5) allowing the Cauchy theorem to apply (details in Sections 5.3 and 5.5). For each DE (the original one and all these homographic transforms), perform Step 4.

Example: (P5) has the exceptional points $u = 1$ and $u = 0$ (poles of A; see Section 5.5.2). These points are regular for the ODE in $(u - 1)^{-1}$ and u^{-1}.

Example: The reduced three-wave interaction dynamical system

$$x' = -2y^2 + z + \gamma x + \delta y, \quad y' = 2xy + \gamma y - \delta x, \quad z' = -2xz - 2z$$

has the exceptional point $y = \delta/2$ [11], not so evident in the system itself but easily unveiled by considering the equivalent third-order ODE for $y(t)$.

Step 4. Find all the families $\mathbf{u} \sim \mathbf{u}_0^{(0)} \chi^\mathbf{p}$ $(\mathbf{u}_0^{(0)} \neq \mathbf{0})$ (details in Section 5.4). Discard those families that are also families of the ODE for singular solutions established at Step 2. Require all components of remaining $\mathbf{p}$'s to be integral. Discard all families having all components of $\mathbf{p}$ positive. For each remaining family, perform Step 5 and at least one of Steps 6 and 7.

Example: (P5) has six families of movable simple pole-like singularities

$$u^2 \sim x^2 \chi^{-2}/(2\alpha), \quad u^{-2} \sim -x^2 \chi^{-2}/(2\beta), \quad (u - 1)^{-2} \sim -\chi^{-2}/(2\delta).$$

Example: The reduced three-wave interaction has the families [11]

$$(x, y, z) \sim (-\chi^{-1}/2, i\chi^{-1}/2, z_0), \quad z_0 \text{ arbitrary}, \quad \text{indices } (-1, 0, 2)$$
$$(x, y, z) \sim (\chi^{-1}, \delta/2, -\chi^{-2}), \quad \text{indices } (-1, 2, 2).$$

The first one will pass the test, while the second will generate at index 2 the conditions $\gamma\delta = 0$, $\gamma(\gamma + 1) = 0$.

In case the DE has too many terms, this step is worth being programmed on a computer, for fear of missing some families.

Warning. If one is unsure about some component u of $\mathbf{u}$ behaving like a positive integer power p of χ, it may be safer to switch to the DE for u^{-1}.

Step 5. From the auxiliary equation of the simplified equation, compute the linear operator $\mathbf{P}(i)$ of (2.35) and the indicial equation (2.35) $\det \mathbf{P}(i) = 0$ (details in Section 5.4). Compute its zeros (the Fuchs indices). Require each index to be integral (details in Section 5.11) and to satisfy the rank condition (5.106).

Example [31, example 5.B]: These are two coupled PDEs with a single family whose linear operator $\mathbf{P}(i)$ is

$$\mathbf{P}(i) = \begin{pmatrix} -\frac{1}{3}(i+2)^2 & \frac{1}{3}(i+2) \\ -(i+2) & i^2 \end{pmatrix}. \tag{6.59}$$

The indices are the zeros of its determinant $(-2, -2, -1, 1)$. For the double index $i = -2$, the rank of $\mathbf{P}(i)$ is one, so the system of PDEs is unstable.

Example [15, 22, 52]: The equation $u_{xxx} - 7uu_{xx} + 11u_x^2 = 0$ has only one family with three indices: $p = -1$, $u_0^{(0)} = -2$, indices $(-6, -1, -1)$. The double index -1 immediately proves the instability.

Step 6 (Non-Fuchsian case). If the degree of the indicial polynomial is strictly lower than N, and if a particular solution is known in closed form, apply the non-Fuchsian perturbative method (details in Section 5.8).

Step 7 (Fuchsian case) Denote by ρ the smallest integer Fuchs index, less than or equal to -1. Define two positive integer upper bounds $k_{\max}$ and $n_{\max}$ representing the cost of the computation to come; see advice below. Solve the linear algebraic system (5.111) in the unknown $\mathbf{u}_j^{(n)}$, $(j, n) \neq (0, 0)$, for the successive values $k = 0$ to $k_{\max}$, $n = 0$ to $n_{\max}$ with $j = k + n\rho$; whenever j is an index i of multiplicity $\text{mult}(i)$.

Require the orthogonality condition (5.112) to be satisfied for any value of the previously introduced arbitrary coefficients.

If $n = 0$ or ($n = 1$ and $i < 0$), assign arbitrary values to $\text{mult}(i)$ components of $\mathbf{u}_i^{(n)}$ defining a basis of $\text{Ker}\,\mathbf{P}(i)$.

If ($n = 1$ and $i \geq 0$) or $n \geq 2$, assign the value 0 to $\text{mult}(i)$ components of $\mathbf{u}_i^{(n)}$ defining a basis of $\text{Ker}\,\mathbf{P}(i)$.

Details in Section 5.7.

Advice for choosing $k_{\max}$ and $n_{\max}$: If the order $n = 0$ fails to describe the general solution, take at least $n_{\max} = 2$; take $k_{\max}$ so as to test the greatest Fuchs index for $n = n_{\max}$ (all details in the remarks at the end of Section 5.7).

This ends the test. Step 6 has been put before Step 7 because in all our examples it allows the process to conclude sooner.

Let us again stress that these sets of conditions may not be sufficient: Painlevé gave the counterexample of the second-order ODE whose general solution is $\pm \operatorname{sn}[\lambda \log(c_1 x + c_2); k)]$, with (c_1, c_2) arbitrary, for which no local test can generate the necessary and sufficient stability condition that $2\pi i \lambda$ be a period of the elliptic function sn. For advanced features, see Section 5.9 and [41].

6.7 The Partial Painlevé Test

In the search for the tiniest piece of integrability, the physicist (see Section 3.7) will perform the above Painlevé test to its end, i.e., without stopping even in case of failure of some condition, so as to collect a bunch of necessary conditions.

Turning to sufficiency, these conditions will then be examined separately in the hope of finding some global element of integrability, most often a Darboux eigenvector.

For instance, the Lorenz model (1.3) has two families,

$$x \sim 2i\chi^{-1}, \quad y \sim -(2i/\sigma)\chi^{-2}, \quad z \sim -(2/\sigma)\chi^{-2}, \quad i^2 = -1, \qquad (6.60)$$

with the same indices $(-1, 2, 4)$, which generate the no-log conditions [32, 107]

$$Q_2 \equiv (8/3)(b - 2\sigma)(b + 3\sigma - 1) = 0, \tag{6.61}$$

$$\begin{aligned} Q_4 \equiv &-4i(b - \sigma - 1)(b - 6\sigma + 2)x_2 + (8/3)(b - 1)(b - 3\sigma + 1)S \\ &- 4b\sigma(b - 3\sigma + 5)r/3 + (-4 + 10b + 30b^2 - 20b^3 - 16b^4)/27 \\ &+ (-38b - 56b^2 - (28/3)b^3 + 88\sigma + 86b^2\sigma)\sigma/3 \\ &- (28/9)b^3\sigma + 70b\sigma^2 - 64\sigma^3 - 58b\sigma^3 + 36\sigma^4 = 0, \end{aligned} \tag{6.62}$$

in which x_2 is arbitrary, S is the Schwarzian of the invariant analysis. Performing a logical *or* operation on these conditions instead of the logical *and* of the mathematician, one obtains the condition on (b, σ)

$$(b - 2\sigma)(b + 3\sigma - 1)(b - \sigma - 1)(b - 6\sigma + 2)(b - 1)(b - 3\sigma + 1) = 0.$$

What is remarkable is that *all* known analytic results on this model (first integrals [77], particular solutions [32], Darboux eigenvectors [48, 78]) belong to one of these six cases. Conversely, to each of the six factors there corresponds such a result, although sometimes only for a finite set of values of (b, σ).

Remarks.

1. With the restriction $S = 0$ one would miss two of the six factors.

2. First integrals of the type $P(x, y, x)e^{\lambda t}$, with P polynomial and λ constant, should not be searched for with the assumption that P is the most general polynomial in three variables. Indeed, P must be an entire function of t, i.e., have no singularities at a finite distance. The generating function of such polynomials is built from the singularity degrees of (x, y, z) [82],

$$\frac{1}{(1 - \alpha x)(1 - \alpha^2 y)(1 - \alpha^2 z)}, \tag{6.63}$$

and it provides the basis, ordered by singularity degrees,

$$(1), \quad (x), \quad (x^2, y, z), \quad (x^3, xy, xz), \quad (x^4, x^2 y, x^2 z, yz, z^2, y^2), \quad \dots$$

P_2 should thus be searched for as a linear combination of 1, x, x^2, y, z. All known first integrals are found at the P_4 level [77].

3. The case $b = 1 - 3\sigma$ is on an equal footing with the case $b = 2\sigma$, which admits the first integral $(x^2 - 2\sigma z)e^{2\sigma t}$, but finding its first integral is still an open problem.

7 Sufficiency: Explicit Integration Methods

We review the algorithmic methods that *may* perform the explicit integration, with emphasis on ODEs. The PDE case is handled in Chapter 8 of this volume [85].

We assume that the application of the Painlevé test (necessary conditions for the PP) has led either to no failure or to a minor failure, corresponding respectively to a presumption of integrability in the Painlevé sense or of partial integrability. If perturbative methods have been used, one has to decide to give up at some perturbation order n (remember the counterexample of Painlevé). The goal is then either to prove the sufficiency (integrability) or to build particular solutions (partial integrability).

If the DE belongs to one of the fully studied classes enumerated in Section 4, the question is solved. Indeed, either it is possible, by some homographic transformation (3.5), to bring the DE back to a normalized ("classified") DE, in which case the integration is finished, or this is impossible, in which case the DE has not the PP.

For a DE that has not been classified, if one excludes the case where the DE is an ODE and defines a new function (a quite improbable event, which has not occurred since 1906), the explicit proof of sufficiency amounts to (the cases below are not mutually exclusive)

• either (ODE case) express the general solution as a finite expression of a finite number of elementary functions (solutions of linear equations, the Weierstrass $\wp$ function, the six Painlevé functions),

- or (PDE case) find a Lax pair.

In the partial integrability situation, one tries to obtain degeneracies of these results: a particular solution or a pair of linear operators able to generate a subclass of solutions.

The methods to handle both cases are the same, and they again rely only on the singularity structure. Their basic common idea is that the singular part of the Laurent expansions (of a *local* nature) contains all the information for a *global* knowledge of the solution.

The two existing methods are known as the *singular part transformation* and the *truncation method*. Before describing them, let us give a few definitions and explain how Painlevé proved the sufficiency for the six equations (P1)–(P6).

7.1 Sufficiency for the Six Painlevé Equations

Painlevé introduced the concept of "intégration parfaite" and used it to solve the question of sufficiency for the six equations discovered by himself and Gambier. The idea is to perform a finite (in the sense of Poincaré: finite expression) single-valued transformation from (Pn) to another ODE that has no more movable singularities, although it may still have fixed critical singularities. Such an ODE has qualitatively the same singularities as a linear ODE, and Painlevé says that its integration is then "parfaite" (achieved) [91, p. 205]: Given any initial conditions, its solution can be computed to arbitrary accuracy (by, e.g., the sequence of coefficients of convergent Taylor series), since one knows in advance where the remaining (fixed) singularities are located. The movable singularities of the original ODE are then totally under control. The equations with fixed critical points therefore constitute a natural extension to the linear equations.

Painlevé defined such transformations (nowadays called "singular part transformations") for each of the six equations (P1)–(P6). These transformations, via logarithmic derivatives, transform (P1)–(P6) into equations for ψ without movable singularities ((P1) [93, p. 14], (P2) [93, p. 15], (P3) [93, p. 16], (P4, P5, P6) [95], [96, III p. 120]):

$$\text{(P1)} \qquad u = -\partial_x^2 \log \psi, \tag{7.1}$$

$$\text{(P2)} \qquad u = \partial_x \log \psi_1 - \partial_x \log \psi_2, \tag{7.2}$$

$$\text{(P3)} \qquad u = e^{-x}(\partial_x \log \psi_1 - \partial_x \log \psi_2), \tag{7.3}$$

$$\text{(P4)} \qquad u = \partial_x \log \psi_1 - \partial_x \log \psi_2, \tag{7.4}$$

$$\text{(P5)} \qquad u = xe^{-x}(2\alpha)^{-1/2}(\partial_x \log \psi_1 - \partial_x \log \psi_2), \tag{7.5}$$

$$\text{(P6)} \qquad u = x(x-1)e^{-x}(2\alpha)^{-1/2}(\partial_x \log \psi_1 - \partial_x \log \psi_2). \tag{7.6}$$

The Lax pairs of (P1)–(P6) can be found in [58, 70].

The two methods developed in the next sections, (7.3) and (7.4), rely on this result.

The *logarithmic derivative* plays a privileged role, as generator of a movable simple pole with a residue generically unity. A prerequisite to the algorithmic derivation of a transformation from u to ψ such as (7.1) is the introduction of a free gauge function, which we denote by φ.

Such a gauge naturally arises if one thinks of an ODE as the canonical reduction of a PDE defined by suppressing the dependence upon all independent variables but x. This is the function φ used in the description of the movable singularities by (6.22) rather than $x - x_0 = 0$. Useless at the stage of building necessary conditions (the Painlevé test), this feature is the key to the algorithmic explicit integration methods.

7.2   The Singular Part(s)

Definition 7.1. The *singular part* of one of the families of movable singularities of a given DE is the finite sum of the Laurent series restricted to the nonpositive powers in the method of pole-like expansions:

$$u_T = \sum_{j=0}^{-p} u_j \chi^{j+p}. \tag{7.7}$$

Synonyms are truncation, truncated expansion.

Given φ, the singular part u_T is a one-parameter (φ_0) family of expressions $u_T(\varphi_0)$, and the two particular values $\varphi_0 = 0$ and $\varphi_0 = \infty$ are of special interest. For the example of KdV (6.25),

$$u_T(0) = -2\left[\frac{\varphi_x}{\varphi} - \frac{\varphi_{xx}}{2\varphi_x}\right]^2 + \frac{\varphi_t}{6\varphi_x} - \frac{2}{3}\left(\frac{\varphi_{xxx}}{\varphi_x} - \frac{3}{2}\left[\frac{\varphi_{xx}}{\varphi_x}\right]^2\right), \tag{7.8}$$

$$u_T(\infty) = -2\left[-\frac{\varphi_{xx}}{2\varphi_x}\right]^2 + \frac{\varphi_t}{6\varphi_x} - \frac{2}{3}\left(\frac{\varphi_{xxx}}{\varphi_x} - \frac{3}{2}\left[\frac{\varphi_{xx}}{\varphi_x}\right]^2\right). \tag{7.9}$$

Definition 7.2. The *singular part operator* $\mathcal{D}$ of a family is defined by

$$\log \varphi \mapsto \mathcal{D}\log\varphi = u_T(0) - u_T(\infty). \tag{7.10}$$

Example 7.1 (KdV). The operator $\mathcal{D}$ is linear and equal to $2\partial_x^2$. This linearity is strongly linked with the Darboux transformation [85].

Example 7.2. The single family of (P1) and the two families of (P2) have the singular parts

(P1) :	$u_T = \chi^{-2} + \dfrac{S}{3}, \quad \mathcal{D} = -\partial_x^2,$	(7.11)
(P2) :	$u_T = \pm\chi^{-1}, \quad \mathcal{D} = \pm\partial_x.$	(7.12)

7.3 Method of the Singular Part Transformation

The method of the singular part transformation is the one used by Painlevé and outlined in Section 7.1. It consists in transforming the DE for u into a DE for φ by the nonlinear transformation

$$u = \mathcal{D}\log\varphi, \tag{7.13}$$

where $\mathcal{D}$ is the singular part operator associated to one of the families of the equation for u.

If the transformed equation for φ can be integrated, then so can the original equation.

Example 7.3 (linearization). The unique first-order first-degree ODE with the PP, namely the Riccati equation (1.1), has a $\mathcal{D}$ operator equal to $-a_2^{-1}\partial_x$, computable from the basic formulae (6.39), (6.51), and (7.10). The transformation $u = -a_2^{-1}\partial_x\log\varphi$ from u to φ leads to the second-order linear equation (1.2) for φ. It is then sufficient to know two particular solutions φ_1 and φ_2 (which *are* functions) of this linear equation to have a global knowledge of the general solution of the Riccati equation by the formula

$$u = -a_2^{-1}\partial_x\log(c_1\varphi_1 + c_2\varphi_2). \tag{7.14}$$

Similarly, the transformation $\wp = -\partial_x^2\log\sigma$ associates to the Weierstrass elliptic function $\wp$ a function σ that is an entire function, the solution of a nonlinear ODE.

Example 7.4 (simplified equation of one of the 50 stable ODEs (4.8)). The ODE

$$E \equiv u'' + uu' - u^3 = 0 \tag{7.15}$$

possesses two families of movable simple poles $u_0 = 1$ and $u_0 = -2$, with the one-parameter particular solutions $u_0/(x - x_0)$. The first family operator is $\mathcal{D} = \partial_x$, and it transforms it into

$$u = \partial_x\log\varphi, \quad E \equiv \varphi\left(\frac{\varphi''}{\varphi^2}\right)' = 0, \tag{7.16}$$

which integrates as $\varphi = a\wp(x - x_0, 0, g_3)$ with (a, x_0, g_3) arbitrary and provides the general solution. The two families of movable simple poles for u correspond to the movable simple zeros of $\wp$ (residue $u_0 = 1$) and to the movable double poles of $\wp$ (residue $u_0 = -2$).

Example 7.5 (indirect linearization). The Ermakov–Pinney equation (6.4), after the transformation $u^{-2} = v$ removing its algebraic singularity

$$E \equiv -\frac{1}{2}vv_{xx} + \frac{3}{4}v_x^2 - \alpha^2 v^2 + \beta^2 v^4 = 0, \tag{7.17}$$

has two families $v \sim \pm(2\beta)^{-1}\chi^{-1}$, and the transformed ODE under $v = (2\beta)^{-1}\partial_x \log \varphi$ [30],

$$\frac{\varphi_{xxx}}{\varphi_x} - \frac{3}{2}\left(\frac{\varphi_{xx}}{\varphi_x}\right)^2 = -2\alpha^2, \qquad (7.18)$$

is a Schwarz ODE (6.6). This integrates the Ermakov–Pinney equation via a finite two-valued expression.

Well suited to DEs possessing only one family (Riccati, Weierstrass, (P1), KdV), this transformation must be adapted, following the Painlevé formulae for (P2)–(P6) in Section 7.1, to suit DEs with more than one family (Jacobi elliptic equation, (P2) to (P6)). This is done in the chapter on PDEs, Chapter 8 [85].

7.4 Method of Truncation (Darboux Transformation)

Perfectly adapted to PDEs [85], this method is rather poor for ODEs, for an intrinsic reason, namely the absence in this case of a Bäcklund transformation (link between two different solutions of the same DE introducing at each iteration at least one more arbitrary parameter in the solution). It nevertheless succeeds, at least partially, in many situations.

The idea [113] is to consider the singular part (7.7) of one family (or the sum $u = \sum_f \mathcal{D} \log \psi_f$ of the singular parts of several families) as a *parametric representation* of a solution in terms of one function ψ linked to χ by $\chi^{-1} = \partial_x \log \psi$ (or several functions ψ_f, one per family f). Every function ψ_f, which defines a singular manifold $\psi_f = 0$, is required to be an entire function, and for instance to satisfy the *same* linear system of two PDEs $L_1 = 0$, $L_2 = 0$ with some adjustable coefficients.

The method consists in identifying to zero the lhs $E(u_T)$ considered as a polynomial of ψ_f and its independent derivatives modulo the constraint that each ψ_f satisfy the linear system. This generates an overdetermined set of *determining equations* whose unknowns are the coefficients u_j of (7.7) and the coefficients of the linear system. The remarkable fact is that the determining equations are easy to solve.

The result is some class of exact solutions, and this class is easily interpreted. If the commutator $[L_1, L_2]$ is identically zero (which is always the case if the linear system has constant coefficients), the solutions are particular ones (PDE case) or any kind (particular or general) (ODE case). If this commutator is zero only when some coefficient of (L_1, L_2) satisfies some PDE, quite probably (L_1, L_2) define a Lax pair.

Again, the ODE case to which we restrict is much less rich than the PDE case [85] to which we refer the reader.

7.4.1 One-Family Truncation

One-family truncation is the celebrated WTC truncation procedure [113].

Applicable to any DE with any number of families, it consists in selecting one of the families $\psi = 0$, in which ψ obeys the linear system of the invariant analysis

$$\psi_{xx} + \frac{S}{2}\psi = 0, \tag{7.19}$$

$$\psi_t + C\psi_x - \frac{C_x}{2}\psi = 0. \tag{7.20}$$

The functions S and C are adjustable functions constrained by the cross-derivative condition (6.48). Consider, e.g., the Ermakov–Pinney ODE [49, 98]

$$E \equiv -\frac{1}{2}vv_{xx} + \frac{3}{4}v_x^2 - \alpha^2 v^2 + \beta^2 v^4 = 0. \tag{7.21}$$

The infinite Laurent series is $v = (2\beta)^{-1}\chi^{-1} + v_1 + O(\chi)$ with v_1 arbitrary and β one of the two square roots of β^2. Thanks to the gauge φ, the coefficient v_1 is not a constant but a function.

The method consists in assuming that a solution v can be represented by the truncation [30]

$$v = v_T = \frac{1}{2\beta}\chi^{-1} + v_1 = \mathcal{D}\log\psi + v_1, \tag{7.22}$$

implying for the lhs E of the DE the similar truncated expansion

$$E \equiv \sum_{j=0}^{4} E_j \chi^{j-4}. \tag{7.23}$$

This generates, in this example, five equations $E_j = 0$ in the unknowns (v_1, S). Among them, E_0 is zero, since the coefficient v_0 of the series for v is already the good one. E_1 is zero, since 1 is a Fuchs index whose orthogonality condition is satisfied. Writing $v_1 = -V_1/(2\beta)$, there remain the three equations

$$E_2 \equiv -4\alpha^2 + S + 6V_1^2 + 6V_{1,x} = 0, \tag{7.24}$$

$$E_3 \equiv 8\alpha^2 V_1 + 2SV_1 - 4V_1^3 + S_x + 2V_{1,xx} = 0, \tag{7.25}$$

$$E_4 \equiv \frac{3}{4}S^2 - 4\alpha^2 V_1^2 + V_1^4 - V_1 S_x + 3SV_{1,x} + 3V_{1,x}^2 - 2V_1 V_{1,xx} = 0. \tag{7.26}$$

The algebraic elimination (i.e., without differentiation) of $V_{1,x}$ and $V_{1,xx}$ among these three equations yields $(S - s)^2 = 0$, with $s = -2\alpha^2$. Then V_1 is found to satisfy the Riccati equation

$$-2V_{1,x} - 2V_1^2 = s, \tag{7.27}$$

hence the particular solution

$$v = \frac{1}{2\beta}(\chi^{-1} - V_1), \tag{7.28}$$

in which each variable χ^{-1} and V_1 satisfies the same Riccati equation and depends on one arbitrary parameter. This is the general solution, which can be written as $v = (2\beta)^{-1}(\partial_x \log \psi_1 - \partial_x \log \psi_2)$, in agreement with the structure of singularities, cf. (7.2).

Remark. The particular solutions generically found by this method consist of the class of polynomials in tanh, which correspond to a constant value for S.

Another example is the (P2) equation, for which the one-family truncation $u = \chi^{-1} + u_1$ provides the one-parameter particular solution $u_1 = 0$, $S = x$ on the condition $\alpha = \frac{1}{2}$, i.e., an algebraic transform of the Airy equation.

7.4.2   Two-Family Truncation

When a DE admits two families with opposite principal parts, such as (7.17), it is natural to seek particular solutions described by two singular manifolds [34]

$$v = \frac{1}{2\beta}[\partial_x \log \psi_1 - \partial_x \log \psi_2 + v_0], \tag{7.29}$$

in which (ψ_1, ψ_2) is a basis of the two-dimensional space of solutions of some ODE whose general solution is entire, e.g., the second-order linear equation with constant coefficients

$$\psi_{xx} - \frac{k^2}{4}\psi = 0, \tag{7.30}$$

$$\Psi_2 = C_1 e^{kx/2} + C_2 e^{-kx/2} = C_0 \cosh \frac{k}{2}(x - x_0), \tag{7.31}$$

$$\psi_1(x) = \Psi_2(x + a), \quad \psi_2(x) = \Psi_2(x - a), \quad a \text{ arbitrary.} \tag{7.32}$$

Substituting (7.29) into (7.17) and eliminating any derivative of (ψ_1, ψ_2) of order greater than or equal to two in x results in a polynomial in the two variables $\psi_{1,x}/\psi_1$, $\psi_{2,x}/\psi_2$. Before identifying it with the null polynomial, one must take account of the first integral μ_0, the ratio of two constant Wronskians

$$\frac{\psi_{1,x}}{\psi_1}\frac{\psi_{2,x}}{\psi_2} = \frac{k^2}{4} - \mu_0 \frac{k}{2}\left(\frac{\psi_{1,x}}{\psi_1} - \frac{\psi_{2,x}}{\psi_2}\right), \quad \mu_0 = \coth ka, \tag{7.33}$$

which splits the polynomial of two variables into the sum of two polynomials in one variable:

$$16\beta^2 E$$
$$\equiv (k^2 - 4\alpha^2 + 6v_0^2 + 6k\mu_0 v_0)\left(\left(\frac{\psi_{1,x}}{\psi_1}\right)^2 + \left(\frac{\psi_{2,x}}{\psi_2}\right)^2\right)$$
$$+ (k\mu_0(k^2 - 4\alpha^2 + 6v_0^2) + 2(3k^2\mu_0^2 - k^2 - 4\alpha^2 + 4v_0^2)v_0)\left(\frac{\psi_{1,x}}{\psi_1} - \frac{\psi_{2,x}}{\psi_2}\right)$$
$$+ 2\alpha^2 k^2 - \frac{k^4}{2} - 3k^3\mu_0 v_0 - 4\alpha^2 v_0^2 - 3k^2 v_0^2 + v_0^4. \quad (7.34)$$

This defines three different algebraic equations in (k, v_0, μ_0); their two solutions

$$k^2 = 4\alpha^2, \quad v_0 = 0, \quad \mu_0 \text{ arbitrary}, \quad\quad\quad (7.35)$$
$$k^2 = 4\alpha^2, \quad v_0 = 2\alpha, \quad k\mu_0 = -2\alpha, \quad\quad\quad (7.36)$$

are just two different representations [34] of a solution of (7.17) depending on two arbitrary constants (μ_0, x_0): With this simple assumption, we have obtained the general solution

$$u^{-2} = v = \frac{1}{2\beta}\left[\frac{\psi_{1,x}}{\psi_1} - \frac{\psi_{2,x}}{\psi_2}\right] = \frac{\alpha}{\beta}\frac{\sinh ka}{\cosh k(x - x_0) + \cosh ka}. \quad (7.37)$$

In particular, with $\mu_0 = 0$ one thus obtains immediately the class of solutions polynomial in the two variables tanh and sech [33], thus augmenting the class indicated at the end of previous section. Evidently, if the DE has only one family, no dependence on sech can be found.

8   Conclusion

The solution of an ODE cannot escape the structure of singularities of the ODE. Such a structure can be studied from the equation itself, without any a priori knowledge of the solution, providing a deep insight into the possibility or not of performing the explicit integration.

Two levels of integrability have been defined: the Painlevé property (the most elementary level) and the integrability in the sense of Poincaré (the practical level).

A first series of methods (globally called "the Painlevé test") provide *necessary* conditions for a differential equation to have the Painlevé property, without any guarantee on the sufficiency. In the case of a negative answer from these first methods, there exist other methods (*Leçons* 8, 9, 10, 13, 19), not developed here, to provide necessary conditions for the general solution to have only a finite amount of movable branching, which implies the integrability in the sense of Poincaré, a weaker property than the PP.

In the case of a positive answer, the DE *may* have the PP, i.e., a general solution free from movable critical singularities. Then, a second series of methods are available to perhaps constructively prove the PP by explicitly building the general solution or some equivalent information (Lax pair). In case of failure of these second methods, the only remaining tool is human ability.

There exists another approach to DEs that is not based on the study of singularities. This is the method of infinitesimal symmetries [88, 89]. It provides reductions of PDEs to "smaller" PDEs or to ODEs, and it may provide first integrals of ODEs. However, the PDEs or ODEs left over after its completion still must be integrated, and the only methods to do so are those based on singularities. For instance, with the ODE (P1), the method of symmetries cannot provide any information (existence or not of a first integral, single-valuedness or multivaluedness).

9 REFERENCES

[1] M.J. Ablowitz and P.A. Clarkson, *Solitons, Nonlinear Evolution Equations and Inverse Scattering* (Cambridge Univ. Press, Cambridge, 1991).

[2] M.J. Ablowitz and H. Segur, *Exact linearization of a Painlevé transcendent*, Phys. Rev. Lett. **38** (1977), 1103–1106.

[3] M.J. Ablowitz, A. Ramani, and H. Segur, *Nonlinear evolution equations and ordinary differential equations of Painlevé type*, Lett. Nuovo Cimento **23** (1978), 333–338.

[4] ———, *A connection between nonlinear evolution equations and ordinary differential equations of P-type*. I, J. Math. Phys. **21** (1980), 715–721; II, **21** (1980), 1006–1015.

[5] M. Abramowitz and I.A. Stegun (eds.), *Handbook of Mathematical Functions*, tenth printing (National Bureau of Standards, Washington, 1972).

[6] G.G. Appelrot, *The problem of motion of a rigid body about a fixed point*, Uchebnye zapiski Moskovskogo universiteta, Otdel fizichesko-matematicheskikh nauk **11** (1894).

[7] A. Beauville, *Monodromie des systèmes différentiels linéaires à pôles simples sur la sphère de Riemann*, Séminaire Bourbaki, $45^{mathrme}$ année, 1992-93, exposé 765, Astérisque **216** (1993), 103–119.

[8] V.A. Belinskii, G.W. Gibbons, D.N. Page, and C.N. Pope, *Asymptotically Euclidean Bianchi IX metrics in quantum gravity*, Phys. Lett. A **76** (1978), 433–435.

[9] D. Benest and C. Frœschlé (eds.), Introduction to methods of complex analysis and geometry for classical mechanics and nonlinear waves (Éditions Frontières, Gif-sur-Yvette, 1994). R. Conte, *Singularities of differential equations and integrability*, pages 49–143. M. Musette, *Nonlinear partial differential equations*, pages 145–195.

[10] D. Bessis, *An introduction to Kowalevski's exponents*, Partially Integrable Evolution Equations in Physics, eds. R. Conte and N. Boccara (Kluwer, Dordrecht, 1990), pages 299–320.

[11] T. Bountis, A. Ramani, B. Grammaticos, and B. Dorizzi, *On the complete and partial integrability of non-Hamiltonian systems*, Physica A **128** (1984), 268–288.

[12] N. Bourbaki, Éléments de mathématique, I Théorie des ensembles, Fascicule de résultats (Hermann, Paris, 1958).

[13] F.J. Bureau, *Sur la recherche des équations différentielles du second ordre dont l'intégrale générale est à points critiques fixes*, Bulletin de la Classe des Sciences **XXV** (1939), 51–68.

[14] _____ , *Differential equations with fixed critical points*, Annali di Mat. pura ed applicata **LXIV** (1964), 229–364 [abbreviated as M. I].

[15] _____ , *Differential equations with fixed critical points*, Annali di Mat. pura ed applicata **LXVI** (1964), 1–116 [abbreviated as M. II].

[16] _____ , *Équations différentielles du second ordre en Y et du second degré en Ÿ dont l'intégrale générale est à points critiques fixes*, Annali di Mat. pura ed applicata **XCI** (1972), 163–281 [abbreviated as M. III].

[17] _____ , *Integration of some nonlinear systems of ordinary differential equations*, Annali di Mat. pura ed applicata **XCIV** (1972), 345–360.

[18] _____ , *Sur des systèmes différentiels du troisième ordre et les équations différentielles associées*, Bulletin de la Classe des Sciences **LXXIII** (1987), 335–353.

[19] _____ , *Differential equations with fixed critical points*, Painlevé transcendents, their asymptotics and physical applications, 103–123, eds. D. Levi and P. Winternitz (Plenum, New York, 1992).

[20] _____ , (third order), in preparation.

[21] É. Cartan, *Sur les variétés à connexion projective*, Bull. Soc. Math. France **52** (1924), 205–241.

[22] J. Chazy, *Sur les équations différentielles du troisième ordre et d'ordre supérieur dont l'intégrale générale a ses points critiques fixes*, thesis, Paris (1910); Acta Math. **34** (1911), 317–385. Table des matières commentée avec index, R. Conte (1991), 6 pages.

[23] _____, *Sur la limitation du degré des coëfficients des équations différentielles algébriques à points critiques fixes*, C. R. Acad. Sc. Paris **155** (1912), 132–135.

[24] _____, *Sur la limitation du degré des coëfficients des équations différentielles algébriques à points critiques fixes*, Acta Math. **41** (1918), 29–69.

[25] P.A. Clarkson, *The Painlevé property and a partial differential equation with an essential singularity*, Phys. Lett. **A 109** (1985), 205–208.

[26] _____, References for the Painlevé equations, about 21 pages.

[27] R. Conte, *Painlevé analysis of nonlinear PDE and related topics: a computer algebra program*, preprint (1988), 1–7. Same title, *Computer algebra and differential equations*, ed. E. Tournier (Academic Press, New York, 1989), page 219.

[28] _____, *Universal invariance properties of Painlevé analysis and Bäcklund transformation in nonlinear partial differential equations*, Phys. Lett. **A 134** (1988), 100–104.

[29] _____, *Invariant Painlevé analysis of partial differential equations*, Phys. Lett. **A 140** (1989), 383–390.

[30] _____, *Unification of PDE and ODE versions of Painlevé analysis into a single invariant version*, Painlevé transcendents, their asymptotics and physical applications, eds. D. Levi and P. Winternitz (Plenum, New York, 1992), pages 125–144.

[31] R. Conte, A.P. Fordy, and A. Pickering, *A perturbative Painlevé approach to nonlinear differential equations*, Physica D **69** (1993), 33–58.

[32] R. Conte and M. Musette, *A simple method to obtain first integrals of dynamical systems*, Solitons and chaos (Research Reports in Physics–Nonlinear Dynamics) eds. I.A. Antoniou and F.J. Lambert (Springer, Berlin, 1991), pages 125–128.

[33] _____ *Link between solitary waves and projective Riccati equations*, J. Phys. **A 25** (1992), 5609–5623.

[34] _____, *Linearity inside nonlinearity: exact solutions to the complex Ginzburg-Landau equation*, Physica D **69** (1993), 1–17.

[35] _____, *A new method to test discrete Painlevé equations*, Phys. Lett. A **223** (1996), 439–448.

[36] G. Contopoulos, B. Grammaticos, and A. Ramani, *Painlevé analysis for the mixmaster universe model*, J. Phys. A **25** (1993), 5795–5799.

[37] _____, *The mixmaster universe model, revisited*, J. Phys. A **27** (1994), 5357–5361.

[38] C.M. Cosgrove, *Painlevé classification of all semilinear partial differential equations of the second order I. Hyperbolic equations in two independent variables*, Stud. Appl. Math. **89** (1993), 1–61.

[39] _____, *Painlevé classification of all semilinear partial differential equations of the second order II. Parabolic and higher dimensional equations*, Stud. Appl. Math. **89** (1993), 95–151.

[40] _____, *All binomial-type Painlevé equations of the second order and degree three or higher*, Stud. Appl. Math. **90** (1993), 119–187.

[41] _____, *Painlevé classification problems featuring essential singularities*, Stud. Appl. Math. **98** (1997), 355–433.

[42] _____, *Corrections and annotations to E.L. Ince*, Ordinary differential equations, Chapter 14, on the classification of Painlevé differential equations, unpublished (1993).

[43] C.M. Cosgrove and G. Scoufis, *Painlevé classification of a class of differential equations of the second order and second degree*, Stud. Appl. Math. **88** (1993), 25–87.

[44] G. Darboux, *Sur la théorie des coordonnées curvilignes et des systèmes orthogonaux*, Annales scientifiques de l'École normale supérieure **7** (1878), 101–150.

[45] _____, *Sur les équations aux dérivées partielles*, C. R. Acad. Sc. Paris **96** (1883), 766–769.

[46] H.T. Davis, Introduction to nonlinear differential and integral equations, no. O-556037 (U.S. Government Printing Office, Washington D.C., 1961).

[47] J.-M. Drouffe, Simplex AMP reference manual, version 1.0 (SPhT, CEA Saclay, F–91191 Gif-sur-Yvette Cedex, 1996).

[48] V.S. Dryuma, *Projective duality in the theory of second order differential equations*, Mat. Issled., Kishinev **112** (1990), 93–103.

[49] V.P. Ermakov, *Équations différentielles du deuxième ordre. Conditions d'intégrabilité sous forme finale.* Univ. Izv. Kiev (1880), Ser. 3, No. 9, 1–25. [English translation by A.O. Harin, 29 pages].

[50] A.P. Fordy, *The Hénon–Heiles system revisited*, Physica D **52** (1991), 204–210.

[51] A.P. Fordy and J. Gibbons, *Some remarkable nonlinear transformations*, Phys. Lett. A **75** (1980), 325–325.

[52] A.P. Fordy and A. Pickering, *Analysing negative resonances in the Painlevé test*, Phys. Lett. A **160** (1991), 347–354.

[53] R. Fuchs, *Sur quelques équations différentielles linéaires du second ordre*, C. R. Acad. Sc. Paris **145** (1905), 555–558.

[54] B. Gambier, *Sur les équations différentielles dont l'intégrale générale est uniforme*, C. R. Acad. Sc. Paris **142** (1906), 266–269, 1403–1406, 1497–1500.

[55] _____, *Sur les équations différentielles du second ordre et du premier degré dont l'intégrale générale est à points critiques fixes*, C. R. Acad. Sc. Paris **143** (1906), 741–743; **144** (1907), 827–830, 962–964.

[56] _____, *Sur les équations différentielles du second ordre et du premier degré dont l'intégrale générale est à points critiques fixes*, thesis, Paris (1909); Acta Math. **33** (1910), 1–55.

[57] C.S. Gardner, J.M. Greene, M.D. Kruskal, and R.M. Miura, *Method for solving the Korteweg-de Vries equation*, Phys. Rev. Lett. **19** (1967), 1095–1097.

[58] R. Garnier, *Sur des équations différentielles du troisième ordre dont l'intégrale générale est uniforme et sur une classe d'équations nouvelles d'ordre supérieur dont l'intégrale générale a ses points critiques fixes*, thesis, Paris (1911); Annales scientifiques de l'École normale supérieure **29** (1912), 1–126.

[59] R. Garnier, *Sur des systèmes différentiels du second ordre dont l'intégrale générale est uniforme*, Annales scientifiques de l'École normale supérieure **77** (1960), 123–144.

[60] V.V. Golubev, Lectures on the integration of the equation of motion of a rigid body about a fixed point (Gostechizdat (State publishing house), Moscow, 1953). English (Israel program for scientific translations, 1960).

[61] B. Grammaticos, B. Dorizzi, and A. Ramani, *Solvable integrodifferential equations and their relation to the Painlevé conjecture*, Phys. Rev. Lett. **53** (1984), 1–4.

[62] G.-H. Halphen, *Sur un système d'équations différentielles*, C. R. Acad. Sc. Paris **92** (1881), 1101–1103. Reprinted, *Œuvres*, volume 2, 475–477 (1918).

[63] M. Hénon and C. Heiles, *The applicability of the third integral of motion: some numerical experiments*, Astron. J. **69** (1964), 73–79.

[64] E. Hille, *Ordinary Differential Equations in the Complex Domain* (J. Wiley and Sons, New York, 1976).

[65] P. Hoyer, *Über die Integration eines Differentialgleichungssystems von der Form* $dx_1/dt = a_1 x_2 x_3 + a_2 x_3 x_1 + a_3 x_1 x_2$, $dx_2/dt = b_1 x_2 x_3 + b_2 x_3 x_1 + b_3 x_1 x_2$, $dx_3/dt = c_1 x_2 x_3 + c_2 x_3 x_1 + c_3 x_1 x_2$ *durch elliptische Funktionen*, Dissertation Königl. Friedrich-Wilhelms Univ., Berlin (1879), 1–36.

[66] L. Hsu and N. Kamran, *Classification of second-order ordinary differential equations admitting Lie groups of fiber-preserving point symmetries*, Proc. London Math. Soc. **58** (1989), 387–416.

[67] E.L. Ince, *Ordinary Differential Equations* (Longmans, Green, and Co., London and New York, 1926). Reprinted (Dover, New York, 1956). See errata in [42].

[68] C. Itzykson and J.-M. Drouffe, *Statistical Field Theory*, two volumes (Cambridge University Press, Cambridge, 1989).

[69] M. Jimbo, M.D. Kruskal, and T. Miwa, *Painlevé test for the self-dual Yang–Mills equation*, Phys. Lett. A **92** (1982), 59–60.

[70] M. Jimbo and T. Miwa, *Monodromy preserving deformations of linear ordinary differential equations with rational coefficients. II*, Physica D **2** (1981), 407–448.

[71] N. Joshi and M.D. Kruskal, *A local asymptotic method of seeing the natural barrier of the solutions of the Chazy equation*, Applications of Analytic and Geometric Methods to Nonlinear Differential Equations, ed. P.A. Clarkson (Plenum, New York, 1993), pages 331–340.

[72] D.J. Kaup, *On the inverse scattering problem for cubic eigenvalue problems of the class* $\psi_{xxx} + 6Q\psi_x + 6R\psi = \lambda\psi$, Stud. Appl. Math. **62** (1980), 189–216.

[73] S.V. Kovalevski, *Sur le problème de la rotation d'un corps solide autour d'un point fixe*, Acta Math. **12** (1889), 177–232.

[74] ———, *Sur une propriété du système d'équations différentielles qui définit la rotation d'un corps solide autour d'un point fixe*, Acta Math. **14** (1890), 81–93.

[75] M.D. Kruskal, *Flexibility in applying the Painlevé test*, Painlevé Transcendents, Their Asymptotics and Physical Applications, eds. D. Levi and P. Winternitz (Plenum, New York, 1992), pages 187–195.

[76] M.D. Kruskal and P.A. Clarkson, *The Painlevé–Kowalevski and poly-Painlevé tests for integrability*, Stud. Appl. Math. **86** (1992), 87–165.

[77] M. Ku's, *Integrals of motion for the Lorenz system*, J. Phys. A **16** (1983), L689–L691.

[78] S. Labrunie and R. Conte, *A geometrical method towards first integrals for dynamical systems*, J. Math. Phys. **37** (1996), 6198–6206.

[79] L.D. Landau and E.M. Lifshitz, Théorie classique des champs, chapter *Problèmes cosmologiques* (Mir, Moscow, 1989, 4th edition).

[80] A. Latifi, M. Musette, and R. Conte, *The Bianchi IX (mixmaster) cosmological model is not integrable*, Phys. Lett. A **194** (1994), 83–92; **197** (1995), 459–460.

[81] P.D. Lax, *Integrals of nonlinear equations of evolution and solitary waves*, Comm. Pure Appl. Math. **21** (1968), 467–490.

[82] G. Levine and M. Tabor, *Integrating the nonintegrable: analytic structure of the Lorenz system revisited*, Physica D **33** (1988), 189–210.

[83] R. Liouville, *Sur les invariants de certaines équations différentielles et sur leurs applications*, Journal de l'École Polytechnique **LIX** (1889), 7–76.

[84] G.M. Murphy, *Ordinary Differential Equations and Their Solutions* (Van Nostrand, Princeton, 1960).

[85] M. Musette, Chapter 8 of this volume.

[86] M. Musette and R. Conte, *Non-Fuchsian extension to the Painlevé test*. Phys. Lett. A **206** (1995), 340–346.

[87] P.K. Nekrasov, *The problem of motion of a rigid body about a fixed point*, Matem. Sb. **16** (1892).

[88] P.J. Olver, *Applications of Lie Groups to Differential Equations* (Springer, Berlin, 1986).

[89] L.V. Ovsiannikov, *Group Analysis of Differential Equations* (Academic Press, New York, 1982).

[90] P. Painlevé, *Leçons sur la théorie analytique des équations différentielles* (Leçons de Stockholm, delivered in 1895) (Hermann, Paris, 1897). Reprinted, Œuvres de Paul Painlevé , vol. I (Éditions du CNRS, Paris, 1973).

[91] _____, *Mémoire sur les équations différentielles dont l'intégrale générale est uniforme*, Bull. Soc. Math. France **28** (1900), 201–261.

[92] _____, *Sur les équations différentielles d'ordre quelconque à points critiques fixes*, C. R. Acad. Sc. Paris **130** (1900), 1112–1115.

[93] _____, *Sur les équations différentielles du second ordre et d'ordre supérieur dont l'intégrale générale est uniforme*, Acta Math. **25** (1902), 1–85.

[94] _____, *Observations au sujet de la Communication précédente* (de Arnaud Denjoy), C. R. Acad. Sc. Paris **148** (1902), 1156–1157.

[95] _____, *Sur les équations différentielles du second ordre à points critiques fixes*, C. R. Acad. Sc. Paris **143** (1906), 1111–1117.

[96] *Œuvres de Paul Painlevé* , 3 volumes (Éditions du CNRS, Paris, 1973, 1974, 1976). Order to: La librairie de CNRS-Éditions, 151 bis, rue Saint-Jacques, F-75005 Paris, phone +33-1-53100505, fax +33-1-53100557, e-mail editions@edition.cnrs.fr).

[97] A. Pickering, *Testing nonlinear evolution equations for complete integrability*, Ph.D. thesis, University of Leeds (1992).

[98] E. Pinney, *The nonlinear differential equation $y''(x) + p(x)y(x) + c/y^3(x) = 0$*, Proc. Amer. Math. Soc. **1** (1950), 681.

[99] H. Poincaré, *Sur un théorème de M. Fuchs*, Acta Math. **7** (1885), 1–32.

[100] _____, *Les méthodes nouvelles de la mécanique céleste*, 3 volumes (Gauthier-Villars, Paris, 1892, 1893, 1899).

[101] J.-F. Pommaret, *Lie Pseudogroups and Mechanics* (Gordon and Breach, New York, 1988).

[102] A. Ramani, B. Grammaticos, and T. Bountis, *The Painlevé property and singularity analysis of integrable and nonintegrable systems*, Physics Reports **180** (1989), 159–245.

[103] K.A. Robbins, *Periodic solutions and bifurcation structure at high R in the Lorenz model*, SIAM J. Appl. Math. **36** (1979), 457–472.

[104] K. Sawada and T. Kotera, *A method for finding N-soliton solutions of the KdV equation and KdV-like equation*, Prog. Theor. Phys. **51** (1974), 1355–1367.

[105] F. Ayres Jr., *Theory and Problems of Differential Equations* (McGraw Hill, New York, 1972).

[106] L. Schlesinger, *Über eine Klasse von Differentialsystemen beliebiger Ordnung mit festen kritischen Punkten*, J. für R. und Angew. Math. **141** (1912), 96–145.

[107] H. Segur, *Solitons and the inverse scattering transform*, Topics in Ocean Physics, eds. A.R. Osborne and P. Malanotte Rizzoli (North-Holland Publ. Co., Amsterdam, 1982), pages 235–277.

[108] T. Sen and M. Tabor, *Lie symmetries of the Lorenz model*, Physica D **44** (1990), 313–339.

[109] A.H. Taub, *Empty space-times admitting a three-parameter group of motions*, Annals of Math. **53** (1951), 472–490.

[110] A. Tresse, *Détermination des invariants ponctuels de l'équation différentielle ordinaire du second ordre $y'' = \omega(x, y, y')$*. Leipzig 1896. 87 S. gr. 8°. Fürstl. Jablonowski'schen Gesellschaft zu Leipzig. Nr. **32** (**13** der math.-naturw. Section). Mémoire couronné par l'Académie Jablonowski; S. Hirkel, Leipzig (1896).

[111] H. Umemura, *Birational automorphic groups and differential equations*, Nagoya Math. J. **119** (1990), 1–80.

[112] G. Valiron, *Cours d'analyse mathématique*, (Masson, Paris, 1950, 2nd edition).

[113] J. Weiss, M. Tabor, and G. Carnevale, *The Painlevé property for partial differential equations*, J. Math. Phys. **24** (1983), 522–526.

4

Asymptotic Studies of the Painlevé Equations

Nalini Joshi

ABSTRACT The main aim of this chapter is to explain direct and natural rigorous methods for carrying out local and some global asymptotic studies near fixed singular points of the classical Painlevé equations. Such methods were first developed by Boutroux (around 1913). Here we review these methods and improve Boutroux's results. Moreover, we show that these methods can also be used to obtain asymptotic behavior in other limits, e.g., when a parameter of the equation becomes large. The methods and results are illustrated here for the first and second Painlevé equations.

1 Introduction

Asymptotics is a strong and fundamental theme that runs through much of mathematics and physics. It is both a heuristic art and a robust rigorous field of mathematical analysis that also appears in many algebraic applications, including number theory. Its strength lies in the range of its methods and the depth of its analytical foundations. The remarkable accuracy of asymptotic approximations can appear startlingly magical, when it is, in fact, rigorous. The aim of this chapter is to introduce the reader to some *nonlinear* asymptotic magic and to illuminate its rigorous foundations.

Nonlinear asymptotics here means the asymptotic study of nonlinear ordinary differential equations (ODEs) by methods that rely only on the equations themselves. In particular, we will describe such direct asymptotic methods for the well-known nonlinear second-order ODEs in the complex plane called the Painlevé equations.

1.1 The Painlevé Equations

For conciseness and simplicity, we will focus on the first and second Painlevé equations (P_I and P_{II}) which are given respectively by

$$y'' = 6y^2 + x, \tag{1.1}$$

$$y'' = 2y^3 + xy + \alpha. \tag{1.2}$$

These (and the other four Painlevé) equations have beautiful properties. Here we outline some of these properties.

Their main distinguishing characteristic, by which they were first identified, lies in the (complex) singularity structure of their highly transcendental solutions. This structure arises from the singularities of the equations. Given a regular point $x = a$ and regular initial values $y(a) = y_0$, $y'(a) = y_1$, standard theorems [1,2] show that a unique solution $y(x)$ exists in a neighborhood of a and is analytic there. Here the key word is "regular," which means that the right side of the equation solved for the highest derivative is analytic at (a, y_0, y_1).

It is easy to see that the Painlevé equations are not regular, i.e., they are singular at infinity in the plane of the independent variable x.

Exercise 1.1. Carry out the transformation $x = 1/t$, $y(x) = Y(t)$ on (1.1)–(1.2). Show that the standard theorems giving existence, uniqueness, and analyticity fail at $t = 0$.

Such a singularity is called a *fixed* singularity (since its location, i.e., $x = \infty$, does not change as initial data change). The standard theorems for ODEs also fail wherever y takes on values at which the right side of the equation becomes singular. For P_I and P_{II} these occur when y becomes unbounded. This can occur at finite, arbitrary points $x = x_0$, as shown in the following exercise.

Exercise 1.2. Given a point $x_0 \in \mathbb{C}$, show that (1.1) admits a formal series solution of the form

$$y(x) = \sum_{n=0}^{\infty} a_n (x - x_0)^{n-2}, \quad a_0 = 1,$$

where a_6 is arbitrary. It can be shown that such series converge [3] and represent the solution around every possible point where y is large.

The location x_0 of this singularity changes with initial data. (See Ref. 3 for the relation between x_0 and initial data given near it.) Such a singularity is called *movable*, since x_0 may change (or move) from solution to solution. The Painlevé equations have the characteristic property that all movable singularities of all solutions are poles. This is commonly referred to as the *Painlevé property*.

The Painlevé equations also possess many other deep properties. For example, their solutions can be written in terms of entire functions, functions that are analytic everywhere except at the fixed singular points of the equation.

Exercise 1.3. Transform a solution $y(x)$ of (1.1) to

$$w(x) = \exp\left\{ -\iint y(x)\, dx\, dx \right\}.$$

Show that $w(x)$ is analytic wherever $y(x)$ is and also wherever $y(x)$ has a movable pole.

Painlevé [4], Gambier [5], and R. Fuchs [6] identified the six Painlevé equations (under some mild conditions) as the only ones (of second order and first degree) with the Painlevé property whose general solutions cannot be written in terms of previously known transcendental functions.

Modern interest in these equations arises from their close relationship to integrable partial differential equations (PDEs) [7, 8]. Ablowitz, Ramani, and Segur [9] found extensive evidence that ODE reductions of completely integrable partial differential equations necessarily have the Painlevé property (in some choice of coordinates). Here integrability means that the equations can be solved through an associated (single-valued) linear system. (See the lectures by Mark Ablowitz in this volume, Chapter 9.)

Exercise 1.4. Consider the modified Korteweg–de Vries (mKdV) equation

$$q_t - 6q^2 q_x + q_{xxx} = 0.$$

Show that the similarity reduction given by

$$q(x,t) = (3t)^{-1/3} f(\xi), \quad \xi = x(3t)^{-1/3}$$

gives the ODE

$$f_{\xi\xi\xi} - 6f^2 f_\xi - f - \xi f_\xi = 0.$$

Integrate this once to show that $f(\xi)$ satisfies P_{II}.

Such reductions also act on the linear problem, or *Lax pair*, associated with the integrable PDE to give a linear problem for the Painlevé equation.

Exercise 1.5. The linear problem (or Lax pair) associated with the mKdV equation is

$$\begin{pmatrix} v_{1x} \\ v_{2x} \end{pmatrix} = \begin{pmatrix} -i\zeta & q(x;t) \\ q(x;t) & i\zeta \end{pmatrix} \begin{pmatrix} v_1 \\ v_2 \end{pmatrix}, \tag{1.3}$$

$$\begin{pmatrix} v_{1t} \\ v_{2t} \end{pmatrix} = \begin{pmatrix} A & B \\ C & -A \end{pmatrix} \begin{pmatrix} v_1 \\ v_2 \end{pmatrix}, \tag{1.4}$$

where

$$A = -4i(\zeta^3 + q^2\zeta/2), \tag{1.5}$$

$$B = 4(q\zeta^2 + iq_x\zeta/2 + q^3/2 - q_{xx}/4), \tag{1.6}$$

$$C = 4(q\zeta^2 - iq_x\zeta/2 + q^3/2 - q_{xx}/4). \tag{1.7}$$

Show that the similarity reduction of the previous exercise leads to the following linear problem, if we define $\lambda = \zeta(3t)^{1/3}$:

$$\begin{pmatrix} v_{1\xi} \\ v_{2\xi} \end{pmatrix} = \begin{pmatrix} -i\lambda & f(\xi) \\ f(\xi) & i\lambda \end{pmatrix} \begin{pmatrix} v_1 \\ v_2 \end{pmatrix}, \tag{1.8}$$

$$\begin{pmatrix} v_{1\lambda} \\ v_{2\lambda} \end{pmatrix} = \begin{pmatrix} \tilde{A} & \tilde{B} \\ \tilde{C} & -\tilde{A} \end{pmatrix} \begin{pmatrix} v_1 \\ v_2 \end{pmatrix}, \tag{1.9}$$

where

$$\tilde{A} = -4i\lambda^2 - 2if^2 - 4i\xi, \tag{1.10}$$

$$\tilde{B} = 4f\lambda + 2if_\xi + \alpha/\lambda, \tag{1.11}$$

$$\tilde{C} = 4\lambda f - 2if_\xi + \alpha/\lambda. \tag{1.12}$$

Show that the compatibility of the system (1.8)–(1.9), i.e.,

$$\frac{\partial^2 v_j}{\partial\xi\partial\lambda} = \frac{\partial^2 v_j}{\partial\lambda\partial\xi}, \tag{1.13}$$

holds if and only if $f(\xi)$ satisfies P_{II}.

The system (1.9) is called a monodromy problem. (See the lecture notes by Gilbert Mahoux, Chapter 2 in this volume.) Note that it has singularities at $\lambda = 0, \infty$. If the fundamental matrix of solutions $\mathbf{V}$ is analytically continued around a closed path containing 0 (or ∞) in its interior, then the resultant solution $\tilde{\mathbf{V}}$ is related to $\mathbf{V}$ through

$$\tilde{\mathbf{V}} = M\mathbf{V},$$

where the matrix M (independent of λ) is called a monodromy matrix. The compatibility condition (1.13) is equivalent to the condition that the similarity invariant data of M, i.e., its trace and determinant, remain constant as ξ changes. It can be shown that each Painlevé equation is associated with such an *isomonodromy* problem.

There is an implicit mapping between the initial data given for a Painlevé equation and the monodromy data of its associated linear problem. Conversely, this mapping can be inverted [10, 11]. That is, the isomonodromy property can be used to solve the Painlevé equation implicitly.

Such solvability is a fundamental property of the Painlevé equations. They possess many other properties including exact special solutions in terms of classical special functions and transformations relating solutions to other solutions (see Refs. [7] and 8 for references). Consequently, their solutions play a distinguished role as nonlinear special functions.

Asymptotics is a natural part of their study as special functions. Asymptotic descriptions are valuable because they give explicit descriptions of the highly transcendental general solutions of the Painlevé equations in terms of known classical functions.

1.2 Asymptotics

The aim of this section is to refresh the reader's knowledge of asymptotics. In particular, we define asymptotic notation, define connection problems, and recall the subtle asymptotics of the Stokes phenomena.

Suppose f and g are two (complex-valued) functions of a complex variable x. Assume that x is approaching x_0 along a path γ in a domain containing x_0 or having it as a limit point and assume that $g(x)$ is bounded below along γ. To compare f and g in this limit, we use the following notations.

Definition 1.1.

1. f is said to be *much, much less than* g (or g is much, much greater than f) in the limit, or

$$f(x) \ll g(x) \text{ as } x \to_\gamma x_0,$$

   if

$$\lim_{x \to x_0} \frac{f(x)}{g(x)} = 0.$$

2. f is *asymptotic to* g, or

$$f(x) \approx g(x) \text{ as } x \to_\gamma x_0,$$

   if

$$\lim_{x \to x_0} \frac{f(x) - g(x)}{g(x)} = 0 \iff \lim_{x \to x_0} \frac{f(x)}{g(x)} = 1.$$

3. f is *asymptotically proportional to* g, i.e.,

$$f(x) \sim g(x) \text{ as } x \to_\gamma x_0,$$

   if

$$\exists \text{ a const } c_0 \neq 0 \text{ s.t. } \lim_{x \to x_0} \frac{f(x)}{g(x)} = c_0.$$

4. f is *of the order of* g, i.e.,

$$f(x) = O\big(g(x)\big) \text{ as } x \to_\gamma x_0,$$

   if

$$\exists M \text{ s.t. } \left| \frac{f(x)}{g(x)} \right| < M \text{ as } x \to_\gamma x_0.$$

As an example, consider

$$\sinh x = \frac{e^x - e^{-x}}{2},$$

in the limit as $x \to +\infty$ along the positive real axis. Then we have

$$\sinh x - \frac{e^x}{2} = -\frac{e^{-x}}{2} \iff \frac{\sinh x - e^x/2}{e^x/2} = -e^{-2x} \tag{1.14}$$

$$\iff \lim_{x \to +\infty} \frac{\sinh x - e^x/2}{e^x/2} = 0. \tag{1.15}$$

This proves that

$$\sinh x \approx \frac{e^x}{2} \text{ as } x \to +\infty.$$

Exercise 1.6. Prove the following asymptotic results:

- $\sinh x \approx -e^{-x}/2$ as $x \to -\infty$

- $\sinh x \sim e^x$ as $x \to +\infty$

- $\sinh x \ll xe^x$ as $x \to +\infty$

- $\sinh x \sin x = O(e^x)$ as $x \to +\infty$

Note that the same function may have different asymptotic representations as $x \to \infty$ along different paths (or directions). This is often confused with the much more subtle true Stokes phenomenon (illustrated below). To avoid confusion, we call it the *crude Stokes phenomenon*.

Definition 1.2. If the leading asymptotic behavior of a function f along a path of approach to x_0 differs from that along another path as $x \to x_0$, it is said to suffer from the *crude Stokes phenomenon*.

The above exercises show that $\sinh x$ suffers from the crude Stokes phenomenon near ∞.

Note also that in the last exercise above $\sinh x \sin x$ has no limit as $x \to +\infty$. However, it does have an asymptotic bound. This shows the usefulness of the big-O notation even when the limit may not exist.

A function f can also be compared with reference functions in the limit as $x \to x_0$. A natural set of reference functions is the set of (integer) powers of $x - x_0$. These lead to (standard) asymptotic series.

Definition 1.3. f is said to be *asymptotic to a series* $\sum_{n=0}^{\infty} a_n(x - x_0)^n$ as $x \to x_0$ if for each integer $N \geq 0$ we have

$$\lim_{x \to x_0} \frac{f(x) - \sum_{n=0}^{N} a_n(x - x_0)^n}{(x - x_0)^N} = 0.$$

In this case, we write

$$f(x) \approx \sum_{n=0}^{\infty} a_n (x - x_0)^n, \quad \text{as } x \to x_0.$$

If the set of reference functions includes other functions such as logarithms and exponentials, the asymptotic series is often referred to as a *generalized* asymptotic series. (See [12] for examples and possible paradoxes that may arise in choosing inappropriate reference functions.)

An asymptotic series may or may not converge. An example of a convergent asymptotic series is

$$\exp(1/x) \approx \sum_{n=0}^{\infty} \frac{1}{n! x^n}, \quad \text{as } x \to \infty.$$

This series is, in fact, a Taylor series for $\exp(1/x)$. But not all Taylor series are asymptotic series. For example, consider

$$\exp(x) = \sum_{n=0}^{\infty} \frac{x^n}{n!},$$

which, although convergent for all finite x, is not an asymptotic series in the limit $x \to \infty$ (or in any limit whatsoever except $x \to 0$).

An example of a divergent asymptotic series is provided by the following exercise.

Exercise 1.7. Define the exponential integral

$$\mathrm{Ei}(x) = \fint_{-\infty}^{x} \frac{e^t}{t} \, dt, \tag{1.16}$$

where the dashed integral sign means a principal value integral if the path of integration is real. As $x \to \infty$, an asymptotic expansion of this function is given by integration by parts. The first two steps of this procedure are

$$\mathrm{Ei}(x) = \frac{e^x}{x} + \fint_{-\infty}^{x} \frac{e^t}{t^2} \, dt$$

$$= \frac{e^x}{x} + \frac{e^x}{x^2} + 2 \fint_{-\infty}^{x} \frac{e^t}{t^3} \, dt. \tag{1.17}$$

Show that at the nth step this gives

$$\mathrm{Ei}(x) = \frac{e^x}{x} \left\{ 1 + \frac{1}{x} + \frac{2}{x^2} + \cdots + \frac{x}{e^x} n! \fint_{-\infty}^{x} \frac{e^t}{t^{n+1}} \, dt \right\}.$$

Hence prove that

$$\mathrm{Ei}(x) \approx \frac{e^x}{x} \sum_{n=0}^{\infty} \frac{n!}{x^n} \qquad \text{as } x \to \infty. \tag{1.18}$$

The difference between asymptoticity and convergence of a series is worth emphasizing here. Convergence of a series is equivalent to the convergence of the sequence of partial sums, for each fixed x in some domain, as $N \to \infty$. Asymptoticity of a series is equivalent to the vanishing of the relative difference between the function and the Nth partial sum, for each fixed N, as $x \to x_0$.

The above example also illustrates the phenomenon of *asymptotics beyond all orders*. The lower limit of integration in $\mathrm{Ei}(x)$ can be varied. For example, the function

$$u(x) := \int_c^x \frac{e^t}{t}\, dt,$$

for some number c, differs from $\mathrm{Ei}(x)$ only by the constant

$$\kappa := \mathrm{Ei}(x) - u(x) = \int_{-\infty}^c \frac{e^t}{t}\, dt.$$

However, this constant is smaller than each term of the asymptotic series for $\mathrm{Ei}(x)$ in the limit $x \to +\infty$ and cannot be identified from this series. To identify κ, we would need either to apply Definition 1.3 of an asymptotic series infinitely often or to sum the series. Neither is possible (with conventional tools).

Moreover, if the function $u(x)$ is analytically continued from the first quadrant $(0 < \arg(x) < \pi/2)$ to the fourth quadrant $(-\pi/2 < \arg(x) < 0)$, the value of κ changes. The change is calculated in the exercise below.

Exercise 1.8. Consider $\mathrm{Ei}(x)$ along two directions of approach to infinity: (i) $\arg(x) = \pi/2$; (ii) $\arg(x) = -\pi/2$. For (i), represent $\mathrm{Ei}(x)$ as a Laplace transform by taking

$$e^{-x}\, \mathrm{Ei}(x) = \int_0^{\infty} q(s) e^{-xs}\, ds,$$

where the path of integration Γ in the s-plane lies in the fourth quadrant. Show that

$$q(s) = \frac{1}{1-s}.$$

As the path in the x-plane changes to (ii), Γ must change to ensure that the Laplace integral is convergent. When x lies along (ii), show that Γ must lie in the upper half-plane $(0 < \arg(s) < \pi)$ for convergence. Hence the change in $e^{-x} \operatorname{Ei}(x)$ is given by

$$\frac{1}{2\pi i} \oint_C \frac{e^{-xs}}{1-s} \, ds = e^{-x},$$

where the closed contour C encloses the pole at $s = 1$. That is, the change in κ is unity.

This is an example of the true Stokes phenomenon.

The crude and true Stokes phenomena show the need for relating the possibly different asymptotic behaviors of a function along different directions of approach to a singularity. More generally, the asymptotic behavior of a function near a singularity may need to be related to its possible behaviors near another singularity. These give rise to connection problems.

Definition 1.4. The problem of relating the asymptotic behavior of a solution $f(x)$ of an ODE near a fixed singularity to its possible behaviors near another singularity (or the same singularity along different directions) is called a *connection problem*.

In studying the asymptotics of solutions of (differential or other) equations, we are often led to a limiting form of the equation in which small terms are neglected. Such forms are called *dominant balances*.

Definition 1.5. An asymptotic limiting form of an equation in which only the largest terms remain is called a *dominant balance*. If the set of largest terms is maximal for the original equation, the balance is called *maximal*. Otherwise, it is called *submaximal*.

For example, the ODE

$$u' + \frac{u}{3z} = u^2 + 1$$

has the maximal dominant balance

$$u' \approx u^2 + 1,$$

and a submaximal balance

$$u^2 \approx -1$$

as $z \to \infty$. In Section 2.2 we study this equation as a model for the first Painlevé equation in the limit as $x \to \infty$.

Connection problems have been the main motivation for the asymptotic study of the Painlevé equations in recent times.

1.3 History

In this section we outline a short history of the asymptotic study of the Painlevé equations.

Boutroux [13,14] was the first to study the local asymptotic behaviors of the Painlevé transcendents (generic solutions of the Painlevé equations) in the limit as the independent variable approaches infinity. His methods were direct. He gave complete locally valid asymptotic behaviors near infinity, and moreover, deduced some qualitative properties on how they vary along lines of points at which the solution has the same value (for example, along lines of zeros).

More recent interest in the asymptotic behaviors of the Painlevé transcendents arose from physical applications, at first in statistical mechanics and later in other fields including string theory. In various scaling or physical limits of several exactly solvable models in statistical mechanics, such as the critical limit of the two-dimensional Ising model, the Painlevé transcendents appear as coefficients of the asymptotic expansions of two-point correlation functions (see [15] for a review). A question of physical interest is to relate the behavior of the correlation functions at large scaling limits to the behavior of the same function at very small scaling limits. The answer to this question lies in the solution of the Painlevé connection problem.

Connection results, which are formulae relating the free parameters of a family of solutions near one fixed singularity to the parameters near another (or the same) fixed singularity, have been found in many special and general cases of the Painlevé equations. Possibly the first of these results was obtained by McCoy, Tracy, and Wu [16] for the third Painlevé equation, with a special choice of parameter and behavior near infinity. Their method of solution relied on being able to sum exactly an asymptotic series for a solution near one of the singularities. (This extraordinary result can be explained in terms of the method of solution for the sine–Gordon equation, whose similarity reduction gives rise to the third Painlevé equation.) Ablowitz and Segur [17,18] next found the connection results for P_{II} (again for special choices of parameter and behavior) by using the linear problem for the modified KdV equation under a similarity reduction. These results were made rigorous by Hastings and McLeod [19] and by Clarkson and McLeod [20].

Jimbo, Miwa, and Sato et al. [21] showed that the isomonodromy problem can be used to solve the connection problems for the Painlevé equations without recourse to integrable PDEs. Since Jimbo et al.'s ground-breaking work, many applications of the isomonodromy method of solving the Painlevé connection problem have appeared in the literature (e.g., [22–26]), and the method has undergone many improvements [27,28].

However, the Painlevé connection problem also admits a direct approach [29–31] that dispenses with auxiliary structures (such as the isomonodromy problem) that these equations may possess. The purpose of this chapter is to show that such direct methods give all the local asymptotic behaviors of the Painlevé equations naturally, and rigorously, and moreover extend to give more global results, which can be used to solve their connection problem. In a way, using the isomonodromy approach to obtain local behaviors is like using a sledgehammer to crack a peanut. Some extensions of these local behaviors were already obtained by Boutroux. The Painlevé property of these equations allows the extension to be made complete (around infinity) for the first and second Painlevé equations. Joshi and Kruskal [31] carried out this completion for generic behaviors of the first and second Painlevé equations, based on a suggestion of Hastings and McLeod [19].

1.4 Main Results

Although Boutroux was a pioneer in this field and his work forms a *tour de force*, there remain some gaps in his proofs of asymptotic validity. In this chapter, we explain his methods in more modern terms and fill the gaps in his local asymptotic results rigorously.

The global asymptotic method of Joshi and Kruskal [31] can also be made rigorous. But for conciseness, and to celebrate the historical occasion of this volume, we prove here only those extensions first pointed out by Boutroux. In fact, we improve on his results, by using the Painlevé property of the Painlevé equations.

In particular, we use the following facts, which are given by the direct proof [3] of the Painlevé property of the Painlevé equations. Suppose $\varepsilon \neq 0$, $|\varepsilon| < 1$, $B > 0$, x_0 s.t. $|x_0| > 1/|\varepsilon|$, η, η' are given where (x_0, η, η') is a regular point for the right side of the Painlevé equation of interest. Let a domain $\mathcal{X}$ be defined such that

$$\mathcal{X} := \{x \mid |x| > 1/|\varepsilon|, |x - x_0| < B\}$$

does not contain a fixed singular point of the equation. Then we have the following:

1. The solution $y(x)$ of the equation defined by initial values $y(x_0) = \eta$, $y'(x_0) = \eta'$ exists and is meromorphic in $\mathcal{X}$.

2. The requirement

$$y(x) \in \mathcal{Y} := \{y \mid |y| < |\log \varepsilon|\}$$

gives a connected subset of $\mathcal{C}$ of $\mathcal{X}$.

3. $\mathcal{C}$ may contain "holes" where $y(x)$ becomes unbounded. (Therefore, they contain the poles of $y(x)$.) These holes are contained inside circular disks of radius approximately $1/|\log \varepsilon|^{1/2}$ for P_I and $1/|\log \varepsilon|$ for P_{II}.

4. $\mathcal{C}$ is covered by paths γ that are unions of segments of rays originating from x_0 and circular arcs around its holes. In particular, there exist such paths connecting x_0 to any other point x in $\mathcal{C}$ of length bounded by $2\pi B$.

Boutroux refrained from using the Painlevé property in his arguments because he obtained asymptotic estimates for a more general family of ODEs (containing the Painlevé equations) some of which do not have this property. As a result, his error estimates were cruder than that obtained in the theorems we prove below.

Also, we give different proofs of some of Boutroux's extensions of his local asymptotic results. These rely on an averaging argument first given in [29, 32].

1.5   Outline of the Chapter

The plan of the remainder of the chapter is as follows. In Section 2 we give brief examples to recall key concepts from asymptotics and to provide models of the asymptotics of the Painlevé equations. In particular, we analyze the asymptotic behavior of the classical Airy function $\text{Ai}(x)$, which is a solution of

$$w'' = xw, \tag{1.19}$$

in the limit as $x \to \infty$ and recall its connection problem. We also reproduce Boutroux's results for the asymptotic analysis of a first-order nonlinear model given by a Riccati equation. Boutroux considered a family of such equations. But we restrict our attention to a particular equation whose linearization is the Airy equation.

Local asymptotic results for the first and second Painlevé equations are then described in Section 3. We give rigorous local generic results both for $x \to \infty$ (for P_I and P_{II}) and $\alpha \to \infty$ (for P_{II}). In Section 4 we extend the local generic results by studying their modulation on a curve through successive zeros of the Painlevé transcendent.

2   Linear and Nonlinear Asymptotic Models

In this section we recall the classical asymptotic results for the Airy functions and improve Boutroux's results for a related Riccati equation.

2.1   Linear Irregular-Singular-Point Theory and Connection Problems

The focus of this chapter is the asymptotic analysis of solutions of differential equations in the limit as we approach a fixed singularity. In general, these are irregular singular points.

Definition 2.1. Consider the second-order linear ODE

$$y'' + p(x)y' + q(x)y = 0 \tag{2.1}$$

near a point x_0. Suppose that both $(x - x_0)p(x)$ and $(x - x_0)^2 q(x)$ are analytic at x_0. Then x_0 is called a *regular* singular point. If one of $(x - x_0)p(x)$ or $(x - x_0)^2 q(x)$ is not analytic at x_0, then it is called an *irregular* singular point.

The definition can be applied to the point at infinity by first transforming to $x = 1/t$ and analyzing the resultant ODE near $t = 0$.

A theorem of L. Fuchs (see [33]) shows that if x_0 is a regular singular point, then at least one solution has the form of a Frobenius series

$$y(x) = \sum_{n=0}^{\infty} a_n (x - x_0)^{n+\alpha}$$

convergent in some domain containing x_0, where α, called the inidicial exponent, is to be found from the ODE. (The second solution is similar, but may involve powers of $\log(x - x_0)$.) If no solution has this form, then the point x_0 must be an irregular singular point.

Now suppose the linear ODE (2.1) has an irregular singular point at $x = \infty$. There is a well-known standard approach for finding its formal asymptotic solutions. Write $y(x) = \exp(S(x))$. We get

$$S'' + (S')^2 + p(x)S' + q(x) = 0.$$

Assume

$$S'' \ll S'^2, \quad x \to \infty,$$

for at least one solution. Otherwise, it can be shown that S is logarithmic or small, which implies an algebraic series for both solutions. We assume that this is not the case. Therefore, we get

$$S'^2 \approx -p(x)S' - q(x).$$

In general, this quadratic asymptotic equation gives rise to two asymptotic solutions. (For an nth-order ODE, this would be an nth-degree equation with n asymptotic solutions.)

To be more specific, consider the Airy equation (1.19). The above approach gives

$$S'^2 = x - S''.$$

Recursive substitution for S'', assuming that it is smaller than x, gives

$$\begin{aligned}
S' &= x^{1/2}(1 - S''/x)^{1/2} \\
&= x^{1/2}\left(1 - (S''/2x) + O(S''/x^2)\right) \\
&= x^{1/2}\left(1 - x^{-3/2}/4 + O(x^{-3})\right) \\
&= x^{1/2} - x^{-1}/4 + O(x^{-5/2}),
\end{aligned} \qquad (2.2)$$

which implies

$$S \approx \frac{2}{3}x^{3/2} - \frac{1}{4}\log x + \text{const.}$$

Iteration shows that there are two formal solutions given by linear combinations of

$$y_+(x) \approx e^{2x^{3/2}/3}x^{-1/4}\sum a_n x^{-3n/2}, \tag{2.3}$$

$$y_-(x) \approx e^{-2x^{3/2}/3}x^{-1/4}\sum(-1)^n a_n x^{-3n/2}, \tag{2.4}$$

as $x \to +\infty$, where $a_0 = 1$ and

$$a_n = \frac{3}{4n}(n - 1/6)(n - 5/6)a_{n-1},$$

for $n \geq 1$.

Exercise 2.1. Show that these asymptotic series formally satisfy the Airy equation. By considering

$$\lim_{n=\infty}\left|\frac{a_{n-1}}{a_n}\right|$$

show that the radius of convergence of these formal solutions is zero.

That is, they are divergent for all $x \neq 0$. The following exercise gives another classical example.

Exercise 2.2. Carry out an asymptotic analysis near infinity for Bessel's equation

$$x^2 y'' + xy' + (x^2 - \nu^2)y = 0,$$

by transforming to $y = \exp(S(x))$. Show that along the positive real axis, the asymptotic behavior is given (in general) by

$$y(x) = \frac{a(x)}{\sqrt{x}}\cos(x - (\nu + \tfrac{1}{2})\pi/2) + \frac{b(x)}{\sqrt{x}}\sin(x - (\nu + \tfrac{1}{2})\pi/2) \tag{2.5}$$

(the phase $(\nu + \tfrac{1}{2})\pi/2$ is not essential but is chosen to give the standard Bessel's functions $J_\nu(x)$ and $Y_\nu(x)$), where $a(x)$, $b(x)$ are linear combinations of

$$w_1 \approx \sum_{n=0}^{\infty}(-1)^n c_{2n}x^{-2n}, \tag{2.6}$$

$$w_2 \approx \sum_{n=0}^{\infty}(-1)^n c_{2n+1}x^{-2n-1}, \tag{2.7}$$

$$c_n = \frac{(4\nu^2 - 1^2)(4\nu^2 - 3^2)\cdots(4\nu^2 - (2n - 1)^2)}{8^n n!}, \quad n > 0, \tag{2.8}$$

$$c_0 = 1. \tag{2.9}$$

Show that these asymptotic series are divergent.

The standard theorems (of existence and analyticity of solutions for ODEs) show that the solutions of Airy's equation must be entire functions (because the equation has no singularities in the finite plane). However, the asymptotic behaviors we found above are multivalued functions. The resolution of this apparent paradox lies in the fact that these formal solutions represent the asymptotic behaviors of the Airy functions only in sectors of angular width less than 2π. That is, the analytic continuation of the asymptotic behaviors of $\mathrm{Ai}(x)$ in some sector (near infinity) is not necessarily its asymptotic behavior in other sectors.

Let $\theta = \arg(x)$. Along (or near) the positive real x-axis, the boundary condition

$$y(x) \approx \frac{1}{x^{1/4}} \exp\left(-\frac{2}{3}x^{3/2}\right)$$

defines a unique solution. However, the alternative condition

$$y(x) \approx \frac{1}{x^{1/4}} \exp\left(+\frac{2}{3}x^{3/2}\right)$$

does *not* specify a solution uniquely, because it leaves the coefficient of the small solution $y_-(x)$ undefined. In other words, this behavior suffers from asymptotics beyond all orders. Moreover, it suffers from the true Stokes phenomenon.

However, this problem no longer occurs along directions where the two formal solutions have the same order of magnitude as $|x| \to \infty$. In fact, the two solutions are the same size wherever

$$\left|\exp\left(\frac{4}{3}x^{3/2}\right)\right| = 1.$$

This occurs where

$$\frac{4}{3}\cos(3\theta/2) = 0,$$

i.e., along directions where $3\theta/2 = \pi/2 + j\pi$, for integer j. Such directions are called *anti-Stokes* lines.

The directions where one solution is minimal relative to the other one is called a *Stokes* line. The next Stokes line (traveling in an anticlockwise direction from the positive real axis) occurs along $\theta = 2\pi/3$. Along this direction, it is y_- that is large and y_+ that is hidden beyond all orders.

The connection problem for Airy's equation is to relate all such asymptotic behaviors of a solution valid along different directions of approach to infinity. To solve this problem, we fix our attention on one solution. Consider the standard solution of Airy's equation $\mathrm{Ai}(x)$ defined by initial conditions

$$\text{Ai}(0) = \frac{3^{-2/3}}{\Gamma(2/3)}, \tag{2.10}$$

$$\text{Ai}'(0) = -\frac{3^{-1/3}}{\Gamma(1/3)}. \tag{2.11}$$

It is well known that $\text{Ai}(x)$ has an integral representation given by

$$\text{Ai}(x) = \frac{1}{2\pi} \int_{\Gamma_1} e^{ikx+ik^3/3} \, dk, \tag{2.12}$$

where Γ_1 is a path starting from infinity in the sector $2\pi/3 < \theta < \pi$ and ending at infinity in the sector $0 < \theta < \pi/3$. (The second solution $\text{Bi}(x)$, linearly independent of $\text{Ai}(x)$, also has an integral representation.) The next exercise shows that such integral representations describe all solutions for appropriate choices of path.

Exercise 2.3. Let S_j be a sector of the complex k-plane described by $k \in [2j\pi/3, (2j+1)\pi/3]$ for $j = 0$, 1, 2. For any integer j modulo 3, let Γ be a path starting at infinity in S_j and ending in S_{j-1}.

1. Show that the integral

$$y(x) = \frac{1}{2\pi} \int_{\Gamma} e^{ikx+ik^3/3} \, dk$$

satisfies the Airy equation.

2. Show that there are only two linearly independent choices of path.

The Airy functions also have a symmetry property under rotation of x by a cube root of unity.

Exercise 2.4. Let $\omega = \exp(-2\pi i/3)$. Show by using the integral representation of $\text{Ai}(x)$ that

$$\text{Ai}(x) = -\omega \, \text{Ai}(\omega x) - \omega^2 \, \text{Ai}(\omega^2 x). \tag{2.13}$$

This property together with (2.12) can be used to solve the Airy connection problem.

First, we show that $\text{Ai}(x)$ is asymptotic to a multiple of y_- in an extended sector centered on the positive real x-axis.

Theorem 2.1.

$$\text{Ai}(x) \approx \frac{1}{\sqrt{\pi}} x^{-1/4} e^{-2x^{3/2}/3}, \qquad |x| \gg 1, \quad |\arg x| < \pi.$$

The proof is outlined in the following exercise. (The definition of saddle points and method of steepest descents can be found in, e.g., [33].)

Exercise 2.5. Consider the integral representation of Ai(x).

1. Show that saddle points of $\rho(k) = k + k^3/(3x)$ are given by $k_s = \pm i\sqrt{x}$.

2. Deform the path Γ_1 to pass through the saddle at $i\sqrt{x}$ (transversely to the imaginary k-axis). Show that this yields the leading-order behavior stated in the theorem above.

3. Note that the saddle points move with x. The calculation is valid until the saddle points hit the boundary of the regions in which Γ_1 must lie. Show that this occurs when $|\arg x| = \pi$.

This classical result simultaneously gives the local asymptotic behavior of Ai(x) and connects it across the anti-Stokes lines $\arg x = \pm\pi/3$ and Stokes lines $\arg x = \pm 2\pi/3$, by effectively tracking Ai(x) along a large circular arc in the complex x-plane. To connect to the one remaining direction $|\arg x| = \pi$ we use the relation (2.13).

Exercise 2.6. Use (2.13) and Theorem 2.1 to find the asymptotic behavior of Ai(x) as $x \to -\infty$ along the negative real axis.

This result completes the solution of the connection problem for Ai(x) near infinity.

2.2 A Nonlinear Model

In this section we describe a first-order nonlinear model studied by Boutroux.

Consider the Riccati equation

$$y'(x) = y^2 + x, \tag{2.14}$$

as $x \to \infty$. This equation can be linearized to (a version of) the Airy equation.

Exercise 2.7. Show that the solution $y(x)$ of (2.14) is given by $y = -w'(x)/w(x)$, where $w(x)$ solves

$$w''(x) + xw(x) = 0. \tag{2.15}$$

(This becomes the standard Airy equation under the transformation $x \mapsto -x$.) Asymptotic properties of $w(x)$ could be derived from the asymptotic behaviors of the Airy functions Ai$(-x)$, Bi$(-x)$. However, instead of using such results for Airy functions, we will start from scratch and develop direct methods for the nonlinear equation as given.

To make the largest terms of this nonlinear equation explicit, consider the transformation

$$y = \sqrt{x}u(z), \quad z = \frac{2}{3}x^{3/2}.$$

This maps (2.14) to

$$u'(z) + \frac{u}{3z} = u^2 + 1. \tag{2.16}$$

Boutroux considered a larger class given by

$$u' + 2p\frac{u}{z} = u^2 + 1, \quad |z| \to \infty, \tag{2.17}$$

for some constant p. This Riccati equation can be linearized through the transformation

$$u(z) = -\frac{\psi'(z)}{\psi(z)}$$

to

$$\psi'' + 2p\frac{\psi'}{z} + \psi = 0,$$

which can in turn be transformed via

$$\psi(z) = z^{1/2-p}w(z)$$

to

$$w'' + \frac{w'}{z} + \left(1 - \frac{(p-1/2)^2}{z^2}\right)w = 0,$$

which is Bessel's equation.

It is well known that Riccati equations such as (2.14), (2.17) have the Painlevé property [34]. So $u(z)$ is meromorphic except at $z = 0$ or ∞. This shows that $y(x)$, and therefore $u(z)$, can have poles only as movable singularities. Recall (see the end of Section 1.2) that as $z \to \infty$, the maximal dominant balance of (2.16) is

$$u' \approx u^2 + 1.$$

Given $\varepsilon \neq 0$, η, we take $|z_0| > 1/|\varepsilon|$ and define a solution by the initial condition $u(z_0) = \eta$. This suggests, and we show below, that for generic values of η, the leading-order behavior of $u(z)$, as $|z| \to \infty$, is given by $\eta + \tan(z - z_0)$.

Before we state and prove this generic result, recall that the function tan is implicitly defined by

$$\int_\eta^u \frac{dv}{v^2 + 1} = z - z_0,$$

where $\tan(z_0) = \eta$. Its period π is given by

$$\oint_C \frac{dv}{v^2 + 1},$$

where C is a closed contour enclosing one of the roots of $v^2 + 1$ in the v-plane. In order to define such integrals, the path of the integral must avoid the roots of $v^2 + 1$ where it becomes singular.

Definition 2.2. The following are called *generic conditions*. Assume $\varepsilon \neq 0$, $B > 0$, z_0 are given with $|\varepsilon| < \frac{1}{2}$, $|z_0| > 1/|\varepsilon|$, $|\varepsilon B| < \frac{1}{2}$.

1. z lies in the domain

$$\mathcal{Z} := \{z \mid |z - z_0| < B, |z| > 1/|\varepsilon|\}.$$

2. γ is a path starting at z_0 in $\mathcal{Z}$ of length $l < 2\pi B$, s.t. $u(z)$, $z \in \gamma$, lies in

$$\mathcal{U} := \{u \mid |u| < |\log \varepsilon|, |u^2 + 1| > |\varepsilon|^{1/7}\}.$$

Definition 2.3. Suppose ε, η, are given that satisfy the generic conditions. Assume $u(z_0) = \eta$. Then the solution defined by this initial condition is called a *generic solution*.

Theorem 2.2. *Suppose the generic conditions are satisfied and u is an associated generic solution. Then this solution satisfies*

$$|\arctan(u - \eta) - z + z_0| < \frac{2\pi B}{3} |\varepsilon|^{6/7} |\log \varepsilon|. \tag{2.18}$$

Proof. Let S be defined by

$$S := \int_\gamma \frac{u}{3z(u^2 + 1)} \, dz. \tag{2.19}$$

Dividing (2.16) by $u^2 + 1$ and integrating gives

$$\int_\Gamma \frac{du}{u^2 + 1} = z - z_0 - S,$$

where Γ is the image of γ under u. Here S is bounded above by

$$|S| \leq 2\pi B \frac{|\varepsilon|}{3} \frac{|\log \varepsilon|}{|\varepsilon|^{1/7}} = \frac{2\pi B}{3} |\varepsilon|^{6/7} |\log \varepsilon| \tag{2.20}$$

by the generic conditions. $\qquad\qquad\square$

By taking Γ to be a closed contour starting at 0 and enclosing one of the points $\pm i$, this theorem extends to give the asymptotic spacing between successive zeros of u. If z_0, z_1 are two such zeros, then we get

$$|z_1 - z_0 \mp \pi| < \frac{2\pi B}{3} |\varepsilon|^{6/7} |\log \varepsilon|, \qquad (2.21)$$

where the choice of sign on the left side is dependent on the orientation of Γ.

Note that the power of $|\varepsilon|$ in (2.18) or (2.21) depends on how closely we skirt around the roots of $u^2 + 1$. If $|u^2 + 1|$ is bounded below by a constant (independent of ε) on γ, this error can be made proportional to $|\varepsilon \log \varepsilon|$.

The generic assumptions leave five possible gaps in the (asymptotic) description of the solution space, namely the limits where the path Γ (or equivalently γ) becomes very long, its endpoint becomes very large or becomes very close to $\pm i$, and where η becomes unboundedly large or arbitrarily close to $\pm i$. We will consider extensions of the theorem in turn for each case.

2.2.1   Very Long Γ

Suppose that Γ has a length greater than or equal to $|\log \varepsilon|$ but lies in a bounded region of the u-plane. Recall that Γ occurs in

$$\int_\Gamma \frac{du}{u^2 + 1},$$

where it can be deformed to N copies of a closed contour enclosing $\pm i$ and a ray connecting a point on such a contour with the given endpoint u. (N and the ray can be chosen in such a way that the ray's length is minimal.) The effect of this deformation is simply to add N multiples of the period π of $\tan(z - z_0)$ in the result of the generic case above. Therefore, the result becomes the same as in the theorem with the addition of $N\pi$ inside the modulus signs on the left side of (2.18).

2.2.2   Very Large η or Very Large Endpoint of Γ

By the previous case, the cases of large η and large endpoint of Γ are equivalent (modulo $N\pi$ being added to the left side of (2.18)). Without loss of generality, we can assume that $|u| \geq |\log \varepsilon|$ on the whole of Γ.

Let $v = -1/u$. Then v satisfies

$$\frac{v'}{v^2} = \frac{1}{v^2} + 1 + \frac{1}{3zv},$$

which implies

$$v' = v^2 + 1 + \frac{v}{3z}.$$

This is almost the same as (2.16), and the same theorem holds for v with η replaced by η_1 where $v(z_0) = \eta_1 = 1/\eta$ and v lies in a ball of radius $1/|\log \varepsilon|$ around the origin.

Since v and $v - \tan(z - z_0)$ are both small, z_0 is close to a zero of $\tan(z - z_0)$. That is, u is close to a pole. Moreover, the estimate (2.21) holds for the spacing between successive poles of u.

2.2.3   Nearly Degenerate Solution

Again without loss of generality we can assume that $|u^2 + 1| \le |\varepsilon|^{1/7}$ over all of Γ. (The case where Γ travels from being close to one of the roots $\pm i$ to the other can be divided into three subcases: where it is close to i, $-i$, or bounded away from both. We consider one of the former cases here. The last case is covered by the generic conditions.)

To be specific, suppose that u is very close to i. Write $u =: i + w$. Then (2.16) becomes

$$w' = w(2i + w) - \frac{i + w}{3z}. \tag{2.22}$$

Definition 2.4. Given ε, z_0, $\eta = i + \tilde{\eta}$, assume that the generic assumptions hold with the following modification.

1. $|\varepsilon|^{1/7}|\log \varepsilon| < \frac{1}{2}$.

2. The path Γ is such that $|w + 2i| > \frac{3}{2}$ and $|\varepsilon|^{6/7}|\log \varepsilon| < |w| \le |\varepsilon|^{1/7}$.

These modified assumptions are called *nearly degenerate conditions*. Let $\eta' = \tilde{\eta}/(2i + \tilde{\eta})$.

Note that we have allowed $u - i$ to be smaller than in the generic case, but not arbitrarily small.

Definition 2.5. Under the nearly degenerate conditions, the solution defined by $u(z_0) = i + \tilde{\eta}$ is called a *nearly degenerate solution*.

Theorem 2.3. *Under the nearly degenerate assumptions, the nearly degenerate solution of (2.22) satisfies*

$$\left| \log\left(\frac{w}{2i + w} \right) - \log \eta' - 2i(z - z_0) \right| < \frac{4\pi B |\varepsilon|^{1/7}}{3|\log \varepsilon|}.$$

Proof. Dividing (2.22) by $w(2i + w)$ and integrating, we get

$$\frac{1}{2i}\left\{ \log\left(\frac{w}{2i + w} \right) - \log \eta' \right\} - (z - z_0) = S := -\int_{z_0}^{z} \frac{i + w}{3zw(2i + w)}\, dz.$$

By the hypothesis, we get

$$|S| \le 2\pi B \cdot \frac{|\varepsilon|}{3} \cdot \frac{3}{2} \cdot \frac{1}{(3/2)|\varepsilon|^{6/7} \log \varepsilon|}.$$

From this, we get the desired result.   □

Notice that the lower bound for w could be as low as $\varepsilon \log \varepsilon$, in which case the error estimate in the theorem would become as large as $O(1/|\log \varepsilon|)$.

By the nearly degenerate conditions, there exists a number κ such that

$$\left|\log(2i + w) - \log 2i - \frac{w}{2i}\right| < \kappa |\varepsilon|^{1/7}.$$

Therefore, we get a refinement of the above result, i.e., there exists a constant μ such that

$$|\log((u - i)/(\eta - i)) - 2i(z - z_0)| < \frac{\mu |\varepsilon|^{1/7}}{|\log \varepsilon|}. \tag{2.23}$$

Exercise 2.8. Prove that (2.23) holds.

This asymptotic behavior of u is an almost degenerate form of the generic leading-order behavior, obtained as a limit as $\tan(z - z_0) \to i$. Note that to leading order u oscillates around the value i.

Moreover, note that this theorem does not cover the case where η (or a point on Γ) is *arbitrarily* close to $\pm i$. These cases lead to the degenerate solutions that Boutroux called *intégrales tronquées*. We discuss these solutions below.

2.2.4   Degenerate Solutions

For simplicity, we restrict our attention to the case where u becomes arbitrarily close to i. Take $\eta = i$. Now we need another set of modified assumptions to reflect the small (possibly zero) size of $|u - i|$.

Definition 2.6. Given ε, assume that the generic conditions hold with the following modification.

1. $|\varepsilon|^{1/7}|\log \varepsilon| < \frac{1}{2}$, $|\varepsilon|^{2/7}/|\log \varepsilon| < \frac{1}{2}$.

2. The path Γ is such that $|w + 2i| > \frac{3}{2}$ and $|w| \le |\varepsilon|^{5/7}|\log \varepsilon|$.

These modified assumptions are called *degenerate assumptions*.

Definition 2.7. Under the degenerate assumptions, the solution defined by $u(z_0) = i$ is called *a degenerate solution*.

(Note that the same definitions apply to the case where η is close to $-i$ with the corresponding sign changes in the above.) Then we get the following result.

Theorem 2.4. *Under the degenerate assumptions, the degenerate solution of (2.22) satisfies*

$$\left|w - \frac{1}{6z}\right| < 2|\varepsilon|^{5/7}|\log \varepsilon|.$$

Proof. The hypothesis and (2.22) give

$$|w'| \leq \frac{5}{2}|\varepsilon|^{5/7}|\log \varepsilon| + \frac{1}{2}|\varepsilon| \leq 3|\varepsilon|^{5/7}|\log \varepsilon|.$$

Rewrite (2.22) as

$$w = \frac{1}{6z} + \frac{1}{2i}\left(-w^2 + w' + \frac{w}{3z}\right).$$

We then get

$$\left|w - \frac{1}{6z}\right| \leq \frac{1}{2}\left(3|\varepsilon|^{5/7}|\log \varepsilon| + |\varepsilon|^{10/7}|\log \varepsilon|^2 + |\varepsilon|^{12/7}|\log \varepsilon|/3\right)$$
$$\leq 2|\varepsilon|^{5/7}|\log \varepsilon| \tag{2.24}$$

as desired.   □

Note that this degenerate asymptotic behavior is not explicitly dependent on a parameter to leading order. In fact, it can be shown that no free parameter appears explicitly at any order of its asymptotic series in powers of $1/z$. This case completes the study of all possible extensions of the generic theorem.

2.2.5   An Asymptotic Description of the Solution Space

We show here that the asymptotic behaviors form a connected component of the solution surface. Let ε_0 be a nonzero number with maximal modulus that satisfies the ε-conditions in the generic, nearly degenerate, and degenerate conditions. We assume $0 < |\varepsilon| \leq |\varepsilon_0|$ in the following.

Theorem 2.5. *The solution surface parameterized by z_0, where $u(z_0) = \eta$, for z_0 and z satisfying $|z| > 1/|\varepsilon|$, is connected. Moreover, the generic, nearly degenerate, and degenerate behaviors are the only behaviors possible.*

The proof is formulated as a sequence of lemmas and exercises below.

Exercise 2.9. Consider the limit $z_0 \to \infty$ of the generic behavior when z_0 lies along a purely imaginary direction. Show that $\tan(z - z_0)$ approaches $\pm i$.

Suppose a line of poles of $u(z)$ is given by

$$\ldots, Z_{-2}, Z_{-1}, Z_0, Z_1, Z_2, \ldots. \tag{2.25}$$

This line is asymptotically parallel to the real z-axis by the results of Section 2.2.2.

Lemma 2.1. *$u(z)$ does not have any poles (or zeros) as $z \to \infty$ above or below this line of poles.*

Proof. Suppose we start near a pole on the line (2.25), travel on a path γ, and encounter another pole of u. The image Γ of γ starts near infinity and then travels into the finite domain before becoming infinite again. We have two possibilities:

1. Γ does not enclose $\pm i$. Then we are back at the starting point in the z-plane.

2. Γ does enclose $\pm i$. Then we have traveled approximately $n\pi$ for some integer n.

Both contradict the hypothesis that we have traveled a nonzero distance in a direction transverse to the real direction. A similar argument applies to zeros of $u(z)$. $\square$

Straightforward calculation gives the following results.

Exercise 2.10.

1. Show that if z_0 lies above the line of poles and $z \to \infty$ in a direction that is not purely imaginary, then the limit of the generic behavior is a nearly degenerate behavior.

2. Show that if z_0 lies above the line of poles and $z \to \infty$ in a purely imaginary direction, then the limit of the generic behavior is a degenerate behavior.

Now we connect the degenerate (or nearly degenerate) solution to a generic one.

Lemma 2.2. *Suppose U is a degenerate solution close to i (say). Let $u = U + \hat{u}$, where $\hat{u} \ll U$ as $|z| \to \infty$ in some sector. Then there exists p such that $\hat{u} \sim z^p \exp(2iz)$, and hence u is a degenerate solution in the interior of a half-plane. Moreover, u is nearly degenerate on the boundary of this half-plane and matches a generic solution there.*

Proof. The first assertion is proved by studying the asymptotic solutions of

$$\hat{u}' = 2i\hat{u} - U' + \hat{u}^2 - \frac{\hat{u}}{3z}.$$

The second arises from the fact that $\hat{u}$ remains small in the interior of a half-plane bounded by the real z-axis. On this axis, however, e^{2iz} becomes oscillatory and matches to the behavior given by (2.23). $\square$

These results also show that the degenerate behaviors are free of poles or zeros in their region of validity. Transforming back to the x-plane shows that in general, for $|x| \gg 1$, the solution $y(x)$ has three (half-)lines of poles, aligned asymptotically with the rays of angle 0, $2\pi/3$, $4\pi/3$.

2.2.6 Special Solutions

There are two distinguished special solutions of (2.14). The equation is invariant under

$$y(x) \mapsto \omega y(\omega x),$$

where ω is a cube root of unity. In general, a solution is mapped to another solution under this transformation. However, two special solutions remain invariant under this mapping, namely those defined by either having a zero or a pole at the origin. We will denote these solutions by y_0 and y_∞.

In general, these solutions have lines of poles that coincide exactly with the rays of angle 0, $2\pi/3$, $4\pi/3$. There are limits of these solutions, however, that are degenerate in two contiguous sectors of width $2\pi/3$ (or a half-plane in z).

There are direct analogues of all the results of this section for the Painlevé equations.

3 The First and Second Painlevé Equations

Consider P_I and P_{II}:

$$y'' = 6y^2 + x, \tag{3.1}$$

$$y'' = 2y^3 + xy + \alpha. \tag{3.2}$$

In this section we prove that the generic local asymptotic behavior of $y(x)$ is given by scaled elliptic functions. The proof is given in detail for $x \to \infty$.

We also state an analogous theorem for the case $\alpha \to \infty$ of P_{II}. This large-parameter limit gives rise to three asymptotic regions described by (i) $x \ll \alpha^{2/3}$, (ii) $x \sim \alpha^{2/3}$, and (iii) $x \gg \alpha^{2/3}$. The last region is the same as the usual $x \to \infty$ limit of P_{II}. The intermediate region is, however, much more complicated than the first or third regions. In particular, it possesses three local domains in its interior (called "turning points" by Kawai et al. [35–37]) where P_{II} becomes asymptotically close to P_I. However, in this chapter we concentrate on the first region and give local rigorous asymptotic results.

3.1 Large Independent Variable

Boutroux showed that these equations can be mapped to

$$u_{zz} = 6u^2 + 1 - \frac{u_z}{z} + \frac{4}{25}\frac{u}{z^2}, \tag{3.3}$$

$$u_{zz} = 2u^3 + u + \frac{a}{z} - \frac{u_z}{z} + \frac{u}{9z^2}, \tag{3.4}$$

respectively, under the transformations

$$y(x) = \sqrt{x}u(z), \quad z = 4x^{5/4}/5,$$
$$y(x) = \sqrt{x}u(z), \quad z = 2x^{3/2}/3, \quad a = 2\alpha/3.$$

These changes of variables can be found by asymptotic arguments (see [31, p. 320]). Boutroux also studied more general ODEs such as

$$y'' = \lambda(y^2 + x^\mu),$$

as $x \to \infty$.

The maximal dominant balances of (3.3), (3.4) are solved by elliptic functions. Boutroux integrated (3.3), (3.4) by multiplying first by u_z and integrating as though only the dominant terms

$$u'' \approx 6u^2 + 1, \quad u'' \approx 2u^3 + u,$$

were present (where primes now denote derivatives with respect to z). However, this leads to problems in estimating the contribution of u_z^2/z in the small terms of the integrated equation.

Exercise 3.1. Show that multiplying (3.3) by u' and integrating leads to

$$\frac{u'^2}{2} = 2u^3 + u + c - \int_{z_0}^{z} \left(\frac{u'^2}{z} + \frac{4}{25} \frac{uu'}{z^2} \right) dz,$$

where c is a constant of integration.

Because u'^2 appears on the right as well as the left of this integrated equation, it requires more argument to show that u'^2 must be bounded above (or below) if $2u^3 + u + c$ is. Boutroux did not provide such an argument.

We eliminate this gap by taking a new approach. That is, we keep the term u'/z with the dominant terms and integrate the resultant asymptotic equations:

$$u'' + \frac{u'}{z} \approx 6u^2 + 1, \quad u'' + \frac{u'}{z} \approx 2u^3 + u.$$

The two terms on the left have integrating factor z. To integrate the polynomial in u on the right we need the integrating factor u'. We combine the two and multiply the equation by z^2u' to get

$$((zu')^2/2)' = (2z^2u^3 + z^2u)' - 4zu^3 - 2zu + \frac{2}{25}(u^2)', \tag{3.5}$$

$$((zu')^2/2)' = \frac{(z^2(u^4 + u^2))'}{2} - zu^4 - zu^2 + a(zu)' - au + \frac{1}{18}(u^2)'. \tag{3.6}$$

Integration yields

$$z^2 u'^2 = z^2(4u^3 + 2u) + 2K - 4\int_{z_0}^{z} z(2u^3 + u)\,dz + \frac{4}{25}u^2, \tag{3.7}$$

$$z^2 u'^2 = z^2(u^4 + u^2) + 2azu + 2K - 2\int_{z_0}^{z} z(u^4 + u^2 + au)\,dz + \frac{u^2}{9}, \tag{3.8}$$

where K is a constant of integration. To fix a solution, assume that initial values are given at a (nonzero) point z_0 by

$$u(z_0) = \eta, \quad u'(z_0) = \eta'.$$

Note that K is related to these values respectively by

$$K = \frac{z_0^2}{2}(\eta'^2 - 4\eta^3 - 2\eta) - 2\eta^2/25, \tag{3.9}$$

$$K = \frac{z_0^2}{2}(\eta'^2 - \eta^4 - \eta^2) - az_0\eta - \eta^2/18. \tag{3.10}$$

If $|z_0|$ and $|z|$ were (equally) large, each term in the dominant balances of (3.7), (3.8) would be of $O(z_0^2, z^2)$. Since this will be the case in our treatment, we redefine

$$E := \frac{K}{z_0^2},$$

and rewrite the above equations as

$$u'^2 = 4u^3 + 2u + 2E$$
$$+ 2E\left(\frac{z_0^2}{z^2} - 1\right) - \frac{1}{z^2}\int_{z_0}^{z} 4z(2u^3 + u)\,dz + \frac{4u^2}{25z^2}, \tag{3.11}$$

$$u'^2 = u^4 + u^2 + 2E$$
$$+ 2E\left(\frac{z_0^2}{z^2} - 1\right) + \frac{2au}{z} - \frac{1}{z^2}\int_{z_0}^{z} 2z(u^4 + u^2 + au)\,dz + \frac{u^2}{9z^2}. \tag{3.12}$$

3.2 Leading-Order Behaviors

To leading order, equations (3.11), (3.12) are solved by elliptic functions. To be specific, we concentrate on P_I in the following, but almost identical arguments can be made for P_{II}.

3.2.1 Elliptic Functions

Consider

$$u'^2 = 4u^3 + 2u + 2E, \tag{3.13}$$

where E is a constant parameter. Let

$$P(u) := 4u^3 + 2u + 2E.$$

In general, $P(u)$ has three distinct roots, which we will denote by u_i, $i = 1, 2, 3$, in the complex plane.

Exercise 3.2. Show that the first-order equation (3.13) can be integrated once more (after taking the square root and separating variables) to give

$$\int_\eta^u \frac{dv}{\sqrt{P(v)}} = z - z_0. \tag{3.14}$$

The solution $u(z)$ defined by the inverse of this function is an elliptic function. Show that it has two periods given by

$$\omega_j = \oint_{C_j} \frac{dv}{\sqrt{P(v)}}, \tag{3.15}$$

where C_j, $j = 1, 2$, are two linearly independent closed contours each enclosing two points in the set $\{u_1, u_2, u_3, \infty\}$.

The elliptic function in this case is a Weierstrass elliptic function. (For standard initial values, this is denoted by $\wp(z)$, see [38].)

3.2.2 Degenerate Elliptic Functions

There exist special values of E for which two roots of $P(u)$ coincide. To find these, consider

$$\frac{dP}{du} = 12u^2 + 2 = 0 \implies u = d_k := (-1/6)^{1/2}, \quad k = 1, 2. \tag{3.16}$$

If such points are also zeros of $P(u)$, then we must have

$$E = D_k := -2d_k \left(2d_k^2 + 1\right)$$

$$= -2\left(-\frac{1}{6}\right)^{1/2}\left(-\frac{1}{3} + 1\right) = -\frac{4}{3}\left(-\frac{1}{6}\right)^{1/2}, \quad k = 1, 2. \tag{3.17}$$

For each such value of E, $P(u)$ has a double root. Note, however, that $P(u)$ cannot have a triple root, because $P''(u) = 24u$ cannot vanish simultaneously with $P'(u)$. Now we estimate the spacing between the roots u_i when E is close to a degenerate value D_k for some k.

Lemma 3.1. *Fix $k = 1$, 2. Let δ be a given number such that $0 \leq |\delta| < 1/256$. Let $E = D_k - 2\delta$ and $u = d_k + \varepsilon$. Then*

$$P(u) = 12d_k\varepsilon^2 + 4\varepsilon^3 - 4\delta.$$

Define

$$p(\varepsilon) = 3d_k\varepsilon^2 + \varepsilon^3 - \delta.$$

Let its roots be ε_j, where $j = 0, 1, 2$. Then the following results hold.

1. *There exist two roots of $p(u)$, ε_0, ε_1, say, such that*

$$\frac{2\sqrt{|\delta|}}{5} < |\varepsilon_j| < \frac{5}{2}\sqrt{|\delta|}, \quad j = 0, 1. \tag{3.18}$$

2. *The remaining root ε_2 satisfies*

$$\frac{4}{25} \leq |\varepsilon_2| \leq \frac{25}{4}.$$

3. *Finally, the roots ε_0, ε_1 satisfy*

$$\frac{2}{5}\sqrt{|\delta|} \leq |\varepsilon_0 - \varepsilon_1| < 5\sqrt{|\delta|},$$

i.e., they are distinct for $\delta \neq 0$.

Proof. In the assumed notation,

$$\begin{aligned}
P(u) &= 4\big(d_k^3 + 3d_k^2\varepsilon + 3d_k\varepsilon^2 + \varepsilon^3\big) + 2(d_k + \varepsilon) + 2(D_k - 2\delta) \\
&= 12d_k\varepsilon^2 + 4\varepsilon^3 - 4\delta,
\end{aligned} \tag{3.19}$$

since d_k is a double root corresponding to the value D_k of E.

Claim 1. *At least one root ε_0, say, of this polynomial satisfies $|\varepsilon_0| < 2^{1/3}|\delta|^{1/3}$.*

If not, then all roots ε_j, for $j = 0, 1, 2$, would satisfy $|\varepsilon_j| \geq 2^{1/3}|\delta|^{1/3}$, which implies

$$|\delta| = |\varepsilon_0\varepsilon_1\varepsilon_2| \geq 2|\delta|,$$

which is a contradiction.

Claim 2. *Moreover, this root satisfies*

$$\frac{\sqrt{|\delta|}}{\sqrt{3}} < |\varepsilon_0| \leq 2\sqrt{|\delta|}.$$

The upper bound follows from

$$|3d_k\varepsilon_0^2| = |\delta - \varepsilon_0^3| \leq 3|\delta|, \tag{3.20}$$

$|1/d_k| = \sqrt{6}$, and $6^{1/4} < 2$. The lower bound follows from using this result in

$$|3d_k\varepsilon_0^2| \geq |\delta| - |\varepsilon_0^3|$$

along with the bound $1 - 8\sqrt{|\delta|} > \frac{1}{2}$.

Claim 3. *The bounds (3.18) hold for another root ε_1.*

First, note that we now have

$$|\varepsilon_1 \varepsilon_2| = \frac{|\delta|}{|\varepsilon_0|} \leq \sqrt{3|\delta|},$$

which implies that (at least one root, which we take to be) ε_1 satisfies

$$|\varepsilon_1| \leq (4|\delta|)^{1/4},$$

by the same argument as in Claim 1. We use the technique in (3.20) twice. First, we get

$$|\varepsilon_1| \leq \frac{13}{5}|\delta|^{3/8}.$$

Using this bound in the same way again shows the desired upper bound in (3.18). The lower bound is obtained in the same way as for ε_0.

Claim 4. *The asserted bounds on ε_2 hold.*

This follows from the equation $\varepsilon_0 \varepsilon_1 \varepsilon_2 = \delta$ and the above bounds on ε_0, ε_1.

Claim 5. *The last assertion of the lemma holds.*

The upper bound is clear from (3.18). To prove the lower bound, we use another relation between the roots of p, namely

$$\varepsilon_0 \varepsilon_1 + \varepsilon_0 \varepsilon_2 + \varepsilon_1 \varepsilon_2 = 0.$$

Rewriting this, we get

$$\varepsilon_0 - \varepsilon_1 = -\left(\frac{\varepsilon_0 \varepsilon_1}{\varepsilon_2} + 2\varepsilon_1\right).$$

The result follows from a usage of the above bounds on the right.   □

Note that these show that if δ is bounded below, then the spacing between the closest roots is also bounded below.

When $E = D_k$, one of the periods w_j of $u(z)$ becomes infinite and u becomes a singly periodic function.

Exercise 3.3. Show that if $P(u) = 4(u - d_k)^2(u - \rho)$, then the inverse of the function defined by

$$\int_\eta^u \frac{dv}{\sqrt{P(v)}} = z - z_0$$

is either given by a hyperbolic trigonometric function or is identically constant.

3.3 Generic Solutions

To prove that the generic asymptotic behavior of the Painlevé transcendent u is given by an elliptic function, we need first to assume some conditions on the initial values and the domain in which we integrate the function. In particular, we assume that the local domain in which we integrate the equation lies in a fixed local patch near infinity. The results we get here will be extended to a larger domain in Section 4.

Note also that in the following we use the Painlevé property of P_I. In particular, we base the definition of our domain $\mathcal{U}$ below on arguments similar to those used in the direct proof of the Painlevé property of P_I obtained in [3].

Definition 3.1. Suppose ε, B, z_0, η, η' are given numbers satisfying the conditions below.

1. $0 < |\varepsilon|$ is given such that

$$|\varepsilon| < \min(1/e, 1/B) \tag{3.21}$$

$$|\log \varepsilon| > \max\big(72\pi, 2/(5\sqrt{eB})\big) \tag{3.22}$$

$$8B|\varepsilon|^{6/7}|\log \varepsilon|^4 < \tfrac{1}{2}. \tag{3.23}$$

2. $|z_0| > 1/\varepsilon$ and z lies in the domain $\mathcal{Z}$, defined by

$$\mathcal{Z} := \{z \mid |z - z_0| < B, |z| > 1/\varepsilon\}.$$

3. $|\eta|$, $|\eta'|$ are bounded above by $|\log \varepsilon|$, $|\eta'| > |\varepsilon|^{1/7}$, and E defined by

$$2E = \eta'^2 - 4\eta^3 - 2\eta - \frac{4\eta^2}{25z_0^2}$$

satisfies $|E - D_k| > 2|\varepsilon|^{2/7}$, $|E| < |\log \varepsilon|^4$. Moreover, $u(z)$ satisfies the initial conditions $u(z_0) = \eta$, $u'(z_0) = \eta'$.

4. γ is a path joining z_0 to z in $\mathcal{Z}$ of length $l < 2\pi B$ such that its image Γ under u lies in the domain

$$\mathcal{U} := \{u \mid |u| < |\log \varepsilon|, |P(u)| > |\varepsilon|^{1/7}\}.$$

These conditions are called *generic conditions*, and the solution $u(z)$ defined in 3 above is called a *generic solution*.

Note that ε_0 can always be found such that the upper bounds on $|\varepsilon|$ are satisfied for $0 < |\varepsilon| < \varepsilon_0$ and that a nonempty connected domain $\mathcal{U}$ and a path γ in $\mathcal{Z}$ exist by the Painlevé property of P_I.

Lemma 3.1 shows that the roots u_i of $P(u)$ are separated by a distance of order $O(|\varepsilon|^{1/7})$. The first of the results of the main theorem below shows that $|u'|$ is bounded below by a nonzero number on γ (in fact on $\mathcal{Z}$) if $|P(u)|$ is so. Therefore, $u(z)$ is invertible (by the inverse function theorem). In particular, the mapping $\gamma \mapsto \Gamma$ is invertible.

Our main result is the following theorem.

Theorem 3.1. *Under these conditions, $\exists$ positive ε_0 such that $\forall 0 < |\varepsilon| \le \varepsilon_0$, the generic solution satisfies*

$$|u'^2 - P(u)| < 8B|\varepsilon||\log \varepsilon|^4,$$

and

$$\left| \int_{\eta_\Gamma}^u \frac{dv}{\sqrt{P(v)}} - (z - z_0) \right| < 16\pi B^2 |\varepsilon|^{6/7} |\log \varepsilon|^4,$$

in $\mathcal{Z}$. Moreover, if z_0 and z_{10} are two successive points in $\mathcal{Z}$ where $u = \eta$, then for $j = 1$ or 2,

$$|(z_{10} - z_0) - \omega_j| < 16\pi B^2 |\varepsilon|^{6/7} |\log \varepsilon|^4.$$

Proof. Consider (3.11), which we rewrite as

$$u'^2 = 4u^3 + 2u + 2E + S, \tag{3.24}$$

where

$$S := 2E\left(\frac{z_0^2}{z^2} - 1\right) - \frac{1}{z^2} \int_{z_0}^z 4z(2u^3 + u)\,dz + \frac{4u^2}{25z^2}. \tag{3.25}$$

Note that we have

$$z = z_0 + Be^{i\theta} \implies \frac{|z_0 + z|}{|z|} = \left| \frac{2z - Be^{i\theta}}{z} \right| \le 2 + |\varepsilon|B < 3$$

and

$$\frac{|z - z_0|}{|z|} = \frac{B}{|z|} \le B|\varepsilon|.$$

Similarly, we have for $\zeta \in \gamma$ that

$$\left| \frac{\zeta}{z} \right| \le \frac{|z_0| + B}{|z|} \le \frac{|z| + 2B}{|z|} < 3.$$

Now, by the generic conditions, we get

$$\left| 2E\left(\frac{z_0^2}{z^2} - 1\right) \right| \le 6B|\varepsilon||\log \varepsilon|^4, \tag{3.26}$$

$$\left| \frac{1}{z^2} \int_{z_0}^z 4z(2u^3 + u)\,dz \right| \le 72\pi B|\varepsilon||\log \varepsilon|^3, \tag{3.27}$$

$$\left| \frac{4u^2}{25z^2} \right| \le \frac{4|\varepsilon|^2|\log \varepsilon|^2}{25}. \tag{3.28}$$

Therefore, S can be bounded as

$$|S| \leq 8B|\varepsilon||\log \varepsilon|^4.$$

This gives the first result. Moreover, we get

$$\left|\frac{S}{P(u)}\right| \leq 8B|\varepsilon|^{6/7}|\log \varepsilon|^4. \tag{3.29}$$

Choose a branch of $(1 + S/P(u))^{1/2}$ by demanding that it have positive real part. Now note that (3.24) gives

$$\int_\eta^u \frac{dv}{P(v)} - (z - z_0) = \int_{z_0}^z \left(\sqrt{1 + \frac{S}{P(u)}} - 1\right) dz. \tag{3.30}$$

Define

$$Q := \sqrt{1 + \frac{S}{P(u)}} + 1.$$

Note that

$$\sqrt{1 + \frac{S}{P(u)}} - 1 = Q - 2.$$

Also, squaring shows that

$$Q^2 = 2Q + \frac{S}{P(u)} \Rightarrow Q = 2 + \frac{S}{QP(u)}.$$

However, the choice of branch ensures that Q is bounded below by $|Q| \geq 1$. Therefore, it follows that

$$|Q - 2| \leq \left|\frac{S}{P(u)}\right| \implies \left|\sqrt{1 + \frac{S}{P(u)}} - 1\right| < 8B|\varepsilon|^{6/7}|\log \varepsilon|^4.$$

Integration then gives the desired result. The separation $z_{10} - z_0$ is based on the same result with the path of integration Γ being a closed contour enclosing two roots of $P(u)$.   □

Lemma 3.1 shows that the roots u_i of $P(u)$ are separated by a distance of order $O(|\varepsilon|^{1/7})$. This shows that the closed contours C_i defining the periods ω_i both exist.

Also, the first of the results of the main theorem above shows that $|u'|$ is bounded below by a nonzero number on γ (in fact, on $\mathcal{Z}$) if $|P(u)|$ is so. Therefore, $u(z)$ is invertible (by the inverse function theorem). In particular, the mapping $\gamma \mapsto \Gamma$ is invertible.

Extensions of this theorem to the case when u is larger than $|\log\varepsilon|$ or E is closer to D_k can be obtained in the same way as in the case of the Riccati equation in Section 2.2.

Because B is finite (and fixed), this result is confined to a local *patch* of the domain $\mathcal{Z}$. A different argument is required to extend the result to a sector of the annular region lying outside the large circle of radius $1/|\varepsilon|$. Such an extension would mean that the local patch is smeared to a domain of size proportional to $1/\varepsilon$, and this violates the second of the generic conditions 1.

3.4   Large Parameter

The purpose of this section is to show that the above direct approach also applies to other singular limits of the Painlevé equations. In particular, we derive here the generic local asymptotic behaviors of the second Painlevé equation in the limit as $\alpha \to \infty$ with $z \ll \alpha$.

The maximal dominant balance of P_{II} in this limit is

$$y'' \approx 2y^3 + \alpha.$$

Exercise 3.4. Show that other possible balances are either subcases of this one or are inconsistent.

This shows that an appropriate transformation (to make this balance explicit) is

$$\begin{aligned}(x) &= \alpha^{1/3}w(z),\\ z &= \alpha^{1/3}x,\end{aligned} \tag{3.31}$$

which transforms P_{II} to

$$w'' = 2w^3 + 1 + \frac{z}{\alpha}w, \tag{3.32}$$

where the primes denote z-derivatives.

Suppose $B > 1$ is given. (The arguments below can be easily changed to cover any $B > 0$.) Let $|z| < B$. Assume that in this region we have a solution of (3.32) defined by (bounded) initial conditions $w = \eta$, $w' = \eta'$ given at a point $z = z_0$. By the arguments given in [3], such a solution always exists. Moreover, it is meromorphic in $|z| < B$ with isolated simple poles of residue ± 1. This solution may be extended to $|z| < |\alpha|$ for any finite nonzero α.

We will work with an integrated form of P_{II}. Multiply (3.32) by w' and integrate as though the small term zw/α were not there. Then, we get

$$w'^2 = w^4 + 2w + 2E_0 + \frac{2}{\alpha}\int_{z_0}^{z} zww'\,dz, \tag{3.33}$$

where

$$2E_0 := \eta'^2 - \eta^4 - 2\eta. \qquad (3.34)$$

The leading-order part of equation (3.33) is solved by Jacobian elliptic functions. Let $E_0 = E$ and

$$P(w) := w^4 + 2w + 2E.$$

In general, $P(w)$ has four distinct roots, which we will denote by w_i, $i = 1, \ldots, 4$, in the complex plane. The leading-order equation

$$w'^2 = w^4 + 2w + 2E \qquad (3.35)$$

can be integrated again to give

$$\int_\eta^w \frac{dv}{\sqrt{P(v)}} = z - z_0. \qquad (3.36)$$

The solution $w(z)$ defined by the inverse of this function is an elliptic function with two periods given by

$$\omega_j = \oint_{C_j} \frac{dv}{\sqrt{P(v)}}, \qquad (3.37)$$

where C_j, $j = 1, 2$, are two linearly independent closed contours each enclosing two roots of $P(v)$ in the v-plane.

As above, there exist special values of E for which two roots of $P(w)$ coincide. To find these, consider

$$P'(w) = 4w^3 + 2 = 0 \implies w = d_k := \left(-\frac{1}{2}\right)^{1/3}, \quad k = 1, 2, 3. \qquad (3.38)$$

If such points are also zeros of $P(w)$, then we must have

$$E = D_k := -\frac{3}{4}\left(-\frac{1}{2P}\right)^{1/3} = \left(\frac{27}{128}\right)^{1/3}, \quad k = 1, 2, 3. \qquad (3.39)$$

For each such value of E, $P(w)$ has a double root. Note, however, that $P(w)$ cannot have a triple root, because $P''(w) = 12w^2$ cannot vanish simultaneously with $P'(w)$.

As before it can be shown that if $E - D_k$ is bounded below by $|\delta|$ (for small enough δ), then the roots of $P(w)$ are separated by a distance of order $\sqrt{|\delta|}$. Define ε by taking

$$\frac{z}{\alpha} = \varepsilon \exp(i\theta),$$

where $\varepsilon \to 0$ as $\alpha \to \infty$. Take ε to be sufficiently small to satisfy the upper bounds given below. For the sake of comparison with the previous result, we give a result here that does *not* rely on the Painlevé property of P_{II}.

Definition 3.2. The following assumptions are called *generic conditions*.

1. $|\varepsilon| > 0$ is given such that

$$|\varepsilon|^{4/7} < \frac{1}{2}, \qquad\qquad (3.40)$$

$$|\varepsilon|^{4/7} > 2|\varepsilon|^{6/7}|\log \varepsilon|^2. \qquad\qquad (3.41)$$

2. $|E - D_k| > h|\varepsilon|^{2/7}$, where h is chosen such that the roots v_i of $P(v)$ are separated by a distance of modulus at least $|\varepsilon|^{1/7}$.

3. Γ is a path in the v-plane, starting at η, of length $l < |\log \varepsilon|$, such that it lies outside (small) circles of radius $|\varepsilon|^{1/7}$ centered at v_i.

4. $|E| < |\log \varepsilon|^4$ (to ensure that the solutions remain bounded).

5. On Γ, $|v| < |\log \varepsilon|$.

6. $|\eta'| > |\varepsilon|^{1/14}$ (to be consistent with the above).

Note that ε_0 can always be found such that the upper bounds on $|\varepsilon|$ are satisfied for $0 < |\varepsilon| < \varepsilon_0$.

Definition 3.3. Suppose ε, η, η' are given that satisfy the generic conditions. A generic solution of P_{II} is one that is defined by initial conditions $w(z_0) = \eta$, $w'(z_0) = \eta'$.

Note that a generic solution may be defined equivalently by the pair η', E satisfying the generic conditions.

Theorem 3.2. *For $0 < |\varepsilon| < \varepsilon_0$, the generic solution satisfies*

$$\left| \int_\eta^w \frac{dv}{\sqrt{P(v)}} - (z - z_0) \right| < \sqrt{2}\varepsilon^{1/2}|\log \varepsilon|.$$

Moreover, if z_0 and z_{10} are two successive points where $w = \eta$, then for $j = 1$ or 2,

$$|(z_{10} - z_0) - w_j| < \sqrt{2}\varepsilon^{1/2}|\log \varepsilon|.$$

The method of proof is similar to that given for Theorem 3.1. But the error bound is cruder because conditions are imposed only on the path Γ in the dependent-variable plane, rather than on its inverse γ. This means that the proof has to convert any integral in the z-plane to one in the v-plane. This requires a division by u' as well as $P(u)$, both of which are bounded below by small quantities. Usage of the Painlevé property improves this error bound to $|\varepsilon|^{(m-1)/m}|\log \varepsilon|^p$ for positive integers $m > 2$, p. Again, this result can be extended to cases where w or Γ become large or where $E - D_k$ becomes small.

4   Global Extensions

In this section we show how to extend the generic local results of the last section (for the limit $z \to \infty$) to obtain global asymptotics. For conciseness and simplicity, we focus on the modulation of the energy-like parameter E along paths linking points where u takes on the same value. The method we use is based on the idea of averaging. For complete generic global results found by the method of multiple scales see [31].

4.1   The Slow Variation of E

Suppose that $\eta = 0$, $\eta' =: p \neq 0$. Starting at z_0, define Ω_j to be the next zero of the generic solution $u(z)$. There are two such points given approximately by the two periods of the local leading-order elliptic-function-type behavior of u. Where convenient, we will fix a choice of $j = 1, 2$ and drop the subscript from Ω.

Write

$$u(z) = U(z) + \hat{u}(z), \qquad (4.1)$$

where

$$U(z_0) = 0, \qquad\qquad U'(z_0) = p, \qquad (4.2)$$
$$\hat{u}(z_0) = 0, \qquad\qquad \hat{u}'(z_0) = 0, \qquad (4.3)$$

and U is an elliptic function with the same periods as the locally valid elliptic-function-type behavior of u. We then have the following lemma.

Lemma 4.1. u is analytic at $\Omega + z_0$ and Ω is given by the inversion of $u(z_0 + \Omega) = 0$.

Proof. The Painlevé property of P_I gives the first result. For the second, we use

$$|u'^2(z_0 + \Omega)| \geq \frac{1}{2}|P(u(z_0 + \Omega))| > \frac{|\varepsilon|^{1/7}}{2} > 0,$$

which follows from Theorem 3.1. Hence $u'(z_0 + \Omega) \neq 0$. Therefore, we can use the inverse function theorem to find Ω.   □

Lemma 4.2. *For a fixed j,*

$$|\Omega_j - \omega_j| < 16\pi B^2 |\varepsilon|^{6/7} |\log \varepsilon|^4.$$

This follows from Theorem 3.1.

Lemma 4.3. *There exists a positive number k such that*

$$|\Omega - \omega + \hat{u}(z_0 + \omega)/p| < k|\varepsilon|^{12/7} |\log \varepsilon|^9.$$

Proof. We have

$$
\begin{aligned}
0 &= u(z_0 + \Omega) \\
&= u(z_0 + \omega + \Omega - \omega) \\
&= u(z_0 + \omega) + (\Omega - \omega)u'(z_0 + \omega) + (\Omega - \omega)^2 g(z), \qquad (4.4)
\end{aligned}
$$

where $z = z_0 + \Omega$ and g is analytic there. Note that

$$
g(z) = \frac{1}{2\pi i} \oint_C \frac{u(\zeta)}{(\zeta - z)(\zeta - z_0 - \omega)^2} \, d\zeta,
$$

where C is a closed contour enclosing z and $z_0 + \omega$. Assuming that $|\zeta - z| > \delta$, g may be estimated in the usual way (by the Cauchy estimate under the generic conditions) as

$$
|g| < \frac{\max_C |u|}{\delta B^2}.
$$

Now, using (4.1) we get

$$
0 = U(z_0 + \omega) + \hat{u}(z_0 + \omega) + (\Omega - \omega)U'(z_0 + \omega) + \sigma(z), \qquad (4.5)
$$

where because

$$
U(z_0 + \omega) = 0, \quad U'(z_0 + \omega) = p
$$

(by the initial conditions and the periodicity of U), we have

$$
\sigma = (\Omega - \omega)(u'(z_0 + \omega) - p) + (\Omega - \omega)^2 g(z).
$$

Claim 6.

$$
|\sigma| < k|\varepsilon|^{12/7}|\log \varepsilon|^9
$$

for some number $k > 0$.

This follows from Theorem 3.1 and the above bound for g. Since we have

$$
\Omega - \omega = -\frac{\hat{u}(z_0 + \omega)}{p} - \frac{\sigma(z)}{p},
$$

the proof of the lemma follows from the above estimate for σ. □

Recall that E was defined implicitly by (3.11). Under the assumption that $\eta = 0$, $\eta' = p$, we have

$$
E(z_0) = \frac{p^2}{2}.
$$

We have written $E(z_0)$ to emphasize the fact that in general, E is not constant but varies with z due to the small terms present in (3.11). To find its variation, we define E more generally by

$$E(z) := \frac{1}{2}(u'^2 - 4u^3 - 2u). \tag{4.6}$$

To make the following calculations more concise, we also adopt the notation that $\mathcal{L}$ denotes any term satisfying

$$|\mathcal{L}| < k|\varepsilon|^{1+m/7}|\log \varepsilon|^n$$

for some positive k, m, and n. Note that $\mathcal{L}$ may vary from line to line below. We leave the actual estimate of terms denoted by $\mathcal{L}$ in each case as an exercise.

Lemma 4.4.

$$E(z_0 + \Omega) - E(z_0) = p\hat{u}'(z_0 + \omega) - \hat{u}(z_0 + \omega) + \mathcal{L}.$$

Proof. We have

$$E(z_0 + \Omega) - E(z_0)$$
$$= \frac{1}{2}\big\{ \big(u'(z_0 + \omega + \Omega - \omega)\big)^2 - 4\big(u(z_0 + \omega + \Omega - \omega)\big)^3$$
$$\quad - 2u(z_0 + \omega + \Omega - \omega)\big\} - \frac{p^2}{2}$$
$$= \frac{1}{2}\big\{ \big(u'(z_0 + \omega)\big)^2 + 2(\Omega - \omega)u'(z_0 + \omega)u''(z_0 + \omega) + (\Omega - \omega)^2 h(z)$$
$$\quad - 4\big[\big(u(z_0 + \omega)\big)^3 + 3(\Omega - \omega)u'(z_0 + \omega)\big(u(z_0 + \omega)\big)^2 + (\Omega - \omega)^2 f(z)\big]$$
$$\quad - 2u(z_0 + \omega) - 2(\Omega - \omega)u'(z_0 + \omega) - 2(\Omega - \omega)^2 g(z)\big\} - \frac{p^2}{2}, \quad (4.7)$$

where f, g, h are analytic functions at $z = z_0 + \Omega$. Using the representation (4.1) and estimating f, g, h as before, we get

$$E(z_0 + \Omega) - E(z_0)$$
$$= \frac{1}{2}\big\{ \big(U'(z_0+\omega)\big)^2 + 2U'(z_0+\omega)\hat{u}'(z_0+\omega) + 2(\Omega-\omega)U'(z_0+\omega)U''(z_0+\omega)$$
$$\quad - 4\big(U(z_0+\omega)\big)^3 - 12\big(U(z_0+\omega)\big)^2 \hat{u}(z_0+\omega)$$
$$\quad - 12(\Omega-\omega)U'(z_0+\omega)\big(U(z_0+\omega)\big)^2 - U(z_0+\omega) - 2\hat{u}(z_0+\omega)$$
$$\quad - 2(\Omega-\omega)U'(z_0+\omega)\big\} - \frac{p^2}{2} + \mathcal{L} \quad (4.8)$$

and thus

$$E(z_0 + \Omega) - E(z_0) = \frac{1}{2}\{p^2 + 2p\hat{u}'(z_0 + \omega) - 2\hat{u}(z_0 + \omega)\} - \frac{p^2}{2} + \mathcal{L}$$
$$= p\hat{u}'(z_0 + \omega) - \hat{u}(z_0 + \omega) + \mathcal{L} \tag{4.9}$$

as required, where we have also used $U'' = 6U^2 + 1$ in the second equality.

$\square$

This result can be restated by using the following lemma.

Lemma 4.5.

$$p\hat{u}'(z_0 + w) - \hat{u}(z_0 + w) = -\int_{z_0}^{z_0+w} \frac{U'^2}{z} \, dz + \mathcal{L}.$$

Proof. The result relies on the equation governing $\hat{u}$, namely

$$\hat{u}'' = 12U\hat{u} + 6\hat{u}^2 - \frac{U'}{z} - \frac{\hat{u}'}{z} + \frac{4U}{25z^2} + \frac{4\hat{u}}{25z^2} \qquad (4.10)$$

$$= 12U\hat{u} - \frac{U'}{z} + \mathcal{L}. \qquad (4.11)$$

Multiplying this equation by U' and using

$$U''' = 12UU', \quad \text{and} \quad U'\hat{u}'' - U'''\hat{u} = (U'\hat{u}' - U''\hat{u})',$$

we get

$$U'\hat{u}' - U''\hat{u} = -\int_{z_0}^{z} \frac{U'^2}{z} \, dz + \mathcal{L}.$$

Evaluation of this result at $z_0 + w$ gives the required result. $\square$

Since $U' = (P(U))^{1/2}$, the period integral in the above lemma can be rewritten in terms of an elliptic integral given by

$$\tilde{\omega} := \oint_C \sqrt{P(u)} \, du,$$

where C is one of two linearly independent closed contours either enclosing two of the roots of P or one root and infinity. We have

$$\int_{z_0}^{z_0+w} \frac{U'^2}{z} \, dz = \frac{\tilde{\omega}}{z_0} + \mathcal{L}.$$

In other words, we have

$$E(z_0 + \Omega) - E(z_0) = -\frac{\tilde{\omega}}{z_0} + \mathcal{L}. \qquad (4.12)$$

This gives a discrete picture of the slow variation of E.

4.2 Averaging

As the point of observation z_0 is varied near infinity, the discrete picture we obtained in the previous section changes. To see how E varies as z_0 changes, we use an implicit method of averaging.

Consider an averaged function $E(z)$ defined to have the same values as E at the zeros of the Painlevé transcendent u and to be interpolated smoothly between these discrete values. (See [39] for a method of constructing such a function.) Such an averaged function $E(z)$ is to be thought of as varying significantly only over large z. In this framework, a period ω is very small (relative to z). Hence the difference (4.12) yields an approximate differential quotient

$$\frac{\Delta_j E}{\Omega_j} := \frac{E(z_0 + \Omega_j) - E(z_0)}{\Omega_j}$$

giving the derivative of the averaged function in the limit as $z \to \infty$.

However, this quotient is different in different directions. From equation (4.12) we have

$$\frac{\Delta_j E}{\Omega_j} = -\frac{\widetilde{\omega}_j}{z_0 \omega_j} + \mathcal{L}.$$

These leading-order values of the differential quotient differ even for arbitrarily small ε. Consequently, the averaged function $E(z)$ must be nonanalytic. Writing $z = x + iy$, $\bar{z} = x - iy$, and defining

$$\frac{\partial}{\partial z} = \frac{1}{2}\left(\frac{\partial}{\partial x} - i \frac{\partial}{\partial y}\right), \tag{4.13}$$

$$\frac{\partial}{\partial \bar{z}} = \frac{1}{2}\left(\frac{\partial}{\partial x} + i \frac{\partial}{\partial y}\right), \tag{4.14}$$

we get

$$\Delta_j E = \Omega_j E_z + \overline{\Omega}_j E_{\bar{z}} + \mathcal{L}.$$

(Note that if $E(z)$ were analytic, then $E_{\bar{z}}$ would be zero.) Using (4.12), we get

$$\omega_j E_z + \overline{\omega}_j E_{\bar{z}} = -\frac{\widetilde{\omega}_j}{z_0} + \mathcal{L}. \tag{4.15}$$

4.3 Properties of the Elliptic Integrals

To analyze the behavior of the averaged $E(z)$ we need to deduce the behaviors of the elliptic integrals ω_j and $\widetilde{\omega}_j$ as functions of E. We will outline their properties through exercises.

Exercise 4.1. Show that $\tilde{\omega}(E)$ satisfies the ODE

$$\tilde{\omega}'' = -\frac{5}{36}\frac{\tilde{\omega}}{E^2 + 2/27}, \tag{4.16}$$

where the primes denote differentiation with respect to E by following the steps below.

1. Integrate $\tilde{\omega}$ by parts by writing

$$du\sqrt{P(u)} = d\left(u\sqrt{P(u)}\right) - u\,d\left(\sqrt{P(u)}\right).$$

   Show that the result gives

$$\tilde{\omega} = \frac{4}{5}w\psi + \frac{6}{5}E\omega,$$

   where

$$\psi := \oint \frac{u}{\sqrt{P(u)}}\,du.$$

2. Integrate ψ by parts in a similar way to show that

$$\psi = -\frac{2}{3}w' + 6E\psi'.$$

3. Use the above two equations for $\tilde{\omega}$, ψ and their derivatives to obtain an equation involving only $\tilde{\omega}$ and w'. Use $\tilde{\omega}' = w$ to get the desired result.

   Note that $2/27 = -D_k^2$. So this ODE for $\tilde{\omega}$ is singular at $E = D_1, D_2$, and at infinity.

   The ODE (4.16) can be used to deduce the asymptotic properties of the elliptic integrals $\tilde{\omega}$, w, w' in the limits as $E \to D_k$ or $E \to \infty$.

Exercise 4.2. Consider (4.16) in the limit as $E \to \infty$.

1. Show that $\tilde{\omega}$, w have the asymptotic behaviors

$$\tilde{\omega} \sim E^{5/6}, \quad w \sim E^{-1/6}.$$

2. Show that in polar coordinates, $z = re^{i\theta}$, (4.15) gives

$$rE_r \approx -\frac{6}{5}E.$$

3. Solve this leading-order equation to show that if $|E|$ is large, then it must decrease as r increases.

   This proves the following lemma.

Lemma 4.6. *For all* $|z_0| > 1/|\varepsilon|$, $|z| > 1/|\varepsilon|$, E *remains bounded.*

Many other properties of E can be found. In particular, we can show that on each fixed ray of approach to infinity (in a sector of width $\pi/2$) E approaches a unique value that is a function of θ. For such results and analyses see [31]. We content ourselves here with proving one more result that was pointed out by Boutroux.

Lemma 4.7. *Let the chain of points consisting of the zeros of u with increasing modulus be denoted inductively by*

$$Z_{m+1} = Z_m + \Omega(Z_m).$$

Then as $m \to \infty$, $E(Z_m)$ *approaches a zero of* $\widetilde{\omega}$.

Proof. Without loss of generality, we restrict the chain by taking a fixed $j = 1$ or 2 and choosing, at each Z_m, the direction given by $\Omega_j \approx \omega_j\big(E(Z_m)\big)$. Now the averaged function E satisfies the differential equation

$$\omega_j \frac{\mathrm{d}E}{\mathrm{d}Z} \approx -\frac{\widetilde{\omega}_j}{Z},$$

where Z is the interpolated variable along the above defined chain. This is a separable ODE, which can be integrated (after dividing by $\widetilde{\omega}_j$) to give

$$\log(\widetilde{\omega}_j) \approx -\log(Z).$$

This shows that

$$\widetilde{\omega}_j \approx \frac{c}{Z},$$

for some constant c. This is the desired result. $\qquad\square$

5   Conclusion

The Painlevé equations and their generic solutions called the Painlevé transcendents appear ubiquitously in physical applications, in particular in scaling limits of exactly solvable models of statistical mechanics. They are also symmetry reductions of soliton equations, to which they are closely related. Their remarkable properties make them special functions of key significance in the theory of integrable systems.

We have shown that the asymptotic limit of the Painlevé equations as the independent variable $x \to \infty$ can be studied rigorously by direct natural methods. All the methods we have displayed here can be applied to other limits of interest, such as the singular limit when an energy-like arbitrary constant (similar to E above) becomes unbounded [40]. They can also be extended to deduce limiting behaviors of the discrete Painlevé equations [41].

Recent developments have relied almost exclusively on *indirect* methods that work through associated linear isomonodromy problems. We showed in this chapter that explicit asymptotic information about the Painlevé transcendents can be obtained *directly*. Direct nonlinear asymptotic methods not only yield rigorous asymptotic results, but also illuminate the study of the analytic properties of solutions [3]. The aim of these notes was to rekindle the spirit of mathematics as practiced by Painlevé and Boutroux, who pioneered analytic methods for differential equations from first principles with direct approaches.

Acknowledgments: Research supported by the Australian Research Council.

6 REFERENCES

[1] E. Hille, *Analytic Function Theory* I and II (Chelsea, New York, 1982).

[2] E. Hille, *Ordinary Differential Equations in the Complex Domain* (John Wiley and Sons, New York, 1976).

[3] N. Joshi and M.D. Kruskal, *A direct proof that the solutions of the six Painlevé equations have no movable singularities except poles*, Stud. Appl. Math. **93** (1994), 187–207.

[4] P. Painlevé, *Sur les équations différentielles du second ordre et d'ordre supérieur dont l'intégrale générale est uniforme*, Acta Math. **25** (1902), 1–85.

[5] B. Gambier, *Sur les équations différentielles du second ordre et du premier degré dont l'intégrale générale est à points critiques fixes*, Acta Math. **33** (1910), 1–55.

[6] R. Fuchs, *Über lineare homogene Differentialgleichungen zweiter Ordnung mit drei im Endlichen gelegenen wesentlich singulären Stellen*, Math. Annalen **63** (1907), 301–321.

[7] M.J. Ablowitz and H. Segur, *Solitons and the Inverse Scattering Transform* (SIAM, Philadelphia, 1981).

[8] M.J. Ablowitz and P.A. Clarkson, *Solitons, Nonlinear Evolution Equations and Inverse Scattering*, volume 149 of London Mathematical Society Lecture Notes in Mathematics (Cambridge University Press, Cambridge, 1991).

[9] M.J. Ablowitz, A. Ramani, and H. Segur, *A connection between non-linear evolution equations and ordinary differential equations of P-type I and II,* J. Math. Phys. **21** (1980), 715–721, 1006–1015.

[10] A.S. Fokas and X. Zhou, *On the solvability of Painlevé II and IV,* Comm. Math. Phys. **144** (1992), 601–622.

[11] A.S. Fokas, U. Mugan, and X. Zhou, *On the solvability of Painlevé I, III and V,* Inverse Problems **8** (1992), 757–785.

[12] F.W.J. Olver, *Asymptotics and Special Functions* (Academic Press, London, 1992).

[13] P. Boutroux, *Recherches sur les transcendantes de M. Painlevé et l'étude asymptotique des équations différentielles du second ordre,* Ann. École Norm. **30** (1913), 265–375.

[14] P. Boutroux, *Recherches sur les transcendantes de M. Painlevé et l'étude asymptotique des équations différentielles du second ordre,* Ann. École Norm. **31** (1914), 99–159.

[15] B. McCoy, *Spin systems, statistical mechanics, and Painlevé functions,* Painlevé transcendents, their asymptotics and physical applications, eds. D. Levi and P. Winternitz, NATO ASI Series B, (Plenum, New York, 1992), pages 377–391.

[16] B. McCoy, C. Tracy, and T.T. Wu, *Painlevé functions of the third kind,* J. Math. Phys. **18** (1977), 1058–1092.

[17] M.J. Ablowitz and H. Segur, *Asymptotic solutions of the Korteweg-deVries equation,* Stud. Appl. Math. **57** (1977), 13–44.

[18] H. Segur and M.J. Ablowitz, *Connection results for the second Painlevé equation,* Physica D **3** (1981), 165–184.

[19] S.P. Hastings and J.B. McLeod, *A boundary value problem associated with the second Painlevé transcendent and the Korteweg -de Vries equation,* Arch. Rational Mech. Anal. **73** (1980), 31–51.

[20] P.A. Clarkson and J.B. McLeod, *A connection formula for the second Painlevé transcendent.* Arch. Rational Mech. Anal. **103** (1988), 97–138.

[21] M. Jimbo, T. Miwa, Y. Môri and M. Sato, *Density Matrix of an Impenetrable Bose Gas and the Fifth Painlevé Transcendent,* Physica **1D** (1980), 80–158.

[22] A.R. Its and V.Yu. Novokshenov, *The Isomonodromic Deformation Method in the Theory of Painlevé Equations,* Lecture Notes in Math. **1191** (Springer-Verlag, Berlin, 1986).

[23] A.R. Its, A.S. Fokas, and A.A. Kapaev, *On the asymptotic analysis of the Painlevé equations via the isomonodromy method*, Nonlinearity **7** (1994), 1291–1325.

[24] A.R. Its and A.A. Kapaev, *The method of isomonodromy deformations and connection formulas for the second Painlevé transcendent*, Math. USSR Izvestiya **31** (1988), 193–207.

[25] A.V. Kitaev, *Isomonodromic technique and elliptic asymptotic formulas for the first Painlevé transcendent*, St. Petersburg Math. J. **5** (1994), 577–605.

[26] P. Deift and X. Zhou, *Asymptotics of the Painlevé II equation*, Comm. Pure Appl. Math. **48** (1995), 277–337.

[27] A.V. Kitaev, *Elliptic asymptotics of the first and the second Painlevé transcendents*, Russian Math. Surveys **49** (1994), 81–150.

[28] A. Bassom, P.A. Clarkson, C. Law, and J.B. McLeod, *Applications of uniform asymptotics to the second Painlevé transcendent*, Arch. Rat. Mech. Anal., **143** (1998), 241–271.

[29] N. Joshi and M.D. Kruskal, *An asymptotic approach to the connection problem for the first and the second Painlevé equations*, Phys. Lett. A **130** (1988), 129–137.

[30] N. Joshi and M.D. Kruskal, *Connection results for the first Painlevé equation*, Painlevé Transcendents, Their Asymptotics and Physical Applications, eds. D. Levi and P. Winternitz, NATO ASI Series B, (Plenum, New York, 1992), pages 61–79.

[31] N. Joshi and M.D. Kruskal, *The Painlevé connection problem: an asymptotic approach I*, Stud. Appl. Math. **86** (1992), 315–376.

[32] N. Joshi and M.D. Kruskal, *The connection problem for Painlevé transcendents*, Physica D **18** (1986), 215–216.

[33] C. Bender and S. Orszag, *Advanced Mathematical Methods for Scientists and Engineers* (McGraw-Hill, New York, 1978).

[34] E. Ince, *Ordinary Differential Equations* (Longmans, Green and Co., London and New York, 1926). Reprinted (Dover, New York, 1954).

[35] T. Kawai and Y. Takei, *On the structure of Painlevé transcendents with a large parameter*, Proc. Japan Acad. Ser. A Math. Sci **69** (1993), 224–229.

[36] T. Kawai and Y. Takei, *WKB analysis of Painlevé transcendents with a large parameter I*, Adv. Math. **118** (1996), 1–33.

[37] T. Kawai and Y. Takei, *On the structure of Painlevé transcendents with a large parameter II*, Proc. Japan Acad. Ser. A Math. Sci **72** (1996), 144–147.

[38] M. Abramowitz and I. Stegun (eds.), *Handbook of Mathematical Functions* (Dover, New York, 1972).

[39] M.D. Kruskal, *Asymptotic theory of Hamiltonian and other systems with all solutions nearly periodic*, J. Math. Phys. **3** (1962), 806–828.

[40] P. Doran-Wu and N. Joshi, *Direct asymptotic analysis of the second Painlevé equation: three limits*, J. Phys. A **30**, 4701–4708.

[41] N. Joshi, *Local asymptotics of the first discrete Painlevé equation as the discrete independent variable approaches infinity*, Methods and applications of analysis **4** (1997), 124–133.

5

2-D Quantum and Topological Gravities, Matrix Models, and Integrable Differential Systems

Philippe Di Francesco

Part A 2-D Quantum Gravity

1 Introduction

These lecture notes are partly based on the review [1] on 2-D quantum gravity and random matrix models. The subdivision into two parts corresponds to two very different approaches to the quantization of 2-D gravity, on the one hand through a discretization of the problem (Part A), on the other hand through a mathematical formulation thereof, as intersection theory on the moduli space of Riemann surfaces (Part B).

1.1 Quantum Gravity in Two Dimensions

Four-dimensional gravity, as expressed through Einstein's general relativity equations, incorporates the interactions between matter and space-time. Quantizing these equations means incorporating all the possible fluctuations of matter and space, however improbable, but weighing them with their probabilities. Attempts to describe such a quantization with ordinary field theory fail due to severe divergences. String theory is an attempt to overcome these difficulties, by replacing particles (points) by stringlike one-dimensional loops, which describe a two-dimensional world-sheet when they evolve in time (interactions being encoded in the topology of this surface, holes being interpreted as fusion/disintegration processes). Polyakov [2] has showed that these theories could be interpreted as two-dimensional gravity theories, by exchanging the roles of the world-sheet and the space-time, and considering the string coordinate in 4-D as a matter field defined on the two-dimensional world-sheet. The advantage is that the Einstein action in 2-D reduces to

$$S_\Sigma(\Lambda, \mathcal{N}) = \Lambda \int_\Sigma \sqrt{|g|} + \mathcal{N} \frac{1}{4\pi} \int_\Sigma \sqrt{|g|} R$$
$$= \Lambda A + \mathcal{N}(2 - 2h), \tag{1.1}$$

where we have identified the area A and the Euler characteristic $2 - 2h$ (h the number of holes, R the scalar curvature) of the surface Σ, respectively multiplied by the cosmological constant Λ and the Newton constant $\mathcal{N}$.

We then wish to perform some Feynman integration over all the possible fluctuations of this two-dimensional world-sheet, namely over all the possible metrics and topologies (numbers of holes) of this surface, and also over the matter fields, weighted by a string action accounting for the probability factor. For simple enough matter theories, such as two-dimensional conformal field theories with a small number of degrees of freedom, this sum could be further simplified, leading to the actual computation of scaling exponents and singularities, indicating a very rich structure of the corresponding gravitational theories [3].

1.2   Discretization: The Ising Example

In this chapter we will not address this continuum approach to two-dimensional quantum gravity. We will use a different approach, namely the discretized version of 2-D quantum gravity. The idea is to replace the sum over all possible metrics and topologies of the world-sheet surface by a sum over random tessellations of these surfaces, of arbitrary genus (number of holes). The matter theories have to be discretized too. To bear in mind an archetypical example, let us illustrate this with the two-dimensional Ising model. In flat two-dimensional space (the plane) its configurations are given by the values of a spin variable $\sigma_i = \pm 1$ at each vertex i of the square lattice, with spacing a (in fact, only a finite rectangle of size $N \times M$ is considered). The spins interact only with their nearest-neighbors, through an energy

$$E(\{\sigma\}) = -J \sum_{(ij)} \sigma_i \sigma_j + h \sum_i \sigma_i, \tag{1.2}$$

where the sums extend over respectively all the lattice bonds (ij) and vertices i (the first term gives the highest probability to parallel (if $J > 0$) or antiparallel (if $J < 0$) configurations, and the second term is a magnetic field, which tends to align the spins in its direction). Each configuration $\{\sigma\}$ is weighed by the Boltzmann probability factor $e^{-\beta E}$, and we can write the partition function of the system as

$$Z(K, H) = \sum_{\sigma_i = \pm 1} e^{-\beta E(\{\sigma\})} \tag{1.3}$$

in the reduced variables $K = \beta J$ and $H = \beta h$. For $H = 0$, the model is known [4], in the thermodynamic limit N, $M \to \infty$, to undergo a second-order phase transition at $K = K_c$ between an ordered phase and a disordered one. The transition is characterized by the scale invariance of the model. It is best seen by considering the spin correlation function

$$\langle \sigma_i \sigma_j \rangle = \frac{1}{Z(K,0)} \sum_{\{\sigma\}} \sigma_i \sigma_j e^{-\beta E} \sim \begin{cases} e^{-|i-j|/\xi} & \text{for } K \neq K_c, \\ \frac{1}{|i-j|^\Delta} & \text{for } K = K_c. \end{cases} \quad (1.4)$$

Except at the critical point $K = K_c$, the effective range of the spin interaction is characterized by the correlation length ξ, which diverges at $K = K_c$ to unveil a subleading algebraic behavior, with a scaling exponent $\Delta = \frac{1}{4}$. The disappearance of the length scale ξ is the so-called scale invariance phenomenon, which guarantees that the continuum thermodynamical limit of the model is described by a conformally invariant field theory (the free Majorana fermion, in the case of the Ising model), namely a field theory locally invariant under scale transformations, rotations, and translations.

To describe the coupling of this model to gravity, we simply have to quantize the square lattice, namely sum over all its possible fluctuations, i.e., tessellations with possibly deformed square tiles (the variations of the metric change the lengths of the bonds), each vertex of the tessellation possibly sharing more or fewer than four squares (this accounts for variations of the curvature of the surface), and finally these tessellations may wrap around holes of the surface, to account for the nontrivial topology. For any such fixed tessellation Γ, we simply consider the Ising model defined on the vertices of Γ, and denote by $Z_\Gamma(K, H)$ the corresponding partition function. The partition function of the 2-D quantum-gravitational Ising model reads

$$Z(\Lambda, \mathcal{N}; K, H) = \sum_{h, A \geq 0} \sum_{\substack{\text{random tessell. } \Gamma \\ \text{with area } A, \text{ genus } h}} e^{\Lambda A + \mathcal{N}(2-2h)} Z_\Gamma(K, H), \quad (1.5)$$

where the contribution of the surfaces of fixed area and genus is weighted by the Einstein factor (1.1).

More generally, we can replace the Ising model by some two-dimensional statistical model, with a set of parameters named collectively by K. The 2-D gravitational version is analogously

$$Z(\Lambda, \mathcal{N}; K) = \sum_{h, A \geq 0} \sum_{\substack{\text{random tessell. } \Gamma \\ \text{with area } A, \text{ genus } h}} e^{\Lambda A + \mathcal{N}(2-2h)} Z_\Gamma(K), \quad (1.6)$$

where $Z_\Gamma(K)$ denotes the partition function of the matter theory on the tessellation Γ.

1.3   Gravitational Observables

In general, we will consider matter systems that for some critical values of
the couplings undergo some continuous phase transitions and are described
in the continuum-thermodynamical limit by some 2-D conformally invariant theories. The latter are indexed by a number, the central charge c, and
have a very rich structure due to their infinite symmetry. In particular, all
the $c < 1$ minimal theories (i.e., with a finite number of basic observables,
like the spin in Ising) have been classified and shown to have central charges
(see the book [5] for an extensive exposition, and [6] for the most relevant
papers)

$$c(p,q) = 1 - 6\frac{(p-q)^2}{pq}. \tag{1.7}$$

Moreover, their finite collection of observables $\phi_{r,s}$ have correlations analogous to that of the spin operator in the Ising model (1.4), with scaling
exponents given by the Kac formula

$$\Delta_{r,s} = \frac{(pr - qs)^2 - (p - q)^2}{pq} \tag{1.8}$$

for $1 \leq r \leq q - 1$, $1 \leq s \leq p - 1$.

When gravity is switched on, the observables of the matter theories get
dressed in a particular way. The simplest of these operators is the identity
operator $\phi_1 \equiv \int_\Sigma \phi_{1,1}$, with $\Delta_{1,1} = 0$, coupled to the cosmological constant
Λ (see (1.1)), as the integral over any surface of $\phi_{1,1}$ measures its area.

In the direct continuum approach, one can show that only the order
parameter fields survive as basic fields of the theory, and that moreover they
acquire different scaling dimensions. Recall that with gravity a correlation
of observables does not feel distances, as the points are integrated over
the surface. The only scale of the model is that fixed by the cosmological
constant Λ, with the dimension of an inverse area (cf. (1.1)). The dressed
observables display a scaling behavior as functions of this constant.

1.4   Tessellations and Diagrammatics: Matrix Model for Pure Gravity

Diagrammatics is an essential feature of field theory: The perturbative
computation of amplitudes is typically reduced to a sum over diagrams,
together with some (Feynman) rules attaching some specific weights to the
diagrams. Let us illustrate this for the simplest of all field theories, namely
the φ^4 theory in $D = 0$ dimension (the field is actually reduced to a point).
Let us compute the partition function (vacuum amplitude)

$$Z(g) = \int_{-\infty}^{\infty} \frac{1}{\sqrt{2\pi}} e^{-(\varphi^2/2) + g\varphi^4/(4!)} \, d\varphi \tag{1.9}$$

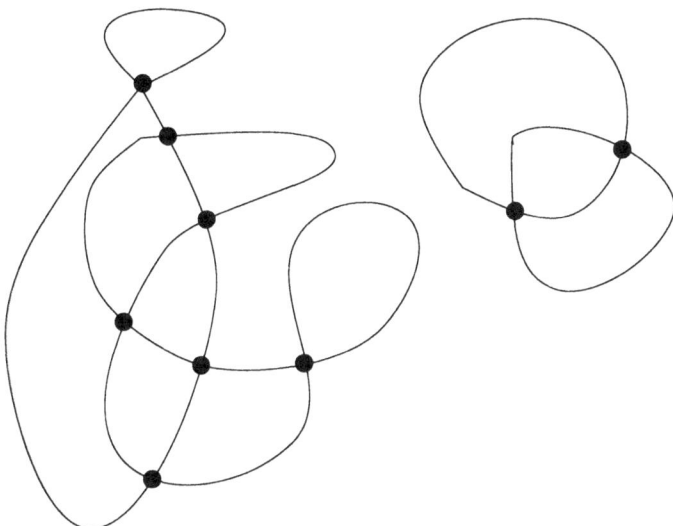

FIGURE 5.1. A typical contribution to $Z_n(g)$. The n $(= 9$ here) vertices are connected through propagators, so as to form a closed not necessarily connected four-valent graph.

as a series expansion in powers of g, in the form $Z(g) = \sum_{n \geq 0} g^n Z_n(g)$, where

$$Z_n(g) = \frac{1}{n!} \int_{-\infty}^{\infty} \frac{1}{\sqrt{2\pi}} \left(\frac{\varphi^4}{4!} \right)^n d\varphi. \qquad (1.10)$$

Such an integral may be computed by using the source representation with

$$Z_s = \int \frac{1}{\sqrt{2\pi}} e^{-(\varphi^2/2) + s\varphi} d\varphi \qquad (1.11)$$

and by taking derivatives with respect to s and taking $s = 0$ in the end:

$$\int \frac{1}{\sqrt{2\pi}} \varphi^{2k} e^{-\varphi^2/2} d\varphi = \frac{d^{2k}}{ds^{2k}} \int \frac{1}{\sqrt{2\pi}} e^{-(\varphi^2/2) + s\varphi} d\varphi \Big|_{s=0}$$

$$= \frac{d^{2k}}{ds^{2k}} e^{s^2/2} \Big|_{s=0}, \qquad (1.12)$$

where we have completed the square and integrated out φ. The derivatives with respect to s must go by pairs, as $(d/ds)e^{s^2/2} = se^{s^2/2}$, and we take $s = 0$ in the end; hence another derivative must hit the s prefactor. Such pairs of derivatives are called propagators, often represented as $(d/ds)^2 Z_s|_{s=0} = \langle \varphi\varphi \rangle = 1$ here.

To evaluate an expression like (1.10), we simply have to draw n small crosses (see Figure 5.1), corresponding to the n monomials φ^4, and to connect them in all possible ways with propagators in such a way as to form a closed graph. Each such connection corresponds in turn to a pair of derivatives with respect to s, acting on Z_s, and receives a weight $\langle \varphi\varphi \rangle = 1$. The symmetry of the crosses is taken into account with the factor 4!, which avoids overcounting the permutations of the legs; hence we are left with a purely combinatorial problem of counting of graphs with labeled legs. Gathering the identical graphs with different labelings, we end up with an overall weight $1/|\mathrm{Aut}(G)|$ per graph G, where $|\mathrm{Aut}(G)|$ is the order of the symmetry group of the graph (it is a certain divisor of the $n!$ term in the denominator of (1.10)). Of course, in this case the answer is easily computable, and we get the simple φ^4 graph-counting formula

$$Z_n(g) = \sum_{\substack{\varphi^4\text{-graphs } G \\ \text{with } n \text{ vertices}}} \frac{g^n}{|\mathrm{Aut}(G)|}. \tag{1.13}$$

To realize a discretization of quantum gravity, we should be able to generate random tessellations of surfaces of fixed genus. The sum (1.13) extends over graphs of arbitrary genus, without distinction. This is why we have to resort to some more sophisticated φ^4 theory, which will distinguish between different genera. This theory is the M^4 matrix theory [7], still a $D = 0$-dimensional field theory, but with a matrix field instead of a scalar one. For technical reasons that will become clear soon, we consider a Hermitian matrix theory, which will give rise to oriented diagrams, realizing the tessellation of orientable surfaces.

The vacuum amplitude of the M^4 Hermitian matrix theory reads

$$Z(g, N) = \int e^{-N\,\mathrm{tr}((M^2/2) - gM^4/4)}\, \mathrm{d}M, \tag{1.14}$$

where the integral extends over Hermitian $N \times N$ matrices M, and the measure is the standard Haar measure over Hermitian matrices

$$\mathrm{d}M = \frac{1}{Z_0(N)} \prod_{i=1}^{N} \mathrm{d}M_{ii} \prod_{i<j} \mathrm{d}\,\mathrm{Re}\,M_{ij}\,\mathrm{d}\,\mathrm{IM}\,M_{ij} \tag{1.15}$$

normalized in such a way that $Z(0, N) = 1$. A prefactor of N has been introduced in the exponential in (1.14), for later counting purposes. As in the scalar case, we expand (1.14) in powers of g as $Z(g, N) = \sum_{n \geq 0} g^n Z_n(N)$, where

$$Z_n(N) = \frac{1}{n!} \int \left(\frac{\mathrm{tr}\,M^4}{4} \right)^n e^{-N\,\mathrm{tr}(M^2/2)}\, \mathrm{d}M. \tag{1.16}$$

We introduce also a source representation, using a fixed Hermitian matrix S':

$$Z_S(N) = \int e^{-N\,\mathrm{tr}(M^2/2) + \mathrm{tr}\,SM}\, \mathrm{d}M, \tag{1.17}$$

and we write

$$\int M_{j_1 i_1} M_{j_2 i_2} \cdots M_{j_k i_k} e^{-N \operatorname{tr}(M^2/2)} \, dM$$

$$= \frac{\partial}{\partial S_{i_1 j_1}} \frac{\partial}{\partial S_{i_2 j_2}} \cdots \frac{\partial}{\partial S_{i_k j_k}} Z_S(N) \Big|_{S=0}$$

$$= \frac{\partial}{\partial S_{i_1 j_1}} \frac{\partial}{\partial S_{i_2 j_2}} \cdots \frac{\partial}{\partial S_{i_k j_k}} e^{(1/N) \operatorname{tr}(S^2/2)} \Big|_{S=0} , \quad (1.18)$$

where we have completed the square and integrated M out. (Notice the interchange of the indices in M and S.) As before, derivatives with respect to S_{ij} should be taken by pairs, more precisely as

$$(\partial/\partial S_{ij}) e^{\operatorname{tr} S^2/(2N)} = (S_{ji}/N) e^{\operatorname{tr} S^2/(2N)};$$

the pair of derivatives must be of the form $(\partial/\partial S_{ij})(\partial/\partial S_{ji})$. We summarize this rule by saying that the propagator is

$$\frac{\partial}{\partial S_{ji}} \frac{\partial}{\partial S_{lk}} Z_S(N)|_{S=0} = \langle M_{ij} M_{kl} \rangle = \frac{1}{N} \delta_{jk} \delta_{jl}. \quad (1.19)$$

The factor $1/N$ comes from the inversion of the quadratic form $N \operatorname{tr} M^2/2$ when completing the square in (1.18). A useful pictorial representation for a matrix element M_{ij} (or equivalently of a derivative with respect to an element S_{ji}) is an oriented double link with a marked end, with the upper line oriented from and the lower line oriented to the marked end.

$$M_{ij} \leftrightarrow \left(\begin{smallmatrix} i \\ j \end{smallmatrix} \, \substack{\longrightarrow \\ \longleftarrow} \right). \quad (1.20)$$

The upper line carries the index i, and the lower line carries the index j. The propagator (1.19) is therefore represented by the continuation of the lines corresponding to the two matrix elements, which therefore have to carry the same indices.

$$\langle M_{ij} M_{ji} \rangle \leftrightarrow \left(\begin{smallmatrix} i \\ j \end{smallmatrix} \, \substack{\longrightarrow \\ \longleftarrow} \, \begin{smallmatrix} i \\ j \end{smallmatrix} \right). \quad (1.21)$$

Let us now show how to compute the integral (1.16). Each monomial $\operatorname{tr} M^4$ can be represented pictorially as

$$\sum_{i,j,k,l} M_{ij} M_{jk} M_{kl} M_{li} \leftrightarrow \sum_{i,j,k,l} \left(\begin{smallmatrix} i & & l \\ & \boxplus & \\ j & & k \end{smallmatrix} \right). \quad (1.22)$$

The integral (1.16) is simply computed by connecting the n vertices (1.22) through propagators of the form (1.21) in all possible ways, so as to build a closed graph with double lines (often called a fatgraph). This generates a collection of (nonconnected) closed graphs, each of which is weighed by a factor

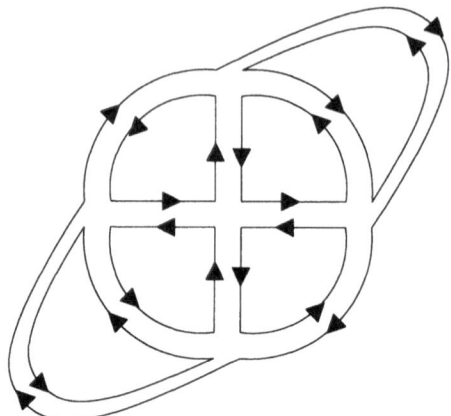

FIGURE 5.2. A sample planar (genus zero) fatgraph with $V = 5$ vertices, $E = 10$ edges, and $L = 7$ loops.

i. Ng per vertex;

ii. N^{-1} per edge (propagator);

iii. N per oriented loop, over which the running matrix index has to be summed from 1 to N.

These are nothing but the Feynman rules for matrix integration. Collecting the Feynman factors above, we see that each graph comes with a total factor (see Figure 5.2)

$$N^{V-E+L} = N^{\chi(G)} = N^{2-2h}, \qquad (1.23)$$

where we have identified the Euler characteristic of the fatgraph with V vertices, E edges, and L loops. In addition to this factor, each graph is weighted by a combinatorial factor, equal to its number of inequivalent labelings, divided by $n!$. Note that the factor $\frac{1}{4}$ accounts for the cyclic symmetry of the trace and avoids overcounting the fatgraphs. Each fatgraph G comes therefore with an overall factor $1/|\operatorname{Aut}(G)|$, the inverse of the order of its symmetry group. We finally get the general expansion for $Z(g, N)$ as a sum over fatgraphs, in the form

$$Z(g, N) = \sum_{A, h \geq 0} \sum_{\substack{\text{fatgraphs with} \\ A \text{vert. and genus } h}} \frac{N^{2-2h} g^A}{|\operatorname{Aut}(G)|}. \qquad (1.24)$$

To identify this expansion with the discrete quantum-gravitational sum (1.6), we still have to note that the dual of a fatgraph in the expansion (1.24) is nothing but a tessellation of some surface with the same genus h, through A squares. Assuming that the square tiles have all unit area, A is identified as the area of the tessellated surface. Upon identifying $\Lambda = \log g$ and $\mathcal{N} = \log N$, the expansions (1.24) and (1.6) are the same!

In conclusion, matrix models are seen to be useful combinatorial tools to generate discretized sums over random surfaces, in which both the area and the genus are fixed. A last remark is in order. The sums (1.24) and (1.13) extend over nonconnected surfaces or graphs. The sum can be reduced to connected objects by simply taking the logarithm of the vacuum amplitudes, which realizes all the necessary subtractions.

So far we have dealt only with pure gravity (no matter). In the following, we will see how matter theories can be represented through some decorations of the initial matrix model, either by adding higher-order terms to the potential (general potential) or by increasing the number of matrices one integrates over (multimatrix models).

1.5 Singularity Structure, Double Scaling Limit

The sum (1.24) is known, from continuum (string) theory, to possess some singularities at some special value of the parameter $g = g_c$. For surfaces of fixed genus h, the singularity reads

$$\log Z_h \sim (g_c - g)^{(5/2)(1-h)}. \tag{1.25}$$

When substituted back in the expansion (1.24), this suggests that we define the following double-scaling limit of the matrix model with quartic potential, namely

$$g \to g_c \quad \text{and} \quad N \to \infty, \tag{1.26}$$

while the parameter

$$x = \frac{g_c - g}{g_c} N^{4/5} \tag{1.27}$$

remains fixed. We expect in such a limit to keep track of the contributions of surfaces of all genera, in a single asymptotic expansion in the variable x:

$$F(x) = \lim \frac{1}{N^2} \log Z(g, N) = \sum_{h \geq 0} f_h x^{(5/2)(1-h)}. \tag{1.28}$$

We will actually show that the string susceptibility $u(x) \propto F''(x)$ satisfies the first Painlevé equation.

More generally, we will have to invent suitable double-scaling limits in gravity + matter theories, and to work out their solutions. The remarkable outcome is that their string susceptibilities will all turn out to obey systems of differential equations, generalizing the pure gravity result.

1.6   Reading Guide

Part A is organized as follows. In Section 2, we recall basic facts on the study of the large N one-matrix model. This corresponds to retaining only the planar (genus 0) graphs in the expansion (1.24). This study will reveal the singularity structure of the vacuum amplitude as a function of the parameter g, and its generalizations for arbitrary potentials. Section 3 will be devoted to the discrete solution of the one-matrix model for all N, which amounts to the discrete Painlevé I equation in the case of pure gravity. In Section 4 we define and take the double scaling limits of these discrete systems, and unveil the underlying integrable differential KdV flow-structure of the models. Section 5 deals with generalizations to multimatrix models, which will relate to generalized KdV flows. We conclude Part A with Section 6.

2   The One-Matrix Model: Large N Limit

2.1   Reduction to a Gas of Eigenvalues

Let us consider the integral over Hermitian matrices of size $N \times N$

$$Z(V, N) = \int e^{-N \operatorname{tr} V(M)} \, \mathrm{d}M, \qquad (2.1)$$

where $V(M)$ is some polynomial of M and $\mathrm{d}M$ denotes the Haar measure (1.15) with some arbitrary but fixed normalization.

The first step in computing such an integral is to reduce it to an integral over N variables, namely the eigenvalues of the diagonalizable matrix M. We write therefore $M = U \Lambda U^\dagger$, where U is a unitary matrix and $\Lambda = \operatorname{diag}(\lambda_1, \lambda_2, \ldots, \lambda_N)$ a real diagonal matrix, and change variables from M to (U, Λ). The Jacobian of this change of variables is best computed, after setting $U = e^{iT}$, T a Hermitian matrix, by evaluating the metric

$$\operatorname{tr}(\mathrm{d}M)^2 = \operatorname{tr}(U(\mathrm{d}\Lambda - i[\Lambda, \mathrm{d}T])^2 U^\dagger)$$

$$= \operatorname{tr}(\mathrm{d}\Lambda)^2 + \sum_{\substack{i,j=1 \\ i \neq j}}^{N} (\lambda_i - \lambda_j)^2 |\mathrm{d}T_{ij}|^2, \qquad (2.2)$$

where the independent variables are λ_i, $i = 1, 2, \ldots, N$, and T_{ij}, $1 \leq i \leq j \leq N$. The Jacobian of the change of variables is simply the square root of the determinant of the metric, namely

$$J = \frac{\partial M}{\partial \Lambda \partial U} = \Delta(\Lambda)^2, \qquad (2.3)$$

where Δ denotes the Vandermonde determinant, defined as

$$\Delta(\Lambda) = \prod_{\substack{i,j=1 \\ i<j}}^{N} (\lambda_i - \lambda_j). \tag{2.4}$$

As the potential term in (2.1) is invariant under unitary transformations, we may write

$$Z(V, N) = \iint \Delta(\Lambda)^2 e^{-N \operatorname{tr} V(\Lambda)} \, d\Lambda \, dU. \tag{2.5}$$

We see that the integration over the unitary group factors out completely, and we may simply set it to 1 by suitably defining the normalization of the Haar measure. So we are finally left with

$$Z(V, N) = \int \Delta(\Lambda)^2 e^{-N \operatorname{tr} V(\Lambda)} \, d\Lambda, \tag{2.6}$$

which is an integral over the N real variables $\lambda_1, \lambda_2, \ldots, \lambda_N$.

2.2   Quartic Potential: Large N Limit

In this subsection we start from the expression (2.6), with the particular quartic potential [7]

$$V(M) = \frac{M^2}{2} + g\frac{M^4}{4}. \tag{2.7}$$

We may now rewrite (2.6) as

$$Z(g, N) = \int e^{-N \sum_i (\lambda_i^2/2 + g\lambda_i^4/4) + \sum_{i \neq j} \log |\lambda_i - \lambda_j|} \, d\Lambda. \tag{2.8}$$

We wish now to take the large N limit of this expression. We remark that both terms in the exponential are of order N^2 as the sums extend over 1, 2, ..., N. Making a saddle-point approximation, the integral $Z(g, N) \sim \int e^{-N^2 S}$ is dominated by the minima of the action S, namely subject to $\partial S/\partial \lambda_i = 0$ for $i = 1, 2, \ldots, N$. This gives for any fixed $i = 1, 2, \ldots, N$

$$\frac{1}{N} \sum_{j \neq i} \frac{2}{\lambda_i - \lambda_j} = \lambda_i + g\lambda_i^3. \tag{2.9}$$

For large N, the distribution of eigenvalues is expected to become a continuous function

$$\rho(\lambda) = \lim_{N \to \infty} \frac{1}{N} \sum_{i=1}^{N} \delta(\lambda - \lambda_i) \tag{2.10}$$

normalized to unity $\int rho(\lambda)\,d\lambda = 1$. In this limit, (2.9) becomes

$$2\int \frac{\rho(\mu)}{\lambda-\mu}\,d\mu = \lambda + g\lambda^3, \tag{2.11}$$

where the integral must be understood as a principal part. Let us introduce the resolvent

$$\omega(z) = \int \frac{\rho(\lambda)}{\lambda - z}\,d\lambda. \tag{2.12}$$

It is related to the eigenvalue distribution ρ through the discontinuity equation across the cut of the integral (2.12), namely

$$\rho(\lambda) = \frac{1}{2\pi}\big(\omega(\lambda + i0) - \omega(\lambda - i0)\big). \tag{2.13}$$

The saddle-point equation (2.11) also takes the simple form

$$\omega(z+i0) + \omega(z-i0) = z + gz^3. \tag{2.14}$$

Instead of directly solving this equation, let us first derive an algebraic equation for $\omega(z)$. Let us multiply both sides of (2.9) by $1/(N(\lambda_i - z))$ and sum over i. We get

$$\begin{aligned}
\frac{1}{N}\sum_i \frac{\lambda_i + g\lambda_i^3}{\lambda_i - z} &= \frac{1}{N^2}\sum_{i,j,i\neq j} \frac{2}{\lambda_i - z}\frac{1}{\lambda_i - \lambda_j} \\
&= \frac{1}{N^2}\sum_{i\neq j} \frac{1}{(\lambda_i - \lambda_j)(\lambda_i - z)} - \frac{1}{(\lambda_i - \lambda_j)(\lambda_j - z)} \\
&= -\frac{1}{N^2}\sum_{i\neq j} \frac{1}{(\lambda_i - z)(\lambda_j - z)}.
\end{aligned} \tag{2.15}$$

In the large N limit, this becomes

$$\omega(z)^2 + (z + gz^3)\omega(z) + \int \rho(\lambda)\frac{\lambda + g\lambda^3 - z - gz^3}{\lambda - z}\,d\lambda = 0, \tag{2.16}$$

where we have used the definition (2.12). The last term in the quadratic equation (2.16) is actually a polynomial of degree 2 in z. Setting

$$P(z) = 4\int \rho(\lambda)\frac{\lambda + g\lambda^3 - z - gz^3}{\lambda - z}\,d\lambda, \tag{2.17}$$

we can solve (2.16) for $\omega(z)$ as

$$\omega(z) = \frac{1}{2}\Big(-z - gz^3 + \sqrt{z^2(1 + gz^2)^2 - P(z)}\Big). \tag{2.18}$$

In principle, the square root may have 6 branch points, resulting in 3 branch cuts on the real axis, hence a disconnected support for ρ. Here and in the following we will always choose the one-cut solution, corresponding to a connected support for ρ (the eigenvalues, interacting through the logarithmic term arising from the Vandermonde determinants, are supposed to fill only one minimum of the potential). This implies that two of the branch points coincide. Writing $\omega(z)$ as

$$\omega(z) = \frac{1}{2}\left(-z - gz^3 + (b_2z^2 + b_1z + b_0)\sqrt{(z - a_1)(z - a_2)}\right), \qquad (2.19)$$

where the a_i and b_i are constants to be determined. Note that with the relation (2.13), the support of the eigenvalue distribution $\rho(\lambda)$ is just $[a_1, a_2]$. As the potential $\lambda^2/2 + g\lambda^4/4$ is an even function of λ, so must be $\rho(\lambda)$; hence the support of $\rho(\lambda)$ must be symmetric with respect to the origin. Let us therefore set $a_2 = -a_1 = a$. The coefficients a, b_0, b_1, b_2 in (2.19) are easily computed by expanding $\omega(z) = -1/z + O(1/z^2)$ for large z, and identifying all the coefficients of the powers of z in the corresponding expansion of (2.19). The result simply reads, power by power of z,

$$
\begin{aligned}
z^3: & \quad b_2 = g, \\
z^2: & \quad b_1 = 0, \\
z^1: & \quad b_0 = 1 + g\frac{a^2}{2}, \\
z^0: & \quad \text{granted by parity}, \\
z^{-1}: & \quad \frac{3}{4}ga^4 + a^2 - 4 = 0.
\end{aligned}
\qquad (2.20)
$$

The last equation yields

$$a^2 = \frac{2}{3g}(\sqrt{1 + 12g} - 1), \qquad (2.21)$$

and the eigenvalue distribution reads

$$\rho(\lambda) = \frac{1}{2\pi}\left(g\lambda^2 + 1 + g\frac{a^2}{2}\right)\sqrt{a^2 - \lambda^2} \qquad (2.22)$$

for $\lambda \in [-a, a]$, with a^2 given by (2.21). Note that for $g = 0$, the distribution reduces to Wigner's semicircle law for the Gaussian matrix model $\rho(\lambda) \sim \sqrt{8 - \lambda^2}$. The saddle-point approximation to the integral (2.8) then reads

$$Z(g, N) \sim e^{-N^2 F_0(g)}, \qquad (2.23)$$

where $F_0(g)$ is the genus $h = 0$ contribution to the connected free energy

$\log Z(g, N)$ and reads

$$F_0(g) = \int \rho(\lambda) \left(\frac{\lambda^2}{2} + g\frac{\lambda^4}{4} \right) d\lambda$$
$$- \int \rho(\lambda) \int \rho(\mu) \log |\lambda - \mu| \, d\mu \, d\lambda \quad (2.24)$$

with $\rho(\lambda)$ as in (2.21)–(2.22). The integrals in (2.24) extend over the support of ρ, namely $[-a, a]$. The logarithmic term in (2.24) is easily disposed of by integrating (2.11) with respect to λ, namely

$$\frac{\lambda^2}{2} + g\frac{\lambda^4}{4} = 2 \int_0^\lambda \int \frac{\rho(\mu)}{t - \mu} \, d\mu \, dt$$
$$= 2 \int \rho(\mu)(\log |\lambda - \mu| - \log |\mu|) \, d\mu. \quad (2.25)$$

This finally yields

$$F_0(g) = F_0(0) + \frac{1}{384}(a^2 - 4)(36 - a^2) - \frac{1}{2} \log \frac{a^2}{4} \quad (2.26)$$

with a^2 given by (2.21). We get an explicit expansion in powers of g around $g = 0$ by just expanding (2.21) and substituting it in (2.26). The coefficient of g^A in this expansion is nothing but the number of planar (genus 0) diagrams with A vertices (weighted by the inverse of the order of their symmetry group).

The expression(2.21) shows the appearance of a singularity at the special value of $g = g_c = -1/12$. Indeed, the expansion of the integral (2.8) in powers of g may be analytically continued from $g > 0$, where it is well-defined, to $g < 0$, where the integration must be taken over an anti-Hermitian matrix M. This continuation, through the expansion (2.26), is valid until one reaches the singularity $g = g_c$. The singular part of the free energy when $g \to g_c$ is easily seen to behave like

$$F_0^{\text{sing}}(g) \sim (g_c - g)^{2+1/2}, \quad (2.27)$$

confirming the announced result (1.25) for genus $h = 0$. (This is actually best seen on the so-called string susceptibility, defined as $u_0(g) = d^2 F_0(g)/dg^2 \sim (g_c - g)^{1/2}$.)

Note also that the distribution of eigenvalues acquires a different behavior on the border of its support when $g \to g_c$. Indeed, one readily sees that

$$\rho(\lambda) \sim (8 - \lambda^2)^{3/2} \quad (2.28)$$

for $g = -1/12$. The generic singularity $(a^2 - \lambda)^{1/2}$ has therefore been smoothened in the critical limit.

2.3   Matter Coupling: The Hard-Dimer Problem

This subsection is a justification for generalizing the problem with quartic potential to a general potential $V(M)$, so as to include some matter theory coupled to the discretized quantum gravity. We illustrate this with the example of the hard-dimer model [12], defined as follows. On a flat square lattice, a configuration of the model consists in occupying arbitrarily the bonds of the lattice by dimers, i.e., small bonds linking two adjacent vertices, with the constraint that a given vertex of the lattice be occupied by at most one such dimer. The dimers therefore avoid each other. The corresponding partition function is obtained as in the Ising case by summing over all configurations of occupied bonds, each weighted by some Boltzmann factor, say K per occupied bond. Defining the model on a random tessellation is straightforward in the dual (matrix fatgraph Γ) formulation, and goes as follows.

Let us define the hard-dimer model on a fatgraph Γ. We simply have to change slightly the Feynman rules we derived in the pure gravity case. First of all, we need two types of edges, to distinguish between empty and occupied ones. This implies the use of two random matrices M and P, with propagators

$$\langle M_{ij} M_{kl} \rangle = \frac{1}{N} \delta_{il} \delta_{jk},$$
$$\langle P_{ij} P_{kl} \rangle = \frac{K}{N} \delta_{il} \delta_{jk}, \qquad (2.29)$$

where the factor K gives the proper weight for occupied bonds. The vertices of Γ can be of two types: empty ($\operatorname{tr} M^4/4$), or with one occupied leg ($\operatorname{tr}(PM^3)/3$). They must be both weighed by Ng, in order to reproduce the total $N^{2-2h} g^A$ factor. The corresponding fatgraphs Γ for the hard-dimer model are obtained by connecting a total of A vertices of either type, through propagators (2.29). The corresponding partition function reads

$$Z(g, N; K) = \int e^{-N \operatorname{tr}(M^2/2 + K^{-1} P^2/2 - gM^3 P/3 - gM^4/4)} \, dP \, dM, \qquad (2.30)$$

where the quadratic piece reproduces the propagators (2.29), upon inversion of the coefficients. Noticing that the dependence of the potential on P is only quadratic, we can perform the Gaussian integration over P, using

$$\int e^{-a \operatorname{tr}(P^2/2) + b \operatorname{tr}(PS)} = e^{-b^2 (\operatorname{tr} S^2)/(2a)} \, dP \qquad (2.31)$$

valid for two arbitrary constants b, $a > 0$, where we simply have completed the square and integrated out P (the normalization of the Haar measure for P is chosen in order for the result of this integral to be 1). Thanks to the relation (2.31) for $a = N/K$ and $b = Ng/3$, the partition function

(2.30) finally reduces to

$$Z(g, N; K) = \int e^{-N \operatorname{tr}(M^2/2 - gM^4/4 - 2Kg^2M^6/18)} \, dM. \qquad (2.32)$$

In conclusion, the coupling of the hard-dimer model to discretized quantum gravity is simply realized through a decoration of the matrix model for pure gravity by a higher-degree potential, whose coefficients are related to the matter parameter K. More generally, we expect higher-degree potentials to describe the coupling of more sophisticated matter theories to discrete quantum gravity. This is certainly a strong enough motivation to study the matrix model (2.8) with a general polynomial potential $V(M)$.

2.4   General Potential: Large N Limit

The analysis of Section 2.2 can be repeated step by step for a general potential $V(\Lambda)$. With the obvious modifications in the various definitions (replace $\lambda + g\lambda^3 \mapsto V'(\lambda)$), we get the quadratic equation for the resolvent

$$\omega(z)^2 + V'(z)\omega(z) + \frac{1}{4}P(z) = 0, \qquad (2.33)$$

where the polynomial $P(z)$ is defined by

$$P(z) = 4 \int \rho(\lambda) \frac{V'(\lambda) - V'(z)}{\lambda - z} \, d\lambda. \qquad (2.34)$$

This leads to the solution

$$\omega(z) = \frac{1}{2}\left(-V'(z) + \sqrt{V'(z)^2 - P(z)}\right). \qquad (2.35)$$

Let us assume that $V(\lambda)$ is an even polynomial of λ, of degree $2m$. The polynomial $V'(z)^2 - P(z)$ of degree $2(2m - 1)$ gives rise in principle to $2(2m - 1)$ branch points for $\omega(z)$, resulting in $2m - 1$ branch cuts for the distribution ρ. Demanding that ρ have a connected support, hence only one cut, amounts to imposing that the polynomial $V'(z)^2 - P(z)$ have $2m - 2$ double zeros, which fixes $P(z)$ completely, together with the fact that $\omega(z) = -1/z + O(1/z^2)$ for large z. This fixes the solution $\omega(z)$ to be

$$\omega(z) = \frac{1}{2}\left(-V'(z) + \alpha_{2m-2}(z)\sqrt{z^2 - a^2}\right), \qquad (2.36)$$

where the remaining cut is again symmetric with respect to the origin (by parity of the potential $V(\lambda)$), resulting in a support $[-a, a]$ for $\rho(\lambda)$, and where the polynomial $\alpha_{2m-2}(z)$, of degree $2m - 2$, is entirely expressible in terms of the coefficients of $V(\lambda)$ (in particular, it is found to be even, thanks to the parity of $V(\lambda)$), and a is also fixed.

The exact relation reads, in fact,

$$\omega(z) = \frac{1}{2\pi} \int_{-a}^{a} \frac{V'(\lambda)}{\lambda - z} \sqrt{\frac{z^2 - a^2}{a^2 - \lambda^2}} \, d\lambda, \qquad (2.37)$$

where a is fixed through the equation

$$\frac{1}{2\pi} \int \frac{\lambda V'(\lambda)}{\sqrt{a^2 - \lambda^2}} \, d\lambda = 1 \qquad (2.38)$$

obtained by identification of the z^{-1} term in $\omega(z) \sim -1/z$ for large z.

With this solution, we finally get the saddle-point expression

$$Z(V, N) \sim e^{-N^2 F_0(V)} \qquad (2.39)$$

in terms of the genus 0 free energy $F_0(V)$, still depending on the coefficients of $V(\lambda)$. We wish now to investigate the singularity structure of this function.

It is by now clear that the singularities of $F_0(V)$, as a function of the coefficients of V, will occur whenever the eigenvalue density

$$\rho(\lambda) = \frac{1}{2\pi} \alpha_{2m-2}(\lambda) \sqrt{a^2 - \lambda^2} \qquad (2.40)$$

displays a smoothened singular behavior at the border of its support, namely when $\rho(\lambda) \sim (a^2 - \lambda^2)^{k+1/2}$ for some $k > 0$. This can be realized by tuning the coefficients of $V(\lambda)$ in such a way as to ensure that some zeros of the polynomial $\alpha_{2m-2}(\lambda)$ coalesce with the borders of the support $\pm a$. By a simple counting of the number of equations and parameters to be fixed, we find that the maximal singularity is obtained by tuning all the coefficients of $V(\lambda)$ so that

$$\rho(\lambda) \sim (a^2 - \lambda^2)^{m-1/2} \qquad (2.41)$$

at the borders of the support of ρ. This is called a multicritical limit. We may introduce an overall factor g for all the coefficients of V (write $V(\lambda) = \sum_i g_i \lambda^{2i}/(2i)$, and take $g_i = g_i^* g^i$), supposed to take their critical relative values (g_i^*), so that the limit $g \to g_c$, g_c some critical value of the parameter g, corresponds to the above tuning. It is then a simple exercise to show that the free energy $F_0(V)$, as a function of g, has the leading singularity

$$F_0(V) \sim (g_c - g)^{2+1/m}. \qquad (2.42)$$

(Again, it is simpler to show that $D^2 F_0(V))/dg^2 \sim (g_c - g)^{1/m}$.)

The singular behavior (2.42) generalizes that of the pure gravity case (2.27) and is characteristic of the coupling of multicritical matter to quantum gravity. This multicriticality will become clear in Section 3.3.

3 The One-Matrix Model: Exact Solution

3.1 Orthogonal Polynomials: The General Method

We now turn to the evaluation of the integral

$$Z(V, N) = \int \Delta(\Lambda)^2 e^{-N \operatorname{tr} V(\Lambda)} \, d\Lambda. \tag{3.1}$$

This is done using the orthogonal polynomial technique. The idea, due to Bessis [8, 9], is to interpret the integral (3.1) as the scalar product of the two Vandermonde determinants with respect to a measure including the exponential term. One therefore wants to represent these determinants in a way that makes the scalar product factorize over the single eigenvalues.

For that purpose, first introduce a set of monic polynomials $p_n(\lambda)$ of degree n in λ (i.e., with $p_n(\lambda) = \lambda^n + O(\lambda^{n-1})$), which are orthogonal with respect to the measure $e^{-NV(\lambda)} \, d\lambda$ over the real numbers, namely

$$(p_n, p_m) = \int_{-\infty}^{\infty} e^{-NV(\lambda)} p_n(\lambda) p_m(\lambda) \, d\lambda = h_n \delta_{n,m}, \tag{3.2}$$

where h_n is a normalization entirely fixed by the measure.

In a second step, we write the Vandermonde determinant as a determinant, namely

$$\Delta(\Lambda) = |1, \lambda, \lambda^2, \ldots, \lambda^{N-2}, \lambda^{N-1}|, \tag{3.3}$$

where the dummy variable λ stands for a column, with λ_i in the ith row. A determinant is unchanged if we replace one of its columns with itself plus a linear combination of the others. As the polynomials p_m are monic, let us replace the last column λ^{N-1} in (3.3) by the column $p_{N-1}(\lambda) = \lambda^{N-1}+$ a linear combination of the other columns. We can repeat this for the column λ^{N-2} and so on, and end up with a representation of the Vandermonde determinant in terms of the orthogonal polynomials only, namely

$$\Delta(\Lambda) = |p_0(\lambda), p_1(\lambda), \ldots, p_{N-1}(\lambda)|. \tag{3.4}$$

The third step consists in evaluating the integral (3.1), by expanding both Vandermonde determinants over the permutations of the symmetric group S_N, namely

$$
\begin{aligned}
Z(V, N) &= \int \sum_{\sigma, \tau \in S_n} \operatorname{sgn}(\sigma\tau) \prod_{i=1}^{N} p_{i-1}(\lambda_{\sigma(i)}) \prod_{j=1}^{N} p_{j-1}(\lambda_{\sigma(j)}) \prod_i e^{-NV(\lambda_i)} \, d\lambda_i \\
&= \sum_{\sigma, \tau \in S_n} \delta_{\sigma, \tau} \prod_{i=1}^{N} h_{i-1} \\
&= N! \prod_{i=0}^{N-1} h_i, \tag{3.5}
\end{aligned}
$$

where, by virtue of the orthogonality relation (3.2), only the p's with $i = j$ and the same variable $\lambda_{\sigma(i)} = \lambda_{\tau(j)}$, i.e., $\sigma = \tau$, have contributed to the sum.

The connected partition function finally reads

$$\log Z(V, N) = \sum_{i=0}^{N-1} \log h_i \tag{3.6}$$

up to an irrelevant additive constant, depending on N only, and which can be absorbed in a redefinition of the Haar measure.

3.2   The Quartic Potential: Discrete Painlevé I

In this subsection we specialize ourselves to the quartic potential case $V(x) = x^2/2 + gx^4/4$ and solve the one-matrix model explicitly [15]. The idea of the solution is that there are relations among the h_i that eventually must enable us to express the resulting free energy (3.6) in terms of g and N. These relations are best exploited by studying two operators acting on the p_m's, namely the multiplication operator by λ and the derivative with respect to λ, respectively denoted by Q and P. They read, as matrices acting on the basis of p's,

$$Q : \quad \lambda p_n(\lambda) = \sum_m Q_{nm} p_m(\lambda),$$

$$P : \quad \frac{\mathrm{d}}{\mathrm{d}\lambda} p_n(\lambda) = \sum_m P_{nm} p_m(\lambda). \tag{3.7}$$

In fact, it is clear that $Q_{nm} = 0$ if $m > n+1$, and that $Q_{n,m+1} = 1$, as the polynomials are monic. We also have $P_{nm} = 0$ if $m \geq n$, and $P_{n,n-1} = n$, as the polynomials are monic.

To evaluate the matrix elements of Q, let us calculate the scalar product

$$(p_r, Qp_n) = Q_{nr}h_r = (Qp_r, p_n) = Q_{rn}h_n, \tag{3.8}$$

where the multiplication by λ is applied to either polynomial. Due to the vanishing of Q_{nm} for $m > n+1$, we conclude that $Q_{nm} = 0$ if $|m - n| > 1$. The remaining elements are $Q_{n,n+1} = 1$ already computed, $Q_{n,n-1} = h_n/h_{n-1}$ thanks to (3.8), and we finally compute

$$(p_n, Qp_n) = \int_{-\infty}^{\infty} e^{-NV(\lambda)} \lambda p_n(\lambda)^2 \, \mathrm{d}\lambda = 0 \tag{3.9}$$

by parity of $V(\lambda) = \lambda^2/2 + g\lambda^4/4$. Hence the Q operator reduces to the simple recursion relation

$$(Qp_n)(\lambda) = \lambda p_n(\lambda) = p_{n+1}(\lambda) + r_n p_{n-1}(\lambda), \tag{3.10}$$

where we have introduced the sequence

$$r_n = \frac{h_n}{h_{n-1}}. \tag{3.11}$$

Note that the p's are entirely specified by the r_n, $n = 0, 1, \ldots, N - 1$, as we have the two initial data $p_0(\lambda) = 1$ and $p_1(\lambda) = \lambda$.

Let us now evaluate the matrix element $P_{n,n-1} = n$ in two different ways:

$$
\begin{aligned}
(p_{n-1}, Pp_n) = nh_{n-1} &= \int_{-\infty}^{\infty} e^{-NV(\lambda)} p_{n-1}(\lambda) \frac{\mathrm{d}}{\mathrm{d}\lambda} p_n(\lambda) \, \mathrm{d}\lambda \\
&= -\int_{-\infty}^{\infty} p_n(\lambda) \frac{\mathrm{d}}{\mathrm{d}\lambda} \left(e^{-NV(\lambda)} p_{n-1}(\lambda) \right) \\
&= N(p_{n-1}, V'(Q)p_n) - (p_n, Pp_{n-1}), \tag{3.12}
\end{aligned}
$$

where we have performed an integration by parts. The last term in the rhs of (3.12) vanishes, as $P_{n-1,n} = 0$, and we are left with

$$nh_{n-1} = N(p_{n-1}, (Q + gQ^3)p_n). \tag{3.13}$$

We must now use (3.10) repeatedly to evaluate the rhs of (3.13), with the result

$$\frac{n}{N} h_{n-1} = h_n + gh_{n-1} r_n (r_{n+1} + r_n + r_{n-1}), \tag{3.14}$$

which can be recast into

$$gn = gr_n \left(1 + g(r_{n+1} + r_n + r_{n-1}) \right), \tag{3.15}$$

which is nothing but the discrete Painlevé I equation. This equation is the main result of this section, and its solutions r_n make it possible to rewrite the free energy (3.6)

$$\log Z(g, N) = \sum_{i=0}^{N-1} (N - i) \log r_i \tag{3.16}$$

up to an irrelevant additive constant depending on N only, which we can dispose of by modifying the normalization of the Haar measure (1.15).

Let us now compute the large N limit of (3.16), using the equation (3.15). The continuum version of gr_n may be viewed as a function $r(z)$ of the scaling variable $z = n/N$, continuously describing the interval $[0, 1]$. Equation (3.15) reduces to

$$gz = r(z) + 3r(z)^2 \equiv W(r(z)), \tag{3.17}$$

whereas the genus 0 free energy is expressed as

$$F_0(g) = \lim_{N \to \infty} \frac{1}{N^2} \log Z(g, N)$$

$$= \lim_{N \to \infty} \frac{1}{N} \sum_{i=0}^{N} \left(1 - \frac{i}{N}\right) \log r_i$$

$$= \int_0^1 (1 - z) \log r(z) \, dz. \qquad (3.18)$$

Let r_c be a critical point of $W(r)$, i.e., a point such that $dW/dr(r_c) = 0$ (hence $r_c = -\frac{1}{6}$). Then, setting $g_c = W(r_c) = -1/12$, we can subtract the equation $g_c = r_c(1 + 3r_c)$ from (3.17) and get

$$gz - g_c = (r(z) - r_c)(1 + 3(r_c + r(z))) = 3(r_c - r(z))^2. \qquad (3.19)$$

Hence $r(z) = r_c(1 - \sqrt{(g_c - gz)/g_c})$, which we substitute in (3.18). With the change of variable $y = (g_c - gz)/g_c$ in the integration, and the definition

$$x = \frac{g_c - g}{g_c}, \qquad (3.20)$$

we finally get

$$F_0(g) = \int_x^1 \frac{y - x}{1 - x} \log\left(r_c(1 - \sqrt{y})\right) dy$$

$$\sim -r_c \int_x^1 (y - x)\sqrt{y} \, dy \sim x^{5/2} \qquad (3.21)$$

for the leading singular behavior of $F_0(g)$ when $g \to g_c$. This reproduces the large N result (2.27) in a totally different way.

The remarkable thing about the above method is that it can be applied to include higher-genus contributions to the free energy. This will be the subject of Section 4.

3.3 General Potentials

We now turn to the case of a general even polynomial potential $V(\lambda)$ of degree $2m$, with the form

$$V(\lambda) = \sum_{p=1}^{m} g_p \frac{\lambda^{2p}}{2p}. \qquad (3.22)$$

The corresponding partition function $Z(V, N)$ is expressed in terms of orthogonal polynomials with respect to the measure $e^{-NV(\lambda)} d\lambda$, through

FIGURE 5.3. A typical term contributing to $(p_n, Q^{2p-1}p_{n-1})$ is represented as a staircase walk of $2p-1$ steps starting at height n and ending at height $n-1$. A step up corresponds to taking the first term in (3.23), and receives the weight 1. A step down corresponds to the second term in (3.23), and receives the weight r_j, where j is the height before the step is made. The total contribution is just the product of these weights along the path. The present term reads $r_{n+1}r_n^2 r_{n-1}r_{n+1}r_n$.

(3.6). We define the operators P and Q as in Section 3.2, and we still have

$$(Qp)_n = p_{n+1} + r_n p_{n-1} \qquad (3.23)$$

with $r_n = h_n/h_{n-1}$. Indeed, the derivation of this recursion relation relies only on the parity of $V(\lambda)$.

The main relation (3.12) reads

$$n = N \frac{1}{h_{n-1}}(p_n, V'(Q)p_{n-1}). \qquad (3.24)$$

Let us write the contribution of a monomial Q^{2p-1} to (3.24). We must use iteratively the recursion (3.23) to go from p_{n-1} to p_n. A simple way of representing this, depicted in Figure 5.3, is to think of each contribution to $(p_n, Q^{2p-1}p_{n-1})$ as attached to a staircase walk of $2p-1$ steps from height n to height $n-1$, with the following weight w_i at each step i:

i. $w_i = 1$ for a step up,

ii. $w_i = r_j$ for a step down, taken from height j,

and the total contribution reads $\prod_{i=1}^{2p-1} w_i$. There are $\binom{2p-1}{p}$ such paths, with p steps up and $p-1$ steps down.

Hence we can write

$$(p_n, Q^{2p-1}p_{n-1}) = \sum_{\substack{\text{staircase} \\ \text{paths}}} \prod_{i=1}^{2p-1} w_i. \qquad (3.25)$$

With the general even potential (3.22), we have to add up all contributions (3.25) of the monomials of V'.

In the large N limit, all the r_n's involved in (3.25), whose indices range from $n-p+1$ to $n+p-1$, with p fixed, can be approximated by the same function $r(z)$, where $z = n/N$. The main relation (3.24) then becomes

$$z = \sum_{p=1}^m g_p \binom{2p-1}{p} r(z)^p \equiv W(r(z)). \qquad (3.26)$$

Note that the polynomial $W(r)$ is also given by the contour integral

$$W(x) = \frac{1}{2i\pi} \oint V'\left(z + \frac{x}{z}\right) dz, \tag{3.27}$$

for any contour around the origin. Let r_c be an $(m-1)$-multicritical point of $W(r)$, namely a point such that $dW/dr(r_c) = d^2W/dr^2(r_c) = \cdots = d^{m-1}W/dr^{m-1}(r_c) = 0$. This is obtained by tuning all the couplings g_p to some particular values

$$g_p = g^p g_{p,c} \tag{3.28}$$

up to an overall free parameter g (indeed, the multicriticality condition consists of $m-1$ algebraic equations, which fix r_c and the g_p's up to an overall scale g, with $g_{1,c} = 1$). Redefining $r(z) \mapsto gr(z)$, (3.26) takes the form $gz = W_c(r)$. Let us set $g_c = W_c(r_c)$ and subtract this equation from (3.26). We get

$$gz - g_c = W_c(r(z)) - W_c(r_c) = (r(z) - r_c)^m \frac{1}{m!} \frac{d^m W}{dr^m}(r_c)$$
$$= \binom{2m-1}{m}(r(z) - r_c)^m. \tag{3.29}$$

Hence $r(z) = r_c\left(1 - \alpha\left((g_c - gz)/g_c\right)^{1/m}\right)$. When substituted in (3.18), this leads to the following singular behavior for the free energy of the $(m-1)$-multicritical potential, after using the same change of variables as in Section 3.2, and with the definition (3.20):

$$F_0(V) = \int_x^1 \frac{y-x}{1-x} \log r_c(1 - \alpha y^{1/m}) \, dy \sim x^{2+1/m}. \tag{3.30}$$

This reproduces the result (2.42) and gives another viewpoint on the multicriticality of the corresponding limit.

4   The Double-Scaling Limit

4.1   Quartic Potential: Painlevé I

Let us go back to the case $V(\lambda) = \lambda^2/2 + g\lambda^4/4$. Introducing $\varepsilon = 1/N$, $z = n/N$, and $r(z) = gr_n$, $r(z + p\varepsilon) = gr_{n+p}$, $p = \pm 1$, we can rewrite the discrete Painlevé equation in the form

$$gz = r(z)\left(1 + r(z+\varepsilon) + r(z) + r(z-\varepsilon)\right) \tag{4.1}$$

and expand it in powers of ε:

$$gz = r(z)\left(1 + 3r(z)\right) + \varepsilon^2 r(z)r''(z) + O(\varepsilon^4). \tag{4.2}$$

Subtracting from this the equation $g_c = W(r_c) = r_c(1 + 3r_c)$, we get

$$gz - g_c = 3(r_c - r(z))^2 + \varepsilon^2 r(z)r''(z) + O(\varepsilon^4). \tag{4.3}$$

Introducing a small parameter a (with the dimension of 1/length in the original physical problem), governing the approach to criticality, we make the change of variables and functions

$$gz = g_c(1 - a^2 y), \quad r(z) = r_c(1 - au(y)). \tag{4.4}$$

This incorporates the large N limit result of Sections 2.2 and 3.2, through the a dependence. The first term in (4.3) is of order a^2. The second term includes the combination $(\varepsilon d/dz)^2 = ((g/g_c)\varepsilon a^{-2} d/dy)^2$. To balance the first term in (4.3), this must be of order a (as it acts on $r(z) \sim au(y)$). Hence ε^2 must be of order $a^{1+2\cdot2} = a^5$. Setting

$$\varepsilon = a^{5/2} = \frac{1}{N} \tag{4.5}$$

we can take the $N \to \infty$ (or $a \to 0$) limit of (4.3), which, after appropriate rescaling of $u(y)$ is nothing but the Painlevé I equation [15]

$$y = u(y)^2 - \frac{1}{3}u''(y). \tag{4.6}$$

Note that we have taken a double limit, in which we have simultaneously sent $N \to \infty$ and $g \to g_c$ ($y = 1$ when $z = 1$; hence $(g_c - g)/g_c \sim a^2$), with

$$x \equiv N^{4/5}\frac{g_c - g}{g_c} \text{ fixed.} \tag{4.7}$$

This is called the double-scaling limit of the matrix model for pure gravity.

The solution $u(y)$ to (4.6) determines the free energy through equation (3.16):

$$F(x) = \lim_{a \to 0} \frac{1}{N^2} \log Z(g, N)$$

$$= \lim_{a \to 0} a^{-5} \int_0^1 (1 - z) \log r(z) \, dz$$

$$= \lim_{a \to 0} \int_{a^{-2}}^{x} (y - x)u(y)dy, \tag{4.8}$$

where we have performed the change of variables and functions (4.4) in the integral and identified the upper bound with $x = (g_c - g)/g_c a^{-2}$ of (4.7). Taking two derivatives with respect to x, we finally get

$$F''(x) = -u(x). \tag{4.9}$$

The function $u(y)$, subject to Painlevé I, is therefore directly interpreted as the string susceptibility of the pure gravity model. We immediately recover the large N behavior $u(x) \sim x^{1/2}$ (cf. (2.27)) from (4.6).

The function $u(x)$ with the behavior $u(x) \sim x^{1/2}$ when $x \to \infty$ solves the pure gravity problem in the following sense. The Painlevé equation can be used as a recursion relation for the coefficients u_h in the asymptotic expansion

$$u(x) = \sum_{h \geq 0} u_h x^{-2+(5/2)(1-h)} \tag{4.10}$$

with the normalization $u_0 = 1$, and u_h (u_h is strictly negative for $h \geq 1$) is the string susceptibility for pure gravity on surfaces of genus h, easily identified as

$$u_h = 5(1-h)\frac{5(1-h)/2 - 1}{2} f_h,$$

for $h \neq 1$, with f_h as in (1.28). In genus 1, the free energy is integrated as $F_1(x) = f_1 \log x$, with $f_1 = -u_1$. The asymptotic solution (4.10) is not convergent; in fact, $|u_h| \sim (2h)!$ for large h. Hence the series is not Borel summable, and the function $u(x)$ is not defined uniquely by the expansion (4.10). Fortunately, we read from (4.10) the perturbative answer to our original problem. To go beyond would require a detailed analysis of the (nonperturbative) singularities of the solution, which is still missing to this day.[1]

4.2 Canonical Commutation Relations: The P, Q Formalism

The idea behind the canonical commutation relations is very simple [16]. We have seen already how to define two operators P and Q within the one-matrix model, representing respectively the derivation with respect to and the multiplication by an eigenvalue. By definition, these operators satisfy the canonical commutation property

$$[P, Q] = 1. \tag{4.11}$$

In this subsection we reconsider the double-scaling limit of Section 4.1 and show that the Painlevé I equation can be recovered from the double scaling limit of the canonical commutation relation (4.11). In particular, we will establish that Q becomes a second-order differential operator, whereas P becomes a third-order differential operator. This will be very useful for later generalizations.

---

[1] Some behaviors can, however, be ruled out: The movable pole singularities of (4.6) can be shown to be unphysical, i.e., incompatible with other requirements from the matrix model, the so-called loop equations.

We start from the definition (3.7) for the Q and P operators in the one-matrix model with quartic potential. Let us rescale the orthogonal polynomials p_n to make them orthonormal, namely, set $\tilde{p}_n = p_n/\sqrt{h_n}$, so that (3.10) becomes more symmetric,

$$(Q\tilde{p})_n = \lambda \tilde{p}_n = \sqrt{r_{n+1}}\tilde{p}_{n+1} + \sqrt{r_n}\tilde{p}_{n-1}, \tag{4.12}$$

or equivalently,

$$Q_{n,m} = (\tilde{p}_m, Q\tilde{p}_n) = \sqrt{r_{n+1}}\delta_{m,n+1} + \sqrt{r_n}\delta_{m,n-1}. \tag{4.13}$$

Let us now take the large N limit. Setting $\varepsilon = 1/N$ as before, we note that the shift operator $\delta_{m,n+1}$ acting on sequences (a_m) can be generated as $e^{\varepsilon d/dz}$, acting on the continuum limit of (a_m), i.e., a function $a(z) = a_m$, for $z = m/N$. Indeed, one just has to write

$$\sum_m \delta_{m,n+1}a_m = a_{n+1} = e^{d/dn}a_n \simeq e^{\varepsilon d/dz}a(z). \tag{4.14}$$

This permits us to rewrite Q as

$$\begin{aligned}
Q &\simeq \sqrt{r(z)}(e^{\varepsilon d/dz} + e^{-\varepsilon d/dz}) \\
&= \sqrt{r_c(1 - au(y))}\left(2 + \varepsilon^2 \frac{d^2}{dz^2} + O(\varepsilon^4)\right) \\
&= \text{const} + \sqrt{-r_c}\left(au - \left(\varepsilon \frac{d}{dz}\right)^2 + O(\varepsilon^4)\right).
\end{aligned} \tag{4.15}$$

The two terms on the rhs of (4.15) are of the same order a, provided that $(\varepsilon d/dz)^2 = (\varepsilon a^{-2}d/dy)^2$ is of order a, and we recover the double-scaling condition $\varepsilon = a^{5/2}$. Retaining only the coefficient of a, we find that

$$Q \to \frac{d^2}{dy^2} - u(y) \tag{4.16}$$

in the double-scaling limit. This limit is a differential operator, acting on functions of the rescaled variable y.

Let us now turn to P. It will be useful to change slightly the definition of the operator P, in the following way:

$$\begin{aligned}
P_{n,m} &= \int_{-\infty}^{\infty} \tilde{p}_m(\lambda)e^{-NV(\lambda)/2}\frac{d}{d\lambda}e^{-NV(\lambda)/2}\tilde{p}_n(\lambda)\,d\lambda \\
&= -\frac{N}{2}(\tilde{p}_m, V'(Q)\tilde{p}_n) + (\tilde{p}_m, \tilde{p}_n') \\
&= -\frac{N}{2}V'(Q)_{n,m} + A_{n,m},
\end{aligned} \tag{4.17}$$

where A is a lower triangular matrix $A_{nm} = 0$ if $n \leq m$. Upon an integration by parts we may as well write

$$P_{n,m} = -\frac{N}{2}V'(Q)_{n,m} + A_{n,m}$$

$$= \frac{N}{2}V'(Q)_{n,m} - A^t_{n,m}, \tag{4.18}$$

where the matrix A^t is upper triangular. Equation (4.18) permits us to compute the matrix elements of P in terms of those of Q only, by using the first equation when $n \leq m$ ($A_{nm} = 0$) and the second one when $n \geq m$ ($A^t_{nm} = 0$). This can be summarized by the following relations between operators:

$$P = \frac{1}{2}(A - A^t), \quad A + A^t = NV'(Q). \tag{4.19}$$

A precise study of the continuum limit enables one to show that

$$P = \text{const} + a\left(\frac{d^3}{dy^3} - \frac{3}{2}u(y)\frac{d}{dy} + \frac{3}{4}u'(y)\right). \tag{4.20}$$

Hence the operator P goes over to a third-order differential operator

$$P \to \frac{d^3}{dy^3} - \frac{3}{4}\left\{u(y), \frac{d}{dy}\right\}, \tag{4.21}$$

where the symbol $\{a, b\} = ab + ba$ stands for the anticommutator.

Substituting the limits (4.16) and (4.20) into the canonical commutation relation (4.11), we immediately get, writing $d \equiv d/dy$ and using the commutation property $[d^n, f(y)] = \sum_{p=1}^{n} \binom{n}{p} d^p f/dy^p d^{n-p}$ (Leibniz rule),

$$[P, Q] = \left[d^3 - \frac{3}{4}\{u, d\}, d^2 - u\right] = -\frac{1}{4}u''' + \frac{3}{4}uu' = 1, \tag{4.22}$$

which is simply, up to a rescaling $y \mapsto y/4$, the first derivative of the Painlevé I equation (4.6).

This gives a natural representation of the solution of the pure gravity model in terms of differential operators. Note that the derivation of (4.20) is not necessary to solve $[P, Q] = 1$ once Q is known to be of the form (4.16) and P is known to be a third-order differential operator. Indeed, the equation $[P, Q] = 1$ fixes entirely the coefficients of P in terms of those of Q, by setting the coefficients of d^2 and d to zero in $[P, Q]$, and we recover (4.21). This remark will be extremely useful in further generalizations.

4.3 General Potential

To derive the double scaling of the one-matrix model with multicritical polynomial potential, we can repeat the successive steps of Section 4.1, namely, express the main equation (3.24) in polynomial terms of the various $r_{n+p} = r(z + p\varepsilon)$, $p = 0, \pm1, \pm2, \ldots$, and expand it in powers of ε. Fixing an $(m-1)$-multicritical point r_c of $W(r)$, we try then to balance the term $(r_c - r)^m$ with some power of ε, which will result in a double-scaling condition. This is, however, tedious, though straightforward in any finite example.

We choose a more elegant route by studying directly the limits of the P and Q operators. The discussion of Section 4.2 for Q did not rely on the particular form of the potential V. We assume the potential is $(m-1)$-multicritical, namely, it depends only on an overall coupling g, with critical value g_c, at which $gz - g_c \sim (r_c - r)^m$. By exactly the same argument as in Section 4.1, we set

$$gz = g_c(1 - a^2y), \quad r(z) = r_c\left(1 - a^{2/m}u(y)\right) \tag{4.23}$$

for a a small parameter governing the approach to criticality $((g_c - g)/g_c \sim a^2)$. Then it is easy to see that the recursion (4.12) is still valid, and leads to

$$\begin{aligned}
Q_{n,m} &= \sqrt{r_{n+1}}\delta_{m,n+1} + \sqrt{r_n}\delta_{m,n-1} \\
&= \sqrt{r_c(1 - a^{2/m}u(y))}\left(e^{\varepsilon d/dz} + e^{-\varepsilon d/dz}\right) \\
&\to \text{const} + \sqrt{r_c}a^{2/m}(d^2 - u) + \cdots
\end{aligned} \tag{4.24}$$

up to higher orders in a, provided that we choose ε in such a way that $(\varepsilon d/dz)^2 \sim (\varepsilon a^{-2}d)$, where $d = d/dy$, is of order $a^{2/m}$. This fixes the scale of ε to be

$$\varepsilon = \frac{1}{N} = a^{(2m+1)/m}. \tag{4.25}$$

This defines the double-scaling limit for the $(m-1)$-multicritical one-matrix model, namely $N \to \infty$ and $g \to g_c$ simultaneously, with

$$x = \frac{g_c - g}{g_c}N^{2m/(2m+1)} \tag{4.26}$$

fixed. In this limit, Q still goes over to a second-order differential operator,

$$Q \mapsto d^2 - u(y). \tag{4.27}$$

The operator P is entirely expressible in terms of Q through (4.18), which is valid for an arbitrary even potential V. In particular, as it is expressed polynomially in terms of Q, it has a finite range, namely $P_{n,m} = 0$ if $|n - m| > B$, B some uniform bound, independent of N (B depends only on the degree of V). This bound ensures that P goes over in the double scaling limit to a differential operator of finite degree p, of the form

$$P \to d^p + v_2 d^{p-2} + v_3 d^{p-3} + \cdots + v_p. \tag{4.28}$$

We must finally write the canonical commutation relation (4.11) $[P, Q] = 1$. To further fix P, we know that we should recover

$$u(y) = y^{1/m} \qquad (4.29)$$

from the second derivative of equation (3.30) in the spherical limit, namely $[u(y)^m]' = 1$ in the differentiated form, which should be reproduced by the relation $[P, Q] = 1$. This further fixes the degree of P to be $2m - 1$ (we use a homogeneity argument, by assigning a weight 1 to the derivative d, hence 2 to $u(y)$, for Q to be homogeneous: The total degree of P is fixed by saying that $[P, Q] = 1$ amounts to $[u^m]' = 1$), and all the coefficients of P are simply expressible in terms of u, by setting the coefficients of d^{2m-2}, d^{2m-3}, ..., d to zero in $[P, Q]$. The last d^0 equation (setting the constant term equal to 1) will be a nonlinear differential equation for $u(y)$, generalizing the differentiated version (4.22) of Painlevé I.

To give the solution in compact form, we need to define the square root of the operator Q. Let

$$L = d + \sum_{j=1}^{\infty} l_j(y) d^{-j} \qquad (4.30)$$

be a pseudodifferential operator of degree 1, namely a formal infinite series in d^{-1} with functional coefficients $l_j(y)$. The operator $d^{-j} = (d^{-1})^j$ commutes with d, and its commutator with functions is computed through the analytic continuation of the Leibniz rule

$$[d^{-j}, f(y)] = \sum_{p \geq 1} (-1)^p \binom{p + j - 1}{p} \frac{d^p f}{dy^p} d^{-j-p}.$$

With these definitions, there exists a unique (up to some irrelevant constants) such operator L such that $L^2 = Q = d^2 - u$. Indeed, this equation can be solved order by order in $l_j(y)$, as it takes a triangular form. Moreover, the operator Q satisfies a parity property (a remnant of the parity property of the potential V), namely that it is invariant under the transformation $*$, defined by $d^* = -d$ for the derivative, $f^* = f$ for functions, and $(AB)^* = B^* A^*$ for any two operators. Therefore, as $Q^* = Q$, its square root L satisfies $L^* = -L$. This permits us to write $L = d + \sum_{j \geq 1} \{l_{2j-1}, d^{-2j+1}\}$ as a manifestly odd operator under $*$. For the first orders of L, we obtain

$$L = Q^{1/2} = d - \frac{1}{4}\{u, d^{-1}\} + \cdots. \qquad (4.31)$$

Any pseudodifferential operator R of degree r can be truncated to its differential part R_+, in which all the negative powers of d are removed. We also denote by $R_- = R - R_+$ the negative part of R. With these definitions, the expression of the operator P in terms of Q for (4.11) to hold is simply

$$P = L_+^{2m-1} = Q_+^{m-1/2}, \qquad (4.32)$$

indeed, expressing the coefficients of P in terms of those of Q. These are polynomials of u, u', u'', ... Note that this expression is manifestly odd under $*$, in agreement with the fact that $[P,Q]^* = [Q^*, P^*] = [-P^*, Q] = [P,Q] = 1$, as $1^* = 1$. We are left with the last equation to be satisfied, namely

$$[Q_+^{m-1/2}, Q] = 1, \tag{4.33}$$

which reduces to an ordinary differential equation for u, generalizing Painlevé I (4.6).

Let us write this equation in the case of the bicritical hard-dimer problem ($m = 3$). We obtain

$$P = (d^2 - u)_+^{5/2} = d^5 - \frac{5}{4}\{u, d^3\} + \frac{5}{16}\{3u^2 + u'', d\}, \tag{4.34}$$

and the canonical commutation relation (4.33) reduces to

$$y = u^3 - uu'' - \frac{1}{2}(u')^2 + \frac{1}{10}u'''' \tag{4.35}$$

after one integration over y and a rescaling $y \mapsto -5y/8$. The equation (4.35) therefore governs the perturbative expansion of the hard-dimer model coupled to quantum gravity in the double-scaling limit.

More generally, equation (4.33) can be recast in terms of the residue of the operator $Q^{m-1/2}$, namely the coefficient of d^{-1} in its formal expansion in powers of d. Denoting by $\operatorname{Res} R$ the residue of the pseudodifferential operator R, and using the fact that $Q^{m-1/2}$ commutes with Q, we rewrite

$$
\begin{aligned}
[Q_+^{m-1/2}, Q] &= [Q, Q_-^{m-1/2}] \\
&= \left[d^2 - u, \frac{1}{2}\{\operatorname{Res}(Q^{m-1/2}), d^{-1}\} + \cdots\right] \\
&= 2[\operatorname{Res}(Q^{m-1/2})]' = 1, \tag{4.36}
\end{aligned}
$$

which permits us to express the differential equation (4.33) in integrated form,

$$4R_m[u, u', u'', \ldots] \equiv 2\operatorname{Res}(Q^{m-1/2}) = y. \tag{4.37}$$

The polynomials $R_m[u]$ (we omit writing the dependence on higher derivatives of u, but it is there) are subject to the recursion [17]

$$R_{m+1}[u]' = \frac{1}{4}R_m[u]''' - uR_m[u]' - \frac{1}{2}u'R_m[u], \tag{4.38}$$

easily obtained by writing $Q^{m+1/2} = QQ^{m-1/2} = Q^{m-1/2}Q$. Equation (4.37) governs the perturbative solution to the $(m-1)$-multicritical matter model coupled to quantum gravity.

In fact, we could have considered the more general solution to $[P, Q] = 1$ made of a linear combination of operators $Q_+^{j-1/2}$ for $j = 1, 2, \ldots, m$ instead of taking only the term $j = m$ (this linear combination is the most general degree-$(2m - 1)$ differential operator P such that $[P, Q] = 1$. The main equation then becomes

$$\left[\sum_{j=1}^{m} \mu_j Q_+^{j-1/2}, Q \right] = 1, \tag{4.39}$$

and it depends on a collection of parameters μ_j of maximal dimension.

4.4   The KdV Hierarchy

More generally, the complete set of equations including all possible multi-critical double-scaling limits of the one-matrix model can be summarized into the single equation (up to some rescalings)

$$y = \sum_{m=2}^{\infty} \mu_m R_m[u] \tag{4.40}$$

with $\mu_m = \delta_{m,k}$ for the $(k - 1)$-multicritical model. With the above-mentioned grading of d (degree 1), y (degree -1), and u (degree 2), and as $R_m[u] \sim u^m$ has degree $2m$, the constant μ_m must have the degree $-(2m + 1)$. Remarkably, a general solution u of (4.40) satisfies the evolution equations of the so-called KdV hierarchy [17], as a function of the variables $t_m = -2\mu_m/(2m + 1)$, $m = 1, 2, \ldots$, and $t_0 = y$. The latter is defined as the set of evolution equations for $L = Q^{1/2}$:

$$\frac{\partial}{\partial t_m} L = [L_+^{2m-1}, L], \tag{4.41}$$

$m = 1, 2, \ldots$, which turns into a set of evolution equations for u as a function of the t_m's. The system (4.41) is integrable, in the sense that $[(\partial/\partial t_m), (\partial/\partial t_n)] = 0$ for all $m, n \geq 0$, as is easily shown by use of the Jacobi identity for nested commutators. The compatibility between the equations (4.40) and (4.41) is left to the reader as an exercise. Note also that as $u = -F''$, the partition function of the matrix model plays the role of τ-function for the hierarchy.

The description through an integrable system is a quite remarkable property of the multicritical solutions to 2-D quantum gravity coupled to matter. We will see that such a correspondence with integrable partial differential systems generalizes beautifully to include many more theories.

4.5   Physical Interpretation

The general equation (4.40) awaits a physical interpretation. This looks like a general equation, interpolating between the various critical points

$\mu_m = \delta_{m,k}$. In fact, the parameters t_k can be interpreted as follows. The fundamental observables of the matter theories we have coupled to gravity (order parameters, cf. the spin σ for the Ising model or the occupation number of the lattice bonds for the hard-dimer model) are naturally expected still to behave as observables in the gravitational theory. As explained in Section 1.3, we have a collection of observables in all these models, corresponding to operators of the original matter theories, dressed by gravity.

The dressed identity operator ϕ_1 is coupled to the renormalized cosmological constant y. With this interpretation, a derivative with respect to y must result in an insertion of the operator ϕ_1 in the sum over surfaces (1.6). More precisely, a derivative with respect to y of the free energy $F(y) \sim \log Z$ amounts to measuring the vacuum expectation value of the operator ϕ_1, namely $F'(y) = \langle \phi_1 \rangle$, and more derivatives amount to measuring higher correlators. In particular, the string susceptibility $u(y) \propto F''(y) = \langle \phi_1 \phi_1 \rangle$ measures the identity two-point function. Note that once gravity is switched on, the identity operator is nontrivial, as it carries the information that a small puncture is made in the surface at the location of the operator, and this cannot be removed without changing the topology.

More generally, in view of (4.39), we have the following picture for the $(m-1)$-multicritical model. Recall that from the continuum point of view, these models are expected to describe the coupling of some critical matter theories, described by conformal field theories, to gravity. For instance, the hard-dimer model lies in the universality class of hard objects on 2-D lattices, known to be described at criticality by a conformal theory with central charge $c(2,5) = -22/5$ (see (1.7)). More generally, the $(m-1)$-multicritical models at hand will describe $c(2, 2m-1)$ conformal theories coupled to 2-D gravity (note the coincidence between the degrees of the operators Q and P, and the two integers labeling the conformal theory). These theories are known to have a collection of m primary operators (order parameters), which will get dressed by gravity.

The gravitational dressing of the observables of the matter theory goes as follows. The parameters t_j, $j = 2, \ldots, m$, are coupled to the other order parameters of the theory, $\phi_2, \phi_3, \ldots, \phi_m$, and a derivative of a correlation function with respect to t_j amounts to the insertion of the operator ϕ_j

$$\frac{\partial}{\partial t_j} \langle \cdots \rangle = \langle \phi_j \cdots \rangle. \tag{4.42}$$

Therefore, the KdV evolution equations (4.41) can be interpreted as rules for computing correlation functions in these models. In particular, these lead to the identification of scaling dimensions for the dressed operators [19], coinciding with those derived from the continuum approach.

The interpolating equation (4.40) describes in addition the flows between various theories, obtained simply by adding new operators to the action: This realizes in a compact way all the possible renormalization group flows between the conformal field theories at the gravitational level.

5   Multimatrix Models

5.1   The Ising Model as a Two-Matrix Model

Let us construct a matrix model [13] to generate the sum (1.5). In the matrix–fatgraph expansion, we wish now to decorate the vertices of the fatgraphs by a spin variable $\sigma = \pm 1$. The interaction between nearest neighbors can be reproduced if we consider a two-matrix model M_+, M_-, each standing for the two possible values of the spin. The vertices of the fatgraph will be either in the $+$ or $-$ state (vertex $\operatorname{tr} M_{\pm}^4/4$), and these vertices must be connected with two types of propagators, with weights

$$\langle (M_+)_{ij}(M_+)_{kl}\rangle = \langle (M_-)_{ij}(M_-)_{kl}\rangle = \frac{e^K}{N}\delta_{il}\delta_{jk},$$

$$\langle (M_+)_{ij}(M_-)_{kl}\rangle = \frac{e^{-K}}{N}\delta_{il}\delta_{jk},$$

(5.1)

where the factors $e^{\pm K}$ reproduce the Boltzmann weights $e^{K\sigma_i\sigma_j}$ for anti-parallel and parallel spins. The vertices are weighted by

$$\operatorname{tr}\frac{(M_+)^4}{4} \mapsto Nge^H,$$

$$\operatorname{tr}\frac{(M_-)^4}{4} \mapsto Nge^{-H},$$

(5.2)

to reproduce the correct magnetic weight $e^{H\sigma_i}$.

The rules (5.1), (5.2) are easily seen to be reproduced by the following two-matrix model:

$$Z(g,N;K,H) =$$
$$\int e^{-N\operatorname{tr}(\alpha M_+^2/2 + \alpha M_-^2/2 + \eta M_+M_- - ge^H M_+^4/4 - ge^{-H}M_-^4/4)}\, dM_+\, dM_-,\quad (5.3)$$

where the quadratic form $\left(\begin{smallmatrix}\alpha & \eta \\ \eta & \alpha\end{smallmatrix}\right)$ is the inverse of $\left(\begin{smallmatrix}e^K & e^{-K} \\ e^{-K} & e^K\end{smallmatrix}\right)$, namely $\alpha = e^K/(2\sinh(2K))$, $\eta = -e^{-K}/(2\sinh(2K))$ (see (1.18)). As in the case of the hard-dimer model, we use this as a motivation to study multimatrix models, the subject of the present section.

5.2   The Itzykson–Zuber Integral and Multimatrix Models

The Itzykson–Zuber integral [10] permits one to reduce coupled multiple matrix integrals to the integration over their eigenvalues, which is a preliminary necessary step toward their complete solution.

It reads as follows: let A and B denote two diagonal matrices of size $N \times N$, with diagonal matrix elements a_i and b_i, respectively. Then we have the integral formula

$$I_N(A, B) = \int e^{\text{tr}(UAU^\dagger B)} \, dU = \alpha_N \frac{\det[e^{a_i b_j}]_{i,j}}{\Delta(A)\Delta(B)}, \tag{5.4}$$

where the integral is taken over the unitary group; α_N is some constant, which can be reabsorbed in the measure; and $\Delta(A)$ stands for the Vandermonde determinant of A (2.4). Let us simply sketch the proof of (5.4). First introduce the Laplacian operator $\mathcal{L} = \text{tr}(d/dM)^2$. When acting on functions of Λ only, where Λ is the diagonal form of M in a unitary change of basis, the Laplacian simply becomes

$$\mathcal{L} = \Delta(\Lambda)^{-2} \sum_k (d/d\lambda_k)\Delta(\Lambda)^2 (d/d\lambda_k).$$

The next step is to prove that $\Delta(\Lambda)\mathcal{L}\Delta(\Lambda)^{-1} = \text{tr}(d/d\Lambda)^2$, i.e., decomposes into a sum of free terms $\sum(d/d\lambda_k)^2$. Finally, the integral $J_N(A, \Lambda) = \Delta(A)\Delta(\Lambda)I_N(A, \Lambda)$ is a solution of the heat equation for $\mathcal{L}$, namely as $\mathcal{L}e^{\text{tr}(AM)} = \text{tr}\,A^2 e^{\text{tr}(AM)}$, we have

$$\sum_k \frac{d^2}{d\lambda_k^2} J_N(A, \Lambda) = \text{tr}\,A^2\, J_N(A, \Lambda). \tag{5.5}$$

Due to the Vandermonde prefactor, J_N is an antisymmetric function of the λ_k, and hence be written as a linear combination of free solutions of the heat equation (5.5), namely of Slater determinants, in the form $J_N = \sum_\mu c_\mu \det[e^{\lambda_i \mu_j}]$, with $\text{tr}\,\mu^2 = \text{tr}\,A^2$. As moreover A and Λ play symmetric roles, only the term $\mu = A$ can survive in this sum, and $J_N(A, \Lambda) = c\det[e^{a_i \lambda_j}]$. This completes the proof of formula (5.4).

Formula (5.4) can be used in many different ways. It provides an effective means of reducing multiple matrix integrals to integrals over eigenvalues. Consider, for instance, the two-matrix integral [11]

$$Z(V, W, N) = \int e^{-N\,\text{tr}(V(M)+W(P)+cMP)} \, dM \, dP \tag{5.6}$$

for two arbitrary polynomial potentials V and W and some coupling constant c. We simply diagonalize P and M through unitary matrices and compute the integral after the change of variables. Recall that $dM = d\Lambda\, dU\,\Delta(\Lambda)^2$ in the change of variables. The integral over P is readily reduced to the eigenvalues by noticing that the integral over M depends only on the eigenvalues μ_j of P. So we rewrite

$$Z(V, W, N) = \int \Delta(\mu)^2 e^{-N\,\text{tr}\,W(\mu)} \int e^{-N\,\text{tr}(V(M)+cM\mu)} \, dM \, d\mu, \tag{5.7}$$

where the integral over the unitary group matrix diagonalizing P has been completely factored out (and been disposed of by changing the normalization of dP). We can now use formula (5.4) to do the integral over the unitary matrix that diagonalizes M,

$$Z(V, W, N) = \int \Delta(\mu)^2 \Delta(\Lambda)^2 e^{-N \operatorname{tr}(V(\Lambda)+W(\mu))} \frac{\det[e^{-Nc\lambda_i \mu_j}]}{\Delta(\Lambda)\Delta(\mu)} \, d\Lambda \, d\mu$$

$$= \int \Delta(\Lambda)\Delta(\mu) e^{-N \operatorname{tr}(V(\Lambda)+W(\mu)+c\Lambda\mu)} \, d\mu \, d\Lambda, \tag{5.8}$$

where we have rearranged the determinant using the antisymmetry of the Vandermonde prefactor.

It is now clear how to apply formula (5.4) to multimatrix models. For instance, we have

$$Z(V_1, \ldots, V_p, N)$$

$$= \int e^{-N \operatorname{tr}(\sum_{i=1}^{p} V_i(M^{(i)}) + \sum_{i=1}^{p-1} c_i M^{(i)} M^{(i+1)})} \, dM^{(1)} \cdots dM^{(p)}$$

$$= \int e^{-N \operatorname{tr}(\sum_{i=1}^{p} V_i(\Lambda^{(i)}) + \sum_{i=1}^{p-1} c_i \Lambda^{(i)} \Lambda^{(i+1)})}$$

$$\times \Delta(\Lambda^{(1)})\Delta(\Lambda^{(p)}) \, d\Lambda^{(1)} \cdots d\Lambda^{(p)} \tag{5.9}$$

by successively applying the Itzykson–Zuber formula (5.4). Note that the coupling between the matrices is simply a chain: The Vandermonde determinants of its ends have not been canceled completely by the denominators of (5.4). It is not difficult to see that the reduction holds for any tree interaction between the matrices. The existence of a loop of interacting matrices prevents us from using (5.4) successively, and is therefore not reducible. We could have had an interaction of the form $(M^{(i)})^k (M^{(j)})^l$, without losing the reducibility property. For our purposes, the integrals of the form (5.9) will be sufficient.

5.3   The Multimatrix Model

Let us now evaluate the integral (5.9). As in the one-matrix case, we introduce a family of monic orthogonal polynomials with respect to the measure $e^{-N(\sum_i V_i(\lambda^{(i)}) + \sum_i c_i \lambda^{(i)} \lambda^{(i+1)})} \, d\lambda^{(1)} \cdots d\lambda^{(p)}$, namely

$$\int e^{-N(\sum_i V_i(\lambda^{(i)}) + \sum_i c_i \lambda^{(i)} \lambda^{(i+1)})} p_n(\lambda_1) q_m(\lambda_p) \, d\lambda^{(1)} \cdots d\lambda^{(p)}$$

$$= h_n \delta_{n,m}. \tag{5.10}$$

By analogy with Section 3.1, we write

$$\Delta(\Lambda_1) = \det[p_{i-1}(\lambda_j^{(1)})]_{ij},$$
$$\Delta(\Lambda_p) = \det[q_{i-1}(\lambda_j^{(p)})]_{ij}. \tag{5.11}$$

Thanks to the orthogonality relation (5.10), by directly expanding the determinants as sums over permutations we end up with

$$Z(V_1, V_2, \ldots, V_p, N) = N! \prod_{i=0}^{N-1} h_i. \tag{5.12}$$

5.4 General Solution: The P, Q Formalism

As in Section 3.1, we will introduce P and Q operators corresponding to differentiation with respect to and multiplication by eigenvalues, and write their canonical commutation relations. For simplicity, we assume that all the potentials $V_1, \ldots, V_p$ are even polynomials. We also rescale the polynomials to have orthonormality, namely, set $\tilde{p}_n = p_n/\sqrt{h_n}$ and $\tilde{q}_m = q_m/\sqrt{h_m}$. In the multimatrix model we can define a pair $(P^{(i)}, Q^{(i)})$ of such operators for each of matrix $M^{(i)}$, $i = 1, 2, \ldots, p$, in the following way. We set

$$Q_{n,m}^{(i)} = (\tilde{p}_m, \lambda^{(i)}\tilde{q}_n),$$

$$P_{n,m}^{(i)} = \int \tilde{p}_m(\lambda^{(1)}) e^{-N(\sum_{j=1}^{i-1} V_j(\lambda^{(j)}) + c_j \lambda^{(j)} \lambda^{(j+1)} + V_i(\lambda^{(i)})/2)}$$

$$\times \frac{\mathrm{d}}{\mathrm{d}\lambda^{(i)}} e^{-N(V_i(\lambda^{(i)})/2 + \sum_{j=i+1}^{p} V_j(\lambda^{(j)}) + c_j \lambda^{(j)} \lambda^{(j+1)})}$$

$$\times \tilde{q}_m(\lambda^{(p)}) \, D\lambda^{(1)} \cdots \mathrm{d}\lambda^{(p)}, \tag{5.13}$$

where we have set $c_p = c_{p+1} = 0$ and used the obvious notation for the scalar product with respect to the multimatrix measure (5.10). These operators obey the canonical commutation relations

$$[P^{(i)}, Q^{(j)}] = \delta_{i,j}. \tag{5.14}$$

Using definition (5.13) we can express $P^{(i)}$ in terms of the operators $Q^{(j)}$ only, through the relations

$$P_{m,n}^{(1)} = -\frac{N}{2} V_1'(Q^{(1)})_{m,n} + (A_1)_{m,n}$$

$$= \frac{N}{2} V_1'(Q^{(1)})_{m,n} + Nc_1 Q_{m,n}^{(2)},$$

$$P_{m,n}^{(j)} = -\frac{N}{2} V_j'(Q^{(j)})_{m,n} + Nc_{j-1} Q_{m,n}^{(j-1)}$$

$$= \frac{N}{2} V_j'(Q^{(j)})_{m,n} + Nc_j Q_{m,n}^{(j+1)}, \tag{5.15}$$

$$P_{m,n}^{(p)} = -\frac{N}{2} V_p'(Q^{(p)})_{m,n} + Nc_{p-1} Q^{(p-1)})_{m,n}$$

$$= \frac{N}{2} V_p'(Q^{(p)})_{m,n} + (A_p)_{m,n},$$

for $j = 2, \ldots, p - 1$. In (5.15), each second line is obtained by integration by parts, and $(A_1)_{n,m} = (\tilde{p}'_m, \tilde{q}_n)$ and $(A_p)_{n,m} = (\tilde{p}_m, \tilde{q}'_n)$. Hence A_1 is an upper triangular matrix and A_p is a lower triangular matrix ($A_p = A_1^t$ if all the V_j's are equal). The system (5.15) is the generalized discrete system governing the solutions of the multimatrix model. Our final aim is to derive its possible (multi)critical limits.

We wish now to argue that the operators $P^{(1)}$ and $Q^{(1)}$ go over to differential operators in a suitable double-scaling limit. The basic ingredient is the fact that both have a finite range, namely, there exists a uniform bound B such that $P^{(1)}_{m,n} = Q^{(1)}_{m,n} = 0$ as soon as $|m - n| > B$. First of all, it is clear that $Q^{(1)}_{m,n} = (\lambda^{(1)} \tilde{p}_n, \tilde{q}_m)$ vanishes if $m > n+1$, and $Q^{(1)}_{n+1,n} = \sqrt{r_{n+1}}$, where $r_n = h_n/h_{n-1}$ as usual. We say that $Q^{(1)}$ has a lower range of 1. A power $(Q^{(1)})^k$ would have lower range k. Let us now evaluate the upper range of $Q^{(1)}$. We first eliminate the $P^{(j)}$ from (5.15) by subtracting every second line from the first. The result is a chain of identities,

$$c_1 Q^{(2)} = \frac{1}{N} A_1 - V'_1(Q^{(1)}),$$
$$c_j Q^{(j+1)} = c_{j-1} Q^{(j-1)} - V'_j(Q^{(j)}), \tag{5.16}$$
$$\frac{1}{N} A_p = c_{p-1} Q^{(p-1)} - V'_p(Q^{(p)}),$$

for $j = 2, \ldots, p - 1$. For $j = 2, \ldots, p - 1$ this gives a recursion, allowing us to express all the $Q^{(j)}$'s as polynomials in $Q^{(1)}$, whose degree depends only on the degrees of the V_j's (hence is independent of N). As $Q^{(1)}$ has a lower range of 1, all the $Q^{(j)}$'s have a finite lower range, depending only on the degrees of the V_i's. But $Q^{(p)}$ has an upper range of 1, as $Q^{(p)}_{m,n} = (\tilde{p}_n, \lambda^{(p)} \tilde{q}_m) = 0$ if $m < n - 1$. Hence $Q^{(p)}$ has a uniformly bound range. Now we reuse the equations (5.16) for $j = p - 1, p - 2, \ldots, 2$ as backward recursions to express all the $Q^{(j)}$'s as polynomials of $Q^{(p)}$, whose degree depends only on the degrees of the V_j's. This induces uniform bounds on their upper range. We conclude that all the $Q^{(j)}$ have a uniform upper and lower range (call it B generically, although some operators may have a smaller range). In principle, equations (5.16) are sufficient to specify the $Q^{(j)}$'s completely, in terms of shift operators $\delta_{n,m+p}$ p ranging from $-B$ to B, with coefficients depending on the r_k's. Using the rest of the equations (5.15), we readily see that the $P^{(j)}$'s also have a uniformly bound range as polynomials in the $Q^{(i)}$'s.

5.5 The Double-Scaling Limit and the Generalized KdV Hierarchy

We can now imagine that a certain double-scaling limit can be taken in which this property guarantees that $Q \equiv Q^{(1)}$ and $P \equiv P^{(1)}$ go over to two differential operators, of respective orders q and p in a rescaled variable y.

We will assume that q and p are coprime integers. We must then write the canonical commutation relations (4.11) $[P,Q] = 1$. The string susceptibility $u(y)$ of the model must appear as the first subleading coefficient of $Q = d^q + \alpha u d^{q-2} + \cdots$, by a simple degree-counting argument.

When we expand $[P,Q]$ in powers of d, we get terms ranging from d^0 to d^{p+q-3}, hence a total of $p+q-2$ equations. The first $p-1$ of them enable one to express the coefficients of P in terms of those of Q, with the result

$$P = Q_+^{p/q}, \qquad (5.17)$$

where the qth root of Q, denoted by $L = Q^{1/q}$, is defined as the degree-1 pseudodifferential operator satisfying $L^q = Q$. The other $q-1$ equations form a coupled differential system for the $q-1$ remaining coefficients of Q, including u. By the usual grade-counting argument, we see that d has degree 1, and u has degree 2; the constant term in Q has degree q, and that of P has degree p. This means that the d^0 term in $[P,Q]$ has the same homogeneity as $[u^{(p+q-1)/2}]'$. Hence, as this is set to 1, we have

$$u \sim y^{2/(p+q-1)}, \qquad (5.18)$$

which generalizes the scaling behavior (4.29), corresponding to $q = 2$, $p = 2m - 1$.

The behavior (5.18) displays a certain ambiguity in the exponent, as different values of p and q may give the same exponent. A simple example of this is the ambiguity between the hard-dimer model, with $q = 2$, $p = 5$, and the Ising model, which turns out to have $q = 3$ and $p = 4$, hence the same susceptibility exponent $\frac{1}{3}$.

In fact, the precise identification of the models is beautifully simple, as the (p,q) model describes a conformal matter theory with central charge $c(p,q)$ (1.7) coupled to quantum gravity. This identification is made by a thorough investigation of the operator content of the (p,q) theory.

The general equation $[Q_+^{p/q}, Q] = 1$ is a system of $q-1$ differential equations for the coefficients of Q. In terms of the qth root L of Q, we have $[L^q, L_-^p] = 1$. Expanding $L_-^p = c_{q-1}L^{-1} + O(d^{-2})$, we get the first equation by equating the coefficient of d^{q-2} to zero in $[P,Q]$, namely $q[c_{q-1}]' = 0$. This gives rise to an integration constant, which we write $qc_{q-1} = (q-1)t_{q-1}$. Then we continue the expansion $L^p = (1-1/q)t_{q-1}L^{-1} + c_{q-2}L^{-2} + O(d^{-3})$. The coefficient of d^{q-3} in $[P,Q]$ reads $q[c_{q-2}]' = 0$, giving rise to another integration constant, which we write $qc_{q-2} = (q-2)t_{q-2}$. Continuing this process we have the equations $qc_j = jt_j$, with integration constants t_j, until we reach the last equation $q[c_1]' = 1$, integrated into $qc_1 = t_1 = y$. These parameters t_j, $j = 1, \ldots, q-1$ are coupled to the order parameters of the theory.

The most general equation for fixed degree q of Q, including all the possible values of p, reads

$$\left[\sum_j \mu_j Q_+^{j/q}, Q\right] = 1. \tag{5.19}$$

It can be shown that if we set $\mu_j = -(j/q + 1)t_{j+q}$ for $j = 1, 2, \ldots$, then equation (5.19) is compatible with the generalized KdV (or reduced KP) flows [20]

$$\frac{\partial}{\partial t_j} L = [L_+^j, L], \tag{5.20}$$

where $L^q = Q$. Identifying the parameters t_j as coupled to the order parameters (for $j = 1, \ldots, q-1$, in which case the t_j are implicit integration constants for (5.19)) and to higher operators (for $j = q + 1, \ldots$, in which case the t_j appear explicitly in the equation (5.19) as some μ's), we see again that the GKdV flows (5.20) provide a means to compute all the correlators of these operators, by the identification (4.42). This is proved by checking the compatibility of the GKdV flows (5.20) with the fundamental equation (5.19), a finite exercise using the Jacobi identity profusely.

To summarize, the operator content of the multimatrix models in the double-scaling limits is simply given by the GKdV flows (5.20). These are again integrable partial differential equations. The partition function of the multimatrix model plays the role of a τ-function for this hierarchy.

5.6   The Ising Model Solution

The solution of the Ising model [13, 14] coupled to 2-D quantum gravity can be proved to be given by $p = 4$ and $q = 3$, starting from the two-matrix model of Section 5.1 and solving it completely by use of the explicit form of (5.15). Let us examine the solution. We set

$$Q = d^3 - \frac{3}{4}\{u, d\} + w \tag{5.21}$$

in all generality, with the requirement that the first subleading coefficient be proportional to u. As argued before, $P = Q_+^{4/3}$ does not exhaust all solutions to $[P, Q] = 1$, with P a fourth-order differential operator. It can actually take the form $P = \mu_4 Q^{4/3} + \mu_2 Q^{2/3}$ (cf. (4.39)).

Setting $\mu_4 = 1$, and $\mu_2 = -5t_5/3$, we get

$$P = d^4 - \{u, d^2\} + \{w, d\} + \frac{1}{2}u^2 + \frac{1}{6}u'' - \frac{5}{3}t_5(d^2 - u), \tag{5.22}$$

and the remaining equations of $[P, Q] = 1$ imply

$$\frac{1}{2}w'' - \frac{3}{2}uw - \frac{5}{2}t_5 w = t_2,$$

$$\frac{1}{12}u'''' - \frac{3}{4}uu'' - \frac{3}{8}(u')^2 + \frac{1}{2}u^3 + \frac{3}{2}w^2 + \frac{5t_5}{12}(3u^2 - u'') = y, \tag{5.23}$$

where we have introduced two integration constants $t_1 = y$ and t_2. These two constants are actually coupled to the two order parameters of the Ising model: The dressed identity ϕ_1 and the dressed spin ϕ_2. With this interpretation, we easily compute $\langle \phi_1 \phi_2 \rangle = w$, whereas $\langle \phi_1 \phi_1 \rangle = u$. The parameter t_5 governs the interpolation between the Ising model and the pure gravity model, with no matter. This parameter is, in fact, coupled to the energy operator ϕ_5 of the Ising model, dressed by gravitation. The coupled system of equations (5.23) governs the perturbative solution to the gravitational 2-D Ising model with arbitrary parameters K and H, of which t_5 and t_2 respectively are some renormalized versions, whereas $y = t_1$ captures the double-scaling limit of the renormalized cosmological constant.

5.7 Beyond Matrix Models

We have experienced the remarkable phenomenon that all the 2-D quantum-gravitational theories we have studied have turned out to be governed by KdV and GKdV flows, together with a canonical commutation relation.

This fact might be much more general, namely that all 2-D quantum-gravitational theories are governed by integrable partial differential systems, supplemented by some canonical commutation relation obeyed by the defining operators.

A first test of this idea is the following. The classification of the $c(p, q)$ conformal theories has, in fact, some fine structure. For given values of p, q there may be several different theories. Those can also be coupled to gravity, hence they should fit into our picture. This means that there are more partial differential systems to be found.

More precisely, this classification [18] associates to each theory a pair of simply laced Lie algebras (of the type A, D, E) with Coxeter numbers p and q. The Coxeter exponents of the two algebras encode the Kac indices (r, s) (1.8) of the actual operators of the theory. The simplest example is the 3-states Potts model, a generalization of the Ising model with a spin taking 3 possible values. This model is described at criticality by the $c(5, 6) = \frac{4}{5}$ conformal theory labeled by the pair of algebras (A_4, D_4), with respective Coxeter exponents $(1, 2, 3, 4)$ and $(1, 3, 3, 5)$ (this means that the operators $\phi_{r,3}$ appear twice). Unfortunately, the matrix model for its coupling to gravity cannot be solved by orthogonal polynomial techniques. Nevertheless, in [19] it is shown that the D-GKdV flows of Drinfeld and Sokolov [20] produce sensible results for the (A, D) theories in general. In the case of the 3-states Potts model, they read as follows. The operator Q is no longer differential, and takes the particular form

$$Q = (d^7 - \{u, d^5\} + \{v, d^3\} + \{w, d\} + u_0 d^{-1} u_0) d^{-1}, \qquad (5.24)$$

where $u(y)$ is the string susceptibility of the model, and v, w, u_0 some functions of y. The operator Q has the symmetry property $Q^* = d^{-1} Q d$. The

D-KdV flows read

$$\frac{\partial}{\partial t_{2k-1}}Q = [Q_+^{(2k-1)/6}, Q], \tag{5.25}$$

for $k = 1, 2, \ldots$, supplemented by a sequence of exceptional flows, the first of which reads

$$\frac{\partial}{\partial t_0}Q = [u_0 d^{-1}, Q]. \tag{5.26}$$

The parameters $t_1, t_3, \ldots, t_0, \ldots$ are coupled to the gravitational operators of the Potts model. A simple degree counting shows that t_3 has the same degree as t_0; therefore, these two constants are coupled to two operators with the same dimension. These are identified as the two copies of $\phi_{3,3}$ of the flat space matter theory.

Moreover, the conjecture is that the 3-states Potts model is governed by the equation $[P, Q] = 1$, where P is a differential operator of order 5. The solution reads $P = Q_+^{5/6}$ (found by equating to 0 the coefficients of d^{10}, d^8, and d^6 in $[P, Q]$, as the odd powers are linked to the even ones by the symmetry property $P^* = -d^{-1}Pd$), and the commutator reduces to a coupled system of differential equations. Due to the infinite tail in Q, this system is a priori infinite. Remarkably, one can show that the infinite set of equations arising from this tail reduces simply to $P(u_0) = 0$, where P acts on the function u_0. But as $P^* = -d^{-1}Pd$, $P = dY$, where $Y = -P^*d^{-1}$, the equation $P(u_0) = 0 = Y(u_0)'$ is a total derivative. We integrate it as $Y(u_0) = t_0/2$. The other equations come from the powers d^4, d^2, and d^0 and give rise to three integration constants, respectively equal to $5t_5/6$, $t_3/2$, $y/6$, with $y = t_1$. The four constants t_1, t_3, t_0, t_5 correspond to the order parameters of the 3-states Potts model. In [19], it was checked that these are compatible with the D-KdV flows (5.25).

6   Conclusion

The above results strongly suggests that, beyond the matrix models, the fundamental structure of 2-D gravity lies in its integrable character, expressed in terms of partial differential systems. The matrix models we have considered so far provide a natural discretization of some integrable systems (KdV, GKdV), but we may think that more general situations, not described by matrix models (or described by matrix models, not solvable with orthogonal polynomial techniques), describing gravity must be linked to other integrable differential systems. We could even dream of a complete classification of integrable differential systems from the consideration of the most general gravitational theories.

Let us finally make a few comments on matrix models. They have appeared in this chapter as the ideal combinatorial tool to generate graphs

tessellating surfaces. As such, they can be applied to many open combinatorial problems. The fundamental fact is that whenever the matrix model
is solvable in terms of orthogonal polynomials, the answer to the combinatorial problem will most probably take the form of a differential system
governing generating functions. Another remark concerns the discrete nature of the matrix models. They indeed generate discrete sets of relations
(like the discrete Painlevé I and its generalizations), of which some scaling
limit has to be taken. But before taking this limit, they provide all the ingredients of discrete integrable systems (i.e., they satisfy discrete integrable
equations, provide discrete tau-functions for these, etc.).

Part B 2-D Topological Gravity

7 Introduction

7.1 Intersection Theory of the Moduli Space of Riemann Surfaces

Another route [21] to quantization of pure gravity consists in directly
considering invariant observables (just like gauge-invariant observables in
gauge theory) of the equivalence classes of all Riemann surfaces under
reparametrizations. The space of such equivalence classes is known to mathematicians as the Deligne–Mumford compactification of the moduli space
of Riemann surfaces

$$\overline{\mathcal{M}}_{h,s} \tag{7.1}$$

with fixed genus h and number s of "punctures," namely points at which
operators can be inserted in order to compute their correlation functions.
These invariants are the so-called intersection numbers of the cohomology
classes of this space, which will play the role of descendent observables
generated by gravity, the fundamental observable being the identity or
puncture operator, denoted by ϕ_0 here.

Typically, these observables are defined as exterior powers of differential
forms

$$\phi_d(x_i) = c_1(\mathcal{L}_{x_i})^{\wedge d}, \tag{7.2}$$

where $\mathcal{L}_x$ denotes some complex line bundle over the surface Σ, whose fiber
is the tangent space $T_x^*\Sigma$ at x, and c_1 denotes the first Chern class. These
are cohomology classes of the moduli space (7.1). The intersection between
various classes is expressed mathematically as the integral over the moduli
space (7.1)

$$I_h(d_1, d_2, \ldots, d_s) = \int_{\overline{\mathcal{M}}_{h,s}} \phi_{d_1} \wedge \phi_{d_2} \wedge \cdots \wedge \phi_{d_s}. \tag{7.3}$$

This number no longer depends on the positions of the insertion points x_i, $i = 1, 2, \ldots, s$, and is a topological invariant. We use the notation

$$I_h(d_1, \ldots, d_s) = \langle \phi_{d_1} \cdots \phi_{d_s} \rangle \tag{7.4}$$

by analogy with a correlator. The numbers I_h (7.3) are nonzero only if the total degree of the integrand is equal to the dimension of the moduli space, namely $\sum_i d_i = 3h - 3 + s$.

Using the definition (7.3), it is possible to derive a number of recursion relations linking these numbers at different values of d_i and h: These are the so-called topological gravity recursions, which encode the cohomological structure of the moduli space. These permit one in principle to compute any "descendent" intersection number in terms of the primary ones, with $d_i = 0$ (the identity ϕ_0 is the only primary gravitational observable in this pure gravity theory).

7.2 Diagrammatics Again: The Kontsevich Matrix Model

The intersection numbers I_h of (7.3) may be rewritten as

$$I_h(d_1, \ldots d_s) = \langle \prod_i \phi_i^{n_i} \rangle, \tag{7.5}$$

where $n_i = |\{j | d_j = i\}|$. Then, the intersection number is nonzero only if $\sum_i i n_i = 3h - 3 + s$. Let $F(s_0, s_1, \ldots)$ denote the generating function of the intersection numbers I_h, for all genera h (implicitly given by the n_i through the relation $h = 1 + (\sum (i - 1)n_i)/3$):

$$F(s_0, s_1, \ldots) = \sum_{n_0, n_1 \cdots \geq 0} \frac{s_0^{n_0} s_1^{n_1}}{n_0! \, n_1!} \cdots \langle \phi_0^{n_0} \phi_1^{n_1} \cdots \rangle \tag{7.6}$$

for arbitrary real parameters s_0, $s_1 \ldots$.

By use of a cell decomposition of the moduli space, Kontsevich [22] has been able to reexpress the generating function F as a purely combinatorial sum over fatgraphs with running loop indices having to be summed over, in the form

$$F(s_0, s_1, \ldots) = \sum_{\substack{\text{fatgraphs } \Gamma \\ \Sigma \text{ indices } i=1,\ldots,N}} \frac{(i/2)^{V}(\Gamma)}{|\operatorname{Aut}(\Gamma)|} \prod_{\text{edges } ij} \frac{2}{\lambda_i + \lambda_j}, \tag{7.7}$$

where $V(\Gamma)$ denotes the number of vertices of Γ, and with the s_i given in terms of the parameters λ_k as

$$s_i = -(2i - 1)!! \sum_{k=1}^{N} \lambda_k^{-2i-1}, \tag{7.8}$$

with $(2i-1)!! = (2i-1)(2i-3)\cdots5\cdot3\cdot1$. This result is, of course, reminiscent of the matrix integrals studied in Part A. They will indeed be compactly expressed in terms of yet another matrix model, known as the Kontsevich model.

7.3   The Main Theorems

The main subject of this Part B will be the detailed proof of the three following theorems, due to Kontsevich and Witten [23].

Theorem 7.1. *The generating function $F(s_0, s_1, \dots)$, equation (7.7), reads $F(s_0, s_1, \dots) = \lim_{N\to\infty} \log \Theta_N(\Lambda)$, where Λ is a diagonal $N \times N$ real matrix with elements $(\lambda_1, \lambda_2, \dots, \lambda_N)$ and $\Theta_N(\Lambda)$ is the following one-Hermitian-matrix integral*

$$\Theta_N(\Lambda) = \frac{\int e^{\mathrm{tr}(iM^3/6 - \Lambda M^2/2)} \, \mathrm{d}M}{\int e^{-\,\mathrm{tr}\,\Lambda(M^2/2)} \, \mathrm{d}M}, \tag{7.9}$$

and the s_i's are expressed in terms of the λ_k's through (7.8).

Theorem 7.2. *The function $\tau(s_0, s_1, \dots) = e^{F(s_0, s_1, \dots)}$ is a tau-function for the KdV hierarchy, when expressed in the variable*

$$t_i = \mathrm{tr}\,\frac{\Lambda^{-2i+1}}{2i-1} = -\frac{s_{i-1}}{(2i-1)!!}, \tag{7.10}$$

$i = 1, 2, \dots$, *namely, if*

$$Q = d^2 - u \quad \text{with } u = -(\partial/\partial t_1)^2 \log \tau, \tag{7.11}$$

then we have the infinite system of partial differential equations

$$\frac{\partial}{\partial t_i} Q = [Q_+^{i-1/2}, Q]. \tag{7.12}$$

Theorem 7.3. *The function u above satisfies the one-matrix canonical commutation relation*

$$[P, Q] = 1 \quad \text{with } P = -\sum_{k\geq1}\left(k + \frac{1}{2}\right)t_{k+1}Q_+^{k-1/2}. \tag{7.13}$$

Theorems 7.1–7.3 establish the equivalence between quantum and topological gravities as described by one-matrix models (cf. (4.39)–(4.41)). It should be noted that this equivalence is a highly nontrivial fact, if only for the very different type of matrix integrals at hand. On the one hand, we have the one-matrix integral (2.1), for which a double-scaling limit has to be taken (V critical and $N \to \infty$ simultaneously); on the other hand, we have the integral (7.9) (also called the matrix Airy function), which serves as a generating function for topological correlators, and in which a simple $N \to \infty$ limit has to be taken.

7.4 Reading Guide

Part B is organized as follows. In Section 9 we show how to compute the Kontsevich matrix integral (7.9), by use of eigenvalue reduction and expansions on the characters of the linear group. The final result takes the simple form of a Wronskian determinant of N functions, related to Airy's function. Section 8 is devoted to the proof of Theorem 7.2, using the previous Wronskian form. In Section 10 we establish Theorem 7.3 and the equivalence between quantum and topological gravities. We also gather more general facts on the generalizations of the intersection theory on the moduli space of Riemann surfaces, known generically as topological field theories, or topological sigma models coupled to topological gravity. We conclude with Section 11.

8 Computing the Kontsevich Integral

8.1 Proof of Theorem 7.1

The proof of Theorem 7.1 is a simple exercise in fatgraph expansion and Feynman rules for matrix models, in order to identify the expansion (7.7) with the fatgraph expansion of (7.9). When expanding (7.9), we draw a collection of vertices with three legs ($\operatorname{tr} M^3/3$ term), receiving a weight $i/2$ per vertex (recall that we need to keep a factor $\frac{1}{3}$ to account for the cyclic symmetry of the trace), and we connect them in all possible ways through vertices. These vertices come with weights given by the inverse of the quadratic form $\operatorname{tr} \Lambda M^2/2 = \sum_{ij}(\lambda_i + \lambda_j)|M_{ij}|^2$, namely

$$\langle M_{ij} M_{kl} \rangle = \delta_{i,l}\delta_{j,k} \frac{2}{\lambda_i + \lambda_j}. \tag{8.1}$$

We still have to sum over all indices running on the loops of the fatgraph. This reproduces exactly the expansion (7.7).

8.2 Explicit Expansion

We wish to compute the Kontsevich integral (7.9) over Hermitian matrices, with the usual Haar measure. For starters, let us evaluate the integral in the denominator,

$$D_N(\Lambda) = \int e^{-\operatorname{tr}\Lambda(M^2/2)}\, dM, \tag{8.2}$$

where the matrix elements λ_k are assumed to be positive to ensure convergence. This integral is Gaussian, and is readily evaluated as

$$D_N(\Lambda) \propto \prod_k \lambda_k^{-1/2} \prod_{k<l} \frac{1}{\lambda_k + \lambda_l} \propto \frac{\Delta(\Lambda)}{\det(\Lambda)^{1/2}\Delta(\Lambda^2)}, \tag{8.3}$$

where the symbol $\propto$ stands for equality up to a multiplicative constant depending only on N.

Let us now turn to the numerator

$$N_N(\Lambda) = \int e^{\mathrm{tr}(iM^3/6 - \Lambda M^2/2)} \, dM = e^{\mathrm{tr}(\Lambda^3/3)} A_N(\Lambda), \qquad (8.4)$$

where we identify

$$A_N(\Lambda) = \int e^{\mathrm{tr}(iM^3/6 + iM\Lambda^2/2)} \, dM \qquad (8.5)$$

after a change of variables $M \mapsto M - i\Lambda$. We now change the integration variable in (8.5) to the unitary and diagonal matrices U and $m = \mathrm{diag}(m_1, m_2, \ldots, m_N)$, where $M = UmU^\dagger$, as in Section 2.1. We get

$$A_N(\Lambda) = \int \Delta(m)^2 e^{i \, \mathrm{tr} \, (m^3/6)} \int e^{(i/2) \, \mathrm{tr}(\Lambda^2 U m U^\dagger)} \, dU \, dm. \qquad (8.6)$$

Fortunately, the integral over the unitary group is of the form (5.4), with $A = \Lambda^2$ and $B = m$, and yields immediately, together with the denominator (8.3),

$$\Theta_N(\Lambda) = \frac{\int \Delta(m)/\Delta(\Lambda^2) \prod_k d\mu_{\lambda_k}(m_k)}{\Delta(\Lambda)/\Delta(\Lambda^2)}$$

$$= \int \prod_{i<j} \frac{m_i - m_j}{i(\lambda_i - \lambda_j)} \prod_k D\mu_{\lambda_k}(m_k), \qquad (8.7)$$

where we have introduced the real measure

$$d\mu_\lambda(x) = \sqrt{\frac{\lambda}{2\pi}} e^{\lambda^3/3 + ix^3/3 + i(\lambda^2/2)x} \, dx \qquad (8.8)$$

(as before, we have rearranged the determinant arising in (5.4) by using the antisymmetry of the prefactor).

To proceed, we use the notation (3.3) introduced in Section 3.1 to denote a determinant by its columns. Then if

$$z(\lambda) = \Theta_1(\lambda) \qquad (8.9)$$

denotes the scalar Kontsevich integral ($z(\lambda)$ is the Airy function) and D denotes the differential operator

$$D = \lambda + \frac{1}{2\lambda^2} - \frac{1}{\lambda} \frac{d}{d\lambda}, \qquad (8.10)$$

then the Kontsevich integral (8.7) takes the form

$$\Theta_N(\Lambda) = \frac{|z, Dz, D^2 z, \ldots, D^{N-1} z|}{|1, \lambda, \lambda^2, \ldots, \lambda^{N-1}|}, \qquad (8.11)$$

where the denominator is the Vandermonde determinant (3.3), and the column $D^p z$ has the element $D^p z(\lambda_i)$ in its ith row. To prove (8.10), we rewrite (8.7) in terms of the Vandermonde determinant of the m's:

$$\Theta_N(\Lambda) = \frac{1}{\Delta(i\Lambda)} \int |1, m, m^2, \ldots, m^{N-1}| \prod_k d\mu_{\lambda_k}(m_k). \qquad (8.12)$$

By multilinearity of the determinant, we can perform the integration row by row, with the result

$$\Theta_N(\Lambda) = \frac{1}{\Delta(i\Lambda)} |\langle 1 \rangle, \langle m \rangle, \langle m^2 \rangle, \ldots, \langle m^{N-1} \rangle|, \qquad (8.13)$$

where $\langle f(m) \rangle$ stands for the one-dimensional integral $\int f(m)\, d\mu_\lambda(m)$, and the columns in (8.13) now depend on λ (with $\lambda \mapsto \lambda_k$ in the kth row). We have

$$z(\lambda) = \langle 1 \rangle = \int \sqrt{\frac{\lambda}{2\pi}} e^{i(m^3/6 + (\lambda^2/2)m - i\lambda^3/3)}\, dm. \qquad (8.14)$$

Let D be the differential operator

$$D = -\sqrt{\lambda} e^{\lambda^3/3} \frac{1}{\lambda} \frac{d}{d\lambda} \frac{1}{\sqrt{\lambda}} e^{-\lambda^3/3}$$

$$= \lambda + \frac{1}{2\lambda^2} - \frac{1}{\lambda}\frac{d}{d\lambda}. \qquad (8.15)$$

Then by definition, we have $\langle m^p \rangle = (iD)^p z(\lambda)$, and (8.11) follows immediately, as the i's cancel between the numerator and denominator.

The function $z(\lambda)$ is a solution of Airy's equation

$$(D^2 - \lambda^2) z(\lambda) = 0 \qquad (8.16)$$

and has the asymptotic expansion

$$z(\lambda) = \sum_{k \geq 0} c_k \lambda^{-3k} \quad \text{with } c_k = -\frac{1}{36^k} \frac{(6k-1)!!}{(2k)!}. \qquad (8.17)$$

We also introduce $\bar{z}(\lambda) = (1/\lambda)Dz(\lambda)$, with the expansion

$$\bar{z}(\lambda) = \sum_{k \geq 0} d_k \lambda^{-3k} \quad \text{with } d_k = \frac{1 + 6k}{1 - 6k}. \qquad (8.18)$$

Thanks to Airy's equation (8.16), we can replace $D^{2p} z(\lambda)$ by $\lambda^{2p} z(\lambda)$ in the numerator of (8.11), up to a linear combination of previous columns of the determinant, which does not affect its value. Analogously, we can replace

$D^{2p+1}z(\lambda)$ by $\lambda^{2p+1}\bar{z}(\lambda)$. Finally, setting $x = 1/\lambda$ (hence $x_k = 1/\lambda_k$), we rewrite

$$\Theta_N(\Lambda) = \frac{|x^{N-1}z, x^{N-2}\bar{z}, x^{N-3}z, \ldots|}{|x^{N-1}, x^{N-2}, \ldots, x, 1|}. \qquad (8.19)$$

Expanding each column in powers of $x^3 = 1/\lambda^3$, we finally get

$$\Theta_N(\Lambda) = \sum_{n_1,\ldots,n_N \geq 0} \prod_{i=1}^{N} a_{n_i}^{(i \bmod 2)} \frac{|x^{3n_1+N-1}, x^{3n_2+N-2}, \ldots, x^{3n_N}|}{|x^{N-1}, x^{N-2}, \ldots, 1|}, \qquad (8.20)$$

where $a_n^{(1)} = c_n$ and $a_n^{(0)} = d_n$ correspond to the expansions of z and $\bar{z}$, respectively.

8.3   Linear Group Character Expression

Remarkably, the expression (8.20) for $\Theta_N(\Lambda)$ involves the Schur functions defined as the characters of the linear group GL(N). The representations of GL(N) are indexed by Young tableaux Y with ℓ_i boxes in the ith row counted from the top, with the constraint $\ell_{i+1} \leq \ell_i$. The classes are represented by the linear invariants of the matrices $X \in$ GL(N), namely the quantities tr X^p, $p = 1, 2, \ldots, N$. Finally, the character of the representation $Y = (\ell_1, \ldots, \ell_N)$ evaluated on the class of X reads

$$\chi_{\ell_1,\ldots,\ell_N}(X) = \frac{|x^{\ell_1+N-1}, x^{\ell_2+N-2}, \ldots, x^{\ell_N}|}{|x^{N-1}, x^{N-2}, \ldots, 1|}, \qquad (8.21)$$

where the numbers $x_1, \ldots, x_N$ encode the linear invariants of X through tr $X^k = \sum_i x_i^k$. The Schur functions (8.21) can be expressed as a single determinant by use of the Schur polynomials p_n, defined by the small-y series

$$\frac{1}{\det(I - yX)} = e^{\sum_{p=1}^{\infty} y^p \, \mathrm{tr}\,(X^p/p)} = \sum_{n \geq 0} y^n p_n(X). \qquad (8.22)$$

The Schur polynomials p_n are usually expressed in the variables $\theta_m(X) = $ tr X^m/m, and they read explicitly

$$p_n(\theta_1, \theta_2, \ldots) = \sum_{\substack{m_k \geq 0 \\ \sum k m_k = n}} \frac{\theta_1^{m_1} \theta_2^{m_2}}{m_1! \, m_2!} \cdots . \qquad (8.23)$$

We extend the definition (8.22) to negative indices by just demanding that $p_n = 0$ when $n < 0$. Then the character (8.21) can be rewritten as

$$\chi_{\ell_1,\ell_2,\ldots,\ell_N}(X) = \det\left[p_{\ell_i+j-i}(X)\right]_{i,j}. \qquad (8.24)$$

This provides a natural way of extending the definition (8.21) to arbitrary (not necessarily ordered) ℓ's.

With this extended definition, the expansion (8.20) is reexpressed as

$$\Theta_N(\Lambda) = \sum_{\substack{n_i \geq 0 \\ i=1,2,\dots,N}} \prod_{i=1}^{N} a_{n_i}^{(i \bmod 2)} \chi_{3n_1,3n_2,\dots,3n_N}(\Lambda^{-1}), \qquad (8.25)$$

where we have identified the matrix argument as $X = \Lambda^{-1}$. Thanks to the formula (8.24), we see that the Kontsevich integral is a function of the variables

$$\theta_m(\Lambda) = \operatorname{tr} \frac{\Lambda^{-m}}{m}. \qquad (8.26)$$

More precisely, the first few terms in the expansion (8.25) read

$$\Theta_N(\Lambda) = c_0 p_0 + c_1 p_3 + c_0 d_1 \left| \begin{pmatrix} p_0 & p_1 \\ p_2 & p_3 \end{pmatrix} \right| + c_0 d_0 c_1 \left| \begin{pmatrix} p_0 & p_1 & p_2 \\ 0 & p_0 & p_1 \\ p_1 & p_2 & p_3 \end{pmatrix} \right| + \cdots$$

$$= 1 - \frac{\operatorname{tr} \Lambda^{-3}}{24} - \frac{(\operatorname{tr} \Lambda^{-1})^3}{6} + \cdots. \qquad (8.27)$$

Note that the terms $\operatorname{tr} \Lambda^{-1} \operatorname{tr} \Lambda^{-2}$ have been automatically canceled.

We are now ready to investigate the large N limit of the expression (8.25). As a function of the variables θ_m (8.26), the integral $\Theta_N(\Lambda)$ has a graded expansion, in that we may attach a degree 1 to each power of Λ^{-1}, hence a term $\chi_{3n_1,\dots,3n_N}(\Lambda^{-1})$ has degree $3(n_1 + \cdots + n_N)$. For instance, the explicit expansion (8.27) has been performed up to degree 3. The large N limit is stable in the following sense. As soon as $N \geq 3k$, all the terms of order less than or equal to $3k$ in the expansion (8.25) depend only on the θ_m, and no longer on N. Indeed, the larger (size N) determinants must have p_0 on the diagonal in all positions $i > 3k$; otherwise, two lines would be equal and the corresponding determinant would vanish. Therefore, their value becomes stable, and the extra lines with p_0 on the diagonal may be removed (this is why only determinants up to size 3 contribute to the order 3 in (8.27). Hence N simply disappears from the series expansion as it grows larger: This makes the large N limit much simpler here than in the matrix models of Part A. We can write

$$\Theta(\theta_1, \theta_2, \dots) = \lim_{N \to \infty} \Theta_N(\Lambda^{-1})$$

$$= \sum_{\substack{n_i \geq 0 \\ i=1,2,\dots}} c_{n_1} d_{n_2} \cdots \chi_{3n_1,3n_2,\dots}(\theta_1, \theta_2, \dots) \qquad (8.28)$$

with an arbitrary number of n_i's.

In the first part of Theorem 7.2, the function $\Theta(\theta_1, \theta_2, \dots)$ must be shown to depend on the odd θ's only, namely $(\partial/\partial\theta_{2i})\Theta = 0$. To prove this, we use the basic property of the Schur polynomials, easily derived from (8.23),

$$\frac{\partial}{\partial\theta_m} p_k = p_{k-m}, \tag{8.29}$$

and the determinant expression (8.24) for the characters χ. The partial derivative $(\partial/\partial\theta_{2i})$ acts on each determinant as the following sum (by multilinearity):

$$\frac{\partial}{\partial\theta_{2i}} \det\left[p_{3n_k - k + l}\right]_{kl}$$

$$= \sum_{s=1,2,\dots} \left(\begin{vmatrix} p_{3n_1} & p_{3n_1+1} & \cdots \\ \vdots & \vdots & \\ p_{3n_s - s + 1 - 2i} & p_{3n_s - s + 2 - 2i} & \cdots \\ \vdots & \vdots & \end{vmatrix} \right), \tag{8.30}$$

where the derivative acts successively on each row s of the determinant. For a fixed s, let us look at the row $s + 2i$: It is identical to the differentiated row s, with simply $n_s \mapsto n_{s+2i}$. Indeed, it reads

$$\left(p_{3n_{s+2i} - (s+2i) + 1}, p_{3n_{s+2i} - (s+2i) + 2}, \dots\right).$$

But in (8.25), with $N \to \infty$, the two indices n_s and n_{s+2i} have to be summed over, with the respective prefactors $a_{n_s}^{(s \bmod 2)}$ and $a_{n_{s+2i}}^{(s+2i \bmod 2)}$, which are both c's or d's. Hence this sum is symmetric under the exchange of n_s and n_{s+2i}, but the determinant is antisymmetric under the exchange of the rows s and $s + 2i$, so the result of the sum is zero. This completes the proof that $\Theta(\theta_1, \theta_2, \dots)$ is a function of only the

$$t_i = \theta_{2i-1}(\Lambda^{-1}) = \operatorname{tr} \frac{\Lambda^{-2i+1}}{2i - 1} \tag{8.31}$$

for $i = 1, 2, \dots$.

9   The Kontsevich Integral as τ-Function of the KdV Hierarchy

Let us now turn to the proof of Theorem 7.2. The first step consists, for finite N, in rewriting the function $\Theta_N(\theta_1, \theta_2, \dots)$ as a single (Wronskian) determinant. This is readily done by use of the multilinearity of the characters χ, to resum each row s of the determinant of p_m's over the integer n_s, with the prefactor $a_{n_s}^{(s \bmod 2)}$. Introducing the function

$$f_s(\theta_1, \theta_2, \dots) = \sum_{n \geq 0} a_n^{(s \bmod 2)} p_{3n - s + N}, \tag{9.1}$$

the resummed sth line of the determinant of the p_m's reads

$$(f_s^{(N-1)}, f_s^{(N-2)}, \ldots, f_s'', f_s', f_s),$$

where we write $f' = (\partial f/\partial \theta_1)$ and where the shifts of the indices of the p's have been translated into derivatives of f_s, using the property $\partial^k p_m/\partial \theta_1^k = p_{m-k}$, a direct consequence of (8.29). This shows that $\Theta_N(\theta_1, \theta_2, \ldots)$ is the Wronskian $W_N(f_1, f_2, \ldots, f_N)$:

$$\Theta_N(\theta_1, \theta_2, \ldots) = W_N(f_1, \ldots, f_N) = \det\left[f_i^{(j-1)} \right]_{1 \leq i,j \leq N}. \qquad (9.2)$$

In a second step, let us introduce the differential operator Δ_N of order N in $d = \partial/\partial \theta_1$ by its action on a function $f(\theta_1)$:

$$\Delta_N(f) = \frac{W_{N+1}(f, f_1, f_2, \ldots, f_N)}{W_N(f_1, f_2, \ldots, f_N)}. \qquad (9.3)$$

We may write $\Delta_N = \sum_{r=0}^{N} w_{r,N} d^{N-r}$, where $w_{r,N}$ is the minor of the Wronskian determinant corresponding to $f^{(N-r)}$, divided by $\Theta_N(\theta_1, \theta_2, \ldots)$. The operator Δ_N is therefore normalized to have $w_{0,N} = 1$. In particular, we have

$$w_{1,N} = -\frac{\partial}{\partial \theta_1} \log \Theta_N(\theta_1, \theta_2, \ldots). \qquad (9.4)$$

In a third step, we derive the following equation for Δ_N:

$$\frac{\partial}{\partial \theta_i} \Delta_N = Q_{i,N} \Delta_N - \Delta_N d^i, \qquad (9.5)$$

where $Q_{i,n} = (\Delta_N d^i \Delta_N^{-1})_+$, a necessary condition for the lhs of (9.5) to be a differential operator. To prove (9.5), it is sufficient to prove it when it acts on the N linearly independent solutions f_1, f_2, $\ldots$, f_N of the differential equation $\Delta_N f = 0$. As the f_s are simple series (9.1) of the Schur polynomials, d^i acts on them as $(\partial/\partial \theta_1)^i = (\partial/\partial \theta_i)$, so that

$$\left(\frac{\partial}{\partial \theta_i} \Delta_N + \Delta_N \frac{\partial^i}{\partial \theta_1^i} \right) f_s = \left\{ \frac{\partial}{\partial \theta_i}, \Delta_N \right\} f_s = \frac{\partial}{\partial \theta_i}(\Delta_N f_s) = 0, \qquad (9.6)$$

which amounts to (9.5) when acting on f_s.

In the fourth, and last step, we let N tend to infinity. Due to the above-mentioned stabilization order by order of the expansion of Θ_N, the same holds for the coefficients $w_{r,N} \mapsto w_r$; hence we finally get a pseudodifferential operator L of order 1, by taking the limit

$$L = \lim_{N \to \infty} \Delta_N d \Delta_N^{-1}. \qquad (9.7)$$

Equation (9.5) for Δ_N is easily translated into the KP flows

$$\frac{\partial}{\partial \theta_i} L = [L_+^i, L].$$

(9.8)

But recall that we have shown that Θ_N depends only on the θ_i with odd index i. In particular, we deduce that $\partial L/\partial \theta_2 = 0$, which implies through (9.8) that $L^2 = L_+^2$ is a differential operator of order 2. More precisely, we have

$$Q = L^2 = (1 - w_1 d^{-1} + \cdots) d^2 (1 + w_1 d^{-1} + \cdots)$$
$$= d^2 + 2 w_1' = d^2 - u$$

(9.9)

with $w_1 = \partial \log \Theta(t_1, t_2, \dots)/\partial t_1$. Hence

$$u(t_1, t_2, \dots) = -2 \frac{\partial^2}{\partial t_1^2} \log \Theta(t_1, t_2, \dots),$$

(9.10)

which amounts to the statement (7.11)–(7.12) of Theorem 7.2.

10   Main Equivalence Theorem Between Topological and Quantum Gravities

10.1   Loop Equations for the Kontsevich Model

There are various ways of getting direct "equations of motion" for the matrix integral Θ_N. One of them is to write that integrals of total derivatives with respect to matrix elements of M vanish (with vanishing boundary conditions at infinity)

$$0 = -2i \int \frac{\partial}{\partial M_{kk}} \left[e^{i \, \text{tr} \, (M^3/6 + M\Lambda^2/2)} \right] dM$$
$$= \int ((M^2)_{kk} \lambda_k^2) e^{i \, \text{tr}(M^3/6 + M\Lambda^2/2)} \, dM,$$

(10.1)

which we write as $\langle (M^2)_{kk} + \lambda_k^2 \rangle = 0$.

Another way is to express the invariance of the matrix integral (say for $A_N(\Lambda)$ (8.5)) under infinitesimal changes of variables. If we perform in (8.5) the change of variable

$$M \mapsto M + i\varepsilon [X, M],$$

(10.2)

where the matrix X is defined by $X_{ij} = \delta_{ik}\delta_{jl}M_{kl}$ and ε is a small parameter, then the change of the integral reads at first order in ε

$$\delta_\varepsilon A_N = i\varepsilon \left\langle M_{ll} - M_{kk} + \frac{i}{2}(\lambda_k^2 - \lambda_l^2) M_{kl} M_{lk} \right\rangle = 0,$$

(10.3)

where the first term comes from the Jacobian of the transformation (10.2)

$$dM \mapsto \left(1 + i\varepsilon(M_{ll} - M_{kk}) + O(\varepsilon^2)\right) dM, \qquad (10.4)$$

and the second term from the variation of the exponential. Dividing (10.3) by $(\lambda_k^2 - \lambda_l^2)$ and summing over $l \neq k$, this can be rewritten as

$$\left\langle \lambda_k^2 + (M_{kk})^2 - 2i \sum_{l \neq k} \frac{M_{kk} - M_{ll}}{\lambda_k^2 - \lambda_l^2} \right\rangle = 0. \qquad (10.5)$$

But the insertion of M_{kk} is generated by acting with $-i(\partial/\lambda_k\partial\lambda_k)$ on $A_N(\Lambda)$, and we finally get a set of constraints on $A_N(\Lambda)$, for $k = 1, 2, \ldots, N$:

$$\left\{ \lambda_k^2 - \left(\lambda_k^{-1} \frac{\partial}{\partial\lambda_k} \right)^2 \right.$$
$$\left. - 2 \sum_{l \neq k} \frac{1}{\lambda_k^2 - \lambda_l^2} \left(\frac{1}{\lambda_k} \frac{\partial}{\partial\lambda_k} - \frac{1}{\lambda_l} \frac{\partial}{\partial\lambda_l} \right) \right\} A_N(\Lambda) = 0. \quad (10.6)$$

This is the matrix generalization of the Airy equation (8.16). With the relations (8.4) and (8.3), this turns into the following equation for $\Theta_N(\Lambda)$:

$$\left\{ \lambda_k^{-2} \left(\sum_l \lambda_l^{-1} \right)^2 + \frac{1}{4} \lambda_k^{-4} + 2 \sum_{l \neq k} \frac{1}{\lambda_k^2 - \lambda_l^2} \left(\lambda_k^{-1} \frac{\partial}{\partial\lambda_k} - \lambda_l^{-1} \frac{\partial}{\partial\lambda_l} \right) \right.$$
$$\left. - 2 \left(1 + \lambda_k^{-2} \sum_l \frac{1}{\lambda_k + \lambda_l} \right) \frac{\partial}{\partial\lambda_k} + \left(\lambda_k^{-1} \frac{\partial}{\partial\lambda_k} \right)^2 \right\} \Theta_N(\Lambda) = 0. \quad (10.7)$$

The differential operator acting on Θ_N is expressed as sums over λ indices $l = 1, 2, \ldots, N$. Letting $N \to \infty$ in (10.7), we may relax this condition and express the operator in terms of invariants of Λ and $d/d\Lambda$, in which the N-dependence disappears. If we expand this large N limit of (10.7) as a power series in $1/\lambda_k$, we get the following constraints:

$$\sum_{m=-1}^{+\infty} \lambda_k^{-2(m+1)} L_m \Theta(\Lambda) = 0, \qquad (10.8)$$

where L_m is a differential operator in the λ_i, expressible in terms of the t_i and $\partial/\partial t_i$ only. We give only the first one:

$$L_{-1} = \frac{t_1^2}{4} - \frac{1}{2} \frac{\partial}{\partial t_1} - \sum_{k=1}^{\infty} \left(k + \frac{1}{2} \right) t_{k+1} \frac{\partial}{\partial t_k}. \qquad (10.9)$$

Differentiating twice with respect to t_1 and dividing by Θ the equation $L_{-1}\Theta = 0$, we obtain

$$\left(-\frac{1}{2} \frac{\partial}{\partial t_1} - \sum_{k=1}^{\infty} \left(k + \frac{1}{2} \right) t_{k+1} \frac{\partial}{\partial t_k} \right) u = 0, \qquad (10.10)$$

where we have used (9.10). Up to a shift $t_2 \mapsto t_2 - \frac{1}{3}$, this is nothing but the canonical relation $[P, Q] = 1$, with Q as in (9.9) and P equal to

$$P = -\sum_{k=1}^{\infty}\left(k + \frac{1}{2}\right)t_{k+1}Q_+^{k-1/2}, \qquad (10.11)$$

which completes the proof of Theorem 7.3.

10.2   Equivalence Between Quantum and Topological Gravities

Theorems 7.2–7.3 above establish a remarkable link between the two following theories. On the one hand, equations (7.12) and (7.13) have been shown, in Part A, to completely determine the perturbative expansions of the gravity + (multicritical)matter theories as defined through (multicritical) double scaling limits of the one-matrix model. On the other hand, here the same equations determine the (perturbative) formal generating function for intersection numbers of the moduli space of Riemann surfaces, with arbitrary genus and punctures.

This provides us with a mapping from the set of all (multicritical) gravity +matter theories defined with one-matrix models to the intersection theory of the moduli space. Precisely, this means that a given $(m-1)$-multicritical theory (with $t_{k+1} = \delta_{k,m}$ in (7.13)) corresponds to a perturbation of the intersection theory of the moduli space around the point $t_{k+1} = \delta_{k,m}$. In this sense, there is a complete correspondence between correlations of observables in the $(m-1)$-critical model and the collection of intersection numbers expressed as the series expansion $F(t_1, t_2, \ldots)$ around the point $t_{k+1} = \delta_{k,m}$.

10.3   Topological Field Theory and Topological Gravity

More generally, Kontsevich has introduced generalizations of his integral (7.9) to describe the other models of 2-D quantum gravity described in Part A, governed by the commutation relation $[P, Q] = 1$, with P and Q two differential operators of orders p and q. Those, in turn, correspond to the intersection theory of coverings of the moduli space.

Even more generally, the topological sigma models provide more candidates for the coupling of matter to quantum gravity. The fields in these models are (stable, holomorphic) maps from the moduli space to some compact variety V (the matter target), usually endowed with a complex structure. The observables of these theories are the so-called Gromov–Witten [25] cohomology classes of this space of maps, and their intersection theory defines the correlation functions of these observables. The remarkable (and probably very general) fact is that these correlators are linked by recursion relations, which may be reinterpreted as some constraints on their generating functions. In many known cases, this generating function turns out to be a tau-function for some integrable partial differential system (e.g., the Toda hierarchy for $V = CP^1$; cf. [26]).

11 Conclusion

In view of all the above, it is tempting to conjecture that 2-D gravity + matter will always be described by some integrable differential systems, hence the importance of the problem of classification of these systems.

For instance, in the case of $c < 1$ conformal models coupled to gravity, the conformal theories based on Lie algebras of E type must be described by E-KdV flows, defined in [20] but never made explicit. In that respect, one should be able to understand what is special about the A, D, E GKdV flows in the set of integrable systems.

Apart from its now obvious physical interest, the framework of 2-D gravity appears also as a beautiful unifying picture for algebraic–geometrical ideas (such as intersection theory on moduli spaces) and differential–geometrical ideas (such as integrability of differential systems).

12 REFERENCES

[1] P. Di Francesco, P. Ginsparg, and J. Zinn-Justin, *2-D quantum gravity and random matrix models*, Phys. Rep. **254** (1995), 1–131.

[2] A. Polyakov, *Quantum geometry of bosonic strings*, Phys. Lett. **B103** (1981), 207–210; *Quantum geometry of fermionic strings*, 211–213.

[3] V. Knizhnik, A. Polyakov, and A. Zamolodchikov, *Fractal structure of 2-D quantum gravity*, Mod. Phys. Lett. **A3** (1988) 819; F. David, *Conformal field theories coupled to 2-D gravity in the conformal gauge*, Mod. Phys. Lett. **A3** (1988), 1651; J. Distler and II. Kawai, *Conformal field theory and 2-D quantum gravity or who's afraid of Joseph Liouville?* Nucl. Phys. **B321** (1989), 509.

[4] L. Onsager, *Crystal statistics. 1. A two-dimensional model with an order disorder transition*, Phys. Rev. **65** (1944) 117–149.

[5] P. Di Francesco, P. Mathieu, and D. Sénéchal, *Conformal Field Theory* (Springer, New York, 1996).

[6] C. Itzykson, H. Saleur, and J.-B. Zuber (eds.), *Conformal Invariance and Applications to Statistical Mechanics*, (World Scientific, 1988).

[7] E. Brézin, C. Itzykson, G. Parisi, and J.-B. Zuber, *Planar diagrams*, Commun. Math. Phys. **59** (1978), 35–51.

[8] D. Bessis, *A new method in the combinatorics of the topological expansion*, Comm. Math. Phys. **69** (1979), 147–163.

[9] D. Bessis, C.Itzykson, and J.-B. Zuber, *Quantum field theory techniques in graphical enumeration*, Adv. in Appl. Math. **1** (1980), 109–157.

[10] C. Itzykson and J.-B. Zuber, *The planar approximation. 2.*, J. Math. Phys. **21** (1980), 411; Harish-Chandra, *Differential operators on a semisimple Lie algebra*, Amer. J. Math. **79** (1957), 87.

[11] S. Chadha, G. Mahoux, and M.-L. Mehta, *A method of integration over matrix variables. 2.*, J. Phys. **A14** (1981), 579.

[12] M. Staudacher, *The Yang–Lee edge singularity on a dynamical planar random surface*, Nucl. Phys. **B336** (1990), 349.

[13] V. Kazakov, *Ising model on a dynamical planar random lattice: exact solution*, Phys. Lett. **A119** (1986), 140–144.

[14] C. Cernkovic, P. Ginsparg, and G. Moore, *The Ising model, the Yang–Lee edge singularity, and 2-D quantum gravity*, Phys. Lett. **B237** (1990), 196.

[15] E. Brézin and V. Kazakov, *Exactly solvable field theories of closed strings*, Phys. Lett. **B236** (1990), 144–150; M. Douglas and S. Shenker, *Strings in less than one dimension*, Nucl. Phys. **B335** (1990), 635; D. Gross and A. Migdal, *Nonperturbative two-dimensional quantum gravity*, Phys. Rev. Lett. **64** (1990), 127; *A nonperturbative treatment of two-dimensional quantum gravity*, Nucl. Phys. **B340** (1990), 333–365.

[16] M. Douglas, *Strings in less than one dimension and generalized KP hierarchies*, Phys. Lett. **B238** (1990), 176.

[17] I. Gelfand and L. Dikii, *Fractional powers of operators, and Hamiltonian systems*, Funct. Anal. Appl. **10:4** (1976), 13 and *The resolvent, and Hamiltonian systems*, Funct. Anal. Appl. **11:2** (1977), 93.

[18] A. Cappelli, C. Itzykson, and J.-B. Zuber, *The ADE classification of minimal and $A_1^{(1)}$ conformal invariant theories*, Comm. Math. Phys. **113** (1987), 1–26.

[19] P. Di Francesco and D. Kutasov, *Unitary minimal models coupled to gravity*, Nucl. Phys. **B342** (1990), 589 and *Integrable models of 2-D quantum gravity*, in *Random Surfaces and Quantum Gravity*, NATO ASI Series **B262** (1991), 35–51.

[20] V. Drinfeld and V. Sokolov, *Lie algebras and equations of Korteweg–De Vries type*, J. Sov. Math. **30** (1985) 1975; Sov. Math. Dokl. **23** No.3 (1981), 457.

[21] E. Witten, *On the structure of the topological phase of two-dimensional gravity*, Nucl. Phys. **B340** (1990), 281–332 and *Two-dimensional gravity and and intersection theory on moduli space*, Surv. in Diff. Geom. **1** (1991), 243–310.

[22] M. Kontsevich, *Intersection theory on the moduli space of curves and the matrix Airy function*, Comm. Math. Phys. **147** (1992), 1–23.

[23] C. Itzykson and J.-B. Zuber, *Combinatorics of the modular group. 2. The Kontsevich integrals*, Int. J. Mod. Phys. **A7** (1992), 5661–5705.

[24] P. Di Francesco, C. Itzykson, and J.-B. Zuber, *Polynomial averages in the Kontsevich model*, Comm. Math. Phys. **151** (1993), 193–219.

[25] M. Kontsevich and Y. Manin, *Gromov–Witten classes, quantum cohomology, and enumerative geometry*, Comm. Math. Phys. **164** (1994), 525–562.

[26] T. Eguchi and S.-K. Yang, *The topological CP^1 model and the large N matrix integral*, Mod. Phys. Lett. **A9** (1994), 2893–2902.

6

Painlevé Transcendents in Two-Dimensional Topological Field Theory

Boris Dubrovin

ABSTRACT This paper is devoted to the theory of WDVV equations of associativity. This remarkable system of nonlinear differential equations was discovered by E. Witten [85] and R. Dijkgraaf, E. Verlinde, and H. Verlinde [24] in the beginning of the 1990s. It was first derived as equations for the so-called primary free energy of a family of two-dimensional topological field theories. Later it proved to be an efficient tool in the solution of problems of the theory of Gromov–Witten invariants, reflection groups and singularities, and integrable hierarchies.

Here we mainly consider the relationships of WDVV to the theory of Painlevé equations. This is a two-way connection. First, any solution to WDVV satisfying certain semisimplicity conditions can be expressed via Painlevé-type transcendents. Conversely, theory of WDVV works as a source of remarkable particular solutions of the Painlevé equations.

This chapter is an extended version of the lecture notes of a course given at the 1996 Cargèse summer school, "The Painlevé Property: One Century Later." It is organized as follows.

In Section 1 we give a sketch of the ideas of two-dimensional topological field theory, we formulate WDVV, and give the main examples of solutions coming from quantum cohomology and from singularity theory. In Section 2 we give a coordinate-free reformulation of WDVV introducing the notion of a Frobenius manifold. We also construct the first main geometrical object, namely the deformed affine connection on a Frobenius manifold. The monodromy of the deformed connection at the origin gives us the first set of important invariants of Frobenius manifolds. In Section 3 we define the class of semisimple Frobenius manifolds. In physics these correspond to two-dimensional topological field theories with all relevant perturbations. We construct the so-called canonical coordinates on such manifolds. In Section 4 we complete the classification of semisimple Frobenius manifolds in terms of monodromy data of a certain universal linear differential operator with rational coefficients. We give a nontrivial example of computation of the monodromy data in quantum cohomology. In the last section we develop a "mirror construction" representing the principal geometrical objects on a semisimple Frobenius manifold by residues and oscillatory integrals of a family of analytic functions on Riemann surfaces.

1 Algebraic Properties of Correlators in 2-D Topological Field Theories. Moduli of a 2-D TFT and WDVV Equations of Associativity

By definition, a quantum field theory (QFT) on a D-dimensional oriented manifold Σ (in our case $D = 2$) consists of:

1. Local fields $\phi_\alpha(x)$, $x \in \Sigma$. The metric $g_{ij}(x)$ on Σ could be one of the fields. It is called gravity.

2. The Lagrangian

$$L = L(\phi, \partial_x \phi, \dots).$$

   The equations of motion of the classical field theory have the form

$$\frac{\delta S}{\delta \phi_\alpha(x)} = 0,$$

   where

$$S[\phi] = \int_\Sigma L(\phi, \partial_x \phi, \dots)$$

   is the classical action.

3. In the path-integral quantization we are interested in the partition function

$$Z_\Sigma = \int [\mathrm{d}\phi] e^{-S[\phi]}$$

   and, more generally, in the (nonnormalized) correlation functions

$$\langle \phi_\alpha(x)\phi_\beta(y)\cdots\rangle_\Sigma = \int [\mathrm{d}\phi]\phi_\alpha(x)\phi_\beta(y)\cdots e^{-S[\phi]}.$$

   The integration in both cases is over the space of local fields ϕ on Σ with an appropriate measure $[\mathrm{d}\phi]$. In the full theory we also have to take an integration over the space of manifolds Σ.

4. The theory admits topological invariance if an arbitrary change of the metric on Σ preserves the action

$$\frac{\delta S}{\delta g_{ij}(x)} \equiv 0.$$

   In the $D = 2$ case such a theory will be called 2-D topological field theory (TFT). For example, in the 2-D case the total curvature functional

$$S = \frac{1}{2\pi} \int_\Sigma R\sqrt{g}\, \mathrm{d}^2 x$$

is topologically invariant. Indeed, due to the Gauss–Bonnet theorem it is equal to the Euler characteristic of the surface Σ.

For a topological field theory the partition function gives a topological invariant of Σ. The correlation functions depend only on the topology of Σ and the fields (but not on their positions). In particular, in the 2-D case we have

$$\langle \phi_\alpha(x)\phi_\beta(y)\cdots\rangle_\Sigma \equiv \langle \phi_\alpha\phi_\beta\cdots\rangle_g.$$

On the rhs. there are just numbers depending on the genus g of the surface Σ and on the labels α, β, ... of the fields.

5. In the *matter sector* of the QFT we integrate over the space of all fields except the metric $(g_{ij}(x))$. For a TFT the correlators of the matter sector have a nice algebraic description, to be presented in a moment. To describe coupling of the QFT to gravity, one has to integrate over the space of metrics. In TFTs, the coupling to gravity can be reduced to an integration over the space of conformal classes of the metrics on Σ, i.e., over the moduli space of Riemann surfaces of genus $g = g(\Sigma)$. This is a much more complicated procedure, until now understood only for the genera $g = 0, 1$.

We now describe the algebraic properties of the matter sector correlators in a 2-D TFT. We will consider simple theories having a finite number of observables in the matter sector

$$\phi_1,\ldots,\phi_n$$

(the so-called primary chiral fields). One can easily derive all the algebraic properties of the correlators using the general Atiyah axioms of a topological field theory. We present here only a summary of these properties.

Definition 1.1. A *Frobenius algebra* is a pair $(A\langle\,,\,\rangle)$ where A is a commutative associative algebra (over $\mathbb{C}$) with a unity, and $\langle\,,\,\rangle$ stands for a symmetric nondegenerate *invariant* bilinear form on A. The invariance means the validity of the following identity:

$$\langle ab, c\rangle = \langle a, bc\rangle \tag{1.1}$$

for three arbitrary vectors a, b, $c \in A$.

Theorem 1.1 (See [22, 23, 32]). *The matter sector correlators of any 2-D TFT with n observables can be encoded by a Frobenius algebra $(A, \langle\ \rangle)$ with a distinguished basis $e_1,\ \ldots,\ e_n$. The genus g correlators of the observables have the form*

$$\langle \phi_{\alpha_1}\phi_{\alpha_2}\cdots\phi_{\alpha_k}\rangle_g = \langle e_{\alpha_1}\cdot e_{\alpha_2}\cdots\phi_{\alpha_k}, H^g\rangle,$$

where

$$H = \eta^{\alpha\beta} e_\alpha \cdot e_\beta \in A,$$
$$(\eta^{\alpha\beta}) = (\eta_{\alpha\beta})^{-1}, \quad \eta_{\alpha\beta} := \langle e_\alpha, e_\beta \rangle.$$

Physicists call $(A, \langle\,,\,\rangle)$ the primary chiral algebra of the TFT. Observe that the structure of the Frobenius algebra is uniquely determined by the genus zero two-point and three-point correlators

$$\langle e_\alpha, e_\beta \rangle = \langle \phi_\alpha \phi_\beta \rangle_0, \quad \langle e_\alpha \cdot e_\beta, e_\gamma \rangle = \langle \phi_\alpha \phi_\beta \phi_\gamma \rangle_0.$$

Usually, the observables are chosen in such a way that the vector e_1 coincides with the unity of the algebra A. Then

$$\langle e_\alpha, e_\beta \rangle = \langle \phi_1 \phi_\alpha \phi_\beta \rangle_0.$$

We now give the two main "physical" examples of Frobenius algebras.

Example 1.1. Let X be a two-dimensional closed oriented manifold without odd-dimensional cohomologies. Take the full cohomology algebra

$$A = H^*(X)$$

with the bilinear form

$$\langle \omega_1, \omega_2 \rangle = \int_X \omega_1 \wedge \omega_2, \quad \omega_1, \omega_2 \in H^*(X) \tag{1.2}$$

(we realize cohomologies by classes of closed differential forms). Symmetry and invariance of this bilinear form are obvious. Nondegeneracy follows from the Poincaré duality theorem. This Frobenius algebra describes the matter sector of the topological sigma model (X is the target space).

Actually, in this example we have a certain graded structure on $(A, \langle\,,\,\rangle)$. Generalizing, we give the following definition.

Definition 1.2. The Frobenius algebra is called *graded* if a linear operator $Q \colon A \to A$ and a number d are defined such that for any $a, b \in A$,

$$Q(ab) = Q(a)b + aQ(b), \tag{1.3a}$$
$$\langle Q(a), b \rangle + \langle a, Q(b) \rangle = d\langle a, b \rangle. \tag{1.3b}$$

The operator Q is called a *grading operator*, and the number d is called the *charge* of the Frobenius algebra. We will consider only the case of diagonalizable grading operators. Then we may assign degrees to the eigenvectors e_α of Q:

$$\deg(e_\alpha) = q_\alpha \quad \text{if } Q(e_\alpha) = q_\alpha e_\alpha. \tag{1.4a}$$

For the topological example the vectors of a homogeneous basis are chosen in such a way that

$$e_\alpha \in H^{2q_\alpha}(X), \quad \deg(e_\alpha) = q_\alpha. \tag{1.4b}$$

The charge d is equal to the half of the dimension of X.

A particular example: $X = \mathbf{CP}^d$. The full cohomology space has dimension $n = d + 1$. The natural basis in $A = H^*(\mathbf{CP}^d)$ is

$$1, \; \omega, \; \omega^2, \ldots, \omega^d,$$

where ω is the standard Kähler form on the projective space. We normalize it by the condition

$$\int_{\mathbf{CP}^d} \omega^d = 1.$$

Then $(A, \langle \, , \, \rangle)$ is isomorphic to the quotient of the polynomial algebra

$$A = \mathbb{C}[\omega]/(\omega^{d+1}),$$

with the bilinear form

$$\langle \omega^k, \omega^l \rangle = \delta_{k+l,d}.$$

Remark. We will also consider below graded Frobenius algebras $(A, \langle \, , \, \rangle)$ over graded commutative associative rings R. In this case we have two grading operators $Q_R \colon R \to R$ and $Q_A \colon A \to A$ satisfying the properties

$$\begin{align}
Q_R(\alpha\beta) &= Q_R(\alpha)\beta + \alpha Q_R(\beta), & \alpha, \beta &\in R, \tag{1.5a} \\
Q_A(ab) &= Q_A(a)b + a Q_A(b), & a, b &\in A, \tag{1.5b} \\
Q_A(\alpha a) &= Q_R(\alpha)a + \alpha Q_A(a), & \alpha &\in R, a \in A, \tag{1.5c} \\
Q_R\langle a, b \rangle + d\langle a, b \rangle &= \langle Q_A(a), b \rangle + \langle a, Q_A(b) \rangle, & a, b &\in A. \tag{1.5d}
\end{align}$$

The number d is called *the charge* of the graded Frobenius algebra.

Example 1.2. [Example of a Frobenius algebra] Let $f(x)$ be a polynomial of $x \in \mathbb{C}^N$ with an isolated singularity at $x = 0$. This means that

$$df(x)|_{x=0} = 0$$

(we may also assume that $f(0) = 0$),

$$df(x)|_{x \neq 0} \neq 0$$

for x sufficiently close to the origin. Take the quotient of the polynomial algebra

$$A = \mathbb{C}[x] \, \Big/ \, \left(\frac{\partial f}{\partial x_1}, \ldots, \frac{\partial f}{\partial x_N} \right). \tag{1.6}$$

This is called the Jacobi ring, or the local algebra of the singularity. This is a finite-dimensional algebra if the singularity has finite multiplicity n. (The number $n = \dim A$ is also called the Milnor number of the singularity.) We define a bilinear form on A by taking the residue

$$\langle p, q \rangle = \frac{1}{(2\pi i)^N} \int_{\bigcap_i |\partial f/\partial x_i| = \varepsilon} \frac{p(x)q(x)\mathrm{d}^N x}{(\partial f/\partial x_1) \cdots (\partial f/\partial x_N)}. \qquad (1.7)$$

Here ε is a sufficiently small positive number. Again, symmetry and invariance of the bilinear form are trivial. Nondegeneracy is less trivial; see the proof in [4, Volume 1, Section 5],. To obtain a graded Frobenius algebra, one way is to take a quasi-homogeneous polynomial $f(x)$. This Frobenius algebra describes the matter sector of a topological Landau–Ginzburg model. The function $f(x)$ is called a superpotential of the theory.

A particular example: a simple singularity of the A_n type. Here $N = 1$, $f(x) = x^{N+1}$. The local algebra is

$$A = \mathbb{C}[x]/(x^{N+1}) = \mathrm{span}(1, x, x^2, \ldots, x^{n-1}),$$

$$x^k \cdot x^l = \begin{cases} x^{k+l}, & k+l < n, \\ 0, & k+l \geq n, \end{cases}$$

$$\langle x^k, x^l \rangle = \mathrm{res}\, \frac{x^{k+l}}{(n+1)x^n} = \begin{cases} 0, & k+l \neq n-1, \\ 1/(n+1), & k+l = n-1. \end{cases}$$

The grading operator is determined by

$$Q(x) = \frac{1}{n+1} x,$$

the charge is

$$d = \frac{n-1}{n+1}.$$

We have already said that the procedure of coupling a 2-D TFT to gravity is more complicated (not settled in full generality). For the genus-zero case it can still be done in an axiomatic way. It turns out that the axioms of coupling to gravity can be reduced to WDVV equations of associativity. Here WDVV stands for Witten–Dijkgraaf–E. Verlinde–H. Verlinde. In the paper [85] the equations of associativity were derived in the setting of topological sigma models. In [24] they were derived in a more general class of TFTs obtained by the so-called twisting from $N = 2$ supersymmetric QFTs. Basically, the idea was to consider correlators of a particular n-dimensional family of TFTs,

$$S \mapsto S - \sum_{\alpha=1}^{n} \int_{\Sigma} \phi_\alpha^{(2)},$$

as functions of the coupling constants $t = (t^1, \ldots, t^n)$. Here $\phi_1^{(2)}, \ldots, \phi_n^{(2)}$ are certain two-forms on Σ in one-to-one correspondence with the observables $\phi_1, \ldots, \phi_n$. The deformation preserves the topological invariance (not the grading!). So one obtains an n-dimensional deformation $(A_t, \langle \, , \, \rangle_t)$ of the n-dimensional Frobenius algebra $(A, \langle \, , \, \rangle) = (A_0, \langle \, , \, \rangle_0)$. A basis $e_1 = 1$, $e_2, \ldots, e_n$ corresponding to the chosen system of observables $\phi_1, \ldots, \phi_n$ is fixed in all the algebras A_t. The following properties of the family of Frobenius algebras $(A_t, \langle \, , \, \rangle_t)$ were proved by WDVV:

$$\langle e_\alpha, e_\beta \rangle_t \equiv \langle e_\alpha, e_\beta \rangle, \tag{WDVV1}$$

$$c_{\alpha\beta\gamma}(t) := \langle e_\alpha \cdot e_\beta, e_\gamma \rangle_t = \frac{\partial^3 F(t)}{\partial t^\alpha \partial t^\beta \partial t^\gamma}. \tag{WDVV2}$$

Here $F(t) = \log Z_0(t)$ is the genus-zero free energy of the family of TFTs (the so-called primary free energy).

The last property is the quasi-homogeneity condition: The structure constants $c_{\alpha\beta\gamma}(t)$ of the algebras A_t are weighted homogeneous functions of degree $q_\alpha + q_\beta + q_\gamma - d$, where we assign the degree $1 - q_\alpha$ to the variable t^α for each $\alpha = 1, \ldots, n$:

$$c_{\alpha\beta\gamma}(\lambda^{1-q_1} t^1, \ldots, \lambda^{1-q_n} t^n) = \lambda^{q_\alpha + q_\beta + q_\gamma - d} c_{\alpha\beta\gamma}(t^1, \ldots, t^n)$$

for an arbitrary $\lambda \neq 0$. Observe that $q_1 = 0$ if $e_1 = 1$. All the quasi-homogeneity equations can be written as a single one for the primary free energy:

$$F(\lambda^{1-q_1} t^1, \ldots, \lambda^{1-q_n} t^n) = \lambda^{3-d} F(t^1, \ldots, t^n) + \text{quadratic}, \tag{WDVV3}$$

where "quadratic" stands for a polynomial of degree at most two in $t^1, \ldots, t^n$. (Later we will slightly modify the quasi-homogeneity requirement for those t^α where $q_\alpha = 1$; see the beginning of Section 2.)

The WDVV equations of associativity describe the problem of classification of n-dimensional families of n-dimensional Frobenius algebras satisfying the above properties WDVV1–WDVV3. One can consider this problem as the first approximation to the problem of classification of 2-D TFTs, at least of those obtained by twisting from $N = 2$ supersymmetric theories. We do not present here other stories of the whole building of a 2-D TFT (coupling to gravity [28, 32, 86], Zamolodchikov-type Hermitian metric on the space of parameters t [18, 29]). The upper stories, unlike to the basement, cannot be built on an arbitrary solution of WDVV. However, before proceeding to the upper stories, we will study the structure of the eventual basement, i.e., of a solution of WDVV. The present chapter is devoted only to this problem of classification of solutions of WDVV equations of associativity.

We finish this section with a sketch of the construction of the deformed 2-D TFTs for the two above examples. Observe first that for a graded

Frobenius algebra $(A_0, \langle \ , \ \rangle, Q, d)$ one can construct a trivial cubic solution of WDVV

$$F_0 = \frac{1}{6}\langle 1, (t)^3 \rangle, \quad t = t^\alpha e_\alpha \in A_0. \tag{1.8}$$

In all the physical examples the free energy $F(t)$ is constructed as an analytic perturbation of a cubic F_0.

Example 1.3. We will assume the $2d$-dimensional target space X to be Kähler. The deformation of F_0 is defined as the generating function of Gromov–Witten invariants. Let us consider the moduli space of instantons

$$X_{[\beta],l} := \{\text{holomorphic } \beta \colon (S^2, p_1, \dots, p_l) \to X,$$
$$\text{given homotopy class } [\beta] \in H_2(X; \mathbb{Z})\}. \tag{1.9}$$

The holomorphic maps β of the Riemann sphere S^2 with marked points $p_1, \dots, p_l$ are considered up to holomorphic changes of parameter. Under certain assumptions about the manifold X (see [6, 57, 66, 73]), it can be shown that $X_{[\beta],l}$ can be compactified to produce an orbifold whose complex dimension is

$$\dim_{\mathbb{C}} X_{[\beta],l} = d + \int_{S^2} \beta^*\big(c_1(X)\big) + l - 3.$$

Here $c_1(X) \in H^2(X)$ is the first Chern class of X.

Observe that any of the marked points p_i defines the evaluation map that we denote by the same symbol

$$p_i \colon X_{[\beta],l} \to X, \quad (\beta, p_1, \dots, p_l) \mapsto \beta(p_i). \tag{1.10}$$

For an element

$$a_1 \otimes a_2 \otimes \cdots \otimes a_k \in \big(H^*(X)\big)^{\otimes k},$$

define the number

$$\langle a_1 \otimes \cdots \otimes a_k \rangle_{[\beta],l} = \begin{cases} 0, & k \neq l, \\ \int_{X_{[\beta],l}} p_1^*(a_1) \wedge \cdots \wedge p_l^*(a_l), & k = l. \end{cases} \tag{1.11}$$

We extend this symbol linearly to the infinite direct sum

$$\mathbb{C} \oplus H^* \oplus H^* \otimes H^* \oplus (H^*)^{\otimes 3} \oplus \cdots$$

with $H^* := H^*(X)$.

Define now the function $F(t)$,

$$t = (t', t'') \in H^*(X) \tag{1.12a}$$

$$t' \in H^2(X)/2\pi i H^2(X, \mathbb{Z}), \quad t'' \in H^{*\neq 2}(X), \tag{1.12b}$$

$$F(t) = F_0(t) + \sum_{[\beta] \neq 0, l} \langle e^{t''} \rangle_{[\beta],l} e^{\int_{S^2} \beta^*(t')}. \tag{1.12c}$$

Here $F_0(t)$ is the cubic (1.8) for the Frobenius algebra $A_0 = H^*(X)$. The exponential

$$e^t := 1 + \frac{t}{1!} + \frac{1}{2!} t \otimes t + \cdots$$

is considered as an element of the infinite direct sum.

The numbers $\langle a_1 \otimes \cdots \otimes a_k \rangle_{[\beta],l}$ can be nonzero only if the following dimension condition holds true

$$\deg a_1 + \cdots + \deg a_l = \dim X_{[\beta],l} = d + \int_{S^2} \beta^* \big(c_1(X)\big) + l - 3.$$

This can be written in the form

$$\sum_{i=1}^{l} (1 - \deg a_i) = 3 - d - \int_{S^2} \beta^* \big(c_1(X)\big). \tag{1.13}$$

We see from this dimension condition that for any $[\beta], l$ the coefficient

$$\big\langle e^{t''} \big\rangle_{[\beta],l}$$

is a polynomial in $t'' \in H^{*\neq 2}(X)$. The coefficients of these polynomials can be proved to be independent of the complex structure on X [47,66,73] and to depend only on the homotopy class of the symplectic structure on X given by the imaginary part Ω of the Kähler metric. They are called *Gromov–Witten invariants* of (X, Ω). (Actually, one can start with a more general situation to define GW invariants of a compact symplectic manifold (X, Ω). To this end one has to consider pseudoholomorphic maps $\beta \colon S^2 \to X$ with respect to an appropriate almost complex structure on X. See details in [47,66,73].)

The family of algebras A_t with the parameter

$$t \in H^*(X)/2\pi i H^2(X, \mathbb{Z})$$

is called the *quantum cohomology* of X. Sometimes one considers quantum cohomology in the restricted sense where the parameter $t = t'$ of the deformation belongs to

$$t \in H^{1,1}(X)/2\pi i H^2(X, \mathbb{Z}).$$

This restricted quantum cohomology is closely related to the Floer symplectic cohomology of (X, Ω) (see [66,72,75]). In the *point of classical limit* $t' \to -\infty$ (i.e., $\int_{S^2} \beta^*(t') \to -\infty$ for any $[\beta] \neq 0$) $F(t) \to f_0(t)$, so the quantum cohomology goes to the classical one.

A particular example. Quantum cohomology of the projective plane $\mathbf{CP}^2$. For $t = t^1 + t^2 \omega + t^3 \omega^2 \in H^*(\mathbf{CP}^2)$ the cubic function $F_0(t)$ is

$$F_0(t) = \frac{1}{2} \left(t^1\right)^2 t^3 + \frac{1}{2} t^1 (t^2)^2.$$

Here $t' = t^2\omega$, $t'' = t^1 + t^3\omega^2$. The series $F(t)$ has the form [57]

$$F(t) = F_0(t) + \sum_{k=1}^{\infty} \frac{N_k}{(3k-1)!} \left(t^3\right)^{3k-1} e^{kt^2}. \tag{1.14}$$

Here

$N_k = \#\{$rational curves of degree k on $\mathbf{CP}^2$

passing through $3k - 1$ generic points.$\}$

For example, $N_1 = 1$ (one line through 2 points), $N_2 = 1$ (one conic through 5 points). One can see that the quasi-homogeneity condition WDVV3 must be modified: The function $F(t)$ has degree $1 = 3 - 2$ (up to quadratic terms) if t^1 has degree 1, t^3 has degree -1, t^2 has degree 0, but $\exp t^2$ has degree 3. This quasi-homogeneity anomaly comes from the term

$$\int_{S^2} \beta^* \left(c_1(\mathbf{CP}^2)\right)$$

in the dimension condition (1.13).

The series $F(t)$ has the nonempty domain of convergence

$$\mathrm{Re}(t^2 + 3\log t^3) < R \tag{1.15}$$

for some positive R. A numerical estimation for R was obtained by [21]:

$$R \approx 1.981.$$

Actually, the following asymptotic ansatz was proposed in [21]:

$$\frac{N_k}{(3k-1)!} \sim a^k b k^{-7/2}, \quad k \to \infty,$$

with $a \approx 0.138$, $b \approx 6.05$. The exact values of the constants a, b are not known.

The structure constants of the restricted quantum cohomology ring are obtained by triply differentiating $F(t)$ and setting $t^1 = t^3 = 0$. The resulting ring has a very simple structure: This is the quotient of the polynomial ring

$$QH^*(\mathbf{CP}^2) = \mathbb{C}[e_2]/(e_2^3 = q)$$

with

$$q = e^{t^2}.$$

Clearly, at the point of classical limit $q \to 0$ one obtains the classical cohomology ring of the projective plane.

The function $F(t)$ can be proved to solve the WDVV equations of associativity [57]. It was observed by Kontsevich that plugging the ansatz (1.14) with $N_1 = 1$ into the equations of associativity, one can recursively compute all the coefficients N_k. We leave it as an exercise for the reader to derive these recursion relations for N_k.

Remark. Set

$$\phi(x) = \sum_{k=1}^{\infty} \frac{N_k}{(3k-1)!} e^{kx}$$

and

$$\psi(x) = \frac{\phi''' - 27}{8(27 + 2\phi' - 3\phi'')}$$

(the prime stands for the x-derivative). Then the coefficients $N_k^{(1)}$ of the expansion

$$\psi(x) = -\frac{1}{8} + \sum_{k=1}^{\infty} \frac{k\, N_k^{(1)}}{(3k)!} e^{kx}$$

are the elliptic Gromov–Witten invariants of $\mathbf{CP}^2$, i.e., they are the number of elliptic curves of degree k passing through $3k$ generic points on $\mathbf{CP}^2$. This was proved in [36].

Also, in the general situation of quantum cohomology of a manifold X one can prove the validity of WDVV for a vast class of manifolds X [57, 66, 73]. The quasi-homogeneity conditions have the form WDVV3 for the dependence of $F(t)$ on the coordinates of the component $t'' \in H^{*\neq 2}(X)$. For the other component $t' = \sum t'^\alpha e'_\alpha \in H^2(X)$ of $t = (t', t'')$ the coordinates t'^α are dimensionless. We then assign the degrees to the exponentials:

$$\deg e^{t'^\alpha} = r_\alpha \tag{1.16}$$

if

$$c_1(X) = \sum r_\alpha e'_\alpha. \tag{1.17}$$

Clearly, for $X = \mathbf{CP}^2$ we obtain the above condition $\deg \exp t^2 = 3$. For Calabi–Yau (CY) varieties X the exponentials $\exp t'^\alpha$ are also dimensionless, since $c_1(X) = 0$. In particular, for CY 3-folds all the GW polynomials

$$\left\langle e^{t''} \right\rangle_{[\beta],l}$$

are just numbers, as follows from the dimension condition (1.13). This means that, essentially, the full quantum cohomology of a CY 3-fold is reduced to the restricted one (we do not consider here the contributions from the odd-dimensional classes of the CY). According to the mirror conjecture [17], the free energy of a CY 3-fold X can be expressed via certain generalized hypergeometric functions. These hypergeometric functions are periods of the holomorphic three-form on the so-called dual CY 3-fold X^*. The mirror conjecture has been proved in [43–45] for CY complete intersections in projective spaces. A general geometrical setting justifying the mirror conjecture was proposed in [87].

In the opposite case of Fano varieties, where $c_1(X) > 0$, nothing is known about the analytic structure of the free energy (besides the trivial example of the projective line, where the full quantum cohomology is reduced to the restricted one). The restricted quantum cohomology can often be computed (actually, they are computed for all Fano complete intersections in [11]). For many examples of Fano varieties it was shown that, as in the above example of $\mathbf{CP}^2$, one can reconstruct all the GW invariants from the restricted quantum cohomology by just solving recursively the WDVV equations of associativity. The restricted quantum cohomology serves as the initial data to specify uniquely the solution of WDVV.

We suggest that the success of this reconstruction of GW invariants of Fano varieties, unlike that of CY varieties, where WDVV gives essentially no information about the GW invariants, is based on the following conjectural property [82] of quantum cohomology of Fano varieties: The deformed Frobenius algebra A_t is semisimple for generic values of the parameter t. In this chapter we describe the general solution of WDVV satisfying the semisimplicity condition. We will show that it can be expressed via certain Painlevé-type transcendents. We will also discuss the problem of selection of the particular solutions of WDVV corresponding to free energies of physically motivated models of 2-D TFT.

Example 1.4. In the topological Landau–Ginzburg models with the superpotential $f(x)$, the deformed Frobenius algebra is given by formulae similar to (1.6), (1.7), where one must use the versal deformation [3, 4]

$$f_s(x) = f(x) + \sum_{i=1}^{n} s^i p_i(x) \tag{1.18}$$

of the singularity. Here $p_1(x) = 1$, $p_2(x)$, $\ldots$, $p_n(x)$ is a basis of the local algebra of the singularity. (Actually, one must choose properly the volume form $d^N x$ in (1.7). The construction of the needed volume form is given in [77].) The metric

$$\sum \eta_{ij}(s) \mathrm{d}s^i \, \mathrm{d}s^j$$

on the space of parameters $s = (s^1, \ldots, s^n)$ has the form

$$\eta_{ij}(s) = \frac{1}{(2\pi i)^N} \int_{\bigcap_j |\partial f_s(x)/\partial x_j| = \varepsilon} \frac{p_i(x)p_j(x) \, \mathrm{d}^N(x)}{(\partial f_s(x)/\partial x_1) \cdots (\partial f_s(x)/\partial x_n)}. \tag{1.19}$$

Under certain assumptions [77], one can prove that this metric has zero curvature. Thus one can introduce new coordinates $(t^1, \ldots, t^n)$ on the space of parameters such that

$$\eta_{ij} \, ds^i \, ds^j = \eta_{\alpha\beta} \, dt^\alpha dt^\beta$$

with a constant matrix $\eta_{\alpha\beta}$. In these coordinates

$$c_{\alpha\beta\gamma}(t) = \frac{1}{(2\pi i)^N} \int_{\cap_j \left| \frac{\partial f_s(x)}{\partial x_j} \right| = \varepsilon} \frac{\frac{\partial f_s(x)}{\partial t^\alpha} \frac{\partial f_s(x)}{\partial t^\beta} \frac{\partial f_s(x)}{\partial t^\gamma} \, d^N(x)}{\frac{\partial f_s(x)}{\partial x_1} \ldots \frac{\partial f_s(x)}{\partial x_n}}. \tag{1.20}$$

The explicit formulae for the A-D-E simple singularities are given in [14].

A particular case: a simple singularity of A_3 type. Here $p_1 = 1$, $p_2 = x$, $p_3 = x^2$ is a basis in the local algebra. So

$$f_s = x^4 + s_1 + s_2 x + s_3 x^2$$

(I use only lower indices in concrete examples.) The metric (1.7) in the coordinates s_1, s_2, s_3 has a matrix depending on s:

$$\langle p_i, p_j \rangle_s = -4 \operatorname{res}_{x=\infty} \frac{p_i(x) p_j(x)}{4x^3 + 2s_3 x + s_2}.$$

We obtain the following matrix of the metric:

$$\eta_{ij}(s) = \begin{pmatrix} 0 & 0 & 1 \\ 0 & 1 & 0 \\ 1 & 0 & -\frac{1}{2}s_3 \end{pmatrix}.$$

Introducing the new coordinates

$$s_1 = t_1 + \frac{1}{8} t_3^2,$$

$$s_2 = t_2,$$

$$s_3 = t_3,$$

we obtain the constant matrix

$$\eta_{\alpha\beta} = \begin{pmatrix} 0 & 0 & 1 \\ 0 & 1 & 0 \\ 1 & 0 & 0 \end{pmatrix}.$$

The new parameterization of the versal deformation has the form

$$P_t(x) \equiv f_s(x) = x^4 + t_1 + \frac{1}{8} t_3^2 + t_2 x + t_3 x^2.$$

The only nontrivial "three-point functions"

$$c_{\alpha\beta\gamma} = -\operatorname{res}_\infty \frac{\partial_\alpha P_t \partial_\beta P_t \partial_\gamma P_t}{\partial_x P_t} \, dx$$

are

$$c_{113} = c_{122} = 1, \quad c_{223} = -\frac{1}{4}t_3, \quad c_{233} = -\frac{1}{4}t_2, \quad c_{333} = \frac{1}{16}t_3^2.$$

This gives a polynomial solution of WDVV:

$$F(t_1, t_2, t_3) = \frac{1}{2}t_1^2 t_3 + \frac{1}{2}t_1 t_2^2 - \frac{1}{16}t_2^2 t_3^2 + \frac{1}{960}t_3^5. \tag{1.21}$$

We can continue our experiments with WDVV and try to find *all* polynomial solutions $F(t_1, t_2, t_3)$. This simple exercise gives only four polynomial solutions [31, 32]! Besides (1.21) they are

$$F = \frac{1}{2}t_1^2 t_3 + \frac{1}{2}t_1 t_2^2 + \frac{1}{6}t_2^3 t_3 + \frac{1}{6}t_2^2 t_3^3 + \frac{1}{210}t_3^7, \tag{1.22}$$

$$F = \frac{1}{2}t_1^2 t_3 + \frac{1}{2}t_1 t_2^2 + \frac{1}{6}t_2^3 t_3^2 + \frac{1}{20}t_2^2 t_3^5 + \frac{1}{3960}t_3^{11}, \tag{1.23}$$

$$F = \frac{1}{2}t_1^2 t_3 + \frac{1}{2}t_1 t_2^2 + t_2^4. \tag{1.24}$$

The last polynomial does not satisfy the semisimplicity condition. It turns out that the first two can be described in terms of singularities of the type B_3 and H_3, respectively. In Section 5 I will explain the construction of the polynomials (1.21)–(1.23) in terms of invariants of the Coxeter groups of the respective types A_3, B_3, H_3, and the generalization of this construction to higher dimensions. Observe that the Coxeter groups A_3, B_3, H_3 are just all the groups of symmetries of Platonic solids (respectively the tetrahedron, octahedron, and icosahedron). So, not only do the WDVV equations of associativity "know" the enumeration of rational plane curves, but they also "know" the list of Platonic solids! See also the conjecture of Section 5 below regarding polynomial solutions of WDVV.

2   Equations of Associativity and Frobenius Manifolds. Deformed Flat Connection and Its Monodromy at the Origin

We give first the precise formulation of WDVV equations of associativity. Next, we will reformulate them in a coordinate-free form.

We look for a function $F\left(t^1, \ldots, t^n\right) \equiv F(t)$, a constant symmetric non-degenerate matrix $(\eta^{\alpha\beta})$, numbers $q_1, \ldots, q_n, r_1, \ldots, r_n, d$ such that, for any $\alpha, \beta, \gamma, \delta = 1, \ldots, n$,

$$\partial_\alpha \partial_\beta \partial_\lambda F(t) \eta^{\lambda\mu} \partial_\mu \partial_\gamma \partial_\delta F(t) = \partial_\delta \partial_\beta \partial_\lambda F(t) \eta^{\lambda\mu} \partial_\mu \partial_\gamma \partial_\alpha F(t) \quad \text{(WDVV1)}$$

(We set

$$\partial_\alpha := \frac{\partial}{\partial t^\alpha}$$

etc.; the summation over repeated indices is assumed.) Equivalently, the algebra

$$A_t = \text{span}(e_1, \dots, e_n)$$

with the multiplication law

$$\begin{aligned} e_\alpha \cdot e_\beta &= c_{\alpha\beta}^\gamma(t) e_\gamma, \\ c_{\alpha\beta}^\gamma(t) &:= \eta^{\gamma\varepsilon} \partial_\varepsilon \partial_\alpha \partial_\beta F(t) \end{aligned} \tag{2.1}$$

must be associative for any t. The algebra will automatically be commutative.

The symmetric nondegenerate bilinear form $\langle \ , \ \rangle$ on A_t defined by

$$\langle e_\alpha, e_\beta \rangle := \eta_{\alpha\beta}, \tag{2.2}$$

where the matrix $(\eta_{\alpha\beta})$ is the inverse of $(\eta^{\alpha\beta})$, is invariant (in the sense of (1.1)), since the expression

$$\langle e_\alpha \cdot e_\beta, e_\gamma \rangle = \partial_\alpha \partial_\beta \partial_\gamma F(t) \tag{2.3}$$

is symmetric with respect to any permutation of α, β, γ.

The variable t^1 will be singled-out, and we require that

$$\partial_\alpha \partial_\beta \partial_1 F(t) \equiv \eta_{\alpha\beta}. \tag{WDVV2}$$

This means that the first basis vector e_1 will be the unit element of all the algebras A_t. From WDVV1, WDVV2 we conclude that $(A_t, \langle \ , \ \rangle)$ is a Frobenius algebra for any t.

The condition WDVV2 is the quasi-homogeneity condition that we write down in the infinitesimal form using the Euler identity for the quasi-homogeneous functions. Introducing the Euler vector field

$$E = \sum_{\alpha=1}^{n} [(1 - q_\alpha)t^\alpha + r_\alpha], \partial_\alpha, \tag{2.4}$$

we require the function $F(t)$ to satisfy

$$\begin{aligned} \mathcal{L}_E F(t) &:= \sum_{\alpha=1}^{n} [(1 - q_\alpha)t^\alpha + r_\alpha] \partial_\alpha F(t) \\ &= (3 - d)F(t) + \frac{1}{2}A_{\alpha\beta}t^\alpha t^\beta + B_\alpha t^\alpha + C \end{aligned} \tag{WDVV3}$$

for some constants $A_{\alpha\beta}$, B_α, C. The numbers q_α, r_α, d must satisfy the following normalization conditions:

$$q_1 = 0, \quad r_\alpha \neq 0 \quad \text{only if} \quad q_\alpha = 1. \tag{2.5}$$

Loosely speaking, we assign the degree $1 - q_\alpha$ to the variable t^α. But if $q_\alpha = 1$, the degree r_α is assigned to $\exp t^\alpha$. With respect to this assignment the function $F(t)$ has degree $3 - d$ up to quadratic terms.

We will consider the class of equivalence of solutions modulo the addition of quadratic polynomials in t.

Exercise 2.1. For any α, β prove that

$$(q_\alpha + q_\beta - d)\eta_{\alpha\beta} = 0. \tag{2.6}$$

Exercise 2.2. Prove that by adding a quadratic polynomial to $F(t)$, the coefficients $A_{\alpha\beta}$, B_α, C in WDVV3 can be normalized in such a way that

$$
\begin{aligned}
&A_{\alpha\beta} \neq 0 \quad \text{only if} \quad q_\alpha + q_\beta = d - 1, \\
&A_{1\alpha} = \sum_\alpha \eta_{\alpha\varepsilon} r_\varepsilon, \\
&B_\alpha \neq 0 \quad \text{only if} \quad q_\alpha = d - 2, \\
&B_1 = 0, \\
&C \neq 0 \quad \text{only if} \quad d = 3.
\end{aligned}
\tag{2.7}
$$

The coefficients $A_{\alpha\beta}$, B_α, C must also be considered as unknown parameters of the WDVV problem.

Trivial solutions are cubics corresponding to graded Frobenius algebras $(A_0, \langle\,,\,\rangle)$:

$$\text{cubic} = \frac{1}{6} c_{\alpha\beta\gamma} t^\alpha t^\beta t^\gamma = \frac{1}{6} \langle 1, (t)^3 \rangle. \tag{2.8}$$

The needed solutions are analytic perturbations of cubics. This means that

$$F(t) = \text{cubic} + \sum_{k,l \geq 0} a_{k,l} (t'')^l e^{kt'}, \tag{2.9}$$

where the vector argument t is subdivided into two parts:

$$t = (t', t''), \quad \deg t' = 0, \quad \deg t'' \neq 0, \tag{2.10}$$

and k, l are multi-indices with all nonnegative coordinates. For

$$t'' \to 0, \quad t' \to -\infty, \tag{2.11}$$

$F(t)$ goes to the cubic. In quantum cohomology this is called the point of classical limit.

There are two main approaches in the WDVV theory.

Algebraic approach We study the formal series solutions (2.9) to the WDVV problem, analyzing, say, the recursion relations for the coefficients $a_{k,l}$. An example of this algebraic approach is the Kontsevich recursion relations for the number of plane curves, and also our discovery of Platonic solids when classifying polynomial solutions to WDVV. A general approach to constructing solutions of WDVV in the class of formal series was recently proposed in [9]. A certain family of formal power series solutions (not satisfying the quasi-homogeneity WDVV3) was very recently constructed in [63].

Analytic approach First to describe *all* solutions to WDVV and then to select the solutions of the needed class (2.9).

In this chapter we will consider the analytic approach to WDVV. This can be applied to the solutions of the form (2.9) only if the series converges near the point of classical limit (2.11). The convergence can be easily verified in concrete examples of quantum cohomologies. However, the general proof of convergence is still missing.

Let me first be more specific about the explicit form of WDVV.

Exercise 2.3. Let Q be the grading operator in $\mathbb{C}^n = \mathrm{span}(e_1, \ldots, e_n)$ defined by

$$Q(e_\alpha) = q_\alpha e_\alpha, \quad \alpha = 1, \ldots, n.$$

Show that WDVV remains invariant under the linear transformations of the variables t:

$$t \mapsto Mt, \quad t = (t^1, \ldots, t^n)^T,$$
$$(e_1, \ldots, e_n) \mapsto (e_1, \ldots, e_n)M^{-1}$$

(the upper label ^{T} stands for the transpose), where the matrix M satisfies the two conditions

$$Me_1 = e_1,$$
$$MQ = QM.$$

Prove that if $d \neq 0$, then the matrix $\eta = (\eta_{\alpha\beta})$ by a transformation of the above form can be reduced to the antidiagonal form

$$\eta_{\alpha\beta} = \delta_{\alpha+\beta, n+1}. \tag{2.12}$$

Derive from WDVV2 that in these coordinates, the function $F(t)$ can be represented in the form

$$F(t) = \frac{1}{2} t^{1\,2} t^n + \frac{1}{2} t^1 \sum_{\alpha=2}^{n-1} t^\alpha t^{n-\alpha+1} + f(t^2, \ldots, t^n) \tag{2.13}$$

for some function f of $n - 1$ variables. WDVV can be written as a system of differential equations for this function.

We will usually consider only the case $d \neq 0$, although this is not important for the mathematical theory of WDVV.

Example 2.1. $n = 2$. Here $f = f(t_2)$. The equations WDVV1 are empty. The quasi-homogeneity condition WDVV3 gives that

$$f = t_2^{(3-d)/(1-d)}, \quad d \neq 1$$
$$f = e^{2t_2/r}, \quad d = 1, \quad E = t_1 \partial_1 + r \partial_2,$$
$$f = -\frac{1}{2} c \log t_2, \quad d = 3, \quad E = t_1 \partial_1 - 2t_2 \partial_2, \quad \mathcal{L}_E F = F + c.$$

Example 2.2. For $n = 3$ the function $f = f(x, y)$, $x = t_2$, $y = t_3$ must satisfy the following PDE:

$$f_{xxy}^2 = f_{yyy} + f_{xxx} f_{xyy}.$$

For generic d the variables t_1, t_2, t_3 have the scaling dimensions 1, $1 - d/2$, $1 - d$, respectively, and the scaling dimension of the function f is $3 - d$. So

$$f(x, y) = \frac{x^4}{y} \phi(\log(yx^q)), \quad q = 2\frac{d-1}{2-d}.$$

Plugging this into the above PDE, one obtains the following complicated third-order ODE for the function ϕ:

$$- 6\phi + 48\phi^2 + 11\phi' + 88q\phi\phi' - (144 + 144q - 3q^2)\phi'^2$$
$$- 6\phi'' + 48(2 + 2q + q^2)\phi\phi'' - 4q(16 + 16q + q^2)\phi'\phi''$$
$$- (13q^2 + 13q^3 + q^4)\phi''^2 + \phi''' + 8q(3 + 3q + q^2)\phi\,\phi'''$$
$$+ 2q^2(1 + q + q^2)\phi'\phi''' - q^3(1 + q)\phi''\phi''' = 0.$$

The nongeneric values are the integer $-2 \leq d \leq 4$. In this case the ansatz for f must be modified. For example, for $d = 2$, $r_2 = r$,

$$f(x, y) = \frac{1}{y} \phi(x + r \log y),$$

where the function ϕ satisfies the ODE

$$\phi'''[r^3 + 2\phi' - r\phi''] - (\phi'')^2 - 6r^2\phi'' + 11r\phi' - 6\phi = 0. \qquad (2.14)$$

The case of quantum cohomology of $\mathbf{CP}^2$ corresponds to $r = 3$ (see Section 1 above). In this case equation (2.14) has a unique solution $\phi = \phi(x)$ of the form

$$\phi = \sum_{k \geq 1} A_k e^{k\,x} \qquad (2.15)$$

normalized by the condition $A_1 = \frac{1}{2}$. Plugging the series (2.15) into equation (2.14), one obtains the recursion relations for the numbers

$$N_k = (3k - 1)!A_k$$

of rational curves of degree k on $\mathbf{CP}^2$ passing through $3k-1$ generic points.

Exercise 2.4. Derive from the recursion relations that the series (2.15) converges if

$$\mathrm{Re}\, x < \log \frac{6}{5}.$$

Recall that the numerical estimate of [21] guarantees convergence for

$$\mathrm{Re}\, x < 1.981.$$

We conclude that in the first nontrivial case $n = 3$ the general solution of WDVV depends on three arbitrary parameters. However, this parameterization does not say anything about the analytic properties of the solutions. For the next case $n = 4$, the situation looks even worse: The function $f = f(t_2, t_3, t_4)$ must be found from an overdetermined system of six PDEs. With increasing, n the overdeterminacy of the system WDVV1 grows rapidly.

In this chapter we will give a complete classification of the solutions of WDVV satisfying the semisimplicity condition. Recall that this condition means that the algebra A_t is semisimple for generic t. The solution will be expressed via certain Painlevé-type transcendents (via particular transcendents of the Painlevé-VI type in the first nontrivial case $n = 3$).

In this section we will develop some preliminary geometrical constructions of the theory of WDVV. First we will give a coordinate-free reformulation of WDVV equations of associativity. The basic idea is to identify the algebra A_t with the space $T_t M$ tangent to the space of the parameters $t \in M$,

$$A_t \ni e_\alpha \leftrightarrow \partial_\alpha \in T_t M, \quad \alpha = 1, \ldots, n.$$

The space of parameters M acquires a new geometrical structure: The tangent spaces $T_t M$ are Frobenius algebras with respect to the multiplication

$$\partial_\alpha \cdot \partial_\beta = c_{\alpha\beta}^\gamma(t)\partial_\gamma \tag{2.16}$$

and metric

$$\langle \partial_\alpha, \partial_\beta \rangle = \eta_{\alpha\beta}. \tag{2.17}$$

We arrive [30] at the following main definition.

Definition 2.1. A (smooth, analytic) *Frobenius structure* on the manifold M is a structure of a Frobenius algebra on the tangent spaces $T_t M = (A_t, \langle\ ,\ \rangle_t)$ depending (smoothly, analytically) on the point t. This structure must satisfy the following axioms.

FM1 The metric on M induced by the invariant bilinear form $\langle\ ,\ \rangle_t$ is flat. Denote by ∇ the Levi–Civita connection for the metric $\langle\ ,\ \rangle_t$. The unit vector field e must be covariantly constant,

$$\nabla e = 0. \qquad (2.18)$$

We use here the word "metric" as a synonym for a symmetric nondegenerate bilinear form on TM, not necessarily a positive one. Flatness of the metric, i.e., vanishing of the Riemann curvature tensor, means that locally a *system of flat coordinates* $(t^1, \ldots, t^n)$ exists such that the matrix $\langle \partial_\alpha, \partial_\beta \rangle$ of the metric in these coordinates becomes constant.

FM2 Let c be the following symmetric trilinear form on TM:

$$c(u, v, w) := \langle u \cdot v, w \rangle. \qquad (2.19)$$

The four-linear form

$$(\nabla_z c)(u, v, w), \quad u, v, w, z \in TM,$$

must also be symmetric.

Before formulating the last axiom we observe that the space $\mathrm{Vect}(M)$ of vector fields on M acquires a structure of a Frobenius algebra over the algebra $\mathrm{Func}(M)$ of (smooth, analytic) functions on M.

FM3 A linear vector field $E \in \mathrm{Vect}(M)$ must be fixed on M, i.e.,

$$\nabla \nabla E = 0. \qquad (2.20)$$

The operators

$$\begin{aligned} Q_{\mathrm{Func}(M)} &:= E, \\ Q_{\mathrm{Vect}(M)} &:= \mathrm{id} + \mathrm{adj}_E \end{aligned} \qquad (2.21)$$

introduce in $\mathrm{Vect}(M)$ a structure of a graded Frobenius algebra of a given charge d over the graded ring $\mathrm{Func}(M)$ (see above the remark after Definition 1.2).

Lemma 2.1. *Locally, a Frobenius manifold with diagonalizable ∇E is described by a solution of WDVV and vice versa.*

Proof.

1. Starting from a solution of WDVV, define the multiplication (2.16) and the metric (2.17) on the tangent planes to the parameter space. In the original coordinates $(t^1, \ldots, t^n)$ the metric is manifestly flat. In these coordinates the covariant derivatives coincide with the partial ones

$$\nabla_\alpha = \partial_\alpha.$$

Since $e = \partial_1$, we have $\nabla e \equiv 0$. The first axiom FM1 is proved. The tensor c in (2.19) has the components

$$c_{\alpha\beta\gamma}(t) \equiv c(\partial_\alpha, \partial_\beta, \partial_\gamma) = \partial_\alpha \partial_\beta \partial_\gamma F(t).$$

So

$$(\nabla_{\partial_\delta} c)(\partial_\alpha, \partial_\beta, \partial_\gamma) = \partial_\alpha \partial_\beta \partial_\gamma \partial_\delta F(t)$$

is totally symmetric. This proves FM2.

Let us now prove FM3. The equations

$$Q_{\mathrm{Vect}(M)}(a \cdot b) = Q_{\mathrm{Vect}(M)}(a) \cdot b + a \cdot Q_{\mathrm{Vect}(M)}(b),$$
$$Q_{\mathrm{Func}(M)}\langle a, b \rangle + \mathrm{d}\langle a, b \rangle = \langle Q_{\mathrm{Vect}(M)}(a), b \rangle + \langle a, Q_{\mathrm{Vect}(M)}(b) \rangle$$

can be recast in the form

$$\mathcal{L}_E(a \cdot b) - \mathcal{L}_E(a) \cdot b - a \cdot \mathcal{L}_E(b) = a \cdot b, \qquad (2.22)$$
$$\mathcal{L}_E \langle a, b \rangle - \langle \mathcal{L}_E a, b \rangle - \langle a, \mathcal{L}_E b \rangle = (2 - d)\langle a, b \rangle. \qquad (2.23)$$

We will prove the last two equations.

The Euler vector field is clearly a linear one. The gradient ∇E is a diagonal constant matrix

$$\nabla E = \mathrm{diag}(1 - q_1, \ldots, 1 - q_n). \qquad (2.24)$$

Triple differentiation of the quasi-homogeneity equation WDVV3 with respect to t^α, t^β, t^γ gives

$$\sum_\varepsilon [(1 - q_\varepsilon)t^\varepsilon + r_\varepsilon] \partial_\varepsilon (c_{\alpha\beta\gamma}(t)) = (q_\alpha + q_\beta + q_\gamma - d)c_{\alpha\beta\gamma}(t). \qquad (2.25)$$

From this and from (2.6) there easily follow the identities (2.22), (2.23) of the definition of a graded Frobenius algebra over a graded ring of functions.

2. Choose locally flat coordinates $(t^1, \ldots, t^n)$ on a Frobenius manifold. We can choose them in a particular way such that $\partial_1, \ldots, \partial_n$ are the eigenvectors of the linear operator $\nabla E \colon TM \to TM$,

$$(\nabla E)\partial_\alpha = \lambda_\alpha \partial_\alpha,$$

for some constant λ_α (in the flat coordinates the matrix of the covariantly constant tensor ∇E is constant). This will be the homogeneous basis for the grading operator $Q_{\text{Vect}(M)}$,

$$Q_{\text{Vect}(M)}\partial_\alpha = (1 - \lambda_\alpha)\partial_\alpha.$$

So

$$E = \sum_{\alpha=1}^{n}(\lambda_\alpha t^\alpha + r_\alpha)\partial_\alpha$$

for some constants r_α. We can kill by a shift all these constants but those for which $\lambda_\alpha = 0$. This gives the form (2.4) of the Euler vector field with

$$q_\alpha := 1 - \lambda_\alpha.$$

From the obvious equation

$$Q_{\text{Vect}(M)}e = 0$$

we immediately obtain that $\lambda_1 = 1$, i.e., $q_1 = 0$.

From the symmetry with respect to α, β, γ, δ of partial derivatives

$$\partial_\delta c_{\alpha\beta\gamma}(t)$$

of the symmetric tensor

$$c_{\alpha\beta\gamma}(t) = \langle \partial_\alpha \cdot \partial_\beta, \partial_\gamma \rangle$$

we conclude the local existence of a function $F(t)$ such that

$$c_{\alpha\beta\gamma}(t) = \partial_\alpha\partial_\beta\partial_\gamma F(t).$$

For the invariant metric we obtain

$$\eta_{\alpha\beta} = \langle \partial_\alpha, \partial_\beta \rangle = \langle \partial_\alpha \cdot \partial_\beta, \partial_1 \rangle = \partial_1\partial_\alpha\partial_\beta F(t).$$

We have proved WDVV1 and WDVV2.

Spelling the last axiom FM3 out, we obtain the following two formulae:

$$\mathcal{L}_E\eta_{\alpha\beta} = \partial_\alpha E^\varepsilon \eta_{\varepsilon\beta} + \partial_\beta E^\varepsilon \eta_{\alpha\varepsilon} = (2 - d)\eta_{\alpha\beta},$$
$$\mathcal{L}_E c_{\alpha\beta}^\gamma = E^\varepsilon \partial_\varepsilon c_{\alpha\beta}^\gamma - \partial_\varepsilon E^\gamma c_{\alpha\beta}^\varepsilon + \partial_\alpha E^\varepsilon c_{\varepsilon\beta}^\gamma + \partial_\beta E^\varepsilon c_{\alpha\varepsilon}^\gamma = c_{\alpha\beta}^\gamma.$$

From this it follows

$$(q_\alpha + q_\beta - d)\eta_{\alpha\beta} = 0, \tag{2.26}$$
$$E^\varepsilon \partial_\varepsilon c_{\alpha\beta}^\gamma = (q_\alpha + q_\beta - q_\gamma)c_{\alpha\beta}^\gamma. \tag{2.27}$$

Using (2.6) we lower the index γ in the last equation to obtain

$$E^\varepsilon \partial_\varepsilon c_{\alpha\beta\gamma} = (q_\alpha + q_\beta + q_\gamma - d)c_{\alpha\beta\gamma}, \quad \alpha, \beta, \gamma = 1, \ldots, n.$$

A triple integration gives

$$E^\varepsilon \partial_\varepsilon F = (3 - d)F + \text{quadratic}.$$

The lemma is proved.   □

Remark. The definition of a Frobenius manifold can be easily translated into an algebraic language as a graded Frobenius algebra structure on the module of derivations of a graded commutative associative algebra. An important extension of this definition for the case of $\mathbb{Z}_2$-graded algebras was done by Kontsevich and Manin [57]. Such *Frobenius supermanifolds* are necessary to deal with Gromov–Witten invariants of manifolds with nontrivial odd-dimensional cohomologies. In this chapter we will not discuss this extension.

Exercise 2.5. Prove that the direct product $M' \times M''$ of two Frobenius manifolds of the *same* charge d carries a natural structure of a Frobenius manifold of charge d, with unity vector field $e' \oplus e''$ and Euler vector field $E' \oplus E''$.

We now address the problem of the (local) classification of Frobenius manifolds coinciding with a local classification of solutions of WDVV. To be more specific we give the following definition.

Definition 2.2. A (local) diffeomorphism

$$\phi : M \to \widetilde{M}$$

of two Frobenius manifolds is called a (*local*) *equivalence* if the differential

$$\phi_* : T_t M \to T_{\phi(t)} \widetilde{M}$$

is an isomorphism of algebras for any $t \in M$ and if

$$\phi^*\langle \ , \ \rangle_{\widetilde{M}} = c^2 \langle \ , \ \rangle_M,$$

where c is a nonzero constant independent of the point of M.

The corresponding free energies F and $\widetilde{F}$ are related by

$$\widetilde{F}(\phi(t)) = c^2 F(t) + \text{quadratic}.$$

Definition 2.3. A Frobenius manifold is called *reducible* if it is equivalent to the direct product of two Frobenius manifolds (see Exercise 2.5 above).

The first main tool in dealing with Frobenius manifolds is a deformation of the Levi–Civita connection ∇. We put

$$\tilde{\nabla}_u v := \nabla_u v + z u \cdot v. \tag{2.28a}$$

Here u, v are two vector fields on M, and z is the parameter of the deformation. We extend this up to a meromorphic connection on the direct product $M \times \mathbb{C}$ by the formulae

$$\tilde{\nabla}_u \frac{d}{dz} = 0,$$

$$\tilde{\nabla}_{d/dz} \frac{d}{dz} = 0, \tag{2.28b}$$

$$\tilde{\nabla}_{d/dz} v = \partial_z v + E \cdot v - \frac{1}{z} \mu v,$$

where

$$\mu := \frac{2-d}{2} - \nabla E = \mathrm{diag}(\mu_1, \ldots, \mu_n), \tag{2.29a}$$

$$\mu_\alpha := q_\alpha - \frac{d}{2}. \tag{2.29b}$$

Here u, v are tangent vector fields on $M \times \mathbb{C}$ having a zero component along $\mathbb{C}$. Observe that $\tilde{\nabla}$ is a symmetric connection.

Proposition 2.1. *For a Frobenius manifold M the curvature of the connection $\tilde{\nabla}$ equals zero. Conversely, if on the tangent spaces to M a structure of a Frobenius algebra is defined satisfying FM1, and if the Euler vector field E satisfies*

$$\mathcal{L}_E \langle \, , \, \rangle = (2-d) \langle \, , \, \rangle \tag{2.30}$$

with a constant d, then M is a Frobenius manifold.

Proof. For a covector

$$\xi = \xi_\alpha \, dt^\alpha + 0 \, dz$$

one has

$$\tilde{\nabla}_\alpha \xi_\beta = \partial_\alpha \xi_\beta - z c_{\alpha\beta}^\gamma \xi_\gamma,$$

$$\tilde{\nabla}_{d/dz} \xi_\beta = \partial_z \xi_\beta - E^\gamma c_{\gamma\beta}^\alpha \xi_\alpha + \frac{1}{z} M_\beta^\varepsilon \xi_\varepsilon,$$

where we denote by M_β^ε the matrix entries of the linear operator $\mu = \frac{1}{2}(2-d) - \nabla E$. Any solution of the system $\tilde{\nabla} \xi = 0$ is a (local) horizontal section of $T^*(M \times \mathbb{C})$ for the connection $\widetilde{\nabla}$. A basis of horizontal sections is given by dz and by n linearly independent solutions of the system

$$\partial_\alpha \xi_\beta = z c_{\alpha\beta}^\gamma \xi_\gamma, \tag{2.31}$$

$$\partial_z \xi_\beta = E^\gamma c_{\gamma\beta}^\alpha \xi_\alpha - \frac{1}{z} M_\beta^\varepsilon \xi_\varepsilon. \tag{2.32}$$

Such a basis exists iff the compatibility conditions

$$\partial_\alpha \partial_\gamma = \partial_\gamma \partial_\alpha, \quad \partial_\alpha \partial_z = \partial_z \partial_\alpha$$

hold true. Differentiating (2.31) with respect to t^γ and subtracting the same expression with α and γ permuted, we obtain the first compatibility condition in the form

$$z(\partial_\gamma c_{\alpha\beta}^\varepsilon - \partial_\alpha c_{\gamma\beta}^\varepsilon)\xi_\varepsilon + z^2(c_{\alpha\beta}^\lambda c_{\lambda\gamma}^\varepsilon - c_{\gamma\beta}^\lambda c_{\lambda\alpha}^\varepsilon)\xi_\varepsilon = 0.$$

This must vanish for arbitrary ξ. We obtain

$$c_{\alpha\beta}^\lambda c_{\lambda\gamma}^\varepsilon = c_{\gamma\beta}^\lambda c_{\lambda\alpha}^\varepsilon$$

(associativity) and

$$\partial_\gamma c_{\alpha\beta}^\varepsilon = \partial_\alpha c_{\gamma\beta}^\varepsilon$$

(local existence of $F(t)$). Similarly, from the compatibility of (2.31) and (2.32) we first obtain

$$\partial_\alpha M_\beta^\varepsilon = 0.$$

So ∇E is a constant matrix. Assume, for simplicity, ∇E to be diagonal, $\nabla E = \mathrm{diag}(1 - q_1, \ldots, 1 - q_n)$. Then we further obtain

$$E^\varepsilon \partial_\varepsilon c_{\alpha\beta}^\gamma = (q_\alpha + q_\beta - q_\gamma) c_{\alpha\beta}^\gamma.$$

As we already know, this together with (2.30) (i.e., with (2.6)) is equivalent to FM3. The proposition is proved. □

Remark. Due to Proposition 2.1, one can alternatively define Frobenius manifolds as those carrying a metric and a linear pencil of affine connections (2.28a) satisfying the above conditions (such a definition was explicitly used in [29]). It is interesting that manifolds with a metric and a linear pencil of affine connections deforming the Levi–Civita connection are known also in mathematical statistics—see the book [1]. This structure appears in the parametric statistics that studies families of probabilistic measures depending on a finite number of parameters. The metric was introduced by Rao about 1945 using the classical Fischer matrix of the family. The deformed Levi–Civita connection was discovered by N.N. Chentsov in 1972. It has the form (2.28a). However, the curvature of the deformed connection does not vanish identically, but it vanishes for two values of the parameter z.

Exercise 2.6. Prove that the solutions of the linear system (2.31), (2.32) are all closed differential forms

$$\xi_\alpha dt^\alpha = d\tilde{t}$$

(the differential along M only).

Choosing a basis of n linearly independent solutions $\xi_\alpha^{(1)}, \ldots, \xi_\alpha^{(n)}$ of the system, we obtain n functions $\tilde{t}_1(t, z), \ldots, \tilde{t}_n(t, z)$. Together with z they give a system of flat coordinates for the connection $\tilde{\nabla}$ on a domain in $M \times \mathbb{C}$. This means that in these coordinates the covariant derivatives coincide with the partial ones.

How to choose a basis of the deformed flat coordinates $\tilde{t}_1(t, z), \ldots, \tilde{t}_n(t, z)$? Let us first forget about the last component (2.28b) of the connection $\tilde{\nabla}$. The first part (2.28a) can be considered as a deformation of the affine structure on M, with z being the parameter of the deformation. We can look for the deformed flat coordinates in the form of the series

$$\tilde{t}_\alpha = \sum_{p=0}^\infty h_{\alpha,p}(t) z^p =: h_\alpha(t; z), \quad \alpha = 1, \ldots, n. \tag{2.33}$$

Lemma 2.2. *The coefficients $h_{\alpha,p}(t)$ can be determined recursively from the relations*

$$\begin{aligned} h_{\alpha,0} &= t_\alpha \equiv \eta_{\alpha\varepsilon} t^\varepsilon, \\ \partial_\beta \partial_\gamma h_{\alpha,p+1} &= c_{\beta\gamma}^\varepsilon \partial_\varepsilon h_{\alpha,p}, \quad p = 0, 1, 2, \ldots, \end{aligned} \tag{2.34}$$

uniquely up to a transformation of the form

$$\tilde{t}_\alpha \mapsto \sum_{\beta=1}^n \tilde{t}_\beta G_\alpha^\beta(z),$$

where the coefficients $G_1, G_2, \ldots$ of the matrix-valued series

$$G(z) = (G_\alpha^\beta(z)) = 1 + z G_1 + z^2 G_2 + \cdots$$

do not depend on t.

Proof. We only have to show that the right-hand sides of (2.34) are second derivatives along t^β and t^γ. This can be proved inductively using the identity

$$\partial_\alpha(c_{\beta\gamma}^\varepsilon \partial_\varepsilon h_{\lambda,p}) - \partial_\beta(c_{\alpha\gamma}^\varepsilon \partial_\varepsilon h_{\lambda,p}) = (c_{\beta\gamma}^\varepsilon c_{\varepsilon\alpha}^\rho - c_{\alpha\gamma}^\varepsilon c_{\varepsilon\beta}^\rho) \partial_\rho h_{\lambda,p-1} = 0$$

due to associativity.   □

The gradients $\nabla h_{\alpha,p}(t)$ and their inner products $\langle \nabla h_{\alpha,p}, \nabla h_{\beta,q} \rangle$ play a very important role in the theory and applications of Frobenius manifolds. Before discussing how to normalize them uniquely I will give here two important identities for these coefficients.

Exercise 2.7. Prove that

$$t_\alpha = \langle \nabla h_{\alpha,0}, \nabla h_{1,1} \rangle, \tag{2.35}$$

$$F(t) = \frac{1}{2}\{\langle \nabla h_{\alpha,1}, \nabla h_{1,1} \rangle \eta^{\alpha\beta} \langle \nabla h_{\beta,0}, \nabla h_{1,1} \rangle \tag{2.36}$$
$$- \langle \nabla h_{1,1}, \nabla h_{1,2} \rangle - \langle \nabla h_{1,3}, \nabla h_{1,0} \rangle\}. \tag{2.37}$$

Exercise 2.8. Prove the identity

$$\nabla \langle \nabla h_\alpha(t; z), \nabla h_\beta(t; w) \rangle = (z + w) \nabla h_\alpha(t; z) \cdot \nabla h_\beta(t; w).$$

Observe that $\langle \nabla h_\alpha(t; z), \nabla h_\beta(t; -z) \rangle$ does not depend on t.

To choose the system of deformed flat coordinates

$$\tilde{t}_1(t; z), \ldots, \tilde{t}_n(t; z)$$

canonically, we will now use the last equation (2.32) of the horizontality of the gradients $\xi_\alpha = \partial_\alpha \tilde{t}(t; z)$:

$$\partial_z \xi_\alpha = \mathcal{U}_\alpha^\beta \xi_\beta - \frac{1}{z}\mu_\alpha \xi_\alpha. \tag{2.38}$$

Here

$$\mathcal{U}_\alpha^\beta(t) := E^\varepsilon c_{\varepsilon\alpha}^\beta \tag{2.39}$$

is the matrix of multiplication by the Euler vector field. The choice of the basis can be made by carefully looking at the behavior of the solutions at $z = 0$, where the connection $\tilde{\nabla}$ has a logarithmic singularity. The analysis of this behavior will provide us with some numerical invariants of the Frobenius manifold.

Let us introduce the numbers

$$\mu_\alpha = q_\alpha - \frac{d}{2}, \quad \alpha = 1, \ldots, n.$$

Recall that they are the entries of the diagonal matrix

$$\mu = \frac{2-d}{2}1 - \nabla E = \mathrm{diag}(\mu_1, \ldots, \mu_n). \tag{2.40}$$

The operator μ is antisymmetric with respect to the metric $\langle \, , \, \rangle$,

$$\langle \mu a, b \rangle + \langle a, \mu b \rangle = 0. \tag{2.41}$$

We say that the operator μ is *resonant* if some of the differences $\mu_\alpha - \mu_\beta$ are nonzero integers. Otherwise, it is called *nonresonant*. We will also use the expressions resonant or nonresonant Frobenius manifold according as the corresponding operator μ is resonant or not. For example, any Frobenius manifold related to quantum cohomology is resonant (all the numbers q_α are integers). The Frobenius manifold on the space of versal deformations of an A_3 singularity is a nonresonant one:

$$\mu = \text{diag}\left(-\frac{1}{4}, 0, \frac{1}{4}\right).$$

We first consider the nonresonant case. In this case the system of deformed flat coordinates can be uniquely chosen in such a way that

$$\tilde{t}_\alpha(t; z) = [t_\alpha + O(z)]z^{\mu_\alpha}, \quad \alpha = 1, \ldots, n$$

(here, as usual, $t_\alpha = \eta_{\alpha\varepsilon}t^\varepsilon$). The coordinates are multivalued analytic functions of z defined for sufficiently small $z \neq 0$. Going along a closed loop around $z = 0$ one obtains the monodromy transformation

$$(\tilde{t}_1, \ldots, \tilde{t}_n) \mapsto (\tilde{t}_1, \ldots, \tilde{t}_n) M_0,$$
$$M_0 = \text{diag}(e^{2\pi i\mu_1}, \ldots, e^{2\pi i\mu_n}).$$

For the resonant case such a choice is not possible. The monodromy matrix M_0 cannot be diagonalized.

We will first rewrite equation (2.32) in matrix form.

The linear change

$$\xi = (\xi_1, \ldots, \xi_n) \mapsto \eta^{-1}\xi^T$$

transforms (2.31), (2.32) as follows:

$$\partial_\alpha \xi = zC_\alpha \xi, \tag{2.42a}$$

$$\partial_z \xi = \left(\mathcal{U} + \frac{1}{z}\mu\right)\xi, \tag{2.42b}$$

where

$$(C_\alpha)^\gamma_\beta = c^\gamma_{\alpha\beta}.$$

The matrix $\mathcal{U}$ is η-symmetric,

$$\mathcal{U}^T\eta = \eta\mathcal{U}, \tag{2.43}$$

and μ is η-antisymmetric,

$$\mu\eta + \eta\mu = 0. \tag{2.44}$$

The solutions of the system (2.42) are gradients of the deformed flat coordinates $\xi = \nabla\tilde{t}$.

Lemma 2.3. *The bilinear form*

$$\langle \xi_1, \xi_2 \rangle := \xi_1^T(-z)\eta \xi_2(z) \tag{2.45}$$

on the space of solutions of (2.42b) *does not depend on z.*

The proof is obvious.

Let us study the classes of equivalence of the system (2.42b) under the gauge transforms

$$\xi \mapsto G(z)\,\xi \tag{2.46a}$$

of the form

$$G(z) = 1 + zG_1 + z^2 G_2 + \cdots, \tag{2.46b}$$

$$G^T(-z)\eta G(z) \equiv \eta. \tag{2.46c}$$

Lemma 2.4. *After an arbitrary gauge transform* (2.46) *the vector function*

$$\tilde{\xi} = G(z)\xi \tag{2.47}$$

satisfies the system

$$\partial_z \tilde{\xi} = \left(\frac{1}{z}\mu + \tilde{U}_1 + z\tilde{U}_2 + \cdots \right)\tilde{\xi},$$

where the matrices $\tilde{U}_{2k+1}$ are η-symmetric and the matrices $\tilde{U}_{2k}$ are η-antisymmetric.

Proof. After the gauge transform the vector function $\tilde{\xi}$ satisfies

$$\partial_z \tilde{\xi} = A(z)\tilde{\xi}$$

with

$$\tilde{\xi} = G(z)\xi A(z) = G(z)\left(\frac{1}{z}\mu + \mathcal{U} \right)G^{-1}(z) + G'(z)G^{-1}(z)$$

$$=: \frac{1}{z}\mu + \tilde{U}_1 + z\tilde{U}_2 + \cdots,$$

where the matrix coefficients $\tilde{U}_k$ are defined by this equation. Using (2.43), (2.44), and (2.46c) one obtains

$$A^T(-z) = \eta \, A(z)\eta^{-1}.$$

This gives

$$\tilde{U}_k^T = (-1)^{k+1}\eta\tilde{U}_k\eta^{-1}, \quad k = 1, 2, \ldots.$$

The lemma is proved. $\qquad\qquad\square$

In the nonresonant case one can choose the gauge transform in such a way that $\tilde{U}_1 = \tilde{U}_2 = \cdots = 0$. The only gauge invariant of the system (2.42b) near the logarithmic singularity $z = 0$ is the diagonal matrix μ.

Let us consider the slightly more general system

$$\partial_z \xi = \left(\frac{1}{z}\mu + U_1 + zU_2 + z^2U_3 + \cdots \right)\xi, \qquad (2.48)$$

whose coefficients satisfy

$$U_k^T = (-1)^{k+1}\eta U_k \eta^{-1}, \quad k = 1, 2, \ldots. \qquad (2.49)$$

Lemma 2.5. *By a gauge transformation of the form* (2.46) *the system* (2.48) *can be reduced to the canonical form*

$$\partial_z \tilde{\xi} = \left(\frac{1}{z}\mu + R_1 + zR_2 + z^2R_3 + \cdots \right)\tilde{\xi}, \qquad (2.50)$$

where the matrices $R_1, R_2, \ldots$ *satisfy*

$$R_k^T = (-1)^{k+1}\eta R_k \eta^{-1}, \qquad (2.51)$$

$$(R_k)_\beta^\alpha \neq 0 \quad only\ if \quad \mu_\alpha - \mu_\beta = k, \quad k = 1, 2, \ldots. \qquad (2.52)$$

Observe that there is only a finite number of nonzero matrices R_k.

Proof. The gauge equivalence of the system (2.48) to a system (2.50) with the matrices R_k satisfying (2.52) is a well-known fact (see, e.g., [41]). Namely, from the recursion relations

$$R_n = U_n + nG_n + [G_n, \mu] + \sum_{k=1}^{n-1}(G_{n-k}U_k - R_kG_{n-k}) \qquad (2.53)$$

we uniquely determine the matrix entries

$$(R_n)_\beta^\alpha \quad \text{for } \mu_\alpha - \mu_\beta = n$$

and

$$(G_n)_\beta^\alpha \quad \text{for } \mu_\alpha - \mu_\beta \neq n,$$

and we put

$$(G_n)_\beta^\alpha = 0 \quad \text{for } \mu_\alpha - \mu_\beta = n.$$

Using induction it can easily be seen that the matrices R_n satisfy the η-symmetry/antisymmetry conditions (2.51) and the matrices G_n satisfy the orthogonality conditions

$$G_n^T = (-1)^{n+1}\eta G_n \eta^{-1} + \sum_{k=1}^{n-1}(-1)^{n+k+1}G_k^T\eta G_{n-k}\eta^{-1}. \qquad (2.54)$$

The lemma is proved. $\square$

We will call (2.50) the *normal form* of the system (2.48). The ambiguity in the choice of the normal form will be described below.

Lemma 2.6. *The matrix solution of the system* (2.50) *is*

$$\xi = z^{\mu} z^{R}, \tag{2.55}$$

where

$$R := R_1 + R_2 + \cdots. \tag{2.56}$$

Proof. From (2.52) we obtain the identity

$$z^{\mu} R_k z^{-\mu} = z^k R_k, \quad k = 1, 2, \ldots. \tag{2.57}$$

So, differentiating the matrix-valued function (2.55), one obtains

$$\partial_z \xi = \frac{\mu}{z} z^{\mu} z^{R} + \frac{1}{z} z^{\mu} R z^{R} = \left(\frac{\mu}{z} + R_1 + R_2 z^2 + \cdots \right) z^{\mu} z^{R}.$$

The lemma is proved.   □

Exercise 2.9. Prove that the monodromy around $z = 0$ of the solution

$$\Xi_0 := z^{\mu} z^{R}$$

has the form

$$\begin{aligned} \Xi_0(z e^{2\pi i}) &= \Xi_0(z) M_0, \\ M_0 &= \exp 2\pi i (\mu + R). \end{aligned} \tag{2.58}$$

We will now represent the parameters of the normal form (2.50) in a geometric way. Let $\mathcal{V}$ be a linear space equipped with a symmetric nondegenerate bilinear form $\langle \, , \, \rangle$ and an antisymmetric operator

$$\mu \colon \mathcal{V} \to \mathcal{V}, \quad \langle \mu a, b \rangle + \langle a, \mu b \rangle = 0.$$

Let us assume, for simplicity, the operator μ to be diagonalizable. Let

$$\operatorname{spec} \mu := \{\mu_1, \ldots, \mu_n\}$$

be the spectrum of μ. Denote by $e_1, \ldots, e_n$ the corresponding eigenvectors. We define a filtration on V:

$$\begin{aligned} 0 = F_0 &\subset F_1 \subset F_2 \subset \cdots \subset \mathcal{V}, \\ F_k &:= \operatorname{span}\{e_\alpha \mid \mu_\alpha + k \notin \operatorname{spec}\}. \end{aligned} \tag{2.59}$$

Obviously, for a nonresonant μ the filtration consists of two terms $0 = F_0 \subset F_1 = \mathcal{V}$.

The associated graded space

$$V_* = \bigoplus_{k \geq 1} V_k,$$
$$V_k = F_k / F_{k-1}$$

is isomorphic to V due to the natural isomorphism

$$V_k \simeq \mathrm{Ker}(\mu + k - 1) \cap F_k \subset V.$$

A linear operator

$$R : V \to V$$

is called μ-*nilpotent* if it commutes with $\exp 2\pi i\mu$,

$$Re^{2\pi i\mu} = e^{2\pi i\mu} R, \tag{2.60}$$

and if

$$R(F_k) \subset F_{k-1}, \quad k = 1, 2, \ldots. \tag{2.61}$$

The associated operator

$$R_* : V_* \to V_*$$

has a natural grading

$$R_* = \bigoplus_{k \geq 1} R_k, \tag{2.62}$$

where the operator R_k shifts the grading by $-k$,

$$R_k(V_m) \subset V_{m-k} \quad \text{for any } m > k. \tag{2.63}$$

Writing all the operators with matrices in the basis of eigenvectors of μ one obtains

$$(R_k)^\alpha_\beta = \begin{cases} R^\alpha_\beta & \text{if } \mu_\alpha - \mu_\beta = k, \\ 0 & \text{otherwise.} \end{cases} \tag{2.64}$$

We say that the μ-nilpotent operator is μ-*skew-symmetric* if

$$\{Rx, y\} + \{x, Ry\} = 0 \quad \text{for any } x, y \in V, \tag{2.65}$$

where

$$\{x, y\} := \langle e^{\pi i\mu} x, y \rangle. \tag{2.66}$$

The corresponding graded components R_k satisfy the following conditions:

$$\langle R_k x, y \rangle = (-1)^{k+1} \langle x, R_k y \rangle \quad \text{for any } x, y \in \mathcal{V}. \tag{2.67}$$

We conclude that the normal form of the system (2.48) is a quadruplet

$$(\mathcal{V}, \langle \, , \, \rangle, \mu, R), \tag{2.68}$$

where

$$\mathcal{V} = \text{span}(e_1, \ldots, e_n)$$

is the n-dimensional space with a bilinear symmetric form

$$\langle e_\alpha, e_\beta \rangle = \eta_{\alpha\beta},$$

an antisymmetric operator

$$\mu = \text{diag}(\mu_1, \ldots, \mu_n),$$

and a μ-nilpotent μ-skew-symmetric operator $R \colon \mathcal{V} \to \mathcal{V}$.

Let us now describe the ambiguity in the choice of the normal-form data. We say that

$$G \colon \mathcal{V} \to \mathcal{V} \tag{2.69}$$

is a μ-*parabolic orthogonal operator* if

$$G = 1 + \Delta, \tag{2.70}$$

where Δ is a μ-nilpotent operator and G satisfies the following orthogonality condition with respect to the bilinear form (2.66):

$$\{G\,x, G\,y\} = \{x, y\}. \tag{2.71}$$

Representing Δ as a sum of the graded components

$$\Delta \simeq \Delta_* = \bigoplus_{k \geq 1} \Delta_k, \tag{2.72}$$

one rewrites the orthogonality condition in the form

$$(1 - \Delta_1^T + \Delta_2^T - \cdots)\eta(1 + \Delta_1 + \Delta_2 + \cdots) = \eta. \tag{2.73}$$

Exercise 2.10. Prove that the monodromy operator (2.58) is orthogonal with respect to the bilinear form (2.66).

Clearly, all μ-parabolic orthogonal operators G form a group, denoted by $\mathcal{G}(\mu, \langle \ , \ \rangle)$. The space of all μ-nilpotent μ-skew-symmetric operators R coincides with the Lie algebra of the nilpotent group. The group $\mathcal{G}(\mu, \langle \ , \ \rangle)$ acts on this space by conjugation:

$$R \mapsto G^{-1} R G. \tag{2.74}$$

In the grading components one has

$$
\begin{aligned}
R_1 &\mapsto R_1, \\
R_2 &\mapsto R_2 + [R_1, \Delta_1], \\
R_3 &\mapsto R_3 + [R_2, \Delta_1] - \Delta_1 R_1 \Delta_1 + [R_1, \Delta_2] + \Delta_1^2 R_1,
\end{aligned}
\tag{2.75}
$$

etc. Two μ-nilpotent μ-skew-symmetric operators related by conjugation (2.74) will be called *equivalent*.

Lemma 2.7. *The set of all normal forms at $z = 0$ of a given system (2.48) is in one-to-one correspondence with the orbit of one normal form with respect to the action (2.74) of the group $\mathcal{G}(\mu, \langle \ , \ \rangle)$.*

Proof. Let us consider two normal forms of the same system (2.48):

$$\partial_z \xi = \left(\frac{\mu}{z} + R_1 + R_2 z + \cdots \right) \xi,$$

$$\partial_z \tilde{\xi} = \left(\frac{\mu}{z} + \tilde{R}_1 + \tilde{R}_2 z + \cdots \right) \tilde{\xi}.$$

They must be related by a gauge transformation

$$\tilde{\xi} = G(z) \xi,$$

where

$$G(z) = 1 + \Delta_1 z + \Delta_2 z^2 + \cdots$$

satisfies (2.46c). Explicitly, we obtain a system of relations identical to (2.54),

$$\tilde{R}_n = R_n + n\, \Delta_n + [\Delta_n, \mu] + \sum_{k=1}^{n-1} (\Delta_{n-k} R_k - \tilde{R}_k \Delta_{n-k}), \quad n = 1, 2, \cdots .$$

From this system we recursively prove that

$$(\Delta_n)_\beta^\alpha = 0 \quad \text{unless} \quad \mu_\alpha - \mu_\beta = n.$$

So

$$G := G(1) = 1 + \Delta_1 + \Delta_2 + \cdots$$

is a μ-parabolic orthogonal operator $G: \mathcal{V} \to \mathcal{V}$. From (2.46c) we derive the orthogonality condition (2.71) in the form (2.73).

Let us derive the relation (2.74) for the operators

$$R = R_1 + R_2 + \cdots, \quad \widetilde{R} = \widetilde{R}_1 + \widetilde{R}_2 + \cdots.$$

Since $\xi = z^\mu z^R$ is a solution to (2.50) (see Lemma 2.4),

$$G(z)z^\mu z^R$$

must be a solution to the system with a tilde. So we must have

$$G(z)z^\mu z^R = z^\mu z^{\widetilde{R}} C$$

for an invertible matrix C. Let us rewrite the last equation in the form

$$z^{-\mu} G(z) z^\mu z^R = z^{\widetilde{R}} C.$$

Using the identities

$$z^{-\mu} \Delta_k z^\mu = z^{-k} \Delta_k$$

we finally obtain

$$(1 + \Delta_1 + \Delta_2 + \cdots)z^R = Gz^R = z\widetilde{R}.$$

Expanding

$$G z^R = G\left(1 + R \log z + \frac{R^2}{2!} \log^2 z + \cdots \right)$$

$$= z^{\widetilde{R}} C = \left(1 + \widetilde{R} \log z + \frac{\widetilde{R}^2}{2!} \log^2 z + \cdots \right) C$$

and equating the coefficients in front of various powers of $\log z$, we obtain

$$C = G,$$
$$GR = \widetilde{R}G.$$

The lemma is proved. $\square$

Definition 2.4. A quadruplet

$$(\mathcal{V}, \langle\,,\,\rangle, \mu, [R]), \tag{2.76}$$

where $\mathcal{V}$ is an n-dimensional linear space with a bilinear symmetric form $\langle\,,\,\rangle$, an antisymmetric diagonalizable operator μ, and an equivalence class $[R]$ of normal forms (2.50) of the system (2.48) will be called *monodromy data at $z = 0$* for this system.

Lemma 2.8. *Two systems of the form*

$$\partial_z \xi^{(i)} = \left(\frac{\mu}{z} + \sum_{k \geq 1} U_k^{(i)} z^{k-1} \right) \xi^{(i)}, \quad i = 1, 2, \tag{2.77}$$

satisfying

$$U_k^{(i)^T} = (-1)^{k+1} \eta U_k^{(i)} \eta^{-1}$$

are equivalent with respect to a gauge transformation of the form (2.46) *iff they have the same monodromy data* (2.76).

Proof. Let the gauge transformations

$$\tilde{\xi}^{(i)} = G^{(i)}(z) \xi^{(i)}, \quad i = 1, 2,$$

reduce the systems (2.77) to the normal forms

$$\partial_z \tilde{\xi}^{(i)} = \left(\frac{\mu}{z} + \sum_{k \geq 1} R_k^{(i)} z^{k-1} \right) \tilde{\xi}^{(i)}, \quad i = 1, 2.$$

If

$$R^{(1)} = G R^{(2)} G^{-1}$$

with

$$G = 1 + \Delta_1 + \cdots \in \mathcal{G}(\mu, \langle \ , \ \rangle),$$

then the gauge transformation

$$\tilde{\xi}^{(2)} = (1 + z\Delta_1 + z^2 \Delta_2 + \cdots) \tilde{\xi}^{(1)}$$

establishes a gauge equivalence of the systems (2.77) for $i = 1$ and $i = 2$. Thus the systems (2.77) are gauge equivalent with

$$\xi^{(2)} = G^{(2)^{-1}}(z)(1 + z\Delta_1 + z^2 \Delta_2 + \cdots) G^{(1)}(z) \, \xi^{(1)}.$$

Conversely, from Lemma 2.7 it follows that gauge equivalent systems have the same monodromy data. The lemma is proved. □

We will now return to Frobenius manifolds. The last component (2.42b) of the system determining horizontal sections of the connection $\tilde{\nabla}$ is a linear system of ODEs with rational coefficients of the form (2.48). The coefficients $\mathcal{U}_\beta^\alpha(t) = E^\varepsilon(t) c_{\varepsilon\beta}^\alpha(t)$ depend parametrically on the point t of the Frobenius manifold. The solutions ξ take values in the space $\mathcal{V} = T_t M$. We may identify the tangent planes at different points t using the Levi–Civita connection ∇ on M. Actually, the space $\mathcal{V}$ is equipped with an additional structure, namely a distinguished vector $e \in \mathcal{V}$. This is an eigenvector of the linear operator μ with the eigenvalue $-d/2$.

Theorem 2.1 (Isomonodromy Theorem (first part)).   *The monodromy data at $z = 0$ of the system (2.42b) do not depend on $t \in M$.*

Proof. The matrix μ and the bilinear form $\langle \, , \, \rangle$ are t-independent by construction. We will now construct a t-independent representative R of the normal form (2.50) of the equation (2.42b)

$$\partial_z \xi = \left(\mathcal{U} + \frac{\mu}{z} \right) \xi,$$

$$\mathcal{U}_\beta^\alpha(t) = E^\varepsilon(t) c_{\varepsilon\beta}^\alpha(t).$$

Let us choose a basis $h_1(t; z), \ldots, h_n(t; z)$ of solutions of the system

$$\partial_\alpha \partial_\beta h_\gamma(t; z) = z c_{\alpha\beta}^\varepsilon(t) \partial_\varepsilon h_\gamma(t; z)$$

of the form (2.33), (2.34). Multiplying, if necessary, the series

$$h_\alpha(t; z) = t_\alpha + \sum_{p \geq 1} h_{\alpha,p}(t) z^p$$

by a matrix-valued series

$$M(z) \equiv \left(M_\beta^\alpha(z) \right) = 1 + \sum z^k M_k,$$

$$h_\alpha(t; z) \mapsto \sum_\varepsilon h_\varepsilon(t; z) M_\alpha^\varepsilon(z)$$

with t-independent coefficients $M_1, M_2, \ldots$, we obtain the identity

$$\langle \nabla h_\alpha(t; -z), \nabla h_\beta(t; z) \rangle = \eta_{\alpha\beta}$$

(see Exercise 2.8). Let us build an $n \times n$-matrix series $G(t; z) = \left(G_\beta^\alpha(t; z) \right)$,

$$G_\beta^\alpha(t; z) = \eta^{\alpha\varepsilon} \partial_\varepsilon h_\beta(t; z),$$

$$G(t; z) = 1 + \sum_{k \geq 1} G_k(t) z^k.$$

The identity (2.46c) reads

$$G^T(t; -z) \eta G(t; z) \equiv \eta.$$

We now perform a gauge transformation on the system (2.42)

$$\xi = G(t; z) \tilde{\xi}.$$

Since G is a matrix solution of the equations

$$\tilde{\nabla}_\alpha G = 0, \quad \alpha = 1, \ldots, n,$$

we obtain

$$\partial_\alpha \tilde{\xi} = 0, \quad \alpha = 1, \ldots, n.$$

The system (2.42b) after the gauge transformation will take the form (2.48) with the matrices U_1, U_2, ... satisfying the symmetry or antisymmetry conditions (2.49)

$$\partial_z \tilde{\xi} = \left(\frac{\mu}{z} + U_1 + z U_2 + \cdots \right) \tilde{\xi}.$$

The full system of the last two equations must still be compatible after the gauge transformation. This implies the t-independence of the coefficients $U_1, U_2, \ldots$. Hence the normal form of the system (2.42b) does not depend on t. The theorem is proved. $\qquad\square$

Definition 2.5. The monodromy data (2.76) of the system (2.42b) are called *monodromy data of the Frobenius manifold at $z = 0$*. Explicitly,

$$R_{1\beta}^\alpha = \mathcal{U}_\beta^\alpha \qquad\qquad \text{for } \mu_\alpha - \mu_\beta = 1, \qquad (2.78a)$$

$$R_{2\beta}^\alpha = \sum_{\mu_\alpha - \mu_\gamma \neq 1} \frac{\mathcal{U}_\gamma^\alpha \mathcal{U}_\beta^\gamma}{\mu_\alpha - \mu_\gamma - 1} \qquad \text{for } \mu_\alpha - \mu_\beta = 2, \qquad (2.78b)$$

etc.

Exercise 2.11. Using formula (2.37), prove that the normalized coefficients $A_{\alpha\beta}$, B_α, C in the quasi-homogeneity equation WDVV3 for the free energy $F(t)$ have the form

$$A_{\alpha\beta} = R_1{}_\alpha^\varepsilon \eta_{\varepsilon\beta}, \qquad (2.79)$$

in particular,

$$r_\alpha = R_1{}_1^\alpha, \qquad (2.80)$$

$$B_\alpha = R_2{}_\alpha^\varepsilon \eta_{\varepsilon 1}, \qquad (2.81)$$

$$C = -\frac{1}{2} R_3{}_1^\varepsilon \eta_{\varepsilon 1}. \qquad (2.82)$$

Remark. We defined our monodromy data as *formal invariants*, i.e., all the gauge transforms (2.46) were defined by formal power series $G(z) = 1 + G_1 z + \cdots$. It is well known, however, that at a regular singularity of the system (2.42b) formal invariants coincide with analytic ones [16]. In other words, all the normalizing transformations are convergent series for sufficiently small $|z|$.

We obtain the following theorem.

Theorem 2.2. *For any Frobenius manifold with the monodromy data* $(\mathcal{V}, \langle\ \rangle, \mu, [R])$, *there exists a fundamental matrix*

$$\Xi^0(t; z) = H(t; z)z^\mu z^R,$$
$$H(t; z) = 1 + H_1(t)z + H_2(t)z^2 = \cdots \tag{2.83}$$

for the system (2.42) defining horizontal sections of $\widetilde{\nabla}$. *The power series converges for sufficiently small* $|z|$. *Here* R *is a representative of the class* $[R]$.

A change of the representative

$$R \mapsto \widetilde{R} = G^{-1}RG,$$

where

$$G = 1 + \Delta_1 + \Delta_2 + \cdots$$

is a μ-*parabolic orthogonal operator, transforms as follows:*

$$\Xi^0 \mapsto \widetilde{\Xi}^0 = \Xi^0 G.$$

The power series transforms as follows

$$\widetilde{H}(t; z)(1 + z\Delta_1 + z^2\Delta_2 + \cdots) = H(t; z).$$

A choice of the fundamental matrix $\Xi^0(t; z)$ *determines a system of deformed flat coordinates* $\tilde{t}_1(t; z), \ldots, \tilde{t}_n(t; z)$ *such that the gradients* $\nabla\tilde{t}_1, \ldots, \nabla\tilde{t}_n$ *are the columns of the matrix* Ξ^0:

$$\big(\tilde{t}_1(t; z), \ldots, \tilde{t}_n(t; z)\big) = \big(h_1(t; z), \ldots, h_n(t; z)\big)z^\mu z^R. \tag{2.84}$$

The functions $\tilde{t}_\alpha(t; z)$ *are uniquely determined up to* t-*independent shifts and* μ-*parabolic transformations.*

We will call the functions $\tilde{t}_\alpha(t; z)$ the *normalized deformed flat coordinates* on the Frobenius manifold. Recall that the columns of the matrix $H(t; z)$ are gradients of the functions $h_1(t; z), \ldots, h_n(t; z)$. The Taylor expansions of these functions for small $|z|$ have the form

$$h_\alpha(t; z) = t_\alpha + \sum_{p=1}^{\infty} h_{\alpha,p}(t)z^p$$

in which the coefficients $h_{\alpha,p}(t)$ satisfy the system of recursion relations (2.34). But now the coefficients are uniquely determined within the ambiguity defined by the action of the group of μ-parabolic orthogonal transformations and up to a t-independent shift.

Exercise 2.12. For a μ-nilpotent operator

$$R = R_1 + R_2 + \cdots$$

define the operators $R_{k,l}$ putting

$$
\begin{aligned}
R_{0,0} &= 1, \\
R_{k,0} &= 0, \quad k > 0, \\
R_{k,l} &= \sum_{i_1 + \cdots + i_l = k} R_{i_1} \cdots R_{i_l}.
\end{aligned}
\tag{2.85}
$$

Prove that the normalized deformed flat coordinates have the following expansion near $z = 0$:

$$
\tilde{t}_\alpha(t; z) = \sum_{k,l \geq 0} \sum_{p=0}^{k} \sum_\varepsilon h_{\varepsilon,p}(t)(R_{k-p,l})_\alpha^\varepsilon z^{k+\mu_\alpha} \frac{\log^l z}{l!}.
\tag{2.86}
$$

Exercise 2.13. Derive the following quasi-homogeneity conditions for the gradient of the functions $h_{\alpha,k}(t)$:

$$
\mathcal{L}_E \nabla h_{\alpha,k} = \left(k + \frac{1}{2}(d - 2) + \mu_\alpha \right) \nabla h_{\alpha,k} + \sum_{\varepsilon,p} \nabla h_{\varepsilon,k-p}(R_p)_\alpha^\varepsilon.
\tag{2.87}
$$

Together with the relations (2.34), these quasi-homogeneity conditions can serve as the recursive definition of the functions $h_{\alpha,k}$, starting from $h_{\alpha,0} = t_\alpha$.

We conclude this section with a description of the monodromy data of Frobenius manifolds with good analytic properties (in particular, of quantum cohomologies).

Proposition 2.2. *For a Frobenius manifold of the form (2.9), all the matrices R_2, R_3, ... vanish and*

$$
(R_1)_\beta^\alpha = \sum_\varepsilon r_\varepsilon c_{\varepsilon\beta}^\alpha.
\tag{2.88}
$$

Here the numbers r_ε enter into the Euler vector field (2.4), and $c_{\alpha\beta}^\gamma$ are the structure constants of the cubic part of $F(t)$ (2.9).

Proof. Due to the isomonodromy theorem it is sufficient to compute the monodromy data of the operator (2.42b) at the point $t = t_0$ of the "classical limit." We have

$$
\mathcal{U}_\beta^\alpha(t_0) = \sum_\varepsilon r_\varepsilon c_{\varepsilon\beta}^\alpha.
$$

The algebra A_{t_0} is graded by the degree

$$
\deg e_\alpha = q_\alpha.
$$

On the other hand, the vector

$$\sum_\varepsilon r_\varepsilon e_\varepsilon$$

has degree one. Thus the operator (2.39) of multiplication in A_{t_0} by this vector increases degrees by one. Hence

$$\mathcal{U}_\beta^\alpha \neq 0 \quad \text{only if} \quad q_\alpha - q_\beta = 1,$$

and the system

$$\partial_z \xi = \left(\mathcal{U}(t_0) + \frac{\mu}{z}\right)\xi$$

is already in the normal form (2.50) with $R = R_1 = \mathcal{U}(t_0)$. The proposition is proved.   □

Corollary 2.1. *In the quantum cohomology of a manifold X, the monodromy data at $z = 0$ is the operator $R = R_1$ of multiplication by the first Chern class $c_1(X)$ acting in the classical cohomologies $H^*(X)$.*

Example 2.3. Let us explain the deformed connection (2.28) in the case of the quantum cohomology of a "sufficiently good" (for example, smooth projective) $2d$-dimensional manifold X (the assumptions and notations are as in Section 1 above). Let $\phi \in H^*(X; \mathbb{C})$ be an arbitrary element. We will construct the function $\tilde{t}_\phi(t; z)$, $t = (t', t'')$, as in (2.10), such that for any ϕ it satisfies

$$\tilde{\nabla} d\tilde{t}_\phi = 0. \tag{2.89}$$

Taking $\phi = \phi_1, \ldots, \phi = \phi_n$ for a basis in $H^*(X, \mathbb{C})$ we will obtain a system of flat coordinates of the deformed connection.

Denote by Q the grading operator (1.4a). We introduce a line bundle $\mathcal{L}$ on the moduli space $X_{[\beta],l}$. The fiber of this bundle at the point $(\beta, p_1, \ldots, p_l) \in X_{[\beta],l}$ is the cotangent line to the Riemann sphere at the first marked point p_1. Let

$$\sigma_1 := c_1(\mathcal{L}) \in H^2(X_{[\beta],l}).$$

Put

$$\tilde{t}_\phi(t; z) = z^{-d/2} \sum_{[\beta],l} \left\langle \frac{z^Q z^{c_1(X)} \phi}{1 - z\sigma_1} \otimes 1 \otimes e^{t''} \right\rangle_{[\beta],l} e^{\int_{S^2} \beta^*(t')}. \tag{2.90}$$

Here we define the symbols

$$\left\langle \frac{a_1}{1 - z\sigma_1} \otimes a_2 \otimes \cdots \otimes a_k \right\rangle_{[\beta],l}$$

as the formal series in z using the expansion

$$\frac{1}{1 - z\sigma_1} = 1 + z\sigma_1 + z^2\sigma_1^2 + \cdots ,$$

and

$$\langle a_1\sigma_1^m \otimes a_2 \otimes \cdots \otimes a_k\rangle_{[\beta],l} =$$

$$\begin{cases} 0, & k \neq l \\ \int_{X_{[\beta],l}} \sigma_1^m \wedge p_1^*(a_1) \wedge p_2^*(a_2) \wedge \cdots \wedge p_l^*(a_l), & k = l. \end{cases} \quad (2.91)$$

These are particular *gravitational descendants* arising in the description of the coupling of the topological sigma-model (= quantum cohomology) to topological gravity [22, 23, 28, 86].

Theorem 2.3. *The function* $\tilde{t}_\phi(t; z)$ *for any* $\phi \in H^*(X)$ *satisfies the equation* (2.89).

Proof. Let us choose some basis $\phi_1, \ldots, \phi_n$ in $H^*(X)$. The formula (2.90) for the functions $(\tilde{t}_{\phi_1}(t; z), \ldots, \tilde{t}_{\phi_n}(t; z))$ can be rewritten in the form

$$(\tilde{t}_{\phi_1}(t; z), \ldots, \tilde{t}_{\phi_n}(t; z)) = (h_1(t; z), \ldots, h_n(t; z)) z^\mu z^R,$$

where the formal series $h_\alpha(t; z)$ have the form

$$h_\alpha(t; z) = t_\alpha + \sum_{p=1}^{\infty} h_{\alpha,p}(t) z^p, \quad \alpha = 1, \ldots, n,$$

$$h_{\alpha,p}(t) = \sum_{[\beta],l} \langle \sigma_1^p \phi_\alpha \otimes 1 \otimes e^{t''}\rangle_{[\beta],l} e^{\int_{S^2} \beta^*(t')}, \quad (2.92)$$

$$\mu = \text{diag}(\mu_1, \ldots, \mu_n), \quad \mu_\alpha = q_\alpha - \frac{d}{2}, \quad \phi_\alpha \in H^{2q_\alpha}(X),$$

and R is the matrix of the operator of multiplication by the first Chern class $c_1(X)$ (cf. (2.84) above). To demonstrate (2.89) it is sufficient to prove that the coefficients $H_{\alpha,p}(t)$ satisfy the recursion relations

$$\partial_\lambda \partial_\mu h_{\alpha,p}(t) = c_{\lambda\mu}^\nu \partial_\nu h_{\alpha,p-1}(t), \quad p \geq 1 \quad (2.93)$$

(see (2.34)), and

$$\mathcal{L}_E \nabla h_{\alpha,p} = \left(p + \frac{d-2}{2} + \mu_\alpha\right) h_{\alpha,p} + \nabla h_{\varepsilon,p-1}(R)_\alpha^\varepsilon. \quad (2.94)$$

The last equation is the particular case of (2.87) for the case of quantum cohomology, where $R = R_1$ is the matrix of multiplication by the first Chern class and $R_2 = R_3 = \cdots = 0$ (see Corollary 2.1). The first relation (2.93) follows from the genus-0 topological recursion relations of Dijkgraaf and Witten [25] (see the derivation in [28, 32]). The second one follows from the recursion relations of Hori [50]. The theorem is proved. $\quad \square$

In general, we know nothing about the analytic properties of the series (2.90) for large $|z|$. But if the quantum cohomology algebra is semisimple (conjecturally, this is the case for Fano varieties X; see below, Section 3), then the asymptotic behavior of the series (2.90) for large $|z|$ is under control. The Stokes parameters of this asymptotic behavior will give us in Section 4 additional parameters of the Frobenius manifold to determine it uniquely.

Example 2.4. We will now compute the flat coordinates of the deformed connection $\tilde{\nabla}$ for the Frobenius manifolds arising in the singularity theory. We consider here only the case of simple singularities $f(x)$ (see the definition in [3,4]). (The formulation of K. Saito's theory of primitive forms in the setting of Frobenius manifolds for more general singularities can be found in [69,74,81].) Simple singularities are labeled by the simply laced Dynkin diagrams A_n, D_n, E_6, E_7, E_8. Denote by $f_t(x)$ the corresponding versal deformation. The variable x is one-dimensional for A_n, or $x = (x_1, x_2)$ for other simple singularities. The parameters are $t = (t^1, \ldots, t_n)$ for A_n, D_n, E_n. Explicitly,

$$A_n: \quad f_t(x) = x^{n+1} + a_n x^{n-1} + \cdots + a_1,$$

$$D_n: \quad f_t(x) = x_1^{n-1} + x_1 x_2^2 + a_{n-1} x_1^{n-2} + \cdots + a_1 + bx_2,$$

$$E_6: \quad f_t(x) = x_1^4 + x_2^3 + a_6 x_1^2 x_2 + a_5 x_1 x_2 + a_4 x_1^2 + a_3 x_2 + a_2 x_1 + a_1,$$

$$E_7: \quad f_t(x) = x_1^3 x_2 + x_2^3 + a_1 + a_2 x_2 + a_3 x_1 x_2 + a_4 x_1 x_2^2 + a_5 x_1$$
$$\qquad + a_6 x_1^2 + a_7 x_2^2,$$

$$E_8: \quad f_t(x) = x_1^5 + x_2^3 + a_8 x_1^3 x_2 + a_7 x_1^2 x_2 + a_6 x_1^3 + a_5 x_1 x_2 + a_4 x_1^2$$
$$\qquad + a_3 x_2 + a_2 x_1 + a_1.$$

The coefficients a_i are some polynomials of the flat coordinates $t^1, \ldots, t^n$. The dependence on the flat coordinates satisfies the following two remarkable identities [40]:

$$\phi_\alpha(x; t)\phi_\beta(x; t) = c_{\alpha\beta}^\gamma(t)\phi_\gamma(x; t) + K_{\alpha\beta}^a(x; t)\frac{\partial f_t(x)}{\partial x^a}, \qquad (2.95)$$

$$\partial_\alpha \phi_\beta(x; t) = \frac{\partial K_{\alpha\beta}^a(x; t)}{\partial x^a}. \qquad (2.96)$$

Here

$$\phi_\alpha(x; t) = \frac{\partial f_t(x)}{\partial t^\alpha}, \quad \alpha = 1, \ldots, n,$$

$K_{\alpha\beta}^a(x; t)$ are some polynomials, the index a takes only one value for A_n or two values for D_n, E_n. The coefficients $c_{\alpha\beta}^\gamma(t)$ coincide with the structure constants of the corresponding Frobenius manifold.

Theorem 2.4. *The oscillatory integrals*

$$\tilde{t}_C(t;z) = z^{(N-2)/2} \int_C e^{zf_t(x)} \, \mathrm{d}^N x \qquad (2.97)$$

($N = 1$ for A_n and $N = 2$ for D_n, E_n) are flat coordinates of the deformed connection $\tilde{\nabla}$. Here C is any N-dimensional cycle in $\mathbb{C}^N$ that goes to infinity along the directions where $\mathrm{Re}\, z f_t(x) \to -\infty$.

Proof. We have to prove that the functions

$$\xi_\alpha = \partial_\alpha \tilde{t}(t;z) = z^{N/2} \int_C \phi_\alpha(x;t) e^{zf_t(x)} \, \mathrm{d}^N x$$

satisfy the system (2.89). Using (2.95), (2.96) we obtain

$$\begin{aligned}
\partial_\alpha \xi_\beta &= z^{N/2} \int_C \frac{\partial K^a_{\alpha\beta}(x;t)}{\partial x^a} e^{zf_t(x)} \mathrm{d}^N x \\
&\quad + z^{(N+2)/2} \int_C \left[c^\gamma_{\alpha\beta}(t)\phi_\gamma(x;t) + K^a_{\alpha\beta}(x;t)\frac{\partial f_t(x)}{\partial x^a} \right] e^{zf_t(x)} \, \mathrm{d}^N x \\
&= z c^\gamma_{\alpha\beta}(t)\xi_\gamma + z^{N/2} \int_C \frac{\partial}{\partial x^a}\left[K^a_{\alpha\beta}(x;t)e^{zf_t(x)} \right] \mathrm{d}^N x \\
&= z c^\gamma_{\alpha\beta}(t)\xi_\gamma
\end{aligned}$$

(we used Stokes's formula

$$\int_C \frac{\partial}{\partial x^a} v^a \, \mathrm{d}^N x = 0$$

for any vector field v^a vanishing on the boundary of C at infinity).

To demonstrate the second equation (2.42b), it is sufficient to prove that

$$z \partial_z \tilde{t}_C = \mathcal{L}_E \tilde{t}_C + \frac{d-2}{2} \tilde{t}_C.$$

Here the Euler vector field and d have the form

$$E = \sum_{\alpha=1}^n \frac{d_\alpha}{h} t^\alpha \partial_\alpha, \quad d_\alpha = m_\alpha + 1,$$

$$d = 1 - \frac{2}{h},$$

where h is the Coxeter number and m_α are the exponents of the corresponding Weyl group $W(A_n)$, $W(D_n)$, $W(E_n)$ (see Section 5 below). One can assign some degrees r_1, r_2 to the variables x_1, x_2 (r_1 only for A_n)

in such a way that the whole deformation $f_t(x)$ is a quasi-homogeneous function of $t^1, \ldots, t^n, x_1, x_2$ of degree one. Explicitly,

$$A_n : \quad r_1 = \frac{1}{n+1},$$

$$D_n : \quad r_1 = \frac{1}{n-1}, \quad r_2 = \frac{n-2}{2n-2},$$

$$E_6 : \quad r_1 = \frac{1}{4}, \quad r_2 = \frac{1}{3},$$

$$E_7 : \quad r_1 = \frac{2}{9}, \quad r_2 = \frac{1}{3},$$

$$E_8 : \quad r_1 = \frac{1}{5}, \quad r_2 = \frac{1}{3}.$$

The coefficients a_α of the versal deformations are quasi-homogeneous polynomials of $t^1, \ldots, t^n$ of degrees $d_\alpha / h = \deg t^\alpha$. The quasi-homogeneity can be recast in the form of the following Euler identity:

$$\sum_a r_a x^a \frac{\partial f_t(x)}{\partial x^a} + \sum_\alpha \frac{d_\alpha}{h} t^\alpha \frac{\partial f_t(x)}{\partial t^\alpha} = f_t(x).$$

Using this identity we obtain

$$\partial_z \tilde{t}_C$$
$$= \frac{N-2}{2z} \tilde{t}_C + z^{(N-2)/2} \int_C f_t(x) e^{z f_t(x)} d^N x$$
$$= \frac{N-2}{2z} \tilde{t}_C + z^{(N-2)/2} \int_C \left[\sum_a r_a x^a \frac{\partial f_t(x)}{\partial x^a} + \sum_\alpha \frac{d_\alpha}{h} t^\alpha \frac{\partial f_t(x)}{\partial t^\alpha} \right] e^{z f_t(x)} d^N x$$
$$= \frac{N-2}{2z} \tilde{t}_C + \frac{1}{z} \mathcal{L}_E \tilde{t}_C + z^{(N-4)/2} \int_C \sum_a \frac{\partial}{\partial x^a} \left(r_a x^a e^{z f_t(x)} \right) d^N x - \frac{r_1 + r_2}{z} \tilde{t}_C$$
$$= \frac{N-2}{2z} \tilde{t}_C + \frac{1}{z} \left(r_1 + r_2 - \frac{N-2}{2} \right) \tilde{t}_C.$$

It remains to check that in all these cases

$$r_1 + r_2 - \frac{N-2}{2} = \frac{2-d}{2} = \frac{h+2}{2h}.$$

The theorem is proved.   □

Let us show that for some cycles $C_1, \ldots, C_n$, the oscillatory integrals $\tilde{t}_{C_1}(t; z), \ldots, \tilde{t}_{C_n}(t; z)$ give independent flat coordinates of the deformed connection $\tilde{\nabla}$. First we will rewrite, following [4], the integral (2.97) as a Laplace-type transform of an appropriate function $p_C(\lambda; t)$:

$$\tilde{t}_C(t; z) = z^{(N-2)/2} \int_0^\infty e^{z\lambda} p_C(\lambda; t) d\lambda, \qquad (2.98)$$

where the integration must be taken along any ray in the half-plane $\operatorname{Re} z\lambda < 0$. Put

$$p_C(\lambda; t) := \oint_{C(\lambda)} \frac{\mathrm{d}^N x}{\mathrm{d} f_t(x)}. \tag{2.99}$$

Here the Gelfand–Leray form $\mathrm{d}^N x / \mathrm{d} f_t(x)$ is defined by the equation

$$\mathrm{d}^N x = \mathrm{d} f_t(x) \wedge \frac{\mathrm{d}^N x}{\mathrm{d} f_t(x)},$$

and the $(N-1)$-cycle $C(\lambda)$ is the intersection of C with the level surface

$$V_\lambda(t) = \{x \mid f_t(x) = \lambda\}. \tag{2.100}$$

Fixing a noncritical value λ_0 for $f_t(x)$, we obtain the *period map*

$$t \mapsto \left[\frac{\mathrm{d}^N x}{\mathrm{d} f_t(x)} \right] \in H^{N-1}(V_{\lambda_0}(t)), \tag{2.101}$$

where the square brackets denote the cohomology class of the form. The map is defined for those t for which λ_0 is not a critical value of $f_t(x)$. In coordinates the map reads

$$t \mapsto \left(p_{\sigma_1}(\lambda_0; t), \ldots, p_{\sigma_n}(\lambda_0; t) \right) \tag{2.102}$$

for a basis

$$\sigma_1, \ldots, \sigma_n \in H_{N-1}(V_{\lambda_0}(t); \mathbb{Z}).$$

The period map is known to be a local diffeomorphism [62]. Now, choosing a basis of N-cycles $C_1, \ldots, C_n$ such that the $(N-1)$-cycles $C_1(\lambda), \ldots, C_n(\lambda)$ are linearly independent, we obtain independent flat coordinates $\tilde{t}_{C_1}(t; z)$, $\ldots, \tilde{t}_{C_n}(t; z)$.

Finally, we note that the nondegeneracy of the period map was not essential to prove the independence of the oscillatory integrals. One could use, instead, the analysis of the asymptotic behavior of the integrals when λ goes to one of the critical values of $f_t(x)$ and C is the corresponding vanishing cycle. We hope, however, that this digression into the singularity theory will help the reader to understand the constructions of Section 5.

Remark. The two main classes of examples of Frobenius manifolds look quite different. There are, however, some unexpected relationships between these two classes of two-dimensional topological field theories. That is, the main characters of a two-dimensional topological field theory constructed from quantum cohomology turn out to coincide with those coming from singularity theory. This phenomenon was first discovered in the quantum cohomologies of Calabi–Yau varieties [17]. It was called the *mirror conjecture* (now partially proved [43–45]). In the last section we will present our version of the mirror construction for semisimple Frobenius manifolds. In particular, we will express the deformed flat coordinates of $\widetilde{\nabla}$ on any semisimple Frobenius manifold satisfying some nondegeneracy condition by oscillatory integrals, and we will also obtain an analogue of the residue formulae (1.19), (1.20). Some general approaches to the mirror conjecture were recently proposed in [9, 44, 61, 87].

3 Semisimplicity and Canonical Coordinates

In this section we introduce the class of semisimple Frobenius manifolds and obtain the main geometrical tool for dealing with them: the canonical coordinates [28].

We recall that a commutative associative algebra A is called *semisimple* if it contains no nilpotents, i.e., nonzero vectors $a \in A$ such that

$$a^m = 0$$

for some positive integer m.

Lemma 3.1. *Any semisimple finite-dimensional Frobenius algebra over $\mathbb{C}$ is isomorphic to the orthogonal direct sum*

$$A \simeq \mathbb{C} \oplus \cdots \oplus \mathbb{C}. \tag{3.1}$$

Proof. For any $a \in A$ denote by L_a the operator of multiplication by a. Since the algebra has a unity, the operator $L_a = 0$ iff $a = 0$. Let $\lambda_1, \ldots, \lambda_k \in A^*$ be the pairwise distinct roots of the commutative algebra A, i.e., linear functions $\lambda_i \colon A \to \mathbb{C}$ such that for any $a \in A$, the eigenvalues of the operator L_a are $\lambda_1(a), \ldots, \lambda_k(a)$. Let

$$A_j := \bigcap_a \mathrm{Ker}\big(L_a - \lambda_j(a)\big)^n, \quad j = 1, \ldots, k,$$

be the corresponding root subspaces. Reducing the commuting operators L_a simultaneously to triangular form, we obtain a decomposition of the algebra into the orthogonal direct sum of the root subspaces

$$A = \bigoplus_{j=1}^{k} A_j, \tag{3.2}$$

$$A_j \cdot A_i = 0 \quad \text{for } i \neq j,$$

and, thus,

$$\lambda_i(A_j) = 0, \quad \text{for } i \neq j.$$

Let $0 \neq v_i \in A_i$ be an eigenvector for a given i, i.e., a vector such that for any $v \in A$,

$$vv_i = \lambda_i(v)v_i.$$

If $\lambda_i(v_i) = 0$, then

$$v_i^2 = 0.$$

This is not possible due to absence of nilpotents. Hence $\lambda_i(v_i) \neq 0$. Put

$$\pi_i = \frac{v_i}{\lambda_i(v_i)}.$$

We obtain

$$\pi_i^2 = \pi_i.$$

Let us prove that each A_i is one-dimensional. If $w_i \neq 0$ is another eigenvector in A_i then

$$w_i \pi_i = \lambda_i(w_i)\pi_i = \lambda_i(\pi_i)w_i = w_i.$$

So w_i is proportional to π_i. If $a \in A_i$ is a vector not proportional to π_i, then the operator of multiplication by $a - \lambda_i(a)\pi_i$ is a nilpotent one. This contradicts the semisimplicity assumption.

So, every A_i for $i = 1, \ldots, k$ is a one-dimensional subalgebra in A generated by the vector π_i such that

$$\pi_i^2 = \pi_i. \tag{3.3}$$

From (3.2) it follows that

$$\pi_i \pi_j = 0 \quad \text{for } i \neq j. \tag{3.4}$$

So $k = n$ and $\pi_1, \ldots, \pi_n$ are the basic idempotents of A. The lemma is proved.   □

Definition 3.1. A Frobenius manifold M is called *semisimple* if for any generic $t \in M$, the algebras $T_t M$ are semisimple.

The semisimplicity of an algebra is an open property. So, if at some point $t = t_0 \in M$ the algebra $T_t M$ is semisimple, it remains semisimple in some neighborhood of t_0.

Exercise 3.1. Prove that the function

$$F(t_1, t_2, t_3, t_4) = \frac{1}{2}t_1^2 t_4 + t_1 t_2 t_3 + f(t_2)$$

and the Euler vector field

$$E = t_1 \partial_1 - t_3 \partial_3 - 2t_4 \partial_4$$

give a solution of WDVV (with $d = 3$) for an arbitrary function $f(t_2)$.

So, nonsemisimple Frobenius manifolds may depend on functional parameters. This fact is well known to experts in mirror symmetry: WDVV equations of associativity provide no information about GW invariants of Calabi–Yau three-folds. For a Calabi–Yau three-fold the Frobenius structure is identically nilpotent.

In the opposite case, semisimple Frobenius manifolds depend on a finite number of parameters. Some of these parameters were described in Section 2: They are the monodromy data at $z = 0$ of the connection $\tilde{\nabla}$. In Section 4, for semisimple Frobenius manifolds we will also define monodromy data at $z = \infty$. We will show that the full list of the monodromy data is a complete local invariant of a semisimple Frobenius manifold. We will also describe the global structure of these manifolds (i.e., the analytic continuation of the local structure) in terms of the monodromy data.

Conjecturally, semisimplicity holds true for quantum cohomology of Fano varieties (see below the example for $\mathbf{CP}^2$). This conjecture is partially supported by the results of [82].

Theorem 3.1. *Let $u_1(t), \ldots, u_n(t)$ be the eigenvalues of the operator of multiplication by the Euler vector field*

$$\det(\mathcal{U}(t) - \lambda \cdot 1) = (-1)^n \prod_{i=1}^{n} (\lambda - u_i(t)), \tag{3.5}$$

$$\mathcal{U}_\beta^\alpha(t) = E^\varepsilon(t) c_{\varepsilon\beta}^\alpha(t). \tag{3.6}$$

Near a semisimple point $t_0 \in M$ they can serve as local coordinates. In these coordinates

$$\frac{\partial}{\partial u_i} \cdot \frac{\partial}{\partial u_j} = \delta_{ij} \frac{\partial}{\partial u_i}, \tag{3.7}$$

$$e = \sum_{i=1}^{n} \frac{\partial}{\partial u_i}, \tag{3.8}$$

$$E = \sum_{i=1}^{n} u_i \frac{\partial}{\partial u_i}, \tag{3.9}$$

$$\langle \, , \, \rangle = \sum_{i=1}^{n} \eta_{ii}(u) du_i^2 \quad \text{where } \eta_{ii}(u) = \frac{\partial t_1}{\partial u_i}, \ t_1 := \eta_{i\varepsilon} t^\varepsilon. \tag{3.10}$$

Lemma 3.2 (Main Lemma). *Let M be a complex-analytic manifold with a structure of Frobenius algebras on the tangent planes $T_t M$ depending analytically on t and satisfying FM1 and FM2 (the quasi-homogeneity FM3 not included). Then local coordinates $u_1, \ldots, u_n$ exist near a semisimple point $t_0 \in M$ such that*

$$\frac{\partial}{\partial u_i} \cdot \frac{\partial}{\partial u_j} = \delta_{ij} \frac{\partial}{\partial u_i}. \tag{3.11}$$

Proof. Near a semisimple point t_0 one can choose a frame of basic idempotents $\pi_1, \dots, \pi_n$,

$$\pi_i \cdot \pi_j = \delta_{ij} \pi_i,$$

depending analytically on the point. It is sufficient to show that the Lie brackets $[\pi_i, \pi_j]$ of these vector fields vanish. Let us use the deformed flat connection

$$\tilde{\nabla}_u v = \nabla_u v + z u \cdot v$$

(no $\tilde{\nabla}_{d/dz}$ component, since we do not assume quasi-homogeneity). Let us introduce the coefficients Γ_{ij}^k and f_{ij}^k from the expansions

$$\nabla_{\pi_j} \pi_i = \sum_k \Gamma_{ij}^k \pi_k,$$

$$[\pi_i, \pi_j] = \sum_k f_{ij}^k \pi_k.$$

Computing those terms of the curvature that are linear in z,

$$\tilde{\nabla}_{\pi_i} \tilde{\nabla}_{\pi_j} - \tilde{\nabla}_{\pi_j} \tilde{\nabla}_{\pi_i} - \tilde{\nabla}_{[\pi_i, \pi_j]} = 0,$$

we obtain the equation

$$\Gamma_{kj}^l \delta_i^l + \Gamma_{ki}^l \delta_{kj} - \Gamma_{ki}^l \delta_j^l - \Gamma_{kj}^l \delta_{ki} = f_{ij}^l \delta_k^l,$$

valid for arbitrary values of the four indices i, j, k, l (no summation with respect to repeated indices in these formulas!). For $l = k$ we obtain

$$f_{ij}^k = 0.$$

The lemma is proved. $\square$

Proof of Theorem 3.1. Let us prove that

$$\mathcal{L}_E \left(\frac{\partial}{\partial u_i} \right) = -\frac{\partial}{\partial u_i}, \tag{3.12}$$

where $u_1, \dots, u_n$ are the local coordinates constructed in the main lemma. We use equation (2.23) of the axiom FM3. This reads

$$\mathcal{L}_E(a \cdot b) - \mathcal{L}_E(a) \cdot b - a \cdot \mathcal{L}_E b = a \cdot b \tag{3.13}$$

for any vector fields a and b. Applying this to the basis idempotents $\pi_i = \partial/\partial u_i$ for $a = \pi_i$, $b = \pi_j$, $i \neq j$, we obtain

$$\mathcal{L}_E(\pi_i) \cdot \pi_j + \pi_i \cdot \mathcal{L}_E(\pi_j) = 0.$$

Hence

$$\mathcal{L}_E(\pi_i) \cdot \pi_j = 0 \quad \text{for } i \neq j.$$

So $\mathcal{L}_E(\pi_i) = \lambda_i \pi_i$ with some factor λ_i. Applying now (3.13) to the case $a = b = \pi_i$ we obtain

$$\lambda_i \pi_i - 2\lambda_i \pi_i = \pi_i.$$

So $\lambda_i = -1$. This proves (3.12). Writing the vector field E in the coordinates $(u_1, \ldots, u_n)$,

$$E = \sum_{i=1}^{n} E^i(u) \frac{\partial}{\partial u_i},$$

we obtain from (3.12)

$$\frac{\partial E^i}{\partial u_j} = 0, \quad i \neq j, \qquad \frac{\partial E^i}{\partial u_i} = 1.$$

Doing, if necessary, a shift of the coordinates $(u_1, \ldots, u_n)$, we arrive at the formula (3.9). The eigenvectors of the operator of multiplication by this vector field are $\partial/\partial u_1, \ldots, \partial/\partial u_n$. The corresponding eigenvalues are u_1, $\ldots$, u_n. To complete the proof of the theorem, we observe that the basis idempotents of a Frobenius algebra are pairwise orthogonal,

$$\langle \pi_i, \pi_j \rangle = \langle e, \pi_i \cdot \pi_j \rangle = 0 \quad \text{for } i \neq j.$$

Consider the 1-form dt_1. By definition, for any vector v,

$$\partial_v t_1 = dt_1(v) = \langle e, v \rangle,$$

where e is the unity of the algebra. Hence

$$\langle \pi_i, \pi_i \rangle = \langle e, \pi_i \cdot \pi_i \rangle = \left\langle e, \frac{\partial}{\partial u_i} \right\rangle = \frac{\partial t_1}{\partial u_i}.$$

The theorem is proved. $\square$

Corollary 3.1. *All the points $t \in M$ where the eigenvalues of $(E(t)\cdot)$ are pairwise distinct are semisimple.*

Definition 3.2. The coordinates $(u_1, \ldots, u_n)$ constructed in Theorem 3.1 are called *canonical coordinates* of the Frobenius manifold.

The canonical coordinates near any point are uniquely defined up to permutations. We will use Latin indices for canonical coordinates, and we put

$$\partial_i := \frac{\partial}{\partial u_i}.$$

We will also show explicitly all the sums with respect to Latin indices, not distinguishing between upper and lower indices. Let us recall that Greek indices are used for flat coordinates, and

$$\partial_\alpha = \frac{\partial}{\partial t^\alpha}.$$

The rules of tensor algebra (raising and lowering indices using $\eta^{\alpha\beta}$ and $\eta_{\alpha\beta}$, the Einstein summation rule, etc.) will be applied only to Greek indices.

We now make an algebraic digression about semisimple Frobenius algebras over $\mathbb{C}$. Let $(A, \langle \ , \ \rangle)$ be such an algebra with a basis $e_1 = e$, e_2, $\dots$, e_n and the multiplication table

$$e_\alpha \cdot e_\beta = c^\gamma_{\alpha\beta} e_\gamma.$$

Let $\pi_1, \dots , \pi_n$ be the idempotents of A. Introduce the basis of *normalized* idempotents

$$f_i = \frac{\pi_i}{\sqrt{\langle \pi_i, \pi_i \rangle}}, \quad i = 1, \dots, n,$$

choosing arbitrary signs of the square roots. Let us introduce the matrix $\Psi = (\psi_{i\alpha})$ putting

$$e_\alpha = \sum_{i=1}^n \psi_{i\alpha} f_i, \quad \alpha = 1, \dots, n.$$

Exercise 3.2. Prove the following formulae:

$$\Psi^T \Psi = \eta, \tag{3.14}$$

$$\psi_{i1} = \sqrt{\langle \pi_i, \pi_i \rangle}, \tag{3.15}$$

$$\pi_i = \sum_{\alpha,\beta=1}^n \psi_{i1} \psi_{i\beta} \eta^{\beta\alpha} e_\alpha, \tag{3.16}$$

$$c_{\alpha\beta\gamma} = \sum_{i=1}^n \frac{\psi_{i\alpha} \psi_{i\beta} \psi_{i\gamma}}{\psi_{i1}}. \tag{3.17}$$

On a semisimple Frobenius manifold the matrix Ψ depends on the point. The above formulae give

$$\langle \ , \ \rangle = \sum_{i=1}^n \psi_{i1}^2(u) du_i^2, \tag{3.18}$$

$$\partial_\alpha = \sum_{i=1}^n \frac{\psi_{i\alpha}(u)}{\psi_{i1}(u)} \partial_i, \tag{3.19}$$

$$\partial_i = \sum_{\alpha, \varepsilon} \eta^{\alpha\varepsilon} \psi_{i\varepsilon}(u) \psi_{i1}(u) \partial_\alpha, \tag{3.20}$$

or, equivalently,

$$dt^\alpha = \sum_{i=1}^{n} \psi_i^\alpha(u)\psi_{i1}(u)du_i, \text{ where } \psi_i^\alpha := \eta^{\alpha\varepsilon}\psi_{i\varepsilon}. \qquad (3.21)$$

We will now rewrite the connection $\widetilde{nabla}$ in the frame of normalized idempotents

$$f_i = \frac{\partial_i}{\sqrt{\langle \partial_i, \partial_i \rangle}}. \qquad (3.22)$$

We recall that the horizontal sections ξ satisfy the compatible system

$$\partial_\alpha \xi = zC_\alpha \xi, \quad (C_\alpha)_\gamma^\beta := c_{\alpha\gamma}^\beta, \qquad (3.23)$$

$$\partial_z \xi = \left(\mathcal{U} + \frac{\mu}{z}\right)\xi, \quad \mathcal{U}_\gamma^\beta := E^\varepsilon c_{\varepsilon\gamma}^\beta, \quad \mu = \text{diag}(\mu_1, \ldots, \mu_n). \qquad (3.24)$$

The operator $\mathcal{U}$ of multiplication by the Euler vector field becomes diagonal in the basis $f_1, \ldots, f_n$:

$$\Psi\mathcal{U}\Psi^{-1} =: U = \text{diag}(u_1, \ldots, u_n). \qquad (3.25)$$

We also introduce the matrix

$$V := \Psi\mu\Psi^{-1} \qquad (3.26)$$

of the operator μ in the same basis. From antisymmetry

$$\langle \mu a, b \rangle + \langle a, \mu b \rangle = 0$$

there follows the antisymmetry of the matrix V,

$$V^T + V = 0. \qquad (3.27)$$

Lemma 3.3. *After the gauge transformation*

$$y = \Psi\xi \qquad (3.28)$$

the system (2.42) reads

$$\partial_i y = (zE_i + V_i)y, \quad i = 1, \ldots, n, \qquad (3.29)$$

$$\partial_z y = \left(U + \frac{V}{z}\right)y. \qquad (3.30)$$

Here E_i are the unit matrices

$$(E_i)_{ab} = \delta_{ia}\delta_{ib}, \qquad (3.31)$$

and V_i are skew-symmetric matrices uniquely determined by the equations

$$[U, V_i] = [E_i, V]. \tag{3.32}$$

The matrices V and Ψ satisfy the differential equations

$$\partial_i \Psi = V_i \Psi, \tag{3.33}$$
$$\partial_i V = [V_i, V]. \tag{3.34}$$

Observe that the matrices V_i are defined at those points where the canonical coordinates are pairwise distinct. Symbolically, (3.32) can be recast in the form

$$V_i = \mathrm{adj}_{E_i}\, \mathrm{adj}_U^{-1}(V). \tag{3.35}$$

Proof. Using (3.25) one obtains

$$\partial_i \xi = z\Pi_i \xi,$$

where Π_i is the operator of multiplication by π_i. By definition of Ψ,

$$\Psi \Pi_i \Psi^{-1} = E_i.$$

So

$$\partial_i y = z E_i y + \widetilde{V}_i y,$$

where

$$\widetilde{V}_i := \partial_i \Psi \cdot \Psi^{-1}. \tag{3.36}$$

Using the orthogonality (3.14), we obtain the antisymmetry of $\widetilde{V}_i$. From the compatibility condition

$$\partial_i \partial_j y = \partial_j \partial_i y$$

it follows that

$$[E_i, \widetilde{V}_j] = [E_j, \widetilde{V}_i]$$

for any i, j. This implies the existence of a symmetric matrix Γ such that

$$\widetilde{V}_i = [E_i, \Gamma], \quad i = 1, \ldots, n.$$

The off-diagonal entries of Γ are uniquely determined. At the points of M where $u_i \neq u_j$ for any $i \neq j$, we thus obtain a uniquely determined skew-symmetric matrix $\widetilde{V}$ such that

$$[U, \widetilde{V}_i] = [E_i, \widetilde{V}], \quad i = 1, \ldots, n.$$

Doing the gauge transformation (3.28) in the system (2.42b), we obtain

$$\partial_z y = \left(U + \frac{V}{z}\right) y.$$

The compatibility condition $\partial_i \partial_z = \partial_z \partial_i$ implies

$$V = \widetilde{V},$$
$$\partial_i V = [V_i, V], \quad i = 1, \ldots, n.$$

The definition (3.36) of the matrix $\widetilde{V}_i = V_i$ reads

$$\partial_i \Psi = V_i \Psi.$$

The lemma is proved. $\square$

Exercise 3.3. Let us consider $V = (V_{ij}(u))$ as a function of

$$u = (u_1, \ldots, u_n)$$

with the values in the Lie algebra so(n). Prove that the equations (3.34) can be considered as time-dependent Hamiltonian systems

$$\frac{\partial V}{\partial u_i} = \{V, H_i(V; u)\}, \quad i = 1, \ldots, n, \tag{3.37}$$

with the quadratic Hamiltonians

$$H_i(V; u) = \frac{1}{2} \sum_{j \neq i} \frac{V_{ij}^2}{u_i - u_j}, \quad i = 1, \ldots, n, \tag{3.38}$$

with respect to the standard linear Poisson bracket on so(n),

$$\{V_{ij}, V_{kl}\} - V_{il}\delta_{jk} - V_{jl}\delta_{ik} + V_{jk}\delta_{il} - V_{ik}\delta_{jl}. \tag{3.39}$$

The canonical coordinates $u_1, \ldots, u_n$ play the role of the time variables of these Hamiltonian systems.

Exercise 3.4. Prove that $\{H_i, H_j\} = 0$ for any i, j. From this and from the commutativity of the flows (3.37), derive that the form

$$\sum_{i=1}^{n} H_i(V; u) \, du_i$$

is closed for any solution $V(u)$ of the system (3.34). This means that (locally) there exists a function $\tau(u)$ such that

$$\frac{\partial \log \tau(u)}{\partial u_i} = H_i(V(u); u), \quad i = 1, \ldots, n. \tag{3.40}$$

This is called a *tau-function* of the solution of the system $V(u)$. In the next section we will show that the system (3.33), (3.34) can be solved by reducing it to some linear Riemann–Hilbert boundary value problem. The tau-function will coincide with the Fredholm determinant of the corresponding system of integral equations (see [52, 71]).

The importance of the tau-function in topological field theory is clear from the following theorem.

Theorem 3.2 ([36]). *Let X be a smooth projective manifold such that the quantum cohomology of X is semisimple. Then the generating function $F^{(1)}(t)$ of the elliptic Gromov–Witten invariants of X is given by the formula*

$$F^{(1)}(t) = \log \left. \frac{\tau(u)}{J^{1/24}} \right|_{u=u(t)},$$

where $\tau(u)$ is the above tau-function and

$$J = \det\left(\frac{\partial t^\alpha}{\partial u_i}\right) = \psi_{11} \cdots \psi_{n1}.$$

Particularly, from this theorem follows the validity of Conjectures 0.1 and 0.2 of recent paper of Givental [46].

We now prove the converse to Lemma 3.3.

Let $V(u)$, $\Psi(u)$ be a solution of the system (3.33), (3.34) with a diagonalizable matrix $V(u)$. We first observe that the product

$$\Psi^{-1}(u)V(u)\Psi(u)$$

does not depend on u. We can therefore find a constant matrix C in such that

$$\Psi^{-1}(u)V(u)\Psi(u) = C\mu C^{-1},$$

where

$$\mu = \mathrm{diag}(\mu_1, \ldots, \mu_n)$$

is a constant diagonal matrix. Doing a change

$$\Psi(u) \mapsto \Psi(u)C$$

we obtain another solution of the linear system (3.33) such that

$$\Psi^{-1}(u)V(u)\Psi(u) = \mathrm{diag}(\mu_1, \ldots, \mu_n). \tag{3.41}$$

After these preliminaries we formulate the following lemma.

Lemma 3.4. *Let* $V(u) = (V_{ij}(u))$, $\Psi(u) = (\psi_{i\alpha}(u))$ *be a solution of the system* (3.33), (3.34) *satisfying* (3.41). *Then the formulae* (3.18), (3.21), (3.17) *define a Frobenius structure on the domain*

$$u_i \neq u_j \quad for \ i \neq j, \qquad \psi_{11}(u) \cdots \psi_{n1}(u) \neq 0. \tag{3.42}$$

Proof. From the antisymmetry of the matrices V_i it follows that

$$\partial_i(\Psi^T \Psi) = 0, \quad i = 1 \ldots, n.$$

Put

$$\eta = (\eta_{\alpha\beta}) = \Psi^T \Psi, \quad (\eta^{\alpha\beta}) = \eta^{-1}. \tag{3.43}$$

The next step is to prove that the 1-forms

$$\sum_{i=1}^{n} \psi_i^{\alpha} \psi_{i1} \, du_i, \quad \text{where } \psi_i^{\alpha} = \eta^{\alpha\varepsilon}\psi_{i\varepsilon},$$

are closed. From (3.33) we obtain

$$\partial_j \psi_{i\alpha} = \frac{V_{ij}}{u_j - u_i} \psi_{j\alpha} \quad \text{for any } i \neq j, \text{ any } \alpha.$$

From this the identity

$$\partial_j(\psi_i^{\alpha}\psi_{i1}) = \partial_i(\psi_j^{\alpha}\psi_{j1})$$

follows. This proves the local existence of functions t^{α} such that

$$dt^{\alpha} = \sum_{i=1}^{n} \psi_i^{\alpha}\psi_{i1} du_i.$$

The differentials dt^1, ... , dt^n are independent on the domain (3.42). So $t^1, \ldots, t^n$ serve as local coordinates on the domain. From the orthogonality relation (3.14) we obtain

$$\partial_{\alpha} = \sum_{i=1}^{n} \frac{\psi_{i\alpha}}{\psi_{i1}} \partial_i.$$

The last step is to prove the symmetry

$$\partial_{\delta}\left(\sum_{i=1}^{n} \frac{\psi_{i\alpha}\psi_{i\beta}\psi_{i\gamma}}{\psi_{i1}}\right) = \partial_{\gamma}\left(\sum_{i=1}^{n} \frac{\psi_{i\alpha}\psi_{i\beta}\psi_{i\delta}}{\psi_{i1}}\right).$$

To prove this we have to use another consequence of (3.33),

$$\partial_i \psi_{i\alpha} = -\sum_{k \neq i} \partial_k \psi_{i\alpha},$$

valid for any i, any α. We leave this computation as an exercise for the reader. The lemma is proved. $\square$

Corollary 3.2. *The classes of local equivalence of semisimple Frobenius manifolds such that 1 is an eigenvalue of ∇E with multiplicity k depend on*

$$k - 1 + \frac{n(n-1)}{2}$$

parameters.

Proof. Take the initial data

$$V(u^0) = \left(V_{ij}(u^0) \right) \tag{3.44}$$

of the antisymmetric matrix V at a point $u^0 = (u_1^0, \dots, u_n^0)$ with

$$u_i^0 \neq u_j^0 \text{ for } i \neq j.$$

Solving the system (3.34) of commuting ODEs, we locally obtain the unique matrix-valued function $V(u)$ and therefore the matrices $V_i(u)$. The solution $\Psi(u)$ of the linear system (3.33) such that

$$\Psi^{-1}(u)V(u)\Psi(u) = \mathrm{diag}(\mu_1, \dots, \mu_n) = \mu$$

is uniquely determined up to the multiplication by a matrix

$$\Psi(u) \mapsto \Psi(u)C,$$
$$C^{-1}\mu C = \mu.$$

The matrices C that preserve the direction of the eigenvector e of μ with the eigenvalue $\mu_1 = -d/2$ produce equivalences of the Frobenius manifolds. The assumption about the multiplicity of the eigenvalue 1 of ∇E means that the eigenvalue $\mu_1 = -d/2$ of μ also has multiplicity k. The vectors Ce considered up to rescalings must be eigenvectors of μ with the same eigenvalue μ_1. The directions of these vectors give $k - 1$ parameters additional to the initial data (3.44). The corollary is proved. □

4 Stokes Matrices and Classification of Semisimple Frobenius Manifolds

In the previous section we parameterized semisimple Frobenius manifolds M by the initial data of the system (3.33), (3.34) of differential equations at a point $t \in M$ such that $u_i(t) \neq u_j(t)$ for $i \neq j$. Typically, however, one has no "natural" point in the Frobenius manifold to specify the initial data (3.44). For example, for Frobenius manifolds with good analytic properties, the "natural" point would be $t_0 = (t'' = 0, t' = -\infty)$. But at this point the Frobenius algebra $T_{t_0}M$ is typically nilpotent (one can keep in mind the example of quantum cohomology where t_0 is the point of classical limit).

So at this point $u_1(t_0) = u_2(t_0) = \cdots = u_n(t_0)$. This is a singular point for the system (3.33), (3.34).

Instead, we will use [28, 32] the monodromy data of the system (3.30) as parameters. Recall that the system is gauge equivalent to the equations (2.42), which determine the horizontal sections of the connection ∇. Some of the monodromy data have already been defined in Section 2. Namely, this part is the monodromy at $z = 0$ of the system (3.30) gauge equivalent to (2.42b). Recall that for a system (3.30) with the monodromy data at $z = 0$ of the form $(\mathcal{V}, \langle\ ,\ \rangle, \mu, [R])$, a fundamental matrix solution $Y_0(z)$ exists such that

$$Y_0 = \Phi(z) z^\mu z^R, \tag{4.1}$$

where

$$\Phi(z) = \Psi + z\Phi_1 + z^2\Phi_2 + \cdots \tag{4.2}$$

is an invertible matrix holomorphic for small $|z|$ satisfying

$$\Phi^T(-z)\Phi(z) = \eta. \tag{4.3}$$

Let us describe the ambiguity in the choice of the normalized solution (4.1).

Let $C_0(\mu, R)$ be the group of all invertible matrices C such that

$$z^\mu z^R C z^{-R} z^{-\mu} = C_0 + zC_1 + \cdots \tag{4.4}$$

is a matrix-valued polynomial in z.

Lemma 4.1. *Two solutions $Y(z)$, $\widetilde{Y}(z)$ of the system (3.30) have the same form (4.1) iff they are related by right multiplication by a matrix $C \in C_0(\mu, R)$.*

Proof. If

$$Y = \Phi(z) z^\mu z^R, \quad \widetilde{Y} = \tilde{\Phi}(z) z^\mu z^R$$

satisfy (3.30), then

$$\widetilde{Y}(z) = Y(z)C$$

for a constant matrix C. We have

$$\Phi^{-1}(z)\tilde{\Phi}(z) = z^\mu z^R C z^{-R} z^{-\mu}.$$

Hence the rhs. must be a polynomial. The converse statement is obvious. The lemma is proved. $\square$

Exercise 4.1. Show that the matrices in $C_0(\mu, R)$ commute with $e^{2\pi i \mu}$ and that they must have the form

$$C = C_0 + C_1 + C_2 + \cdots$$

with

$$(C_k)_\beta^\alpha \neq 0 \quad \text{only if} \quad \mu_\alpha - \mu_\beta = k, \quad k = 0, 1, \ldots.$$

In particular, the matrix C_0 commutes with μ.

Remark. In the case of a Frobenius manifold we have an additional structure of the monodromy data of (3.30) at $z = 0$. Namely, an eigenvector e of the matrix V with the eigenvalue $\mu_1 = -d/2$ must be chosen. It corresponds to the unity of M. We must therefore impose an additional constraint on the matrix C in (4.4): The component C_0 must preserve the chosen vector. Observe that the chosen vector corresponds to the first column ψ_{i1} of the matrix Ψ.

The second part is the monodromy data at $z = \infty$, which we are now going to define.

We first describe the monodromy data at $z = \infty$ of the system

$$\frac{dy}{dz} = \left(U + \frac{1}{z}V\right) * y \tag{4.5}$$

with arbitrary $n \times n$ matrices of the form

$$U = \mathrm{diag}(u_1, \ldots, u_n), \quad u_i \neq u_j, \tag{4.6}$$

and V with the property

$$V^T = -V \tag{4.7}$$

being a diagonalizable matrix

$$\Psi^{-1}V\Psi = \mu = \mathrm{diag}(\mu_1, \ldots, \mu_n). \tag{4.8}$$

The point $z = \infty$ is an irregular singularity for the system (3.30). So, in the problem of finding the normal form of the system (3.30), we have to distinguish formal gauge equivalences

$$y \mapsto G\left(\frac{1}{z}\right)y,$$

$$G\left(\frac{1}{z}\right) = 1 + \frac{G_1}{z} + \frac{G_2}{z^2} + \cdots$$

and analytic ones, where the series converges for sufficiently large $|z|$.

Definition 4.1. Two systems

$$\frac{dy^{(i)}}{dz} = \left(U + \frac{1}{z}V^{(i)}\right)y^{(i)}, \quad i = 1, 2, \tag{4.9}$$

are called *analytically equivalent at $z = \infty$* if there exists a gauge transformation

$$y^{(2)} = G(z)y^{(1)} \tag{4.10}$$

with the matrix-valued function $G(z)$ analytic at $z = \infty$ satisfying $G(\infty) = 1$ and the orthogonality condition

$$G^T(-z)G(z) = 1.$$

The *monodromy at $z = \infty$ of the system* (3.30) is the class of analytic equivalence of this system.

Below we will explain how one can parameterize the monodromy at infinity by Stokes matrices of the system (3.30). But first we will show that the system is, to some extent, uniquely determined by the monodromy at $z = 0$ and $z = \infty$.

Lemma 4.2. *Let* (4.9) *be two systems analytically equivalent at $z = \infty$. Then the matrix G establishing the gauge equivalence is a rational function of z of the form*

$$G = 1 + \frac{G_1}{z} + \frac{G_2}{z^2} + \cdots + \frac{G_m}{z^m}. \tag{4.11}$$

Proof. Let the given gauge transformation (4.10) be analytic for $|z| > M$ for some constant M. Choose a point z_0 with $|z_0| > M$ and the fundamental matrix solutions $Y^{(i)}(z)$ of the systems (4.9) with the initial data

$$Y^{(i)}(z_0) = 1, \quad i = 1, 2.$$

For any z with $|z| > M$ we must have

$$G(z)Y^{(1)}(z) = Y^{(2)}(z)C$$

for some constant nondegenerate matrix C. The solutions $Y^{(1,2)}(z)$ can be analytically continued along any path in $\mathbb{C} \setminus 0$. The formula

$$G(z) = Y^{(2)}(z)CY^{(1)}(z)^{-1}$$

gives the analytic continuation of $G(z)$ (recall that $\det Y^{(i)}(z) = \sum_i u_i \times \exp(z - z_0) \neq 0$). We obtain a single-valued analytic function in $\overline{\mathbb{C}} \setminus 0$ such that $G(\infty) = 1$. Near the point of regular singularity $z = 0$ the entries of the matrices $Y^{(1,2)}(z)$ do not grow faster than some power of $|z|$. Therefore, $G(z)$ has also at most a polynomial growth at $z = 0$. So it must be a rational function having a pole only at $z = 0$. The lemma is proved. $\qquad\square$

Exercise 4.2. Prove that the determinant of the matrix (4.11) is identically equal to 1.

Remark. Gauge transformations with rational $G(z)$ are called *Schlesinger transformations* [52]. For the case of Frobenius manifolds they induce certain symmetries of WDVV, i.e., changes of variables

$$t \mapsto \hat{t},$$
$$F \mapsto \hat{F},$$

mapping solutions to solutions. We give here the explicit form [32] of such symmetries for the case $m \leq 1$ in (4.11).

Type 1. $G = \text{const}$, $G\mu = \mu G$, G permutes the two eigenvectors of μ with the numbers 1 and κ. Then

$$\hat{t}_\alpha = \partial_\alpha \partial_\kappa F(t),$$

$$\frac{\partial^2 \hat{F}}{\partial \hat{t}^\alpha \partial \hat{t}^\beta} = \frac{\partial^2 F}{\partial t^\alpha \partial t^\beta}, \tag{4.12}$$

$$\hat{\eta}_{\alpha\beta} = \eta_{\alpha\beta}.$$

Type 2.

$$G = 1 + \frac{A}{z},$$

where

$$A_{ij} = \frac{\psi_{i1}\psi_{j1}}{t_1}.$$

Then

$$\hat{t}^1 = \frac{1}{2}\frac{t_\sigma t^\sigma}{t_1},$$

$$\hat{t}^\alpha = \frac{t^\alpha}{t_1}, \quad \alpha \neq 1, n,$$

$$\hat{t}^n = -\frac{1}{t_1}, \tag{4.13}$$

$$\hat{F} = t_1^{-2}\left[F - \frac{1}{2}t^1 t_\sigma t^\sigma\right],$$

$$\hat{\eta}_{\alpha\beta} = \eta_{\alpha\beta}.$$

One may also take the superposition of (4.13) with any transformation of the form (4.12)

We now classify the systems of the form (3.30) having the same monodromies at $z = 0$ and $z = \infty$. We will show that, generically, these systems must coincide. There remain, however, some subtleties in the nongeneric situation. The ambiguity of the reconstruction of the system (3.30) starting from the monodromy data at $z = 0$ and $z = \infty$ will be completely described in terms of the monodromy at $z = 0$.

Let us choose a representative R in the class of equivalence $[R]$ of the monodromy data at $z = 0$ of the system (3.30). Let us consider the centralizer of the monodromy matrix

$$M_0 = \exp 2\pi i(\mu + R) \tag{4.14}$$

in the group of invertible matrices, i.e., the matrices C commuting with M_0:

$$C^{-1}M_0C = M_0. \tag{4.15}$$

For any such matrix C the product

$$z^\mu z^R C z^{-R} z^{-\mu} = \sum_k A_k z^k \tag{4.16}$$

is a matrix-valued Laurent polynomial in z. In particular, for a matrix $C \in C_0(\mu, R)$, the rhs. of (4.16) contains only nonnegative powers of z. Denote by $\mathcal{C}(\mu, R)$ the group quotient of the centralizer (4.15) by the subgroup $C_0(\mu, R)$.

Example 4.1. For a nonresonant μ the group $\mathcal{C}(\mu, R)$ consists of one element.

Example 4.2. The group $\mathcal{C}(\mu, R)$ with a resonant μ and $R = 0$ is not trivial. It is isomorphic to the subgroup of "upper triangular" parabolic matrices in the centralizer of $\exp 2\pi i\mu$:

$$C = \cdots + C_{-2} + C_{-1} + 1, \tag{4.17}$$

where

$$(C_k)^\alpha_\beta \neq 0 \quad \text{only if} \quad \mu_\alpha - \mu_\beta = k, \quad k = -1, -2, \dots. \tag{4.18}$$

Let the two systems of the form (4.9) have the same monodromy data at $z = 0$ and $z = \infty$. We will associate with such a pair a matrix $C \in \mathcal{C}(\mu, R)$ where μ, R are the monodromy data of the systems (4.9) at $z = 0$. Let $Y^{(1)}(z)$ be the matrix solution of the system

$$\partial_z Y^{(1)} = \left(U + \frac{1}{z}V^{(1)}\right)Y^{(1)} \tag{4.19}$$

of the form

$$Y^{(1)}(z) = \Phi(z)z^\mu z^R.$$

Let $G_0(z) = 1 + O(z)$ and $G_\infty(z) = 1 + O(1/z)$ be the gauge transformations of the system (4.19) to another system of the same form

$$\partial_z Y^{(2)} = \left(U + \frac{1}{z} V^{(2)} \right) Y^{(2)}. \tag{4.20}$$

The matrix-valued functions $G_0(z)$ and $G_\infty(z)$ are assumed to be analytic near $z = 0$ and $z = \infty$, respectively. Near $z = 0$ we obtain a solution

$$Y_0^{(2)}(z) = G_0(z) Y^{(1)}(z)$$

of (4.20). Continuing $Y^{(1)}(z)$ analytically along a ray ρ in a neighborhood of infinity we produce another solution of the system (4.20),

$$Y_\infty^{(2)}(z) = G_\infty(z) Y^{(1)}(z).$$

Continuing $Y_\infty^{(2)}(z)$ back along the same ray ρ, we obtain two matrix solutions of (4.20) defined in a neighborhood of $z = 0$. They must be related by a multiplication by an invertible matrix C_{12}

$$Y_\infty^{(2)}(z) = Y_0^{(2)}(z) C_{12}.$$

We rewrite the last equation in the form

$$G_0^{-1}(z) G_\infty(z) = \Phi(z) z^\mu z^R C_{12} z^{-R} z^{-\mu} \Phi^{-1}(z). \tag{4.21}$$

The rhs. must be a meromorphic function near $z = 0$. This means, in particular, that the matrix C_{12} commutes with the monodromy matrix M_0. We arrive at the following result.

Theorem 4.1. *The set of all systems*

$$\partial_z \tilde{Y} = \left(U + \frac{1}{z} \tilde{V} \right) \tilde{Y}$$

of the form (3.30) whose monodromy data at $z = 0$ and $z = \infty$ coincide with those of the given system

$$\partial_z Y = \left(U + \frac{1}{z} V \right) Y$$

is in one-to-one correspondence with the elements of the group $C(\mu, R)$.

Proof. The above construction associates with a pair of these systems an element $C = C_{12}$ of the centralizer of M_0. It remains to show that the two systems coincide iff

$$z^\mu z^R C z^{-R} z^{-\mu}$$

is a polynomial in z. Indeed, if this is the case, the rhs. of (4.21) is analytic at $z = 0$. Therefore, $G_\infty(z)$ is analytic at $z = 0$. Using the normalization $G_\infty(\infty) = 1$ we conclude that $G_\infty(z) \equiv 1$. The converse statement is obvious. The theorem is proved. $\qquad \square$

We now proceed to a "quantitative" description of the monodromy at infinity of systems of the form (3.30). We first show that all the systems (3.30) with given pairwise distinct values of $u_1, \ldots, u_n$ are gauge equivalent at $z = \infty$ with respect to *formal* gauge transformations. It is sufficient to construct a gauge transformation

$$\widetilde{Y} = G(z)Y \tag{4.22}$$

of the system (3.30) to the system with constant coefficients

$$\partial_z \widetilde{Y} = U\widetilde{Y}. \tag{4.23}$$

Lemma 4.3. *For any system* (3.30) *there exists a unique formal series*

$$G(z) = 1 + \frac{A_1}{z} + \frac{A_2}{z^2} + \cdots \tag{4.24}$$

satisfying

$$G^T(-z)G(z) = 1 \tag{4.25}$$

such that (4.22) *transforms* (3.33) *to the system* (4.23) *with constant coefficients.*

Proof. For the coefficients of the formal series (4.24), one obtains the recursion relations

$$[U, A_1] = V,$$
$$[U, A_{k+1}] = A_k V - k A_k, \quad k = 1, 2, \ldots.$$

Representing

$$A_k = B_k + D_k$$

as the sum of an off-diagonal matrix B_k and a diagonal one D_k, we obtain

$$B_1 = \mathrm{adj}_U{}^{-1}(V),$$
$$D_k = \frac{1}{k}\,\mathrm{diag}(B_k V),$$
$$B_{k+1} = \mathrm{adj}_U{}^{-1}(A_k V - k A_k),$$

where "diag" stands for the diagonal part of the matrix. This proves the existence and uniqueness of the series $G(z)$.

Let us choose a fundamental matrix $Y(z)$ for the system (3.30) such that

$$Y^T(-z)Y(z) \equiv 1.$$

Then

$$G(z)Y(z)$$

is a formal solution of the system (4.23). Therefore, for an appropriate constant invertible matrix C,

$$G(z)Y(z) = e^{zU}C.$$

Computing the product

$$(G^{-1}(z))^T G^{-1}(-z) = e^{zU}(CC^T)^{-1}e^{-zU}$$

we conclude that $CC^T = 1$, since the lhs. is a formal series in inverse powers of z of the form $1 + O(1/z)$. This proves the orthogonality relation (4.25). The lemma is proved.   □

The series $G(z)$ typically diverges. However, in certain sectors of the complex z-plane near $z = \infty$ it serves as the asymptotic development of an actual solution of the original system.

We recall that a series

$$a_0 + \frac{a_1}{z} + \frac{a_2}{z^2} + \cdots$$

is an asymptotic expansion of the function $f(z)$ for $|z| \to \infty$ in the sector

$$\alpha < \arg z < \beta$$

if for any n and any sufficiently small positive ε

$$z^n \left[f(z) - \sum_{k=0}^n \frac{a_k}{z^k} \right] \to 0$$

as $|z| \to \infty$ uniformly in the sector

$$\alpha + \varepsilon < \arg z < \beta - \varepsilon.$$

This fact will be denoted for brevity by

$$f(z) \sim a_0 + \frac{a_1}{z} + \frac{a_2}{z^2} + \cdots, \quad |z| \to \infty, \quad \alpha < \arg z < \beta.$$

Let us set

$$Y_{\text{formal}}(z) = \left(1 + \frac{A_1}{z} + \frac{A_2}{z^2} + \cdots \right) e^{zU}, \tag{4.26}$$

where the coefficients of the formal series are defined in Lemma 4.3. We say that a matrix solution $Y(z)$ of the system (3.30) has the asymptotic expansion

$$Y(z) \sim Y_{\text{formal}}(z), \quad |z| \to \infty, \quad \alpha < \arg z < \beta,$$

if in the same sector

$$Y(z)e^{-zU} \sim Y_{\text{formal}}(z)e^{-zU} = 1 + \frac{A_1}{z} + \frac{A_2}{z^2} + \cdots.$$

Definition 4.2. A line ℓ through the origin in the complex z-plane is called *admissible* for the system (3.30) if

$$\operatorname{Re} z(u_i - u_j)|_{z \in \ell \backslash 0} \neq 0 \quad \text{for any } i \neq j. \tag{4.27}$$

Let us fix an admissible line ℓ and an orientation on it. According to the orientation the line splits into a negative and a positive part, respectively ℓ_- and ℓ_+. Let the parts have the equations

$$\begin{aligned}
\ell_+ &= \{z \mid \arg z = \phi\}, \\
\ell_- &= \{z \mid \arg z = \phi - \pi\}.
\end{aligned} \tag{4.28}$$

We construct the two sectors

$$\begin{aligned}
\Pi_{\text{right}} &: \quad \phi - \pi - \varepsilon < \arg z < \phi + \varepsilon, \\
\Pi_{\text{left}} &: \quad \phi - \varepsilon < \arg z < \phi + \pi + \varepsilon,
\end{aligned} \tag{4.29}$$

for some sufficiently small positive ε.

Theorem 4.2. *There exist unique solutions* $Y_{\text{right/left}}(z)$ *of* (3.30), *analytic in the sectors* $\Pi_{\text{right/left}}$, *respectively, having the asymptotic expansion*

$$Y_{\text{right/left}}(z) \sim Y_{\text{formal}}(z) \tag{4.30}$$

as $|z| \to \infty$ *in these sectors.*

See the proof in [7].

We are now ready to define Stokes matrices of the system (3.30). In the narrow sector

$$\Pi_+ : \quad \phi - \varepsilon < \arg z < \phi + \varepsilon \tag{4.31}$$

we have two solutions. They must be related by the multiplication by a matrix:

$$Y_{\text{left}}(z) = Y_{\text{right}}(z)S, \quad z \in \Pi_+. \tag{4.32}$$

Similarly, in the opposite narrow sector Π_-,

$$Y_{\text{left}}(z) = Y_{\text{right}}(z)S_-, \quad z \in \Pi_-. \tag{4.33}$$

Definition 4.3. The matrices S, S_- are called *Stokes matrices of the system* (3.30).

Lemma 4.4. *Two systems with equal Stokes matrices with respect to the same admissible oriented line* ℓ *are analytically equivalent near* $z = \infty$.

Proof. Let $Y^{(1)}_{\text{left/right}}(z)$, $Y^{(2)}_{\text{left/right}}(z)$ be the solutions of the corresponding systems with the needed asymptotic expansions in the sectors $\Pi_{\text{left/right}}$, respectively. Let us consider the piecewise analytic, matrix-valued function $G(z)$ defined for sufficiently large $|z|$ as

$$G(z) = \begin{cases} Y_{\text{right}}^{(2)}(z)Y_{\text{right}}^{(1)}{}^{-1}(z), & z \in \Pi_{\text{right}}, \\ Y_{\text{left}}^{(2)}(z)Y_{\text{left}}^{(1)}{}^{-1}(z), & z \in \Pi_{\text{left}}. \end{cases}$$

In the sectors Π_+, Π_- we have

$$Y_{\text{left}}^{(1,2)}(z) = Y_{\text{right}}^{(1,2)}(z)S, \quad z \in \Pi_+,$$

$$Y_{\text{left}}^{(1,2)}(z) = Y_{\text{right}}^{(1,2)}(z)S_-, \quad z \in \Pi_-.$$

So $G(z)$ is a single-valued analytic function for $|z| > M$ for some large constant M. In the sectors $\Pi_{\text{right/left}}$,

$$G(z) \sim 1 + O\left(\frac{1}{z}\right).$$

Therefore, $z = \infty$ is a removable singularity for this function, and

$$G(\infty) = 1.$$

This function $G(z)$ establishes the needed gauge transformation between the systems. The lemma is proved.   □

We will now describe the algebraic properties of the Stokes matrices. We first describe explicitly all nonadmissible lines. Each of them consists of two *Stokes rays*

$$R_{ij} := \{z \mid z = -ir(\bar{u}_i - \bar{u}_j), r \geq 0\}, \quad i \neq j \tag{4.34}$$

(some of them may coincide). We explain: For $z \in R_{ij}$

$$|e^{zu_i}| = |e^{zu_j}|;$$

on the right of R_{ij}

$$|e^{zu_i}| < |e^{zu_j}|,$$

and on the left of R_{ij}

$$|e^{zu_i}| > |e^{zu_j}|.$$

The ray R_{ji} is the opposite one to R_{ij}. An admissible line ℓ must contain no Stokes rays. The sectors $\Pi_{\text{right/left}}$ can be extended up to the first nearest Stokes ray (see [7]).

Theorem 4.3. *The Stokes matrices* $S = (s_{ij})$, S_- *of the system* (3.30) *satisfy the following properties:*

$$S_- = S^T, \tag{4.35}$$

$$s_{ii} = 1, \quad i = 1, \ldots, n, \tag{4.36a}$$

$$s_{ij} \neq 0 \quad only \ if \quad R_{ij} \subset \Pi_{\text{left}}. \tag{4.36b}$$

Proof. We know that for any two matrix solutions $Y_1(z)$, $Y_2(z)$ of the system (3.30), the product

$$Y_1^T(-z)Y_2(z)$$

does not depend on z. Let us choose for $z \in \Pi_{\text{right}}$, $Y_2(z) = Y_{\text{right}}(z)$, $Y_1(z) = Y_{\text{left}}(z)$. Using the asymptotic expansions

$$Y_{\text{right}}(z) \sim G(z)e^{zU},$$
$$Y_{\text{left}}(-z) \sim G(-z)e^{-zU},$$

valid for $z \in \Pi_+$ with $G(z)$ defined in Lemma 4.3, and the orthogonality condition (4.25), we obtain

$$Y_{\text{left}}^T(-z)Y_{\text{right}}(z) \equiv 1, \quad z \in \Pi_+.$$

Let us analytically continue this formula in the counterclockwise direction through the ray ℓ_+. We obtain after analytic continuation

$$Y_{\text{right}}(z) \mapsto Y_{\text{left}}(z)S^{-1},$$
$$Y_{\text{left}}(-z) \mapsto Y_{\text{right}}(-z)S_-.$$

So

$$S_-^T Y_{\text{right}}^T(-z)Y_{\text{left}}(z)S^{-1} \equiv 1, \quad z \in \Pi_-.$$

As above, we show that

$$Y_{\text{right}}^T(-z)Y_{\text{left}}(z) \equiv 1, \quad z \in \Pi_-.$$

Therefore,

$$S_-^T = S.$$

Let us now prove (4.36). Comparing the asymptotic expansions of both sides of (4.32) for $z \in \Pi_+$, we conclude that

$$e^{zU} S e^{-zU} \sim 1, \quad |z| \to \infty, \quad z \in \Pi_+.$$

This means that

$$e^{z(u_i - u_j)}s_{ij} \sim \delta_{ij}, \quad |z| \to \infty, \quad z \in \Pi_+.$$

For the diagonal terms this implies $s_{ii} = 1$. For the off-diagonal terms we have

$$|e^{z(u_i - u_j)}| \to \infty \quad \text{for } |z| \to \infty, \ z \in \Pi_+,$$

if $R_{ij} \subset \Pi_{\text{right}}$. So, for those pairs $i \neq j$ for which $R_{ij} \subset \Pi_{\text{right}}$, we must have $s_{ij} = 0$. The opposite ray satisfies $R_{ji} \subset \Pi_{\text{left}}$, and

$$|e^{z(u_j - u_i)}| \to 0 \quad \text{for } |z| \to \infty, \ z \in \Pi_+.$$

So s_{ji} need not be zero. The lemma is proved.   □

We see that the Stokes matrix S contains $n(n-1)/2$ independent parameters.

To complete the list of the monodromy data we define the central connection matrix

$$Y_0(z) = Y_{\text{right}}(z)C, \quad z \in \Pi_+ \tag{4.37}$$

(observe: the branch cut in the definition of $Y_0(z)$ must be chosen along ℓ_-).

The monodromy (μ, R) at $z = 0$, the monodromy S at $z = \infty$, and the central connection matrix C are not independent. First of all, we have the following *cyclic relation*:

$$C^{-1}S^T S^{-1}C = M_0 = \exp 2\pi i(\mu + R). \tag{4.38}$$

This expresses a simple topological fact: On the punctured plane $\mathbb{C} \setminus 0$, a loop around infinity is homotopic to a loop around the origin. Another property comes from the orthogonality relations

$$\begin{aligned}
S &= Ce^{-\pi iR}e^{-\pi i\mu}\eta^{-1}C^T, \\
S^T &= Ce^{\pi iR}e^{\pi i\mu}\eta^{-1}C^T.
\end{aligned} \tag{4.39}$$

We leave the proof of these identities as an exercise for the reader.

The matrix C is defined up to transformations of the form

$$C \mapsto BC, \quad BSB^T = B, \tag{4.40a}$$

preserving the relations (4.38), (4.39), and

$$C \mapsto CC_0, \quad C_0 \in \mathcal{C}_0(\mu, R), \tag{4.40b}$$

corresponding to a change of the solution $Y_0(z)$.

Exercise 4.3. Prove that the equivalence classes (4.40) of the central connection matrices of systems (3.30) with a given monodromy (μ, R) at the origin and a given monodromy S at infinity are in one-to-one correspondence with the group $\mathcal{C}(\mu, R)$.

The properties (4.38) and (4.39) typically specify the central connection matrix C of the system with given μ, R, S, essentially uniquely with an ambiguity (4.40) that does not affect the Frobenius structure. This reflects the claim of Theorem 4.1 (here "typically" means triviality of the group $\mathcal{C}(\mu, R)$). Anyhow, the following uniqueness theorem holds.

Lemma 4.5. *If two systems*

$$\partial_z Y^{(1,2)} = \left(U + \frac{1}{z}V^{(1,2)}\right)Y^{(1,2)}$$

have the same matrices μ, R, S (with respect to the same admissible oriented line ℓ), C, then $V^{(2)} = V^{(1)}$.

The proof is similar to that of Lemma 4.4. We leave it as an exercise.

Let us return to semisimple Frobenius manifolds. Starting from a point $t_0 \in M$ such that the eigenvalues $u_1(t_0), \dots, u_n(t_0)$ of the operator $\mathcal{U}(t_0) = (E(t_0)\cdot)$ are pairwise distinct, ordering these eigenvalues, choosing signs of the square roots of $\langle \partial_i, \partial_i \rangle$, and fixing an oriented line ℓ on the complex z-plane admissible for the points $u_1(t_0), \dots, u_n(t_0)$, we define the Stokes matrix $S = S(t_0)$ and the central connection matrix $C = C(t_0)$. We will now prove that these matrices do not change under small variations of t_0. Observe that the property of admissibility of the line ℓ is stable under small perturbations of t_0.

Theorem 4.4 (Isomonodromy Theorem (second part)). *The Stokes matrix S and the central connection matrix C do not depend on the point of a semisimple Frobenius manifold.*

Proof. The coefficients A_1, A_2, ... of the solution $Y_{\text{formal}}(z; u)$ are analytic functions on u due to Lemma 4.3. From the uniqueness of $Y_{\text{formal}}(z; u)$ it easily follows that

$$\partial_i Y_{\text{formal}}(z; u) = (zE_i + V_i)Y_{\text{formal}}(z; u), \quad i = 1, \dots, n.$$

The same statements are true for the solutions $Y_{\text{right/left}}(z; u)$ and, as we already know from Section 2, for the solution $Y_0(z; u)$. Using the definitions

$$S = Y_{\text{right}}^{-1}(z; u)Y_{\text{left}}(z; u), \quad z \in \Pi_+,$$
$$C = Y_{\text{right}}^{-1}(z; u)Y_0(z; u), \quad z \in \Pi_{\text{right}},$$

we obtain

$$\partial_i S = 0, \quad \partial_i C = 0.$$

The theorem is proved. $\qquad\qquad\qquad\qquad\qquad\qquad\qquad\qquad\qquad\quad\square$

Together with the results of Section 2, we conclude that the monodromy data μ, R, S, C do not depend on the point of the Frobenius manifold.

We will now show how to reconstruct the semisimple Frobenius manifold starting from the monodromy data.

To reconstruct the operator (3.30) and the solutions $Y_{\text{right/left}}$, Y_0 for given u_1, ..., u_n (μ, R, S, C), one must solve a certain Riemann–Hilbert boundary value problem. Let D be the disk

$$|z| < \rho$$

for some $\rho > 0$, and let P_{right} and P_{left} be the two components of $\mathbb{C} \setminus \ell$ intersected with the external parts of the disk. We must construct a piecewise-analytic function

$$\Phi(z) = \begin{cases} \Phi_{\text{right}}(z), & z \in P_{\text{right}}, \\ \Phi_{\text{left}}(z), & z \in P_{\text{left}}, \\ \Phi_0(z), & z \in D, \end{cases}$$

respectively continuous in the closures of P_{right}, P_{left}, D, such that:

1. On the positive (i.e., the one belonging to ℓ_+) part of the common boundary of P_{right} and P_{left}, the boundary values of the functions are related by

$$\Phi_{\text{left}}(z) = \Phi_{\text{right}}(z) e^{zU} S e^{-zU}. \tag{4.41}$$

2. On the negative part of the common boundary of P_{right} and P_{left}, the boundary values of the functions are related by

$$\Phi_{\text{left}}(z) = \Phi_{\text{right}}(z) e^{zU} S^T e^{-zU}. \tag{4.42}$$

3. On the common boundary of D and P_{right}, the boundary values of the functions are related by

$$\Phi_0(z) = \Phi_{\text{right}}(z) e^{zU} C z^{-R} z^{-\mu}. \tag{4.43}$$

4. On the common boundary of D and P_{left}, the boundary values of the functions are related by

$$\Phi_0(z) = \Phi_{\text{left}}(z) e^{zU} S^{-1} C z^{-R} z^{-\mu}. \tag{4.44}$$

5. For $|z| \to \infty$ within $P_{\text{right/left}}$,

$$\Phi_{\text{right/left}}(z) \to 1. \tag{4.45}$$

Theorem 4.5. *If the Riemann–Hilbert boundary value problem 1–5 has a unique solution at a point $u^0 = (u_1^0, \ldots, u_n^0)$, $u_i^0 \neq u_j^0$ for $i \neq j$, then a unique solution $\Phi = \Phi(z; u_1, \ldots, u_n)$ exists for u sufficiently close to u^0 and it is an analytic function of u. It can be continued analytically to a meromorphic function on the universal covering of the space*

$$\mathbb{C}^n \setminus \operatorname{diag} := \{(u_1, \ldots, u_n) \mid u_i \neq u_j \text{ for } i \neq j\}. \tag{4.46}$$

The proof follows from the general theory of Riemann-Hilbert boundary value problems (see [65, 71]).

Having a solution $\Phi = \left(\Phi_{\text{right}}(z; u), \Phi_{\text{left}}(z; u), \Phi_0(z; u)\right)$ of the Riemann–Hilbert boundary value problem we can reconstruct the solutions

$$\begin{aligned} Y_{\text{right/left}}(z; u) &= \Phi_{\text{right/left}}(z; u) e^{zU}, \\ Y_0(z; u) &= \Phi_0(z; u) z^\mu z^R. \end{aligned} \tag{4.47}$$

Let us introduce notation for the coefficients of the expansion of the matrix $\Phi_0(z; u) = \left(\Phi_{0i\alpha}(z; u)\right)$ near $z = 0$:

$$\Phi_{0i\alpha}(z; u) = \sum_{p=0}^{\infty} \phi_{i\alpha,p}(u) z^p. \tag{4.48}$$

Observe that

$$\phi_{i\alpha,0}(u) = \psi_{i\alpha}(u). \tag{4.49}$$

Theorem 4.6 (Isomonodromy Theorem (third part)). *For given μ, R, S, C satisfying (4.36), (4.38), (4.39), let the Riemann–Hilbert boundary problem (4.41)–(4.45) have a unique solution $\Phi = \Phi(z; u^0)$ at a point $u^0 = (u_1^0, \ldots, u_n^0)$, $u_i^0 \neq u_j^0$ for $i \neq j$ such that*

$$\prod_{i=1}^{n} \phi_{i1,0}(u^0) \neq 0. \tag{4.50}$$

Then the formulae

$$\eta_{\alpha\beta} = \sum_{i=1}^{n} \phi_{i\alpha,0}(u)\phi_{i\beta,0}(u), \tag{4.51}$$

$$e = \sum_{i=1}^{n} \partial_i, \tag{4.52}$$

$$E = \sum_{i=1}^{n} u_i \partial_i, \tag{4.53}$$

$$t_\alpha = \sum_{i=1}^{n} \phi_{i1,1}(u)\phi_{i\alpha,0}(u), \tag{4.54}$$

$$c_{\alpha\beta\gamma} = \sum_{i=1}^{n} \frac{\psi_{i\alpha}\psi_{i\beta}\psi_{i\gamma}}{\psi_{i1}}, \tag{4.55}$$

$$F = \frac{1}{2}\sum_{i=1}^{n}\left[\eta^{\alpha\beta}\phi_{i\alpha,1}\phi_{i\beta,0}\phi_{i1,1}^{2} - \phi_{i1,2}\phi_{i1,1} - \phi_{i1,0}\phi_{i1,3}\right], \tag{4.56}$$

define a semisimple Frobenius structure on a small neighborhood of u^0.

Proof. Let us define the matrix-valued functions $Y_{\text{right/left}}(z;u)$, $Y_0(z;u)$ by the formulae (4.47) and prove that they satisfy the linear system (3.29), (3.30) with

$$V(u) = [U, A_1(u)], \tag{4.57}$$
$$V_i(u) = [E_i, A_1(u)], \tag{4.58}$$

where the matrix $A_1(u)$ is defined from the asymptotic expansion

$$A_1(u) := \lim_{|z|\to\infty, z\in\Pi_+} z\left(\Phi_{\text{right}}(z;u) - 1\right). \tag{4.59}$$

Let us consider the piecewise-analytic function

$$Y(z;u) = \begin{cases} Y_{\text{right}}(z;u), & z \in \Pi_{\text{right}} \\ Y_{\text{left}}(z;u), & z \in \Pi_{\text{left}}, \\ Y_0(z;u), & z \in D. \end{cases}$$

We first prove that the matrix $Y(z;u)$ is invertible for any z, u. Indeed, $\det Y(z;u)e^{-z\sum u_i}$ is a piecewise-analytic function of z having no jumps on the intersections of the domains Π_{right}, Π_{left}, D and going to 1 when $|z| \to \infty$. Thus

$$\det Y(z;u) \equiv e^{z(u_1+\cdots+u_n)}.$$

We now introduce piecewise-analytic functions

$$G_i(z;u) := \partial_i Y(z;u) \cdot Y^{-1}(z;u).$$

By construction of S, C, it follows that $G_i(z;u)$ has no jumps on the intersections of the domains Π_{right}, Π_{left}, D. So it is an analytic matrix-valued function on $\mathbb{C} \setminus 0$. At $|z| \to \infty$ it has the asymptotic expansion

$$G_i(z;u) = \partial_i\left[\left(1 + \frac{A_1}{z} + \cdots\right)e^{zU}\right]e^{-zU}\left(1 - \frac{A_1}{z} + \cdots\right)$$
$$\sim zE_i + V_i + O\left(\frac{1}{z}\right).$$

At $z = 0$ the function $G_i(z; u)$ has a finite limit

$$G_i(z; u) = \partial_i \left[(\Psi(u) + O(z)) \, z^\mu z^R \right] z^{-R} z^{-\mu} \left[\Psi^{-1}(u) + O(z) \right]$$
$$= \partial_i \Psi(u) \cdot \Psi^{-1}(u) + O(z)$$

due to the constant value of μ, R. Hence

$$G_i(z; u) = z E_i + V_i$$

and

$$\partial_i Y = (z E_i + V_i) Y, \quad i = 1, \ldots, n.$$

In particular,

$$\partial_i \Psi = V_i \Psi.$$

Similarly, considering the piecewise-analytic function

$$G_z := \partial_z Y(z; u) \cdot Y^{-1}(z; u)$$

we obtain that

$$G_z = U + \frac{V}{z},$$

where the matrix $V = V(u)$ is defined in (4.57).

To prove the orthogonality conditions

$$\Phi_{\text{right/left}}^T(-z; u) \Phi_{\text{right/left}}(z; u) \equiv 1,$$
$$\Phi_0^T(-z; u) \Phi_0(z; u) \equiv \eta$$

we will consider the piecewise-analytic matrix-valued function

$$G(z) := \begin{cases} Y_{\text{right}}(z; u) Y_{\text{left}}^T(-z; u), & z \in \Pi_{\text{right}}, \\ Y_{\text{left}}(z; u) Y_{\text{right}}^T(-z; u), & z \in \Pi_{\text{left}}. \end{cases}$$

For $z \in \Pi_+ \cap \Pi_{\text{right}}$,

$$G(z) = Y_{\text{right}}(z; u) Y_{\text{left}}^T(-z; u) = Y_{\text{right}}(z; u) S Y_{\text{right}}^T(-z; u).$$

For $z \in \Pi_+ \cap \Pi_{\text{left}}$,

$$G(z) = Y_{\text{left}}(z; u) Y_{\text{right}}^T(-z; u) = Y_{\text{right}}(z; u) S Y_{\text{right}}^T(-z; u).$$

So, $G(z)$ has no jumps on ℓ_+. Similarly, it has no jumps on ℓ_-. Using (4.39) one obtains that for $z \in \Pi_{\text{right}}$ near $z = 0$,

$$G(z) = Y_0(z; u) e^{\pi i R} e^{\pi i \mu} \eta^{-1} Y_0^T(-z; u)$$
$$= \Phi_0(z; u) \eta^{-1} \Phi_0^T(-z; u) = 1 + O(z).$$

A similar computation gives the same behavior of $G(z)$ at $z \to 0$, $z \in \Pi_{\text{left}}$. So $G(z) \equiv 1$. This proves the orthogonality conditions.

The equations (4.54), (4.56) simply express (2.35), (2.37). Note that the functions $t_1(u)$, ..., $t_n(u)$ are independent coordinates at the points u where the product

$$\prod_{i=1}^{n} \psi_{i\alpha}(u)$$

is nonzero. The theorem is proved.   □

Exercise 4.4. Show that the product (4.50) does not vanish identically unless the matrix

$$e^{zU} S e^{-zU}$$

is independent in one of the variables $(u_1, \ldots, u_n)$.

The isomonodromy theorem gives a structure of semisimple Frobenius manifolds on small domains in the space of isomonodromic deformations of the operator

$$L = \frac{\mathrm{d}}{\mathrm{d}z} - \left(U + \frac{V}{z} \right)$$

with rational coefficients. The parameters of this Frobenius manifold are the monodromy data

$$(\mu, e, R, S, C) \tag{4.60}$$

of the operator satisfying the above properties (4.6), (4.7). Here e is a selected eigenvector of the matrix V with the eigenvalue μ_1 (μ_1 being the corresponding diagonal entry of the matrix μ). The choice of e corresponds to the choice of the first column of the matrix Ψ in the formulae (4.50)–(4.56). (We need not fix the bilinear form $\langle \, , \, \rangle$. It is given by (4.51).) This also demonstrates that, locally, any semisimple Frobenius manifold can be realized in such a way.

Exercise 4.5. We say that the Stokes matrix S is *reducible* if it has the form $S = S' \oplus S''$ with respect to some decomposition of the set of indices $\{1, \ldots, n\} = I' \cup I''$ into a union of two nonempty nonintersecting subsets. Prove that a reducible matrix S can make up a part of the monodromy data only if $\exp 2\pi i \mu_1$ is the eigenvalue of both the matrices $S'^T S'^{-1}$ and $S''^T S''^{-1}$. Prove that the Stokes matrix of a reducible Frobenius manifold is reducible (see Exercise 2.5).

We will now describe the structure of the analytic continuation of semisimple Frobenius manifolds. According to Theorem 4.5 and due to the formulae (4.54), (4.56), the functions t_α and F can be analytically continued to meromorphic functions on the universal covering of $\mathbb{C}^n \setminus \text{diag}$. Since the canonical coordinates are defined up to a reordering, the structure of analytic continuation of the Frobenius manifold with given monodromy data (4.60) is described by the action of the fundamental group

$$\pi_1\left((\mathbb{C}^n \setminus \text{diag})/S_n, (u_1^0, \ldots, u_n^0)\right) = \mathcal{B}_n$$

(the braid group) on the monodromy data computed at a given point u^0. The global structure of the Frobenius manifold is described by the stationary subgroup $\mathcal{B}_n{}^0 \subset \mathcal{B}_n$ of the given monodromy data (4.60).

To compute the action of the braid group $\mathcal{B}_n$ on the monodromy data, and also to describe the dependence of the monodromy data on the admissible oriented line ℓ, we will briefly present here the theory of Stokes factors (see [7]).

Let us label all the Stokes rays (4.34) of the system (3.30) in counter-clockwise order starting from the first one in Π_{right}. We obtain the rays

$$
\begin{aligned}
R^{(1)}, \ldots, R^{(m)} &\text{ in } \Pi_{\text{right}}, \\
R^{(m+1)}, \ldots, R^{(2m)} &\text{ in } \Pi_{\text{left}}.
\end{aligned}
\tag{4.61}
$$

We will use the cyclic labeling $R^{(k \pm 2m)} = R^{(k)}$. Observe that the narrow sectors Π_+ and Π_- contain no Stokes rays. For generic $(u_1, \ldots, u_n)$ one has

$$m = \frac{n(n-1)}{2},$$

but some coincidences of the Stokes rays may happen in the nongeneric situation when there are three u_i, u_j, u_k on a line or two pairs u_i, u_j and u_k, u_l on two parallel lines. Let us consider the sector of the z-plane from $R^{(k)}e^{-i\varepsilon/2}$ to $R^{(m+k)}e^{-i\varepsilon}$. According to Theorem 4.2 there exists a unique solution $Y^{(k)}(z)$ of (3.30) such that

$$Y^{(k)}(z) \sim Y_{\text{formal}}(z), \quad |z| \to \infty, \tag{4.62}$$

within the above sector. This solution can be extended while preserving the asymptotics into the open sector

$$\Pi_k: \quad \text{from } R^{(k-1)} \text{ to } R^{(m+k)}. \tag{4.63}$$

On the intersection of two subsequent sectors one has a constant matrix K_j defined by

$$Y^{(j+1)}(z) = Y^{(j)}(z)K_j, \quad z \in \Pi_j \cap \Pi_{j+1}. \tag{4.64}$$

Lemma 4.6.

$$Y_{\text{right}} = Y^{(1)}, \quad Y_{\text{left}} = Y^{(m+1)}, \tag{4.65}$$

$$S = K_1 \cdots K_m. \tag{4.66}$$

The proof is obvious.

Definition 4.4. The matrices K_j are called *Stokes factors* of the matrix S.

Exercise 4.6. Prove that

$$K_{m+j} K_j^T = 1. \tag{4.67}$$

How to find the Stokes factors knowing the Stokes matrix S and the configuration of pairwise distinct complex numbers $u_1, \ldots, u_n$? The clue is in the following property of Stokes factors (see [7]).

Lemma 4.7. *All the diagonal entries of K_j equal 1. Out of the off-diagonal entries $(K_j)_{ab}$, all equal zero but those for which the Stokes ray R_{ba} coincides with $R^{(j)}$.*

Proof. On $z \in \Pi_j \cap \Pi_{j+1}$ one must have

$$e^{zU} K_j e^{-zU} \to 1 \quad \text{as } |z| \to \infty.$$

Hence $(K_j)_{aa} = 1$ (as in the proof of Theorem 4.3). On the intersection the absolute values

$$|e^{z(u_a - u_b)}|$$

can go to either $+\infty$ or 0 for any pair $a \neq b$ but those for which R_{ab} or R_{ba} coincides with $R^{(j)}$. Indeed, the whole intersection $\Pi_j \cap \Pi_{j+1}$ lies on the right of the oriented line

$$R^{(m+j)} \cup \left(-R^{(j)}\right).$$

If

$$R_{ab} = R^{(m+j)}, \quad R_{ba} = R^{(j)},$$

then on the right from the oriented line one has

$$|e^{z(u_a - u_b)}| \to 0 \quad \text{as } |z| \to \infty.$$

The lemma is proved.   □

Theorem 4.7. *Any Stokes matrix S with the above properties can be uniquely factorized into the product $S = K_1 \cdots K_m$ of Stokes factors of the above form.*

See the proof in [7].

From the factorization (4.66), it follows that the Stokes matrix does not change if one deforms the admissible line ℓ without intersecting any of the Stokes rays. We now describe what happens if the admissible oriented line $\ell = \ell_+ \cup (-\ell_-)$ passes through the Stokes ray R moving counterclockwise. Instead, one may consider a deformation of one of the Stokes rays R passing through ℓ_+ moving clockwise.

Lemma 4.8. *After the above deformation, the new solutions $Y'_{\text{right/left}}$, the new Stokes matrix S', and the new connection matrix C' have the form*

$$Y_{\text{right}} = Y'_{\text{right}} K_R^T, \tag{4.68a}$$

$$Y'_{\text{left}} = Y_{\text{left}} K_R, \tag{4.68b}$$

$$S' = K_R^T S K_R, \tag{4.68c}$$

$$C' = K_R^T C \tag{4.68d}$$

(the last formula holds true modulo the ambiguity (4.40)*). Here K_R is the Stokes factor corresponding to the Stokes ray R.*

The proof follows from Lemma 4.6 and from Exercise 4.6.

We now are ready to compute the action of the braid group $\mathcal{B}_n$ on the monodromy data describing the analytic continuation of the Frobenius manifold. First, the action of $\mathcal{B}_n$ on the monodromy at $z = 0$ is trivial. We then compute the action on the Stokes matrix S. Let us assume that the canonical coordinates $(u_1, \ldots, u_n)$ are ordered in such a way that S is an upper triangular matrix. We choose the standard generators $\sigma_1, \ldots, \sigma_{n-1}$ of the braid group $\mathcal{B}_n$. The generator σ_i is given by a deformation of $(u_1, \ldots, u_n)$ such that:

1. u_k remains fixed for $k \neq i,\, i + 1$.

2. u_i and u_{i+1} are permuted moving counterclockwise.

Let us deform $(u_1, \ldots, u_n)$ in the coefficients of the operator

$$L = \frac{\mathrm{d}}{\mathrm{d}z} - \left(U + \frac{V(u)}{z} \right).$$

Due to isomonodromy, the matrices S and C remain unchanged until some of the Stokes rays pass through ℓ. After this we must reorder the canonical coordinates to preserve upper triangularity of the Stokes matrix and then to compute the new matrices S' and C'' using Lemma 4.8. We must recall here that the operator L for a given ordering of the canonical coordinates $(u_1, \ldots, u_n)$ is determined up to a transformation

$$L \mapsto JLJ, \tag{4.69}$$

where J is an arbitrary diagonal matrix of the form

$$J = \mathrm{diag}(\pm 1, \ldots, \pm 1). \tag{4.70}$$

Thus the matrices

$$S \text{ and } JSJ, \quad C \text{ and } JC \tag{4.71}$$

must be identified. So, what we need is actually an action of $\mathcal{B}_n$ on the equivalence classes of the matrices S and C with respect to the identifications (4.71).

The result is given by the following theorem

Theorem 4.8. *The analytic continuation of a semisimple Frobenius manifold is described by the action*

$$S \mapsto \beta(S),$$
$$C \mapsto \beta(C) \qquad\qquad (4.72)$$

of the braid group $\mathcal{B}_n \ni \beta$ *on the Stokes matrix* $S = (s_{ij})$ *and the central connection matrix* C. *For the standard generator* $\beta = \sigma_i$, *the action has the form*

$$\sigma_i(S) = K^{(i)}(S)SK^{(i)}(S),$$
$$\sigma_i(C) = K^{(i)}(S)C, \qquad\qquad (4.73)$$

$$\left(K^{(i)}(S)\right)_{kk} = 1, \quad k = 1,\ldots,n, \quad k \neq i, i+1,$$
$$\left(K^{(i)}(S)\right)_{i+1,i+1} = -s_{i,i+1}, \qquad\qquad (4.74)$$
$$\left(K^{(i)}(S)\right)_{i,i+1} = \left(K^{(i)}(S)\right)_{i+1,i} = 1.$$

All other entries of the matrix $K^{(i)}(S)$ *are equal to zero.*

Proof. Let us assume that during the deformation σ_i, the coordinates u_i and u_{i+1} remain sufficiently close to each other. Then all the Stokes rays but $R_{i,i+1}$ and $R_{i+1,i}$ will be only slightly deformed, and they will return to their original positions (with renumbering $i \leftrightarrow i+1$) after the end of the deformation. But the rays $R_{i,i+1}$ and $R_{i+1,i}$ interchange their positions, rotating clockwise. In particular, it is the ray $R = R_{i+1,i}$ that passes through the positive half-line ℓ_+ rotating clockwise. At the very last moment before the collision, the configuration of the Stokes rays is such that $R^{(1)} = R_{i+1,i}$ and $R^{(m+1)} = R_{i,i+1}$, and we may assume that $R^{(1)}$ and $R^{(m+1)}$ contain no other Stokes rays. From Theorem 4.7 we obtain a factorization of S into the product of upper triangular Stokes factors

$$S = K_1 K_2 \cdots K_m,$$

where the only nonzero off-diagonal entry of the matrix K_1 sits in the $(i, i+1)$ box, and all the factors $K_2, \ldots, K_m$ have zero in the $(i, i+1)$ place. From this we obtain that

$$(K_1)_{i,i+1} = s_{i,i+1}.$$

We must now apply the formulae (4.68) to compute the new matrices S', C' with

$$K_R = K_{m+1} = (K_1^T)^{-1}.$$

After applying the permutation $i \leftrightarrow i+1$, we arrive at the formulae (4.74). The theorem is proved. $\qquad\qquad \square$

Example 4.3. For $n = 3$ the generators σ_1, σ_2 of $\mathcal{B}_3$ act as follows in the space of Stokes matrices:

$$S = \begin{pmatrix} 1 & x & y \\ 0 & 1 & z \\ 0 & 0 & 1 \end{pmatrix},$$

$$\sigma_1(x, y, z) = (-x, z, y - x\,z), \quad \sigma_2(x, y, z) = (y, x - y\,z, -z). \tag{4.75}$$

Exercise 4.7. Prove that the braid

$$\zeta = (\sigma_1 \ldots \sigma_{n-1})^n \tag{4.76}$$

acts trivially on Stokes matrices.

The braid ζ generates the center of $\mathcal{B}_n$ (see [12]). So the quotient of $\mathcal{B}_n$ by the center acts on the space of Stokes matrices. For $n = 3$ the quotient is isomorphic to the modular group $\mathrm{PSL}_2(\mathbb{Z})$ [12].

Let $\mathcal{B}_n(S, C) \subset \mathcal{B}_n$ be the stationary subgroup of the equivalence class (4.71) of the pair S, C. We realize it as a subgroup in the fundamental group

$$\pi_1\big([\mathbb{C}^n \setminus \mathrm{diag}]/S_n, (u_1^0, \ldots, u_n^0)\big)$$

and construct the corresponding covering

$$M(S, C) \to [\mathbb{C}^n \setminus \mathrm{diag}]/S_n,$$

i.e., a covering such that the group of deck transformations of the fiber is isomorphic to $\mathcal{B}_n(S, C)$. From Theorem 4.8 we derive the following theorem.

Theorem 4.9.

1. *For given monodromy data (μ, e, R, S, C), the Frobenius structure extends from a small neighborhood of u^0 to a dense open subset in the manifold $M(S, C)$. We denote this Frobenius structure on $M(S, C)$ by $\mathrm{Fr}(\mu, e, R, S, C)$.*

2. *Let (μ, e, R, S, C) be the monodromy data of a semisimple Frobenius manifold M computed at the point $u^0 = (u_1^0, \ldots, u_n^0)$ with respect to an admissible oriented line ℓ. Let M^0 be the open part of the Frobenius manifold M consisting of all points $t \in M$ such that all the eigenvalues $u_1(t), \ldots, u_n(t)$ of the operator of multiplication by the Euler vector field are pairwise distinct. Then the map*

$$M^0 \mapsto \mathrm{Fr}(\mu, e, R, S, C)$$

is well defined, and it is an equivalence of Frobenius manifolds.

Example 4.4. Let us compute the monodromy data of quantum cohomology of $\mathbf{CP}^2$, i.e., of the solution (1.14) of WDVV equations of associativity. The monodromy at $z = 0$ is completely determined by the classical cohomology $H^*(\mathbf{CP}^2)$ together with the first Chern class $c_1(\mathbf{CP}^2)$ (see Section 2). We obtain

$$\mu = \operatorname{diag}(-1, 0, 1), \quad R = \begin{pmatrix} 0 & 0 & 0 \\ 3 & 0 & 0 \\ 0 & 3 & 0 \end{pmatrix}.$$

Let us compute the Stokes matrix at the semisimple point

$$t_1 = t_3 = 0, \quad \text{arbitrary } t_2 \text{ with } \operatorname{Re} t_2 < R. \tag{4.77}$$

Here R is the radius of convergence (1.15). Let us write $q = \exp t_2$. The system (2.42) for horizontal sections $(\xi_1, \xi_2, \xi_3) = (\partial_1 \tilde{t}, \partial_2 \tilde{t}, \partial_3 \tilde{t})$ of the connection $\widetilde{nabla}$ can be reduced to two third-order equations,

$$\begin{aligned} \partial_2^3 \phi &= z^3 q \phi, \\ (z \partial_z)^3 \phi &= 27 z^3 q \phi, \end{aligned} \tag{4.78}$$

for the function

$$\phi = \phi(t_2 z) = \frac{\xi_1}{z},$$

$$(\xi_1, \xi_2, \xi_3) = \left(z\phi, \frac{1}{3} z \partial_z \phi, \frac{1}{9} \partial_z (z \partial_z \phi) \right).$$

The system (4.78) is equivalent to the single equation

$$(z \partial_z)^3 \Phi = 27 z^3 \Phi \tag{4.79}$$

using the quasi-homogeneity

$$\phi(t_2, z) = \Phi(z q^{1/3}). \tag{4.80}$$

The problem is reduced to the computation of the Stokes matrix of the generalized hypergeometric equation (see [34]). We must carefully select the basis of formal solutions of (4.79) at $z \to \infty$ corresponding to the basis of columns of $Y_{\text{formal}}(z)$ of the solution (4.26) of the gauge-equivalent system (3.30).

The operator $\mathcal{U}$ of multiplication by the Euler vector field in the basis $e_1 = \partial_1$, $e_2 = \partial_2$, $e_3 = \partial_3$ has the matrix

$$\mathcal{U}(t) = \begin{pmatrix} 0 & 0 & 3q \\ 3 & 0 & 0 \\ 0 & 3 & 0 \end{pmatrix}, \quad t = (0, t_2, 0), \quad q = e^{t_2}. \tag{4.81}$$

The canonical coordinates (i.e., the eigenvalues of $\mathcal{U}$) at the point (4.77) take the values

$$u_1 = 3q^{1/3}, \quad u_2 = 3\bar{\varepsilon}^2 q^{1/3}, \quad u_3 = 3\varepsilon^2 q^{1/3}, \tag{4.82}$$

where

$$\varepsilon = \exp\frac{\pi i}{3}.$$

The corresponding idempotents of the quantum cohomology algebra are

$$\pi_1 = \frac{1}{3}(e_1 + q^{-1/3}e_2 + q^{-2/3}e_3),$$

$$\pi_2 = \frac{1}{3}(e_1 + \varepsilon^2 q^{-1/3}e_2 + \bar{\varepsilon}^2 q^{-2/3}e_3),$$

$$\pi_3 = \frac{1}{3}(e_1 + \bar{\varepsilon}^2 q^{-1/3}e_2 + \varepsilon^2 q^{-2/3}e_3).$$

The invariant metric is given by

$$\langle \pi_1, \pi_1 \rangle = \frac{1}{3}q^{-2/3}, \quad \langle \pi_2, \pi_2 \rangle = \frac{1}{3}\bar{\varepsilon}^2 q^{-2/3}, \quad \langle \pi_3, \pi_3 \rangle = \frac{1}{3}\varepsilon^2 q^{-2/3}.$$

Evaluating the square root, we obtain the normalized idempotents

$$f_1 = \frac{1}{\sqrt{3}}(q^{1/3}e_1 + e_2 + q^{-1/3}e_3),$$

$$f_2 = \frac{1}{\bar{\varepsilon}\sqrt{3}}(q^{1/3}e_1 + \varepsilon^2 e_2 + \bar{\varepsilon}^2 q^{-1/3}e_3),$$

$$f_3 = \frac{1}{\varepsilon\sqrt{3}}(q^{1/3}e_1 + \bar{\varepsilon}^2 e_2 + \varepsilon^2 q^{-1/3}e_3).$$

This gives the matrix $\Psi = (\psi_{i\alpha})$:

$$\Psi = \frac{1}{\sqrt{3}}\begin{pmatrix} q^{-1/3} & 1 & q^{1/3} \\ \bar{\varepsilon}q^{-1/3} & -1 & \varepsilon q^{1/3} \\ \varepsilon q^{-1/3} & -1 & \bar{\varepsilon}q^{1/3} \end{pmatrix}. \tag{4.83}$$

We can easily compute the matrix V at the point of interest (cf. [70]). But what we need is to determine the asymptotic structure of the solutions of (4.79) at $z \to \infty$. We must choose the basis $\tilde{t}_1^\infty$, $\tilde{t}_2^\infty$, $\tilde{t}_3^\infty$ of the coordinates $\tilde{t}$ such that the matrix

$$Y_{ij} := \frac{\partial_i \tilde{t}_j^\infty}{\psi_{i1}}$$

has the expansion (4.26), i.e.,

$$Y_{ij} \sim \left(\delta_{ij} + O\left(\frac{1}{z}\right)\right) e^{zu_j}, \quad i,j = 1,2,3. \tag{4.84}$$

This gives the three solutions ϕ_1, ϕ_2, ϕ_3 of the system (4.78) such that

$$\phi_j = \frac{1}{z}\frac{\partial}{\partial t^1}\tilde{t}_j^\infty = \frac{1}{z}\sum_{i=1}^{3}\partial_i\tilde{t}_j^\infty = \frac{1}{z}\sum_{i=1}^{3}\psi_{i1}Y_{ij}.$$

For the corresponding basic solutions of (4.79), we obtain the required expansions

$$\Phi_1 \sim \frac{1}{\sqrt{3}}\frac{e^{3z}}{z}\left(1+O\left(\frac{1}{z}\right)\right),$$

$$\Phi_2 \sim \frac{\bar{\varepsilon}}{\sqrt{3}}\frac{e^{3\bar{\varepsilon}^2 z}}{z}\left(1+O\left(\frac{1}{z}\right)\right), \tag{4.85}$$

$$\Phi_3 \sim \frac{\varepsilon}{\sqrt{3}}\frac{e^{3\varepsilon^2 z}}{z}\left(1+O\left(\frac{1}{z}\right)\right).$$

We must now compute the Stokes matrix of the equation (4.79) with respect to the bases of solutions having the asymptotic expansions (4.85) in the right/left half-planes $\Pi_{\text{right/left}}$ with some admissible oriented line ℓ.

The Stokes rays of equation (4.79) have the form

$$R_{12} = \{-\rho\varepsilon \mid \rho \geq 0\},$$
$$R_{13} = \{\rho\bar{\varepsilon} mid\rho \geq 0\}, \tag{4.86}$$
$$R_{23} = \{\rho \mid \rho \geq 0\},$$

the rays R_{21}, R_{31}, R_{32} are opposite to the above. We choose the admissible line

$$\ell = \{r\,e^{i\alpha} \mid -\infty < r < \infty\} \tag{4.87}$$

for a fixed small $\alpha > 0$ oriented according to the positive direction of r. We will now use a suitable Meijer function [64] to compute the Stokes matrix.

Lemma 4.9. *The function*

$$g(z) = \frac{1}{(2\pi)^2 i}\int_{-c-i\infty}^{-c+i\infty}\Gamma^3(-s)e^{\pi i s}z^{3s}\,ds \tag{4.88}$$

defined for $z \neq 0$,

$$-\frac{5\pi}{6} < \arg z < \frac{\pi}{6}, \tag{4.89}$$

where c is any positive number, satisfies (4.79). The analytic continuation of this function has the asymptotic expansion

$$g(z) \sim \frac{1}{\sqrt{3}}\bar{\varepsilon}\frac{e^{3\bar{\varepsilon}^2 z}}{z} = \Phi_2(z), \quad |z| \to \infty, \tag{4.90}$$

in the sector

$$-\frac{5\pi}{3} < \arg z < \pi. \tag{4.91}$$

It satisfies the identity

$$g(ze^{2\pi i}) - 3g(ze^{4\pi i/3}) + 3\,g(ze^{2\pi i/3}) - g(z) = 0. \tag{4.92}$$

Proof (cf. [64]. Using the Stirling formula

$$\log \Gamma(z) = \left(z - \frac{1}{2}\right)\log z - z + \frac{1}{2}\log(2\pi) + O\!\left(\frac{1}{z}\right)$$

and

$$\lim_{|y|\to\infty} |\Gamma(x+iy)|e^{\pi y/2}|y|^{1/2-x} = \sqrt{2\pi}, \quad x,y \text{ real},$$

we prove the uniform convergence of the integral in the domain (4.89) and its independence on c. Differentiation gives

$$(z\partial_z)^3 g = \frac{27}{(2\pi)^2 i}\int_{-c-i\infty}^{-c+i\infty} s^3\Gamma^3(-s)e^{\pi i s}z^{3s}\,\mathrm{d}s.$$

Using the property of the Γ function

$$s\Gamma(-s) = -\Gamma(1-s)$$

and doing a shift $s \mapsto s+1$, we obtain for the rhs. the integral

$$\frac{27}{(2\pi)^2 i}\int_{-c-i\infty}^{-c+i\infty} \Gamma^3(-s)e^{\pi i s}z^{3(s+1)}\,\mathrm{d}s = 27z^3 g.$$

To derive the asymptotic expansion we use Laplace's method. Representing the integrand in the form exp phase and using the Stirling's formula, one obtains the following asymptotic expansion for the phase:

$$3\log\Gamma(-s) + \pi i s + 3s\log z \sim -3\left(s+\frac{1}{2}\right)\log s + 3s\log z + (3-2\pi i)s$$

valid for

$$-\frac{3\pi}{2} < \arg s < -\frac{\pi}{2}. \tag{4.93}$$

For large $|z|$ the phase has a critical point at

$$s \sim ze^{-2\pi i/3} - \frac{1}{2}.$$

This critical point is in the domain (4.89) if

$$-\frac{5\pi}{6} < \arg z < \frac{\pi}{6}.$$

At the critical point, the value of the phase is

$$\text{phase}_0 \sim -\frac{3}{2}\log z + 3ze^{-2\pi i/3} + \frac{3}{2}\log 2\pi + \pi i,$$

and the second s-derivative taken at this point is

$$\text{phase}_0'' \sim -\frac{3e^{2\pi i/3}}{z}.$$

Applying Laplace's formula for the integral

$$g(z) \sim \frac{1}{(2\pi)^2 i}\frac{1}{\sqrt{2\pi}}\frac{e^{\text{phase}_0}}{\sqrt{\text{phase}_0''}}$$

we obtain the asymptotics (4.90). The asymptotics remains valid in a wider sector

$$-\frac{5\pi}{3} < \arg z < \pi.$$

Indeed, during this analytic continuation, i.e., counterclockwise until R_{32} and clockwise until R_{21}, the exponential e^{zu_2} remains dominant.

To derive the identity (4.92) we observe that the equation is invariant under the rotation

$$z \mapsto ze^{2\pi i/3}.$$

This generates a linear operator, A, in the 3-dimensional space of solutions of (4.79). Let us prove that all the eigenvalues of A are equal to 1. Indeed, near $z = 0$, all solutions have the form

$$\Phi(z) = \sum_{m=0}^{\infty} \frac{z^{3m}}{(m!)^3}[a_m + b_m \log z + c_m \log^2 z], \qquad (4.94)$$

where a_0, b_0, c_0 are arbitrary parameters and the coefficients a_m, b_m, c_m for $m > 0$ are uniquely determined from the recursion relations

$$c_m = c_{m-1},$$

$$b_m + \frac{2}{m}c_m = b_{m-1},$$

$$a_m + \frac{1}{m}b_m + \frac{2}{3m^2}c_m = a_{m-1}.$$

The operator

$$(A\phi)(z) = \Phi(ze^{2\pi i/3})$$

in the basis of solutions of the form (4.94), with only one of the three parameters a_0, b_0, c_0 nonzero, is given by a triangular matrix with all 1's on the diagonal. Writing the Cayley–Hamilton theorem as

$$(A - 1)^3 = 0,$$

we obtain

$$A^3g - 3A^2g + 3Ag - g = 0.$$

This gives the identity (4.92). The lemma is proved. $\square$

Let us construct the three solutions

$$\Phi^{\text{right}}(z) = \left(\Phi_1^{\text{right}}(z), \Phi_2^{\text{right}}(z), \Phi_3^{\text{right}}(z)\right),$$

which have an asymptotic behavior of the form (4.85):

$$\Phi_j^{\text{right}}(z) \sim \Phi_j(z), \quad |z| \to \infty, \quad -\pi < \arg z < \frac{\pi}{3}, \quad j = 1, 2, 3.$$

We can take

$$\Phi^{\text{right}}(z) = \left(-g(e^{2\pi i/3}z), g(z), g(e^{-2\pi i/3}z)\right). \tag{4.95}$$

Similarly, the components of the vector function $\Phi^{\text{left}}(z)$ must have the asymptotics

$$\Phi_j^{\text{left}}(z) \sim \Phi_j(z), \quad |z| \to \infty, \quad 0 < \arg z < 4\pi/3, \quad j = 1, 2, 3.$$

We take

$$\Phi^{\text{left}}(z) = \left(-g(e^{-4\pi i/3}z), g(e^{-2\pi i}z) - 3g(e^{-4\pi i/3}z), g(e^{-2\pi i/3}z)\right). \tag{4.96}$$

The only novelty to be proved is the formula for Φ_2^{left}. Indeed, it follows from Lemma 4.9 that

$$\Phi_2^{\text{left}}(z) = g(e^{-2\pi i}z) - 3g(e^{-4\pi i/3}z) \sim \Phi_2(z), \quad |z| \to \infty, \quad \frac{\pi}{3} < \arg z < \frac{4\pi}{3}.$$

Using the identity (4.92) we may rewrite this function as

$$\Phi_2^{\text{left}}(z) = g(z) - 3g(e^{-2\pi i/3}z) \sim \Phi_2(z), \quad |z| \to \infty, \quad 0 < \arg z < \frac{\pi}{3}.$$

Applying again the identity (4.92) we obtain that in the sector

$$0 < \arg z < \frac{\pi}{3},$$

one has

$$\left(\Phi_1^{\text{left}}(z), \Phi_2^{\text{left}}(z), \Phi_3^{\text{left}}(z)\right) = \left(\Phi_1^{\text{right}}(z), \Phi_2^{\text{right}}(z), \Phi_3^{\text{right}}(z)\right) S$$

with

$$S = \begin{pmatrix} 1 & 0 & 0 \\ 3 & 1 & 0 \\ -3 & -3 & 1 \end{pmatrix}. \tag{4.97}$$

This is the Stokes matrix of the quantum cohomology of $\mathbf{CP}^2$. Changing the sign of the normalized idempotent f_3, we can reduce S to the form

$$S = \begin{pmatrix} 1 & 0 & 0 \\ 3 & 1 & 0 \\ 3 & 3 & 1 \end{pmatrix}. \tag{4.98}$$

The matrix (4.97) was obtained from physical considerations in [19]. The main argument was that in Landau–Ginzburg models of 2-D topological field theory, the entries of the Stokes matrix must be integers. Then, since the eigenvalues of $S^T S^{-1}$ must all be 1, one arrives at the following Diophantine equation for the entries:

$$x^2 + y^2 + z^2 - xyz = 0,$$

where

$$S = \begin{pmatrix} 1 & x & y \\ 0 & 1 & z \\ 0 & 0 & 1 \end{pmatrix}.$$

All the integer solutions to the equation have the form

$$x = 3x_1, \quad y = 3y_1, \quad z = 3z_1,$$

where x_1, y_1, z_1 are integer solutions to Markov equations

$$x_1^2 + y_1^2 + z_1^2 - 3x_1 y_1 z_1 = 0.$$

The solutions of Markov equation are known to be all equivalent to $(1, 1, 1)$ modulo the action (4.75) of the braid group. This solution of Markov equation just corresponds to the Stokes matrix (4.98).

In the next section we will construct polynomial Frobenius manifolds starting from an arbitrary finite Coxeter group. In particular, for the Coxeter groups with simply laced Dynkin diagrams, these coincide with the Frobenius manifolds of singularity theory. It can be shown, using this construction, that in this example, S is the variation operator of the singularity computed in the so-called distinguished basis of vanishing cycles [4].

Here we define a remarkable operation of the *tensor product* of Frobenius manifolds. We are motivated by the results of Kaufmann, Kontsevich, and Manin [53, 57] describing the quantum cohomology of the direct product of two varieties.

Let M', M'' be two Frobenius manifolds of respective dimensions n' and n''. We say that a Frobenius manifold M of the dimension $n'n''$ is the *tensor product* $M = M' \otimes M''$ if it has the following structure:

1. The tangent planes TM with the bilinear form $\langle \, , \, \rangle$ and the unit vector field e are represented as

$$(TM, \langle \, , \, \rangle, e) = TM' \otimes TM'', \langle \, , \, \rangle' \otimes \langle \, , \, \rangle'', e' \otimes e'')$$

(as usual, we identify the tangent planes at different points using the Levi-Civita flat connection). Thus, the flat coordinates on M have double labels,

$$t = (t^{\alpha' \alpha''}), \qquad 1 \le \alpha' \le n', \quad 1 \le \alpha'' \le n''.$$

The unit vector field is

$$e = \frac{\partial}{\partial t^{1'1''}}.$$

The matrix of $\langle \, , \, \rangle$ has the form

$$\eta_{\alpha'\alpha''\beta'\beta''} = \eta_{\alpha'\beta'}\eta_{\alpha''\beta''}.$$

2. At the points

$$t \in M, \quad t^{\alpha'\alpha''} = 0 \quad \text{for } \alpha' > 1, \alpha'' > 1, \tag{4.99}$$

the algebra $T_t M$ is the tensor product

$$T_t M = T_{t'} M' \otimes T_{t''} M'',$$
$$t' = (t^{2'1''}, \ldots, t^{n'1''}),$$
$$t'' = (t^{1'2''}, \ldots, t^{1'n''}),$$

i.e.,

$$c^{\gamma'\gamma''}_{\alpha'\alpha''\beta'\beta''}(t) = c^{\gamma'}_{\alpha'\beta'}(t')c^{\gamma''}_{\alpha''\beta''}(t'').$$

In these formulae, $\eta_{\alpha'\beta'}$, $c^{\gamma'}_{\alpha'\beta'}$, $\eta_{\alpha''\beta''}$, and $c^{\gamma''}_{\alpha''\beta''}$ are respectively the invariant bilinear form and the structure constants of the Frobenius manifolds M' and M''.

3. The charge is

$$d_M = d_{M'} + d_{M''},$$

and the Euler vector field on M has the form

$$E = \sum_{\alpha',\alpha''} t^{\alpha'\alpha''}(1 - q_{\alpha'} - q_{\alpha''})\frac{\partial}{\partial t^{\alpha'\alpha''}}$$

$$+ \sum r_{\alpha'}\frac{\partial}{\partial t^{\alpha'1''}} + \sum r_{\alpha''}\frac{\partial}{\partial t^{1'\alpha''}}. \quad (4.100)$$

Here

$$E' = \sum_{\alpha'=1}^{n'}[(1 - q_{\alpha'})t^{\alpha'} + r_{\alpha'}]\partial_{\alpha'},$$

$$E'' = \sum_{\alpha''=1}^{n''}[(1 - q_{\alpha''})t^{\alpha''} + r_{\alpha''}]\partial_{\alpha''}$$

are the Euler vector fields on M' and M'', respectively.

For any two semisimple Frobenius manifolds M', M'', we will now describe their tensor product $M = M' \otimes M''$ in terms of the monodromy data of the factors.

Lemma 4.10.

1. *If $M = M' \otimes M''$ with semisimple M' and M'', then M is semisimple.*

2. *Let $t'_0 \in M'$, $t''_0 \in M''$ be two points such that (a) $t_0^{1'} = t_0^{1''}$, and (b) the values of the canonical coordinates $u_{i'} = u_{i'}(t'_0)$, $i' = 1, \ldots, n'$, $u_{i''} = u_{i''}(t''_0)$, $i'' = 1, \ldots, n''$, satisfy the properties*

$$u_{i'} \neq u_{j'}, \qquad i' \neq j',$$
$$u_{i''} \neq u_{j''}, \qquad i'' \neq j'',$$
$$u_{i'} + u_{i''} \neq u_{j'} + u_{j''}, \quad (i', i'') \neq (j', j'').$$

Let ℓ be a line in the z-plane such that for any $z \in \ell \setminus 0$

$$\mathrm{Re}[z(u_{i'} - u_{j'})] \neq 0, \quad i' \neq j'$$
$$\mathrm{Re}[z(u_{i''} - u_{j''})] \neq 0, \quad i'' \neq j''$$
$$\mathrm{Re}[z(u_{i'} - u_{j'})] + \mathrm{Re}[z(u_{i''} - u_{j''})] \neq 0, \quad (i', i'') \neq (j', j'').$$

Then the Stokes matrix S of M at the point t_0 with the coordinates

$$t^{\alpha'1''} = t_0^{\alpha'}, \quad \alpha' = 1,\ldots,n',$$
$$t^{1'\alpha''} = t_0^{\alpha''}, \quad \alpha'' = 1,\ldots,n'', \quad (4.101)$$
$$t^{\alpha'\alpha''} = 0, \quad \alpha' > 1, \; \alpha'' > 1,$$

is the tensor product of the Stokes matrices S' of M' at the point t_0' and S'' of M'' at the point t_0'':

$$S = S' \otimes S''.$$

Proof. If $t_0' \in M'$, $t_0'' \in M''$ are semisimple points of the Frobenius manifolds, the point (4.101) will be a semisimple point of M. The idempotents of the algebra

$$T_{t_0'} M' \otimes T_{t_0''} M''$$

are tensor products $\pi_{i'} \otimes \pi_{i''}$, $i' = 1, \ldots, n'$, $i'' = 1, \ldots, n''$. The operator of multiplication by the Euler vector field (4.100) at the point (4.101) has the form

$$\mathcal{U} = 1' \otimes \mathcal{U}'' + \mathcal{U}' \otimes 1'' - t^1 1' \otimes 1'',$$

where $t^1 = t_0^{1'} = t_0^{1''}$. The eigenvalues of this operator are

$$u_{i'} + u_{i''} - t^1, \quad 1 \leq i' \leq n', \quad 1 \leq i'' \leq n''.$$

These are the values of the canonical coordinates on M at the points of the $(n' + n'' - 1)$-dimensional locus (4.99).

Let $Y'_{\text{right/left}}(z; t_0')$, $Y''_{\text{right/left}}(z; t_0'')$ be the solutions of the system (3.30) for M' and M'', respectively, with the asymptotic behavior (4.26) in the right/left half-planes with respect to the admissible line ℓ. Then the solutions of the system (3.30) for M with the needed asymptotic expansion (4.26) are

$$Y_{\text{right}}(z; t_0) = e^{-zt^1} Y'_{\text{right}}(z; t_0') \otimes Y''_{\text{right}}(z; t_0''), \quad z \in \Pi_{\text{right}},$$

$$Y_{\text{left}}(z; t_0) = e^{-zt^1} Y'_{\text{left}}(z; t_0') \otimes Y''_{\text{left}}(z; t_0''), \quad z \in \Pi_{\text{left}}.$$

The lemma is proved. $\qquad\qquad\qquad\qquad\qquad\qquad\qquad\qquad\qquad\qquad$ □

Theorem 4.10 (Theorem–Definition). *Let*

$$M = \text{Fr}(\mu' \otimes 1 + 1 \otimes \mu'', e' \otimes e'', R' \otimes 1 + 1 \otimes R'', S' \otimes S'', C' \otimes C''),$$
$$M' = \text{Fr}(\mu', e', S', C'),$$
$$M'' = \text{Fr}(\mu'', e'', S'', C'').$$

Then

$$M = M' \otimes M''.$$

Proof. Let $u' = (u_{1'}, \ldots, u_{n'}) \in M'$, $u'' = (u_{1''}, \ldots, u_{n''}) \in M''$ be two regular points of these Frobenius manifolds, i.e., points such that the Riemann–Hilbert boundary value problem of the form (4.41)–(4.45) for each of the manifolds has a unique solution $(Y_0', Y_{\text{right}}', Y_{\text{left}}')$ and $(Y_0'', Y_{\text{right}}'', Y_{\text{left}}'')$, respectively. Performing, if necessary, a diagonal shift

$$u_{i'} \mapsto u_{i'} + c, \quad i' = 1, \ldots, n'$$

we may also assume that

$$t^{1'}(u') = t^{1''}(u'') =: t^1.$$

Then the functions

$$Y_0 = e^{-zt^1} Y_0' \otimes Y_0'',$$
$$Y_{\text{right}} = e^{-zt^1} Y_{\text{right}}' \otimes Y_{\text{right}}'',$$
$$Y_{\text{left}} = e^{-zt^1} Y_{\text{left}}' \otimes Y_{\text{left}}'',$$

will give the solution of the Riemann–Hilbert boundary value problem for the manifold M. It follows that the matrix Ψ at the points (4.101) is also a tensor product:

$$\Psi = \left(\psi_{i'\alpha'}(u') \psi_{i''\alpha''}(u'') \right).$$

Using the formulae of the isomonodromy theorem we conclude that $M = M' \otimes M''$. The theorem is proved. $\qquad\square$

Example 4.5. Let M be the Frobenius manifold corresponding to the quantum cohomology of $\mathbf{CP}^1$, i.e.,

$$F = \frac{1}{2} t_1^2 t_2 + e^{t_2},$$
$$E = t_1 \partial_1 + 2\partial_2.$$

The monodromy data are

$$\mu = \text{diag}\left(-\frac{1}{2}, \frac{1}{2}\right), \quad R = \begin{pmatrix} 0 & 0 \\ 2 & 0 \end{pmatrix}, \quad S = \begin{pmatrix} 1 & 2 \\ 0 & 1 \end{pmatrix}$$

(the computation of S is similar to the above computation of the Stokes matrix of the quantum cohomology of $\mathbf{CP}^2$, but it is simpler). The tensor square of this Frobenius manifold computed according to Theorem 4.10 describes the quantum cohomology of $\mathbf{CP}^1 \times \mathbf{CP}^1$.

Example 4.6. Let M_h be the polynomial two-dimensional Frobenius manifolds of the form

$$F = \frac{1}{2} t_1^2 t_2 + t_2^{h+1}, \quad h \in \mathbb{Z}, \quad h \geq 3.$$

The tensor product of the form $M_{h'} \otimes M_{h''}$ is a polynomial 4-dimensional Frobenius manifold only in the following three cases: $M_3 \otimes M_3$, $M_3 \otimes M_4$, $M_3 \otimes M_5$.

In the next section we will establish a relation between polynomial Frobenius manifolds and finite Coxeter groups. We will see that the manifolds M_h correspond to the groups $I_2(h)$ of symmetries of the regular h-gon in the plane. In particular, for $h = 3$ we obtain that $I_2(3)$ is the Weyl group of type A_2; for $h = 4$, $I_2(4)$ is the Weyl group of type B_2. Their tensor products also correspond to certain finite Coxeter groups. Namely,

$$M_{A_2} \otimes M_{A_2} = M_{D_4}, \tag{4.102}$$

$$M_{A_2} \otimes M_{B_2} = M_{F_4}, \tag{4.103}$$

$$M_{A_2} \otimes M_{I_2(5)} = M_{H_4}, \tag{4.104}$$

with the notation for finite Coxeter groups as in [15]; see also the next section). Besides these, there are only two more cases where the tensor product of two polynomial Frobenius manifolds is again a polynomial Frobenius manifold. They correspond to the following Coxeter groups:

$$M_{A_2} \otimes M_{A_3} = M_{E_6}, \tag{4.105}$$

$$M_{A_2} \otimes M_{A_4} = M_{E_8}. \tag{4.106}$$

More generally, in singularity theory our operation of tensor product of the Frobenius structures on the parameter space of versal deformation of an isolated quasi-homogeneous singularity corresponds to the operation of the direct sum of singularities. Denoting by $M_{f(x)}$ the Frobenius structure on the parameter space of versal deformations of the singularity of a function $f(x)$, we obtain

$$M_{f(x)+g(y)} = M_{f(x)} \otimes M_{g(y)}.$$

Indeed, according to Deligne (see [4]), the variation operator of the direct sum of the singularities is the tensor product of the variation operators of the summands. From this point of view the identifications (4.102), (4.105), (4.106) become obvious. The equalities (4.103) and (4.104) seem not to admit a simple explanation within the framework of singularity theory. However, they are in agreement with the embeddings of Frobenius manifolds obtained by the folding of Dynkin diagrams explained in the next section (I am thankful to J.-B. Zuber for bringing this point to my attention).

5   Monodromy Group and Mirror Construction for Semisimple Frobenius Manifolds

We will introduce a new metric [30, 32] on an open subset of a Frobenius manifold M. The inverse of this metric will be a symmetric bilinear form on the cotangent bundle T^*M defined everywhere.

Definition 5.1. The *intersection form* of the Frobenius manifold M is the bilinear form on T^*M defined by the formula

$$(\omega_1, \omega_2) := i_{E(t)}(\omega_1 \cdot \omega_2), \quad \omega_1, \omega_2 \in T_t^* M. \tag{5.1}$$

On the rhs. the product of one-forms $T_t^* M \times T_t^* M \to T_t^* M$ is defined using the algebra structure on $T_t M$ and the isomorphism

$$\langle \, , \, \rangle : T_t M \to T_t^* M.$$

In flat coordinates, the components of the intersection form are given by the formula

$$g^{\alpha\beta}(t) := (dt^\alpha, dt^\beta) = E^\varepsilon(t) c_\varepsilon^{\alpha\beta}(t)$$
$$= (d + 1 - q_\alpha - q_\beta) F^{\alpha\beta}(t) + A^{\alpha\beta}. \tag{5.2}$$

Here

$$c_\varepsilon^{\alpha\beta}(t) = \eta^{\alpha\gamma} c_{\gamma\varepsilon}^\beta(t),$$
$$F^{\alpha\beta}(t) = \eta^{\alpha\lambda} \eta^{\beta\mu} \frac{\partial^2 F(t)}{\partial t^\lambda \partial t^\mu},$$
$$A^{\alpha\beta} = \eta^{\alpha\lambda} \eta^{\beta\mu} A_{\lambda\mu},$$

where the constant matrix $A_{\lambda\mu}$ was defined in WDVV3.
From (5.2) one obtains

$$g^{\alpha\beta}(t) = t^1 \eta^{\alpha\beta} + \tilde{g}^{\alpha\beta}(t^2, \ldots, t^n)$$

with

$$\tilde{g}^{\alpha\beta}(t^2, \ldots, t^n) = \sum_{\varepsilon=2}^n E^\varepsilon(t) c_\varepsilon^{\alpha\beta}(t).$$

So the bilinear form does not identically degenerate.

Definition 5.2. The locus $\Sigma \subset M$,

$$\Sigma = \{ t \in M \mid \det(g^{\alpha\beta}(t)) = 0 \}, \tag{5.3}$$

is called *discriminant* of the Frobenius manifold M.

Exercise 5.1. Prove that the discriminant is specified by the equation

$$\det \mathcal{U}(t) = 0,$$

where $\mathcal{U}(t)$ is the operator of multiplication by the Euler vector field.

The inverse

$$(g_{\alpha\beta}) = (g^{\alpha\beta})^{-1} \tag{5.4}$$

defines a metric on the open subset $M \setminus \Sigma$.

Lemma 5.1.

1. *The Christoffel coefficients of the Levi–Civita connection for the metric* (5.4) *in the flat coordinates* t^α *are uniquely determined from the equation*

$$\Gamma^{\alpha\beta}_\gamma := -g^{\alpha\varepsilon}\Gamma^\beta_{\varepsilon\gamma} = \left(\frac{d+1}{2} - q_\beta\right)c^{\alpha\beta}_\gamma. \tag{5.5}$$

2. *The metric* (5.4) *on* $M \setminus \Sigma$ *is flat.*

Te proof can be found in [32].

For brevity we will call the bilinear form $g^{\alpha\beta}(t)$ on $T^*_t M$ the *contravariant metric* and the expressions $\Gamma^{\alpha\beta}_\gamma := -g^{\alpha\varepsilon}\Gamma^\beta_{\varepsilon\gamma}$ the *contravariant Christoffel coefficients* of the Levi–Civita connection for the metric.

We make a digression about linear pencils of contravariant metrics.

Let $\left(g_1^{ij}(x), \Gamma_1{}^{ij}_k(x)\right)$ and $\left(g_2^{ij}(x), \Gamma_2{}^{ij}_k(x)\right)$ be two contravariant metrics invertible on an open subset of a manifold M together with the corresponding contravariant Christoffel coefficients.

Definition 5.3. We say that the two contravariant metrics form a *linear quasi-homogeneous pencil* of charge d if:

1. For any $\lambda \in \mathbb{C}$ the metric

$$g_1^{ij}(x) - \lambda g_2^{ij}(x)$$

does not degenerate on an open subset in M.

2. The functions

$$\Gamma_1{}^{ij}_k(x) - \lambda\Gamma_2{}^{ij}_k(x)$$

are the contravariant Christoffel coefficients of the metric (5.4).

3. There exists a function $\varphi(x)$ on M such that the vector fields

$$E^i(x) := g_1^{ij}(x)\frac{\partial\varphi}{\partial x^j}, \quad e^i(x) := g_2^{ij}(x)\frac{\partial\varphi}{\partial x^j} \tag{5.6}$$

have the following properties:

$$[e, E] = e, \tag{5.7}$$

$$\mathcal{L}_E g_1^{ij}(x) = (d-1)\, g_1^{ij}(x), \quad \mathcal{L}_E g_2^{ij}(x) = (d-2)g_2^{ij}(x),$$
$$\mathcal{L}_e g_1^{ij}(x) = g_2^{ij}(x), \qquad\qquad \mathcal{L}_e g_2^{ij}(x) = 0. \tag{5.8}$$

Theorem 5.1. *The intersection form of a Frobenius manifold together with the flat metric $\langle\,,\,\rangle$ form a flat pencil of charge d.*

The proof can be derived from Lemma 5.1 (see [32]). The function $\varphi(t)$ equals $\varphi = t_1 = \eta_{1\varepsilon} t^\varepsilon$.

It can be shown [33] that conversely a manifold with a flat pencil of contravariant metrics carries a natural Frobenius structure such that in the flat coordinates for g_2^{ij}, the metric g_1^{ij} has the form (5.2) (cf. [32, 35]). A detailed proof will be published elsewhere.

Definition 5.4. A function $x = x(t)$ is called the *flat coordinate* of a metric if the differential dx is covariantly constant with respect to the Levi–Civita connection for the metric.

The flat coordinates of the intersection form on a Frobenius manifold are determined from the system of linear differential equations

$$g^{\alpha\varepsilon} \partial_\beta \xi_\varepsilon + \sum_\varepsilon \left(\frac{1}{2} - \mu_\varepsilon \right) c_\beta^{\alpha\varepsilon} \xi_\varepsilon = 0, \tag{5.9}$$

where $\xi_\beta = \partial_\beta \xi$.

Definition 5.5. The equations (5.9) are called the *Gauss–Manin system of the Frobenius manifold.*

Exercise 5.2. Prove that the flat coordinates of the linear pencil

$$g^{\alpha\beta}(t) - \lambda \eta^{\alpha\beta}$$

have the form

$$x(t^1 - \lambda, t^2, \ldots, t^n), \tag{5.10}$$

where $x(t^1, t^2, \ldots, t^n)$ are flat coordinates of the intersection form. Prove that the gradients $\xi^\alpha = \eta^{\alpha\beta} \partial_\beta x(t^1 - \lambda, t^2, \ldots, t^n)$ satisfy the system of equations

$$(\mathcal{U} - \lambda) \partial_\beta \xi + C_\beta \left(\frac{1}{2} + \mu \right) \xi = 0, \tag{5.11}$$

$$(\mathcal{U} - \lambda) \partial_\lambda \xi = \left(\frac{1}{2} + \mu \right) \xi. \tag{5.12}$$

This is an extension of the Gauss–Manin system (5.9) onto $M \times \mathbb{C}_\lambda$. The second equation (5.12) has rational coefficients in λ. As above, the compatibility of the full system will imply the isomonodromy of the Fuchsian system (5.12).

Digression. One can see by a straightforward computation that the system

$$(\mathcal{U} - \lambda) \partial_\beta \phi + C_\beta \mu \phi = 0, \tag{5.13a}$$

$$(\mathcal{U} - \lambda) \partial_\lambda \phi = \mu \phi \tag{5.13b}$$

is also compatible. We will use it to reduce WDVV for $n = 3$, $d \neq 0$ to a particular case of the Painlevé VI equation (semisimplicity is assumed). For $n = 3$ the matrix μ degenerates:

$$\mu = \mathrm{diag}(\mu_1, 0, -\mu_1).$$

So, the equations (5.13b) for the vector function $\phi = (\phi_1, \phi_2, \phi_3)^T$ splits into a 2×2 subsystem for $\chi = (\phi_1, \phi_3)^T$ and a quadrature for ϕ_2:

$$\frac{d\chi}{d\lambda} = -\mu_1 A(\lambda) \chi, \tag{5.14}$$

$$A(\lambda) = \frac{A_1}{\lambda - u_1} + \frac{A_2}{\lambda - u_2} + \frac{A_3}{\lambda - u_3}. \tag{5.15}$$

Here u_1, u_2, u_3 are the eigenvalues of $U(t)$ (i.e., the canonical coordinates), the 2×2-matrices have the form

$$A(\lambda) = \mu_1 \begin{pmatrix} v_1^1(\lambda; t) & -v_3^1(\lambda; t) \\ v_1^3(\lambda; t) & -v_3^3(\lambda; t) \end{pmatrix},$$

where the matrix $\left(v_\beta^\alpha(\lambda; t) \right)$ is defined to be $(\mathcal{U}(t) - \lambda)^{-1}$, and

$$A_i = \begin{pmatrix} \psi_{i1}\psi_{i3} & -\psi_{i3}^2 \\ \psi_{i1}^2 & -\psi_{i1}\psi_{i3} \end{pmatrix}, \quad i = 1, 2, 3. \tag{5.16}$$

Clearly, the matrices satisfy the conditions

$$\det A_i = \mathrm{tr}\, A_i = 0, \quad i = 1, 2, 3 \tag{5.17a}$$

$$A_1 + A_2 + A_3 = \begin{pmatrix} 1 & 0 \\ 0 & -1 \end{pmatrix}. \tag{5.17b}$$

Following [52], we introduce coordinates p, q, k on the space of matrices A_1, A_2, A_3 satisfying (5.17). The coordinate q is the root of the linear equation

$$[A(q)]_{12} = 0; \tag{5.18a}$$

the coordinate p is the value

$$p = [A(q)]_{11}. \tag{5.18b}$$

Explicitly,

$$q = \frac{g^{11}g^{22} - g^{12^2}}{g^{11}},$$

$$p = \mu_1 \frac{g^{11}g^{22}}{g^{12^3} + g^{11}g^{12}g^{13} - g^{11}g^{12}g^{22} - g^{11^2}g^{23}}. \tag{5.19}$$

The entries of the matrices A_i can be expressed via the coordinates p, q, k as follows:

$$\psi_{i1}\psi_{i3} = -\frac{q-u_i}{2\mu_1^2 P'(u_i)}\left[P(q)p^2+2\mu_1\frac{P(q)}{q-u_i}p+\mu_1^2\left(q+2u_i-\sum u_j\right)\right],$$

$$\psi_{i3}^2 = -k\frac{q-u_i}{P'(u_i)}, \tag{5.20}$$

$$\psi_{i1}^2 = -k^{-1}\frac{q-u_i}{4\mu_1^4 P'(u_i)}\left[P(q)p^2+2\mu_1\frac{P(q)}{q-u_i}p+\mu_1^2\left(q+2u_i-\sum u_j\right)\right]^2,$$

where the polynomial $P(\lambda)$ has the form

$$P(\lambda) := (\lambda-u_1)(\lambda-u_2)(\lambda-u_3). \tag{5.21}$$

The compatibility of the system (5.13) implies

$$\partial_i q = \frac{P(q)}{P'(u_i)}\left[2p+\frac{1}{q-u_i}\right],$$

$$\partial_i p = -\frac{P'(q)p^2 + (2q+u_i-\sum u_j)p + \mu_1(1-\mu_1)}{P'(u_i)}, \tag{5.22}$$

and it gives a quadrature for the function $\log k$:

$$\partial_i \log k = (2\mu_1-1)\frac{q-u_i}{P'(u_i)}. \tag{5.23}$$

Eliminating p from the system, we obtain a second-order differential equation for the function $q = q(u_1, u_2, u_3)$:

$$\partial_i^2 q = \frac{1}{2}\frac{P'(q)}{P(q)}(\partial_i q)^2 - \left[\frac{1}{2}\frac{P''(u_i)}{P'(u_i)} + \frac{1}{q-u_i}\right]\partial_i q$$
$$+\frac{1}{2}\frac{P(q)}{(P'(u_i))^2}\left[(2\mu_1-1)^2 + \frac{P'(u_i)}{(q-u_i)^2}\right], \quad i = 1,2,3.$$

The system (5.22) is invariant with respect to transformations of the form

$$u_i \mapsto au_i + b,$$
$$q \mapsto aq + b.$$

Introducing the invariant variables

$$x = \frac{u_3 - u_1}{u_2 - u_1},$$

$$y = \frac{q}{u_2 - u_1} - \frac{u_1}{u_2 - u_1},$$

we obtain for the function $y = y(x)$ the following particular Painlevé VI equation:

$$y'' = \frac{1}{2}\left[\frac{1}{y} + \frac{1}{y-1} + \frac{1}{y-x}\right](y')^2 - \left[\frac{1}{x} + \frac{1}{x-1} + \frac{1}{y-x}\right]y'$$
$$+ \frac{1}{2}\frac{y(y-1)(y-x)}{x^2(x-1)^2}\left[(2\mu_1 - 1)^2 + \frac{x(x-1)}{(y-x)^2}\right]. \quad \text{(PVI}(\mu))$$

Conversely, for a solution $y(x)$ of the equation PVI(μ), we construct functions $q = q(u_1, u_2, u_3)$ and $p = p(u_1, u_2, u_3)$ by putting

$$q = (u_2 - u_1)y\left(\frac{u_3 - u_1}{u_2 - u_1}\right) + u_1,$$

$$p = \frac{1}{2}\frac{P'(u_3)}{P(q)}y'\left(\frac{u_3 - u_1}{u_2 - u_1}\right) - \frac{1}{2}\frac{1}{q - u_3}.$$

Then we compute the quadrature (5.23), which determines the function k (this provides us with one more arbitrary integration constant). After this we are able to compute the matrix $(\psi_{i\alpha}(u))$ from the equations (5.20) and

$$(\psi_{12}, \psi_{22}, \psi_{32})$$
$$= \pm i(\psi_{21}\psi_{33} - \psi_{23}\psi_{31}, \psi_{13}\psi_{31} - \psi_{11}\psi_{33}, \psi_{11}\psi_{23} - \psi_{13}\psi_{21}).$$

The last step is to reconstruct the flat coordinates $t = t(u)$ and the tensor $c_{\alpha\beta\gamma}$ using the formulae (3.21) and (3.17).

Example 5.1. Applying the above procedure to the three polynomial solutions (1.21)–(1.23) of WDVV, we obtain the following three algebraic solutions of PVI(μ) with $\mu = -\frac{1}{4}, -\frac{1}{3}, -\frac{2}{5}$ [32,34], represented in parametric form by

$$y = \frac{(s-1)^2(1+3s)(9s^2-5)^2}{(1+s)(25 - 207s^2 + 1539s^4 + 243s^6)},$$

$$x = \frac{(s-1)^3(1+3s)}{(s+1)^3(1-3s)}, \qquad (5.24)$$

$$y = \frac{(2-s)^2(1+s)}{(2+s)(5s^4 - 10s^2 + 9)},$$

$$x = \frac{(2-s)^2(1+s)}{(2+s)^2(1-s)}, \qquad (5.25)$$

$$y = \frac{(s-1)^2(1+3s)^2(-1+4s+s^2)(7-108s^2+314s^4-588s^6+119s^8)^2}{(1+s)^3(-1+3s)P(s^2)},$$

$$x = \frac{(-1+s)^5(1+3s)^3(-1+4s+s^2)}{(1+s)^5(-1+3s)^3(-1-4s+s^2)}, \qquad (5.26a)$$

where

$$P(z) = 49 - 2133z + 34308z^2 - 259044z^3 + 16422878z^4 - 7616646z^5$$
$$+ 13758708z^6 + 5963724z^7 - 719271z^8 + 42483z^9. \quad (5.26b)$$

Some other particular solutions of Painlevé VI in relation to Frobenius manifolds were constructed in [80].

Let us return to the intersection form of a Frobenius manifold. Due to Lemma 5.1, in a neighborhood of a point $t_0 \in M \setminus \Sigma$ one can choose n independent flat coordinates $x^1(t), \ldots, x^n(t)$ of the intersection form. In these coordinates the matrix

$$g^{ab} = (dx^a, dx^b) = \frac{\partial x^a}{\partial t^\alpha} \frac{\partial x^b}{\partial t^\beta} g^{\alpha\beta}(t) \quad (5.27)$$

becomes constant, and the Christoffel coefficients vanish. The flat coordinates are determined uniquely up to shifts and orthogonal transformations in $O(n, g^{ab})$.

Exercise 5.3. Show that for $d \neq 1$ the flat coordinates $x(t)$ can be chosen in such a way that

$$\mathcal{L}_E x(t) = \frac{1-d}{2} x(t).$$

So, for $d \neq 1$, the flat coordinates of the intersection form satisfying the quasi-homogeneity condition of Exercise 5.3 are determined uniquely up to a transformation from $O(n, g^{ab})$.

The solutions of the Gauss–Manin system can be analytically continued along any path in $M \setminus \Sigma$. We obtain a multivalued *period map*

$$t \mapsto \left(x^1(t), \ldots, x^n(t)\right) \quad (5.28)$$

defined on $M \setminus \Sigma$ (cf. the end of Section 2 above). The multivaluedness of the period map is described by a representation of the fundamental group of the complement to the discriminant

$$\pi_1(M \setminus \Sigma; t_0) \to O(n, g^{ab}) \quad (5.29)$$

(for $d = 1$, instead of the orthogonal group, we obtain a representation into the group of affine isometries of the metric g^{ab}).

Definition 5.6. The image $W(M)$ of the representation (5.29) is called the *monodromy group* of the Frobenius manifold.

Our main aim now is to compute the monodromy group of a semisimple Frobenius manifold in terms of the Stokes matrix of the manifold.

In the semisimple case, doing the gauge transform

$$\phi = \Psi\xi, \quad (5.30)$$

we rewrite the extended Gauss–Manin system (5.11), (5.12) in the form

$$(U - \lambda)\frac{\mathrm{d}\phi}{\mathrm{d}\lambda} = \left(\frac{1}{2} + V\right)\phi, \tag{5.31a}$$

or, equivalently,

$$\frac{\mathrm{d}\phi}{\mathrm{d}\lambda} = \sum_{i=1}^{n} \frac{B_i}{\lambda - u_i}\phi \tag{5.31b}$$

with

$$B_i = -E_i\left(\frac{1}{2} + V\right), \tag{5.31c}$$

where E_i is the unit matrix (3.31),

$$\partial_i\phi = -\frac{B_i}{\lambda - u_i}\phi + V_i\phi, \quad i = 1, \ldots, n. \tag{5.32}$$

We obtain a Fuchsian system (5.31b) with the matrix residues B_1, ... , B_n of a particular form (5.31c). The compatibility of (5.31) provides the isomonodromy of the dependence of the coefficients of the system on the position of the poles u_1, ... , u_n. We will now relate the structure of the monodromy of the Fuchsian system (5.31b) to the Stokes matrix of the Frobenius manifold.

Lemma 5.2. *Let $\phi^{(1)}$, $\phi^{(2)}$ be two solutions of the system (5.31). Then the bilinear form*

$$\left(\phi^{(1)}, \phi^{(2)}\right) := \phi^{(1)}{}^{T}(U - \lambda)\phi^{(2)} \tag{5.33}$$

depends neither on λ nor on $u_1, \ldots, u_n$.

The proof can be obtained by straightforward differentiation.

Remark. We recall that the solutions $\phi = (\phi_1, \ldots, \phi_n)^T$ of the system (5.31) correspond to flat coordinates $x(t)$ of the intersection form

$$\phi_i = \sum \psi_{i\alpha}\eta^{\alpha\beta}\partial_\beta x\left(t^1 - \lambda, t^2, \ldots, t^n\right). \tag{5.34}$$

If $\phi^{(1)}$ corresponds to $x_1(t)$, and $\phi^{(2)}$ to $x_2(t)$, then the bilinear form (5.33) equals

$$\left(\phi^{(1)}, \phi^{(2)}\right) = (\mathrm{d}x_1, \mathrm{d}x_2) - \lambda\langle \mathrm{d}x_1, \mathrm{d}x_2\rangle. \tag{5.35}$$

We will now construct, essentially following [7], a particular system of solutions of the Fuchsian system (5.31b). Let us choose an argument φ in such a way that

$$\arg(u_i - u_j) \neq \frac{\pi}{2} + \varphi \pmod{2\pi} \quad \text{for any } i \neq j. \tag{5.36}$$

We make n distinct parallel branch cuts $L_1, \ldots, L_n$ in the complex λ-plane of the form

$$L_j = \{\lambda = u_j + i\rho e^{-i\varphi}, \rho \geq 0\}, \quad j = 1, \ldots, n. \qquad (5.37)$$

Each branch cut has positive and negative sides

$$L_j^+ = \left\{\lambda \,\Big|\, \arg(u_j - \lambda) = -\frac{\pi}{2} - \varphi + 0\right\},$$

$$L_j^- = \left\{\lambda \,\Big|\, \arg(u_j - \lambda) = -\frac{\pi}{2} - \varphi + 2\pi - 0\right\}.$$

On the complement

$$\mathbb{C} \setminus \bigcup_j L_j \qquad (5.38)$$

the single-valued functions $\sqrt{u_1 - \lambda}, \ldots, \sqrt{u_n - \lambda}$ are well defined. We specify them uniquely by requiring that on L_j^+,

$$\arg \sqrt{u_j - \lambda} = -\frac{\pi}{4} - \frac{\varphi}{2} + 0. \qquad (5.39)$$

Let us choose small loops $\gamma_1, \ldots, \gamma_n$ going around the points $u_1, \ldots, u_n$ in the counterclockwise direction. Let $R_1^*, \ldots, R_n^*$ be the monodromy transformations in the space of solutions of (5.31b) corresponding to the loops $\gamma_1, \ldots, \gamma_n$.

Lemma 5.3.

1. *There exist unique solutions $\phi^{(1)}(\lambda), \ldots, \phi^{(n)}(\lambda)$ of (5.31b), analytic in (5.38), such that*

$$R_j^* \phi^{(j)} = -\phi^{(j)}, \quad j = 1, \ldots, n, \qquad (5.40)$$

$$\phi_a^{(j)}(\lambda) = \frac{\delta_a^j}{\sqrt{u_j - \lambda}} + O\left(\sqrt{u_j - \lambda}\right), \quad \lambda \to u_j. \qquad (5.41)$$

2. *Introduce a symmetric matrix $G = (G_{ij})$,*

$$G_{ij} = \left(\phi^{(i)}, \phi^{(j)}\right). \qquad (5.42)$$

The monodromy transformations R_j^* are the reflections

$$R_j^* \phi^{(i)} = \phi^{(i)} - 2G_{ij}\phi^{(j)}, \quad i, j = 1, \ldots, n, \qquad (5.43)$$

in the hyperplanes orthogonal to $\phi^{(j)}$ with respect to the bilinear form (5.33).

Proof. The matrix residue B_j in (5.31b) has one eigenvalue $-\frac{1}{2}$ and $n-1$ eigenvalues 0. So one can construct a fundamental system of solutions $\phi^{(j)}(\lambda)$, $r_2(\lambda)$, ... , $r_n(\lambda)$ such that $R_j^* \phi^{(j)} = -\phi^{(j)}$, $R_j^* r_k = r_k$, $k = 2$, ... , n. This means that the last $n-1$ solutions are analytic at $\lambda = u_j$. The solution $\phi^{(j)}(\lambda)$ is determined uniquely up to a nonzero factor. From this, the first part of the lemma easily follows.

To prove the second part, let us represent $\phi^{(i)}(\lambda)$ as a linear combination of $\phi^{(j)}(\lambda)$ and of the solutions analytic at $\lambda = u_j$,

$$\phi^{(i)}(\lambda) = C_{ij}\phi^{(j)}(\lambda) + r_{ij}(\lambda).$$

Here C_{ij} is some constant, and the solution $r_{ij}(\lambda)$ is analytic at $\lambda = u_j$. Computing the bilinear form (5.42) and using Lemma 5.2, we obtain

$$G_{ij} = \lim_{\lambda \to u_j} \sum_{a=1}^{n} (u_a - \lambda)\phi_a^{(i)}(\lambda)\phi_a^{(j)}(\lambda) = C_{ij}.$$

We obtain

$$C_{ij} = G_{ij} \quad \text{for } i \neq j.$$

A similar computation gives

$$G_{ii} = 1.$$

We obtain a representation

$$\phi^{(i)}(\lambda) = G_{ij}\phi^{(j)}(\lambda) + r_{ij}(\lambda). \tag{5.44}$$

So

$$
\begin{aligned}
R_j^* \phi^{(i)} &= -G_{ij}\phi^{(j)}(\lambda) + r_{ij}(\lambda) \\
&= \phi^{(i)}(\lambda) - 2G_{ij}\phi^{(j)}(\lambda) \\
&= \phi^{(i)}(\lambda) - 2\frac{(\phi^{(i)}, \phi^{(j)})}{(\phi^{(j)}, \phi^{(j)})}\phi^{(j)}(\lambda).
\end{aligned}
$$

The lemma is proved.   □

We will now establish, using the technique of [7], a simple relation between the matrix G (5.42) for the system (5.31b) and the Stokes matrix of the operator (3.30).

Let us assume that the angle φ is chosen in such a way that the order of the rays L_1, ... , L_n in the complex λ-plane corresponds to the order of the complex numbers u_1, ... , u_n in the following sense: Looking along the ray L_j from the endpoint $\lambda = u_j$, we must see L_{j-1} as the nearest ray on the left and L_{j+1} as the nearest one on the right, $2 \leq j \leq n-1$.

Lemma 5.4. *The oriented line $\ell = \ell_+ \cup (-\ell_-)$,*

$$\ell_+ = \{z \mid \arg z = \varphi\},$$

is admissible for the operator (3.30). The corresponding Stokes matrix S is upper triangular. It satisfies the relation

$$S + S^T = 2G, \tag{5.45}$$

where G is the matrix (5.42).

Proof. The admissibility is obvious from (5.36). Let us construct the fundamental matrices $Y_{\text{right}}(z)$, $Y_{\text{left}}(z)$ having the needed asymptotic expansion (5.26) in the half-planes Π_{right}, Π_{left}. We will construct them by taking an appropriate inverse Laplace transform of the solutions $\phi^{(j)}(\lambda)$ defined in Lemma 5.3. Put

$$Y_{aj}(z) = -\frac{\sqrt{z}}{2\sqrt{\pi}} \int_{C_j} \phi_a^{(j)}(\lambda) e^{\lambda z} \mathrm{d}\lambda, \quad a, j = 1, \ldots, n. \tag{5.46}$$

Here C_j is an infinite contour coming from infinity along the positive side of the branch cut L_j, then encircling the point $\lambda = u_j$ and finally returning to infinity along the negative side of the branch cut L_j. Since $\lambda = \infty$ is a regular singularity of the system (5.31b), the solutions $\phi^{(j)}(\lambda)$ do not grow at $\lambda \to \infty$ faster than some power of $|\lambda|$. We conclude that the integral converges absolutely for $z \in \Pi_{\text{left}}$. Using (5.31b) and integrating by parts, we prove that the matrix $Y(z) = \left(Y_a^j(z)\right)$ satisfies the equation (3.30). To obtain the asymptotic expansion of this solution as $|z| \to \infty$ we can, due to Watson's lemma [84], integrate the terms of the convergent expansion (5.41) of the solution $\phi^{(j)}(\lambda)$ near $\lambda = u_j$. By so doing, we easily see that the solution $Y_{\text{left}}(z) := Y(z)$ has the needed asymptotic expansion

$$Y_{aj}(z) \sim \left(\delta_{aj} + O\left(\frac{1}{z}\right)\right) e^{z u_j}$$

as $|z| \to \infty$, $z \in \Pi_{\text{left}}$.

Let us now construct the fundamental matrix $Y_{\text{right}}(z)$. We must choose the system of the opposite branch cuts

$$L_j' = \{\lambda = u_j - i\rho e^{-i\varphi}, \rho \geq 0\}, \quad j = 1, \ldots, n, \tag{5.47}$$

to construct the corresponding solutions $\phi^{(j)\prime}(\lambda)$ and to define

$$Y_{\text{right}\,aj}(z) = -\frac{\sqrt{z}}{2\sqrt{\pi}} \int_{C_j'} \phi_a^{(j)\prime}(\lambda) e^{\lambda z} \mathrm{d}\lambda, \quad a, j = 1, \ldots, n. \tag{5.48}$$

Here the contour C_j' goes around the branch cut L_j'. As above, we prove that the solution $Y_{\text{right}}(z) := \left(Y_{\text{right}\,a}^j(z)\right)$ of (3.30) has the needed asymptotic expansion in Π_{right} as $|z| \to \infty$. It remains to establish a relation between the integrals (5.46) and (5.48). To analytically continue $Y_{\text{left}}(z)$ through ℓ_+ in the clockwise direction into Π_{right}, we must rotate the branch cuts L_j in the counterclockwise direction until they take the places of L'_j, $j = 1, \ldots,$ n. For $j = 1$, such a deformation does not meet any obstruction. So

$$\phi^{(1)'} = \phi^{(1)}.$$

To deform L_2 to L'_2 we must pass through the branch cut L_1. This is equivalent to the action of the monodromy transformation R_1^*. So

$$\phi^{(2)'} = R_1^* \phi^{(2)}.$$

Continuing this process, we obtain that

$$\phi^{(k)'} = R_1^* R_2^* \cdots R_{k-1}^* \phi^{(k)}, \quad k = 2, \ldots, n.$$

Using the computation in the proof of the Coxeter identity (see [15]) we obtain

$$\phi^{(k)} = 2G_{k1}\phi^{(1)'} + 2G_{k2}\phi^{(2)'} + \cdots + 2G_{kk-1}\phi^{(k-1)'} + \phi^{(k)'}, \quad k = 1, \ldots, n.$$

The lemma is proved. $\qquad\qquad\qquad\qquad\qquad\qquad\qquad\qquad\qquad\qquad$ $\square$

Corollary 5.1. *If*

$$\det(S + S^T) \neq 0, \tag{5.49}$$

then the functions $\phi^{(1)}(\lambda), \ldots, \phi^{(n)}(\lambda)$ define a basis of the space of solutions of (5.31b).

Exercise 5.4. Prove that the Stokes matrix of the quantum cohomology of a manifold X of an even complex dimension (assuming semisimplicity of the quantum cohomology) satisfies the nondegeneracy condition (5.49).

In the rest of this section I will assume that $d \neq 1$ and that the Stokes matrix satisfies the nondegeneracy condition (5.49).

All the above constructions of the basis $\phi^{(1)}(\lambda), \ldots, \phi^{(n)}(\lambda)$ were done for a given fixed point $(u_1, \ldots, u_n)$ of the Frobenius manifold. Since the solutions $\phi^{(j)}(\lambda)$ are uniquely determined, they become locally well defined analytic functions of $(u_1, \ldots, u_n)$.

Lemma 5.5. *The above-mentioned solutions $\phi^{(1)}(\lambda)$ also satisfy the equations* (5.32).

Proof. Let us consider the vector function

$$\tilde{\phi}^{(j)} := \partial_i \phi^{(j)} + \frac{B_i}{\lambda - u_i} \phi^{(j)} - V_i \phi^{(j)}$$

for some i between 1 and n. Because of the compatibility of (5.31) and (5.32), the vector function $\tilde{\phi}^{(j)}$ satisfies (5.31). It is easy to see that this solution is regular near the points $\lambda = u_1, \ldots, \lambda = u_n$. Hence $\tilde{\phi}^{(j)} = 0$. The lemma is proved. $\square$

Let $\widetilde{M} = \widetilde{\mathrm{Fr}}(e, \mu, R, S, C; u^0)$ be the Frobenius structure on the universal covering of $\mathbb{C}^n \setminus \mathrm{diag}$ defined by the given monodromy data (e, μ, R, S, C) with $\mu_1 \neq -\frac{1}{2}$, $\det(S + S^T) \neq 0$. The discriminant $\tilde{\sigma}$ of this Frobenius manifold consists of the lifts of the coordinate hyperplanes $u_i = 0$, $i = 1$, $\ldots, n$.

Let $\mathcal{E}$ be n-dimensional linear space equipped with a symmetric nondegenerate bilinear form $(\ ,\ )$ on the dual space $\mathcal{E}^*$ having in some basis e^1, $\ldots, e^n$ the Gram matrix

$$(e^i, e^j) = (S + S^T)_{ij}. \tag{5.50}$$

For any $1 \leq i \leq n$ denote by

$$R_i : \mathcal{E} \to \mathcal{E} \tag{5.51}$$

the transformation dual to the reflection $R_i^* : \mathcal{E}^* \to \mathcal{E}^*$ in the hyperplanes orthogonal to e^i:

$$R_i^*(x) = x - (x, e^i)e^i. \tag{5.52}$$

Theorem 5.2. *The image of the monodromy representation in the group $O(\mathcal{E}, (\ ))$ of orthogonal transformations of the space $\mathcal{E}$,*

$$\pi_1(\widetilde{M}; u^0) \to O(\mathcal{E}, (\ )), \tag{5.53}$$

is the group generated by the reflections $R_1, \ldots, R_n$.

Proof. According to Lemma 5.3, locally we have a basis $\phi^{(1)}(\lambda; u), \ldots,$ $\phi^{(n)}(\lambda; u)$ of solutions of (5.31), (5.32). The formula

$$x_j(\lambda; u) = \frac{2\sqrt{2}}{1 - d} \sum_a (u_a - \lambda)\psi_{a1}(u)\phi_a^{(j)}(\lambda; u) \tag{5.54}$$

gives flat coordinates of the linear pencil $(\ ) - \lambda \langle\ ,\ \rangle$. Due to Lemma 5.4 we have

$$(dx_i, dx_j) - \lambda \langle dx_i, dx_j \rangle = (S + S^T)_{ij}. \tag{5.55}$$

We obtain a locally well defined isometry (the period map of the Frobenius manifold, cf. (2.102))

$$\widetilde{M} \setminus \widetilde{\Sigma} \to \mathcal{E}, \tag{5.56a}$$

$$u \mapsto (x_1(u), \ldots, x_n(u)), \tag{5.56b}$$

where

$$x_j(u) := x_j(\lambda; u)|_{\lambda=0}.$$

The monodromy around the branch $u_i = 0$ of $\widetilde{\Sigma}$ in the given chart of the universal covering of $\mathbb{C}^n \setminus \text{diag}$ is equivalent to the monodromy of the vector function

$$\big(x_1(\lambda; u), \ldots, x_n(\lambda; u)\big)$$

corresponding to a small loop around $\lambda = u_i$ in the λ-plane. We obtain the transformation

$$x_k(u) \mapsto x_k(u) - (S + S^T)_{ki} x_i(u), \quad k = 1, \ldots, n, \tag{5.57}$$

where we identify the coordinates in $\mathcal{E}$ with the dual basis in $\mathcal{E}^*$. This means that locally the monodromy group is generated by the reflections (5.57).

What happens with the analytic continuation into another chart of $\widetilde{M}$? We arrive in another chart when some of the Stokes rays (4.34) pass through the admissible line ℓ. Simultaneously, two of the branch cuts $L_1, \ldots, L_n$ pass one through the other. The Stokes matrix changes according to the rule (4.68). It is sufficient to understand what happens with the flat coordinates $x_1(\lambda; u), \ldots, x_n(\lambda; u)$ with an elementary transformation (4.74) of the braid group.

Lemma 5.6. *The elementary braid σ_i permuting the points $\lambda = u_i$ and $\lambda = u_{i+1}$ in the complex λ-plane gives the following transformation of the solutions $\phi^{(1)}(\lambda), \ldots, \phi^{(n)}(\lambda)$:*

$$\sigma_i(\phi^{(k)}) = \phi^{(k)}, \quad k \neq i, i+1,$$
$$\sigma_i(\phi^{(i)}) = \phi^{(i+1)}, \tag{5.58}$$
$$\sigma_i(\phi^{(i+1)}) = R_{i+1}^* \phi^{(i)} = \phi^{(i)} - S_{i,i+1} \phi^{(i+1)}.$$

We leave the proof as an exercise to the reader.

We obtain, after the analytic continuation along the braid σ_i, that the new monodromy transformations in $\mathcal{E}^*$ are reflections in the hyperplanes orthogonal to the vectors

$$(e^1, \ldots, e^{i-1}, e^{i+1}, R_{i+1}^*(e^i), e^{i+2}, \ldots, e^n).$$

But a reflection with respect to the hyperplane orthogonal to $R_{i+1}^*(e^i)$ is equal to $R_{i+1}^* R_i^* R_{i+1}^*$. This transformation belongs to the group generated by $R_1^*, \ldots, R_n^*$. The theorem is proved. $\square$

To complete our description of an arbitrary semisimple Frobenius manifold in terms of an appropriate "singularity theory," we must construct an analogue of versal deformation. This can be done (at least, under the nondegeneracy assumption (5.49)) in the following way [32, Appendix I]. We will construct a family of functions $\lambda(p; u)$, $u = (u_1, \ldots, u_n)$, of the complex variable p defined in an open domain $\mathcal{D}$ of a Riemann surface $\mathcal{R}$ realized as a branched covering of the complex plane with a finite number of sheets. The Riemann surface may depend on u. However, the projection of the domain $\mathcal{D}$ on the complex plane will be fixed. These functions depend on complex pairwise distinct parameters $u_1, \ldots, u_n$ belonging to a sufficiently small domain $\Omega \subset \mathbb{C}^n$.

The first main property is that $\lambda(p; u)$, as function of $p \in \mathcal{D}$, has critical values just $u_1, \ldots, u_n$. The corresponding critical points must not be degenerate. The second condition that we require from the function $\lambda(p; u)$ is that for any two points $p_i^{(1,2)} \in \mathcal{D}$ with the same critical value u_i, we have

$$\lambda''(p_i^{(1)}; u) = \lambda''(p_i^{(2)}; u).$$

Here the prime denotes the p-derivative.

Definition 5.7. The function $\lambda(p; u)$ on $\mathcal{D} \times \Omega$ satisfying the above two properties is called the *superpotential* of some domain M_Ω in the Frobenius manifold M if:

1. The canonical coordinates $(u_1, \ldots, u_n)$ map M_Ω to $\Omega \subset \mathbb{C}^n$.

2. For any critical points $p_1, \ldots, p_n \in \mathcal{D}$ of $\lambda(p; u)$ with the respective critical values $u_1, \ldots, u_n$, the following expressions for the flat metric $\langle \, , \, \rangle$ on $T_t M$, the intersection form $(\, , \,)$ (outside the discriminant Σ), and the multiplication of tangent vectors hold true:

$$\langle \partial', \partial'' \rangle_t = - \sum_{i=1}^n \operatorname{res}_{p=p_i} \frac{\partial'(\lambda(p; u(t))\mathrm{d}p)\partial''(\lambda(p; u(t))\mathrm{d}p)}{\mathrm{d}\lambda(p; u(t))}, \tag{5.59}$$

$$(\partial', \partial'')_t$$
$$= - \sum_{i=1}^n \operatorname{res}_{p=p_i} \frac{\partial'(\log \lambda(p; u(t))\mathrm{d}p)\partial''(\log \lambda(p; u(t))\mathrm{d}p)}{\mathrm{d}\log \lambda(p; u(t))}, \tag{5.60}$$

$$\langle \partial' \cdot \partial'', \partial''' \rangle_t$$
$$= - \sum_{i=1}^n \operatorname{res}_{p=p_i} \frac{\partial'(\lambda(p; u(t))\mathrm{d}p)\partial''(\lambda(p; u(t))\mathrm{d}p)\partial'''(\lambda(p; u(t))\mathrm{d}p)}{\mathrm{d}p\,\mathrm{d}\lambda(p; u(t))}.$$
$$\tag{5.61}$$

In these formulae ∂', ∂'', ∂''' are any three vector fields on M,

$$\mathrm{d}\lambda(p; u) := \frac{\partial \lambda(p; u)}{\partial p}\mathrm{d}p, \quad \mathrm{d}\log \lambda(p; u) := \frac{\partial \log \lambda(p; u)}{\partial p}\mathrm{d}p.$$

3. For some 1-cycles $Z_1, \ldots, Z_n$ in $\mathcal{D}$ the integrals

$$\tilde{t}_j(u; z) = \frac{1}{\sqrt{z}} \int_{Z_j} e^{z\lambda(p;u)} dp, \quad j = 1, \ldots, n, \qquad (5.62)$$

converge and give a system of independent flat coordinates of the connection $\tilde{\nabla}$.

Example 5.2. For the polynomial Frobenius manifold corresponding to the singularity A_n, the versal deformation

$$\lambda = p^{n+1} + a_n p^{n-1} + \cdots + a_1$$

gives the required superpotential. The variables $u_1, \ldots, u_n$ are the critical values of this function. Locally one can express the coefficients of the polynomial as single-valued functions of u_j. For other simple singularities, the versal deformation is a family of polynomials of two variables. However, one can reduce double integrals for the residues (1.19), (1.20) and for the oscillatory integrals (2.97) to one-dimensional residues and single integrals of the above form. For the D_n case this was done in [24] (the superpotential becomes a rational function). For the case of E_6, the singularity of the superpotential is algebraic. This was found in [60].

Example 5.3. For the case where the Riemann surfaces $\mathcal{R}$ can be compactified at infinity in such a way that there is exactly one branch point on the Riemann surface over any of the critical values u_j, the Frobenius manifold can be identified with a Hurwitz space of branched coverings. The Frobenius structure on the Hurwitz spaces was constructed in [26, 27] (see also [32]). In [59], the method of [26, 27] was extended to as well produce some algebro-geometric solutions satisfying WDVV1 and WDVV2 but not WDVV3.

We will now construct a superpotential for semisimple Frobenius manifolds satisfying the nondegeneracy assumption (5.49).

Let $u^0 = (u_1^0, \ldots, u_n^0)$, $u_i^0 \neq u_j^0$ for $i \neq j$, be any point of M (written in the canonical coordinates). We choose the branch cuts $L_1^0, \ldots, L_n^0$ as above. For $M \ni u$ sufficiently close to u^0, we will choose the branch cuts $L_1, \ldots, L_n$ coinciding with $L_1^0, \ldots, L_n^0$ outside some small neighborhoods of the points $u_1^0, \ldots, u_n^0$, respectively. This allows us to construct solutions $\phi^{(1)}(\lambda; u), \ldots, \phi^{(n)}(\lambda; u)$ as in Lemma 5.3. Set

$$G^{ij} = \left(\phi^{(i)}, \phi^{(j)} \right).$$

Let (G_{ij}) be the inverse matrix. Let us consider the solution to (5.31) (cf. [8])

$$\phi(\lambda; u) = \sum_{i,j=1}^{n} G_{ij} \phi^{(j)}(\lambda; u). \qquad (5.63)$$

Lemma 5.7. *The solution $\phi(\lambda; u) = (\phi_a(\lambda; u))$ for $\lambda \to u_j$ has the behavior*

$$\phi_a(\lambda; u) = \frac{\delta_{aj}}{\sqrt{u_j - \lambda}} + O(1), \quad a, j = 1, \ldots, n. \tag{5.64}$$

The proof follows from (5.44).

Denote by $p = p(\lambda; u)$ the corresponding flat coordinate of the intersection form

$$p(\lambda; u) = \frac{\sqrt{2}}{1 - d} \sum_{a=1}^{n} (u_a - \lambda) \psi_{a1}(u) \phi_a(\lambda; u), \tag{5.65}$$

i.e.,

$$\frac{\partial p(\lambda; u)}{\partial u_a} = \frac{1}{\sqrt{2}} \psi_{a1}(u) \phi_a(\lambda; u). \tag{5.66}$$

It is analytic in the domain

$$\mathbb{C} \setminus \bigcup_j L_j. \tag{5.67}$$

For $\lambda \to u_j$ it behaves as follows:

$$p(\lambda; u) = p_j + \sqrt{2} \psi_{j1} \sqrt{u_j - \lambda} + O(u_j - \lambda), \tag{5.68}$$

where $p_j = p(u_j; u)$. For $\lambda \to \infty$ the function $p(\lambda; u)$ has a regular singularity. Hence $dp(\lambda; u)/d\lambda$ has at most a finite number of zeros $r_1, \ldots, r_N$ in (5.67). Without loss of generality we may assume that all these zeros are simple and that they do not belong to the branch cuts L_j.

Let $\mathcal{D}_0$ be the image of the domain

$$\lambda \in \mathbb{C} \setminus \bigcup_j L_j^0$$

with respect to the map $p(\lambda; u^0)$. In particular, the two sides of the branch cut L_j^0 open to produce a smooth boundary curve of $\mathcal{D}_0$ passing through p_j^0. Write

$$\zeta_j^0 = p(r_j; u^0), \quad j = 1, \ldots, N.$$

Let us consider the inverse $\lambda = \lambda(p; u^0)$ to the function $p(\lambda; u^0)$. It lives on a certain branched covering $\hat{\mathcal{D}}_0$ of the domain $\mathcal{D}_0$ obtained by cutting $\mathcal{D}_0$ along some paths going from $\zeta_1^0, \ldots, \zeta_N^0$ to infinity and by subsequent gluing of a finite number of copies of $\mathcal{D}_0$ with the cuts. Near a point of the boundary of $\hat{\mathcal{D}}_0$ passing through p_j^0, we have

$$\lambda = u_j^0 - \frac{1}{2\psi_{j1}^2(u^0)} (p - p_j^0)^2 + O(p - p_j^0)^3.$$

Thus we can analytically continue $\lambda(p; u^0)$ through the boundary of $\mathcal{D}_0$ near a domain $\mathcal{D}$ containing p_j^0 as its internal point. We can repeat this construction for all u sufficiently close to u^0 (actually, it is sufficient to require the points u_j not to intersect the branch cuts L_i^0). In this way we will produce a family of Riemann surfaces with the branch points $\zeta_1, \ldots, \zeta_N$. We may also assume that the image of (5.67) with respect to the map $p(\lambda; u)$ for any u close to u^0 belongs to the projection of the domain $\mathcal{D}$ onto the complex p-plane. This completes the construction of the family of functions $\lambda(p; u)$.

The cycles Z_j that we need to compute the integrals (5.62) have the form

$$Z_j = p(C_j; u) \tag{5.69}$$

(more precisely, an arbitrary lift of this cycle on the Riemann surface), where C_j was defined in (5.46).

Theorem 5.3. *The function $\lambda(p; u)$ is a superpotential of the Frobenius manifold for a sufficiently small neighborhood of the point u^0.*

Proof. By construction, the function $\lambda(p; u)$ has critical values $u_1, \ldots, u_n$. For any critical point $p_j \in \mathcal{D}$ on the Riemann surface with the critical value u_j, we obtained

$$\lambda = u_j - \frac{1}{2\psi_{j1}^2(u)}(p - p_j)^2 + O(p - p_j)^3.$$

So the second derivatives of $\lambda(p; u)$ do not depend on the choice of the critical point.

Let us prove the formulae (5.59)–(5.61). We take $\partial' = \partial_a$, $\partial'' = \partial_b$, $\partial''' = \partial_c$ as the vector fields along the canonical coordinates. Then

$$\partial_a(\lambda(p; u)dp) = [\delta_{aj} + O(p - p_j)]dp, \quad p \to p_j,$$

$$d\lambda(p; u) = -\left[\frac{p - p_j}{\psi_{j1}^2} + O(p - p_j)^2\right]dp, \quad p \to p_j.$$

So

$$\mathrm{res}_{p=p_j} \frac{\partial_a(\lambda(p; u)dp)\partial_b(\lambda(p; u)dp)}{d\lambda(p; u)} = -\psi_{j1}^2 \delta_{aj}\delta_{bj}.$$

Thus the formula (5.59) gives

$$\langle \partial_a, \partial_b \rangle = \delta_{ab}\psi_{a1}^2.$$

This coincides with (3.15).

Similarly, (5.60) for $u_a \neq 0$ gives

$$(\partial_a, \partial_b) = \delta_{ab}\frac{\psi_{a1}^2}{u_a}.$$

This coincides with the definition of the intersection form written in the canonical coordinates. Finally, the last formula (5.61) gives

$$\langle \partial_a \cdot \partial_b, \partial_c \rangle = \delta_{ab}\delta_{ac}\psi_{a1}^2.$$

This is equivalent to (3.15) together with the definition of the canonical coordinates $\partial_a \cdot \partial_b = \delta_{ab}\partial_a$.

To prove (5.62), we use the fact that the integrals

$$\tilde{t}_j = -\sqrt{z}\int_{C_j} p(\lambda; u)e^{z\lambda}\,\mathrm{d}\lambda, \quad j = 1,\dots,n, \qquad (5.70)$$

give flat coordinates of the deformed connection. Let us first prove their independence. Indeed, the Jacobi matrix

$$Y_{aj}(z; u) = \frac{1}{\psi_{a1}}\frac{\partial \tilde{t}_j}{\partial u_a} = -\sqrt{z}\int_{C_j}\phi_a(\lambda; u)e^{z\lambda}\mathrm{d}\lambda$$

coincides with the fundamental matrix $Y^{\mathrm{left}}(z; u)$ (up to a factor $2\sqrt{\pi}$) due to Lemma 5.7 and Lemma 5.4. The final step of the derivation is an integration by parts and a change of the integration variable $\lambda \mapsto p$:

$$\tilde{t}_j = -\frac{1}{\sqrt{z}}\int_{C_j} p(\lambda; u)\,de^{\lambda z}$$

$$= \frac{1}{\sqrt{z}}\int_{C_j} e^{\lambda z}\frac{dp(\lambda; u)}{d\lambda}\,\mathrm{d}\lambda$$

$$= \frac{1}{\sqrt{z}}\int_{Z_j} e^{z\lambda(p; u)}\mathrm{d}p.$$

The theorem is proved.   □

Example 5.4. Let the reflections $R_1^*, \dots, R_n^*$ generate a finite group W acting in the Euclidean space $\mathcal{E}$ of dimension n. Recall [15] that finite groups generated by reflections (5.43) are called *Coxeter groups*. Let us assume the group W to be an irreducible one. We will construct, following [31], a Frobenius manifold M_W with the monodromy group W.

The underlying manifold of M_W will be the orbit space

$$M_W = \mathcal{E}/W. \qquad (5.71)$$

The coordinate ring of M_W is, by definition, the ring of W-invariant polynomials

$$\mathbb{C}[x_1,\dots,x_n]^W,$$

where $x_1,\dots,x_n$ are Euclidean coordinates on $\mathcal{E}$. Due to Chevalley's theorem [15], M_W has a natural structure of a graded affine algebraic variety:

$$\mathbb{C}[x_1,\dots,x_n]^W \simeq \mathbb{C}[y^1,\dots,y^n], \qquad (5.72)$$

where

$$y^i = y^i(x_1, \ldots, x_n), \quad i = 1, \ldots, n,$$

are some homogeneous W-invariant polynomials of degrees

$$d_i := \deg y^i(x) = m_i + 1, \tag{5.73}$$

and $m_1, \ldots, m_n$ are the exponents of the Coxeter group. The basic invariant polynomials determine a coordinate system on the orbit space M_W. They are uniquely determined up to invertible transformations of the form

$$y^i(x) \mapsto y^{i'}\big(y^1(x), \ldots, y^n(x)\big), \quad i = 1, \ldots, n,$$

with quasi-homogeneous polynomials $y^{i'}(y^1, \ldots, y^n)$ of the same degree d_i.

We will construct a polynomial Frobenius structure on M_W. This means that the structure functions $c^\gamma_{\alpha\beta}$ will be elements of the ring (5.72). Important ingredients of this construction will be the Arnold's construction of the convolution of invariants [2, 42] and also the flat coordinates on the orbit space M_W discovered by K. Saito et al. in [76, 78].

Let $y^1(x)$ be the invariant polynomial of the maximal degree $h = d_1$. The number h is called *Coxeter number* of the group W. We define the unit vector field

$$e := \frac{\partial}{\partial y^1} \tag{5.74}$$

and the Euler vector field

$$E := \frac{1}{h} \sum_a x_a \frac{\partial}{\partial x_a}. \tag{5.75}$$

The unit vector field is defined up to a constant factor.

The construction of the metric $\langle \ , \ \rangle$ and of the multiplication law of tangent vectors is more complicated. Let $(\ , \ )$ denote the W-invariant Euclidean metric on the space $\mathcal{E}$. We will use the orthonormal coordinates $x_1, \ldots, x_n$ with respect to this metric.

Let us define a bilinear symmetric form on T^*M_W. In the coordinates $y^1, \ldots, y^n$, it has the matrix

$$(dy^i, dy^j) = \sum_{a=1}^n \frac{\partial y^i}{\partial x_a} \frac{\partial y^j}{\partial x_a} = g^{ij}(y) \tag{5.76}$$

for some polynomials $g^{ij}(y)$, $y = (y^1, \ldots, y^n)$ (these exist due to Chevalley's theorem). The matrix $\big(g^{ij}(y)\big)$ is invertible on $M_W \setminus \Sigma$, where the discriminant Σ consists of all singular orbits.

Theorem 5.4. *There exists a unique, up to an equivalence, polynomial Frobenius structure on the space of orbits of a finite Coxeter group with the unity vector field (5.74), the Euler vector field (5.75), and the intersection form (5.76).*

Sketch of the proof. We put

$$\langle \, , \, \rangle := \mathcal{L}_e(\, , \,) \tag{5.77}$$

(cf. (5.8)). This gives the *Saito metric* on the orbit space M_W. According to [76, 78], this metric is flat, and there exists a distinguished system of basic homogeneous W-invariant polynomials $t^1(x), \ldots, t^n(x)$ such that

$$\eta^{\alpha\beta} := \langle dt^\alpha, dt^\beta \rangle$$

is a constant nondegenerate matrix. Our main observation is that the metrics (,) and $\langle \, , \, \rangle$ form a flat pencil (see the definition above). This allows us to reconstruct the Frobenius structure inverting the formula (5.2) (in the present case $A^{\alpha\beta} = 0$ in (5.2)). Namely, we define a W-invariant homogeneous polynomial F of degree $2h + 2$ from the equations

$$\eta^{\alpha\lambda}\eta^{\beta\mu}\frac{\partial^2 F}{\partial t^\lambda \partial t^\mu} = \frac{h\,(dt^\alpha, dt^\beta)}{\deg t^\alpha + \deg t^\beta - 2}, \quad \alpha, \beta = 1, \ldots, n \tag{5.78}$$

(cf. (5.2)). Such a polynomial exists, and it satisfies WDVV. The theorem is proved. □

Observe that for the Frobenius structure on M_W

$$d = 1 - \frac{2}{h}, \quad q_\alpha = 1 - \frac{\deg t^\alpha}{h}, \quad \alpha = 1, \ldots, n. \tag{5.79}$$

Exercise 5.5. Prove that the monodromy group of the Frobenius manifold M_W is isomorphic to W. (Hint: prove that the flat coordinates of the intersection form coincide with the Euclidean coordinates in the space W.)

Exercise 5.6. Prove that the Frobenius manifolds M_W satisfy the semisimplicity condition.

In particular, for $n = 2$, the polynomial Frobenius manifold corresponding to the group $I_2(k)$ of symmetries of the regular k-gon has the form

$$F = \frac{1}{2}t_1^2 t_2 + t^{k+1}.$$

For $n = 3$ there are three irreducible finite Coxeter groups $W(A_3)$, $W(B_3)$, and $W(H_3)$. They are the groups of symmetries of, respectively, the regular tetrahedron, octahedron, and icosahedron. The corresponding polynomial Frobenius manifolds have respectively the form (1.21), (1.22), and (1.23). We also give here the list of all our polynomial Frobenius manifolds in dimension 4.

Group $W(A_4)$:

$$F = \frac{1}{2}t_1^2 t_4 + t_1 t_2 t_3 + \frac{1}{2}t_2^3 + \frac{1}{3}t_3^4 + 6t_2 t_3^2 t_4 + 9t_2^2 t_4^2 + 24t_3^2 t_4^3 + \frac{216}{5}t_4^6.$$

Group $W(B_4)$:

$$F = \frac{1}{2}t_1^2 t_4 + t_1 t_2 t_3 + t_2^3 + \frac{t_2 t_3^3}{3} + 3t_2^2 t_3 t_4 + \frac{t_3^4 t_4}{4}$$
$$+ 3t_2 t_3^2 t_4^2 + 6t_2^2 t_4^3 + t_3^3 t_4^3 + \frac{18t_3^2 t_4^5}{5} + \frac{18t_4^9}{7}.$$

Group $W(D_4)$:

$$F = \frac{1}{2}t_1^2 t_4 + t_1 t_2 t_3 + t_2^3 t_4 + t_3^3 t_4 + 6t_2 t_3 t_4^3 + \frac{54}{35}t_4^7.$$

Group $W(F_4)$:

$$F = \frac{1}{2}t_1^2 t_4 + t_1 t_2 t_3 + \frac{t_2^3 t_4}{18} + \frac{3t_3^4 t_4}{4} + \frac{t_2 t_3^2 t_4^3}{2}$$
$$+ \frac{t_2^2 t_4^5}{60} + \frac{t_3^2 t_4^7}{28} + \frac{t_4^{13}}{2^4 \cdot 3^2 \cdot 11 \cdot 13}.$$

Group $W(H_4)$:

$$F = t_1 t_2 t_3 + \frac{t_1^2 t_4}{2} + \frac{2t_2^3 t_4}{3} + \frac{t_3^5 t_4}{240} + \frac{t_2 t_3^3 t_4^3}{18} + \frac{t_2^2 t_3 t_4^5}{15} + \frac{t_3^4 t_4^7}{2^3 \cdot 3^3 \cdot 5}$$
$$+ \frac{t_2 t_3^2 t_4^9}{2 \cdot 3^4 \cdot 5} + \frac{8t_2^2 t_4^{11}}{3^4 \cdot 5^2 \cdot 11} + \frac{t_3^3 t_4^{13}}{2^2 \cdot 3^6 \cdot 5^2} + \frac{2t_2^2 t_4^{19}}{3^8 \cdot 5^3 \cdot 19} + \frac{32t_4^{31}}{3^{13} \cdot 5^6 \cdot 29 \cdot 31}.$$

As was shown in [13], these are all the semisimple polynomial solutions of WDVV for $n = 4$ satisfying the conditions

$$0 < q_\alpha \le d < 1, \quad \alpha = 2, 3, 4.$$

Other examples of polynomial solutions of WDVV associated with finite Coxeter groups can be found in [90].

Remark. There are some inclusions between the polynomial Frobenius manifolds of the form

$$M_W := \mathbb{C}^n / W$$

(the orbit spaces) for a finite Coxeter group W acting in n-dimensional Euclidean space. These inclusions correspond to the operation of folding Dynkin graphs [4]. As shown in [88], if the Dynkin graph of a Coxeter group W' is obtained by folding the Dynkin graph of another Coxeter group W, then the corresponding orbit space $M_{W'}$ is a (graded) linear subspace in M_W with respect to the Saito linear structure. From our construction we immediately conclude that the inclusion

$$M_{W'} \subset M_W$$

is also an embedding of Frobenius manifolds. We obtain the following list of embeddings (they were obtained in [90] by a straightforward computation):

$$M_{B_n} \subset M_{A_{2n-1}},$$
$$M_{I_2(k)} \subset M_{A_{k-1}},$$
$$M_{F_4} \subset M_{E_6},$$
$$M_{H_3} \subset M_{D_6},$$
$$M_{H_4} \subset M_{E_8}.$$

(The group W_{G_2} coincides with $W_{I_2(6)}$, and therefore $M_{G_2} \subset M_{A_5}$.) The inclusions mean that, for example,

$$F_{E_8}(t_1, 0, t_3, 0, 0, t_6, 0, t_8) = F_{H_4}(t_1, t_3, t_6, t_8).$$

Conjecture. *Any irreducible semisimple polynomial Frobenius manifold is equivalent to M_W for some finite irreducible Coxeter group W.*

Remark. According to our construction, the Frobenius structure depends not only on the monodromy group W but also on the equivalence class of the ordered system of generating reflections $R_1^*, \ldots, R_N^*$. The equivalence is established by simultaneous conjugations of the generators by (,)-orthogonal transformations and by the following action of the braid group:

$$\begin{aligned}
\sigma_i(R_k^*) &= R_k^*, \quad k \neq i, i+1, \\
\sigma_i(R_i^*) &= R_{i+1}^*, \\
\sigma_i(R_{i+1}^*) &= R_{i+1}^* R_i^* R_{i+1}^*.
\end{aligned} \qquad (5.80)$$

Here, as above, the σ_i's are the standard generators of the braid group $\mathcal{B}_n$. Any such equivalence class is determined by the orbit of the Stokes matrix $S = (S_{ij})$:

$$S_{ii} = 1, \qquad S_{ij} = (e_i^*, e_j^*) \quad \text{for } i < j \qquad (5.81)$$

with respect to the $\mathcal{B}_n$-action (5.80). Here e_i^* is the basis of normals to the mirrors of the reflections normalized by the condition

$$(e_i^*, e_i^*) = 2$$

for any i. For example, for an algebraic Frobenius manifold, the orbit of the given Stokes matrix S must be finite. For the first nontrivial case $n = 3$ the classification of finite orbits of the action of $\mathcal{B}_3$ on the space of 3×3 Stokes matrices satisfying the nondegeneracy condition (5.49) was obtained in [34]. Namely, there are only five finite orbits. Three of them correspond to the standard system of generating reflections in the groups $W(A_3)$, $W(B_3)$, $W(H_3)$ of symmetries of the regular tetrahedron, octahedron, and icosahedron, respectively. Recall the construction of a standard system of generating reflections in the group of symmetries of a regular polyhedron. Let O be the center of the polyhedron, M the center of its face, A a vertex of the face, H the center of an edge of the face having A as an endpoint. Then the reflections with respect to the planes OMA, OMH, and OAH generate the group of symmetries of the polyhedron [20]. A reordering of these generators gives the same equivalence class.

One can repeat this construction with the cube (the reciprocal of the octahedron) or with the dodecahedron (the reciprocal of the icosahedron) just to obtain the same system of generators in $W(B_3)$ and in $W(H_3)$, respectively. We are now able to describe the remaining two finite orbits of the action of $\mathcal{B}_3$. The corresponding mirrors of the reflections are obtained by applying the above construction to the great icosahedron and great dodecahedron. The description of these regular Kepler–Poinsot star polyhedra can be found in Coxeter's book [20]. As above, their reciprocals give the same equivalence class. All these regular star polyhedra have icosahedral symmetry. Thus, we obtain three classes of triplets of generating reflections in the group $W(H_3)$.

The above classification was applied in [34] to the problem of the classification of algebraic solutions of $(\text{PVI}(\mu))$. One can show that the standard systems of generators in $W(A_3)$, $W(B_3)$, $W(H_3)$ correspond to the polynomial solutions (1.21)–(1.23) of WDVV (thus, to the algebraic solutions (5.24)–(5.26b) of $(\text{PVI}(\mu))$). The last two finite orbits give algebraic (nonpolynomial) Frobenius manifolds.

The problem of classification for any n of the finite orbits of the action of the braid group $\mathcal{B}_n$ in the space of $n \times n$ Stokes matrices remains open. The solution of this problem could be useful to prove the above conjecture. For $n = 4$ one can prove that all finite orbits of irreducible Stokes matrices satisfying the nondegeneracy condition (5.49) correspond to a system of generating reflections of a finite Coxeter group acting in $\mathbb{R}^4$. In the groups $W(A_4)$ and $W(B_4)$ there is only one equivalence class of systems of generating reflections, namely, the standard one. In the groups $W(D_4)$ and $W(F_4)$ there are two classes. Finally, in the group $W(H_4)$ there are 10 classes of systems of generating reflections. One of them corresponds to the standard system of generators in the group of symmetries of the regular 600-cell, 6 others to 4-dimensional regular star polyhedra considered modulo reciprocity (see the definitions in [20]), but the remaining 3 classes do not have a clear geometrical meaning. Out of the full list of 16 finite orbits, we give here Stokes matrices of representatives in the finite $\mathcal{B}_4$-orbits corresponding to only nonstandard systems of generators:

$$D_4: \quad S = \begin{pmatrix} 1 & -1 & 0 & 1 \\ 0 & 1 & -1 & -1 \\ 0 & 0 & 1 & 1 \\ 0 & 0 & 0 & 1 \end{pmatrix},$$

$$F_4: \quad S = \begin{pmatrix} 1 & -1 & 0 & \sqrt{2} \\ 0 & 1 & -\sqrt{2} & -\sqrt{2} \\ 0 & 0 & 1 & 1 \\ 0 & 0 & 0 & 1 \end{pmatrix},$$

$$H_4: \quad S = \begin{pmatrix} 1 & -1 & 0 & \frac{1+\sqrt{5}}{2} \\ 0 & 1 & -1 & -\frac{1+\sqrt{5}}{2} \\ 0 & 0 & 1 & \frac{-1+\sqrt{5}}{2} \\ 0 & 0 & 0 & 1 \end{pmatrix},$$

$$S = \begin{pmatrix} 1 & -1 & 0 & \frac{1+\sqrt{5}}{2} \\ 0 & 1 & -\frac{1+\sqrt{5}}{2} & -\frac{1+\sqrt{5}}{2} \\ 0 & 0 & 1 & 1 \\ 0 & 0 & 0 & 1 \end{pmatrix},$$

$$S = \begin{pmatrix} 1 & -1 & 0 & \frac{1-\sqrt{5}}{2} \\ 0 & 1 & -\frac{1-\sqrt{5}}{2} & -\frac{1-\sqrt{5}}{2} \\ 0 & 0 & 1 & 1 \\ 0 & 0 & 0 & 1 \end{pmatrix}.$$

The construction of Theorem 5.4 was generalized in [35] to produce quasi-polynomial solutions of WDVV, i.e., solutions with $d = 1$ of the form

$$F(t^1, \ldots, t^n) = \text{cubic} + f(t^2, \ldots, t^{n-1}, \exp t^n)$$

with a polynomial f. The monodromy group of these Frobenius manifolds are some extensions of affine Weyl groups. In particular, the monodromy group of the quantum cohomology of $\mathbf{CP}^1$ is given by this construction (see [32]).

Example 5.5. Let us compute the monodromy group of the quantum cohomology of $\mathbf{CP}^2$. The Stokes matrix S (4.97) of this Frobenius manifold satisfies the nondegeneracy condition (5.49). The basic reflections in the monodromy group in the basis of the flat coordinates (x, y, z) corresponding to the basis $\Phi^{\text{left}} = (\Phi_1^{\text{left}}, \Phi_2^{\text{left}}, \Phi_3^{\text{left}})$ of the solutions (4.96) have the matrices

$$R_1^* = \begin{pmatrix} -1 & -3 & 3 \\ 0 & 1 & 0 \\ 0 & 0 & 1 \end{pmatrix}, \quad R_2^* = \begin{pmatrix} 1 & 0 & 0 \\ -3 & -1 & 3 \\ 0 & 0 & 1 \end{pmatrix}, \quad R_3^* = \begin{pmatrix} 1 & 0 & 0 \\ 0 & 1 & 0 \\ 3 & 3 & -1 \end{pmatrix}.$$

The full monodromy group of the Frobenius manifold is obtained by adding the monodromy transformation corresponding to the only nontrivial loop

$$t_2 \mapsto t_2 + 2\pi i.$$

This corresponds to the rotation

$$z \mapsto z e^{2\pi i/3}.$$

Using the explicit formulae (4.96) and the identity (4.92), we immediately obtain the required transformation

$$\Phi^{\text{left}}(z e^{2\pi i/3}) = \Phi^{\text{left}}(z)T,$$

or, equivalently,

$$(x, y, z) \mapsto (x, y, z)T \tag{5.82}$$

with

$$T = \begin{pmatrix} 0 & -1 & 0 \\ 0 & 0 & 1 \\ -1 & -3 & 3 \end{pmatrix}. \tag{5.83}$$

For the matrix

$$T_0 := TR_1^* = \begin{pmatrix} 0 & -1 & 0 \\ 0 & 0 & 1 \\ 1 & 0 & 0 \end{pmatrix} \tag{5.84}$$

we have the identities

$$T_0^3 = -1, \quad R_2^* = T_0^{-1} R_1^* T_0, \quad R_3^* = T_0^{-1} R_2^* T_0. \tag{5.85}$$

So, we introduce the new system of generators A, B, C in the full monodromy group by putting

$$A = R_1^*, \quad B - T_0^4 = -T_0, \quad C = T_0^3 = -1. \tag{5.86}$$

All the transformations of the group preserve the integer lattice in $\mathbb{R}^3$. They also preserve the indefinite quadratic form with the Gram matrix $S + S^T$:

$$q(x, y, z) = 2(x^2 + y^2 + z^2 + 3xy - 3xz - 3yz). \tag{5.87}$$

The group acts discretely on the complexification of the cone $q(x, y, z) > 0$. Introducing the coordinates r, τ, $\bar{\tau}$,

$$
\begin{aligned}
x &= \frac{ir}{2} \frac{2\tau\bar{\tau} - 3(\tau + \bar{\tau}) + 2}{\tau - \bar{\tau}}, \\
y &= \frac{ir}{2} \frac{2\tau\bar{\tau} + \tau + \bar{\tau} - 2}{\tau - \bar{\tau}}, \\
z &= \frac{ir}{2} \frac{2\tau\bar{\tau} - (\tau + \bar{\tau}) - 2}{\tau - \bar{\tau}},
\end{aligned}
\tag{5.88}
$$

we obtain the action of the generating transformations

$$A: \qquad \left(r \mapsto r, \tau \mapsto -\frac{1}{\tau}, \bar{\tau} \mapsto -\frac{1}{\bar{\tau}}\right),$$

$$B: \quad \left(r \mapsto r, \tau \mapsto \frac{1}{1-\tau}, \bar{\tau} \mapsto \frac{1}{1-\bar{\tau}}\right), \qquad (5.89)$$

$$C: \qquad\qquad \left(r \mapsto -r, \tau \mapsto \tau, \bar{\tau} \mapsto \bar{\tau}\right).$$

We have proved the following theorem

Theorem 5.5. *The monodromy group of the quantum cohomology of* $\mathbf{CP}^2$ *is isomorphic to* $\mathrm{PSL}_2(\mathbb{Z}) \times \{\pm\}$.

It would be interesting to develop an appropriate theory of invariants for this action of the modular group. This could help to obtain analytic formulae for the quantum cohomology of $\mathbf{CP}^2$.

Acknowledgments: I would like to thank the organizers of the Cargèse summer school for this invitation and generous support. I thank A.B. Givental for fruitful discussions on Theorem 3.2.

6   References

[1] S.-I. Amari, *Differential-Geometric Methods in Statistics,* Lecture Notes in Statistics **28** (Springer, Berlin, 1985).

[2] V.I. Arnol'd, *Wave front evolution and equivariant Morse lemma,* Comm. Pure Appl. Math. **29** (1976), 557–582.

[3] ———, *Singularities of Caustics and Wave Fronts* (Kluwer, Dordrecht, 1990).

[4] V.I. Arnol'd, S.M. Gusein-Zade, and A.N. Varchenko, *Singularities of Differentiable Maps,* volumes I, II (Birkhäuser, Boston, 1988).

[5] M.F. Atiyah, *Topological quantum field theories,* Publ. Math. I.H.E.S. **68** (1988), 175.

[6] M. Audin, *An introduction to Frobenius manifolds, moduli spaces of stable maps and quantum cohomology,* Preprint (1997).

[7] W. Balser, W.B. Jurkat, and D.A. Lutz, *Birkhoff invariants and Stokes multipliers for meromorphic linear differential equations,* J. Math. Anal. Appl. **71** (1979), 48–94.

[8] _____ , *On the reduction of connection problems for differential equations with an irregular singular point to ones with only regular singularities*, SIAM J. Math. Anal. **12** (1981), 691–721.

[9] S. Barannikov, M. Kontsevich, *Frobenius manifolds and formality of Lie algebras of polynomial vector fields*, alg-geom/9710032.

[10] K. Behrend, *Gromov–Witten invariants in algebraic geometry*, Inv. Math. **127** (1997), 601–627.

[11] A. Beauville, *Quantum cohomology of complete intersections*, alg-geom/9501008.

[12] J. Birman, *Braids, Links, and Mapping Class Groups* (Princeton Univ. Press, Princeton, 1974).

[13] A. Blanco Cedeño, *On polynomial solutions of equations of associativity*, Preprint ICTP IC/97/152 (1997).

[14] B. Blok and A. Varchenko, *Topological conformal field theories and the flat coordinates*, Int. J. Mod. Phys. **A7** (1992), 1467.

[15] N. Bourbaki, *Groupes et algèbres de Lie*, Chapitres 4, 5 ct 6 (Masson, Paris, 1981).

[16] E.A. Coddington and N. Levinson, Theory of Ordinary Differential Equations (McGraw-Hill, New York, 1955).

[17] P. Candelas, X.C. de la Ossa, P.S. Green, and L. Parkes, *A pair of Calabi–Yau manifolds as an exactly soluble superconformal theory*, Nucl. Phys. **B359** (1991), 21–74.

[18] S. Cecotti and C. Vafa, *Topological-antitopological fusion*, Nucl. Phys. **B367** (1991), 359–461.

[19] _____ , *On classification of N = 2 supersymmetric theories*, Comm. Math. Phys. **158** (1993), 569–644.

[20] H.S.M. Coxeter, *Regular Polytopes* (Macmillan, New York, 1963).

[21] P. Di Francesco and C. Itzykson, *Quantum intersection rings*, The Moduli Space of Curves, eds. R. Dijkgraaf, C. Faber, and G. van de Geer (Birkhäuser, Basel, 1995), pages 81–148.

[22] R. Dijkgraaf, *Intersection theory, integrable hierarchies and topological field theory*, New symmetry principles in quantum field theory, ed. C. Itzykson, NATO ASI series (Plenum, New York, 1991), pages 95–158. hep-th/9201003.

[23] _____, *Notes on topological string theory and 2-D quantum gravity*, String theory and quantum gravity, ed. A.W. Smith (World Scientific, River Edge NJ, 1990), pages 91–156.

[24] R. Dijkgraaf, E. Verlinde, and H. Verlinde, *Topological strings in d < 1*, Nucl. Phys. **B 352** (1991), 59–86.

[25] R. Dijkgraaf and E. Witten, *Mean field theory, topological field theory, and multimatrix models*, Nucl. Phys. **B 342** (1990), 486–522.

[26] B. Dubrovin, *Differential geometry of moduli spaces and its application to soliton equations and to topological field theory*, Preprint No. 117, Scuola Normale Superiore, Pisa (1991).

[27] _____, *Hamiltonian formalism of Whitham-type hierarchies and topological Landau–Ginsburg models*, Comm. Math. Phys. **145** (1992), 195–207.

[28] _____, *Integrable systems in topological field theory*, Nucl. Phys. **B 379** (1992), 627–689.

[29] _____, *Geometry and integrability of topological–antitopological fusion*, Comm. Math. Phys. **152** (1993), 539–564.

[30] _____, *Integrable systems and classification of 2-dimensional topological field theories*, Integrable Systems, eds. O. Babelon, O. Cartier, and Y. Kosmann-Schwarbach (Birkhäuser, Basel, 1993), pages 313–359.

[31] _____, *Differential geometry of the space of orbits of a Coxeter group*, Preprint SISSA-29/93/FM (February 1993).

[32] _____, *Geometry of 2-D topological field theories*, Integrable Systems and Quantum Groups, eds. M. Francaviglia and S. Greco, Lecture Notes in Mathematics **1620** (Springer, Berlin, 1996), pages 120–348.

[33] _____, *Flat pencils of metrics and Frobenius manifolds*, Integrable Systems and Algebraic Geometry, eds. M.-H. Saito, Y. Shimizu, and K. Ueno, (World Scientific, Singapore, 1998), pages 47–72. math.DG/9803106.

[34] B. Dubrovin and M. Mazzocco, *Monodromy of certain Painlevé-VI transcendents and reflection groups*, Preprint SISSA 149/97/FM (1997).

[35] B. Dubrovin and Zhang Y., *Extended affine Weyl groups and Frobenius manifolds*, Compositio Math. **111** (1998), 167–219.

[36] _____, *Bihamiltonian hierarchies in 2-D topological field theory at one-loop approximation*, Comm. Math. Phys. **198** (1998), 311–361. hep-th/9712232.

[37] A. Duval and C. Mitschi, *Matrices de Stokes et groupe de Galois des équations hypergéométriques confluentes généralisées*, Pacific J. Math. **138** (1989), 25–56.

[38] T. Eguchi, H. Kanno, Y. Yamada, and Yang S.-K, *Topological strings, flat coordinates and gravitational descendants*, Phys. Lett. **B305** (1993), 235–241.

[39] T. Eguchi, Y. Yamada, and Yang S.-K., *Topological field theories and the period integrals*, Mod. Phys. Lett. A **8** (1993), 1627–1638.

[40] _____, *On the genus expansion in the topological string theory*, Rev. Math. Phys. **7** (1995), 279.

[41] F. R. Gantmacher, *The Theory of Matrices* (Chelsea, New York, 1960).

[42] A.B. Givental, *Convolution of invariants of groups generated by reflections, and connections with simple singularities of functions*, Funct. Anal. **14** (1980), 81–89.

[43] _____, *Equivariant Gromov–Witten invariants*, Internat. Math. Res. Notices **13** (1996), 613–663. alg-geom/9603021 (1996).

[44] _____, *Stationary phase integrals, quantum Toda lattice, flag manifolds, and the mirror conjecture*, alg-geom/9612001 (1996).

[45] _____, *A mirror theorem for toric complete intersections*, Topological field theory, primitive forms and related topics, Progr. Math. **160** (Birkhäuser, Boston, 1998), pages 141–175.

[46] _____, *Elliptic Gromov–Witten invariants and the generalized mirror conjecture*, math.AG/9803053 (1998).

[47] M. Gromov, *Pseudo-holomorphic curves in symplectic manifolds*, Invent. Math. **82** (1985), 307–347.

[48] N. Hitchin, *Poncelet polygons and the Painlevé equations*, Geometry and Analysis, ed. Ramanan (Tata Inst. of Fundamental Research, Bombay, 1995), pages 151–185.

[49] _____, *Frobenius manifolds*, With notes by D. Calderbank. NATO Adv. Sci. Inst. Ser. C Math. Phys. Sci. **12**, Gauge Theory and Symplectic Geometry (Montreal, 1995), pages 69–112

[50] K. Hori, *Constraints for topological strings in $D \geq 1$*, Nucl. Phys. **B 439** (1995), 395–420.

[51] A.R. Its and V.Yu. Novokshenov, *The Isomonodromic Deformation Method in the Theory of Painlevé Equations*, Lecture Notes in Mathematics **1191** (Springer, Berlin, 1986).

[52] M. Jimbo and T. Miwa, *Monodromy preserving deformations of linear ordinary differential equations with rational coefficients.* II. Physica D **2** (1981), 407–448.

[53] R. Kaufmann, The geometry of moduli spaces of pointed curves, the tensor product in the theory of Frobenius manifolds and the explicit Künneth formula in quantum cohomology (Ph.D. thesis, Max Planck Institut in Bonn, 1997).

[54] M. Kontsevich, *Intersection theory on the moduli space of curves,* Funktsional'nyi Analiz i Ego Prilozheniya **25** (1991), 50–57; Funct. Anal. Appl. **25** (1991), 123–129.

[55] ———, *Intersection theory on the moduli space of curves and the matrix Airy function,* Comm. Math. Phys. **147** (1992), 1–23.

[56] ———, *Enumeration of rational curves via torus action,* Comm. Math. Phys. **164** (1994), 525–562.

[57] M. Kontsevich and Yu.I. Manin, *Gromov–Witten classes, quantum cohomology and enumerative geometry,* Comm. Math. Phys. **164** (1994), 525–562.

[58] ———, *Quantum cohomology of a product (with Appendix by R. Kaufmann),* Inv. Math. **124** (1996), 313–339.

[59] I.M. Krichever, *The τ-function of the universal Whitham hierarchy, matrix models and topological field theories,* Comm. Pure Appl. Math. **47** (1994), 437–475.

[60] W. Lerche and N. Warner, *Exceptional Seiberg–Witten Geometry from ALE Fibrations,* hep-th/9608183 (1996).

[61] B.-H. Lian, K. Liu, and S.-T. Yau, *Mirror principle I,* alg-geom/9712011 (1997).

[62] E. Looijenga, *A period mapping for certain semi-universal deformations,* Compositio Math. **30** (1975), 299–316.

[63] A. Losev, *"Hodge strings" and elements of K. Saito's theory of the primitive forms,* hep-th/9801179 (1998).

[64] Y. L. Luke, Mathematical functions and their approximations (Academic Press, New York, 1975).

[65] B. Malgrange, *Équations différentielles à coefficients polynomiaux* (Birkhäuser, Basel, 1991).

[66] D. McDuff and D. Salamon, *J-holomorphic curves and quantum cohomology* (American Mathematical Society, Providence RI, 1994).

[67] Yu.I. Manin, *Frobenius manifolds, quantum cohomology, and moduli spaces*, Preprint MPI 96–113 (1996).

[68] _____, *Sixth Painlevé equation, universal elliptic curve, and mirror of P^2*, Amer. Math. Soc. Transl. (2) **186** (1998), 131–151. alg-geom/9605010,

[69] _____, *Three constructions of Frobenius manifolds: a comparative study*, math.QA/9801006.

[70] Yu.I. Manin and S.A. Merkulov, *Semisimple Frobenius (super)manifolds and quantum cohomology of P^r*, alg-geom/9702014 (1997).

[71] T. Miwa, *Painlevé property of monodromy presereving equations and the analyticity of τ-functions*, Publ. RIMS **17** (1981), 703–721.

[72] S. Piunikhin, *Quantum and Floer cohomology have the same ring structure*, Preprint MIT (March 1994).

[73] Y. Ruan and G. Tian, *A mathematical theory of quantum cohomology*, Math. Res. Lett. **1** (1994), 269–278.

[74] C. Sabbah, *Frobenius manifolds: isomonodromic deformations and infinitesimal period mappings*, Exposition. Math. **16** (1998), 1–57.

[75] V. Sadov, *On equivalence of Floer's and quantum cohomology*, Preprint HUTP-93/A027 (1993).

[76] K. Saito, *On a linear structure of a quotient variety by a finite reflection group*, Publ. RIMS, Kyoto Univ., **29** (1993), 535–579.

[77] _____, *Period mapping associated to a primitive form*, Publ. RIMS **19** (1983), 1231–1264.

[78] K. Saito, T. Yano, and J. Sekeguchi, *On a certain generator system of the ring of invariants of a finite reflection group*, Comm. in Algebra **8(4)** (1980), 373–408.

[79] I. Satake, *Flat structure for the simple elliptic singularity of type $\tilde{E}_6$ and Jacobi forms*, Proc. Japan Acad. Ser. A Math. Sci. **69** (1997), no. 7. hep-th/9307009.

[80] J. Segert, *Frobenius manifolds from Yang–Mills instantons*, Math. Res. Lett. **5** (1998), 327–344. dg-ga/9710031.

[81] A. Takahashi, *Primitive forms, topological LG models coupled to gravity and mirror symmetry*, math.AG/9802059 (1998).

[82] Tian Gang, Xu Geng, *On the semisimplicity of the quantum cohomology algebra of complete intersections*, alg-geom/9611035 (1996).

[83] W. Wasow, Asymptotic expansions for ordinary differential equations (Wiley, New York, 1965).

[84] E.T. Whittaker and G.N. Watson, *A Course of Modern Analysis* (American Mathematical Society Press, New York, 1979).

[85] E. Witten, *On the structure of the topological phase of two-dimensional gravity*, Nucl. Phys. **B 340** (1990), 281–332.

[86] ———, *Two-dimensional gravity and intersection theory on moduli space*, Surv. Diff. Geom. **1** (1991), 243–210.

[87] ———, *Lectures on mirror symmetry*, Essays on Mirror Manifolds, ed.S.-T Yau (International Press Co., Hong Kong, 1992).

[88] T. Yano, *Free deformation for isolated singularity*, Sci. Rep. Saitama Univ. **A 9** (1980), 61–70.

[89] S.-T. Yau, ed., *Essays on Mirror Manifolds* (International Press Co., Hong Kong, 1992).

[90] J.-B. Zuber, *On Dubrovin's topological field theories*, Mod. Phys. Lett. **A 9** (1994), 749–760.

[91] ———, *Graphs and reflection groups*, Comm. Math. Phys. **179** (1996), 265–294.

[92] ———, *Generalized Dynkin diagrams and root systems and their folding*, Topological field theory, primitive forms and related topics, eds. C.M. Kashiwara, A. Matsuo, K. Saito, and I. Satake (Birkhäuser, Basel, 1998), pages 453–493. hep-th/9707046.

7

Discrete Painlevé Equations

Basile Grammaticos, Frank W. Nijhoff, and Alfred Ramani

Introduction

The Cargèse summer school celebrated the 100th anniversary of the Painlevé property, the property that was introduced by Painlevé and subsequently by Gambier and their school to classify ordinary differential equations (ODEs) according to the singularity behavior of their solutions, [1–4]. In the other contributions to this volume the various implications of the Painlevé property as well as of the particular *differential* equations in which this property is manifested are explained in detail.

In these lectures we will concentrate on *difference* equations rather than differential equations.

When in school we are introduced to the notions of differentiation we learn first about differences, derivatives being obtained a posteriori by a limit that is—in fact—a reductive procedure: We lose information. So, somehow, in spite of the century-long domination of mathematical analysis by notions of differentiation and integration, the latter concepts are born out of what is actually a "poor man's approach." Wouldn't it be more beautiful if everything could be done on the basis of exact and finite manipulations, i.e., working with finite differences rather than with differential calculus?

Of course, one objection would be that the continuum is the state of matter as it appears to us in our observation of nature, in particular in physics. Thus, in order to capture the smoothness of all movement and the coherence of all material we need the smooth operations of calculus. This is undoubtedly true, but when we delve deeper and investigate Nature on a deeper level where the fundamental laws of physics are no longer governed by classical mechanics and instead quantum mechanics is starting to dominate, this natural smoothness disappears and we find ourselves faced again with a discrete state of affairs (either appearing in the guise of discrete quantum numbers, noncommutative operations, or otherwise). This insight was already spoken about by the people who stood at the birth of quantum mechanics. Let us cite:

> To be sure, it has been pointed out that the introduction of a
> space–time continuum may be considered as contrary to nature

in view of the molecular structure of everything which happens
on a small scale. It is maintained that perhaps the success of the
Heisenberg method points to a purely algebraic description of
nature, that is to the elimination of continuum functions from
physics. Then, however, we must also give up, by principle,
the space–time continuum. It is not unimaginable that human
ingenuity will some day find methods which will make it possible
to proceed along such a path. At the present time, however, such
a program looks like an attempt to breathe in empty space.

A. Einstein, *Physics and Reality*,[1] 1936.

Maybe the time has finally come to embark on the program alluded
to in these prophetic words, and maybe we should no longer let us be
dominated by the *horror vacui*. In fact, in more recent years, voices have
been raised in favor of a fundamentally discrete description of nature, and
serious attempts have been made to develop tools to this end, cf., e.g., [5–9].
Although the resulting models are still in their infancy, there is more to it
than just toys and games. So, what is holding us back?

When it comes to mathematical modeling in terms of discrete systems
described by difference equations there seems to be a far greater ambiguity
in writing down the basic equations of a model. Whereas with continuous models described in terms of differential equations a great deal of
know-how exists, for instance on how certain terms may represent certain
phenomena, that know-how is greatly lacking in dealing with difference
equations. Furthermore, there is the problem of the complexity of the theory. In comparison with the theory of differential equations, for which we
have at our disposal a large arsenal of methods, approaches, and theories, the theory of difference equations is highly underdeveloped. This was
recognized by Birkhoff already at the beginning of the twentieth century,
who embarked on a systematic program of developing the analytic theory
of difference equations. Unfortunately, since the Second World War these
early works [10, 11] have been generally forgotten. And this still involves in
principle only the linear theory. When we come to the theory of nonlinear
difference equations, no coherent framework exists to date.

To improve the situation we need a guiding principle that will help us
effectively to deal with these two drawbacks: On the one hand it will guide
us around the ambiguity in mathematical modeling, while on the other
hand it will lead to models for which effectively a great deal of theory can
be developed. Fortunately, there is a very good and effective criterion that
one can maintain as such a guiding principle: *integrability!*

Why integrability? First of all, integrable systems are the ones that,
albeit nonlinear and highly nontrivial, can be dealt with by systematic and
rigorous approaches. It can be argued that they are too special to represent

---

[1]Printed in *Essays in Physics*, Philosophical Library, New York, 1950.

the wealth of physical phenomena, but wait: Isn't that precisely what we need that integrable systems are so very special that we lose all ambiguity in searching for the governing equations for our models?

This chapter deals with discrete Painlevé equations [12]. Their connection with the Painlevé property has been discussed at length in the other chapters of this volume. What the present chapter is about is to find an answer to the following questions: (1) *What are the proper discrete analogues of the Painlevé transcendents* P_I–P_{VI}, and (2) *What is the defining property generalizing to the discrete case the Painlevé property?*

To comment on the latter question let us make the following remarks: The Painlevé property requires that the singularities of the solutions of differential equations involving multivaluedness (so excluding poles) do not depend on the particularity of the solution, i.e., on the initial data. The phenomenon of continuous systems that we can get "close" to a singularity is absent in discrete systems in view of their nonlocal nature, which implies that the usual methods to investigate singularities fail. Even the very concept of a singularity needs to be revised in dealing with discrete systems. So, here lies the difficulty in formulating a proper discrete analogue of the Painlevé property that does not somehow refer back to an underlying continuous structure. All these questions have opened up new avenues of investigation.

The nature of innovative research is never a straight-line trajectory that one can readily follow; it is rather a twisted curve with many bends and self-intersections. The fundamental research in discrete systems is still in its infancy, and many of the concepts have to be developed by trial and error. This is very much the case in the area of discrete Painlevé equations. It would be misleading to suggest that we are ready to present a full-blown theory of discrete Painlevé equations, starting from unambiguous definitions and axioms. Unlike the situation of the continuous theory, in the discrete domain we have to work by developing examples and using combinations of different methods and approaches, working often by analogy. At present, we are gradually reaching a stage where we seem to be gaining a beginning of an understanding, having obtained a rather representative family of difference equations that can be used as a starting point for developing the concepts we need in the discrete domain. In this chapter we hope to persuade the reader that many of the desirable properties of the Painlevé equations hold for the examples at our disposal, and that by the end of the chapter the reader will be convinced that this is indeed an extremely rich area of research in which many new results await us in the future.

The organization of this chapter is as follows. In Section 1 we will discuss the issue of integrability of discrete systems in general, and focus in particular on integrable mappings and correspondences, i.e., finite-dimensional integrable discrete systems. This allows us to exhibit where the subtle points are by which discrete systems differ from continuous ones. In Section 2 we will focus on the relation between integrable systems of KdV type, i.e., integrable *partial difference equations* and their similarity reductions leading to examples of discrete Painlevé equations. In Section 3 another method of obtaining discrete Painlevé equations (dPs) is presented: the singularity confinement approach that plays the role of the discrete Painlevé property. The bilinear formalism for discrete equations will also be developed. The singularity structure of the dPs is used to bilinearize them. In Section 4 we shall present the special properties of the dPs that will convince us that these equations are indeed the proper analogues of the continuous Painlevé transcendents. Finally, in Section 5 we will develop the isomonodromic deformation formulation for the various dPs that may form the starting point for the systematic asymptotic analysis of the discrete transcendental functions that solve the dPs. The latter, however, opens up a large body of work for the future, to which, we hope, this chapter may form the inspiration.

1   Integrable Discrete Systems

In this first section we will familiarize the reader with integrability in the discrete domain. We will mainly discuss finite-dimensional mappings and correspondences, i.e., systems that can more or less be regarded as the counterparts of a finite-dimensional (possibly Hamiltonian) system with a discrete rather than continuous time-flow. So instead of defining the flows, e.g., by

$$\dot{\mathbf{y}} \equiv \frac{d\mathbf{y}}{dt} = \mathbf{f}(\mathbf{y}; t) \tag{1.1}$$

(for some (vector-valued) variable $\mathbf{y} \in \mathbb{R}^n$), we can, for example, think of a (system of) difference equation(s) of the type

$$\mathbf{y}(t + \delta) = \mathbf{f}(\mathbf{y}; t). \tag{1.2}$$

In the theory of dynamical systems such difference equations often appear in the guise of an iterated map of the form

$$\mathbf{y} = \mathbf{y}_0 \mapsto \mathbf{y}_1 = \varphi(\mathbf{y}) \mapsto \mathbf{y}_2 = \varphi(\varphi(\mathbf{y})) \mapsto \cdots. \tag{1.3}$$

One circumstance in which such maps arise is from the Poincaré section of the orbit of a continuous system, see Figure 7.1. We will be interested in discrete systems, for instance of the form of (1.2) but not necessarily so, that have nice regularity properties that we would like to call *integrable*. What we precisely mean by integrability in the discrete arena is a bit tricky to explain right now, but we will come back to that later, giving rather precise definitions.

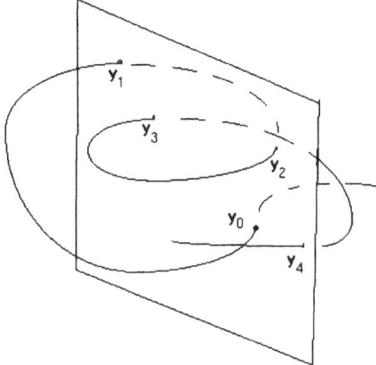

FIGURE 7.1. Dynamical mapping arising from the Poincaré section of an orbit.

In fact, integrable discrete systems, in whatever form, contrast as a matter of course with the nonintegrable ones, which we expect to exhibit complex and random behavior such as chaos and strange attractors. In the context of chaos theory, dynamical maps have been extensively studied over the last few decades; see many of the standard textbooks such as the one by Devaney [13]. A typical discrete system that has the symptomatic behavior of a chaotic system is the logistic map

$$x_{n+1} = f(x_n), \quad f(x) = ax(1-x). \tag{1.4}$$

Such systems exhibit the Feigenbaum period-doubling bifurcation scheme and all the characteristics of a system with chaotic behavior and consequently, such systems are *not* the ones in which we will be interested in this chapter.

In contrast to the enormous amount of research that has been undertaken into systems of the type (1.4), the investigation of nontrivial *integrable* discrete systems has been lagging greatly behind. In a sense this is a strange situation, since one would expect that the integrable dynamical systems would, by their supposed regularity, be much easier to study. Somehow, the investigations of integrable and nonintegrable discrete systems should be interlinked, because we could think of a scenario where integrable discrete systems would serve as a starting point for perturbation approaches. So far, very little work has been done in these directions.

In order to get a handle on the discrete Painlevé equations, it is necessary first to get a feeling of what integrable discrete systems are in general. Therefore, in this first section we will be mainly dealing with *autonomous* discrete systems, i.e., those for which in (1.2) $\mathbf{f}(\mathbf{y};t) = \mathbf{f}(\mathbf{y})$, $\mathbf{f}$ does not depend explicitly on the independent (time) variable. The discrete Painlevé equations, our main topic, are typically *not* of this type and are more complicated. However, in order to get some understanding of the latter equations we have to gain some familiarity with autonomous discrete systems first.

1.1   What Is an Integrable Discrete System?

The question of constructing integrable maps goes back to to the beginning of the twentieth century to the work of Julia, Ritt, and Fatou. (For an excellent review of the developments of integrable maps we would like to refer to the review paper by Veselov in [14, 15], from which some of the material of this section is taken.)

The question asked by Julia and Ritt in the 1920s was the following: *Is it possible to find polynomial mappings f, g : $\mathbf{C} \to \mathbf{C}$ such that they commute, i.e.,*

$$f \circ g = g \circ f,$$

and such that the orbits of f and g do not intersect?

Why is this commutativity property related to some kind of integrability? In terms of dynamical systems we can view the map f as defining some iterative time-flow, while viewing g as a symmetry of this dynamical system, i.e.,

$$x_{n+1} = f(x_n), \quad x'_n = g(x_n).$$

Of course, for the symmetry to be nontrivial we need somehow that the symmetry generator g act transitively on the orbit given by the flow f. This is a reasonable definition of a (complex) one-dimensional map with a nontrivial symmetry, and hence we can use this to define some form of integrability for the discrete system.

The question is thus to find all possible nontrivial pairs of polynomial maps (we will soon defer the requirement of polynomiality) (f, g), up to change of variables, of course, which means up to an invertible polynomial change of coordinates $f \mapsto \varphi^{-1} \circ f \circ \varphi$.

It turns out that the answer to the above question is rather simple, and was given already by Julia, Ritt, and Fatou [16–18]: *Up to affine transformations, the only polynomial maps f, apart from the linear ones, that allow for a commutative map whose orbit does not intersect with that of g are either the pure power $f(z) = z^k$ or the Chebyshev polynomials $f(z) = \pm T_k(z)$ for some natural number k.*

Exercise 1.1. Recalling the definition of the Chebyshev polynomials,

$$T_k(z) = \cos(k \arccos z),$$

show that a commutative map is given by one of the Chebyshev polynomials itself.

Exercise 1.2. Show that there do not exist any nontrivial commutative maps for the quadratic map $f(z) = z^2 + c$, when $c \neq 0$, -2.

So, this exhausts all (polynomial) discrete integrable discrete systems in one (complex) dimension! What about higher dimensions?

In principle, we can ask exactly the same question as above for any dimension. In that case we have in mind polynomial mappings[2] $f, g \colon \mathbf{C}^n \to \mathbf{C}^n$. Thus, our f and g are given by $f = (f_1(z), \ldots, f_n(z))$, where $f_i \in \mathbf{C}[z_1, \ldots, z_n]$. The degree of the map can be defined to be the number of preimages of a generic point. Again, we are interested in finding nontrivial commutative maps $g \circ f = f \circ g$ up to sensible changes of variables. Thus, we have to factor out the group called the *Cremona group*, denoted by $\mathrm{GA}_n(\mathbf{C})$, which consists of all invertible polynomial changes of coordinates of the form $f \mapsto \varphi^{-1} \circ f \circ \varphi$. This problem has been investigated by Veselov in [19], and he has shown that to each simple complex Lie algebra one can associate an infinite series of integrable polynomial mappings, generalizing the Chebyshev polynomial case.

Let us focus on the situation in dimension 2. Mappings of the form we are interested in are now automorphisms of the (complex) plane:

$$(x, y) \mapsto (\tilde{x}, \tilde{y}) = \Phi(x, y) = (f(x, y), g(x, y)). \tag{1.5}$$

Again, we shall speak of commutative mappings as being *symmetries* of the dynamical system generated by (1.5). So, as in the one-dimensional case; a map $\Psi = (\varphi, \psi)$ is called a symmetry of the map $\Phi = (f, g)$ if it commutes with Φ such that Φ and Ψ do not have common iterations.

The *affine Cremona group* $\mathrm{GA}_2(\mathbf{C})$ is formed by the polynomial mappings of $\mathbf{C}^2$ to itself whose inverses are also polynomial. Within this setting we have a number of results, obtained much later than the early work by Julia, Ritt, and Fatou. Using results by Jung [20] and Wright [21], who gave a description of all abelian subgroups of $\mathrm{GA}_2(\mathbf{C})$, Veselov obtained the following result:

Proposition 1.1. *A dynamical system of the form* (1.5) *possesses a nontrivial symmetry iff it is possible to transform* Φ *to affine or triangular form via a change of coordinates, or (equivalently) if the degree of the polynomials defining the iterations of* Φ *is bounded.*

In the special case of symplectic (area-preserving maps), i.e., maps for which

$$J = \frac{\partial(\tilde{x}, \tilde{y})}{\partial(x, y)} = 1, \tag{1.6}$$

a corollary of this result can be formulated as follows:

Proposition 1.2. *A symplectic map* Φ *in* $\mathrm{GA}_2(\mathbf{C})$ *that possesses a nontrivial symmetry has always a nonconstant polynomial integral* $I(x, y)$, *i.e.,*

$$I(\tilde{x}, \tilde{y}) = I(x, y). \tag{1.7}$$

---

[2]In the case of rational mappings we consider $f, g \colon \mathbf{C}P^n \to \mathbf{C}P^n$.

The latter statement is, of course, what we somehow expect: integrability as the existence of a first integral of the map. The systematic search for discrete maps of the plane possessing a first integral is curiously enough something that has been studied only in recent years. One first example was found by McMillan in 1971 [22], and can be found below. This, however, is not a polynomial map but a rational map. Integrability somehow seems to defy polynomial maps, and early work led to more or less negative results. To give a few examples:

1. Invertible *quadratic* maps defined by second-degree polynomials. By an appropriate change of variables they can always be cast into the form

    $$\Phi(x, y) = (\alpha_0 + \alpha_1 x + \alpha_2 y + \alpha x^2, \beta_0 + \beta_1 x + \beta_2 y + \beta x^2),$$

    such that $\alpha_1 \beta_2 - \alpha_2 \beta_1 \neq 0$, $\alpha \beta_2 - \alpha_2 \beta = 0$. The following can be shown: The dynamical system given by Φ possesses a nontrivial symmetry iff $\alpha = \alpha_2 = 0$ or $\alpha = \beta = 0$, so only if the map is linear. A special subcase is the Hénon map [23] with $\Phi(x, y) = (1 + y - ax^2, bx)$ $(b \neq 0)$, which possesses a nontrivial symmetry iff $a = 0$.

2. Maps of standard type, i.e., mappings that can be given as the second-order difference equation

    $$\tilde{x} - 2x + \underset{\sim}{x} = f(x), \tag{1.8}$$

    the undertilde denoting the preimage of the map. Of course, (1.8) can be cast into the first-order form

    $$\tilde{x} = y,$$
    $$\tilde{y} = 2y - x + f(y).$$

    This map exhibits necessarily exponential growth of the degree of the polynomial f, unless the polynomial f is linear, $f(x) = ax + b$, which implies from a result of Veselov that it cannot possess a nontrivial symmetry for any polynomial f with degree > 1. So (nonlinear) polynomial maps of standard type are never integrable!

These examples demonstrate already how difficult it is actually to find nontrivial examples of integrable discrete systems. We should necessarily go beyond polynomial maps to find examples, and it is only in recent years that systematically a search for such systems has been undertaken.

1.2 Integrable Maps of the Plane

The first integrable map of the plane to itself was probably found by McMillan in [22]. He searched for area-preserving rational maps carrying an invariant. The family of mappings he found were basically the class of maps given under (a) in the list below.

A more systematic approach was adopted in a short paper [24] by Suris, who derived a series of integrable mappings of standard type, i.e., of the form

$$x_{n+1} - 2x_n + x_{n-1} = f(x_n). \tag{1.9}$$

From the requirement that these maps should carry an invariant, i.e., an analytic function of two variables $\Phi(x, y)$ such that under the map (1.9) we have

$$\Phi(x_{n+1}, x_n) = \Phi(x_n, x_{n-1}), \tag{1.10}$$

he obtained the following list:
(a) Rational Maps:

$$f(x) = \frac{A + Bx + Cx^2 + Dx^3}{1 - \varepsilon(E + \frac{1}{3}Cx + \frac{1}{2}Dx^2)};$$

$$\Phi(x, y) = \frac{1}{2}(x - y)^2 + \varepsilon\left[-\frac{1}{2}A(x + y) - \frac{1}{2}Bxy - \frac{1}{6}Cxy(x + y)\right.$$
$$\left. - \frac{1}{4}Dx^2y^2 - \frac{1}{2}E(x - y)^2\right].$$

(b) Trigonometric Maps:

$$f(x) = \frac{2}{\omega\varepsilon} \arctan \alpha, \tag{1.11}$$

$$\alpha = \left[\frac{(\omega\varepsilon/2)(A \sin \omega x + B \cos \omega x + C \sin 2\omega x + D \cos 2\omega x)}{1 - (\omega\varepsilon/2)(A \cos \omega x - B \sin \omega x + C \cos 2\omega x - D \sin 2\omega x + E)}\right]; \tag{1.12}$$

$$\Phi(x, y) = \frac{1}{\omega^2}(1 - \cos \omega(x - y))$$
$$+ \frac{\varepsilon}{2\omega}[A(\cos \omega x + \cos \omega y) - B(\sin \omega x + \sin \omega y)C \cos \omega(x + y)$$
$$- D \sin \omega(x + y) + E \cos \omega(x - y)].$$

(c) Hyperbolic (Exponential) Maps:

$$f(x) = \frac{1}{\alpha\varepsilon} \ln \left[\frac{1 + \alpha\varepsilon(Be^{-\alpha x} + De^{-2\alpha x} - E)}{1 - \alpha\varepsilon(Ae^{\alpha x} + Ce^{2\alpha x} + E)}\right];$$

$$\Phi(x, y) = \frac{1}{\alpha^2} \cosh(x - y)$$
$$+ + \frac{\varepsilon}{2\alpha}[-A(e^{\alpha x} + e^{\alpha y}) + B(e^{-\alpha x} + e^{-\alpha y})$$
$$- Ce^{\alpha(x+y)} + De^{-\alpha(x+y)} - 2E \cosh \alpha(x - y)].$$

In the above A, B, C, D, E as well as ω and α are constants, ε is a formal parameter, and the way in which these maps were obtained is by searching for a (biholomorphic in some neighborhood $|x - y| < \delta$) invariant of the form

$$\Phi(x, y) = \Phi_0(x, y) + \varepsilon \Phi_1(x, y)$$

that truncates at first order in ε, while the map is given in terms of the formal expansion

$$f(x) = \sum_{k=1}^{\infty} \varepsilon^k f_k(x).$$

Suris has proven that under natural analyticity conditions on $f(x)$ as well as by imposing the symmetry condition $\Phi(x, y) = \Phi(y, x)$, the above classification is exhaustive if one imposes truncation at first order in ε on the invariant.

Exercise 1.3. Introducing the variable

$$u_n \equiv x_n + \frac{1}{2} f(x_n) - x_{n-1} = x_{n+1} - x_n - \frac{1}{2} f(x_n)$$

derive the functional equation

$$\Phi\left(x, x + \frac{1}{2} f(x) - u\right) = \Phi\left(x, x + \frac{1}{2} f(x) + u\right).$$

Introducing also $\varphi(x - y) = \Phi_0(x, y)$ and expanding in orders of ε, ε^2, ε^3, show by elimination of the coefficients f_1, f_2, f_3 that φ obeys the equation

$$\varphi'''(u) = c\varphi'(u),$$

and deduce the various possibilities (a)–(c).

The largest family of integrable mappings of the plane known so far was constructed by Quispel, Roberts, and Thompson in [25], and we will refer to this family as the QRT-family of mappings. The construction is very beautiful and proceeds as follows.

The QRT-family is given by the map

$$x \mapsto \tilde{x} = \frac{f_1(y) - x f_2(y)}{f_2(y) - x f_3(y)}, \qquad y \mapsto \tilde{y} = \frac{g_1(\tilde{x}) - y g_2(\tilde{x})}{g_2(\tilde{x}) - y g_3(\tilde{x})}, \qquad (1.13)$$

in which $f_1, f_2, f_3, g_1, g_2, g_3$ are fourth-order polynomials given by the following rule:

$$\begin{pmatrix} f_1(x) \\ f_2(x) \\ f_3(x) \end{pmatrix} = \begin{pmatrix} \alpha_0 x^2 + \beta_0 x + \gamma_0 \\ \delta_0 x^2 + \varepsilon_0 x + \zeta_0 \\ \kappa_0 x^2 + \lambda_0 x + \mu_0 \end{pmatrix} \times \begin{pmatrix} \alpha_1 x^2 + \beta_1 x + \gamma_1 \\ \delta_1 x^2 + \varepsilon_1 x + \zeta_1 \\ \kappa_1 x^2 + \lambda_1 x + \mu_1 \end{pmatrix},$$

$$\begin{pmatrix} g_1(x) \\ g_2(x) \\ g_3(x) \end{pmatrix} = \begin{pmatrix} \alpha_0 x^2 + \delta_0 x + \kappa_0 \\ \beta_0 x^2 + \varepsilon_0 x + \lambda_0 \\ \gamma_0 x^2 + \zeta_0 x + \mu_0 \end{pmatrix} \times \begin{pmatrix} \alpha_1 x^2 + \delta_1 x + \kappa_1 \\ \beta_1 x^2 + \varepsilon_1 x + \lambda_1 \\ \gamma_1 x^2 + \zeta_1 x + \mu_1 \end{pmatrix}.$$

The mapping (1.13) constitutes an 18-parameter family of mappings of the plane $\mathbb{R}^2 \to \mathbb{R}^2$. The mapping is the most general one that allows for a corresponding K-family of invariant curves filling the plane, which is the general biquadratic curve

$$
(\alpha_0 + K\alpha_1)x^2y^2 + (\beta_0 + K\beta_1)x^2y
$$
$$
+ (\gamma_0 + K\gamma_1)x^2 + (\delta_0 + K\delta_1)xy^2(\varepsilon_0 + K\varepsilon_1)xy + (\zeta_0 + K\zeta_1)x
$$
$$
+ (\kappa_0 + K\kappa_1)y^2 + (\lambda_0 + K\lambda_1)y + (\mu_0 + K\mu_1) = 0. \quad (1.14)
$$

Let us give a brief proof of the statement. It is easily noted that the mapping (1.13) can be written in the form

$$
\begin{pmatrix} x^2 \\ x \\ 1 \end{pmatrix} \cdot \left[\begin{pmatrix} f_1 \\ f_2 \\ f_3 \end{pmatrix} \times \begin{pmatrix} \tilde{x}^2 \\ \tilde{x} \\ 1 \end{pmatrix} \right] = 0,
$$

where $f_i = f_i(y)$, $i = 1, 2, 3$. Similarly, for $g_i = g_i(\tilde{x})$, $i = 1,\ 2,\ 3$, we have

$$
\begin{pmatrix} y^2 \\ y \\ 1 \end{pmatrix} \cdot \left[\begin{pmatrix} g_1 \\ g_2 \\ g_3 \end{pmatrix} \times \begin{pmatrix} \tilde{y}^2 \\ \tilde{y} \\ 1 \end{pmatrix} \right] = 0.
$$

The special choice for f and g amounts to

$$
\mathbf{f} = (\mathbf{A}_0 \cdot \mathbf{Y}) \times (\mathbf{A}_1 \cdot \mathbf{Y}), \quad \mathbf{g} = (\mathbf{A}_0^T \cdot \tilde{\mathbf{X}}) \times (\mathbf{A}_1^T \cdot \tilde{\mathbf{X}}),
$$

where $\mathbf{f}$ and $\mathbf{g}$ are the vectors with components f_i, g_i, respectively, and the vectors $\mathbf{X}$ are given by $(x^2, x, 1)^T$, and similarly for $\mathbf{Y}$, the superscript T denoting matrix transposition, and where the matrices $\mathbf{A}_i$, $i = 0$, 1, are given by

$$
\mathbf{A}_i = \begin{pmatrix} \alpha_i & \beta_i & \gamma_i \\ \delta_i & \varepsilon_i & \zeta_i \\ \kappa_i & \lambda_i & \mu_i \end{pmatrix}.
$$

The argument is now very simple: Noting on the one hand that

$$
\mathbf{X} \cdot \left[(\mathbf{A}_0 \cdot \mathbf{Y}) \times (\mathbf{A}_1 \cdot \mathbf{Y})) \times \tilde{\mathbf{X}} \right]
$$
$$
= (\mathbf{X}^T \cdot \mathbf{A}_1 \cdot \mathbf{Y}) \left(\tilde{\mathbf{X}}^T \cdot \mathbf{A}_0 \cdot \mathbf{Y} \right) - (\mathbf{X}^T \cdot \mathbf{A}_0 \cdot \mathbf{Y}) \left(\tilde{\mathbf{X}}^T \cdot \mathbf{A}_1 \cdot \mathbf{Y} \right),
$$

and on the other hand that

$$
\mathbf{Y} \cdot \left[(\mathbf{A}_0^T \cdot \tilde{\mathbf{X}}) \times (\mathbf{A}_1^T \cdot \tilde{\mathbf{X}})) \times \tilde{\mathbf{Y}} \right]
$$
$$
= \left(\tilde{\mathbf{X}}^T \cdot \mathbf{A}_0 \cdot \tilde{\mathbf{Y}} \right) \left(\tilde{\mathbf{X}}^T \cdot \mathbf{A}_1 \cdot \mathbf{Y} \right) - \left(\tilde{\mathbf{X}}^T \cdot \mathbf{A}_0 \cdot \mathbf{Y} \right) \left(\tilde{\mathbf{X}}^T \cdot \mathbf{A}_1 \cdot \tilde{\mathbf{Y}} \right),
$$

both being equal to zero, one gets a determinant relation of the form

$$\left(\mathbf{X}^T \cdot \mathbf{A}_1 \cdot \mathbf{Y}\right)\left(\widetilde{\mathbf{X}}^T \cdot \mathbf{A}_0 \cdot \widetilde{\mathbf{Y}}\right) = \left(\mathbf{X}^T \cdot \mathbf{A}_0 \cdot \mathbf{Y}\right)\left(\widetilde{\mathbf{X}}^T \cdot \mathbf{A}_1 \cdot \widetilde{\mathbf{Y}}\right),$$

which can be integrated to obtain the invariant

$$\frac{\mathbf{X}^T \cdot \mathbf{A}_0 \cdot \mathbf{Y}}{\mathbf{X}^T \cdot \mathbf{A}_1 \cdot \mathbf{Y}} = K. \qquad (1.15)$$

It is easy to see that the latter relation precisely yields the biquadratic (1.14).

Exercise 1.4. Show that the mapping (1.13) also carries the invariant

$$\frac{\widetilde{\mathbf{X}}^T \cdot \mathbf{A}_0 \cdot \mathbf{Y}}{\widetilde{\mathbf{X}}^T \cdot \mathbf{A}_1 \cdot \mathbf{Y}} = K. \qquad (1.16)$$

As is mentioned in the appendix of the review paper [26] (cf. also [27]) the QRT-mappings are not, in general, area-preserving (i.e., symplectic) but measure-preserving, meaning that the Jacobian of the map is not exactly one but there exists a function $m(x, y)$ such that

$$J = \frac{\partial(\tilde{x}, \tilde{y})}{\partial(x, y)} = \frac{m(x, y)}{m(\tilde{x}, \tilde{y})}.$$

A crucial step in the proof of [27] is provided by the following exercise.

Exercise 1.5. Show that the map (1.13) can be written as a composition $\iota_2 \circ \iota_1$ of two involutions

$$\iota_1 : \begin{cases} x \longmapsto \tilde{x} = \dfrac{f_1(y) - x f_2(y)}{f_2(y) - x f_3(y)}, \\ y \longmapsto \tilde{y} = y, \end{cases}$$

and

$$\iota_2 : \begin{cases} x \longmapsto \tilde{x} = x, \\ y \longmapsto \tilde{y} = \dfrac{g_1(x) - y g_2(x)}{g_2(x) - y g_3(x)}, \end{cases}$$

where $\iota_1^2 = \iota_2^2 = \mathrm{id}$.

This property indicates that the family (1.13) of mappings of the plane belongs, in fact, to the class of *reversible mappings*. This class, which does not entirely overlap the class of integrable mappings, has been studied in great generality in the literature; see [26] for a review. Integrability of a mapping in this context was defined in [28] requiring that the phase plane of a reversible mapping is foliated by curves invariant under the mapping and its reversing symmetries (involutions).

Exercise 1.6. Show that the QRT-map (1.13) is measure-preserving with $m(x, y)$ given by

$$m(x, y) = \left(\mathbf{X}^T \cdot \mathbf{A}_1 \cdot \mathbf{Y}\right)^{-1}$$

in the notation above.

The measure-preservation of the QRT-map (1.13) means that according to a result of Moser [29] there exists a local change of variables such that the map is carried over to an area-preserving one. In terms of the geometry of the curve (1.14) this means that we have an elliptic surface rather than an elliptic curve on which the map is parametrized.[3] The integrability of the QRT-map is not in contradiction with the fact that 2-2 correspondences associated with the full biquadratic equation

$$\Phi(x, y) = \sum_{0 \le i, j \le 2} a_{ij} x^i y^j = 0 \tag{1.17}$$

have in general exponential growth, which in itself would be a sign of nonintegrable behavior. As was shown in [30], there are distinct subcases of such biquadratic mappings for which slow (polynomial) growth occurs, namely cases in which images under the map are "glued," the most important ones being the following:

Symmetric Case: $\Phi(x, y) = \Phi(y, x)$, leading to the symmetric biquadratic

$$\Phi(x, y) = a_{22} x^2 y^2 + a_{12} xy(x + y) + a_{02}(x^2 + y^2)$$
$$+ a_{11} xy + a_{01}(x + y) + a_{00} = 0, \quad (1.18)$$

in which case the QRT-map (1.13) is symmetric as well, i.e., $f = g$.

Factorizable Case: The biquadratic factorizes, e.g.,

$$\Phi(x, y) = (\alpha_1 xy + \beta_1 x + \gamma_1 y + \delta_1)(\alpha_2 xy + \beta_2 x + \gamma_2 y + \delta_2),$$

leading to a "double" discrete Riccati map

$$x \mapsto y = -\frac{\beta_i x + \delta_i}{\alpha_i x + \gamma_i}, \quad i = 1, 2.$$

Exercise 1.7. Show that in the symmetric case, i.e., $f_i = g_i$ in (1.13), the mapping reduces to the three-point map

$$\tilde{x} = \frac{f_1(x) - \underline{x}\, f_2(x)}{f_2(x) - \underline{x}\, f_3(x)}, \tag{1.19}$$

namely by introducing an odd/even staggering of the time updates of the original mapping.

---

[3] A.P. Veselov, private communication.

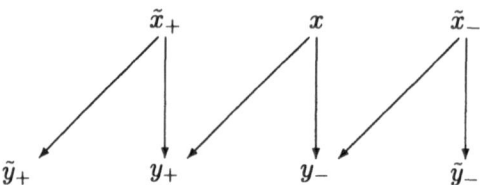

We can explain the integrability of the general QRT-map, even in the case that the 2-2 correspondence defined by the invariant is not among the cases in which gluing of images reduces the growth, because of the special form (1.13) of the map. In fact, if we insist on viewing the QRT-map as a correspondence, we must realize that we need to take into account both (1.15) as well as (1.16), i.e., for the QRT-map we have

$$\Phi(x, y) = \Phi(\tilde{x}, y) = \Phi(\tilde{x}, \tilde{y}) = 0.$$

Thus, if we describe the dynamics of the QRT map *viewed as a correspondence* coming from the biquadratics, we easily see that given at any iterate of the map a value x and solving the biquadratic to yield two images y_+ and y_-, the second condition above implies that among the two preimages of each image y_+ or y_- there must be both x as well as another preimage $\tilde{x}_+$, respectively $\tilde{x}_-$, that form the iterates of the map, i.e., one for each image. Thus, as indicated in the diagram below, there is no room for growth. Obviously, the argument holds only for biquadratics and is no longer valid if the invariant is of higher order in either x or y. Recently, the singularity confinement (which will be explained in Section 3), was established for the full 18-parameter QRT-map, providing another proof of integrability.

In Section 3 and 4 we will often refer to the QRT-family as a starting point for the construction of nonautonomous mappings, which will turn out to be discrete analogues of Painlevé equations. It has to be pointed out, nonetheless, that the transition from autonomous to nonautonomous is a very fundamental one by which the character of solutions is essentially changed.

1.3   Discrete Liouville-Integrable Systems

The notion of a *Lagrangian correspondence* goes back to the papers by Logan [31] and Maeda [32]. A more modern exposition can be found in [14, 15] with applications to integrability, cf. also [33–36].

Suppose that we have a phase space given by some smooth variety $\mathcal{M}$, and a function $L : \mathcal{M} \times \mathcal{M} \to \mathbb{R}R$, which we will consider to be a *discrete Lagrangian*. Similar to the continuous case, the corresponding discrete dynamical system is generated by stationarity of the action functional

$$S(x) = \sum_{n \in \mathbf{Z}} L(x_n, x_{n+1}), \tag{1.20}$$

i.e., the equation $\delta S(x) = 0$, the infinitesimal variation of the action, is equivalent to the discrete Euler–Lagrange equations

$$\left(\frac{\partial L(x,y)}{\partial x}\right)_{\substack{x=x_n \\ y=x_{n+1}}} + \left(\frac{\partial L(x,y)}{\partial y}\right)_{\substack{x=x_{n-1} \\ y=x_n}} = 0. \qquad (1.21)$$

The corresponding map $\Phi : (x_{n-1}, x_n) \mapsto (x_n, x_{n+1})$ is in general multivalued and maybe even not everywhere defined. It is given by its graph

$$\Gamma_\Phi \subset \mathcal{M} \times \mathcal{M}.$$

If $\mathcal{M}$ is a symplectic manifold, with symplectic form ω, we can naturally define a symplectic form on its Cartesian product in the standard way:

$$\Omega = \pi_1^* \omega - \pi_2^* \omega,$$

in which $\pi_i : \mathcal{M} \times \mathcal{M} \to \mathcal{M}(i = 1, 2)$ denote the natural projections. We can introduce the following definition [15]:

Definition 1.1. A relation $\Gamma_\Phi \subset \mathcal{M} \times \mathcal{M}$ is called a symplectic correspondence if the relation $\Phi : \mathcal{M} \to \mathcal{M}$ preserves the natural symplectic form, in other words, if the restriction of $(\pi_1^* \ominus \pi_2^*)\omega$ to Γ_Φ is zero.

If we have a Lagrangian map on a symplectic manifold $\mathcal{M}$, then we have automatically a symplectic correspondence as follows. Suppose L is nondegenerate, i.e., the (Hessian) matrix of second derivatives of L is regular. Then we can introduce the form

$$\omega = \sum_{i,j} \frac{\partial^2 L}{\partial x_i \partial y_j} dx_i \wedge dy_j. \qquad (1.22)$$

Definition 1.2. A symplectic relation $\Phi : \mathcal{M} \to \mathcal{M}$ is said to be *integrable* if it has exactly $N = \frac{1}{2} \dim \mathcal{M}$ independent integrals $I_1, \ldots, I_N$ in involution.

Noting that here involutivity is the usual notion implying the commutativity of the Hamiltonian vector field associated with the invariants, this corresponds exactly to the definition of integrability in the usual Liouville sense. Later on, in dealing with discrete Painlevé equations, our notion of integrability has to change, since the Liouville sense no longer applies to those equations.

In the continuum case we know the implications of Liouville integrability: There is a transformation to action-angle variables in terms of which the flow is linear and the orbits wind in a rational way over a set of tori (in the compact case). This is the well-known Arnold'd–Liouville–Darboux theorem. It turns out, as proven by Veselov (see, e.g., [15]) that in the discrete case we have exactly the same situation. Thus, we have the following result.

Theorem 1.1 (Discrete Theorem à la Liouville). *If a symplectic relation $\Phi : \mathcal{M} \to \mathcal{M}$ has N independent integrals in involution in accordance with Definition 1.2, then any compact nonsingular level $\mathcal{M}_c \equiv \{x \in \mathcal{M} \mid I_k(x) = c_k, i = 1, \ldots, N\}$ is a disconnected union of tori on which the dynamics take place according to regular shifts.*

As we have seen in the previous section, in a multivalued map (correspondence) the growth of the number of copies is a measure of the complexity of the underlying dynamical system. This idea, going back to Arnold'd [37], has been exploited in a body of recent work connecting these copies to the actions of certain Coxeter groups on matrices and associated birational transformations; cf., e.g., [38, 39]. For us it suffices to bring across the intuitive idea that slow growth is a sign of regularity, possibly of integrability. In its simplest form we can think of dynamical systems where the iterations are obtained at each step by solving an algebraic system and where generically one expects exponential growth of the number of iterates. This can be quantified by defining for a map $\Phi : \mathcal{M} \to \mathcal{M}$

$$N_x^{\Phi}(k) \equiv \#\Phi^k(x),$$

in which Φ^k denotes the kth iterate of the map; one would for a 2-2 correspondence expect

$$N_x^{\Phi}(k) = 2^k,$$

which can be pictorially represented by the *tree* diagram of Figure 7.2. However, iterates may overlap, and if there is "enough" symmetry in the system (as one expects in the integrable cases) one may have a situation in which gluing of images occurs as depicted in Figure 7.3. This is actually what happens, and Veselov has proven in [19] that the Liouville integrability implies (at most) polynomial growth.

Theorem 1.2 (Veselov [19]). *The number of different images of a point under the kth iteration of an integrable (in the sense of Definition 1.1) symplectic correspondence grows not faster than some polynomial in k, i.e., if p is the number of components of the level $\mathcal{M}_c$ containing a point x and m is the maximal number of images at each iteration step, then there exists a constant C such that*

$$N_x^{\Phi}(k) < C k^{mp-1}.$$

Thus, Liouville-integrable mappings have at most polynomial growth, and by reversing the argument one may expect that slow growth is an indication of the integrability of a discrete system. Discrete Painlevé equations are not integrable in the Liouville sense, but nonetheless the growth phenomena and branching aspects play a role there as well. One may argue that these matters are closely linked to the notion of single-valuedness of a

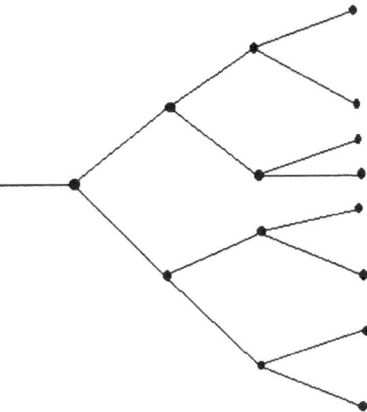

FIGURE 7.2. Exponential growth in a 2-2 correspondence.

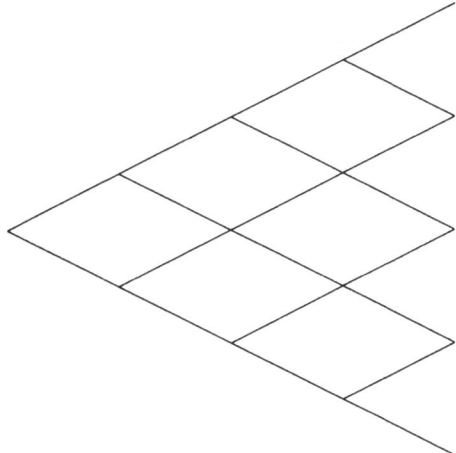

FIGURE 7.3. Gluing of images for an integrable multivalued map.

solution to a discrete dynamical system. These issues and their implications for the singularity analysis of discrete systems will play an important role in the investigation of a discrete version of the Painlevé property, and they will be investigated in more detail in Section 3.

1.4   Multidimensional Integrable Maps

A systematic construction of multidimensional integrable mappings, generalizing the McMillan family, was undertaken in [40]. The construction starts from the investigation of initial value problems for a partial difference version of the Korteweg–de Vries (KdV) equation. We will mention extensively such *integrable lattice equations* in the next section. Without going into the details, we mention that the construction of [40], which proceeds by imposing *periodic* initial conditions on the lattice, leads to a finite-dimensional reduction of the lattice KdV given by an integrable rational map $mathbbR^{2P} \to mathbbR^{2P}$, ($P$ corresponding to the period of the initial data on the lattice), of the form

$$\tilde{x}_j = y_j,$$

$$\tilde{y}_j = x_{j+1} - \frac{\varepsilon\delta}{1+y_{j+1}} + \frac{\varepsilon\delta}{1+y_j}, \quad j = 1,\ldots,P, \tag{1.23}$$

under the condition

$$\sum_{j=1}^{P} x_j = C_1, \sum_{j=0}^{P-1} y_j = C_2.$$

Unlike what obtains in the two-dimensional case, volume preservation does not imply symplecticity, and the proof of Liouville integrability in these cases is more involved and relies on tools like Lax pairs and r-matrices. The proof of integrability of this family of multidimensional maps was given in [41]; cf. also [42]. In the simplest case (period $P = 2$ initial data) the map reduces to a special case in the McMillan family, namely

$$x_{n+1} + x_{n-1} = \frac{2ax_n}{1 - x_n^2}. \tag{1.24}$$

We will single out this particular equation, since a nonautonomous version of it will turn out to be exactly one of the first found cases of a discrete Painlevé transcendent.

However, the autonomous case is much simpler, and its invariant curve

$$I(x,y) = (1 - x^2)(1 - y^2) + 2axy = \text{const} \tag{1.25}$$

can actually be parametrized in terms of Jacobi elliptic functions. Recall the addition formula for the elliptic sine function:

$$\text{sn}(u + v) = \frac{\text{sn}\,u\,\text{cn}\,v\,\text{dn}\,v + \text{sn}\,v\,\text{cn}\,u\,\text{dn}\,u}{1 - k^2\,\text{sn}^2\,u\,\text{sn}^2\,v}.$$

Exercise 1.8. Use the addition formula for the elliptic sine function to show that the map (1.24) has a solution of the form $x_n = A\,\text{sn}(\alpha n + \theta_0; k)$ and give expressions for the parameter α, the amplitude A, and the modulus k in terms of the parameter a and the value of the invariant I.

A *Lax pair* for this map was first found in the papers [40, 41], and it can be written in the following form. Consider the eigenvalue (spectral) problem

$$L_n(h)\varphi_n(h) = \lambda\varphi_n(h), \tag{1.26a}$$

with

$$L_n(h) = \begin{pmatrix} a & 1 + x_{n+1} & 1 & 0 \\ 0 & 0 & 1 - x_n & 1 \\ h & 0 & a & 1 - x_{n+1} \\ h(1 + x_n) & h & 0 & 0 \end{pmatrix}. \tag{1.26b}$$

The parameter h plays the role of a Floquet parameter similar to that of Lax pairs for the periodic Toda chain; cf., e.g., [43]. Imposing on the eigenvector $\varphi_n(h)$ the discrete evolution in terms of the variable n of the form

$$\varphi_{n+1}(h) = M_n(h)\varphi_n(h), \tag{1.27a}$$

with

$$M_n(h) = \begin{pmatrix} \dfrac{a}{1 + x_{n+1}} & 1 & 0 & 0 \\ 0 & 0 & 1 & 0 \\ 0 & 0 & \dfrac{a}{1 - x_{n+1}} & 1 \\ h & 0 & 0 & 0 \end{pmatrix}, \tag{1.27b}$$

we have an overdetermined system leading to the discrete Lax equation

$$L_{n+1}(h)M_n(h) = M_n(h)L_n(h), \tag{1.28}$$

which basically tells us that the evolution in discrete time n is obtained from a similarity transformation on the matrix $L_n(h)$ at each step of the iteration.

Exercise 1.9. Show that the McMillan map (1.24) arises from (1.28) with the matrices $L_n(h)$ and $M_n(h)$ given by (1.26b) and (1.27b).

It is easy to see that the invariant (1.25) arises from the characteristic equation

$$P(h, \lambda) = \det(\lambda - l_n(h)) = 0, \tag{1.29}$$

which describes the invariant elliptic curve under the McMillan map.

More generally, the symmetric QRT-mapping (1.19) can be solved in terms of elliptic functions parametrizing the symmetric biquadratic curve (cf., e.g., the final section of [44]). However, a Lax pair is to our knowledge not known for the general symmetric case.

The generalized McMillan map, i.e., the extension to higher-dimensional maps rather than maps form the plane to itself, was found in [40] as the periodic reduction of a lattice KdV system. A 4-dimensional map is given by the following coupled system of second-order equations:

$$x_{n+1} - y_{n-1} = \frac{a}{1 + x_n} - \frac{a}{1 + y_n}, \tag{1.30a}$$

$$y_{n+1} + x_{n-1} + y_{n-1} = \frac{a}{1 + y_n} - \frac{a}{1 - x_n - y_n}. \tag{1.30b}$$

The parametrization is now more complicated and requires the implementation of the technique of finite-gap integration; cf., e.g., [45]. In this case we need functions carrying particular addition formulae defined on a hyperelliptic curve of genus $g = 2$ that is obtained from a Lax representation of the form (1.28) but now involving 6×6 matrices.

Exercise 1.10. Show that the 4-dimensional mapping (1.30) can be obtained from a discrete Lax pair of the form (1.26) with $L_n(h)$ and $M_n(h)$ given by 6×6 matrices forming the natural extensions of (1.26b) and (1.27b). Use the Lax pair to calculate two independent invariants from the spectral curve (1.29).

The McMillan map can thus be generalized to any dimension by carrying further the generalization from the two-dimensional map to the four-dimensional to any dimension in a natural way. Another generalization consists of a hierarchy of integrable maps of multiplicative type coming from periodic reductions of the lattice-modified KdV equation; cf. [40]. The lowest member in this hierarchy takes the form

$$x_{n+1}x_{n-1} = \kappa \frac{(1 + ax_n)(x_n + b)}{(1 + cx_n)(x_n + d)}, \tag{1.31}$$

sometimes referred to as the Hirota equation. (It is also closely related to a discrete version of the sine–Gordon equation.) Generalizations of this multiplicative map were given in [40, 41]. The connection between autonomous maps and their nonautonomous extensions that form the discrete versions of the Painlevé equations will be exploited at great length in the following sections.

2   Similarity Reduction and Direct Linearization

In the previous section we have explored to a certain degree the notion of integrability in the discrete domain. So far, the examples were limited to *autonomous* systems in which case we have a clear-cut definition of an integrable dynamical map, either in the sense of the existence of nontrivial commutative maps or in the sense of Liouville (existence of invariants in involution). When we come to nonautonomous systems, the situation becomes less clear. Ideally, we would like to have a defining property like the Painlevé property by which one could attempt to classify discrete systems. However, difference systems being essentially nonlocal, the local properties of singularities in the complex plane do not naturally extend to lattices on which the discrete systems live. We will, therefore, have to postpone the question of what could be a discrete version of the Painlevé property to the later sections, developing first a number of canonical examples of systems that we would like to consider to be the natural discrete analogues of the Painlevé equations.

In order to arrive at such examples we are inspired by the well-known connection between soliton systems and Painlevé equations. We know that similarity reductions of some of the famous soliton equations (modified KdV, Boussinesq, derivative NLS, Heisenberg ferromagnet, etc.) lead to ordinary differential equations (ODEs) that are exactly the Painlevé transcendents or near relatives of them. If somehow we can discretize the soliton equations (which we can), thereby retaining the integrability aspects, and if next we can devise a procedure for implementing the similarity reduction on the discrete equation, we will have a construction that *should* lead to discrete analogues of the corresponding Painlevé transcendents. So, our approach consists of two steps:

- Find proper discrete analogues of soliton equations.

- Develop on the discrete lattice a way of obtaining their similarity solutions.

Both steps are highly nontrivial, since brute-force discretization of a differential equation, for example by doing simple replacements of the form

$$u_x(x,t) \mapsto \Delta_x u(x,t) = u(x + \delta, t) - u(x,t),$$
$$u_t(x,t) \mapsto \Delta_t u(x,t) = u(x, t + \varepsilon) - u(x,t),$$

or something similar, generally would destroy the integrability of the equation. Experience shows that such simple recipes almost always fail and will therefore not lead to integrable difference equations. However, genuinely integrable difference equations have been constructed over the years (cf., e.g., [46–57]; cf. also [58, 59] for reviews), and the discretizations involved are much more sophisticated than the simple replacements described above. How to find such discrete counterparts of soliton equations in a systematic way is a story in itself, and we will give some ingredients in this section. In principle the case of ordinary difference equations, being the analogues of ordinary differential equations, the situation is somewhat simpler in that we have only one independent variable to play around with. Even in that case it is highly nontrivial to find exact discretizations, and only recently has a more or less systematic approach been developed (see the lectures of R. Conte and of M. Musette in this volume; cf. also [60]). However, one should not discard the possibility of the existence of integrable discrete systems that do not have *any* continuous counterpart as well, examples of which we will encounter in Section 3. In this sense again the discrete integrable systems are more fundamental than the continuous ones.

The next step: Once we have a good discrete soliton equation, i.e., a partial difference equation, how to obtain a reduction to an ordinary difference equation is even less clear. Simple scaling arguments, like the ones that apply to the continuous case and that we summarize in the next section, do not apply to these discrete soliton equations. More dramatically, to date there does not yet exist a satisfactory theory of (continuous or discrete) symmetries for partial difference equations of the type that we would like to consider, even though there is a growing body of work directed toward the aim of setting up such a theory; cf., e.g., [32, 61–66]. In spite of these difficulties, in [67] we were able to deploy a method, sometimes referred to as the *direct linearization approach*, to tackle the problem of obtaining the similarity reduction of soliton systems on the lattice. This method, which is particularly well suited to exploit the rich algebraic structure of soliton systems, goes back to an early paper by Fokas and Ablowitz [68], and was further developed in [69–71]. It is quite powerful in providing the interrelations between the various relevant equations and as such can be effectively used to obtain both the relevant lattice equations and the basic equations governing the similarity reduction. In this section we will focus on this approach to obtain discrete Painlevé equations, while in the later sections we will introduce alternative methods including a discrete analogue of the Painlevé property.

2.1   The Method of Similarity Reduction

One of reasons why the Painlevé equations have reemerged during the seventies is the connection with soliton equations. As has been mentioned in many of the earlier chapters, it is well known that Painlevé transcendents have emerged from similarity reduction of various soliton equations [72, 73]. There is a rather large literature on this subject; cf., e.g., [74, 75]. This connection, in its turn, has been the starting point of many of the analytic approaches to study the transcendental solutions. As a consequence of this connection various treatments of solving the Painlevé transcendents have been given; see [68, 73, 76, 78–81].

Also in physics the Painlevé equations have appeared in various guises, particularly in connection with exactly solvable models in statistical mechanics, where they have arisen in the calculation of correlation functions in the scaling limit; cf., e.g., [82–85]. The reasons as to why Painlevé transcendents arise as solutions of the correlation functions is still far from understood. The solvability of these models has been conjectured to be in some sense deeply rooted in the underlying analytic property of the Painlevé equations: that of the absence of movable critical points (other than poles). Another intriguing appearance of the Painlevé transcendents was in the discovery some years ago that the partition function of matrix models believed to describe two-dimensional quantum gravity obeys the Painlevé I equation; cf. [86–90]. Here, and in the earlier work [91], there is a connection already with discrete models, since in fact, P_I appears as the continuum limit (large N-limit) of a nonlinear recursion relation known in this context as the "string equation." Actually, these recursion relations were well known in the theory of orthogonal polynomials and what is now known under the name *discrete Painlevé I* (see Section 3) was already found in 1939 by Shohat [92] and rediscovered by Freud in [93] in connection with the asymptotics of orthogonal polynomials; cf. [94] for a more modern perspective. From an isomonodromic point of view this equation has been investigated only very recently in [95].

Since we will need the connection between soliton systems and Painlevé equations for our derivation of *discrete* Painlevé transcendents, let us give a short summary of how the similarity reduction of a soliton equation works in practice.

Consider the modified KdV (MKdV) equation

$$v_t + v_{xxx} - 6v^2 v_x = 0, \tag{2.1}$$

which carries the Lax pair

$$\partial_x \phi(x, t; \lambda) = L(x, t; \lambda) \phi(x, t; \lambda), \tag{2.2a}$$
$$\partial_t \phi(x, t; \lambda) = M(x, t; \lambda) \phi(x, t; \lambda), \tag{2.2b}$$

in which

$$L = \begin{pmatrix} \lambda & v \\ v & -\lambda \end{pmatrix},$$

$$M = \begin{pmatrix} -4\lambda^3 + 2\lambda v^2 & -4\lambda^2 v - 2\lambda v_x - v_{xx} + 2v^3 \\ -4\lambda^2 v + 2\lambda v_x - v_{xx} + 2v^3 & 4\lambda^3 - 2\lambda v^2 \end{pmatrix}.$$

Exercise 2.1. Show that the compatibility conditions of (2.2), i.e.,

$$L_t - M_x + [L, M] = 0,$$

is valid iff the potential v obeys the MKdV equation (2.1).

The MKdV as well as its Lax representation are invariant under a scaling transformation of the form

$$v(x, t) \mapsto \frac{1}{\rho} v(\rho x, \rho^3 t),$$

$$\phi(x, t; \lambda) \mapsto \phi\left(\rho x, \rho^3 t; \frac{1}{\rho}\lambda\right).$$

Requiring that the solution itself be scaling-invariant leads to the following conditions of self-similarity

$$v(x, t) = \frac{1}{\rho} v(\rho x, \rho^3 t) = \frac{1}{t^{1/3}} V(\xi), \tag{2.3}$$

in which $\xi = x t^{-1/3}$ is the *similarity variable*. The condition (2.3) can also be reformulated in terms of a differential constraint to which we shall refer as the *similarity constraint*. For the MKdV it reads

$$v + x v_x + 3t v_t = 0. \tag{2.4}$$

On the level of the Lax representation the reduction arises from

$$\lambda \phi_\lambda = x\phi_x + 3t\phi_t. \tag{2.5}$$

Thus, we obtain a new function $\phi(x, t; \lambda) = \Phi(\xi; k)$, where $k = t^{1/3}\lambda$, for which we have the system

$$\partial_\xi \Phi(\xi, k) = A(\xi, k)\Phi(\xi, k), \tag{2.6a}$$

$$k\frac{\mathrm{d}}{\mathrm{d}k}\Phi(\xi, k) = B(\xi, k)\Phi(\xi, k), \tag{2.6b}$$

with matrices A and B arising from the Lax pair (2.2) together with (2.5) and (2.3). Equation (2.6) forms what is nowadays referred to as the monodromy problem for the second Painlevé transcendent P_{II}.

Exercise 2.2. Find the explicit form of the matrices A and B of (2.6) in terms of $V(\xi)$.

Exercise 2.3. Show that the compatibility condition of the system (2.6) gives rise to

$$k\frac{\mathrm{d}}{\mathrm{d}k}A - B_\xi + [A, B] = 0,$$

and that this leads to P_{II} in the form

$$V_{\xi\xi} = 2V^3 - \frac{1}{3}\xi V + \mu.$$

Below we give a few more examples of Painlevé transcendents arising from similarity reductions of continuous soliton equations.

There is a large number of soliton equations that can be reduced to Painlevé or Painlevé-type ODEs. The reduction is not always easy to find. A general hypothesis that this connection between PDEs and ODEs is actually very fundamental for the integrability of the nonlinear evolution equations was stated in a paper by Ablowitz, Ramani, and Segur, [73] postulating that *"Every ordinary differential equation that arises as a similarity reduction of a nonlinear evolution equation which is solvable by the inverse scattering transform is of Painlevé type."*

So far, however, no exhaustive classification exists of integrable nonlinear evolution equations according to their reductions to Painlevé types of equations. (It should be mentioned that different reductions on one and the same NLEE can yield different ODEs, all of Painlevé type.) We will restrict ourselves here by mentioning a number of special cases that will turn out to be useful for the connection with discrete problems later.

Another fundamental equation is the Schwarzian KdV (SKdV)

$$\psi_t = \psi_x S(\psi), \quad S(\psi) \equiv \frac{\psi_{xxx}}{\psi_x} - \frac{3}{2}\frac{\psi_{xx}^2}{\psi_x^2}, \tag{2.7}$$

in which $S(\psi)$ denotes the Schwarzian derivative of ψ, which is a specialization of the Krichever–Novikov equation [96], which is invariant under Möbius transformations

$$\psi \mapsto \frac{a\psi + b}{c\psi + d}, \tag{2.8}$$

i.e., under the group PSL(2). Within the KdV family, (2.7) is probably the most fundamental equation. It seems that similarity reductions of Schwarzian equations such as (2.8) have not received much attention, although it is an easy exercise to obtain them. In [97] it was noted that for example the similarity reduction of the Schwarzian KdV is obtained very straightforwardly by considering (2.7) together with the similarity constraint

$$3\mu\psi = x\psi_x + 3t\psi_t, \tag{2.9}$$

the solution of which leads to the following *ansatz* for the solutions of the SKdV equation:

$$\psi(x, t) = t^\mu \Psi(\xi), \quad \xi = \frac{x}{t^{1/3}}. \tag{2.10}$$

Imposing (2.10) we obtain a reduction to a third-order ODE of the form

$$\frac{\Psi'''}{\Psi'} - \frac{3}{2}\frac{\Psi''^2}{\Psi'^2} = \mu\frac{\Psi}{\Psi'} - \frac{1}{3}\xi. \tag{2.11}$$

Equation (2.11) is no longer Möbius invariant, as the scaling symmetry breaks the invariance under PSL(2). However, (2.11) is directly related to the Painlevé II equation, using the well-known correspondence

$$y = -\frac{1}{2}\frac{\Psi''}{\Psi'} \implies S(\Psi) = -2(y' + y^2), \tag{2.12}$$

thus leading to the Painlevé II (P_{II}) equation

$$y'' = 2y^3 - \frac{1}{3}\xi y + \left(\frac{1}{6} - \frac{1}{2}\mu\right). \tag{2.13}$$

It is curious that the third-order equation (2.11), which we could loosely baptize the Schwarzian P_{II} (SP_{II}) equation, doesn't seem to occur as such in the class of Chazy equations of [3]; cf. also [4]. There exist other Schwarzian equations that have a similar relation with Painlevé equations, cf. [97], and which we will encounter below.

The Boussinesq (BSQ) family of equations is another family of NLEEs whose similarity reductions are particularly interesting. They were investigated for the first time in [98,99]. In [98] it was shown that the (potential) BSQ equation

$$u_{tt} + \frac{1}{3}u_{xxxx} + 4u_x u_{xx} = 0, \tag{2.14}$$

which is the integrated version of the BSQ equation, leads to an equation of Painlevé type of the form

$$\gamma U' = U''^2 + 4U'^3 - \frac{9}{4}(\eta U' - U)^2 + \beta, \qquad (2.15)$$

by considering the reduction

$$u(x,t) = \frac{1}{\sqrt{t}}(U(\eta) - \frac{1}{12}\eta^3), \quad \eta = \frac{x}{\sqrt{t}}. \qquad (2.16)$$

Equation (2.15) is a second-order second-degree ODE obeying the Painlevé property; cf. [3]. The complete classification of such equations was achieved only fairly recently by Cosgrove in [100]; cf. also [101].

Exercise 2.4. Show that the reduction (2.16) when applied to (2.14) leads to (2.15). (Note that β and γ are integration constants.)

Equation (2.15) is related via a Miura transformation to P_{IV}, which is more directly obtained from the modified Boussinesq (MBSQ) equation, cf. [98], which reads

$$q_{tt} + \frac{1}{3}q_{xxxx} + 2q_t q_{xx} - 2q_x^2 q_{xx} = 0. \qquad (2.17)$$

Exercise 2.5. Show that the similarity reduction

$$q_x(x,t) = V(\eta) + \frac{1}{2}\eta \qquad (2.18)$$

leads to P_{IV}.

Many examples of reductions from soliton equations to Painlevé equations or equations of Painlevé type exist; cf., e.g., [102–105]. Sometimes these reductions can be quite complicated and a general recipe cannot be given. In Chapter 10 in this volume more systematic methods for finding these reductions on the basis of symmetry approaches are developed. What interests us in the present chapter is to extend the similarity reduction approach to the discrete situation to obtain (hopefully) discrete versions of the Painlevé equations. To achieve that, however, some major difficulties have to be overcome. First we need a systematic method to obtain discrete soliton systems, and then we need to implement scaling reductions. Even the latter in the absence of genuine symmetry approaches on the lattice is a major task. In this section we show how to do it using the direct linearization approach, which we will develop next, and this approach indeed gives us proper analogues of discrete Painlevé equations in a more or less systematic way.

2.2 Infinite-Matrix Scheme

The general scheme from which we obtain the discrete equations as well as the similarity constraints is one that uses infinite matrices. It has been very successful in yielding insights into integrable hierarchies for various settings (cf., e.g., [70,71], and [59] for a review and guide to the literature). In particular, it incorporates the various interrelations between equations (such as Miura and Bäcklund transformations) and allows for a treatment of the equations without fixing a priori preferred variables (such as in the Grassmannian approach of Sato and others).

For the sake of brevity, let us formulate the structure in a rather abstract way. We basically need three ingredients to formulate the entire scheme:

- An infinite ($\mathbb{Z} \times \mathbb{Z}$) matrix $\mathbf{C}$ tht we can take of the form

$$\mathbf{C} = \int_\Gamma \rho_\ell \mathbf{c}_\ell {}^t\mathbf{c}_{\ell'} \, d\lambda, \tag{2.19}$$

in which $\mathbf{c}_\ell$ and ${}^t\mathbf{c}_{\ell'}$ are infinite vectors with components $(\mathbf{c}_\ell)_j = ({}^t\mathbf{c}_{\ell'})_j = \ell^j$, and ρ_ℓ depends on additional variables that are to be determined later. The integrations over the contour Γ and measure $d\lambda$ need not be specified at this point, but we we will loosely assume that they can be chosen such that the objects to be introduced below are well defined. From the definition it is clear that $\mathbf{C}$ is symmetric: $\mathbf{C} = {}^t\mathbf{C}$ (the left superscript t denotes the adjoint of the $\mathbb{Z} \times \mathbb{Z}$ matrix).

- Matrices $\mathbf{\Lambda}$ and ${}^t\mathbf{\Lambda}$ that define the operations of index-raising when multiplied at the left, respectively the right, as well as matrices $\mathbf{I}$ and ${}^t\mathbf{I}$ that count the index label, namely by $(\mathbf{I} \cdot \mathbf{c}_\ell)_j = j(\mathbf{c}_\ell)_j$, $(\mathbf{I} \cdot {}^t\mathbf{c}_{\ell'})_j = j({}^t\mathbf{c}_{\ell'})_j$. The interrelation between the variables $\mathbf{\Lambda}$, ${}^t\mathbf{\Lambda}$, $\mathbf{I}$ and ${}^t\mathbf{I}$ is given by

$$\mathbf{\Lambda}^j \cdot \mathbf{I} = (\mathbf{I} + j) \cdot \mathbf{\Lambda}, \tag{2.20a}$$

$$ {}^t\mathbf{I} \cdot {}^t\mathbf{\Lambda}^j = {}^t\mathbf{\Lambda} \cdot (\mathbf{I} + j). \tag{2.20b}$$

- An infinite matrix $\mathbf{\Omega}$ obeying the equations

$$\mathbf{\Omega}\mathbf{\Lambda}^j - (-{}^t\mathbf{\Lambda})^j \mathbf{\Omega} = \mathbf{O}_j, \tag{2.21a}$$

$$ {}^t\mathbf{I} \cdot \mathbf{\Omega} + \mathbf{\Omega} \cdot \mathbf{I} + \mathbf{\Omega} = 0, \tag{2.21b}$$

in which

$$\mathbf{O}_k = \sum_{j=0}^{k-1} (-{}^t\mathbf{\Lambda})^j \cdot \mathbf{O} \cdot \mathbf{\Lambda}^{k-1-j}, \tag{2.22}$$

and where $\mathbf{O}$ is a projection matrix on the central element, i.e., $(\mathbf{O} \cdot \mathbf{C}_k)_{i,j} = \delta_{i,0} \mathbf{C}_{0,j}$, etc., obeying also

$$ {}^t\mathbf{I} \cdot \mathbf{O} = \mathbf{O} \cdot \mathbf{I} = 0. \tag{2.23}$$

There are linear equations for the object $\mathbf{C}$, the form of which depends on the choice of the factor ρ_ℓ. If we choose ρ_ℓ to depend on continuous variables x_1, x_2, x_3, $\ldots$, then this factor takes the form

$$\rho_\ell = \prod_{\nu \text{ odd}} e^{\ell^\nu x_\nu}, \qquad (2.24a)$$

whereas if ρ_ℓ depends on discrete variables, we can take

$$\rho_\ell = \prod_\nu \left(\frac{p_\nu + \ell}{p_\nu - \ell} \right)^{n_\nu}. \qquad (2.24b)$$

Thus we can impose for $\mathbf{C}$ one of the following:

Continuous Case: $\partial_{x_j} \mathbf{C} = \mathbf{\Lambda}^j \cdot \mathbf{C} - \mathbf{C} \cdot (-{}^t\mathbf{\Lambda})^j.$ (2.25a)

Discrete Case: $\widetilde{\mathbf{C}} \cdot (p - {}^t\mathbf{\Lambda}) = (p + \mathbf{\Lambda}) \cdot \mathbf{C}.$ (2.25b)

These equations are linear. However, using all the tools, we can now derive either continuous or nonlinear equations for the relevant objects. These objects are the following:

- The infinite matrix

$$\mathbf{U} = \mathbf{C} \cdot (1 + \mathbf{\Omega} \cdot \mathbf{C})^{-1}. \qquad (2.26)$$

- An infinite determinant

$$\tau \equiv \det{}_{Z \times Z} (1 + \mathbf{\Omega} \cdot \mathbf{C}). \qquad (2.27)$$

To make sense of the latter we can use the expansion

$$\det(1 + A) = 1 + \sum_i A_{ii} + \sum_{i<j} \begin{vmatrix} A_{ii} & A_{ij} \\ A_{ji} & A_{jj} \end{vmatrix} + \cdots,$$

and impose that the integrations in (2.19) be such that the expansions, which are all of the form

$$\mathrm{tr}_Z \left((\mathbf{\Omega} \cdot \mathbf{C})^k \right),$$

terminate. The relevant quantities in terms of which one can derive closed-form nonlinear equations are actually the individual entries of the infinite matrix $\mathbf{U}$, which, by the way, is symmetric: ${}^t\mathbf{U} = \mathbf{U}$ as a consequence of the symmetry of $\mathbf{C}$.

Let us now derive some of the basic equations from this scheme. Let us investigate what happens if we apply a shift on τ according to (2.25b)

$$\begin{aligned}
\widetilde{\tau} &= \det\left(1 + \mathbf{\Omega} \cdot \widetilde{\mathbf{C}} \right) = \det\left\{ 1 + \mathbf{\Omega} \cdot (p + \mathbf{\Lambda}) \cdot \mathbf{C} \cdot (p - {}^t\mathbf{\Lambda})^{-1} \right\} \\
&= \det\left\{ 1 + \left[(p - {}^t\mathbf{\Lambda}) \cdot \mathbf{\Omega} + \mathbf{O} \right] \cdot \mathbf{C} \cdot (p - {}^t\mathbf{\Lambda})^{-1} \right\} \\
&= \det\left\{ 1 + \mathbf{\Omega} \cdot \mathbf{C} + (p - {}^t\mathbf{\Lambda})^{-1} \cdot \mathbf{O} \cdot \mathbf{C} \right\} \\
&= \det(1 + \mathbf{\Omega} \cdot \mathbf{C}) \det\left\{ 1 + (p - {}^t\mathbf{\Lambda})^{-1} \cdot \mathbf{O} \cdot \mathbf{C} \cdot (1 + \mathbf{\Omega} \cdot \mathbf{C})^{-1} \right\},
\end{aligned}$$

from which it follows that

$$\frac{\widetilde{\tau}}{\tau} = \det\left(\mathbf{1} + (p - {}^t\mathbf{\Lambda})^{-1} \cdot \mathbf{O} \cdot \mathbf{U}\right) = 1 + \left(\mathbf{U} \cdot (p - {}^t\mathbf{\Lambda})^{-1}\right)_{0,0}. \qquad (2.28)$$

The right-hand side is the central element of an infinite matrix obtained as a formal series of entries of the infinite matrix $\mathbf{U}$. This special combination plays a particularly important role in what follows. Equation (2.28) is reminiscent of relations between the τ-function in the approach of the Japanese school and eigenfunctions of the linear problems associated with integrable hierarchies; cf. also [106]. The object τ, being defined as an infinite determinant, can be shown to obey bilinear relations, which we will not study here. In Sections 3 and 4 we will come back to bilinear forms in connection with the discrete Painlevé equations.

The relation (2.28) also has a continuous counterpart, which can be derived by using (2.25) instead of (2.25b).

Exercise 2.6. Use (2.25) to prove that

$$\partial_{x_j} \log \tau = \mathrm{tr}_{\mathbf{Z}}\left(\mathbf{O}_j \cdot \mathbf{U}\right).$$

For the infinite matrix $\mathbf{U}$ we can derive relations using the definition (2.26) and again (2.25b), as follows:

$$\widetilde{\mathbf{U}} \cdot (p - {}^t\mathbf{\Lambda}) = \left(\mathbf{1} + \mathbf{\Omega} \cdot \widetilde{\mathbf{C}}\right)^{-1} \cdot \widetilde{\mathbf{C}} \cdot (p - {}^t\mathbf{\Lambda}) = \left(\mathbf{1} + \mathbf{\Omega} \cdot \widetilde{\mathbf{C}}\right)^{-1} \cdot (p + \mathbf{\Lambda}) \cdot \mathbf{C}$$

$$= \left(\mathbf{1} - \widetilde{\mathbf{U}} \cdot \mathbf{\Omega}\right) \cdot (p + \mathbf{\Lambda}) \cdot \mathbf{C}$$

$$= (p + \mathbf{\Lambda}) \cdot \mathbf{C} - \widetilde{\mathbf{U}} \cdot \left[(p - {}^t\mathbf{\Lambda}) \cdot \mathbf{\Omega} + \mathbf{O}\right] \cdot \mathbf{C}.$$

Thus,

$$\widetilde{\mathbf{U}} \cdot (p - {}^t\mathbf{\Lambda}) \cdot (\mathbf{1} + \mathbf{\Omega} \cdot \mathbf{C}) = (p + \mathbf{\Lambda}) \cdot \mathbf{C} - \widetilde{\mathbf{U}} \cdot \mathbf{O} \cdot \mathbf{C},$$

which, after dividing by $(\mathbf{1} + \mathbf{\Omega} \cdot \mathbf{C})$, leads to the matrix Riccati equation

$$\widetilde{\mathbf{U}} \cdot (p - {}^t\mathbf{\Lambda}) = (p + \mathbf{\Lambda}) \cdot \mathbf{U} - \widetilde{\mathbf{U}} \cdot \mathbf{O} \cdot \mathbf{U}. \qquad (2.29)$$

Equation (2.29) forms the starting point for the construction of integrable lattice equations. By combining different shifts associated with different parameters p one can actually derive all relevant discrete equations within the KdV family from this single relation. Equation (2.29) has its continuous counterpart as well, which can be derived by a similar set of manipulations starting from (2.25).

Exercise 2.7. Use the definition (2.26) of $\mathbf{U}$ and the relation (2.25) to derive the relation

$$\partial_{x_j} \mathbf{U} = \mathbf{\Lambda}^j \cdot \mathbf{U} - \mathbf{U} \cdot (-{}^t\mathbf{\Lambda})^j - \mathbf{U} \cdot \mathbf{O}_j \cdot \mathbf{U}. \qquad (2.30)$$

It is clear that since the relations (2.25a) are linear, involving commuting factors, we can impose on $\mathbf{C}$ simultaneously all possible relations of that form. Thus by imposing (2.25) for any subset of values $j \in \mathbb{Z}$ we necessarily have a solution $\mathbf{C}$ depending on all $\ldots, x_{-1}, x_0, x_1, \ldots$, implying that all relations of the type (2.30) hold for $\mathbf{U}$ simultaneously, involving the variables x_j that we have selected. This is, in fact, the origin of the integrable *hierarchy* of nonlinear equations for $\mathbf{U}$ leading to the KdV hierarchy for its central entry, the MkdV hierarchy for another one of its entries and the SKdV hierarchy for yet another one, etc.; cf., e.g., [70, 71].

Once we have selected a particular set of continuous or discrete variables, we have yet another type of relation that we can consistently impose on the quantity $\mathbf{C}$, involving the operation generated by the matrices $\mathbf{I}$ and ${}^t\mathbf{I}$. These relations are a consequence of the behavior of the plane wave-factor ρ_ℓ under the application of the differential operator $\ell d/d\ell$. In fact, noting that

Continuous Case:
$$\ell \frac{\partial}{\partial \ell} \rho_\ell = \sum_j j x_j \partial_{x_j} \rho_\ell, , \tag{2.31}$$

Discrete Case:
$$\ell \frac{\partial}{\partial \ell} \rho_\ell = \sum_\nu n_\nu p_\nu \left(\frac{1}{p_\nu - \ell} - \frac{1}{p_\nu + \ell} \right) \rho_\ell, \tag{2.32}$$

we are motivated from (2.19) to the choice

$$\mathbf{C} + \mathbf{I} \cdot \mathbf{C} + \mathbf{C} \cdot {}^t\mathbf{I} = \sum_\nu p_\nu n_\nu \left((p_\nu + \mathbf{\Lambda})^{-1} \cdot \mathbf{C} - \mathbf{C} \cdot (p_\nu - {}^t\mathbf{\Lambda})^{-1} \right) \tag{2.33}$$

in the discrete case, in which the n_ν form the collection of discrete variables on which we allow $\mathbf{C}$ to depend on with corresponding parameters p_ν. To derive the corresponding equation for $\mathbf{U}$ we need to multiply (2.33) from the left by $1 + \mathbf{\Omega} \cdot \mathbf{C} = 1 - \mathbf{\Omega} \cdot \mathbf{U}$ and use (2.21) to derive the relation

$$\mathbf{U} + \mathbf{I} \cdot \mathbf{U} + \mathbf{U} \cdot {}^t\mathbf{I} = \sum_\nu n_\nu p_\nu \left(\frac{1}{p_\nu + \mathbf{\Lambda}} \cdot \mathbf{U} - \mathbf{U} \cdot \frac{1}{p_\nu - {}^t\mathbf{\Lambda}} \right.$$
$$\left. + \mathbf{U} \cdot \frac{1}{p_\nu - {}^t\mathbf{\Lambda}} \cdot \mathbf{O} \cdot \frac{1}{p_\nu + \mathbf{\Lambda}} \cdot \mathbf{U} \right). \tag{2.34}$$

Exercise 2.8. Prove (2.34).

Exercise 2.9. Prove from (2.33) that the continuous counterpart of (2.34) is given by the same equation as the one for $\mathbf{C}$ itself.

Here we observe already a crucial difference between the continuous and discrete cases: Whereas in the continuous case the equation involving the matrices $\mathbf{I}$ and ${}^t\mathbf{I}$ are linear (both for $\mathbf{C}$ as well as for $\mathbf{U}$), in the discrete case we obtain a *nonlinear* equation for the latter. In what follows the interpretation of (2.33) and (2.34) is that they represent the similarity constraints on the solutions of the lattice equations derived from (2.29), thereby corresponding to a "hidden" scaling-invariance. In the next sections we will employ the various relations that we have derived from the direct linearization framework to derive for the entries of the infinite matrix $\mathbf{U}$ on the one hand integrable lattice equations and on the other hand their similarity constraints in closed form. From those two ingredients the derivation of discrete Painlevé equations can be immediately implemented, as we shall see.

2.3 Lattice Equations

Now that we have the framework in place from which the similarity reduction on the lattice can be derived, we can now exploit the relations derived earlier to get closed-form discrete equations, as well as their similarity reduction. We will work within the KdV family, which means that we impose the symmetry

$$\mathbf{C} = {}^{t}\mathbf{C} \implies \mathbf{U} = {}^{t}\mathbf{U}.$$

Without imposing this symmetry we would end up with 3-dimensional equations, thus leading to KP-type of lattice systems rather than to the KdV family. In fact, the infinite set of differential relations (2.30) without further constraints leads to the continuous KP hierarchy, whereas imposing the above symmetry constraints implies an additional set of algebraic relations of the form

$$\mathbf{U} \cdot (-{}^{t}\boldsymbol{\Lambda})^{j} = \boldsymbol{\Lambda}^{j} \cdot \mathbf{U} - \mathbf{U} \cdot \mathbf{O}_{j} \cdot \mathbf{U}, \quad j \text{ even}, \tag{2.35}$$

arising from (2.30) for variables x_j carrying an *even* label. This set of relations reduces the system from the 2+1-dimensional KP hierarchy to the 1+1-dimensional KdV hierarchy. The same is true in the discrete case.

In order to obtain the lattice equations in the KdV family let us now introduce a number of objects in terms of which one can derive from the basic system (2.29) closed-form equations, namely

$$u \equiv U_{0,0}, \quad s_{\alpha,\beta} \equiv \left(\frac{1}{\alpha + \boldsymbol{\Lambda}} \cdot \mathbf{U} \cdot \frac{1}{\beta + {}^{t}\boldsymbol{\Lambda}} \right)_{0,0}, \tag{2.36a}$$

as well as

$$v_{\alpha} \equiv 1 - \left(\frac{1}{\alpha + \boldsymbol{\Lambda}} \cdot \mathbf{U} \right)_{0,0}. \tag{2.36b}$$

In (2.36) α and β are arbitrary parameters that we can tune to our liking. The KdV case is distinguished by the fact that the variable $s_{\alpha,\beta}$ is invariant under the interchange of the parameters α and β. It is obvious from the definitions that v_α and w_β can be obtained from $s_{\alpha,\beta}$ in the limits $\beta \to \infty$, respectively $\alpha \to \infty$, and that u is recovered in the simultaneous limit. However, we shall be interested as well in cases where either $\alpha = 0$ or $\beta = 0$ or both.

From the basic equations (2.29) we can now derive the following set of relations for the objects defined in (2.36):

$$\tilde{s}_\alpha = (p + u)\tilde{v}_\alpha - (p - \alpha)v_\alpha, \tag{2.37a}$$

$$s_\beta = (p + \beta)\tilde{v}_\beta - (p - \tilde{u})v_\beta, \tag{2.37b}$$

as well as

$$1 - (p + \beta)\tilde{s}_{\alpha,\beta} + (p - \alpha)s_{\alpha,\beta} = \tilde{v}_\alpha v_\beta. \tag{2.38}$$

Closed-form equations can be obtained by considering two different transformations,

$$\rho_k \mapsto \tilde{\rho}_k, \quad \rho_k \mapsto \hat{\rho}_k,$$

both of the same type (2.54), but for different parameter values p, respectively q. Since the dependence on the two variables n and m will enter multiplicatively in the central object $\mathbf{C}$, it is clear that $\mathbf{C}$ depends linearly on these variables implying the permutability of these transformations. As a consequence, there is another set of relations that one can derive. Thus, from the system of relations (2.37) for both translation types we can then derive the Miura-type transformation

$$p - q + \hat{u} - \tilde{u} = (p - \alpha)\frac{\hat{v}_\alpha}{\tilde{\hat{v}}_\alpha} - (q - \alpha)\frac{\tilde{v}_\alpha}{\tilde{\hat{v}}_\alpha}, \tag{2.39a}$$

$$p + q + u - \hat{\tilde{u}} = (p - \alpha)\frac{v_\alpha}{\tilde{v}_\alpha} + (q + \alpha)\frac{\hat{\tilde{v}}_\alpha}{\hat{v}_\alpha}. \tag{2.39b}$$

Next, the transformations $u \mapsto \tilde{u}$ and $u \mapsto \hat{u}$ are identified with translations on a two-dimensional lattice with grid points labeled by integers n, resp. m. Thus, all objects (2.36) and (2.36b) are then interpreted as functions of the lattice sites (n, m) and the shifts by translations

$$(n, m) \mapsto (n + 1, m), \quad (n, m) \mapsto (n, m + 1),$$

respectively. Thus, as we shall show, by simple elimination procedures we obtain the various lattice versions of the KdV system from the relations (2.38), (2.39).

In a similar way the similarity constraints are derived from the basic relation (2.29), and the corresponding monodromy problems will de obtained from (2.34), as we shall explain in Section 5. In order to implement these relations, let us first give the "crude" forms of the similarity constraints.

To obtain closed-form constraints for single objects we need the cases that the parameters α and β go either to zero or infinity (otherwise, we will get constraints containing derivatives with respect to the variables α and β). So, we will focus primarily on similarity constraints for the objects u, v_0, and $s_{0,0}$, but we need some other objects along the way as well. The similarity constraint for u is obtained from

$$u = np\left(1 - \frac{v_p}{\tilde{v}_p}\right) + mq\left(1 - \frac{v_q}{\hat{v}_q}\right), \tag{2.40}$$

whereas the similarity constraint for v_0 is obtained from

$$0 = n\left(\frac{v_p}{\tilde{v}_p}\frac{\tilde{v}_0}{v_0} - 1\right) + m\left(\frac{v_q}{\hat{v}_q}\frac{\hat{v}_0}{v_0} - 1\right). \tag{2.41}$$

Various expressions for the fractions $\tilde{v}_p/v_p$, $\hat{v}_p/v_p$ follow from (2.37) and (2.38). The similarity constraint for $z = s_{0,0} + \frac{n}{p} + \frac{m}{q}$ is given by

$$z = \frac{n}{p}\frac{v_p}{\tilde{v}_p}v_0\,\tilde{v}_0 + \frac{m}{q}\frac{v_q}{\hat{v}_q}v_0\hat{v}_0. \tag{2.42}$$

We will present the final closed-form expressions for the similarity constraints below.

Exercise 2.10. Derive (2.40)–(2.42) from (2.34).

2.4   The Lattice KdV Family and Lattice Similarity Constraints

Let us now write down the explicit formulae for the similarity reductions of the lattice KdV family. Starting from the relations of the previous section we will obtain lattice equations and similarity constraints for the quantities $u =: u_{n,m}$ in the case of the lattice KdV equation, $v_0 =: v_{n,m}$ in the case of the lattice MKdV equation, and $z = z_{n,m}$ for the lattice SKdV equation. Let us start with the last case, in which we actually obtain a very simple equation, which was first given in [59] and which reads

$$\frac{(z_{n,m} - z_{n+1,m})(z_{n,m+1} - z_{n+1,m+1})}{(z_{n,m} - z_{n,m+1})(z_{n+1,m} - z_{n+1,m+1})} = \frac{q^2}{p^2}. \tag{2.43}$$

On the left-hand side of (2.43) we recognize the canonical conformally-invariant cross ratio of four points in the complex plane, which indicates invariance under Möbius transformations

$$z \mapsto \frac{az + b}{cz + d}, \quad ad - bc \neq 0.$$

In fact, taking into account that the cross ratio constitutes the discrete analogue of the Schwarzian derivative (cf., e.g., [107]) it is not hard to establish that (2.43) in a proper continuum limit indeed yields the SKdV equation (2.7). Let us mention that (2.43) is a special case of the more general equation for $s_{\alpha,\beta} =: s_{n,m}$ with free parameters α and β that can be derived from (2.38) and that was first given in [51,52], namely

$$\frac{1 - (p + \beta)s_{n+1,m} + (p - \alpha)s_{n,m}}{1 - (q + \beta)s_{n,m+1} + (q - \alpha)s_{n,m}}$$

$$= \frac{1 - (q + \alpha)s_{n+1,m+1} + (q - \beta)s_{n+1,m}}{1 - (p + \alpha)s_{n+1,m+1} + (p - \beta)s_{n,m+1}}, \quad (2.44)$$

equation (2.43) being obtained in the special case that $\alpha = \beta = 0$.

The integrability aspect of the lattice SKdV (2.43) is manifested in the existence of a Lax pair of the form

$$\phi_{n+1,m}(k) = L_{n,m}(k^2)\phi_{n,m}(k), \quad (2.45a)$$

$$\phi_{n,m+1}(k) = M_{n,m}(k^2)\phi_{n,m}(k), \quad (2.45b)$$

in which the 2×2 matrices L and M are given by

$$L_{n,m}(k) = \begin{pmatrix} 1 & z_{n,m} - z_{n+1,m} \\ \dfrac{k^2/p^2}{z_{n,m} - z_{n+1,m}} & 1 \end{pmatrix},$$

$$M_{n,m}(k) = \begin{pmatrix} 1 & z_{n,m} - z_{n,m+1} \\ \dfrac{k^2/q^2}{z_{n,m} - z_{n,m+1}} & 1 \end{pmatrix}. \quad (2.46)$$

Exercise 2.11. Show that the compatibility of the two different ways of calculating $\phi_{n+1,m+1}$ from (2.45) leads to the condition

$$L_{n,m+1}M_{n,m} = M_{n+1,m}L_{n,m}, \quad (2.47)$$

and show that in the case of the matrices (2.45a) and (2.45b) this condition gives rise to the lattice Schwarzian KdV, equation (2.43).

The more general equation (2.44) is very important, too, since by tuning the free parameters α and β we obtain a number of different lattice equations related to the KdV system. In fact, in the limit $\beta \to \infty$, taking $1 - \beta s_{n,m} \to v_{n,m}$ we obtain the equation

$$(p - \alpha)\frac{v_{n,m+1}}{v_{n+1,m+1}} - (q - \alpha)\frac{v_{n+1,m}}{v_{n+1,m+1}}$$

$$= (p + \alpha)\frac{v_{n+1,m}}{v_{n,m}} - (q + \alpha)\frac{v_{n,m+1}}{v_{n,m}}, \quad (2.48)$$

which for $\alpha = 0$ constitutes a lattice version of the (potential) modified KdV (MKdV) equation. Next, by taking the limit $\alpha, \beta \to \infty$ such that $\alpha\beta s_{n,m} \to u_{n,m}$, we obtain the equation

$$(p - q + u_{n,m+1} - u_{n+1,m})(p + q - u_{n+1,m+1} + u_{n,m}) = p^2 - q^2, \quad (2.49)$$

which is the lattice (potential) KdV equation. All these variables are not unrelated, and various Miura-type relations can be established on the basis of (2.38) and (2.39).

Let us now consider the similarity reduction of the different equations in the KdV family. From (2.40)–(2.42) these are easily calculated using also the relations (2.38) and (2.39). In this way we obtain for the lattice KdV equation (2.49) the constraint

$$u_{n,m} + np\frac{u_{n+1,m} - u_{n-1,m}}{2p + u_{n-1,m} - u_{n+1,m}} + mq\frac{u_{n,m+1} - u_{n,m-1}}{2q + u_{n,m-1} - u_{n,m+1}} = 0, \qquad (2.50)$$

which was first given in [67]. As mentioned earlier, the nonlinear constraint (2.50) represents the hidden scaling invariance of the corresponding solutions of the lattice KdV equation, and as such it is exactly the discrete analogue of the corresponding linear continuous constraint.

In a similar fashion we can find discrete analogues of (2.4) and (2.9). Indeed, for the lattice MKdV equation (2.48), equation (2.41) leads immediately to the similarity constraint

$$n\frac{v_{n+1,m} - v_{n-1,m}}{v_{n+1,m} + v_{n-1,m}} + m\frac{v_{n,m+1} - v_{n,m-1}}{v_{n,m+1} + v_{n,m-1}} = 0. \qquad (2.51)$$

Both (2.50) and (2.51) represent the similarity constraint for the special parameter case $\mu = \frac{1}{3}$ in the associated Painlevé II equation. The direct linearization approach so far has not provided the information on other values of the parameter. So we have to bring the parameter-dependence in by hand in the similarity constraints. To see how to do this let us turn to the similarity constraint for the lattice Schwarzian KdV (2.43), namely (cf. [97]),

$$3\mu z_{n,m} = 2n\frac{(z_{n+1,m} - z_{n,m})(z_{n,m} - z_{n-1,m})}{z_{n+1,m} - z_{n-1,m}}$$
$$+ 2m\frac{(z_{n,m+1} - z_{n,m})(z_{n,m} - z_{n,m-1})}{z_{n,m+1} - z_{n,m-1}}. \qquad (2.52)$$

For $\mu = \frac{1}{3}$ this equation is a direct consequence of (2.42) however, it is not hard to see that the extra freedom provided by the factor μ in the term on the left-hand side is exactly analogous to the corresponding term in the continuous constraint (2.9). That (2.52) is indeed the correct constraint for any value of μ is not a priori evident, but follows from the investigation of the consistency of (2.52) with the lattice equation (2.43) from the point of view of initial value problems on the lattice. We will give no details of these arguments here but refer to the forthcoming publication [108]. Once we have established (2.52) for general μ, the corresponding general parameter constraints for the lattice KdV and MKdV follow by exploiting the relations (2.38) and (2.39) between the quantities $z_{n,m}$, $v_{n,m}$, and $u_{n,m}$. This leads, in fact, to an additional free constant (up to an alternating term) on the right-hand side of (2.51), which we will use also in the following sections. Further details can be found in [108].

2.5 Semicontinuous Limits

Let us now investigate what happens under a continuum limit, bringing us eventually back to the original SKdV equation. Since there are two discrete variables in the lattice equation, n and m, we have to perform the continuum limit in two steps: one letting the variable m become continuous, reducing our equation to a *differential–difference equation*, i.e., an equation with one discrete and one continuous variable, and a second step in which the remaining discrete variable will become continuous. Both steps are achieved by shrinking the corresponding lattice step (encoded in the parameters p and q) to zero. Such limits can be performed in various ways, and we will highlight two particular continuum limits in what follows, leading to different branches of discrete Painlevé equations. It must be remarked that the similarity reduction of lattice equations within the KdV class allows for a direct derivation of discrete equations (i.e., before taking continuum limits) as well. This gives rise to a complicated but very rich equation that is a discrete form of the Painlevé III equation (cf. [108]) and that contains various subcases of lower discrete Painlevé equations. However, many aspects of this rich equation still remain to be investigated, and we will not present these results here at this stage.

"Skew" Continuum Limit

This limit involves first a change of variables on the lattice, namely $u_{n,m} =: u_{n'}(m)$, and then taking the limit by the transformation

$$\delta \equiv p - q \mapsto 0, \quad m \mapsto \infty, \quad \delta m \mapsto \tau, \qquad (2.53)$$

where $n' = n + m$ is to remain fixed. This limit is motivated from the behavior of discrete plane-wave factors

$$\rho_k(n, m) = \left(\frac{p+k}{p-k}\right)^n \left(\frac{q+k}{q-k}\right)^m, \qquad (2.54)$$

which govern the linear dispersion of the lattice equations; cf. [51]. Under (2.53) these plane-wave factors behave like

$$\left(\frac{p+k}{p-k}\right)^n \left(\frac{q+k}{q-k}\right)^m \mapsto \left(\frac{p+k}{p-k}\right)^{n'} e^{2k\tau/(p^2-k^2)}; \qquad (2.55)$$

cf. [51, 52].

The continuum limit applied to the general lattice KdV (2.44) yields

$$\partial_\tau s_n = \frac{s_{n+1} - s_{n-1} + 2p(s_n^2 + s_{n+1}s_{n-1}) - (2p + \alpha + \beta)s_n s_{n+1}}{2p - (p+\alpha)(p+\beta)s_{n+1} + (p-\alpha)(p-\beta)s_{n-1}}$$
$$- \frac{(2p - \alpha - \beta)s_n s_{n-1}}{2p - (p+\alpha)(p+\beta)s_{n+1} + (p-\alpha)(p-\beta)s_{n-1}}. \qquad (2.56)$$

Analogous to the fully discrete case, (2.56) has various interesting specifications for special choices of the variables α and β. For instance, the differential-difference analogue of the lattice KdV (2.49) is given by

$$1 + \partial_\tau u_n = \frac{2p}{2p - u_{n+1} + u_{n-1}}, \qquad (2.57)$$

which is related to the Kac–van Moerbeke–Volterra equation [109], whereas the continuum limit of (2.48) is given by

$$\partial_\tau \log v_n = \frac{v_{n+1} - v_{n-1}}{(p+\alpha)v_{n+1} + (p-\alpha)v_{n-1}}, \qquad (2.58)$$

which for $\alpha = 0$ is a differential–difference version of the MKdV equation, while for $\alpha = p$ it is a potential version of the Toda lattice equation; cf. [51]. For $\alpha = \beta = 0$, (2.56) reduces to the following differential–difference version of the Schwarzian KdV (cf. also [107, 110]):

$$\frac{p}{2}\partial_\tau z_n = \frac{(z_{n+1} - z_n)(z_n - z_{n-1})}{z_{n+1} - z_{n-1}}, \qquad (2.59)$$

i.e., (2.56) for $\alpha = \beta = 0$.

Let us now show how in the semicontinuum limit we can actually use the similarity constraint to obtain a reduction to a single ordinary nonautonomous difference equation that can be identified to be a difference version of the Painlevé II equation; cf. [67]. In fact, starting from the similarity constraint for the MKdV equation, in the limit (2.53), namely

$$\frac{1}{2}(3\mu - 1) - a_0(-1)^n = \frac{v_{n+1} - v_{n-1}}{v_{n+1} + v_{n-1}}\left[n + 2\tau\frac{\partial_\tau(v_{n+1}v_{n-1})}{v_{n+1}^2 - v_{n-1}^2}\right], \qquad (2.60)$$

c_0, c_1 being constants, one can use (2.58) to eliminate the derivatives with respect to τ to obtain the discrete Painlevé II (dP$_{\text{II}}$) equation

$$x_{n+1} + x_{n-1} = \frac{2p}{\tau}\frac{b_n - nx_n}{1 - x_n^2}, \qquad (2.61)$$

in terms of the variable

$$x_n \equiv \frac{v_{n+1} - v_{n-1}}{v_{n+1} + v_{n-1}}. \qquad (2.62)$$

Considering now the similarity reduction of the semicontinuous SKdV equation, (2.52) in the limit (2.53) reduces to

$$3\mu z_n = 2n\frac{(z_{n+1} - z_n)(z_n - z_{n-1})}{z_{n+1} - z_{n-1}} + 2\tau\frac{(z_{n+1} - z_n)\dot{z}_{n-1} + (z_n - z_{n-1})\dot{z}_{n+1}}{z_{n+1} - z_{n-1}}$$
$$- 2\tau\frac{(z_{n+1} - z_n)(z_n - z_{n-1})}{(z_{n+1} - z_{n-1})^2}(\dot{z}_{n+1} + \dot{z}_{n-1}). \quad (2.63)$$

Eliminating again the derivatives with respect to the variable τ from (2.63) one obtains, after one "integration," the following nonautonomous third-order nonlinear difference equation:

$$\frac{4\tau}{p}\frac{(z_{n+2} - z_{n+1})(z_n - z_{n-1})}{(z_{n+2} - z_n)(z_{n+1} - z_{n-1})} = 3\mu\frac{z_n}{z_{n+1} - z_n} - a_n, \quad (2.64)$$

in which $a_n \equiv n + (-1)^n a_0 - \frac{1}{2}(3\mu - 1)$, a_0 being an integration constant. We will refer to (2.64) as the discrete Schwarzian Painlevé II (dSP$_{\text{II}}$) equation, since in a continuum limit it reduces to the Schwarzian Painlevé II equation (2.11). Equation (2.64) is related to the dP$_{\text{II}}$ equation (2.61) via the discrete Hopf–Cole transformation

$$x_n = \frac{z_{n+1} + z_{n-1} - 2z_n}{z_{n+1} - z_{n-1}}. \quad (2.65)$$

Exercise 2.12. Show that (2.65) follows from (2.62) together with the relation

$$v_{n+1}v_n = p(z_n - z_{n+1}), \quad (2.66)$$

where the latter equation follows (2.38) for $\alpha = \beta = 0$.

Using (2.64) we now obtain

$$a_n + \frac{\tau}{p}(1 + x_{n+1})(1 - x_n) = 3\mu\frac{z_n}{z_{n+1} - z_n}, \quad (2.67)$$

which will subsequently yield (2.61). Thus the similarity reduction of the SKdV to the second Painlevé transcendent outlined in the introduction holds perfectly well also for the discrete case.

We note that from the equations given above one can also straightforwardly derive a Miura transformation connecting the dP$_{\text{II}}$ equation (2.61) to an equation called discrete P34 (dP34), which is the discrete analogue of the 34th equation in the Painlevé–Gambier classification; cf. [12]. In fact, returning to the continuum situation, introducing the KdV variable

$$w = -y' - y^2 + \frac{1}{6}\xi, \quad (2.68)$$

where $y = y(\xi)$ obeys the P_{II} equation (2.13), we obtain P34 in the form

$$w''w = \frac{1}{2}w'^2 - 2w^3 + \frac{1}{3}\xi w^2 - \frac{\mu^2}{8}. \tag{2.69}$$

In the discrete case, interpreting (2.67) as a Miura transformation between the dP_{II} (2.61) and an equation for the variable

$$w_n \equiv \frac{z_{n+1} + z_n}{z_{n+1} - z_n}, \tag{2.70}$$

we obtain from the dSP_{II} equation (2.64) the following equation for w_n:

$$(w_{n+1} + w_n)(w_n + w_{n-1}) = \frac{8\tau}{p} \frac{w_n^2 - 1}{3\mu w_n - (2a_n + 3\mu)}, \tag{2.71}$$

which is the dP34 equation that was first given in [111]. The Miura transformation

$$3\mu w_n = \frac{2\tau}{p}(1 + x_{n+1})(1 - x_n) + 2a_n + 3\mu,$$
$$a_n = n + a_0(-1)^n - \frac{1}{2}(3\mu - 1), \tag{2.72}$$

was established in [112]. The inverse Miura transformation reads

$$x_n = \frac{w_{n-1} - w_n + 2}{w_{n-1} + w_n}. \tag{2.73}$$

Exercise 2.13. Derive (2.73).

Another equation playing an important role in this scheme is the Painlevé-type equation that one obtains from the similarity constraint for the semicontinuous KdV equation (2.57), which reads

$$0 = u_n + \frac{(np - \tau)(u_{n+1} - u_{n-1})}{2p + u_{n-1} - u_{n+1}}$$
$$+ 2p\tau \frac{u_{n+1} - u_{n-1} + p\partial_\tau(u_{n+1} + u_{n-1})}{(2p + u_{n-1} - u_{n+1})^2}. \tag{2.74}$$

Using equation (2.57) itself in order to eliminate as before the derivatives with respect to τ, we can obtain the following equation for the variable

$$R_n \equiv \frac{2p}{h_n h_{n-1}}, \quad h_n \equiv 2p + u_{n-1} - u_{n+1}, \tag{2.75}$$

namely,

$$1 = ((n+1)p - 2\tau) R_{n+1} - ((n-1)p - 2\tau) R_n$$
$$+ p\tau \left(R_{n+2} R_{n+1} + R_{n+1}^2 - R_n^2 - R_n R_{n-1} \right). \tag{2.76}$$

Equation (2.76) is a third-order difference equation, which is the discrete analogue of the similarity reduction of the KdV equation, which is a third-order differential equation. It is related to the dP$_{\text{II}}$ equation via a Miura transformation that can be derived using the relation

$$h_n = p\frac{v_{n+1} + v_{n-1}}{v_n}, \tag{2.77}$$

which follows from (2.39). Using this relation and (2.66) we obtain the following expression for R_n in terms of the Schwarzian KdV variable:

$$R_n = \frac{2}{p}\frac{(z_{n-1} - z_n)^2}{(z_{n+1} - z_{n-1})(z_n - z_{n-2})}. \tag{2.78}$$

It is interesting to note that the variable R_n itself is the variable that obeys the Kac–van Moerbeke–Volterra equation; cf. also [59]. Thus we see that all these equations are interconnected on the discrete level even in the similarity reduced case.

The Miura transformation (2.72) together with the discrete Hopf-Cole transformation (2.65) constitutes on the level of the similarity reduced system a sequence of "difference substitutions"

$$\text{dSP}_{\text{II}} \xrightarrow{\text{Hopf Cole}} \text{dP}_{\text{II}} \xrightarrow{\text{Miura}} \text{dP}_{34}.$$

The existence of such a sequence of difference substitutions was already demonstrated for the classical Liouville theory in [107]; cf. also [110]. The persistence of the above scheme on the Painlevé level suggests that it would be useful to investigate the Hamiltonian aspect of the (discrete) Painlevé equations within a unified framework that includes the Schwarzian equations.

"Straight" Continuum Limit

The semicontinuous limit (2.53), which leads to the differential–difference Schwarzian equation (2.59), is obviously not the only way to obtain a semidiscrete Schwarzian equation. In fact, one can also perform a "straight" continuum limit, not involving a change of lattice variables, of the form

$$m \mapsto \infty, \quad q \mapsto \infty \quad \text{such that} \quad \frac{m}{q} \mapsto \xi, \tag{2.79}$$

together with the expansions ($z_n' \equiv \partial_\xi z_n$)

$$z_{n,m} \mapsto z_n(\xi) + \frac{1}{q}z_n'(\xi) + \frac{1}{2q^2}z_n''(\xi) + \cdots, \tag{2.80}$$

which leads to various other differential–difference KdV-type equations. For example, in this limit we obtain the differential–difference SKdV in the form

$$z_n' = -v_n^2 \implies z_{n+1}'z_n' = p^2(z_{n+1} - z_n)^2, \tag{2.81}$$

using also (2.66). Note that in comparison with (2.59) this equation is of lower order in the lattice shifts, but involves two derivatives with respect to the continuous variable. In this case the semicontinuous similarity constraint reads

$$3\mu z_n = \xi z_n' + 2n \frac{(z_{n+1} - z_n)(z_n - z_{n-1})}{z_{n+1} - z_{n-1}},\qquad (2.82)$$

from which one easily derives the following alternative Schwarzian P_{II} equation:

$$p^2\xi^2 \frac{(z_{n+2} - z_n)(z_{n+1} - z_{n-1})}{(z_{n+2} - z_{n+1})(z_n - z_{n-1})} = \left(\frac{3\mu z_n}{z_{n+1} - z_n} + \frac{3\mu z_n}{z_n - z_{n-1}} - 2n \right)$$
$$\times \left(\frac{3\mu z_{n+1}}{z_{n+2} - z_{n+1}} + \frac{3\mu z_{n+1}}{z_{n+1} - z_n} - 2(n+1) \right). \quad (2.83)$$

Equation (2.83) is a different discrete Schwarzian P_{II} equation. By construction both discrete equations (2.64) and (2.83) lead to the SP_{II} equation in an appropriate subsequent continuum limit on the remaining discrete variable n (see the next remark). However, as is suggested by the treatment in [67] the two discrete SP_{II} equations have different asymptotic behavior, and we expect, therefore, that their solutions will be given in terms of different transcendental functions.

Similar to the previous semicontinuous limit, there is a Miura chain connecting (2.83) to a discrete version of P_{II} itself and to another discrete P34. In fact, introducing the same variable w_n as before in (2.70) we can derive the following alternative version of dP34 [97]:

$$p^2\xi^2 \frac{(w_{n+1} + w_n)(w_n + w_{n-1})}{[3\mu(w_{n+1} + w_n) - 4(n+1)][3\mu(w_n + w_{n-1}) - 4n]}$$
$$= \frac{1}{4}(w_n^2 - 1), \quad (2.84)$$

which we shall refer to as alt-dP34. An alternative discretization of P_{II} is obtained from the limit of (2.48), which reads

$$\partial_\xi \log(v_{n+1} v_n) = p \left(\frac{v_{n+1}}{v_n} - \frac{v_n}{v_{n+1}} \right),\qquad (2.85)$$

together with the similarity constraint

$$\frac{1}{2}(3\mu - 1) - a_0(-1)^n = \xi \partial_\xi (\log v_n) + n \frac{v_{n+1} - v_{n-1}}{v_{n+1} + v_{n-1}},\qquad (2.86)$$

(c being a constant). Combining (2.85) and (2.86) we obtain for the variable $f_n \equiv v_{n+1}/v_n$ an equation of the form

$$3\mu - 1 = n \frac{f_n f_{n-1} - 1}{f_n f_{n-1} + 1} + (n+1) \frac{f_n f_{n+1} - 1}{f_n f_{n+1} + 1} + p\xi \left(f_n - \frac{1}{f_n} \right),\qquad (2.87)$$

to which we shall refer as *alternative* discrete Painlevé II (alt-dP$_{II}$), which is different from (2.61). Again there is a Miura transformation between the alternative dP34, (2.84), and alt-dP$_{II}$, (2.87), which is given by

$$2p\xi f_n = (w_n + 1)\left(3\mu - \frac{4(n+1)}{w_{n+1} + w_n}\right). \tag{2.88}$$

However, its inverse is nonlocal, namely[4]

$$f_n f_{n-1} = \frac{w_{n-1} + 1}{w_n - 1}. \tag{2.89}$$

We mention that the alt-dP$_{II}$ (2.87) was first given in a slightly different form in [111]. A systematic study of this equation was recently undertaken in [113].

Exercise 2.14. Derive (2.88) and (2.89) from (2.81) and (2.82) using also the definitions of w_n and f_n.

Finally, we mention that also in this limit there exists a third-order Painlevé-type of equation associated with the KdV lattice. In fact, the differential–difference equation arising from the lattice KdV (2.49) in the straight limit reads

$$u_n' + u_{n+1}' + (p + u_n - u_{n+1})^2 = p^2, \tag{2.90}$$

whereas the associated similarity constraint now reads

$$0 = u_n + pn\frac{u_{n+1} - u_{n-1}}{2p + u_{n+1} - u_{n-1}} + \xi u_n'. \tag{2.91}$$

Equation (2.91) is directly related to the dressing chain (cf., e.g., [114]) in terms of the variable $g_n \equiv p + u_n - u_{n+1}$, namely[5]

$$g_{n+1}' + g_n' = g_{n+1}^2 - g_n^2, \tag{2.92}$$

and by eliminating the derivatives with respect to the variable ξ by using (2.92) in conjunction with the similarity constraint (2.91) we obtain the third-order equation

$$1 + \frac{2p^2(n-1)}{(g_n + g_{n-1})(g_{n-1} + g_{n-2})}$$
$$- \frac{2p^2(n+1)}{(g_{n+1} + g_n)(g_n + g_{n-1})} = \xi(g_{n-1} - g_n). \tag{2.93}$$

---

[4]It can be argued that (2.89) is actually the Miura transformation, and (2.88) its inverse. However, we take the point of view that the member that contains the free parameter μ should be considered to be the actual Miura transformation as is the case in (2.72).

[5]The full dressing chain equation including shift parameters is obtained by allowing the lattice parameter p to depend on n.

Equation (2.93), which is the straight-limit analogue of (2.76), is related to (2.87) via the Miura-type of transformation

$$g_n + g_{n-1} = p \left(f_n + \frac{1}{f_{n-1}} \right). \tag{2.94}$$

In conclusion, we can maintain that the similarity approach to obtain discrete Painlevé equations is very fruitful in obtaining a clear insight into the many interrelations among the various forms in which discrete Painlevé equations arise. This approach is currently being used to obtain a deeper understanding of the higher discrete Painlevé transcendents, notably in connection with the Boussinesq family of integrable lattices; cf. [115]. However, in the next two sections we will not continue on this path but rather focus on other derivations of the discrete Painlevé equations. In Section 5 we will come back to the similarity approach in order to give a systematic derivation of Lax pairs (monodromy problems) for the discrete Painlevé equations.

3   The Painlevé Property for Discrete Systems

One of the characteristic properties of continuous integrable systems that has become over the years a powerful integrability criterion is the Painlevé property [116]. Put in a nutshell, the Painlevé property is the requirement that the movable (i.e., initial-condition-dependent) singularities be single-valued. The extension of this property to discrete systems has been elusive for years. Still, it was clear that singularities had to play a role in the integrability of discrete systems. Consider, for instance, the following rational mapping (which, as it will turn out, includes the discrete P_I):

$$x_{n+1} + x_n + x_{n-1} = a + \frac{b}{x_n}. \tag{3.1}$$

If we choose the appropriate initial conditions, it may well happen that $x_n = 0$ for some n. Since x_n appears at the denominator of the rhs of (3.1), this will lead to the divergence of x_{n+1}. Iterating the mapping, we observe that the divergence propagates. In fact, as Kruskal pointed out [117], it is not so much the fact that we find $x_{n+1} = \infty$ that is troublesome. The main difficulty comes from the fact that in iterating further we hit upon indeterminate forms of the type $\infty - \infty$, $0 \cdot \infty$, and so on. This is where single-valuedness is violated. Thus, if we wish to restore single-valuedness, we must somehow resolve the indeterminacies. The way to do this is through a continuity (with respect to the initial condition) argument. We assume that $x_n = \varepsilon$ instead of strictly zero, and expand the x_{n+1}, x_{n+2}, $\ldots$ in Laurent series of ε. Our essential requirement is that at some iteration step the singularity must disappear. This is the property we have dubbed the "singularity confinement" [118]. It plays the role of the Painlevé property for discrete systems and has turned out to be an integrability criterion of great efficiency.

In this section we will study the discrete Painlevé equations, some of which we have already encountered in the previous section, all over again from the point of view of the singularity analysis. Not only will we recover most of the results of Section 2 from a different point of view, but the singularity confinement hypothesis has been instrumental as a tool for finding new discrete Painlevé equations as well as a number of its properties; cf. the earlier review [119]. We believe that ultimately the singularity patterns one encounters in these discrete systems will give a handle on the problem of classification of ordinary difference equations much in the spirit of the original work of Painlevé and his school.

3.1   Discrete Painlevé Equations from Singularity Confinement

How can one derive a discrete Painlevé equation through the use of the singularity confinement criterion? The first all-important ingredient is a functional form for the mapping to be analyzed. There are two ways to obtain such a form. The first is to start from a straightforward discretization of a given continuous equation. One must then introduce sufficient freedom, as far as the functional form is concerned, leaving some parameters to be determined by the integrability criterion. The second method is to start from an autonomous mapping known to be integrable in terms of elliptic functions (which are the autonomous limits of the Painlevé transcendents), like the mappings we have encountered in the first lecture, and to de-autonomize it. Both methods have been used with success in the derivation of discrete Painlevé equations [120].

In order to illustrate this approach we shall perform the singularity confinement analysis on the mapping (3.1), where we consider that a and b are functions of n. Our assumption is that at some iteration step n, x_n is regular, while x_{n+1} vanishes. Following the ideas we sketched in the introduction to this section, we set $x_{n+1} = \varepsilon$. We obtain thus the following sequence of values:

$$x_{n+2} = \frac{b_{n+1}}{\varepsilon} + a_{n+1} - x_n + \mathcal{O}(\varepsilon), \tag{3.2}$$

$$x_{n+3} = -\frac{b_{n+1}}{\varepsilon} + a_{n+2} - a_{n+1} + x_n + \mathcal{O}(\varepsilon). \tag{3.3}$$

(Note that there is no way one can make the divergence of x_{n+3} disappear.) Computing x_{n+4}, we find that it diverges unless $a_{n+3} - a_{n+2}=0$. Thus for confinement a must be constant. Implementing this constraint we obtain

$$x_{n+4} = \frac{b_{n+1} - b_{n+2} - b_{n+3}}{b_{n+1}} \varepsilon + \mathcal{O}(\varepsilon^2). \tag{3.4}$$

We ask next for x_{n+5} to be finite, and we obtain the second condition

$$b_{n+1} - b_{n+2} - b_{n+3} + b_{n+4} = 0. \tag{3.5}$$

The solution of (3.5) is $b_n = \alpha n + \beta + \gamma(-1)^n$, i.e., b_n is linear in n, up to a parity-dependent constant. We can for the time being ignore this even-odd dependence (we shall come back to it in the next section). Putting $b_n = z \equiv \alpha n + \beta$ we have

$$x_{n+1} + x_n + x_{n-1} = a + \frac{z}{x_n}. \tag{3.6}$$

This is a discrete form of P_I. This last statement needs some explanation. How does one attribute a name (a number?) to a *discrete* Painlevé equation (dP)? The simplest way is to take the continuous limit of the mapping and base the naming on the continuous equation obtained. This is the method that is used traditionally. Still, it is not free of ambiguities. Another, more reasonable, method would be to name a dP on the basis of its number of effective parameters. The complication stems from the fact that, contrary to the continuous case, the dPs do not have a unique canonical form. As we shall see in what follows, alternative forms exist for all the dPs.

Let us now turn to another discrete equation,

$$x_{n+1} + x_{n-1} = \frac{a + bx_n}{1 - x_n^2}, \tag{3.7}$$

which is the McMillan map (1.24) that we studied in Section 1, and determine its integrable nonautonomous form, based on the singularity confinement property. We assume that for some n we have a regular x_n and $x_{n+1} = \sigma + \varepsilon$ where $\sigma = \pm 1$. (In this way we cover the two possibilities of x going through a root of the denominator of the rhs.). Iterating further we obtain

$$x_{n+2} = -\frac{b_{n+1} + \sigma a_{n+1}}{2\varepsilon} + \frac{a_{n+1} - \sigma b_{n+1}}{4} - x_n + \mathcal{O}(\varepsilon), \tag{3.8}$$

$$x_{n+3} = -\sigma + \frac{2b_{n+2} - b_{n+1} - \sigma a_{n+1}}{b_{n+1} + \sigma a_{n+1}} \varepsilon + \mathcal{O}(\varepsilon^2). \tag{3.9}$$

The condition for x_{n+4} to be finite reads

$$b_{n+1} - 2b_{n+2} + b_{n+3} + \sigma(a_{n+1} - a_{n+3}) = 0, \tag{3.10}$$

which leads to $a_{n+1} = a_{n+3}$ and $b_{n+1} - 2b_{n+2} + b_{n+3} = 0$. Thus we have $b_n = \alpha n + \beta (\equiv z)$ and $a_n = \delta + \gamma(-1)^n$. Again we shall, for the time being, ignore the even–odd dependence and take a as a strict constant. We obtain finally

$$x_{n+1} + x_{n-1} = \frac{a + zx_n}{1 - x_n^2}, \tag{3.11}$$

which is the discrete P_{II}, (2.61), that we have already encountered in the previous section.

Before presenting the derivation of dP_{III} let us introduce the procedure we have used in order to fix the functional form. The autonomous form of the general symmetric three-point mapping (1.19) can be written as [25]

$$f_1(x_n) - f_2(x_n)(x_{n+1} + x_{n-1}) + f_3(x_n)x_{n+1}x_{n-1} = 0, \tag{3.12}$$

where the f_i's are polynomials in x. In fact, we recall from the first section that when the f_i's are particular n-independent quartic polynomials, the mapping (3.12) is solvable in terms of elliptic functions. Next, we assume that the continuous equation is in terms of x. Expanding x_{n+1} and x_{n-1} we have

$$x_{n\pm 1} = x \pm \varepsilon x' + \frac{\varepsilon^2}{2}x'' + \mathcal{O}(\varepsilon^3). \tag{3.13}$$

Since we look for a second-order equation, we have

$$(xf_3(x) - f_2(x))x'' = f_3(x)x'^2 + \cdots, \tag{3.14}$$

where all lower order terms are made to vanish identically through an appropriate choice of the parameters. Thus, if we discretize a specific Painlevé equation, we must start by choosing f_2 and f_3 in such a way as to get $f_3(x)/(xf_3(x) - f_2(x))$ to coincide with the factor multiplying x'^2 in that equation. In the case of P_{III} we have $x'' = x'^2/x + \cdots$, which leads to $f_2 = 0$ (which has as a consequence that f_1 and f_3 are quadratic instead of quartic). We look thus for a form of dP_{III}

$$x_{n+1}x_{n-1} = \frac{g(x_n - a)(x_n - b)}{(x_n - c)(x_n - d)}, \tag{3.15}$$

where a, b, c, d, and g a priori depend on n. When does a divergence appear in the iteration of (3.15)? This happens whenever x passes through one of the roots of the denominator. Let us first consider the case where $x_{n-1} = c_{n-1}$ while x_{n-2} is free. We find thus that x_n diverges and $x_{n+1}c_{n-1} = g_n$. Iterating further we would have found that the x's obtained do not, in general, depend on the free quantity x_{n-2}. The only way to restore this dependence, i.e., to confine the singularity, at this stage is to balance the singularity of x_n by a singularity of the rhs. of (3.15). Thus x_{n+1} must be equal to a root of the denominator. The assumption $x_{n+1} = c_{n+1}$ is not acceptable because it corresponds to a periodic singularity and is thus in contradiction with the requirement that the singularity be movable. We are left with the choice $x_{n+1} = d_{n+1}$, and we have $c_{n-1}d_{n+1} = g_n$. In a symmetric way, starting from $x_{n-1} = d_{n-1}$ we obtain $c_{n+1}d_{n-1} = g_n$. Thus the first condition for confinement is

$$c_{n+1}/d_{n+1} = c_{n-1}/d_{n-1}, \qquad (3.16)$$

and c/d is periodic of period two.

This is not, however, the only source of divergence: x_n may be equal to one of the roots of the numerator. This would lead to a divergent x_{n+2} unless the divergence is balanced by an appropriate vanishing factor. Let us start with the case $x_{n-1} = a_{n-1}$. This leads to $x_n = 0$. In order to obtain a finite x_{n+2} we must have $x_{n+1} = b_{n+1}$. (As previously, the confining condition $x_{n+1} = a_{n+1}$ is not acceptable because it corresponds to a nonmovable singularity.) We find that we need $a_{n-1}b_{n+1} = g_n a_n b_n/c_n d_n = a_{n+1}b_{n-1}$, where the last equation is obtained from the case $x_{n-1} = b_{n-1}$. Thus the second confinement condition is

$$a_{n+1}/b_{n+1} = a_{n-1}/b_{n-1}. \qquad (3.17)$$

So a/b is also periodic of period two. In addition, using the expression of g above, we also have

$$a_{n-1}b_{n+1}c_n d_n = a_n b_n c_{n+1}d_{n-1}. \qquad (3.18)$$

We remark at this stage that the two conditions (3.16)–(3.17) relate n's of the same parity. Thus, the parameters with even and odd indices are related only by (3.18). If one introduces the simplifying assumption that the ratios c/d and a/b are not just periodic but strictly constant, one obtains simply dP$_{\mathrm{III}}$. A dependent variable transformation $x_n \mapsto x_n c_n/C$ allows one to replace c_n by the constant C (which we shall denote simply by c from now on). Since c/d is a constant, d is now also a constant, and from (3.18) we find that a and b are both proportional to λ^n for some λ.

We are not going to proceed further to the derivation of the higher dPs. Details can be found in [120].

Exercise 3.1. Derive the singularity confinement constraints that lead to the exact n-dependence of the parameters of the dP$_{\mathrm{IV}}$ and dP$_{\mathrm{V}}$ mappings. (Hint: The starting point for dP$_{\mathrm{IV}}$ is the expression

$$(x_{n+1} + x_n)(x_n + x_{n-1}) = \frac{Q_4(x_n)}{P_4(x_n)} \qquad (3.19)$$

and for dP$_{\mathrm{V}}$

$$(x_{n+1}x_n - 1)(x_n x_{n-1} - 1) = \frac{Q_5(x_n)}{P_5(x_n)}, \qquad (3.20)$$

where the Q's are quartic polynomials and the P's are quadratic ones.)

3.2 Continuous Limits of Discrete Painlevé Equations

How can one find a continuous limit of a discrete Painlevé equation? First, one must, through the appropriate scaling, introduce a small step in the variation of the independent variable (one recalls that n is an integer and thus varies by steps of one). Thus the independent variable at the continuous limit becomes $t = \varepsilon n$, where ε is a quantity that goes to zero. Next, we expand $x = x_n$ in a power series: $x = w_0 + \varepsilon w_1 + \varepsilon^2 w_2 + \mathcal{O}(\varepsilon^3)$. The variables x_{n+1} and x_{n-1} are further expanded including derivatives of w. We have thus

$$w_i(n \pm 1) = w_i \pm \varepsilon w_i' + \frac{\varepsilon^2}{2} w_i'' + \mathcal{O}(\varepsilon^3)). \qquad (3.21)$$

Similarly, we have $a = a_0 + \varepsilon a_1 + \varepsilon^2 a_2 + \cdots$ and also $z = z_0 + \varepsilon z_1 + \varepsilon^2 z_2 + \cdots$. The difficult part is to decide which of the z_i's is the independent variable t and which of the w_i's is the dependent variable w of the continuous equation. This requires some experience (and also quite a bit of trial-and-error experimentation). One must, of course, keep the lower terms in the expansion of x, while assuming that they are constant or that they have a given expression in terms of t.

In the case of dP$_I$, it is clear, given its form, that w_0 cannot be the variable of the continuous equation (an unbalanced w^2 would appear at lowest order, while the derivative enters only at order ε^2). Similarly, the assumption that w_1 is the right variable must be soon abandoned due to inconsistencies. We are thus led to assume that w_2 is the variable of the continuous P$_I$. To make a long story short, we find that the proper expansion is

$$x = 1 + \varepsilon^2 w, \quad z = -3 - \varepsilon^4 t, \quad a = 6, \qquad (3.22)$$

and at the limit $\varepsilon \to 0$, equation (3.6) becomes $w'' + 3w^2 + t = 0$.

The continuous limit of dP$_{II}$ can be obtained through the ansatz

$$x = \varepsilon w, \quad z = 2 + \varepsilon^3 t, \quad a = \varepsilon^3 \mu, \qquad (3.23)$$

leading to the continuous P$_{II}$: $w'' = 2w^3 + tw + \mu = 0$. At this point, a caveat is in order. While computing the continuous limit, one must be particularly cautious not to put unnecessarily to zero parameters that must be present in the final equation. It may even happen that one performs unwittingly the coalescence limit (a notion to be expended in the next section) on top of the continuous limit resulting in a "lower" equation, for instance finding P$_I$ instead of the expected P$_{II}$.

Another remark concerning the continuous limits is that a *Taylor* expansion in ε may not always be sufficient. In this case one must introduce negative powers of ε through the adequate scaling, or, equivalently, introduce a Laurent expansion of the variables and parameters. Still, we insist that finding the continuous limit is a delicate procedure, and one may miss an important finding due to an inadequate computation of a continuous limit.

For the continuous limit of dP_{III} we start with the form

$$x_{n+1}x_{n-1} = \frac{cd(x-az)(x-bz)}{(x-c)(x-d)}, \tag{3.24}$$

where $z = \lambda^n$. As expected, the continuous variable is $w(=w_0) = x$. We find, moreover, that $\lambda = 1 + \varepsilon$, $c = 1/\varepsilon - \alpha/2$, $d = -1/\varepsilon - \alpha/2$, $a = \beta\varepsilon - \gamma\varepsilon^2/2$, $b = -\beta\varepsilon - \gamma\varepsilon^2/2$, leading to the equation

$$w'' = \frac{w'^2}{w} + w^3 + \alpha w^2 + \gamma t - \frac{\beta^2 t^2}{w}, \tag{3.25}$$

which is P_{III}, albeit in a noncanonical form.

Exercise 3.2. Prove that the dP_{IV} and dP_V equations have P_{IV} and P_V, respectively, as continuous limits, as expected.

3.3 Asymmetric Forms of Discrete Painlevé Equations

In Section 3.1 we have seen that the application of the singularity confinement method leads naturally to discrete Painlevé equations. We have provided simple forms for the latter by ignoring the even–odd dependence. However, this is an unnecessary simplification that results in loss of degrees of freedom. We shall here reconsider these parity-dependent terms, starting with the example of dP_I. We have, in full generality (with $Z_n = \alpha n + \beta$),

$$X_{n+1} + X_n + X_{n-1} = \frac{Z_n + \gamma(-1)^n}{X_n} + a. \tag{3.26}$$

Let us write explicitly the mapping for even and odd indices:

$$X_{2m+1} + X_{2m} + X_{2m-1} = \frac{Z_{2m} + \gamma}{X_{2m}} + a, \tag{3.27}$$

$$X_{2m+2} + X_{2m+1} + X_{2m} = \frac{Z_{2m+1} - \gamma}{X_{2m+1}} + a. \tag{3.28}$$

Next we introduce two variables $x_m = X_{2m}$, $y_m = X_{2m+1}$, one for each parity. We can now rewrite this system as

$$y_m + x_m + y_{m-1} = \frac{z_m + \gamma}{x_m} + a, \tag{3.29}$$

$$x_{m+1} + y_m + x_m = \frac{z_m + \alpha - \gamma}{y_m} + a, \tag{3.30}$$

with $z_m = 2\alpha m + \beta$.

Thus in the new variables the mapping becomes a system of two two-point mappings. The important remark is that this mapping has now one more genuine parameter than the symmetric dP_I. A careful computation of the continuous limit of this asymmetric dP_I leads to P_{II}. In fact, putting $x = 1 + \varepsilon w + \varepsilon^2 u$, $y = 1 - \varepsilon w + \varepsilon^2 u$, $z = 1 - \varepsilon^3 m$, $a = 2$, $\gamma = -\varepsilon^3 c/4$ we obtain a first relation

$$u = \frac{1}{4}(w^2 - w' + t) \tag{3.31}$$

with $t = \varepsilon m$, leading to

$$w'' = 2w^3 + 2tw + c. \tag{3.32}$$

This shows that the richness of the parity-dependent terms should not be discarded by an argument "$(-1)^n$ does not possess a continuous limit." Rather, this even–odd dependence points to the fact that there exists another form of the discrete Painlevé equations, and it turns out that the latter is more general and has interesting properties.

In the same spirit as for dP_I we can present the asymmetric form of dP_{II},

$$y_m + y_{m-1} = \frac{z_m x_m + \gamma}{1 - x_m^2}, \tag{3.33}$$

$$x_{m+1} + x_m = \frac{(z_m + \alpha)y_m + \delta}{1 - y_m^2}, \tag{3.34}$$

where $z_m = 2m\alpha + \beta$. As a matter of fact, this form was directly obtained in [121], where we were looking for possible forms of dP_{III}. The continuous limit of (3.33)–(3.34) confirms this fact, namely that the asymmetric form of dP_{II} is indeed dP_{III}.

The most interesting result, however, in the domain of the asymmetric forms of the dP's, is the asymmetric dP_{III}. Let us consider again the final steps of the singularity confinement analysis for dP_{III}. Instead of making the unnecessary simplification that c/d and a/b are constant, we proceed in full generality. We first use the dependent variable transformation (for even and odd indices separately) and impose $c_{2n} = c_e$, $c_{2n+1} = c_o$ and thus $d_{2n} = d_e$, $d_{2n+1} = d_o$. We now find that (3.18) becomes $a_{n-1}b_{n+1}c_n d_n = a_n b_n c_{n-1}d_{n-1}$, and the same equation at step $n+1$ is $a_n b_{n+2}c_{n-1}d_{n-1} = a_{n+1}b_{n+1}c_n d_n$, i.e.,

$$a_{n+1}/a_{n-1} = b_{n+2}/b_n. \tag{3.35}$$

Combining the latter with (3.17) implies that the ratio

$$a_{n+1}/a_{n-1} = b_{n+1}/b_{n-1}$$

must be a *parity-independent* constant. Calling this ratio q, we find that both a and b are proportional to q^n. We have, in fact, $a_{2n} = q^n a_e$, $a_{2n+1} =$

$q^n a_o$ and the analogous expression for b. Moreover, (3.18) now becomes

$$qa_e b_e c_o d_o = a_o b_o c_e d_e. \tag{3.36}$$

If $q = \lambda^2$ and $a_o = \lambda a_e$, $b_o = \lambda b_e$, we recover the dP$_{\text{III}}$ case. As a last step we separate the even and odd x's, calling, for instance, the odd ones y with the redefinition $x_{2n} \mapsto x_n$, $x_{2n+1} \mapsto y_n$, and similarly for (a, b, c, d), which at odd n's will now be called (p, r, s, t). Because of the redefinition of n, what was previously called x_{n+2} is now just x_{n+1}. We obtain two coupled equations for x and y,

$$y_n y_{n-1} = \frac{st(x_n - a)(x_n - b)}{(x_n - c)(x_n - d)}, \tag{3.37}$$

$$x_{n+1} x_n = \frac{cd(y_n - p)(y_n - r)}{(y_n - s)(y_n - t)}, \tag{3.38}$$

where c, d, s, t are constants and a, b, p, r are proportional to q^n. The constraint (3.18) now becomes

$$qabst = prcd. \tag{3.39}$$

This system was derived in our work [121] and identified by Jimbo and Sakai [122] as a discrete form of P$_{\text{VI}}$. This can be confirmed through the continuous limit

$$q = 1 + \varepsilon, \quad z = q^n, \quad y = \frac{x - z}{x - 1} + \varepsilon x' z \frac{z - 1}{2(x - 1)^2} + \varepsilon^2 u, \tag{3.40a}$$

$$a = z(1 + \varepsilon A), \quad b = \frac{z}{q1 + \varepsilon A}, \quad c = 1 + \varepsilon C, \quad d = \frac{1}{c}, \tag{3.40b}$$

$$p = z(1 + \varepsilon P), \quad r = \frac{z}{1 + \varepsilon P}, \quad s = 1 + \varepsilon S, \quad t = \frac{1}{s}. \tag{3.40c}$$

This automatically satisfies (3.39). We obtain, for $\varepsilon \to 0$,

$$x'' = \frac{x'^2}{2}\left(\frac{1}{x} + \frac{1}{x-1} + \frac{1}{x-z}\right) - x'\left(\frac{1}{z} + \frac{1}{z-1} + \frac{1}{x-z}\right)$$
$$+ \frac{x(x-1)(x-z)}{z^2(z-1)^2}\left(\alpha + \frac{\beta z}{x^2} + \frac{\gamma(z-1)}{(x-1)^2} + \frac{\delta z(z-1)}{(x-z)^2}\right) \tag{3.41}$$

with $\alpha = 2S^2$, $\beta = -2P^2$, $\gamma = 2C^2$, and $\delta = -2A(A + 1)$. Thus dP$_{\text{VI}}$, for which no symmetric form has been obtained yet, can be expressed as a system that is nothing but the asymmetric dP$_{\text{III}}$.

We leave it to the courageous reader to show that dP$_{\text{IV}}$ and dP$_{\text{V}}$ also possess asymmetric forms that at the continuous limit lead again to P$_{\text{VI}}$.

The even–odd dependence of some parameters is not the only one that can result from the application of singularity confinement. One can very well find ternary or higher symmetries. We shall not enter into these details. The treatment of such cases is analogous to that of the cases with even–odd dependence. One introduces a sufficient number of components and converts the mapping to a system (where all equations but two are just local, algebraic relations).

3.4   Alternative Forms of dPs: Limits and Degeneracies

We have already pointed out that the discrete systems are more fundamental objects than the continuous ones. The latter can always be considered as just limiting cases of the former, but the discrete systems have more freedom. The same is evidently true for the discrete Painlevé equations. One domain where this difference becomes manifest is that of particular cases of the discrete Painlevé equations, by taking limits of the parameters. While the initial dP has some continuous Painlevé equation as its continuous limit, the continuous limit of the discrete equation where some of the parameters are taken to specific values may not be the same continuous Painlevé equation. In this sense, if one starts from a dP (with a given name) and takes special limits, one can end with a dP with a different name. (This is admittedly a semantic difference: It is based essentially on the fact that we are naming the discrete equations after their continuous limits).

Before examining realistic examples based on dP_{II} or higher equations, let us point out that we can also consider the limiting cases of dP_I. If we put $a = 0$ in (3.6) we obtain a nonautonomous integrable equation

$$x_{n+1} + x_n + x_{n-1} = z/x_n, \tag{3.42}$$

which *does not* have a continuous limit. (In [123] we have dubbed this equation dP_0, but we must not seek a particular meaning to this name, other that it is an equation that has less freedom than dP_I).

Let us now turn to dP_{II} and introduce a scaling of all variables and parameters so as to transform the equation to

$$x_{n+1} + x_{n-1} = \frac{zx_n + a}{\rho^2 - x_n^2}. \tag{3.43}$$

By taking the $\rho \to 0$ limit we obtain the equation

$$x_{n+1} + x_{n-1} = \frac{z}{x_n} + \frac{a}{x_n^2}, \tag{3.44}$$

which is a well known form of dP_I.

Exercise 3.3. Find the continuous limit of (3.44).

Similarly, one can study limits of dP$_{\text{III}}$, which we shall rewrite here for convenience as

$$x_{n+1}x_{n-1} = -\frac{\gamma x_n^2 + \zeta x_n + \mu}{\alpha x_n^2 + \beta x_n + \gamma}. \tag{3.45}$$

The full dP$_{\text{III}}$ corresponds to $\gamma \neq 0$. We find (through application of the singularity confinement criterion) that $\zeta = \zeta_0 \lambda^n$ and $\mu = \mu_0 \lambda^{2n}$. The interesting limits correspond to $\gamma = 0$. We first obtain the equation (still for $\zeta \propto \lambda^n, \mu \propto \lambda^{2n}$)

$$x_{n+1}x_{n-1} = \frac{\zeta x_n + \mu}{(x_n + \beta)x_n}. \tag{3.46}$$

This is a novel form of dP$_{\text{II}}$. Its continuous limit can be obtained through $x = 1 + \varepsilon w$, $\beta = -2 + \varepsilon^3 g$, $\zeta = -2\lambda^n$, $\mu = \lambda^{2n}$, where $\lambda = 1 + \varepsilon^3/2$, leading to $w'' = 2w^3 + wt + g$. A further limit can be obtained, starting from (3.30), by taking $\beta = 0$, in addition to $\gamma = 0$. In this case we obtain

$$x_{n+1}x_{n-1} = \frac{\zeta}{x_n} + \frac{\mu}{x_n^2}, \tag{3.47}$$

where, by the gauge $x \mapsto x\lambda^{n/2}$, μ can be taken as a constant and ζ of the form $\zeta_0 \lambda^{n/2}$. Equation (3.47) is a discrete P$_\text{I}$, as can be seen from the continuous limit obtained through $x = 1 + \varepsilon^2 w$, $\zeta = 4\kappa^n$, $\mu = -3$, and $\kappa (\equiv \lambda^{1/2}) = 1 - \varepsilon^5/4$, leading to $w'' = 6w^2 + t$. More limiting cases based on equations dP$_{\text{IV}}$ and dP$_{\text{V}}$ can be found in [124].

The cases we have named degenerate need some explanation. The way we have obtained the discrete Painlevé equations was to assume a functional form and apply singularity confinement. The important assumption at that stage is that the functional form is fixed. Moreover, our analysis was limited to the generic case without considering nongeneric ones, which can lead to different results. Let us illustrate this in the case of dP$_{\text{II}}$. The starting point for the application of singularity confinement is an autonomous mapping of the form

$$x_{n+1} + x_{n-1} = \frac{a + bx_n}{\rho^2 - x_n^2}. \tag{3.48}$$

Now, suppose that the numerator exactly divides the denominator. In this case the autonomous starting point is

$$x_{n+1} + x_{n-1} = \frac{\kappa}{x_n + \rho}. \tag{3.49}$$

In this case, the singularity confinement analysis we have performed for the full dP$_{\text{II}}$ is no longer valid: The singularity structure of (3.49) is completely different. What we obtain (after translating x so as to put ρ to zero) is a dP of the form

$$x_{n+1} + x_{n-1} = \frac{z}{x_n} + a, \tag{3.50}$$

where again z is linear in n, and a is a constant. The continuous limit is obtained through $x = 1 + \varepsilon^2 w$, $a = 4$, $z = -2 - \varepsilon^5 n$, leading at $\varepsilon \to 0$ to $w'' + 2w^2 + t = 0$, with $t = \varepsilon n$.

Cases like the one just analyzed, obtained through some common factor cancellation in the autonomous form, are called (by us) degenerate. The main idea is first to reduce the fraction on the rhs of the autonomous mapping and then use singularity confinement to deautonomize it properly.

The same procedure as above has been used in the case of dP_{III}. One obtains

$$x_{n+1}x_{n-1} = -\frac{ax_n + b}{cx_n + d}, \tag{3.51}$$

and the deautonomization of this equation yields $a = a_0\lambda^n$ and $d = d_0\lambda^n$. Unless $c = 0$, we can always take $c = 1$, through division, and a proper gauge allows us to take $b = 1$. Equation (3.51) in its nonautonomous form is a novel form of discrete P_{II}.

It goes without saying that one can combine degeneracies and limits. For instance, the limit $d = 0$ in (3.51) leads to the equation ($c = 1$)

$$x_{n+1}x_{n-1} = a + \frac{1}{x_n}, \tag{3.52}$$

where $a = a_0\lambda^n$. This is another form of dP_I. The continuous limit is obtained through $x = x_0(1 + \varepsilon^2 w)$, where $x_0^3 = -\frac{1}{2}$, $a = 3x_0^2\lambda^n$, with $\lambda = 1 - \varepsilon^5/3$, leading to $w'' + 3w^2 + t = 0$. An equivalent equation can be obtained from (3.51) by taking $a = 0$:

$$x_{n+1}x_{n-1} = \frac{1}{x_n + d}. \tag{3.53}$$

Equation (3.53) is transformed into (3.52) by taking $x \mapsto 1/x$ and exchanging a, d. Another limit, leading to another dP_I, is $c = 0$. We obtain

$$x_{n+1}x_{n-1} = ax_n + b, \tag{3.54}$$

where a, here, is a constant and $b = b_0\lambda^n$ with continuous limit $w'' + 6w^2 + t = 0$ obtained through $x = 1 + \varepsilon^2 w$, $a = 2$, $b_0 = -1$, and $\lambda = 1 + \varepsilon^5$. An equivalent equation can also be obtained by taking $b = 0$ in (3.51). We obtain

$$x_{n+1}x_{n-1} = \frac{ax_n}{x_n + d}. \tag{3.55}$$

Equations (3.54) and (3.55) are related through the transformation $x \mapsto 1/x$, with the appropriate relations of the parameters.

We leave to the courageous reader the task of investigating the limits and degeneracies of the dP_{IV} and dP_V families. To the less courageous but still interested one we indicate that a detailed analysis of these cases has been presented in [124]. Still, since this work has one omission, we present, here another exercise based on the dP_V family. It was shown in [124] that the limit of dP_V that reads

$$(x_{n+1}x_n - 1)(x_n x_{n-1} - 1) = \frac{\kappa(x_n^2 + 1) + \mu x_n}{\alpha x_n + \beta} \qquad (3.56)$$

is, in fact, a dP_{IV}, while the $\beta = 0$ limit of (3.56) is a dP_{34} (This curious numbering will be explained in the next section). Now, (3.56) has a degenerate form whenever the denominator divides the numerator:

$$(x_{n+1}x_n - 1)(x_n x_{n-1} - 1) = px_n + q. \qquad (3.57)$$

Singularity confinement leads to $p = p_0 \lambda^{3n}$, $q = q_0 \lambda^{2n}$. The equation thus obtained is a dP_{II} with continuous limit $w'' = 6w^3 + 12tw + \mu$ obtained through $x = 1/\sqrt{3} + w$, $\lambda^n = 1 + \varepsilon^2 t$, $q_0 = \frac{4}{3}$, $p_0 = -8\sqrt{3}/9 - 2\varepsilon^3\mu/3$.

3.5   Other Methods to Derive Discrete Painlevé Equations

In the previous sections we have seen two of the main methods for the derivation of discrete Painlevé equations: similarity reductions of lattices and singularity confinement. In the last section we shall encounter another method, that of the direct derivation of the linear problem associated to a dP. This Lax-pair approach is also intimately related to the orthogonal polynomial method, discrete AKNS approach, and so on. We shall study the Lax pairs for discrete Painlevé equations extensively in the next section. In this section we shall present still other methods that lead to discrete Painlevé equations.

A first method is through the discrete analogues of Miura transforms. We have encountered Miura transformations in connection with discrete Painlevé equations already in Section 2. Historically, the Miura transform was found as the transformation relating the KdV to modified KdV, and we have seen that this connection persists on the level of the lattice equations as well. Thus, we could use the term "modified" also for the dP's, namely starting with a given dP and using a discrete Miura one obtains a different modified dP. This was demonstrated for the first time in [112] where the discrete Miura between dP_{II} and its modified version, namely dP_{34}, (cf. (2.71)), was given in the form[6]

---

[6]It should be pointed out, however, that as a similarity reduction, following the

$$y_n = (1 + x_n)(1 - x_{n+1}) - \frac{\bar{z}}{2} \qquad (3.58)$$

(where $\bar{z} = (z_n + z_{n+1})/2$), which brings dP$_{\mathrm{II}}$, in the form (3.11), over into

$$(y_{n+1} + y_n)(y_n + y_{n-1}) = \frac{4y^2 - m^2}{y + \bar{z}/2}, \qquad (3.59)$$

which is again the discrete form of P$_{34}$, as is confirmed by its continuum limit.

Exercise 3.4. Find the continuum limit of (3.59).

We must point out that this presentation, although following the traditional approach, is not fully satisfactory. We prefer a more detailed one that starts from a Miura transform being given *as a system*, i.e., (3.58) complemented by

$$x_n = \frac{m + y_n - y_{n-1}}{y_n + y_{n-1}}. \qquad (3.60)$$

Eliminating either of the x or y variables, one obtains the corresponding equation for the other one, i.e., dP$_{34}$ (3.59) or P$_{\mathrm{II}}$ (3.11), respectively. Thus the two equations are treated in a unified way. Still, from the scope of this section (providing new dPs) the asymmetric treatment consisting of (3.58) on top of (3.11) is sufficient.

Discrete Miuras exist, of course, for the other forms of dP$_{\mathrm{II}}$. Let us show this for the form (3.57) of dP$_{\mathrm{II}}$ we encountered in the last section. In this case the Miura system reads

$$y_n = \frac{\lambda^n(rx_n + \lambda^n)}{x_n x_{n+1} - 1}, \qquad (3.61)$$

$$x_n = \frac{\lambda^n(r^2 y_n + \lambda)}{r\lambda(y_n y_{n-1} - 1)}. \qquad (3.62)$$

Eliminating y between (3.61)–(3.62) leads to dP$_{\mathrm{II}}$ (3.57) with $p_0 = r/\lambda^2$, $q_0 = r^2/\lambda$, while elimination of x leads to the dP$_{34}$

$$(y_n y_{n+1} - 1)(y_n y_{n-1} - 1) = \frac{\lambda^{2n-1}(r^2 y_n + \lambda)(r^2 + y_n \lambda)}{r^2(y_n + \lambda^{2n})}. \qquad (3.63)$$

(This last equation can be obtained from equation (6.7) of [124] as a special limit, $a = 0$.)

A second method that has produced several new dPs is the use of auto-Bäcklund transforms of continuous Painlevé equations [111]. Here is the

---

scheme that was elaborated at length in Section 2, the continuous P$_{34}$ is not the modified P$_{\mathrm{II}}$, but the reverse: P$_{\mathrm{II}}$ is the modified P$_{34}$. This is perhaps an explanation as to why the "modified" terminology has not caught on in the case of the Painlevé equations.

general principle. The auto-Bäcklund transformation of a given Painlevé equation is a relation of the form

$$w(t, \tilde{\alpha}) = F(w'(t, \alpha), w(t, \alpha), t), \tag{3.64}$$

where $w(t, \alpha)$ is a solution of a Painlevé equation corresponding to the parameter α, and $w(t, \tilde{\alpha})$ a solution of the same equation corresponding to some other value $\tilde{\alpha}$ of the parameter. It suffices to find still another value $\underset{\sim}{\alpha}$ of the parameter for which a relation of the form

$$w(t, \underset{\sim}{\alpha}) = G(w'(t, \alpha), w(t, \alpha), t) \tag{3.65}$$

exists and eliminate $w'(t, \alpha)$ between the two. This task is considerably facilitated by the fact that, as was shown in [125], the only auto-Bäcklund transform linear in w' preserving the Painlevé property is the one involving Riccati forms

$$w(t, \tilde{\alpha}) = \frac{w' + aw^2 + bw + c}{dw^2 + ew + f} \tag{3.66}$$

with $a, b, \dots, f$ functions of t alone. Once w' is eliminated, one obtains a discrete equation relating $w(\tilde{\alpha})$, $w(\underset{\sim}{\alpha})$, and $w(\alpha)$, where t enters as a simple parameter, the mapping's independent variable being α. Let us just illustrate here this approach in the case of P_{II}. If $w(t, \alpha)$ is a solution of P_{II} $w'' = 2w^3 + tw + \alpha$, then

$$w(t, -\alpha) = -w(t, \alpha), \tag{3.67}$$

$$w(t, \alpha + 1) = -w(t, \alpha) - \frac{1 + 2\alpha}{2w^2(t, \alpha) + 2w'(t, \alpha) + t}, \tag{3.68}$$

are also solutions of the same equation with the appropriate α. Combining these two auto-Bäcklund transforms, we construct $w(t, \alpha - 1)$ and eliminate w'. We thus obtain

$$\frac{z_n}{x_{n+1} + x_n} + \frac{z_{n-1}}{x_{n-1} + x_n} = -2x_n^2 - t, \tag{3.69}$$

where $x_n = w(t, \alpha)$ and $z_n = \alpha + \frac{1}{2}$. This equation is another discrete form of P_I.

Let us give a further example obtained from the auto-Bäcklund of P_{III}. It is known that the P_{III}

$$w'' = \frac{w'^2}{w} - \frac{w'}{t} + \frac{1}{t}(\alpha w^2 + \beta) + \gamma w^3 + \frac{\delta}{w} \tag{3.70}$$

has the following auto-Bäcklund (we limit ourselves here to the $\gamma\delta \neq 0$ case, which can be scaled to $\gamma = 1$, $\delta = -1$):

$$w(-\alpha, -\beta) = -w(\alpha, \beta), \tag{3.71}$$

$$w(-\beta, -\alpha) = w^{-1}(\alpha, \beta), \tag{3.72}$$

$$w(-\beta - 2, -\alpha - 2) = w \left(1 + \frac{2 + \alpha + \beta}{t(w' + w^2 + 1)/w - 1 - \beta} \right) \tag{3.73}$$

(where in the last relation the w in the rhs must be understood as $w(\alpha, \beta)$). We consider the case $\alpha \neq \beta$. Using (3.72)–(3.73) (and the analogue of (3.73) starting from $w(-\beta, -\alpha)$, which leads to $w(\alpha - 2, \beta - 2)$), we can eliminate w' and obtain a relation between $w(\alpha - 2, \beta - 2)$, $w(\alpha, \beta)$, and $w(\alpha + 2, \beta + 2)$, i.e., a one-dimensional 3-point mapping on the (α, β)-plane. We introduce the independent variable $z = (\alpha + \beta + 2)/4$ (so as to have $\Delta z = 1$) and the two parameters μ, κ through $\mu = (\beta - \alpha - 2)/4$, $\kappa = -it/2$. We choose $x = i/w$ as the mapping variable and obtain

$$\frac{z_n}{x_{n+1}x_n + 1} + \frac{z_{n-1}}{x_n x_{n-1} + 1} = \kappa\left(-x_n + \frac{1}{x_n}\right) + z_n + \mu. \tag{3.74}$$

It is, moreover, straightforward to put $\kappa = 1$ through scaling. So the final form of this equation is

$$\frac{z_n}{x_{n+1}x_n + 1} + \frac{z_{n-1}}{x_n x_{n-1} + 1} = -x_n + \frac{1}{x_n} + z_n + \mu, \tag{3.75}$$

which is the alt-dP$_{\text{II}}$ equation (2.87) of the previous section. The continuous limit is obtained through $x = 1 + \varepsilon u$, $z = 2 - \varepsilon^2 t/2$ with $t = \varepsilon n$, $\mu = \varepsilon^3(\frac{1}{4} - m/2)$. At the limit $\varepsilon \to 0$ we obtain $u'' = 2u^3 + tu + m$, i.e., the P$_{\text{II}}$ equation. This means that (3.75) is an alternative form of the dP$_{\text{II}}$ equation: Its properties have been studied in great detail in [113]. In conclusion, we observe that Bäcklund transformations for continuous Painlevé equations yield discrete Painlevé equations belonging to the "alternative family" that also belong to the second continuum limit exhibited in Section 2.5. For completeness we mention that such Bäcklund transformations were studied extensively in the past by the Minsk school; cf., e.g., [126].

The last method we shall present here is the discrete dressing approach. It is well known in the continuous case that proper reductions of the dressing chain lead to Painlevé equations; cf. [114,127]. The same holds true for the discrete dressing. (The details of this approach are unpublished as yet, but we do not despair of having them ready for publication in some foreseeable future.) To make a long story short, we find the discrete dressing equation to be

$$\frac{R_{n+1,m}}{R_{n+2,m-1}} + \frac{R_{n+2,m}}{R_{n+1,m}} = \frac{R_{n+1,m}}{R_{n,m}} + \frac{R_{n,m+1}}{R_{n+1,m}} + \Lambda(n), \tag{3.76}$$

where Λ is a function depending on n only.

Periodicity assumptions on R may lead to discrete Painlevé equations.

Exercise 3.5. Prove that (3.76), under the hypothesis $R_{n+2,m} = R_{n,m}$ (and thus $\Lambda(n + 2) = \Lambda(n)$), reduces to the alternative dP$_{\text{II}}$ (3.75).

Even more interesting is to prove that under the assumption $R_{n+3,m} = R_{n,m+1}$ one can find another dP. We shall not present here any of the tedious details but give the final result

$$(y_{n-1} - 1)\left(\frac{z_n}{x_n} + 1 - y_n\right) = x_n(y_n - 1) + a, \qquad (3.77)$$

$$(x_{n+1} - 1)\left(\frac{z_n - c}{y_n} + 1 - x_n\right) = y_n(x_n - 1) - b, \qquad (3.78)$$

with $a + b + c = z_n - z_{n+1}$. This equation is an alternative dP$_{\text{IV}}$. Its continuous limit can be obtained if we start with $x = 1 + \varepsilon w$, $y = 1 + \varepsilon u$, $z = 1 - 2\varepsilon t$, $a = \varepsilon^2 \alpha$, $b = \varepsilon^2 \beta$, and thus $c = \varepsilon^2(2 - \alpha - \beta)$.

We do not believe that the methods we have enumerated here exhaust the possibilities for the derivation of discrete Painlevé equations. Still, their systematic use has allowed us over the past few years to establish a fairly complete list of discrete analogues to the Painlevé equations.

3.6   Discrete Painlevé Equations and Beyond

Our analysis in this section has focused on discrete Painlevé equations given as three-point mappings with some extension to systems of two two-point mappings. In fact, this analysis of dPs is part of a vast program of ours in which we have established the equivalent of the Gambier classification for discrete systems in the form of three-point mappings [128].

Several questions remain open at this moment concerning the discrete analogues of integrable differential equations. In this study we have limited ourselves to equations of the first and second order explicit in the highest derivative. However, there exists a class of equations where the highest derivative enters through some power larger than one: the Briot–Bouquet first order equations [12] $(w')^n = f(w, z)$ and the equations studied by Bureau [129] and Cosgrove [100] at second order, of the form $(w'')^n = f(w', w, z)$. Their discrete analogues have never been studied (at least to the authors' knowledge). Even without resorting to nonlinearities of the highest derivative, equations in the form of a system of two two-point mappings have not been studied in any generality. Their study is all the more interesting (and difficult), in that several examples of integrable mappings $x_{n+1} = f(x_n, y_n, n)$, $y_{n+1} = g(x_n, y_n, n)$ in non-QRT form do exist [111]. Moreover, the guide of the continuous limit does not exist here, since the study of continuous systems $w' = f(w, u, z)$, $u' = g(w, u, z)$ is far from complete. The same is true for third-order equations $w''' = f(w'', w', w, z)$. The study of the corresponding 4-point mappings is only schematic at this moment, although it presents considerable interest: In fact, in [130] it was shown that the natural form of a q-deformed dP$_{\text{I}}$ is the 4-point equation

$$\beta q^{-n}(x_{n+1} - q^2 x_{n-2}) = q\frac{(n)_q + \alpha q^n}{x_n} - \frac{(n-1)_q + \alpha q^{n-1}}{x_{n-1}}, \qquad (3.79)$$

where $(n)_q = (q^n - 1)/(q - 1)$. The discrete analogue of higher-order equations would also be of great interest, since it would allow us to obtain discrete forms for well-known integrable Hamiltonian systems.

Moving away from the purely discrete systems, one encounters further interesting problems. In the domain of differential–difference equations we have already proposed in [131] an integrability criterion that blends singularity confinement with the Painlevé singularity analysis. A first application of this method has revealed the existence of the delay–differential equivalents to the Painlevé equations. Here also, there exist several possibilities combining higher-order derivatives with higher-order mappings. We expect the equations that will result from such studies (and the ones mentioned in the previous paragraphs) to be much richer than the ones discovered one century ago by Painlevé and Gambier.

3.7   Bilinearization of the Discrete Painlevé Equations

As an application of the discrete singularity analysis let us now study the bilinearization of the discrete Painlevé equations. The Hirota bilinear formalism has been one of the most powerful tools for the study of integrable evolution equations. Curiously, its use in the case of the Painlevé equations has been, at the least, hesitant. This is all the more curious in that the solutions of these equations are meromorphic in the complex plane of the independent variable and should thus possess simple expressions in terms of ratios of entire functions. (This is precisely what the Hirota formalism does: Introducing a dependent-variable transformation, it allows the expression of the latter in terms of the τ-functions, which are assumed to be entire). However, the systematic application of the bilinear approach to the Painlevé equations had to wait till very recently: Only in 1992 did Hietarinta and Kruskal [132] give the standard bilinear forms of the first five Painlevé transcendents. The bilinearization of P_{VI} had to wait even longer, since we have produced it only this year [133].

In this section we shall present the bilinearization of the first three dPs. Before proceeding to the actual computations let us examine the singularity structure of some simple cases and try to put it to use in the choice of the dependent variable transformation. The two cases that we will examine here are the standard dP_I (3.6) and dP_{II} (3.11). Here z is linear in the independent variable n, i.e., $z = \alpha n + \beta$. For the needs of the present section a schematic singularity structure will suffice. In the case of dP_I we have just seen that a singularity appears whenever the x_n in the denominator happens to vanish. This has as a consequence that both x_{n+1} and x_{n+2} diverge, whereupon x_{n+3} vanishes again and x_{n+4} is finite (i.e., the singularity is indeed confined). Thus the singularity pattern is $\{0, \infty, \infty, 0\}$. In the case of dP_{II} a singularity appears whenever x_n in the denominator takes the value $+1$ or -1. Thus we have two singularity patterns, which in this case turn out to be $\{-1, \infty, +1\}$ and $\{+1, \infty, -1\}$.

How can we use this information in order to express x in terms of τ-functions? Let us start with dP$_I$. Since we have only one singularity pattern, we expect one τ-function to suffice. As τ-functions are entire, x must be a ratio of products of such functions. Hence, let us assume that x_n contains a τ-function F_n in the numerator and that F_n passes through zero, so, $x_n = 0$. Since x_{n+1} and x_{n+2} are infinite, the denominator of x must contain F_{n-1} and F_{n-2} (which ensures that F_n appears in the denominators of x_{n+1} and x_{n+2}, respectively). Finally, since x_{n+3} vanishes, x_n must contain F_{n-3} in the numerator. Thus, the expression of x, dictated by the singularity pattern, is

$$x_n = \frac{F_n F_{n-3}}{F_{n-1} F_{n-2}}. \tag{3.80}$$

As we shall see in the following sections, this expression suffices for the bilinearization (in fact, trilinearization) of dP$_I$. That the choice (3.80) is a reasonable one can also be seen through the continuous limit of this expression. We know, for dP$_I$, that the continuous limit is obtained through $x = 1 + \varepsilon^2 w$ at $\varepsilon \to 0$. Implementing this limit on (3.80) we obtain $w = 2\partial_z^2 \log F$, a transformation that is at the base of the (continuous) Hirota bilinear formalism.

In the case of dP$_{II}$ we have two singularity patterns, and so we expect two τ-functions to appear in the expression of x. Let us start with the pattern $\{-1, \infty, +1\}$. The divergent x may be related to a vanishing τ-function, say F, in the denominator. In order to ensure that x_{n-1} and x_{n+1} are respectively -1 and $+1$, we choose x_n in the form $x_n = -1 + pF_{n+1}/F_n = 1 + qF_{n-1}/F_n$, where p, q must be expressed in terms of the second τ-function G. We turn now to the second pattern $\{+1, \infty, -1\}$ related to the vanishing of the τ-function G. We obtain in this case $x_n = 1 + rG_{n+1}/G_n = -1 + sG_{n-1}/G_n$, where r, s are expressed in terms of F. Combining the two expressions in terms of F and G we obtain the following simple expression for x,

$$x_n = -1 + \frac{F_{n+1} G_{n-1}}{F_n G_n} = 1 - \frac{F_{n-1} G_{n+1}}{F_n G_n}, \tag{3.81}$$

which satisfies both singularity patterns. The continuous limit of (3.81) reads $w = \partial_z \log F/G$, which is the expected transformation in the case of P$_{II}$. Expressions (3.81) can be written in a way that recalls the bilinear formalism for continuous systems. The Hirota D operator plays an important part here also. Starting with D, defined through its action on the dot product $D(f \cdot g) = (\partial_x - \partial_{x'})f(x)g(x')|_{x=x'} = f'(x)g(x) - f(x)g'(x)$, we introduce the shift operator e^D. Its action on a dot product is $e^{\lambda D} f \cdot g = f(x + \lambda)g(x - \lambda)$, and thus $F_{n+1} G_{n-1}$ can be obtained simply as $e^D F \cdot G$, where e^D operates on the discrete variable n.

We start our systematic construction of bilinear expressions for dPs with the second discrete Painlevé equation. As explained in the previous section, two τ-functions are needed here, related to the nonlinear variable through (3.81), and thus we expect dP$_{II}$ to be given as a system of two bilinear relations. Equation (3.81) does indeed provide the first equation of the system. By eliminating the denominator $F_n G_n$ we obtain

$$F_{n+1}G_{n-1} + F_{n-1}G_{n+1} - 2F_n G_n = 0. \qquad (3.82)$$

In order to obtain the second equation we rewrite dP$_{II}$ as $(x_{n+1} + x_{n-1}) \times (1 - x_n)(1 + x_n) = zx_n + a$. We use the two possible definitions of x_n in terms of F, G in order to simplify the expressions $1 - x_n$ and $1 + x_n$. Next, we obtain two equations by using each of these two definitions for x_{n+1} combined with the alternative definition for x_{n-1}. We obtain thus

$$F_{n+2}F_{n-1}G_{n-1} - F_{n-2}F_{n+1}G_{n+1} = F_n^2 G_n(zx_n + a), \qquad (3.83)$$

$$G_{n-2}G_{n+1}F_{n+1} - G_{n+2}G_{n-1}F_{n-1} = G_n^2 F_n(zx_n + a). \qquad (3.84)$$

Finally, we add equation (3.83) multiplied by G_{n+2}, and (3.84) multiplied by F_{n+2}. Up to the use of the upshift of (3.82), a factor $F_{n+1}G_{n+1}$ appears in both sides of the resulting expression. After simplification, the remaining equation is indeed bilinear,

$$F_{n+2}G_{n-2} - F_{n-2}G_{n+2} = z(F_{n+1}G_{n-1} - F_{n-1}G_{n+1}) + 2aF_n G_n, \qquad (3.85)$$

where a symmetric expression was used for x_n in the rhs., obtained as the arithmetic mean of the two rhs. of (3.81). Equations (3.82) and (3.85), taken together, are the bilinear form of dP$_{II}$.

In the case of dP$_I$ we shall examine several of its forms starting with

$$x_{n+1} + x_{n-1} = z_n/x_n + a. \qquad (3.86)$$

Here, the singularity pattern is $\{0, \infty, a, \infty, 0\}$. This pattern suggests the transformation

$$x = \frac{F_{n+2}F_{n-2}}{F_{n+1}F_{n-1}}. \qquad (3.87)$$

Substituting back into (3.86) we obtain readily

$$F_{n+3}F_{n-1}F_{n-2} - F_{n-3}F_{n+1}F_{n+2}$$
$$= z_n F_{n+1}F_n F_{n-1} + aF_{n+2}F_n F_{n-2}. \qquad (3.88)$$

This trilinear form cannot be reduced further. Let us now go back to the "standard" dP$_I$ given by

$$x_{n+1} + x_n + x_{n-1} = z_n/x_n + a. \qquad (3.89)$$

As we have seen above, the proper dependent-variable transform, dictated by the singularity patterns of dP$_I$, $\{0, \infty, \infty, 0\}$, is

$$x_n = \frac{F_{n+1}F_{n-2}}{F_n F_{n-1}}.$$

Instead of substituting into (3.89) we use the discrete derivative of the latter,

$$x_{n+2} - x_{n-1} = z_{n+1}/x_{n+1} - z_n/x_n. \tag{3.90}$$

By reducing to common denominator we obtain the trilinear form

$$F_{n+3}F_{n-1}F_{n-2} - F_{n-3}F_{n+1}F_{n+2}$$
$$= z_{n+1}F_{n+1}^2 F_{n-2} - z_n F_{n-1}^2 F_{n+2}, \tag{3.91}$$

which, again, cannot be reduced further. So the standard dP_I, too, does not possess a bilinear form but rather a trilinear one.

The last form of dP_I we shall treat here is a multiplicative form of dP_I, which reads

$$x_{n+1}x_{n-1} = z_n/x_n + a/x_n^2. \tag{3.92}$$

Here z_n is not an affine function of n, but rather $z_n = z_0\lambda^n$. The singularity pattern of this equation is $\{0, \infty, 0\}$, and thus the transformation $x_n = F_{n+1}F_{n-1}/F_n^2$ should be used here. We find that a full F^2 factor drops out, and we obtain

$$F_{n+2}F_{n-2} = z_n F_{n+1}F_{n-1} + aF_n^2. \tag{3.93}$$

Thus the dP_I (3.92) has a bilinear expression.

The final example we shall present is dP_{III}. Here two singularity patterns exist, $\{c, \infty, d\}$ and $\{d, \infty, c\}$. The existence of these two singularity patterns suggests the introduction of two τ-functions F, G:

$$x_n = c\left(1 + \frac{F_{n+1}G_{n-1}}{F_n G_n}\right) = d\left(1 + \frac{F_{n-1}G_{n+1}}{F_n G_n}\right). \tag{3.94}$$

The singularity patterns described above correspond to the cases where either F or G passes through zero. Starting from this ansatz the bilinear form of dP_{III} was given in [5]:

$$cF_{n+1}G_{n-1} - dF_{n-1}G_{n+1} + (c-d)F_n G_n = 0, \tag{3.95}$$

$$\frac{cd}{c-d}(cF_{n+2}G_{n-2} - dG_{n+2}F_{n-2}) + (c-a)(d-b)F_n G_n$$
$$+ c(d-b)F_{n+1}G_{n-1} + d(c-a)G_{n+1}.F_{n-1} = 0 \tag{3.96}$$

The first equation comes just from equating the two expressions of x_n. However, this is not the simplest form. In fact, as we have seen, two "potential" singularities exist when x passes through a root of the numerator. The patterns are $\{a_{n-1}, 0, b_{n+1}\}$ and $\{b_{n-1}, 0, a_{n+1}\}$, and we remark that no singularity actually develops. This suggests the introduction of two more auxiliary τ-functions J, K in the following manner:

$$x_n = c\left(1 + \frac{F_{n+1}G_{n-1}}{F_n G_n}\right) = d\left(1 + \frac{F_{n-1}G_{n+1}}{F_n G_n}\right) = \frac{J_n K_n}{F_n G_n}, \qquad (3.97)$$

$$\frac{1}{x_n} = \frac{1}{a}\left(1 + \frac{J_{n+1}K_{n-1}}{J_n K_n}\right) = \frac{1}{b}\left(1 + \frac{J_{n-1}K_{n+1}}{J_n K_n}\right) = \frac{F_n G_n}{J_n K_n}. \qquad (3.98)$$

These equations, in turn, imply the dP$_{\text{III}}$ for x. By equating the expressions of x and $1/x$ we have four equations for the four τ-functions, and thus the bilinearization is trivially obtained:

$$c(F_n G_n + F_{n+1}G_{n-1}) = d(F_n G_n + F_{n-1}G_{n+1}) = J_n K_n, \qquad (3.99)$$

$$\frac{1}{a}(J_n K_n + J_{n+1}K_{n-1}) = \frac{1}{b}(J_n K_n + J_{n-1}K_{n+1}) = F_n G_n. \qquad (3.100)$$

Exercise 3.6. Compute the continuous limits of the bilinear forms of the first three dPs and compare them with the existing bilinear results of [134].

4   Properties of the Discrete Painlevé Equations

Why are the discrete Painlevé equations called *Painlevé* equations? As we have seen in the previous sections, they are integrable nonautonomous discrete systems that at the continuous limit go over to the usual, continuous Painlevé equations. In this sense, they constitute discretizations of the Painlevé equations. Is that enough in order to justify their name? The key word here is "integrability." It turns out that once we have integrability, the systems obtained have a host of properties that allow us to draw a close parallel between the continuous and discrete systems. Thus thanks to their integrable character, the discrete Painlevé equations have the look and feel of the continuous ones, and thus their name is amply justified. Still, one must keep in mind that the discrete systems live at a more fundamental level than the continuous ones. This leads to noteworthy differences between the discrete and continuous systems. First, we have many more discrete equations than continuous ones: One cannot give a simple canonical form for each of the dPs as was done for the continuous ones. Second, there exist dPs that do *not* have a continuous limit (dP$_0$ is such an example). Moreover, the continuous limit, in some cases, washes out subtle differences that may exist between dPs. For instance, both asymmetric dP$_{\text{IV}}$ and asymmetric dP$_{\text{V}}$ give P$_{\text{VI}}$ as a continuous limit, but asymmetric dP$_{\text{V}}$ is richer than asymmetric dP$_{\text{IV}}$. Despite these differences, the continuous and discrete Painlevé equations are closely related objects and the following sections will establish this through a systematic study of the properties of the dPs.

4.1   Coalescence Cascades

The six continuous Painlevé equations are known to form a coalescence cascade. This means that by taking the appropriate limits of the dependent and independent variables (w, t) as well as the parameters of the equation, we can recover a "lower" equation starting from a "higher" in the following pattern:

$$\mathrm{P_{VI}} \to \mathrm{P_V} \to \{\mathrm{P_{IV}}, \mathrm{P_{III}}\} \to \mathrm{P_{II}} \to \mathrm{P_I}.$$

We shall show below that this is true for the discrete equations as well. At this point we must make clear what we mean by *coalescence cascade*. The coalescence is a limiting procedure performed on the parameters of the equation but also on the dependent variable as well as on the explicitly n-dependent ones. In that way one gets an equation that has fewer parameters than the equation one starts with. Thus through this procedure one can obtain "lower" ones starting from a "higher" one (provided that we take the appropriate limits of dependent variables as well as of the parameters and also the explicitly n-dependent variables). The analogy with the continuous Painlevé equations is perfect. In this case the coalescence chain is

$$d\mathrm{P_{VI}} \to d\mathrm{P_V} \to \{d\mathrm{P_{IV}}, d\mathrm{P_{III}}\} \to d\mathrm{P_{II}} \to d\mathrm{P_I}.$$

In what follows we will present the result for the five standard forms. Thus the first limit from $d\mathrm{P_{VI}}$ to $d\mathrm{P_V}$ will not be given here. (Comments on this limit will be given at the end of the section). The following conventions will be used. The variables and parameters of the "higher" equation will be given in capital letters (X, Z, P, Q, A, B, C), while those of the "lower" equation are given in lowercase letters (x, z, p, q, a, b, c). The small parameter that will introduce the coalescence limit will be denoted by δ.

In order to illustrate the process, let us work out in full detail the case $d\mathrm{P_{II}} \to d\mathrm{P_I}$. We start with the equation

$$X_{n+1} + X_{n-1} = \frac{Z X_n + A}{1 - X_n^2}. \tag{4.1}$$

We put $X = 1 + \delta x$, whereupon the equation becomes

$$4 + 2\delta(x_{n+1} + x_{n-1} + x_n) = -\frac{Z(1 + \delta x_n) + A}{\delta x_n}. \tag{4.2}$$

Now, clearly, Z must cancel A up to order δ, and this suggests the ansatz $Z = -A - 2\delta^2 z$. Moreover, the $\mathcal{O}(\delta^0)$ term in the rhs must cancel the 4 of the lhs, and we are thus led to $A = 4 + 2\delta a$. Using these values of Z and A, we obtain (at $\delta \to 0$)

$$x_{n+1} + x_{n-1} + x_n = \frac{z}{x_n} + a, \tag{4.3}$$

i.e., precisely dP_I.

The coalescence dP_{III} to dP_{II} requires a more delicate limit, since the independent variable of dP_{III} enters in an exponential way. In order to perform the limit we take $\lambda = 1 + \gamma \delta^r$ for some r, whereupon λ^n becomes $1 + n\gamma\delta^r + \mathcal{O}(\delta^{2r})$, and thus, at the limit, p, q are of the form $\alpha + \beta n + \mathcal{O}(\delta^{2r})$ with $\beta = \alpha\gamma\delta^r$. We start from

$$X_{n+1}X_{n-1} = \frac{AB(X_n - P)(X_n - Q)}{(X_n - A)(X_n - B)}. \tag{4.4}$$

The ansatz for X is here, too, $X = 1 + \delta x$. For the remaining quantities we obtain

$$A = 1 + \delta, \quad B = 1 - \delta, \tag{4.5a}$$
$$P = 1 + \delta + \delta^2(z + a)/2 + \mathcal{O}(\delta^3), \tag{4.5b}$$
$$Q = 1 - \delta + \delta^2(z - a)/2 + \mathcal{O}(\delta^3), \tag{4.5c}$$

so in fact, $r = 2$, and at the limit $\delta \to 0$, dP_{III} reduces exactly to dP_{II}:

$$x_{n+1} + x_{n-1} = \frac{zx_n + a}{1 - x_n^2}. \tag{4.6}$$

As we saw above, dP_{IV} also reduces to dP_{II}. Here we start from

$$(X_{n+1} + X_n)(X_n + X_{n-1}) = \frac{(X_n^2 - A^2)(X_n^2 - B^2)}{(X_n - Z)^2 - C^2} \tag{4.7}$$

and put $X = 1 + \delta x$. We take

$$A = 1 + \delta, \qquad B = 1 - \delta, \tag{4.8}$$
$$C = \delta - \delta^2 a/2, \quad Z = 1 - \delta^2 z/4. \tag{4.9}$$

The result at $\delta \to 0$ is precisely dP_{II} given by (4.6).

In the case of dP_V,

$$(X_{n+1}X_n - 1)(X_n X_{n-1} - 1)$$
$$= \frac{PQ(X_n - A)(X_n - 1/A)(X_n - B)(X_n - 1/B)}{(X_n - P)(X_n - Q)}, \tag{4.10}$$

two different limits exist. In order to obtain dP_{IV} we put $X = 1 + \delta x$ and take

$$A = 1 + \delta a, \qquad B = 1 - \delta b, \tag{4.11}$$
$$P = 1 + \delta(z + c), \quad Q = 1 + \delta(z - c), \tag{4.12}$$

i.e., $\lambda = 1 + \alpha \delta$, such that $z = \alpha n + \beta$. At the limit $\delta \to 0$ we obtain dP$_{\text{IV}}$ (4.7) in terms of the variable x. The case of the coalescence dP$_{\text{V}}$ to dP$_{\text{III}}$ requires a different ansatz. Here we put $X = x/\delta$. Moreover, we take

$$P = \frac{p}{\delta}, \quad Q = \frac{q}{\delta}, \quad A = \frac{a}{\delta}, \quad B = \frac{b}{\delta}. \tag{4.13}$$

We obtain then at the limit $\delta \to 0$

$$x_{n+1}x_{n-1} = \frac{pq(x_n - a)(x_n - b)}{(x_n - p)(x_n - q)}. \tag{4.14}$$

While this is not exactly the form of dP$_{\text{III}}$ (4.4), it is very easy to reduce it to the latter. We introduce y through $x = y\lambda^n$ (recall $p = p_0\lambda^n$, $q = q_0\lambda^n$) and obtain with $\mu = 1/\lambda$

$$y_{n+1}y_{n-1} = \frac{p_0 q_0 (x_n - a\mu^n)(x_n - b\mu^n)}{(x_n - p_0)(x_n - q_0)}, \tag{4.15}$$

which is obviously of the form (4.4).

Two remarks are necessary here. The first is that our cascade is incomplete because the form of dP$_{\text{VI}}$ as a three-point mapping that would belong to this "standard" cascade is not yet known. All the other forms of dP$_{\text{VI}}$ known belong to different cascades, which brings us to our second remark. Since there are many more dPs than continuous Painlevé equations, it is natural that they can organize themselves into more than one cascade.

Exercise 4.1. Starting from alternative-dP$_{\text{II}}$ (3.75), show that there exist a coalescence limit to the alternative dP$_{\text{I}}$ (3.69).

In [128] it was shown that the coalescence procedure works not only for the Painlevé equations but for all the equations of the Painlevé–Gambier classification. On the basis of this result we can surmise that this property is characteristic of integrable systems in general and points to the existence of some generic equation that encompasses a whole class. In the case of the Painlevé equations this is the role of P$_{\text{VI}}$, but for the dPs, several such equations seem to exist.

4.2 Particular Solutions of the Discrete Painlevé Equations

How can one construct particular solutions to the discrete Painlevé equations? First we must point out that (in analogy with the continuous Painlevé equations) dPs possess two kinds of solutions: rational ones and solutions in terms of special functions. Both exist *only* for some special values of the parameters of the equation. Moreover, the codimension of the rational and special-functions solutions of the dPs need not be the same. Let us make all this clear through specific examples.

We start with the rational solutions. The simplest approach for finding an elementary rational solution is to assume a form, for instance $x_n = \alpha n + \beta$, and substitute it into the equation. The parameters of the equation for which such a solution exists as well as the exact values of the α, β can be easily obtained. Starting with P_{II},

$$x_{n+1} + x_{n-1} = \frac{a + zx_n}{1 - x_n^2}, \tag{4.16}$$

we find readily that for $a = 0$, $x_n \equiv 0$ is a solution. The next rational solutions are easily obtained. For $a = \pm\Delta z$ we have $x = \mp\Delta z/(z - 2)$, where $\Delta z = z_{n+1} - z_n$. The procedure can be extended to higher rational solutions: Substitute a rational form $P(z)/Q(z)$ and determine the coefficients. However, the process is tedious, and since there exists a much simpler approach for the construction of these higher solutions, we shall not pursue it further. Let us give rather the *elementary* rational solution of dP_{III}. We start with

$$x_{n+1}x_{n-1} = \frac{ab(x_n - p)(x_n - q)}{(x_n - a)(x_n - b)}, \tag{4.17}$$

where a, b are constants and p, q proportional to λ^n. We look for a solutions $x \approx \lambda^{n/2}$. Substituting into (4.17) we find that the condition for the existence of this solution is $aq = bp$ and then $x = \sqrt{aq}$. Rational solutions can also be obtained for the higher dPs, as well as for the alternative forms [135]. The interested reader can find the details in [136, 137].

Let us now turn to the solutions in terms of special functions. It is well known that in the continuous case one can obtain solutions by assuming that the solution of the Painlevé equation is also the solution of an auxiliary equation of Riccati form. In the discrete case all we have to do is to assume a form

$$x_{n+1} = -\frac{\alpha_n x_n + \beta_n}{\gamma_n x_n + \delta_n}, \tag{4.18}$$

which leads to

$$x_{n-1} = -\frac{\delta_{n-1} x_n + \beta_{n-1}}{\gamma_{n-1} x_n + \alpha_{n-1}}, \tag{4.19}$$

and substitute back into the equation that must be identically satisfied for all x_n. This gives an overdetermined set of equations for the $\alpha, \beta, \gamma, \delta$, which can be satisfied only when some conditions hold among the parameters of the equation. The discrete Riccati equation is subsequently linearized by the standard Cole–Hopf substitution $x = P/Q$, leading to

$$P_{n+1} = \frac{\alpha_n}{\gamma_n} P_n + \frac{\beta_n}{\gamma_n} Q_n, \tag{4.20}$$

$$Q_{n+1} = -P_n - \frac{\delta_n}{\gamma_n} Q_n, \tag{4.21}$$

and finally a linear equation for Q,

$$\gamma_n Q_{n+2} - \left(\alpha_n - \frac{\delta_{n+1}\gamma_n}{\gamma_{n+1}}\right) Q_{n+1} + \left(\beta_n - \frac{\alpha_n \delta_n}{\gamma_{n+1}}\right) Q_n = 0. \qquad (4.22)$$

Let us make this more specific by applying the method to dP_{II}. First, it is clear that through scaling of x and a simple division we can set two of the unknowns α, β, γ, δ to any finite value. Let us take $\gamma = 1$ and $\delta = -1$. We find readily that for dP_{II} to be satisfied we must have $\alpha = 1$, $\beta = z/2 - 1 - \Delta z/4$, and $a = \Delta z/2$. In this case equation (4.22) becomes

$$Q_{n+2} - 2Q_{n+1} + \frac{1}{4}(2z - \Delta z)Q_n = 0, \qquad (4.23)$$

which was shown in [112] to be a discrete analogue of the Airy equation.

The method described above for the construction of the special-function-type solutions is quite efficient, but in most cases a shortcut can be used in order to speed up calculations. Given the forms of the dPs the idea is that the special solutions should be related to some factorization (or separation). Let us illustrate this in the case of dP_{III} (4.17). We assume a separation of the form

$$x_{n+1} = \frac{b(x_n - p)}{(x_n - a)}, \qquad (4.24)$$

$$x_{n-1} = \frac{a(x_n - q)}{(x_n - b)}. \qquad (4.25)$$

The system (4.24)–(4.25) is consistent, *provided* that $bp = \lambda aq$. Its linearization (through $x = P/Q$) leads directly to

$$Q_{n+2} + (a - b)Q_{n+1} + (p_0 \lambda^n - ab)Q_n = 0, \qquad (4.26)$$

which was shown in [121] to be a q-discrete form of Bessel's equation.

Once we obtain the elementary solutions of the dPs for some set of values of the parameters of the equation, the following question arises: How does one obtain the analogous solutions for other values of the parameters? For this, two approaches exist. The first is based on the auto-Bäcklund or the Schlesinger transformations of the equations. This will be the object of the next section. The other method for the construction of "higher" solutions is the direct one based on the bilinear formalism. In this case the equation is expressed in terms of a τ-function (through a dependent variable transformation). It can be shown that for special-function-type solutions, the τ-function assumes the form of a Casorati determinant involving the discrete special functions.

For instance, we have shown in [138] that dP_{II} has Casorati-type solutions that can be expressed in terms of (discrete) Airy functions. Starting from

(4.16) we introduce the dependent-variable transformation

$$x_n = \frac{\tau_{N+1}^{n+1} \tau_N^n}{\tau_{N+1}^n \tau_N^{n+1}} - 1. \tag{4.27}$$

The τ-function is characterized by a parameter N, which in the case of Casorati solutions takes integer values and is directly related to the size of the Casorati determinant. The precise form of the τ-function for these solutions is

$$\tau_N^n = \begin{vmatrix} A_n & A_{n+2} & \cdots & A_{n+2N-2} \\ A_{n+1} & A_{n+3} & \cdots & A_{n+2N-1} \\ \vdots & \vdots & \ddots & \vdots \\ A_{n+N-1} & A_{n+N+1} & \cdots & A_{n+3N-3} \end{vmatrix}, \tag{4.28}$$

where A_n is the discrete Airy function, solutions of the equation

$$A_{n+2} = 2A_{n+1} - (pn + q)A_n. \tag{4.29}$$

The constants p, q, are related to the independent variable z and the parameter a of dP$_{\mathrm{II}}$ (4.16) through $z = 2pn + 2q + (2N-1)p$, $a = -(2N+1)p$. Results of this form exist to date only for dP$_{\mathrm{II}}$ and dP$_{\mathrm{III}}$ [139] but there exist no fundamental difficulties to extend them to the higher dPs.

4.3 The Auto-Bäcklund and Schlesinger Transforms

As we have seen in the last section, one way to obtain "higher" solutions for a given dP is through the use of auto-Bäcklund transformations. The auto-Bäcklund transformations are relations that allow one to relate the solution of a given dP to the solution of the same dP with different values of the parameter. We must make clear at this point that the use of these transformations is not limited to the special solutions of the Painlevé equations. The auto-Bäcklund transformations can relate *any* solution of a discrete Painlevé equation to a solution of the same dP, provided that the parameters of the two equations are related in the appropriate way. The Schlesinger transformations are just particular auto-Bäcklund transformations. As such they relate solutions of the same equation. However, the Schlesinger transformations relate solutions corresponding to the same monodromy data except for *integer* differences in the monodromy exponents. In the discrete case the relation of the Schlesinger transformations to monodromy exponents is not always clear. However, we can use an analogy with the continuous case. If one uses the proper parametrization of the equation, the Schlesinger transformations can be shown to be associated to simple changes of the parameters. The discrete case can be analyzed in the same spirit. By using the proper parametrization, one can identify, among the auto-Bäcklund transformations, those that correspond to simple changes of the parameters and that can thus be dubbed Schlesinger transformations.

How can one find the auto-Bäcklund (and Schlesinger) transformations for a given dP? The general principle is the following. First, obtain a Miura transformation that transforms the equation into a new one (the "modified" one). Second, find the invariance of the latter, usually associated to some discrete transformation. Third, implement these discrete transformations and return to the initial equation through the inverse of the Miura transformation. In the process we find that the parameters of the initial equation have been modified, and thus the chain of transformations defines indeed an auto-Bäcklund transformation. It may turn out that the Miura transformation itself is already an auto-Bäcklund transformation (in the sense that it transforms a given dP to the same dP) but the "modified" equation is defined at different lattice points than the initial one. Obtaining the Miura transformation can be facilitated once we remark that all known Miura tranformations have the form of a discrete Riccati equation, i.e., a homographic mapping.

Let us illustrate the construction of the auto-Bäcklund transformation to dP_{II}. We have shown in the previous section that dP_{II} (4.16) and dP_{34} (3.59) are related through a system of Miura transformations given by (3.58) and (3.60), where the parameters of dP_{II} and dP_{34} are related through $m = a + \Delta z/2$, where $\Delta z = z_{n+1} - z_n$.

Thus we start with a given x corresponding to a (or equivalently to $m - \Delta z/2$). We transform with (3.58) to y satisfying a dP_{34} with this m, and since dP_{34} depends only on m^2, we can change the sign of m. Using (3.60) with $m' = -m$, we come back to an x' satisfying dP_{II} with parameter $a' = -m - \Delta z/2$. Finally, by changing the sign of x', we obtain a new $\tilde{x}$ that satisfies dP_{II} with $\tilde{a} = -a' = m + \Delta z/2 = a + \Delta z$. Combining these transformations in one expression, we obtain

$$\tilde{x}_n = -x_n + \frac{(x_n - 1)(a + \Delta z/2)}{(1 + x_{n-1})(x_n - 1) + (a + \Delta z)/2}, \qquad (4.30)$$

where x_n is a solution of dP_{II} at the point z_n and parameter a, and $\tilde{x}_n$ a solution at the same point and parameter $\tilde{a} = a + \Delta z$. We must remark here that given the simple form of dP_{II}, the auto-Bäcklund transformation (4.30) is in fact a Schlesinger transformation, and no further transformations are needed.

A precious help in the derivation of auto-Bäcklund transformations of dPs is the bilinear formalism. As we have seen in the previous section, the variable x of dP_{II} can be expressed in terms of two τ-functions F and G as $x_n = -1 + F_{n+1}G_{n-1}/F_nG_n = 1 - F_{n-1}G_{n+1}/F_nG_n$. Using this substitution in the Miura transformation (4.30) we obtain $y_n + (z_n + \Delta z/2)/2 = G_{n+2}G_{n-1}/G_{n+1}G_n$, i.e., the variable of the dP_{34} involves only one τ-function. Let us rephrase the construction of the auto-Bäcklund transformation in the language of the bilinear formalism. We start with the bilinear dP$_{II}$ with τ-functions F, G:

$$F_{n+1}G_{n-1} + F_{n-1}G_{n+1} - 2F_nG_n = 0, \qquad (4.31)$$

$$F_{n+2}G_{n-2} - F_{n-2}G_{n+2} = z(F_{n+1}G_{n-1} - F_{n-1}G_{n+1}) + 2aF_nG_n. \quad (4.32)$$

For the compatibility of (4.31)–(4.32) one can obtain a hexalinear dP$_{34}$ on G alone with parameter $m^2 = (a+\Delta z/2)^2$. The important point here is that at the level of dP$_{34}$ only m^2 is fixed. We can thus change the sign of m without changing the equation dP$_{34}$. The fact that G satisfies the hexalinear dP$_{34}$ means precisely that the two equations (4.31) and (4.32) are compatible, with a such that $m = a + \Delta z/2$. But then two equations of the same form as (4.31), (4.32) with the very same G but with F replaced by some other τ-function H and a replaced by a' will also be compatible, provided that $-m = a' + \Delta z/2$. This corresponds to dP$_{II}$ for the ordered pair (H, G) with parameter $a' = -m - \alpha/2$, or, by inspection of the symmetry properties of (4.31)–(4.32) upon interchanging H and G, to a dP$_{II}$ for the ordered pair (G, H) with parameter $\tilde{a} \equiv -a' = m + \Delta z/2 = a + \Delta z$. Thus starting from (F, G) with parameter a we can construct a pair (G, H) with parameter $a + \Delta z$. This is precisely the auto-Bäcklund transformation of dP$_{II}$. The details of the construction of the function H can be found in [138]. We give here just the result of this calculation:

$$H_nF_{n+1} = G_{n+1}G_n(z + a + \Delta z) - 2G_{n+2}G_{n-1}, \qquad (4.33)$$

$$H_{n+1}F_n = G_{n+1}G_n(z - a) - 2G_{n+2}G_{n-1}, \qquad (4.34)$$

or, combining (4.33)–(4.34),

$$H_nF_{n+1} - H_{n+1}F_n = (2a + \Delta z)G_{n+1}G_n. \qquad (4.35)$$

Thus starting from a given (F, G) at parameter a we can construct H (4.33 suffices), and iterating further (also backwards) we obtain the solution at any $a + n\Delta z$, $n \in \mathbb{Z}$.

In order to find the Schlesinger transformations of dP$_{III}$ we will rely again on the bilinear formalism. We start from the variable x expressed in terms of *two* τ-functions F and G and introduce a new variable u, which can be expressed in terms of only *one* τ-function, say G. We remark that $(x_n - c)$ is proportional to F_{n+1}/F_n, while $(x_n - d)$ contains F_{n-1}/F_n. Upshifting the last object and multiplying by the first allows us to get rid of F entirely. We thus find a first relation between x and u,

$$u_n = (x_n - c)(x_{n+1} - d), \qquad (4.36)$$

where u is given by $u_n = cdG_{n-1}G_{n+2}/G_nG_{n+1}$. This is the first half of the Miura transformation. Eliminating F between the two bilinear equations (3.95)–(3.96) leads to a (hexalinear) equation for G that can also be expressed in u. The latter would be the "modified" equation of dP$_{\text{III}}$. However, it is simpler to obtain the second half of the Miura transformation and proceed from there. As we have shown in [140], the complement of the Miura involves a rational expression that must be homographic in both u_n and u_{n-1}. We readily obtain

$$x_n = \frac{u_n u_{n-1}/(cd) - u_n - u_{n-1} + cd - ab}{-u_n/d - u_{n-1}/c + c + d - a - b}. \tag{4.37}$$

Eliminating x_n and x_{n+1} between (4.36), (4.37), and its upshift leads to an equation for u_{n-1}, u_n, u_{n+1}. This equation is, after a change of variables, a discrete form of the dP$_{\text{V}}$ equation (although not all the parameters of a dP$_{\text{V}}$ are present). Introducing $U = u - cd$, we obtain

$$(U_n U_{n+1} - \lambda^2 abcd)(U_n U_{n-1} - abcd)$$
$$= \frac{cd(U_n + bd)(U_n + \lambda ac)(U_n + ad)(U_n + \lambda bc)}{U_n + cd}. \tag{4.38}$$

In order to define a different Miura transformation we can introduce the quantity

$$w_n = (1/x_n - 1/a)(1/x_{n+1} - 1/(\lambda b)), \tag{4.39}$$

which depends only on the τ-function K but not J. The second half of the Miura transformation, analogous to (4.37), is

$$x_n = \frac{-aw_{n-1}/\lambda - bw_n \lambda + 1/a + 1/b - 1/c - 1/d}{abw_n w_{n-1} + 1/(ab) - 1/(cd) - w_n \lambda - w_{n-1}/\lambda}. \tag{4.40}$$

Again eliminating x leads to an equation for w. Introducing $W_n = w_n - 1/(\lambda ab)$, we obtain for this equation

$$\left(W_n W_{n+1} - \frac{1}{\lambda^2 abcd}\right)\left(W_n W_{n-1} - \frac{1}{abcd}\right)$$
$$= \frac{(W_n + 1/(\lambda bd))(W_n + 1/(ac))(W_n + 1/(\lambda bc))(W_n + 1/(ad))}{\lambda ab W_n + 1}. \tag{4.41}$$

The quantity $\widetilde{W}_n = U/(\lambda abcd)$ satisfies an equation, obtained from (4.38), namely,

$$\left(\widetilde{W}_n \widetilde{W}_{n+1} - \frac{1}{\lambda^2 abcd}\right)\left(\widetilde{W}_n \widetilde{W}_{n-1} - \frac{1}{abcd}\right)$$
$$= \frac{(\widetilde{W}_n + 1/(\lambda ac))(\widetilde{W}_n + 1/(bd))(\widetilde{W}_n + 1/(\lambda bc))(\widetilde{W}_n + 1/(ad))}{\lambda ab \widetilde{W}_n + 1}. \tag{4.42}$$

Equation (4.41) has the same form as (4.42), provided that one introduces the parameters

$$\tilde{a} = a\sqrt{\lambda}, \quad \tilde{b} = b/\sqrt{\lambda}, \quad \tilde{c} = c\sqrt{\lambda}, \quad \tilde{d} = d/\sqrt{\lambda}. \qquad (4.43)$$

We define

$$\tilde{w}_n = \widetilde{W}_n + 1/(\lambda \tilde{a}\tilde{b}) = u_n/(\lambda abcd) \qquad (4.44)$$

and

$$\tilde{x}_n = \frac{-\tilde{a}\tilde{w}_{n-1}/\lambda - \tilde{b}\tilde{w}_n\lambda + 1/\tilde{a} + 1/\tilde{b} - 1/\tilde{c} - 1/\tilde{d}}{\tilde{a}\tilde{b}\tilde{w}_n\tilde{w}_{n-1} + 1/(\tilde{a}\tilde{b}) - 1/(\tilde{c}\tilde{d}) - \tilde{w}_n\lambda - \tilde{w}_{n-1}/\lambda}. \qquad (4.45)$$

Given this definition of $\tilde{x}$ and since $\widetilde{W}$ satisfies (4.42), it follows that

$$\tilde{w}_n = (1/\tilde{x}_n - 1/\tilde{a})(1/\tilde{x}_{n+1} - 1/(\lambda \tilde{b})), \qquad (4.46)$$

and therefore $\tilde{x}$ satisfies dP$_{\text{III}}$ with parameters $\tilde{a}, \tilde{b}, \tilde{c}, \tilde{d}$. The transformation from x to $\tilde{x}$ through (4.36), (4.43), (4.44), and (4.45) defines an auto-Bäcklund transformation for dP$_{\text{III}}$. In this case this is indeed a Schlesinger transformation, which we denote by S_c^a. (The convention used here is to give explicitly the parameters associated to x_n, rather than x_{n+1}, in (4.36) and (4.39).) The inverse transformation $(S_c^a)^{-1}$ can be obtained by defining w through (4.39), $\underline{u} = w\lambda abcd$, and finally $\underline{x}$ through the analogue of (4.37).

In a similar way we can introduce the transformations S_c^b, $S_d^a = (S_c^b)^{-1}$, and $S_d^b = (S_c^a)^{-1}$. They correspond to multiplying the two parameters that appear explicitly by $\sqrt{\lambda}$ while dividing the two others by the same quantity. These are the most elementary Schlesinger transformations. Using them we can construct further Schlesinger transformations that act separately on $\{a, b\}$ or $\{c, d\}$. For instance, the product $S_c^a S_d^a$ corresponds to $a \to a\lambda$, $b \to b/\lambda$, $c \to c$, $d \to d$.

Following the same general procedure one can construct the Schlesinger transformations of the higher dPs as well as those of the alternative forms.

5   Monodromy Problems and q-Difference Equations

In this final section we will report on how far the isomonodromic deformation theory for the discrete Painlevé equations has been developed. The immediate answer is: Not very far! This is a rich area that waits to be exploited. The starting point is the establishment of isomonodromic deformation problems, which form the Lax pairs for the corresponding discrete Painlevé equations. As was explained in great detail in some of the other chapter in this volume, notably the one by M. Ablowitz et al., Chapter 9, for the Painlevé equations the isospectral theory (i.e., the inverse scattering scheme) that is used to find solutions of the initial value problem of nonlinear evolution equations of soliton type is replaced by the isomonodromy theory. This approach goes back to the early works of R. Fuchs [141] and L. Schlesinger [142], and a more modern approach was initiated by the Kyoto school [143, 144], which is explained in Chapter 2, by Mahoux. In the isomonodromic deformation theory the direct problem starts with the extraction of monodromy data (consisting of connection matrices, Stokes matrices etc.) from the linear differential equation, which we will loosely refer to as the monodromy problem, and examples of which in the continuous case we have seen already in Section 2. In fact, as we have seen, these monodromy problems can be obtained by implementing the similarity reduction to the Lax pair of the nonlinear evolution equation (NLEE) that we investigated. The inverse procedure, namely the reconstruction of the potential, which in our case is the Painlevé transcendent, is then obtained in terms of solutions of a Riemann–Hilbert (RH) problem, which albeit not in closed form, can be investigated asymptotically and from which in principle information on the *connection problem*, i.e., how the asymptotic behaviors around different singular points are related, can be obtained. This is a large program, which is well underway for the continuous Painlevé transcendents and to which many authors have contributed in a substantial way; cf., e.g., [77–81].

For the discrete Painlevé equations we are only at the very beginning of this ambitious program. So far, the only place where a serious effort was undertaken to perform the asymptotic analysis of an isomonodromy problem for a discrete Painlevé equation was in [95] for dP$_I$ in connection with the theory of orthogonal polynomials and two-dimensional quantum gravity. For all other discrete Painlevé equations the work still has to be started. What does exist, however, is a collection of monodromy problems for a number of dPs, which gives us good hope that some day the program of studying discrete Painlevé transcendents can be brought into shape. Furthermore, these monodromy problems might give us a handle on the issue of obtaining a classification of discrete Painlevé equations, which in view of their great abundance is an urgent problem. Since in the discrete case we have not yet something like a *classifying group*, similar to the group of Möbius transformations in the continuous case, nor a good way of dividing nonlinear difference equations into classes according to their "shape," other means of classification are needed. One possible characteristic would be to classify discrete Painlevé equations according to their singularity patterns, which have been explained in the previous lecture. Another way would be to identify classes of monodromy problems that might lead to a group-theoretic classification. In Section 5.2 below we will give a scheme of isomonodromic deformation problems which could form a starting point for such an approach.

We will show in this section that there are indeed systematic ways of obtaining monodromy problems for the discrete Painlevé equations and that even though we have not yet obtained them for all dPs (notably, dP_{IV} and dP_V have remained elusive) we have good hope of obtaining a complete list of Lax pairs in the near future.

5.1 Derivation from Infinite-Matrix Structure

We will now show how isomonodromic deformation problems can be derived systematically on the basis of the infinite-matrix structure that we developed in Section 2. Using the scheme of Section 2.2 we can introduce a linear set of relations for infinite vectors. In the same spirit of the derivation of the lattice equations one can then extract from this infinite linear system a set of closed-form linear equations in terms of a finite number of components only, which then leads to the Lax pairs as well as to the monodromy problem (the latter by implementing the similarity constraints).

Let us start by introducing the infinite vector

$$\mathbf{u}_k = \rho_k \left(\mathbf{1} + \mathbf{U} \cdot \frac{1}{k + {}^t\mathbf{\Lambda}} \cdot \mathbf{O} \right) \cdot \mathbf{c}_k, \tag{5.1}$$

having components u_k^j $(j \in \mathbb{Z})$ with ρ_k given in (2.24b) and $\mathbf{c}_k$ denoting, as before, the vector with entries $(\mathbf{c}_k)_j = k^j$. Recalling that the action of the operators $\mathbf{\Lambda}$ and $\mathbf{I}$ on $\mathbf{c}_k$ is given by

$$\mathbf{\Lambda} \cdot \mathbf{c}_k = k\mathbf{c}_k, \quad \mathbf{I} \cdot \mathbf{c}_k = k\frac{d}{dk}\mathbf{c}_k,$$

and making use of the relations (2.20), we can derive from the matrix Riccati equation (2.29) for $\mathbf{U}$ the following *linear* relations for $\mathbf{u}_k$:

$$(p - k)\tilde{\mathbf{u}}_k = (p + \mathbf{\Lambda} - \tilde{\mathbf{U}} \cdot \mathbf{O}) \cdot \mathbf{u}_k, \tag{5.2a}$$

$$(p + k)\mathbf{u}_k = (p - \mathbf{\Lambda} + \mathbf{U} \cdot \mathbf{O}) \cdot \tilde{\mathbf{u}}_k. \tag{5.2b}$$

Exercise 5.1. Prove (5.2).

Exercise 5.2. Use the symmetry valid for the KdV family, $\mathbf{U} = {}^t\mathbf{U}$, to prove the following identity:

$$k^2\mathbf{u}_k = \mathbf{\Lambda}^2 \cdot \mathbf{u}_k - \mathbf{U} \cdot (\mathbf{O} \cdot \mathbf{\Lambda} - {}^t\mathbf{\Lambda} \cdot \mathbf{O}) \cdot \mathbf{u}_k. \tag{5.3}$$

From (5.2) one can derive Lax pairs for the various lattice equations within the KdV family, and the relation (5.3) can, in fact, be used to establish the gauge transformations between the various Lax pairs, as will be shown below. Next we need to construct the similarity constraint for $\mathbf{u}_k$, which can be obtained in a similar way, now using the similarity constraint

(2.34) as a starting point. The result is the following:

$$
k\frac{\mathrm{d}}{\mathrm{d}k}\mathbf{u}_k
$$

$$
= \mathbf{I}\cdot\mathbf{u}_k + \sum_\nu n_\nu p_\nu \left[\frac{1}{p-k} - \left(1 + \mathbf{U}\cdot\frac{1}{p_\nu - {}^t\mathbf{\Lambda}}\cdot\mathbf{O}\right)\cdot\frac{1}{p_\nu + \mathbf{\Lambda}}\right]\cdot\mathbf{u}_k.
\tag{5.4}
$$

Exercise 5.3. Prove (5.4).

The next step is to derive concrete Lax pairs and monodromy problems from the infinite set of equations encoded in (5.2)-(5.4). This can be achieved by singling out specific components, the choice of which will determine which equation within the lattice KdV family we are dealing with. Let us illustrate this with two specific examples.

Example 5.1 (Lattice (potential) KdV). In this case we take the components u_k^0 and u_k^1 to form the two-component vector

$$
\phi_k \equiv (p-k)^n (q-k)^m \begin{pmatrix} u_k^0 \\ u_k^1 \end{pmatrix},
\tag{5.5}
$$

for which it is immediate to derive the linear relation

$$
\widetilde{\phi}_k = L_k \phi_k, \, L_k \equiv \begin{pmatrix} p - \tilde{u} & 1 \\ k^2 - p^2 + * & p + u \end{pmatrix},
\tag{5.6}
$$

in which $u = U_{0,0}$ as before and $*$ stands for the product of the diagonal entries of the matrix L_k. Equation (5.6) is one part of the Lax pair of the potential lattice KdV equation. The other part is identical, apart from the replacements $\tilde{u} \mapsto \hat{u}$, $p \mapsto q$, leading to

$$
\widehat{\phi}_k = M_k \phi_k, \, M_k \equiv \begin{pmatrix} q - \hat{u} & 1 \\ k^2 - q^2 + * & q + u \end{pmatrix}.
\tag{5.7}
$$

In this way we obtain the Lax representation for (2.49) from the compatibility condition

$$
\widehat{L}_k M_k = \widetilde{M}_k L_k.
$$

In order to obtain the monodromy problem that gives us the similarity constraint we have to use (5.4) and write down what this gives us for the vector ϕ_k. Applying the constraint for u_k^0 and u_k^1, respectively, we obtain

$$
k\frac{\mathrm{d}}{\mathrm{d}k}u_k^0 = np\left(\frac{1}{p-k}u_k^0 - v_{-p}u_k^{(p)}\right) + mq\left(\frac{1}{q-k}u_k^0 - v_{-q}u_k^{(q)}\right),
\tag{5.8a}
$$

$$
k\frac{\mathrm{d}}{\mathrm{d}k}u_k^1 = u_k^1 + np\left(\frac{1}{p-k}u_k^1 - u_k^0 - s_{-p}u_k^{(p)}\right)
$$

$$
+ mq\left(\frac{1}{q-k}u_k^1 - u_k^0 - s_{-q}u_k^{(q)}\right),
\tag{5.8b}
$$

where $u_k^{(\alpha)}$ is defined by

$$u_k^{(\alpha)} \equiv \left(\frac{1}{\alpha + \Lambda} \cdot \mathbf{u}_k \right)_0, \tag{5.9}$$

and where the objects v_α and s_α were defined in (2.36b). In order to bring this set of equations into shape we need a number of relations, namely (2.37) and (2.39) for $\alpha = -p, -q$ together with

$$\tilde{v}_p v_{-p} = 1, \quad 2p + \underline{u} - \tilde{u} = p \frac{\tilde{v}_0 + \underline{v}_0}{v_0} = 2p \frac{\tilde{v}_p}{v_p} \tag{5.10}$$

(cf. also (2.77)) (and similar equations with $\tilde{}$ replaced by $\hat{}$ and p replaced by q), as well as from (5.2) the relations

$$(p - k)\tilde{u}_k^{(\alpha)} = (p - \alpha)u_k^{(\alpha)} + \tilde{v}_\alpha u_k^0, \tag{5.11}$$

applied to $\alpha = p$ (and similarly for the replacements). To make a long story short, we end up with the following differential equation for the vector ϕ_k:

$$k \frac{d}{dk} \phi_k = \begin{pmatrix} n + m & 0 \\ -np - mq & n + m + 1 \end{pmatrix} \phi_k - \frac{2np^2}{2p + \underline{u} - \tilde{u}} \begin{pmatrix} 1 & 0 \\ -p + \tilde{u} & 0 \end{pmatrix} \phi_k$$

$$- \frac{2mq^2}{2q + \underline{u} - \hat{u}} \begin{pmatrix} 1 & 0 \\ -q + \hat{u} & 0 \end{pmatrix} \phi_k. \tag{5.12}$$

In order to make (5.12) into a monodromy problem in the usual sense we should use the Lax representation (2.45), (2.46) to express $\tilde{\phi}_k$ and $\hat{\phi}_k$ in terms of ϕ_k, thus leading to a differential equation with regular singularities at $k^2 = 0$, ∞ and $k^2 = p^2$, q^2.

Exercise 5.4. Prove that the compatibility of (5.12) with (2.45a), or covariantly with (2.45b), yields both the lattice KdV equation (2.49) and the singularity constraint (2.50).

Example 5.2 (Lattice (potential) MKdV). To obtain the Lax representation for the lattice MKdV equation, a similar calculation needs to be performed. In this case we single out the components u_k^0 together with $u_k^{(\alpha)}$ for some fixed value of α, to construct the two-component vector

$$\psi_k \equiv (p - k)^n (q - k)^m \begin{pmatrix} u_k^{(\alpha)} \\ u_k^0 \end{pmatrix}. \tag{5.13}$$

First, using the relations (2.37) we obtain from (5.4) the Lax matrices now in the form

$$\tilde{\psi}_k = L_k \phi_k, \quad L_k \equiv \begin{pmatrix} p - \alpha & \tilde{v}_\alpha \\ \dfrac{k^2 - \alpha^2}{v_\alpha} & (p + \alpha)\dfrac{\tilde{v}_\alpha}{v_\alpha} \end{pmatrix}, \tag{5.14}$$

and a similar expression for M_k by making the usual replacements.

Exercise 5.5. Derive (5.14) and show that the compatibility of the Lax pair gives rise to the lattice MKdV equation (2.48).

Exercise 5.6. Use (2.37) to show that the Lax representations (5.6) and (5.14) are related via the following gauge transformation (cf. [57]):

$$L_k^{(MKDV)} = \tilde{U}_k L_k^{(KDV)} U_k^{-1}, \quad U_k = \begin{pmatrix} -s_\alpha & v_\alpha \\ k^2 - \alpha^2 & 0 \end{pmatrix} \tag{5.15}$$

(and a similar formula relating the matrices M_k).

In order to derive a monodromy problem from (5.4) we need to restrict ourselves to the case $\alpha = 0$, because otherwise we would obtain from the term with the operator **I** derivatives with respect to α, hence equations that are no longer closed-term. Using for $u_k^{(0)} \equiv u_k^{(\alpha=0)}$ the additional relation

$$k\frac{\mathrm{d}}{\mathrm{d}k} u_k^{(0)} = -u_k^{(0)} + n\left(\frac{k}{p-k} u_k^{(0)} + (1 + p s_{0,-p}) u_k^{(p)}\right)$$

$$+ m\left(\frac{k}{q-k} u_k^{(0)} + (1 + q s_{0,-q}) u_k^{(q)}\right), \tag{5.16}$$

in addition to (5.8b) and the relation $1 + p s_{0,-p} = \tilde{v}_0 v_{-p}$ that follows from (2.38), we easily obtain the following monodromy problem:

$$k\frac{\mathrm{d}}{\mathrm{d}k} \psi_k = \begin{pmatrix} -1 & 0 \\ 0 & n+m \end{pmatrix} \psi_k$$

$$+ \frac{2n v_0}{\tilde{v}_0 + \underset{\sim}{v}_0} \begin{pmatrix} 0 & \tilde{v}_0 \\ 0 & -p \end{pmatrix} \psi_k + \frac{2m v_0}{\hat{v}_0 + \underset{\wedge}{v}_0} \begin{pmatrix} 0 & \hat{v}_0 \\ 0 & -q \end{pmatrix} \underset{\wedge}{\psi}_k. \tag{5.17}$$

Exercise 5.7. Prove similarly as in the KdV case that the compatibility of (5.17) with (5.14) yields both the lattice KdV equation (2.48) and the singularity constraint (2.51) for the special value $\mu = \frac{1}{3}$.

Exercise 5.8. Infer the Lax representation (2.45) from (5.14) for the lattice MKdV equation by performing a gauge transformation with the matrix

$$\begin{pmatrix} 1 & 0 \\ 0 & 1/v_0 \end{pmatrix}$$

and using the relations $\tilde{v}_0 v_0 = p(z - \tilde{z})$, $\hat{v}_0 v_0 = q(z - \hat{z})$. Show that when applied to the monodromy problem (5.17) this gauge leads to the following monodromy problem for the lattice SKdV equation (2.43):

$$\left(J_+ + k\frac{\mathrm{d}}{\mathrm{d}k}\right)\phi_{n,m} = (n+m) J_- \phi_{n,m} + \frac{2n}{p}\frac{z-\underset{\sim}{z}}{z-\tilde{z}}\begin{pmatrix} 0 & \tilde{z}-z \\ 0 & 1 \end{pmatrix}\phi_{n-1,m}$$

$$+ \frac{2m}{q}\frac{z-\underset{\wedge}{z}}{z-\hat{z}}\begin{pmatrix} 0 & \hat{z}-z \\ 0 & 1 \end{pmatrix}\phi_{n,m-1}, \tag{5.18}$$

where

$$J_+ = \begin{pmatrix} 1 & 0 \\ 0 & 0 \end{pmatrix}, \quad J_- = \begin{pmatrix} 0 & 0 \\ 0 & 1 \end{pmatrix}.$$

5.2 Deautonomization of Maps

An alternative approach for obtaining monodromy problems for dPs was developed in [145]. The idea is very simple, and is stated as follows: We have noted at many instances in the previous sections that if we take the autonomous limit of the discrete Painlevé equations, namely by replacing the factors containing the explicit dependence on the independent discrete variable by constants, we recover the underlying integrable maps of the plane, all of which are special cases of the 18-parameter QRT-mapping of [25]. In fact, that is what was underlying the philosophy of the approach of the Section 3: to start from a known autonomous integrable map and then, somehow, to introduce an n-dependence in the free parameters in such a way that singularity confinement is retained to obtain a discrete Painlevé equation.

It is natural to investigate whether this approach will also yield a deeper insight into the equations, e.g., produce for us the corresponding Lax pairs of the dPs. The answer is positive, and in [145] we obtained a number of novel Lax pairs for the known dPs at that moment. A very strong confirmation of the singularity confinement hypothesis was produced in that paper by the fact that we were able to produce a Lax pair for dP$_{\text{III}}$ for which at that point in time singularity confinement was the *only* indication of its integrability.

In the search for monodromy problems for the dPs we thus start from Lax pairs for autonomous integrable mappings. Unfortunately, we do not know (yet) a Lax pair for the general QRT-mapping even in the symmetric case given by (1.19). However, a large family of integrable mappings was constructed in [57], namely by period reduction of a general class of integrable lattice equations forming the lattice analogue of the Gel'fand–Dikii (GD) hierarchy. We will not present any details of the construction of [57] here. For our present purpose it is sufficient to mention that the mappings coming from the lattice GD hierarchy fall into two general classes. Both are given by a spectral problem of the form

$$L(h)\phi = \lambda\phi, \widehat{\phi} = M(h)\phi, \tag{5.19}$$

where h has the interpretation of the Floquet parameter corresponding to the quasi-periodicity of the solution, and λ is the spectral parameter. For the two classes the Lax matrices $L(h)$ and $M(h)$ take on the following forms:

Mappings of GD Type. In this case we have

$$L(h) = \Sigma_h^N + \Sigma_h^{N-1} X^{(N-1)} + \cdots + \Sigma_h X^{(1)} + X^{(0)}, \qquad (5.20a)$$

$$M(h) = \Sigma_h + D, \quad N = 2, 3, \ldots . \qquad (5.20b)$$

These mappings are related to the lattice versions of the KdV ($N = 2$), Boussinesq ($N = 3$), and higher-order equations in the GD hierarchy.

Mappings of MGD Type. Taking in this case

$$L(h) = \Sigma_h^N Y^{(N)} + \cdots + \Sigma_h Y^{(1)} + Y^{(0)}, \qquad (5.21a)$$

$$M(h) = \Sigma_h Z^{(1)} + Z^{(0)}, \qquad (5.21b)$$

subject to the linear restrictions

$$\sum_{j=0}^{N} Y_{n+j}^{(j)} = 0, Z_{n+1}^{(1)} + Z_n^{(0)} = \text{const}, \qquad (5.21c)$$

the corresponding mappings are reductions of the lattice analogue of the *modified* Gel'fand–Dikii hierarchy (MGD), namely for $N = 2$ the lattice MKdV and for $N = 3$ the lattice-modified BSQ (MBSQ) equation.

In (5.20) and (5.21) the matrix Σ_h is the periodic shift matrix

$$\Sigma_h = \begin{pmatrix} 0 & 1 & & & \\ & 0 & 1 & & \\ & & \ddots & \ddots & \\ & & & 0 & 1 \\ h & & & & 0 \end{pmatrix},$$

and all coefficients $X^{(j)}$, $Y^{(j)}$ as well as W and D are diagonal $2P \times 2P$ (in the case of *even* reduction) or $(2P - 1) \times (2P - 1)$ (in the case of *odd* reduction) matrices. We will refer to P as the *dimension* of the map and to N as the *rank* of the map. It is noted that the forms (5.20a), resp. (5.21a), are the precise discrete analogues of the Nth-order differential spectral problems that are the basis of the continuous GD-hierarchy; cf., e.g., [106].

In order to obtain nonautonomous versions of the equations that arise from the Lax pairs (5.20) and (5.21) we have to replace the spectral problem by a differential equation. Naively, this is done by making the replacement

$$\lambda \mapsto h \frac{d}{dh},$$

i.e., the Euler operator with respect to the Floquet parameter h. Thus, we obtain from (5.20) the isomonodromic deformation system

$$h \frac{d}{dh} \phi_n(h) = L_n(h)\phi_n(h), \quad \phi_{n+1}(h) = M_n(h)\phi_n(h), \qquad (5.22)$$

the compatibility of which leads to the system

$$\frac{d}{dh}M_n(h) = L_{n+1}(h)M_n(h) - M_n(h)L_n(h), \tag{5.23}$$

which for appropriate choices of the matrices $L_n(h)$ and $M_n(h)$ should lead to discrete Painlevé equations.

However, in the case of (5.21) the replacement is not always so obvious. Since the autonomous mappings are mostly of multiplicative type, we may expect that in this case the corresponding nonautonomous equations have an explicit n-dependence that enters in an exponential way. This has led us to postulate in [145] replacement of the form

$$\lambda \mapsto q^{hd/dh},$$

i.e., in terms of a q-shift operation. Thus we are led to the q-difference system

$$\phi_n(qh) = L_n(h)\phi_n(h), \quad \phi_{n+1}(h) = M_n(h)\phi_n(h) \tag{5.24}$$

rather than to the differential system (5.22). In fact, as we shall see below, this scheme whose compatibility now reads

$$M_n(qh)L_n(h) = L_{n+1}(h)M_n(h) \tag{5.25}$$

gives rise to the monodromy problem for dP$_{\text{III}}$!

In the following we shall discuss in more detail the results coming from the above scheme yielding isomonodromic deformation problems for a number of the dPs.

5.2.1   dP$_{\text{I}}$

In this case we take odd dimension, $P = 2$, and rank 2, meaning that we deal with 3×3 matrix reduction of the lattice KdV equation. Denoting the Lax matrices as

$$L_n(h) = \begin{pmatrix} \lambda_1 & v_2 & 1 \\ h & \lambda_2 & v_3 \\ hv_1 & h & \lambda_3 \end{pmatrix}, \quad M_n(h) = \begin{pmatrix} d_1 & 1 & 0 \\ 0 & d_2 & 1 \\ h & 0 & d_3 \end{pmatrix}, \tag{5.26}$$

we obtain from (5.23) the set of relations

$$\begin{cases} d_1 + \hat{v}_3 = v_1 + d_2, \\ d_2 + \hat{v}_1 = v_2 + d_3, \\ d_3 + \hat{v}_2 = v_3 + d_1, \end{cases} , \quad \begin{cases} \hat{\lambda}_3 + d_1\hat{v}_1 = d_3 v_1 + \lambda_1 + 1, \\ \hat{\lambda}_1 + d_2\hat{v}_2 = d_1 v_2 + \lambda_2, \\ \hat{\lambda}_2 + d_3\hat{v}_3 = d_2 v_3 + \lambda_3, \end{cases} \tag{5.27}$$

together with $(\hat{\lambda}_i - \lambda_i)d_i = 0$, $i = 1$, 2, 3. In the autonomous case the λ_i are constant, whereas in the nonautonomous case we want to recover some nontrivial dependence on the discrete time-variable n. In order to achieve this we have to "tune" the diagonal entries d_i: Some of them need to be equal to zero. There are two distinct cases: Either

$$d_2 = d_3 = 0, \; d_1 \neq 0 \implies \hat{\lambda}_1 = \lambda_1,$$

or

$$d_3 = 0, d_1, \; d_2 \neq 0 \implies \hat{\lambda}_1 = \lambda_1, \hat{\lambda}_2 = \lambda_2.$$

In the former case we have

$$d_1 = \frac{\lambda_1 - \lambda_2}{v_2} = \frac{1 + \lambda_1 - \hat{\lambda}_3}{\hat{v}_1} \implies \hat{\lambda}_3 = 1 + \lambda_2, \qquad (5.28)$$

using also $\hat{v}_1 = v_2$. Taking into account that there is a Casimir $v_1 + v_2 + v_3 = C -$ const, it is easy to see that we obtain from the set of relations (5.27) the dP$_I$ equation in terms of $x_n \equiv v_2 = v_2(n)$, namely in the form

$$x_{n+1} + x_n + x_{n-1} = C + \frac{\lambda_1 - \lambda_2}{x_n}, \qquad (5.29a)$$

$$\lambda_2 = \frac{1}{2}n + (-1)^n \lambda_0, \quad \lambda_0, \quad \lambda_1 = \text{const.} \qquad (5.29b)$$

In the latter case we have from the second set of relations (5.27)

$$d_1 = \frac{1 + \lambda_1 - \hat{\lambda}_3}{\hat{v}_1}, \quad d_2 = \frac{\hat{\lambda}_2 - \hat{\lambda}_3}{v_3}, \qquad (5.30)$$

together with the condition

$$\hat{\lambda}_1 + \hat{\lambda}_2 + \hat{\lambda}_3 + d_1 \hat{v}_1 + d_2 \hat{v}_2 = 1 + \lambda_1 + \lambda_2 + \lambda_3 + d_1 v_2 + d_2 v_3,$$

which yields $\hat{\lambda}_3 = 1 + \lambda_3$. Using again the Casimir $C = v_1 + v_2 + v_3$ we arrive at a coupled system in terms of $y_n \equiv v_1 = v_1(n)$ and $z_n \equiv v_3 = v_3(n)$, namely

$$y_{n+1} + y_n + z_n = C - \frac{\lambda_2 - \lambda_0 - n}{z_n}, \qquad (5.31a)$$

$$y_{n+1} + z_{n+1} + z_n = C - \frac{\lambda_1 - \lambda_0 - n}{y_{n+1}}, \qquad (5.31b)$$

which is the asymmetric form of dP$_I$ (cf. (3.29) and (3.30)), and thus, as we have seen in Section 3, is actually an alternative discrete form of P$_{II}$.

5.2.2   Alt-dP$_I$

In this case we take even smallest period $P = 1$ reduction of the lattice BSQ, thus rank $N = 3$ and 2×2 matrices. The Lax matrices (5.20) are now of the form

$$L_n(h) = \begin{pmatrix} hu_1 + \lambda_1 & h + v_2 \\ h^2 + hv_1 & hu_2 + \lambda_2 \end{pmatrix}, \quad M_n(h) = \begin{pmatrix} d_1 & 1 \\ h & d_2 \end{pmatrix}. \qquad (5.32)$$

We obtain from (5.23) the set of relations

$$
\begin{cases}
d_1 + \hat{u}_2 = u_1 + d_2, \\
d_2 + \hat{u}_1 = u_2 + d_1, \\
d_1 \hat{u}_1 + \hat{v}_2 = d_1 u_1 + v_1, \\
d_2 \hat{u}_2 + \hat{v}_1 = d_2 u_2 + v_2,
\end{cases}
\qquad
\begin{cases}
\hat{\lambda}_2 + d_1 \hat{v}_1 = d_2 v_1 + \lambda_1 + 1, \\
\hat{\lambda}_1 + d_2 \hat{v}_2 = d_1 v_2 + \lambda_2,
\end{cases}
\tag{5.33}
$$

together with $(\hat{\lambda}_i - \lambda_i)d_i = 0$, $i = 1, 2$. Again we need to tune the diagonal entries d_i in order to get a nontrivial dependence of the λ_i on the discrete variable n. Thus, choosing

$$
d_2 = 0, \ d_1 \neq 0 \implies \hat{\lambda}_1 = \lambda_1,
$$

we obtain from the second set in (5.33)

$$
d_1 = \frac{\hat{\lambda}_1 - \hat{\lambda}_2}{v_2} = \frac{1 + \lambda_1 - \hat{\lambda}_2}{\hat{v}_1},
\tag{5.34}
$$

leading to $\hat{\lambda}_2 = \lambda_2 + 1$. Furthermore, from the first set of relations in (5.33) we have a Casimir $u_1 + u_2 = C_1 = \text{const.}$, as well as a relation of the form $v_1 + v_2 - u_1 u_2 = C_2 = \text{const.}$ Thus, noting that

$$
d_1 = u_1 + \hat{u}_1 - C_1 \implies v_2 = \frac{\lambda_1 - \lambda_2}{u_1 + \hat{u}_1 - C_1},
$$

and choosing as our variable $x_n + \frac{1}{2}C_1 \equiv u_1 = u_1(n)$, we obtain the equation

$$
\frac{\lambda_1 - \lambda_0 - n}{x_{n+1} + x_n} + \frac{\lambda_1 - \lambda_0 - n + 1}{x_n + x_{n-1}} = C_2 + \frac{1}{4}C_1^2 - x_n^2,
\tag{5.35}
$$

which is the alt-dP$_\text{I}$ equation; cf. (3.69). This discrete Painlevé equation arose in the context of continuous Painlevé equations already in [144].

5.2.3   dP$_\text{II}$

At the end of Section 1 we have seen that the autonomous version of the dP$_\text{II}$ equation is exactly the McMillan map. Thus we can start from the Lax pair (1.26b), (1.27b) and deautonomize it to obtain a monodromy problem for dP$_\text{II}$. this Lax pair corresponds to the even $P = 2$ mapping reduction of the lattice KdV, which means that we have rank $N = 2$. Thus we have in this case 4×4 Lax matrices, which we write as

$$
L_n(h) = \begin{pmatrix} \lambda_1 & v_2 & 1 & 0 \\ 0 & \lambda_2 & v_3 & 1 \\ h & 0 & \lambda_3 & v_4 \\ h v_1 & h & 0 & \lambda_4 \end{pmatrix}, \quad
M_n(h) = \begin{pmatrix} d_1 & 1 & 0 & 0 \\ 0 & d_2 & 1 & 0 \\ 0 & 0 & d_3 & 1 \\ h & 0 & 0 & d_4 \end{pmatrix}.
\tag{5.36}
$$

From (5.23) we now obtain the set of relations

$$
\begin{cases}
d_1 + \hat{v}_4 = v_1 + d_3, \\
d_2 + \hat{v}_1 = v_2 + d_4, \\
d_3 + \hat{v}_2 = v_3 + d_1, \\
d_4 + \hat{v}_3 = v_4 + d_2,
\end{cases}
\quad
\begin{cases}
\hat{\lambda}_4 + d_1 \hat{v}_1 = d_4 v_1 + \lambda_1 + 1, \\
\hat{\lambda}_1 + d_2 \hat{v}_2 = d_1 v_2 + \lambda_2, \\
\hat{\lambda}_2 + d_3 \hat{v}_3 = d_2 v_3 + \lambda_3, \\
\hat{\lambda}_3 + d_4 \hat{v}_4 = d_3 v_4 + \lambda_4,
\end{cases}
\tag{5.37}
$$

together with $(\hat{\lambda}_i - \lambda_i)d_i = 0$, $i = 1, 2, 3$. Tuning the diagonal entries by taking

$$
d_2 = d_4 = 0, \ d_1, \ d_3 \neq 0 \implies \hat{\lambda}_1 = \lambda_1, \ \hat{\lambda}_3 = \lambda_3,
$$

we can deduce

$$
d_1 = \frac{1 + \lambda_1 - \hat{\lambda}_4}{\hat{v}_1} = \frac{\hat{\lambda}_1 - \lambda_2}{v_2},
\tag{5.38a}
$$

$$
d_3 = \frac{\hat{\lambda}_3 - \lambda_4}{\widehat{C} - v_2} = \frac{\lambda_3 - \hat{\lambda}_2}{\widehat{C} - \hat{v}_1},
\tag{5.38b}
$$

having taken into account $\hat{v}_1 = v_2$, $\hat{v}_3 = v_4$, and the Casimirs

$$
v_1 + v_3 = C = C_1 + (-1)^n C_2, v_2 + v_4 = \widehat{C} = C_1 - (-1)^n C_2,
$$

C_1, C_2 being constants. Comparing the two expressions for d_1 and d_3, we have

$$
\hat{\lambda}_4 = \lambda_2 + 1, \quad \hat{\lambda}_2 = \lambda_4 \implies \lambda_2 = \frac{1}{2}n + \lambda_0(-1)^n, \quad \lambda_0 = \text{const.}
$$

Thus, taking for our variable, e.g., $\frac{1}{2}\widehat{C} - x_n \equiv v_2 = v_2(n)$, we obtain dP$_{\mathrm{II}}$ in the form

$$
x_{n+1} + x_{n-1}
+ \frac{(\lambda_1 + \lambda_3 - n - \frac{1}{2})x_n + \frac{1}{2}\widehat{C}\left[\lambda_1 - \lambda_3 + \frac{1}{2} - 2\lambda_0(-1)^n\right]}{\frac{1}{4}\widehat{C}^2 - x_n^2} = 0.
\tag{5.39}
$$

Exercise 5.9. Show that by tuning the diagonal entries in (5.36) differently, namely by taking $d_3 = d_4 = 0$, $d_1, d_2 \neq 0$, we arrive again at a monodromy problem for dP$_{\mathrm{II}}$.

Yet another choice of the diagonal entries d_i does produce another system, namely by taking $d_4 = 0$, $d_1, d_2, d_3 \neq 0$, leading to $\lambda_1, \lambda_2, \lambda_3 = \text{const.}$ In that case we have from (5.37) still the same (alternating) Casimirs given above, but in (5.38) the second expressions for d_1 and d_3 no longer hold true. Furthermore, for d_2 we have now

$$
d_2 = \frac{d_1 v_2 + \lambda_2 - \lambda_1}{\hat{v}_2},
\tag{5.40}
$$

whereas from the second set in (5.37) we can deduce that $\hat{\lambda}_4 = \lambda_4 + 1$, implying that $\lambda_4 = n + \lambda_0$, $\lambda_0 = \text{const}$. Taking in this case $\frac{1}{2}C - y_n \equiv v_1 = v_1(n)$, we obtain the coupled system

$$x_n + y_{n+1} = \frac{\lambda_1 - \lambda_2}{\frac{1}{2}C - x_{n+1}} - \frac{(\frac{1}{2}\widehat{C} - x_n)(\lambda_1 - \lambda_0 - n)}{(\frac{1}{2}C - x_{n+1})(\frac{1}{2}\widehat{C} - y_{n+1})}, \tag{5.41a}$$

$$y_n - x_{n+1} = \frac{\lambda_1 - \lambda_0 - n}{\frac{1}{2}\widehat{C} - y_{n+1}} - \frac{\lambda_3 - \lambda_0 - n}{\frac{1}{2}\widehat{C} - x_n}, \tag{5.41b}$$

which is an as to yet unidentified discrete Painlevé equation, which we expect in the continuum limit to be related to the variant form of $\mathrm{dP_{III}}$. Another asymmetric form of $\mathrm{dP_{II}}$ was given in the paper [121] and carries a monodromy problem of Ablowitz–Ladik type (see the next section).

5.2.4 Alt-$\mathrm{dP_{II}}$

It was noted in [113] that the autonomous alt-$\mathrm{dP_{II}}$ equation arises from the lattice version of the MBSQ equation. This indicates which type of monodromy problem to take into consideration for the alt-$\mathrm{dP_{II}}$ case. In fact, MBSQ requires a rank-3 multiplicative form, i.e., in the MGD class, and the mapping reduction takes place for smallest even period $P = 1$, thus leading to a 2×2 matrix system. Thus taking the Lax matrices in the form

$$L_n(h) = \begin{pmatrix} hu_1 + \lambda_1 & hw_2 + v_2 \\ h^2 w_1 + hv_1 & hu_2 + \lambda_2 \end{pmatrix}, \quad M_n(h) = \begin{pmatrix} d_1 & z_2 \\ hz_1 & d_2 \end{pmatrix}, \tag{5.42}$$

the only difference with (5.32) residing in the highest order terms in the off-diagonal entries, we obtain from (5.23) the set of relations

$$\begin{cases} d_1\widehat{w}_1 + \hat{u}_2 z_1 = u_1 z_1 + d_2 w_1, \\ d_2\widehat{w}_2 + \hat{u}_1 z_2 = u_2 z_2 + d_1 w_2, \\ d_1\hat{u}_1 + \hat{v}_2 = d_1 u_1 + v_1, \\ d_2\hat{u}_2 + \hat{v}_1 = d_2 u_2 + v_2, \end{cases} \tag{5.43a}$$

$$\begin{cases} \hat{\lambda}_2 + d_1\hat{v}_1 = d_2 v_1 + \lambda_1 + 1, \\ \hat{\lambda}_1 + d_2\hat{v}_2 = d_1 v_2 + \lambda_2, \end{cases} \tag{5.43b}$$

in addition to

$$\widehat{w}_2 z_1 = w_1 z_2, \widehat{w}_1 z_2 = z_1 w_2. \tag{5.44}$$

Furthermore, this being within the MGD class, we impose the constraints

$$w_2 + u_1 + v_2 + \lambda_1 = 0, \quad w_1 + u_2 + v_1 + \lambda_2 = 0, \tag{5.45}$$

as well as

$$z_2 + d_1 = \text{const}, \quad z_1 + d_2 = \text{const}. \tag{5.46}$$

We note, importantly, that in this special case we still have a monodromy problem of differential type even though we might have expected here a q-difference system. The latter, however, would not be compatible with some of the relations (5.43). An important remark is that we can still go through the derivation of the alt-dP$_\mathrm{I}$ case and derive (5.35) in terms of the variable u_1. Introducing the variables $w_{1,2}$ and $z_{1,2}$, however, will give extra freedom, and in fact, these variables are not being specified by the compatibility relations (5.43) alone. In turn, they are specified by the constraints (5.45) and (5.46), which turn out to be *compatible* (and therein resides the nontriviality of this reduction) with the relations coming from the system (5.43). So, by eliminations similar to the alt-dP$_\mathrm{I}$ case we obtain from (5.43) in the usual way the set of conditions

$$w_2 \widehat{w_2} = C_1 \frac{z_2}{z_1} = C_1 \frac{v_2}{\hat{v}_1}, \tag{5.47a}$$

$$(\lambda_1 - \lambda_2) w_2 = (\hat{u}_1 + u_1 - C_2) v_2, \tag{5.47b}$$

$$\hat{v}_1 \hat{v}_2 + (\lambda_1 - \lambda_2)(\hat{u}_1 - u_1) = v_1 v_2, \tag{5.47c}$$

in which C_1 and C_2 are integration constants, as well as $\lambda_2 = \lambda_0 + n$, $\lambda_0, \lambda_1 = \text{const}$. Instead of concentrating on u_1, however, we now take as our main variable $x_n \equiv w_2 = w_2(n)$ and use the constraints (5.45) to derive

$$\frac{C_1}{x_n} + C_2 - u_1 + \frac{C_1(\lambda_1 - \lambda_2 + 1)}{C_3 - x_{n-1}x_n} + n - 1 + \lambda_0 = 0, \tag{5.48a}$$

$$x_n + u_1 + \frac{C_1(\lambda_1 - \lambda_2 + 1)}{C_3 - x_n x_{n+1}} + \lambda_1 = 0, \tag{5.48b}$$

where C_3 is the constant in (5.46) and where we have set $z_1 = C_1$ without loss of generality, leading to

$$z_2 = x_{n+1}x_n, \quad v_2 = \frac{(\lambda_1 - \lambda_2)x_n x_{n+1}}{C_3 - x_n x_{n+1}}, \quad \hat{v}_1 = \frac{(\lambda_1 - \lambda_2)x_n x_{n+1}}{C_3 - x_n x_{n+1}}.$$

Obviously, (5.48) leads immediately to

$$x_n + \frac{C_1}{x_n} + C_2 + \frac{C_1(\lambda_1 - \lambda_2 + 1)}{C_3 - x_{n-1}x_n} + \frac{C_1(\lambda_1 - \lambda_2 + 1)}{C_3 - x_n x_{n+1}}$$
$$+ \lambda_1 + \lambda_0 + n - 1 = 0, \tag{5.49}$$

which is the alt-dP$_\mathrm{II}$ equation, provided that $C_3 = C_1$. But in fact, that is precisely the condition we need in order for (5.48) to be compatible with the relations (5.47). Moreover, since it can be shown that in terms of u_1 the system (5.47) gives rise again to the alt-dP$_\mathrm{I}$ equation, we might consider (5.48) to constitute a Miura transformation between this equation and the alt-dP$_\mathrm{II}$, which is somewhat surprising. In fact, there is also a kind of inverse relation of the form

$$\hat{u}_1 + u_1 = C_2 + \frac{C_1}{x_{n+1}} - x_n. \tag{5.50}$$

5.2.5 dP$_\mathrm{III}$

The autonomous version of the dP$_\mathrm{III}$ equation corresponds exactly with the $P = 2$ even reduction of the lattice MKdV equation. This suggests that a monodromy problem for dP$_\mathrm{III}$ would arise from the MGD class of rank $N = 2$ with a 4×4 representation of the form

$$L_n(h) = \begin{pmatrix} \lambda_1 & v_2 & u_3 & 0 \\ 0 & \lambda_2 & v_3 & u_4 \\ hu_1 & 0 & \lambda_3 & v_4 \\ hv_1 & hu_2 & 0 & \lambda_4 \end{pmatrix}, \tag{5.51a}$$

$$M_n(h) = \begin{pmatrix} d_1 & w_2 & 0 & 0 \\ 0 & d_2 & w_3 & 0 \\ 0 & 0 & d_3 & w_4 \\ hw_1 & 0 & 0 & d_4 \end{pmatrix}. \tag{5.51b}$$

Since the dP$_\mathrm{III}$ equation was the first example of a discrete Painlevé equation in which the dependence on the discrete variable n enters via exponents, it was natural to expect that this monodromy problem should be of q-difference type rather than differential type. This turned out to be the case, as was shown in [145]. Let us therefore investigate the compatibility (5.25), from which we obtain now the set of relations

$$\begin{cases} d_1\hat{u}_1 + \hat{v}_4 w_1 = w_4 v_1 + d_3 u_1, \\ d_2\hat{u}_2 + \hat{v}_1 w_2 = qw_1 v_2 + d_4 u_2, \\ d_3\hat{u}_3 + \hat{v}_2 w_3 = w_2 v_3 + d_1 u_3, \\ d_4\hat{u}_4 + \hat{v}_3 w_4 = w_3 v_4 + d_2 u_4, \end{cases} \tag{5.52a}$$

$$\begin{cases} \hat{\lambda}_4 w_1 + d_1\hat{v}_1 = d_4 v_1 + q\lambda_1 w_1, \\ \hat{\lambda}_1 w_2 + d_2\hat{v}_2 = d_1 v_2 + \lambda_2 w_2, \\ \hat{\lambda}_2 w_3 + d_3\hat{v}_3 = d_2 v_3 + \lambda_3 w_3, \\ \hat{\lambda}_3 w_4 + d_4\hat{v}_4 = d_3 v_4 + \lambda_4 w_4, \end{cases} \tag{5.52b}$$

together with

$$\hat{u}_4 = \frac{w_3}{w_1} u_1, \quad \hat{u}_1 = \frac{w_4}{w_2} u_2, \quad \hat{u}_2 = q\frac{w_1}{w_3} u_3, \quad \hat{u}_3 = \frac{w_2}{w_4} u_4$$

from which we have the Casimir

$$u_1 u_3 = \kappa_n = q^{n/2} \alpha^{(-1)^n} C, \quad u_2 u_4 = \kappa_{n+1}. \tag{5.53}$$

Tuning, as in the dP$_{II}$ case, the diagonal entries by taking

$$d_2 = d_4 = 0, \; d_1, \; d_3 \neq 0 \implies \hat{\lambda}_1 = \lambda_1, \; \hat{\lambda}_3 = \lambda_3,$$

and taking without loss of generality $w_1 = w_3 = 1$, we can deduce

$$d_1 = \frac{q\lambda_1 - \hat{\lambda}_4}{\hat{v}_1} = w_2 \frac{\hat{\lambda}_1 - \lambda_2}{v_2}, \tag{5.54a}$$

$$d_3 = \frac{\lambda_3 - \hat{\lambda}_2}{\hat{v}_3} = w_4 \frac{\hat{\lambda}_3 - \lambda_4}{v_4}, \tag{5.54b}$$

having taken into account $\hat{v}_1 = qv_2/w_2$, $\hat{v}_3 = v_4/w_4$. Comparing both expressions for d_1 and d_3, we conclude that

$$\hat{\lambda}_4 = q\lambda_2, \quad \hat{\lambda}_2 = \lambda_4 \implies \lambda_2 = q^{n/2} \beta^{(-1)^n} \lambda_0, \quad \beta, \quad \lambda_0 = \text{const.} \tag{5.55}$$

Since we are dealing with the MGD class, we have the constraints

$$v_1 = u_2 + \lambda_4, \quad v_2 = u_3 + \lambda_1, \quad v_3 = u_4 + \lambda_2, \quad v_4 = u_1 + \lambda_3, \tag{5.56}$$

as well as

$$w_2 - d_1 = \text{const}, \quad w_4 - d_3 = \text{const},$$

where the constant can be taken to be 1 without loss of generality. Let us choose $x_n \equiv u_1 = u_1(n)$ to be our variable. This implies that $u_2 = \kappa_{n+1}/x_{n-1}$, $u_3 = \kappa_n/x_n$, and $u_4 = x_{n-1}$. The variables v_i are obtained in terms of x_n from the constraints (5.56), and using these it is not hard to check that the remaining two equations in the first set of (5.52) are satisfied. The latter fact is indicative of the consistency of the reduction imposed by the constraints. Thus we get the equation for x_n in the form

$$x_{n+1} x_{n-1} = \kappa_{n+1} \frac{(\kappa_n + \lambda_2 x_n)(x_n + \lambda_3)}{(\kappa_n + \lambda_1 x_n)(x_n + \lambda_4)}, \tag{5.57}$$

which is the dP$_{III}$ equation. The Lax pair constructed in this way was to our knowledge the first example of a monodromy problem of q-difference type.

Exercise 5.10. Determine what discrete system arises from the q-difference monodromy problem with Lax matrices (5.51) using a different tuning of the diagonal entries, namely $d_4 = 0$, $d_1, d_2, d_3 \neq 0$.

We finish this section by remarking that we have by far not exhausted yet all possibilities of constructing monodromy problems arising from the GD and MGD classes as considered above. It cannot be excluded that monodromy problems for dP_{IV} and dP_V arise in this way, but we haven't found them yet. These possibilities are currently under study. One advantage of this approach to obtain monodromy problems is that it might ultimately yield a kind of classification of discrete Painlevé equations according to rank and dimension within the (M)GD class. As said before, to date no study of the asymptotic aspects of the monodromy problems has been undertaken. Such a study might possibly yield insight into the nature of the discrete transcendents.

5.3   Other Approaches

The methods described above are obviously not the only ones that yield isomonodromic deformation problems for the discrete Painlevé equations. The advantage of the monodromy problems coming from deautonomization of the GD mappings is that they form a well-defined class of discrete systems in terms of which discrete Painlevé equations can be identified with certain reductions of nonlinear integrable lattice equations. Within this setting one could eventually hope to arrive at a kind of classification of dPs according to their reduction. A disadvantage might be that the dimension of the matrix representation is not fixed. In order to use effectively the resulting monodromy problems in order to extract asymptotic information, it might be advantageous to work with 2×2 matrices only. Without giving any details, we mention that alternative 2×2 monodromy problems for dP_I and dP_{II} have been obtained by Joshi et al. in [146] starting from the Ablowitz–Ladik spectral problem and by Levi et al. in [147] within a framework of nonisospectral flows. It is safe to assume that these two approaches are very closely related to each other and lead to a slightly simpler monodromy problem for dP_{II} than the one given in (5.36) above.

In the important paper by Jimbo and Sakai [122], in which they identified the asymmetric dP_{III} with a discrete Painlevé VI equation, the starting point was also a 2×2 monodromy problem of q-difference type that mimics the classic Fuchsian approach to obtain P_{VI}. Since this discrete Painlevé VI is so far the richest dP for which we know a monodromy problem, let us see how this one fits into our picture. To make a connection between the 4×4 Lax matrices and a 2×2 problem, let us investigate somewhat more closely the Lax representation (5.51) for the dP_{III} equation—a similar argument holds for (5.36) in the case of dP_{II} as well. After application of a permutation of rows and columns, the Lax matrices can be cast into the block form

$$\bar{L}_n(h) = \left(\begin{array}{c|c} \Lambda_1 + \sigma_h U_1 & V_2 \\ \hline \sigma_h V_1 & \Lambda_2 + \sigma_h U_2 \end{array} \right), \quad \overline{M}_n(h) = \left(\begin{array}{c|c} D & W \\ \hline \sigma_h & 0 \end{array} \right), \quad (5.58)$$

in which

$$\Lambda_{1,2} = \begin{pmatrix} \lambda_{1,2} & 0 \\ 0 & \lambda_{3,4} \end{pmatrix}, \quad U_{1,2} = \begin{pmatrix} u_{1,2} & 0 \\ 0 & u_{3,4} \end{pmatrix}, \quad V_{1,2} = \begin{pmatrix} v_{1,2} & 0 \\ 0 & v_{3,4} \end{pmatrix},$$

and similarly

$$D = \begin{pmatrix} d_1 & 0 \\ 0 & d_3 \end{pmatrix}, \quad W = \begin{pmatrix} w_2 & 0 \\ 0 & w_4 \end{pmatrix}, \quad \sigma_h = \begin{pmatrix} 0 & 1 \\ h & 0 \end{pmatrix},$$

and where we have the constraints

$$W = 1 + D, \quad V_1 = U_2 + \sigma_h \Lambda_2 \sigma_h^{-1}, \quad V_2 = \Lambda_1 + \sigma_h U_1 \sigma_h^{-1}, \qquad (5.59)$$

corresponding to the constraints on the entries of the matrices (5.51).

In this shape it is very easy to extract a 4×4 monodromy problem for the qP$_{VI}$ of [122], realizing that qP$_{VI}$ is the asymmetric form of dP$_{III}$, i.e., it is obtained from the alternating form of dP$_{III}$ on the half-integer lattice. Thus, taking into account (5.55), the asymmetric form of dP$_{III}$ can be obtained by introducing variables x_n on the even lattice sites and variables y_n on the odd lattice sites. The only thing one needs to do is to rewrite the Lax pair in terms of "double shifts" on the lattice, which is easily obtained from the M-part of (5.58) iterated twice. In this manner we obtain

$$\Phi_{qh} = \mathcal{L}_h \Phi_h = \left(\begin{array}{c|c} \Lambda_1 + \sigma_h U_1 & V_2 \\ \hline \sigma_h V_1 & \Lambda_2 + \sigma_h U_2 \end{array} \right) \Phi_h, \qquad (5.60a)$$

$$\widetilde{\Phi}_h = \mathcal{M}_h \Phi_h = \left(\begin{array}{c|c} D_2 D_1 + W_2 \sigma_h & D_2 W_1 \\ \hline \sigma_h D_1 & \sigma_h W_1 \end{array} \right) \Phi_h, \qquad (5.60b)$$

in which the tilde $\tilde{}$ denotes in this context the double shift, and where we have the constraints (5.59) as before. Note that we have now matrices D_1, W_1 as well as D_2, W_2, since we need D and W shifted over a half-lattice as well. The conditions on the various matrices in the 2×2 blocks of (5.60) are easily worked out from the compatibility conditions

$$\mathcal{M}_{qh} \mathcal{L}_h = \widetilde{\mathcal{L}}_h \mathcal{M}_h$$

and lead to

$$\widetilde{\Lambda}_1 = \Lambda_1, \quad \widetilde{\Lambda}_2 = q \Lambda_2, \quad \widetilde{U}_1 W_2 = Q \sigma_h U_1 W_2 \sigma_h^{-1},$$
$$\widetilde{U}_2 W_1^{-1} = Q \sigma_h U_2 W_1^{-1} \sigma_h^{-1},$$

in which $Q = \mathrm{diag}(q, 1)$, together with

$$D_1 V_2 = W_1 (\Lambda_1 - \Lambda_2), \quad D_2 \widetilde{V}_1 = Q \Lambda_1 - \sigma_h \widetilde{\Lambda}_2 \sigma_h^{-1}.$$

Introducing Casimirs via $\det U_1 = \kappa_1 = \kappa_1^0 q^n$, $\det U_2 = \kappa_2 = \kappa_2^0 q^n$, where $\kappa_{1,2}^0$ are constants, and choosing $u_1 \equiv y$, $u_2 \equiv x$, we obtain the asymmetric dP_{III} in the form

$$\tilde{x}x = q\kappa_2 \frac{(\kappa_1 + \lambda_1 y)(y + \lambda_4)}{(\kappa_1 + \lambda_2 y)(y + \lambda_3)}, \tag{5.61a}$$

$$y\underset{\sim}{y} = \kappa_1 \frac{(\kappa_2 + \lambda_3 x)(x + \lambda_4)}{(\kappa_2 + \lambda_2 x)(x + q\lambda_1)}, \tag{5.61b}$$

where $\lambda_2 = \lambda_2^0 q^n$, $\lambda_4 = \lambda_4^0 q^n$, λ_2^0, λ_4^0 as well as λ_1, λ_3 are constants. Equation (5.61) is the qP_{VI} equation of [122] and was given above in (3.37), (3.38). Following the scheme outlined above, the next step is to reduce the 4×4 system (5.60) to a 2×2 system following the same derivation as given in the beginning of this section. A comparison of these results with the monodromy problem of [122] would be useful. It should be mentioned that although the derivation of the asymmetric dP_{III} follows straightforwardly from the logic of our previous papers, e.g., [121, 145], it was the great breakthrough of [122] to recognize this system as a proper discrete analogue of P_{VI}. As was mentioned in Section 3.3, there exist other discrete versions of P_{VI} related to asymmetric forms of dP_{IV} and dP_V. However, for these systems the monodromy problems are still unknown.

Exercise 5.11. Derive the form of the monodromy problem (5.60) by applying twice consecutively (5.58) and combining the double shifts into one single shift, i.e., $\hat{\hat{\Phi}}_h = \tilde{\Phi}_h$, and express the entries in the Lax matrices $\mathcal{L}_h$ and $\mathcal{M}_h$ in terms of $x \equiv u_1$ and $y \equiv \hat{u}_1$ involving double shifts only.

Exercise 5.12. Show that the asymmetric form of dP_{II} is obtained along similar lines as outlined above for the asymmetric dP_{III}, namely starting from the 4×4 system (5.36) and considering double shifts. Calculate the corresponding isomonodromy problem and show that the compatibility conditions leads to the asymmetric dP_{II} (3.33), (3.34) in the form

$$\tilde{x} + x = \frac{4}{C_1 C_2} \frac{(\lambda_1 - \lambda_2 - \lambda_3 + \lambda_4) + (\lambda_1 - \lambda_2 + \lambda_3 - \lambda_4)y}{1 - y^2}, \tag{5.62a}$$

$$\tilde{y} + y = \frac{4}{C_1 C_2} \frac{(\lambda_1 + \lambda_2 - \lambda_3 - \lambda_4) + (\lambda_1 - \lambda_2 + \lambda_3 - \lambda_4)\tilde{x}}{1 - \tilde{x}^2}, \tag{5.62b}$$

taking $x = 1 - 2v_1/C_1$, $y = 1 - 2v_2/C_2$, with the Casimirs $C_1 = \operatorname{tr} V_1$, $C_2 = \operatorname{tr} V_2$, and where $\lambda_2 = n + \lambda_2^0$, $\lambda_4 = n + \lambda_4^0$, λ_1, λ_3, λ_2^0, λ_4^0 being constants.

The above results show that there might be a unified framework in which we can embed monodromy problems for all discrete Painlevé equations. Ultimately, these monodromy problems should be used to investigate the transcendental solutions of the discrete equations. This is a major task for the future, and so far very few results exist. If this program is taken up, we can expect many new developments in the near future. With this we want to conclude our rather incomplete introduction to this new field of discrete Painlevé equations in the hope that it will form an inspiration for young researchers who would like to dedicate themselves to this intriguing new field.

Acknowledgments: F.W.N. is grateful to the Laboratoire de Physique Nucléaire de l'Université Paris VII for its hospitality during a visit where the major part of this work was carried out. B.G. and F.W.N. are grateful for support from the Commission of the European Communities under contract no. ERBCHBICT941546.

6 REFERENCES

[1] P. Painlevé, *Mémoire sur les équations différentielles dont l'intégrale générale est uniforme*, Bull. Soc. Math. France **28** (1900), 201–261; *Sur les équations différentielles du second ordre et d'ordre supérieur dont l'intégrale générale est uniforme*, Acta Math. **25** (1902), 1–85.

[2] B. Gambier, *Sur les équations différentielles du second ordre et du premier degré dont l'intégrale générale est à points critiques fixes*, Acta Math. **33** (1909), 1–55.

[3] J. Chazy, *Sur les équations différentielles du troisième ordre et d'ordre supérieur dont l'intégrale générale a ses points critiques fixes*, Acta Math. **34** (1911), 317–385.

[4] M.R. Garnier, *Sur des équations différentielles du troisième ordre dont l'intégrale est uniforme et sur une classe d'équations nouvelles d'ordre supérieur dont l'intégrale a ses points critiques fixes*, Ann. Sci. de l'ENS vol. XXIX, # 3, (1912), 1–126.

[5] R.P. Feynman, *Simulating physics with computers*, Int. J. Theor. Phys. **21** (1982), 467.

[6] Y. Nambu, *Field theory and Galois' fields*, Int. J. Theor. Phys. **21** (1982), 625–636.

[7] T.D. Lee, *Can time be a discrete variable?* Phys. Lett. **122B** (1983), 217–220.

[8] G. 't Hooft, *Quantization of discrete deterministic theories by Hilbert space extension*, Nucl. Phys. **B342** (1990), 471–485.

[9] G. 't Hooft, K. Isler, and S. Kalitzin, *Quantum Field theoretic behavior of a deterministic cellular automaton*, Nucl. Phys. **B386** (1992), 495–519.

[10] G.D. Birkhoff, *General theory of linear difference equations*, Trans. Amer. Math. Soc. **12** (1911), 243–284; *The generalized Riemann problem for linear differential equations and the allied problems for linear difference and q-difference equations*, Proc. Am. Acad. Arts Sci. **49** (1913), 521–568.

[11] N.E. Nörlund, *Vorlesungen über Differenzenrechnung* (Kopenhagen, 1923).

[12] E.L. Ince, *Ordinary Differential Equations* (Longmans, Green and co., London and New York, 1926). Reprinted (Dover, New York, 1956).

[13] R.L. Devaney, *An Introduction to Chaotic Dynamical Systems*, (Addison-Wesley, 1989).

[14] A.P. Veselov, *What is an integrable mapping?*, What is Integrability?, ed. V.E. Zakharov (Springer Verlag, Berlin, 1991), Springer Series in Nonlinear Dynamics, pages 251–272.

[15] A.P. Veselov, *Integrable maps*, Russ. Math. Surv. **46** (1991), 1–51.

[16] G. Julia, *Mémoire sur la permutabilité des fractions rationnelles*, Ann. Sci. ENS **39** (1922), 131–215.

[17] J.F. Ritt, *Permutable rational functions*, Trans. Amer. Math. Soc. **25** (1923), 399–448.

[18] P. Fatou, *Sur l'itération analytique et les substitutions permutables*, J. Math. Pures Appl. **23** (1924), 1–49.

[19] A.P. Veselov, *Integrable mappings and Lie algebras*, Sov. Phys. Dokl. **35** (1987), 211–213.

[20] H. Jung, *Über ganze birationale Transformationen der Ebene*, J. Reine Angew. Math. **184** ((1942), 161–172.

[21] D. Wright, *Abelian subgroups of* $\mathrm{Aut}_k(k[x,y])$ *and applications to actions on the affine plane*, Ill. J. Math. **23** (1979), 579–633.

[22] E.M. McMillan, *A problem in the stability of periodic systems*, Topics in Modern Physics. A Tribute to E.U. Condon, eds. W.E. Brittin and H. Odabasi (Colorado Associated Univ. Press, Boulder, 1971), pages 219–244.

[23] M. Hénon, *A two-dimensional mapping with a strange attractor*, Commun. Math. Phys. **50** (1967), 69–77.

[24] Yu.B. Suris, *Integrable mappings of standard type*, Funct. Anal. Appl. **23** (1987), 74–76.

[25] G.R.W Quispel, J.A.G. Roberts, and C.J. Thompson, *Integrable mappings and soliton equations*, Phys. Lett. **A126** (1988), 419–421; ibid. II, Physica **D34** (1989), 183–192.

[26] J.A.G. Roberts and G.R.W. Quispel, *Chaos and time-reversal symmetry*, Phys. Rep. 216 (1992), 63.

[27] J.A.G. Roberts, *Order and Chaos in Reversible Dynamical Systems*, Ph.D. thesis, Mathematics Department, University of Melbourne, 1990.

[28] V.I. Arnold'd and M.B. Sevryuk, *Oscillations and bifurcations in reversible systems*, Nonlinear Phenomena in Plasma Physics and Hydrodynamics, ed. R. Sagdeev (MIR Publications, Moscow, 1986), pages 31–64.

[29] J. Moser, *On the volume elements on a manifold*, Trans. Am. Math. Soc. **120** (1965), 286–294.

[30] A.P. Veselov, *Growth and integrability in the dynamics of maps*, Commun. Math. Phys. **145** (1992), 181–193.

[31] J.D. Logan, *First integrals in discrete variational calculus*, Aequationes Math. **9** (1973), 210–220.

[32] S. Maeda, *Canonical structure and symmetries for discrete systems*, Math. Japonica **25** (1980) 405–420; *Extension of Noether theorem*, Math. Japon. **26** (1981), 85–90.

[33] A.P. Veselov, *Integrable Lagrangian correspondences and the factorization of matrix polynomials*, Funct. Anal. Appl. **25** (1991), 112–122.

[34] P.A. Deift, L.C. Li, and C. Tomei, *Matrix factorizations and integrable systems*, Commun. Pure Appl. Math. **42** (1989), 443 521.

[35] J. Moser and A.P. Veselov, *Discrete versions of some classical integrable systems and factorization of matrix polynomials*, Commun. Math. Phys. **139** (1991), 217–243.

[36] M. Bruschi, O. Ragnisco, P.M. Santini, and G.-Z. Tu, *Integrable symplectic maps*, Physica **49D** (1991), 273–294.

[37] V.I. Arnold'd, *Dynamics of complexity of intersections*, Bol. Soc. Bras. Mat. **21** (1990), 1–10.

[38] G. Falqui and C.-M. Viallet, *Singularity, complexity, and quasi-integrability of rational maps*, Commun. Math. Phys. **154** (1993), 111–125.

[39] S. Boukraa and J.-M. Maillard, *Factorization properties of birational mappings*, Physica **A220** (1995), 403–470.

[40] V. Papageorgiou, F.W. Nijhoff, and H.W. Capel, *Integrable mappings and nonlinear integrable lattice equations*, Phys. Lett. **147A** (1990), 106–114.

[41] H.W. Capel, F.W. Nijhoff, and V.G. Papageorgiou, *Complete integrability of Lagrangian mappings and lattices of KdV type*, Phys. Lett. **155A** (1991), 377–387.

[42] F.W. Nijhoff, V.G. Papageorgiou, and H.W. Capel, *Integrable time-discrete systems: lattices and mappings*, Quantum Groups, ed. P.P. Kulish, Lecture Notes in Mathematics **1510** (Springer, Berlin, 1992), pages 312–325.

[43] M. Toda, *Theory of Nonlinear Lattices*, Springer Series in Solid State Physics **20** (Springer, Berlin, 1981).

[44] R.J. Baxter, *Exactly Solved Models in Statistical Mechanics*, (Associated Press, London, 1982).

[45] E.D. Belokolos, A.I. Bobenko, V.Z. Enolskii, A.R. Its, and V.B. Matveev, *Algebro-Geometric Approach to Nonlinear Evolution Equations* (Springer Verlag, 1994).

[46] M.J. Ablowitz and F.J. Ladik, *A nonlinear difference scheme and inverse scattering*, Stud. Appl. Math. **55** (1976), 213–229; *On the solution of a class of nonlinear partial difference equations*, ibid. **57** (1977), 1–12.

[47] R. Hirota, *Nonlinear partial difference equations I-III*, J. Phys. Soc. Japan **43** (1977), 1424–1433, 2074–2089.

[48] R. Hirota, *Discrete analogue of generalized Toda equation*, J. Phys. Soc. Japan **50** (1981), 3785–3791.

[49] D. Levi, L. Pilloni, and P.M. Santini, *Integrable three-dimensional lattices*, J. Phys. A: Math. Gen. **14** (1981), 1567–1575.

[50] E. Date, M. Jimbo, and T. Miwa, *Method for generating discrete soliton equations I-V*, J. Phys. Soc. Japan **51** (1982), 4116–4131, **52** (1983), 388–393, 761–771.

[51] F.W. Nijhoff, G.R.W. Quispel, and H.W. Capel, *Direct linearization of nonlinear difference-difference equations*, Phys. Lett. **97A** (1983), 125–128.

[52] G.R.W. Quispel, F.W. Nijhoff, H.W. Capel, and J. van der Linden, *Linear integral equations and nonlinear difference-difference equations*, Physica **125A** (1984), 344–380.

[53] F.W. Nijhoff, H.W. Capel, G.L. Wiersma, and G.R.W. Quispel, *Bäcklund transformations and three-dimensional lattice equations*, Phys. Lett. **105A** (1984), 267–272.

[54] F.W. Nijhoff, H. Capel, and G.L. Wiersma, *Integrable lattice systems in two and three dimensions*, Geometric Aspects of the Einstein Equations and Integrable Systems, ed. R. Martini, Lecture Notes in Physics **239** (Springer, Berlin, 1985), pages 263–302.

[55] G. Wiersma and H.W. Capel, *Lattice equations, hierarchies and Hamiltonian structures*, Physica **142A** (1987), 199–244; ibid. **149A** (1988), 49–74.

[56] F.W. Nijhoff and H.W. Capel, *The direct linearisation approach to hierarchies of integrable PDEs in 2+1 dimensions: I. Lattice equations and the differential-difference hierarchies*, Inv. Problems **6** (1990), 567–590.

[57] F.W. Nijhoff, V.G. Papageorgiou, H.W. Capel, and G.R.W. Quispel, *The lattice Gel'fand-Dikii hierarchy*, Inv. Probl. **8** (1992), 597–621.

[58] H.W. Capel and F.W. Nijhoff, *Integrable lattice equations*, Important Developments in Soliton Theory, eds. A.S. Fokas and V.E. Zakharov, (Springer Lect. Notes in Nonlinear Dynamics, 1993), pages 38–57.

[59] F.W. Nijhoff and H.W. Capel, *The discrete Korteweg-de Vries equation*, Acta Applicandae Mathematicae **39** (1995), 133–158.

[60] R. Conte and M. Musette, *A new method to test discrete Painlevé equations*, Phys. Lett. **223A** (1996), 439–448.

[61] S. Maeda, *The similarity method for difference equations*, IMA Journal of Applied Mathematics **38** (1987), 129–134.

[62] V.A. Dorodnitsyn, *Finite difference models entirely inheriting symmetry of original differential equations*, Modern Group Analysis: Advanced Analytical and Computational Methods in Mathematical Physics, ed. N.H. Ibragimov (Kluwer Acad. Publ., 1993).

[63] V.A. Dorodnitsyn, *Symmetries of finite-difference equations*, in: Handbook of Lie Group Analysis of Differential Equations, eds. N.H. Ibragimov et al. (CRC Press Inc, 1994).

[64] D. Levi and P. Winternitz, *Continuous symmetries of discrete equations*, Phys. Lett. **A152** (1991), 335–338; *Symmetries and Conditional Symmetries of Differential-Difference Equations*, J. Math. Phys. **34** (1993), 3713–3730.

[65] D. Levi, L. Vinet, and P. Winternitz, *Lie group formalism for difference equations*, J. Phys. **A** (1996).

[66] G.R.W. Quispel, *Lie symmetries and the integration of difference equations*, Phys. Lett. **A184** (1993), 64–70; G. Byrnes, G.R.W. Quispel and R. Sahadevan, *Factorizable Lie Symmetries and the Linearization of Difference Equations*, Nonlinearity **8** (1995), 443–459.

[67] F.W. Nijhoff and V.G. Papageorgiou, *Similarity reductions of integrable lattices and discrete analogues of the Painlevé II equation*, Phys. Lett.**153A** (1991), 337–344.

[68] A.S. Fokas and M.J. Ablowitz, *Linearization of the Korteweg-de Vries and Painlevé II equations*, Phys. Rev. Lett. **47** (1981), 1096–1100.

[69] P.M. Santini, A.S. Fokas, and M.J. Ablowitz, *The direct linearization of a class of nonlinear evolution equations*, J. Math. Phys. **25** (1984), 2614–2619.

[70] F.W. Nijhoff, G.R.W. Quispel, J. van der Linden, and H.W. Capel, *On some linear integral equations generating solutions of nonlinear partial differential equations*, Physica **119A** (1983), 101–142.

[71] F.W. Nijhoff, *Linear integral transformations and hierarchies of integrable nonlinear evolution equations*, Physica **D31** (1988), 339–388.

[72] M.J. Ablowitz and H. Segur, *Exact Linearization of a Painlevé Transcendent*, Phys. Rev. Lett. **38** (1977), 1103–1106.

[73] M.J. Ablowitz, A. Ramani, and H. Segur, *A connection between nonlinear evolution equations and ordinary differential equations of P-type, I,II*, J. Math. Phys. **21** (1980), 715–721; 1006–1015.

[74] M.J. Ablowitz and H. Segur, *Solitons and the Inverse Scattering Transform* (SIAM, Philadelphia, 1981).

[75] M.J. Ablowitz and P.A. Clarkson, *Solitons, Nonlinear Evolution Equations and Inverse Scattering*, LMS Lecture Notes **149** (Cambridge University Press, Cambridge, 1991).

[76] H. Flaschka and A.C. Newell, *Monodromy- and spectrum-preserving deformations I*, Commun. Math. Phys. **76** (1980), 65–116.

[77] A.R. Its and V.Y. Novokshenov, *The Isomonodromic Deformation Theory in the Theory of Painlevé Equations*, Lecture Notes in Mathematics **1191** (Springer, Berlin, 1986).

[78] A.S. Fokas and M.J. Ablowitz, *On the initial value problem of the second Painlevé equation*, Commun. Math. Phys. **91** (1983), 381–403.

[79] A.R. Its, *"Isomonodromy" solutions of equations of zero curvature*, Math. USSR Izv. **26** (1986), 497–529.

[80] A.S. Fokas, U. Mugan, and M.J. Ablowitz, *A Method of Linearization for Painlevé IV, V*, Physica **D30** (1988), 247–283.

[81] A.S. Fokas and X. Zhou, *On the solvability of Painlevé II and IV*, Commun. Math. Phys. **144** (1992), 601–622.

[82] T.T. Wu, B.M. McCoy, C.A. Tracy, and E. Barouch, *Spin–spin correlation functions for the two-dimensional Ising model: Exact theory in the scaling limit*, Phys. Rev. **B13** (1976), 316–374.

[83] H.G. Vaidya and C.A. Tracy, *Transverse Time-Dependent Spin Correlation Functions for the One-Dimensional XY Model at Zero Temperature*, Physica **92A** (1978), 1–41.

[84] M. Jimbo, T. Miwa, Y. Môri, and M. Sato, *Density matrix of an impenetrable Bose gas and the fifth Painlevé transcendent*, Physica **1D** (1980), 80–158.

[85] J.H.H. Perk, H.W. Capel, G.R.W. Quispel, and F.W. Nijhoff, *Finite-temperature correlations for the Ising chain in a transverse field*, Physica **123A** (1984), 1–49.

[86] E. Brézin and V.A. Kazakov, *Exactly solvable field theories of closed strings*, phys. lett. **236b** (1990), 144–150.

[87] D.J. Gross and A.A. Migdal, *Non-perturbative two-dimensional quantum gravity*, Phys. Rev. Lett. **64** (1990) 127–130, Nucl. Phys. **B340** (1990), 333–365.

[88] V. Periwal and D. Shevitz, *Unitary-matrix models as exactly solvable string theories*, Phys. Rev. **64** (1990), 1326–1329.

[89] M.R. Douglas and S.H. Shenker, *Strings in less than one dimension*, Nucl. Phys. **B335** (1990), 635–654.

[90] O. Alvarez and P. Windey, *Universality in two-dimensional quantum gravity*, Nucl. Phys. **B348** (1991), 490–506.

[91] D. Bessis, C. Itzykson, and J.-B. Zuber, *Quantum field theory techniques in graphical enumeration*, Adv. Appl. Math. **1** (1980), 109–157.

[92] J.A. Shohat, *A differential equation for orthogonal polynomials*, Duke Math. J. **5** (1939), 401–417.

[93] G. Freud, *On the coefficients in the recursion formulae of orthogonal polynomials*, Proc. Roy. Irish Acad. Sect. A **76** (1976), 1–6.

[94] A.P. Magnus, *Painlevé type differential equations for the recurrence coefficients of semi-classical orthogonal polynomials*, J. Comput. Appl. Math. **57** (1995), 215–237.

[95] A.S. Fokas, A.R. Its, and A.V. Kitaev, *Discrete Painlevé equations and their appearance in quantum gravity*, Comm. Math. Phys. **142** (1991), 313–344.

[96] I.M. Krichever and S.P. Novikov, *Holomorphic bundles over algebraic curves and nonlinear equations*, Russ. Math. Surv. **35** (1980), 53–79.

[97] F.W. Nijhoff, *On some "Schwarzian" equations and their discrete analogues*, Algebraic Aspects of Integrable Systems; In Memory of Irene Dorfman, eds. A.S. Fokas and I.M. Gel'fand (Birkhäuser Verlag, 1996), pages 237–260.

[98] G.R.W. Quispel, F.W. Nijhoff, and H.W. Capel, *Linearization of the Boussinesq equation and of the modified Boussinesq equation*, Phys. Lett. **91A** (1982), 143–145.

[99] T. Nishitani and M. Tajiri, *On similarity solutions of the Boussinesq equation*, Phys. Lett. **89A** (1982), 379–380.

[100] C.M. Cosgrove and G. Scoufis, *Painlevé classification of a class of differential equations of the second order and second degree*, Stud. Appl. Math. **88** (1993), 25–87.

[101] C.M. Cosgrove, *All binomial-type Painlevé equations of the second order and degree three or higher*, Stud. Appl. Math. **90** (1993), 119–187.

[102] M Boiti and F. Pempinelli, *Nonlinear Schrödinger equation, Bäcklund transformations and Painlevé transcendents*, Nuova Cim. **59B** (1980), 40–58.

[103] G.R.W. Quispel and H.W. Capel, *The nonlinear Schrödinger equation and the anisotropic Heisenberg spin chain*, Phys. Lett. **88A** (1982) 371–374; Physica **117A** (1983), 76–102.

[104] A.S. Fokas, R.A. Leo, L. Martina, and G. Soliani, *The scaling reduction of the three-wave resonant system and the Painlevé VI equation*, Phys. Lett. **A115** (1986), 329–332.

[105] L. Martina and P. Winternitz, *Analysis and applications of the symmetry group of the multidimensional three-wave resonant interaction problem*, Ann. Phys. **196** (1989), 231– 277.

[106] L.A. Dickey, *Soliton equations and Hamiltonian systems*, (World Scientific, Signapore, 1991).

[107] L.D. Faddeev and L.A. Tahktajan, *Liouville model on the lattice*, Springer Lect. Notes Phys. **246** (1986), 166–179.

[108] F.W. Nijhoff, A. Ramani, B. Grammaticos, and Y. Ohta, *On discrete Painlevé equations associated with the lattice KdV systems and the Painlevé VI equation*, solv-int/9812011.

[109] M. Kac and P. van Moerbeke, *On an explicitly soluble system on nonlinear differential equations related to certain Toda lattices*, Adv. Math. **16** (1975), 160–169.

[110] A.Yu. Volkov, *Miura transformation on the lattice*, Theor. Math. Phys. **74** (1988), 96–99.

[111] A.S. Fokas, B. Grammaticos, and A. Ramani, *From continuous to discrete Painlevé equations*, J. Math. Anal. Appl. **180** (1993), 342–360.

[112] A. Ramani and B. Grammaticos, *Miura transforms for discrete Painlevé equations*, J. Phys. A: Math. Gen. **25** (1992), L633–637.

[113] F.W. Nijhoff, J. Satsuma, K. Kajiwara, B. Grammaticos, and A. Ramani, *A study of the alternative discrete Painlevé-II equation*, Inv. Probl. **12** (1996), 697–716.

[114] A.P. Veselov and A.B. Shabat, *Dressing chains and the spectral theory of the Schrödinger operator*, Funct. Anal. Appl. **27** (1993), 81–96.

[115] F.W. Nijhoff, *The lattice Boussinesq equation and the discrete Painlevé IV equation*, in preparation.

[116] A. Ramani, B. Grammaticos and T. Bountis, *The Painlevé property and singularity analysis of integrable and nonintegrable systems*, Phys. Rep. **180** (1989), 159.

[117] M.D. Kruskal, *"Completeness" of the Painlevé test—General considerations—Open problems*, Chapter, 14, this volume.

[118] B. Grammaticos, A. Ramani, and V.G. Papageorgiou, *Do integrable mappings have the Painlevé property?*, Phys. Rev. Lett. **67** (1991), 1825–1828.

[119] B. Grammaticos and A. Ramani, *Discrete Painlevé equations: derivation and properties*, Applications of Analytic and Geometric Methods to Nonlinear Differential Equations, ed. P.A. Clarkson, NATO ASI Series C **413** (Kluwer, Dordrecht, 1993), pages 299–314.

[120] A. Ramani, B. Grammaticos, and J. Hietarinta, *Discrete versions of the Painlevé equations*, Phys. Rev. Lett. **67** (1991), 1829–1832.

[121] B. Grammaticos, F.W. Nijhoff, V. Papageorgiou, A. Ramani, and J. Satsuma, *Linearization and solutions of the discrete Painlevé III equation*, Phys. Lett. **185A** (1994), 446–452.

[122] M. Jimbo and H. Sakai, *A q-analog of the sixth Painlevé equation*, Lett. Math. Phys. **38** (1996), 145–154.

[123] B. Grammaticos and B. Dorizzi, *Integrable systems and numerical integrators*, J. Math. Comp. Sim. 37 (1994), 341.

[124] A. Ramani and B. Grammaticos, *Discrete Painlevé equations: coalescences, limits and degeneracies*, Physica A **228** (1996), 160–171.

[125] A.S. Fokas and M.J. Ablowitz, *A unified approach to transformations and elementary solutions of Painlevé equations*, J. Math. Phys. **23** (1982), 2033–2042.

[126] V.I. Gromak and N.A. Lukashevich, *Analytiskie Swoistwa Rechenii Uravnenii Painlevé* (Minsk University Press, 1990).

[127] V.E. Adler, *Nonlinear chains and Painlevé equations*, Physica **D73** (1994), 335–351.

[128] B. Grammaticos and A. Ramani, *Retracing the Painlevé–Gambier classification for discrete systems*, Math. Appl. An. **4** (1997), 196.

[129] F. Bureau, *Equations différentielles du second ordre en $\ddot{y}$ et du second degré en y dont l'intégrale est à points critiques fixes*, Ann. Mat. Pura Appl. (IV) **91** (1972), 163–281.

[130] F.W. Nijhoff, *On a q-deformation of the discrete Painlevé I equation and q-orthogonal polynomials*, Lett. Math. Phys. **30** (1994), 327–336.

[131] B. Grammaticos, A. Ramani, and K. M. Tamizhmani, *An integrability test for differential-difference systems*, J. Phys. A: Math. Gen. **27** (1994), 559.

[132] J. Hietarinta and M.D. Kruskal, *Hirota forms for the six Painlevé equations from singularity analysis*, NATO ASI series B278 (Plenum, 1992), page 175.

[133] Y. Ohta, A. Ramani, B. Grammaticos, and K.M. Tamizhmani, *From discrete to continuous Painlevé equations: a bilinear approach*, Phys. Lett. **216A** (1996), 255–261.

[134] A. Ramani, B. Grammaticos, and J. Satsuma, *Bilinear discrete Painlevé equations*, J. Phys. A: Math. Gen. **28** (195), 4655–4665.

[135] A.P. Bassom and P.A. Clarkson, *New exact solutions of the discrete fourth Painlevé equation*, Phys. Lett. **A194** (1994), 358–370.

[136] K.M. Tamizhmani, B. Grammaticos, and A. Ramani, *Schlesinger transforms for the discrete Painlevé IV equation*, Lett. Math. Phys. **29** (1993), 49–54.

[137] K.M. Tamizhmani, B. Grammaticos, A. Ramani, and Y. Ohta, *A study of the discrete P_V: Miura transformations and particular solutions*, Lett. Math. Phys. **38** (1996), 289–296.

[138] K. Kajiwara, Y. Ohta, J. Satsuma, B. Grammaticos, and A. Ramani, *Casorati determinant solutions for the discrete Painlevé II equation*, J. Phys. A: Math. Gen. **27** (1994), 915–922.

[139] K. Kajiwara, Y. Ohta, and J. Satsuma, *Casorati determinant solutions for the discrete Painlevé III equation*, J. Math. Phys **36** (1995), 4162–4174.

[140] M. Jimbo, H. Sakai, A. Ramani, and B. Grammaticos, *Bilinear structure and Schlesinger transforms of the q-P_{III} and q-P_{VI} equations*, Phys. Lett. **217A** (1996), 111–118.

[141] R. Fuchs, *Über lineare homogene Differentialgleichungen zweiter Ordnung mit drei im Endlichen gelegene wesentlich singuläre Stellen*, Math. Ann. **63** (1907), 301-321.

[142] L. Schlesinger, *Über eine Klasse von Differentialsystemen beliebiger Ordnung mit festen kritischen Punkten*, J. für Math. **141** (1912), 96–145.

[143] M. Jimbo, T. Miwa, and K. Ueno, *Monodromy preserving deformation of linear ordinary differential equations with rational coefficients*, Physica **2D** (1981), 306–352.

[144] M. Jimbo and T. Miwa, *ibid. II*, Physica **D2** (1981) 407–448; *ibid. III*, Physica **D4** (1981), 26–46.

[145] V.G. Papageorgiou, F.W. Nijhoff, B. Grammaticos, and A. Ramani, *Isomonodromic deformation problems for discrete analogues of Painlevé equations*, Phys. Lett. **A164** (1992), 57–64.

[146] N. Joshi, D. Burtonclay, and R.G. Halburd, *Nonlinear nonautonomous discrete dynamical system from a general discrete isomonodromy problem*, Lett. Math. Phys. **26** (1992), 123–131.

[147] D. Levi, O. Ragnisco, and M.A. Rodriguez, *On nonisospectral flows, Painlevé equations, and symmetries of differential and difference equations*, Theor. Math. Phys. **93** (1992), 1409–1414.

8

Painlevé Analysis for Nonlinear Partial Differential Equations

Micheline Musette

ABSTRACT The Painlevé analysis introduced by Weiss, Tabor, and, Carnevale (WTC) in 1983 for nonlinear partial differential equations (PDEs) is an extension of the method initiated by Painlevé and Gambier at the beginning of this century for the classification of algebraic nonlinear differential equations (ODEs) without movable critical points. In this chapter we explain the WTC method in its invariant version introduced by Conte in 1989 and its application to solitonic equations in order to find algorithmically their associated Bäcklund transformations. Many remarkable properties are shared by these so-called integrable equations, but they are generically no longer valid for equations modeling physical phenomena. Belonging to this second class, some equations called "partially integrable" sometimes keep remnants of integrability. In that case, the singularity analysis may also be useful for building closed-form analytic solutions, which necessarily agree with the singularity structure of the equations. We display the privileged role played by the Riccati equation and systems of Riccati equations that are linearizable, as well as the importance of the Weierstrass elliptic function, for building solitary waves or more elaborate solutions.

1 Introduction

During the past thirty years, interest in nonlinear phenomena has been growing in different fields of modern physics, such as optics, fluid dynamics, condensed matter, elementary particle physics, statistical mechanics, and astrophysics. Although the manifestation of those phenomena varies according to the different fields, they present a common feature in their mathematical description. The link comes from their description by *nonlinear evolution equations* (i.e., PDEs) whose solutions represent the propagation of waves with a permanent profile. Moreover, the analytical methods for solving them are directly inspired by the works of the famous mathematicians L. Fuchs, H. Poincaré, and P. Painlevé, as explained in Conte's contribution to this volume, Chapter 3.

The propagation of a bell-shaped solitary wave on water was approximately explained by the mathematical physicists J. Boussinesq [14] and Lord Rayleigh [114] only thirty years after its experimental discovery by

Scott Russell in 1844 [117]. The full explanation was later given in 1895 by Korteweg and de Vries (KdV [73]), who derived the nonlinear dispersive equation

$$u_t + u_{xxx} + 3(u^2)_x = 0, \qquad (1.1)$$

possessing the two-parameter (k, τ) exact solution

$$u^{\mathrm{sw}}(k, \theta) = \frac{k^2}{2} \operatorname{sech}^2 \frac{\theta}{2}, \quad \operatorname{sech} = \frac{1}{\cosh}, \qquad (1.2)$$

$$\theta = k\xi + \tau, \quad \xi = x - ct, \quad c = k^2. \qquad (1.3)$$

The name "soliton" was introduced by Zabusky and Kruskal in 1965 [137] when they solved the initial value problem for the KdV equation (1.1) and discovered solutions describing the elastic collision of several waves (1.2).

In this chapter we shall restrict our study by the method of singularities to nonlinear evolution equations possessing two different levels of integrability: *complete integrability* or *partial integrability*, including some chaotic PDEs that possess explicit analytic solutions in very special circumstances.

Complete integrability means that

- either the nonlinear partial differential equation can be related to a linear partial differential equation by an explicit transformation,

- or the equation passes the Painlevé test and possesses the Painlevé ¡ property (PP) for PDEs, i.e., *firstly*, on every noncharacteristic manifold its general solution has no movable critical singularities in the complex plane of an arbitrary function $\varphi(x, t)$; *secondly*, the PDE possesses an auto-Bäcklund transformation or is related by a Bäcklund transformation to another PDE possessing the PP (PDEs passing the "weak Painlevé" test [1] and related by a hodograph transformation [26] to another equation possessing the PP are outside the scope of this chapter),

- or the equation possesses solitary waves, N-soliton solutions for arbitrary N, an infinite number of conservation laws, bi-Hamiltonian structures, infinite-dimensional Lie algebras, ... ,

- or the equation satisfies the Ablowitz–Ramani–Segur (ARS) [3,4] conjecture on the relationship of *all* its reductions to ODEs without movable critical points.

More explicit definitions concerning the properties of this first class of equations, as well as classical examples, will be given in Section 2.

Partial integrability means that some above-listed properties are not satisfied (in particular, the Painlevé test may be satisfied only with some constraints on the function φ, or may never be satisfied regardless of the choice of φ, and the ARS conjecture is no longer valid) but the equation possesses explicit analytic solutions like, for instance, degenerate solitary waves, N-shock solutions, N-soliton solutions with N bounded [10, 46, 87], or it retains some pieces of integrability like degenerate Bäcklund transformations or a finite number of conservation laws [10, 57, 105]. For equations belonging to this second class, methods for finding particular solutions, which must agree with the singularity structure of the equation, will be developed in Section 4.

Section 3 contains the main subject of our lectures: It is devoted to the WTC [134] method and its extensions in the invariant version introduced by Conte [28], for finding algorithmically the auto-Bäcklund transformation of integrable nonlinear PDEs.

2   Integrable Equations

We present here a few classical examples of nonlinear partial differential equations either explicitly related to linear partial differential equations or characterized by the properties of complete integrability mentioned in the previous section.

2.1   Integration by Direct Linearization

Some equations can be linearized by an explicit transformation:

Example 2.1. The **Burgers** equation

$$u_t + (u_x + u^2)_x = 0 \qquad (2.1)$$

is linearized into the heat equation [51]

$$u = (\log \varphi)_x, \quad ((\varphi_t + \varphi_{xx})/\varphi)_x = 0. \qquad (2.2)$$

Example 2.2. The generalized **Eckhaus** equation [19, 25, 77]

$$iu_t + u_{xx} + (\beta^2 |u|^4 + 2\beta e^{i\gamma}(|u|^2)_x)u = 0, \quad (\beta, \gamma) \in \mathcal{R}, \qquad (2.3)$$

is linearizable into the Schrödinger equation for $\beta \cos \gamma \neq 0$,

$$i\nu_t + \nu_{xx} = 0, \quad u = \sqrt{\frac{1}{2\beta \cos \gamma}} \frac{\nu}{\sqrt{\varphi}} e^{-(i/2) \tan \gamma \log \varphi}, \qquad (2.4)$$

with $\varphi_x = |\nu|^2$ and

$$|u|^2 = \frac{1}{2\beta \cos \gamma} (\log \varphi)_x. \qquad (2.5)$$

If $\gamma = \pi/2$, the Kundu [75, 76] gauge transformation $u = \nu e^{i\beta\varphi}$ transforms the more general higher-order nonlinear Schrödinger equation (HNLS)

$$iu_t + u_{xx} + \delta|u|^2 u + (\beta^2|u|^4 + 2i\beta|u|^2)_x)u = 0, \quad (\beta, \delta) \in \mathcal{R}, \qquad (2.6)$$

into the nonlinear Schrödinger equation (NLS)

$$i\nu_t + \nu_{xx} + \delta|\nu|^2\nu = 0. \qquad (2.7)$$

The natural question is then; Where do these miraculous transformations from u to another field come from? This will be answered in Section 3.2.1.

2.2 Reduction to ODEs with the Painlevé Property

Ablowitz, Ramani, and Segur [3–5] and McLeod and Olver [89] conjectured a link between integrable NLPDEs and the Painlevé ODEs [106]: For the integrable NLPDEs particularly studied in this chapter, all known reductions to ODEs are single-valued algebraic transforms of the Weierstrass or Painlevé equations. Some of them are listed in Table 8.1.

2.3 Construction of Solitary Wave Solutions

In the integrable case, the solitary waves sech and sech^2 are degenerate elliptic functions, obtained by imposing boundary conditions on the general solution of the ODE defining the traveling wave reduction. In the partially or nonintegrable case, the general solution of the reduction may not exist. One then looks for particular solutions, taking advantage of the singularity structure of the ODE by the method of subequations in Section 4.

Example 2.3 (KdV). The reduction $u(x, t) = U(\xi)$, $\xi = x - ct$, of (1.1) yields the ODE

$$(-cU + U'' + 3U^2)' = 0. \qquad (2.8)$$

After two integrations, this equation becomes

$$-cU^2/2 + U^3 + U'^2/2 + K_1 U + K_2 = 0, \qquad (2.9)$$

which can be identified with the Weierstrass elliptic equation

$$\wp'^2 = 4\wp^3 - g_2\wp - g_3, \quad (g_2, g_3) \text{ real constants}, \qquad (2.10)$$

$$u = c/6 - 2\wp(x - ct - x_0, c^2/12 - K_1, K_2/2 + K_1 c/12 - (c/6)^3). \qquad (2.11)$$

The solitary wave (1.2) is found by imposing the boundary conditions $U(\xi) \to 0$, $U'(\xi) \to 0$, $U''(\xi) \to 0$, when $|\xi| \to \infty$. Note that for $K_1 = K_2 = 0$, equation (2.9) is a degenerate elliptic equation and

$$\wp(x - ct - x_0, c^2/12, -(c/6)^3)$$
$$= -(c/4)\,\mathrm{sech}^2\left(\sqrt{c}(x - ct - x_0)/2\right) + c/12. \qquad (2.12)$$

TABLE 8.1. Some reductions of a PDE to an ODE and their solutions. The PDEs $E(u, x, t) = 0$ (KdV, MKdV, sG, Bq, NLS, Tzi) are respectively defined by the equations (1.1), (2.94), (2.118), (3.51), (2.130), (2.64). The reduction to an ODE for $U(\xi)$ is defined by the two expressions of u in terms of (U, x, t) and of ξ in terms of (x, t). The letter K, with or without subscript, denotes an arbitrary constant. The last column indicates the elementary function ($\wp$, (P1)–(P6)), whose general solution of the ODE is a single-valued algebraic transform.

PDE	u	ξ	ODE	$\wp$, (Pn)
KdV	U	$x - ct$	$U'^2 + 2U^3 - cU^2 + 2K_1 U + 2K_2 = 0$	$\wp$
KdV	$U - \lambda t$	$x + 3\lambda t^2$	$U'' + 3U^2 - \lambda\xi + K = 0$	(P1)
MKdV	U	$x - ct$	$U'^2 - U^4 - cU^2 + K_1 U + K_2 = 0$	$\wp$
MKdV	$(3t)^{-1/3} U$	$x(3t)^{-1/3}$	$U'' - 2U^3 - \xi U + K = 0$	(P2)
sG	$-i \log U$	$x - ct$	$cU'^2 + U^3 + KU^2 + U = 0$	$\wp$
sG	$-i \log U$	xt	$U'' - U'^2/U + U'/\xi + (1 - U^2)/(2\xi) = 0$	(P3)
Bq	U	$x - ct$	$(U''/3) + U^2 + c^2 U + K_1\xi + K_2 = 0$	$\wp$, (P1)
Bq	$2(U' + \xi - t^2)$	$x - t^2 + K_1$	$U''^2 + 4U'^3 + 12(\xi U' - U)U'' + K_2 U' + K_3 = 0$	(P2) [21, 113]
Bq	$(U' - \xi^2/2)/t$	$xt^{-1/2}$	$U''^2/2 + U'^3 - (9/8)(U - \xi U')^2 + K_1(U - \xi U') + K_2 U' + K_3 = 0$	(P4) [17, 21, 113]
NLS	(2.20)	$x - ct$	(2.23)	$\wp$
NLS	$e^{i(xt - 4t^3/3)} U$	$x - t^2$	$U'' + 2\varepsilon U^3 - 2\xi U = 0$	(P2) [121]
NLS	$t^{-1/2}\sqrt{U'}\, e^{i\varphi}$	$xt^{-1/2}$	$4U''^2 + 4\varepsilon U'^3 + KU' + (\xi U' - U)^2/4 = 0$	(P4) [11, 18, 21]
Tzi	$\log U$	$x - ct$	$-cU'^2 + 2aU^3 + KU^2 - a_0 = 0$	$\wp$
Tzi	$\log U$	xt	$(\xi U'/U)' + aU + a_0 U^{-2} = 0$	(P3) [24]

Example 2.4. The generalized **Tzitzéica** equation

$$u_{xt} + ae^u + a_1 e^{-u} + a_0 e^{-2u} = 0, \quad a \neq 0, \qquad (2.13)$$

includes Liouville ($a_1 = a_0 = 0$), sinh–Gordon ($a_1 \neq 0, a_0 = 0$), or Tzitzéica [122,123] ($a_0 \neq 0, a_1 = 0$) equations. It is polynomial in the variable $v = e^u$,

$$vv_{xt} - v_x v_t + av^3 + a_1 v + a_0 = 0. \qquad (2.14)$$

Its reduction $(v, x, t) \to (V, \xi = x - ct)$ can be integrated once

$$-cV'^2 + 2aV^3 - 6KV^2 - 2a_1 V - a_0 = 0, \quad K \text{ arbitrary}, \qquad (2.15)$$

and possesses the general two-parameter solution [31]

$$aV = K + 2c\wp(\xi - \xi_0, (3K^2 + aa_1)/c^2, (4K^3 + 2aa_1 K + a^2 a_0)/(4c^3)). \qquad (2.16)$$

Moreover, a linear superposition of two waves with opposite directions,

$$v(x, t) = Af(x - ct) + Bg(x + ct), \qquad (2.17)$$

is compatible with the Tzitzéica equation by assuming that f and g satisfy the following second-order ODE with constant coefficients:

$$f'' = A_1 f^2 + B_1, \quad g'' = A_2 g^2 + B_2. \qquad (2.18)$$

A particular solution of (2.13) for $a_1 = 0$ is then [100]

$$ae^u = 2c\wp(x - ct - x_1, g_2, K + a^2 a_0/(8c^3))$$
$$\qquad\qquad - 2c\wp(x + ct - x_2, g_2, K - a^2 a_0/(8c^3)), \qquad (2.19)$$

c, x_1, x_2, g_2, K arbitrary constants.

Example 2.5 (NLS). The reduction $u(x, t) = \rho(\xi)e^{i[-\Omega t + \varphi(\xi)]}$ of (2.130) yields the coupled ODEs

$$-c\rho' + 2\varphi'\rho' + \varphi''\rho = 0, \qquad (2.20)$$
$$\rho'' + (\Omega - (\varphi')^2 + c\varphi')\rho + 2\varepsilon\rho^3 = 0. \qquad (2.21)$$

Equation (2.20) admits the integrating factor ρ,

$$\varphi' = c + K_1/S, \quad S = \rho^2. \qquad (2.22)$$

Then (2.21) admits the integrating factor ρ'; hence

$$S'^2 = -4\varepsilon S^3 - 4\alpha S^2 + 8K_2 S - K_1^2, \quad \alpha = \Omega + c^2/4, \qquad (2.23)$$

an elliptic equation for S with the general solution

$$S = -\alpha/(3\varepsilon) - \wp(x - ct - x_0, g_2, g_3)/\varepsilon, \qquad (2.24)$$

$$g_2 = 8\varepsilon(K_2 + \alpha^2/(6\varepsilon)), \quad g_3 = (2\alpha/3)^3 + 8K_2\alpha\varepsilon/3 + \varepsilon^2 K_1^2. \qquad (2.25)$$

The one-soliton solution is obtained for the values of K_1, K_2, making the Weierstrass elliptic function degenerate into a trigonometric function:

$$\wp(\xi, g_2, g_3) \to a_1 + a_2 \operatorname{sech}^2 k\xi. \qquad (2.26)$$

This happens in two cases:

1. $a_1 = K_1 = K_2 = 0$, $k^2 = -\alpha$, $\rho^2 = (k^2/\varepsilon)\operatorname{sech}^2 k\xi$,

2. $a_1 K_1 K_2 \neq 0$, $a_1 = -(k^2 + \alpha)/(3\varepsilon)$, $\rho^2 = a_1 + (k^2/\varepsilon)\operatorname{sech}^2 k\xi$.

They respectively correspond for equation (2.130) to the three-parameter (c, k, x_0) solution ("bright" soliton) [138]

$$\varepsilon > 0 : u = \varepsilon^{-1/2} k \operatorname{sech}(k(x - ct - x_0)) e^{icx/2 + i(k^2 - (c/2)^2)t} \qquad (2.27)$$

and the four-parameter (c, k, K, x_0) solution ("dark" soliton) [139]

$$\varepsilon < 0 : u = (-\varepsilon)^{-1/2} \left[(k/2)\tanh(k(x - ct - x_0)/2) - i(K - c/2)\right]$$
$$\times e^{iKx - 2i[k^2/4 + (K - c/2)^2 + K^2/2]t}. \qquad (2.28)$$

2.4 Conservation Laws

Definition 2.1. Given a PDE $E(u; x, t) = 0$, a conservation law is a relation

$$T_t + X_x = 0, \qquad (2.29)$$

where T and X, respectively called *density* and *flux*, depend on x, t, u, and its derivatives. If the total variation of X in the interval $a \leq x \leq b$ is zero, the quantity $I = \int_a^b T \, dx$ is a constant of the motion called a *conserved quantity*. "Integrable" PDEs possess an infinite number of conservation laws [2, 6, 44, 126, 127]. For example, the first three conservation laws are:

(a) for the KdV equation (1.1) [71, 94, 135]

$$T_1 = u, \qquad X_1 = 3u^2 + u_{xx}; \qquad (2.30)$$
$$T_2 = u^2/2, \qquad X_2 = 2u^3 + uu_{xx} - u_x^2/2; \qquad (2.31)$$
$$T_3 = 2u^3 - u_x^2, \qquad X_3 = 9u^4 + 6u^2 u_{xx} - 12uu_x^2 - 2u_x u_{3x} + u_{xx}^2; \qquad (2.32)$$

(b) for the MKdV equation (2.94) [94]

$$T_1 = u, \qquad\qquad X_1 = 2u^3 + u_{xx}; \tag{2.33}$$
$$T_2 = u^2/2, \qquad\qquad X_2 = 3u^4/2 + uu_{xx} - u_x^2/2; \tag{2.34}$$
$$T_3 = u^4/4 - u_x^2/4, \quad X_3 = u^6 + u^3u_{xx} - 3u^2u_x^2 - u_xu_{3x}/2 + u_{xx}^2/4; \tag{2.35}$$

(c) for the sG equation (2.118) [42, 80, 120]

$$T_1 = u_x^2/2, \qquad\qquad\qquad\qquad X_1 = \cos u; \tag{2.36}$$
$$T_2 = u_x^4/4 - u_{xx}^2, \qquad\qquad\qquad X_2 = u_x^2 \cos u; \tag{2.37}$$
$$T_3 = 3u_x^6 - 12u_x^2u_{xx}^2 + 16u_x^3u_{3x} + 72u_{3x}^2,$$
$$X_3 = (2u_x^4 - 24u_{xx}^2) \cos u. \tag{2.38}$$

(d) for the NLS equation (2.130), we reproduce three of the five conservation laws given by Zakharov and Shabat [138, 139] for $\varepsilon = \pm 1$:

$$\varepsilon = +1, \quad I_1 = \int_{-\infty}^{+\infty} |u|^2 \, dx, \quad I_2 = \int_{-\infty}^{+\infty} (\bar{u}u_x - u\bar{u}_x) \, dx,$$

$$I_3 = \int_{-\infty}^{+\infty} (|u_x|^2 - \frac{1}{2}|u|^4) \, dx, \tag{2.39}$$

$$\varepsilon = -1, \quad I_1 = \int_{-\infty}^{+\infty} (1 - |u|^2) \, dx, \quad I_2 = -\int_{-\infty}^{+\infty} (\bar{u}u_x - u\bar{u}_x) \, dx,$$

$$I_3 = \int_{-\infty}^{+\infty} (|u|^4 + |u_x|^2 - 1) \, dx \tag{2.40}$$

(where $\bar{u}$ denotes the complex conjugate of u).

For the Tzitzéica equation (2.64), Dodd and Bullough [42] first obtained two nontrivial conservation laws; then Mikhailov [93] gave a recursion formula for an infinite set of nontrivial polynomial conserved densities.

2.5 Bäcklund Transformations

2.5.1 Definition

A Bäcklund [8] transformation (BT) between two given PDEs

$$E_1(u; x, t) = 0, \quad E_2(v; x', t') = 0 \tag{2.41}$$

is a set of four relations [41, vol. III, Chap. XII])

$$F_j(u, v, u_x, v_{x'}, u_t, v_{t'}, \ldots; x, t, x', t') = 0, \quad j = 1, 2, \tag{2.42}$$
$$x' = X(x, t, u, u_x, u_t, v), \quad t' = T(x, t, u, u_x, u_t, v), \tag{2.43}$$

such that the elimination of u (resp. v) between (F_1, F_2) implies

$$E_2(v; x', t') = 0 \quad (\text{resp. } E_1(u; x, t) = 0).$$

In case the two PDEs are the same, the BT is called an auto-BT.

Bäcklund theory originates from the work of Lie and Bäcklund for the study of surfaces in differential geometry. The subject was subsequently developed by Goursat [56] and Clairin [23]. Bäcklund transformations represent an extension of Lie contact transformations. They were first obtained for second-order PDEs in two independent variables, linear in the highest derivatives (i.e., a special type of Monge–Ampère equation).

For more details on BTs, the reader is advised to consult the book by Rogers and Shadwick [115] and the classical book of Goursat [55].

2.5.2 Examples: Second-Order PDEs

Example 2.6 (Burgers and heat equations). Given the two equations

$$E_1 \equiv u_t + (u_x + u^2)_x = 0, \quad E_2 \equiv v_t + v_{xx} = 0, \tag{2.44}$$

the two relations defining the BT are

$$F_1 \equiv v_x - uv = 0, \quad F_2 \equiv v_t + u^2 v + v u_x = 0. \tag{2.45}$$

Indeed, the elimination of v (resp. u) yields the identities

$$(F_2/v)_x - (F_1/v)_t \equiv E_1, \quad v \neq 0, \quad \text{and} \quad F_2 + F_{1,x} + u F_1 \equiv E_2. \tag{2.46}$$

Example 2.7 (Liouville and d'Alembert). Given the two equations

$$E_1 \equiv u_{xt} - e^u = 0, \quad E_2 \equiv v_{xt} = 0, \tag{2.47}$$

the two relations

$$F_1 \equiv u_x - v_x + \lambda e^{(u+v)/2} = 0, \tag{2.48}$$

$$F_2 \equiv u_t + v_t + (2/\lambda)e^{(u-v)/2} = 0, \tag{2.49}$$

where λ is an arbitrary real constant called a Bäcklund parameter, define a Bäcklund transformation as shown by the elimination of v (resp. u):

$$F_{1,t} + F_{2,x} - (1/\lambda)e^{(u-v)/2}F_1 - (\lambda/2)e^{(u+v)/2}F_2 \equiv 2E_1, \tag{2.50}$$

$$F_{1,t} - F_{2,x} + (1/\lambda)e^{(u-v)/2}F_1 - (\lambda/2)e^{(u+v)/2}F_2 \equiv -2E_2. \tag{2.51}$$

Thus, the general solution of the d'Alembert equation

$$v = f(x) + g(t), \quad (f, g) \text{ arbitrary functions,} \tag{2.52}$$

provides, by integration of the ODEs (2.48)–(2.49), a solution of (2.47),

$$e^u = 2\varphi_x\varphi_t/\varphi^2, \quad \varphi = \frac{\lambda}{2}\int^x e^f \mathrm{d}x + (1/\lambda)\int^t e^{-g}\mathrm{d}t, \tag{2.53}$$

which is the general solution. Traveling waves are built by the choice

$$\varphi = \coth(\alpha x) - \tanh(\beta t) \Longrightarrow e^u = 2\alpha\beta/\cosh^2(\alpha x - \beta t). \tag{2.54}$$

Remark. The auto-Bäcklund transformation

$$F_1 \equiv (u + U)_x + 2\lambda\sinh((u - U)/2) = 0, \tag{2.55}$$
$$F_2 \equiv (u - U)_t - (2/\lambda)e^{(u+U)/2} = 0, \tag{2.56}$$

where u and U are two solutions of the Liouville equation, has been given in 1987 [110] and later recovered [35] in the framework of the singularity analysis.

Example 2.8 (Sine–Gordon). Given two solutions u and U of the sine–Gordon equation

$$E_1 \equiv u_{xt} - \sin u = 0, \quad E_2 \equiv U_{xt} - \sin U = 0, \tag{2.57}$$

the auto-Bäcklund transformation is defined by

$$F_1 \equiv (u + U)_x - 2\lambda\sin((u - U)/2) = 0, \tag{2.58}$$
$$F_2 \equiv (u - U)_t - (2/\lambda)\sin((u + U)/2) = 0, \quad \lambda \text{ arbitrary constant,} \tag{2.59}$$

as can easily be checked quite similarly to the Liouville and d'Alembert case, by elimination of U (resp. u) between these two relations:

$$F_{1,t} + F_{2,x} + (1/\lambda)\cos((u + U)/2)F_1 + \lambda\cos((u - U)/2)F_2 \equiv 2E_1, \tag{2.60}$$

$$F_{1,t} - F_{2,x} - (1/\lambda)\cos((u + U)/2)F_1 + \lambda\cos((u - U)/2)F_2 \equiv 2E_2. \tag{2.61}$$

Lamb [79] built from (2.58)–(2.59) infinite families of solutions, e.g., the N-soliton solution: At the first iteration, one starts from the solution $U = 0$ ("vacuum"), and the integration of the ODEs (2.58)–(2.59) yields

$$\tan(u/4) = e^{\lambda x + \lambda^{-1}t + \delta}, \quad \delta \text{ arbitrary constant,} \tag{2.62}$$

i.e., the one-soliton solution

$$u_x = 2\lambda\,\mathrm{sech}(\lambda x + \lambda^{-1}t + \delta), \quad u_t = 2\lambda^{-1}\,\mathrm{sech}(\lambda x + \lambda^{-1}t + \delta). \tag{2.63}$$

Example 2.9 (Tzitzéica). For the Tzitzéica equation (Tzi)

$$u_{xt} = e^u - e^{-2u} \tag{2.64}$$

the auto-BT [35] is explicitly written in Section 3.2.5. This result is based on an equivalent representation of the matrix Lax pair (given by Tzitzéica [122, 123] and rediscovered by Mikhailov [92, 93])

$$\frac{\partial}{\partial x} \begin{pmatrix} \varphi \\ \partial_x \varphi \\ \partial_t \varphi \end{pmatrix} = \begin{pmatrix} 0 & 1 & 0 \\ 0 & U_x & \lambda e^{-U} \\ e^U & 0 & 0 \end{pmatrix} \begin{pmatrix} \varphi \\ \partial_x \varphi \\ \partial_t \varphi \end{pmatrix}, \tag{2.65}$$

$$\frac{\partial}{\partial t} \begin{pmatrix} \varphi \\ \partial_x \varphi \\ \partial_t \varphi \end{pmatrix} = \begin{pmatrix} 0 & 0 & 1 \\ e^U & 0 & 0 \\ 0 & \lambda^{-1} e^{-U} & U_t \end{pmatrix} \begin{pmatrix} \varphi \\ \partial_x \varphi \\ \partial_t \varphi \end{pmatrix}, \tag{2.66}$$

and of the Moutard [96] transformation between two solutions U, u:

$$e^u = -e^U + 2\varphi_x \varphi_t / \varphi^2 \tag{2.67}$$

in terms of two components $Y_1 = \varphi_x / \varphi$, $Y_2 = \varphi_t / \varphi$.

Since 1973, BTs have been found for PDEs of order greater than two. Different approaches have been used for deriving those transformations:

1. the method of Clairin [23, 81],

2. the method of differential forms developed by Wahlquist and Estabrook [45, 128, 129]),

3. the method of bilinear transformations of Hirota [59, 61, 90],

4. the method of gauge transformations developed by Boiti et al. [12, 13] and Levi et al. [83, 84].

   In the last two methods, the BT results from the elimination of the wave function between the Lax pair and the DT. In next sections these two main concepts of complete integrability are briefly recalled; then the principle of the method of gauge transformations is presented. Lax pairs and DTs are explicitly given for the PDEs of the AKNS scheme (KdV, MKdV, sine–Gordon, NLS) and for some fifth-order PDEs, respectively in Sections 2.6.3 and 2.6.4. But let us first give some definitions.

2.6   Darboux Transformation and Lax Pair

2.6.1   Definitions

Crum–Darboux transformation. This transformation is a key in the theory of nonlinear integrable evolution equations for building soliton solutions and understanding their "asymptotically linear" superposition rules. It is based on a result obtained by the French mathematician Gaston Darboux in the special case of the Sturm–Liouville equation (also called Schrödinger equation in quantum mechanics). We briefly recall this old theorem [40] and its generalization due to Crum [39].

Theorem 2.1 (Darboux). *The linear Schrödinger equation*

$$\psi_{xx} + (u + \lambda)\psi = 0 \tag{2.68}$$

is invariant under

$$\psi \mapsto \tilde{\psi} = (\partial_x - \frac{\psi_{0,x}}{\psi_0})\psi, \tag{2.69}$$

$$u \mapsto \tilde{u} = u + 2(\log \psi_0)_{xx}, \tag{2.70}$$

where $\psi_0 \equiv \psi(x, \lambda_0)$ is an eigenfunction of (2.68) with parameter λ_0.

The essential point is that the new potential $\tilde{u}$ depends only on ψ_0 and not on ψ. This transformation can then be iterated to obtain the following result.

Theorem 2.2 (Crum). *The function*

$$\tilde{\psi} = \frac{W(\psi_1, \psi_2, \ldots, \psi_N, \psi)}{W(\psi_1, \psi_2, \ldots, \psi_N)}, \tag{2.71}$$

where ψ_1, ψ_2, ... , ψ_N are eigenfunctions of (2.68) associated with parameters λ_1, λ_2, ... , λ_N and the symbol W represents the Wronskian determinant, solves the equation (2.68) for the potential

$$\tilde{u} = u + 2\left(\log W(\psi_1, \psi_2, \ldots, \psi_N)\right)_{xx}. \tag{2.72}$$

Lax pair. In 1968 Lax [82] explained in a very transparent way the greater part of the result of Gardner et al. [54] by introducing the operators

$$L = -\partial_x^2 - u(x, t), \quad A = -4\partial_x^3 - 6u\partial_x - 3u_x \tag{2.73}$$

such that the KdV equation (1.1) may be represented as

$$\partial_t L = [A, L], \tag{2.74}$$

called the Lax representation. Equation (2.74) expresses the compatibility between the two partial differential equations of the system

$$\begin{rcases} L\psi = \lambda\psi \\ \psi_t = A\psi \end{rcases} \iff \begin{cases} \psi_{xx} + (u + \lambda)\psi = 0 \\ \psi_t + (2u - 4\lambda)\psi_x - u_x\psi = 0 \end{cases}$$

called a Lax pair. This equivalence results from the identity

$$\psi_{xxt} - \psi_{txx} \equiv \mathrm{KdV}(u)\psi. \tag{2.75}$$

The system (2.6.1) is invariant under the Darboux transformation (2.69)–(2.70) with the compatibility condition

$$(\partial_t \tilde{L})\tilde{\psi} = [\tilde{A}, \tilde{L}]\tilde{\psi}, \tag{2.76}$$

where $(\tilde{L}, \tilde{A})$ results from the substitution of u by $\tilde{u}$ in (L, A).

2.6.2 Bäcklund Gauge Transformation

A general procedure to obtain BT for nonlinear PDEs derived as compatibility conditions between a given generalized Lax pair of operators was simultaneously considered by Boiti et al. and Levi et al. in 1982. It has provided new results for multidimensional nonlinear PDEs. Here we report only the principle of the method. Let us consider the Lax pair

$$\psi_x = L\psi, \quad \psi_t = M\psi, \tag{2.77}$$

where ψ is an $N \times N$ matrix as well as L, M which have a preassigned dependence on a matrix "potential" $Q(x,t)$ and on a constant parameter λ. The compatibility condition between the two equations of the system (2.77) implies the following nonlinear equation:

$$L_t - M_x + [L, M] = 0. \tag{2.78}$$

To construct the BT for this nonlinear partial differential equation one has to consider two different systems of type (2.77) corresponding to two different "potentials," say $Q(x,t)$ and $\tilde{Q}(x,t)$:

$$\psi_x = L(Q(x,t);\lambda)\psi, \quad \psi_t = M(Q(x,t);\lambda)\psi, \tag{2.79}$$

$$\tilde{\psi}_x = \tilde{L}(\tilde{Q}(x,t);\lambda)\tilde{\psi}, \quad \tilde{\psi}_t = \tilde{M}(\tilde{Q}(x,t);\lambda)\tilde{\psi}. \tag{2.80}$$

One assumes that the following generalized DT holds between the wave functions ψ and $\tilde{\psi}$:

$$\tilde{\psi} = B\psi, \tag{2.81}$$

where B is a matrix function of $Q, \tilde{Q}, x, t$, and λ. The compatibility between (2.81) and the system (2.79)–(2.80) gives the auto-BT

$$B_x = \tilde{L}B - BL, \quad B_t = \tilde{M}B - BM. \tag{2.82}$$

By cross-differentiating these two relations one gets

$$(\tilde{L}_t - \tilde{M}_x + [\tilde{L}, \tilde{M}])B - B(L_t - M_x + [L, M]) = 0, \tag{2.83}$$

which implies that if $Q(x,t)$ satisfies the nonlinear PDE (2.78), then $\tilde{Q}(x,t)$ satisfies the same equation. This exactly coincides with the definition of the BT previously given in Section 2.5.1.

Let us also mention the book of Matveev and Salle [91] as a basic reference on the Darboux transformation and its development in soliton theory.

In the extension of Painlevé analysis to NLPDEs [134], if a PDE fulfills the necessary conditions of integrability ("Painlevé test"), one tries to determine a Lax pair and a Darboux transformation relating two solutions of the same PDE in order to constructively prove the sufficiency of these conditions. A method (truncation procedure) leading to such a Lax pair and DT will be explained in Section 3.2.2. In this formalism, the link with the notion of "general solution" is that the knowledge of the BT a priori allows one to build wide classes of solutions. In one space dimension the "good" Lax pair of a given nonlinear PDE must depend on the solution of this equation and an arbitrary constant λ. In the next section we show by means of examples how to derive the Lax pair and DT from the associated BT. In each case it will be our aim to show in Section 3 how these two pieces of information can be found algorithmically by singularity analysis.

2.6.3   Examples: AKNS Scheme

Example 2.10 (Korteweg–de Vries). The conservative form of Kdv is (1.1), and we define the potential form as

$$u = w_x, \quad F(w) \equiv w_t + w_{xxx} + 3w_x^2 = 0. \tag{2.84}$$

Given two solutions w and W of (2.84), the auto-BT is defined by [81]

$$(w + W)_x = 2\lambda - (w - W)^2/2, \tag{2.85}$$
$$(w + W)_t = -2(w_x^2 + w_x W_x + W_x^2) - (w - W)(w - W)_{xx}, \tag{2.86}$$

where λ is the Bäcklund parameter. After changing variables w, W to $W, Y = (w - W)/2$, the gradient of Y is defined by the Riccati equations

$$Y_x = \lambda - U - Y^2, \ U = W_x, \tag{2.87}$$
$$Y_t = (U_x - (2U + 4\lambda)Y)_x. \tag{2.88}$$

The transformation

$$Y = \partial_x \log \psi \tag{2.89}$$

linearizes these Riccati equations into one second-order ODE and one first-order PDE:

$$\psi_{xx} + (U - \lambda)\psi = 0, \tag{2.90}$$
$$\psi_t + (2U + 4\lambda)\psi_x - (U_x + G(t))\psi = 0, \quad G \text{ an arbitrary function.} \tag{2.91}$$

The Lax pair of KdV is defined by these two linear equations, which satisfy the compatibility condition

$$\psi_{xxt} - \psi_{txx} = E(U)\psi, \tag{2.92}$$

while the DT for KdV is defined by the x-derivative of (2.89),

$$u - U = 2\partial_x^2 \log \psi. \tag{2.93}$$

Example 2.11 (Modified Korteweg–de Vries). The conservative form of MKdV is

$$E(u) \equiv u_t + u_{xxx} - 2a^{-2}(u^3)_x = 0, \qquad (2.94)$$

and we define the potential form as

$$u = w_x, \quad F(w) \equiv w_t + w_{xxx} - 2a^{-2}w_x^3 = 0. \qquad (2.95)$$

Given two solutions w and W of (2.95), the auto-BT is given by [81]

$$(w + W)_x = -2a\lambda \sinh((w - W)/a), \qquad (2.96)$$

$$(w + W)_t = 8\lambda^2 W_x - 4\lambda W_{xx} \cosh((w - W)/a)$$
$$+ 4a(2\lambda^3 - \lambda W_x^2/a^2) \sinh((w - W)/a), \qquad (2.97)$$

where λ is the Bäcklund parameter. The change of variables

$$(w, W) \mapsto (W, Y = e^{(w-W)/a}) \qquad (2.98)$$

maps these equations into the two Riccati equations for Y,

$$Y_x = -2(U/a)Y + \lambda(1 - Y^2), \quad U = W_x, \qquad (2.99)$$

$$Y_t = 2A_1 Y + B_1(1 + Y^2) + C_1(1 - Y^2)$$
$$= (-4\lambda U/a + (2(U/a)^2 - 4\lambda^2 + 2(U_x/a))Y)_x, \qquad (2.100)$$

$$A_1 = \frac{U_{xx}}{a} - 2\frac{U^3}{a^3} + 4\lambda^2 \frac{U}{a}, \quad B_1 = -2\lambda \frac{U_x}{a}, \quad C_1 = 2\lambda \frac{U^2}{a^2} - 4\lambda^3. \qquad (2.101)$$

The compatibility condition of this "Riccati pseudopotential" Y is

$$Y_{xt} - Y_{tx} = -(2/a)E(U)Y. \qquad (2.102)$$

The Lax pair is obtained by linearizing these two Riccati equations by the transformation

$$Y = \psi_1/\psi_2, \qquad (2.103)$$

$$\begin{pmatrix} \psi_1 \\ \psi_2 \end{pmatrix}_x = \begin{pmatrix} -U/a & \lambda \\ \lambda & U/a \end{pmatrix} \begin{pmatrix} \psi_1 \\ \psi_2 \end{pmatrix}, \qquad (2.104)$$

$$\begin{pmatrix} \psi_1 \\ \psi_2 \end{pmatrix}_t = \begin{pmatrix} A_1 & B_1 + C_1 \\ B_1 - C_1 & -A_1 \end{pmatrix} \begin{pmatrix} \psi_1 \\ \psi_2 \end{pmatrix}, \qquad (2.105)$$

while the Darboux transformation is defined by

$$u - U = a\partial_x \log Y, \qquad (2.106)$$

which, by elimination of Y_x with (2.99), is identical to

$$u + U = a\lambda(Y^{-1} - Y). \qquad (2.107)$$

The homographic transformation with $\alpha = U/a$,

$$Y = \lambda\chi/(1 + \alpha\chi), \tag{2.108}$$

maps the Riccati system (2.99)–(2.100) into the simpler form

$$\chi_x = 1 + (S/2)\chi^2, \tag{2.109}$$
$$\chi_t = -C + C_x\chi - (1/2)(CS + C_{xx})\chi^2, \tag{2.110}$$
$$S = 2\left(\frac{U_x}{a} - \left(\frac{U}{a}\right)^2 - \lambda^2\right), \quad C = 2\left(\frac{U_x}{a} - \left(\frac{U}{a}\right)^2 + 2\lambda^2\right). \tag{2.111}$$

We shall see that the relation between the two functions S and C,

$$S - C + 6\lambda^2 = 0, \tag{2.112}$$

corresponds to the singular manifold (SM) equation of the KdV equation [134] and can be found algorithmically [109] when one performs the Painlevé analysis of the MKdV equation. In the variable

$$f = a(Y - 1)/(Y + 1), \tag{2.113}$$

the system (2.99)–(2.100) and the DT (2.106) or (2.107) become [125, 127]

$$u - U = 2a^2 f_x/(a^2 - f^2), \tag{2.114}$$
$$u + U = -4a^2\lambda f/(a^2 - f^2), \tag{2.115}$$
$$af_x = -(U/a)(a^2 - f^2) - 2\lambda af, \tag{2.116}$$
$$af_t = A_1(a^2 - f^2) + B_1(a^2 + f^2) - 2C_1 af. \tag{2.117}$$

Example 2.12 (Sine-Gordon).

$$E(u) \equiv u_{xt} - \sin u = 0. \tag{2.118}$$

Given two solutions u and U of (2.118), the auto-BT is given by [79]

$$(u + U)_x = -4\lambda\sin((u - U)/2), \tag{2.119}$$
$$(u - U)_t = -\lambda^{-1}\sin((u + U)/2), \tag{2.120}$$

where λ is the Bäcklund parameter. The change of variables

$$(u, U) \mapsto (U, Y = e^{-i(u-U)/2}) \tag{2.121}$$

maps these equations into the two Riccati equations for Y,

$$Y_x = iU_x Y + \lambda(1 - Y^2), \tag{2.122}$$
$$Y_t = ((1 - Y^2)\cos U + i(1 + Y^2)\sin U)/(4\lambda). \tag{2.123}$$

The compatibility condition of the Riccati pseudopotential Y is

$$Y_{xt} - Y_{tx} = iE(U)Y. \tag{2.124}$$

The Lax pair is obtained by linearizing the Riccati system

$$Y = \frac{\psi_1}{\psi_2}, \tag{2.125}$$

$$\begin{pmatrix} \psi_1 \\ \psi_2 \end{pmatrix}_x = \begin{pmatrix} iU_x/2 & \lambda \\ \lambda & -iU_x/2 \end{pmatrix} \begin{pmatrix} \psi_1 \\ \psi_2 \end{pmatrix},$$

$$\begin{pmatrix} \psi_1 \\ \psi_2 \end{pmatrix}_t = \frac{1}{4\lambda} \begin{pmatrix} 0 & e^{iU} \\ e^{-iU} & 0 \end{pmatrix} \begin{pmatrix} \psi_1 \\ \psi_2 \end{pmatrix},$$

while the Darboux transformation for sine–Gordon is defined by

$$u - U = 2i \log Y, \tag{2.126}$$

$$(u + U)_x = 2i\lambda(Y^{-1} - Y). \tag{2.127}$$

The homographic transformation (2.108) with $\alpha = -iU_x/2$ maps the Riccati system (2.122)–(2.123) into the simpler form (2.109)–(2.110), with

$$S = -iU_{xx} + U_x^2/2 - 2\lambda^2, \quad C = -e^{iU}/(4\lambda^2). \tag{2.128}$$

The relation between S and C,

$$S + C_{xx}/C - (1/2)(C_x/C)^2 + 2\lambda^2 = 0, \tag{2.129}$$

represents the SM equation obtained by Conte [28] when performing the invariant Painlevé analysis of the sine–Gordon equation.

Example 2.13 (Nonlinear Schrödinger).

$$E(u) \equiv iu_t + u_{xx} + 2\varepsilon|u|^2 u = 0, \quad \varepsilon = \pm 1. \tag{2.130}$$

Given two solutions u and u' of (2.130), the auto-BT can be written as [22, 72, 81, 85]

$$(u + U)_x = (u - U)\sqrt{4\lambda^2 - \varepsilon|u + U|^2}, \tag{2.131}$$

$$(u + U)_t = i(u - U)_x\sqrt{4\lambda^2 - \varepsilon|u + U|^2}$$
$$+ i\varepsilon(u + U)(|u + U|^2 + |u - U|^2)/2. \tag{2.132}$$

The extension to NLS of the transformation (2.115) is

$$u + U = -4\lambda f/(1 + \varepsilon|f|^2). \tag{2.133}$$

Therefore, the change of variables $(u, U) \mapsto (U, f)$ transforms (2.131) into

$$-f_x + \varepsilon f^2 \bar{f}_x = (1 - \varepsilon|f|^2)(U(1 + \varepsilon|f|^2) + 2\lambda f). \tag{2.134}$$

The elimination of $\bar{f}$ between this equation and its complex conjugate, assuming $1 - |f|^4 \neq 0$, provides

$$f_x = -2\lambda f - U - \varepsilon \bar{U} f^2, \qquad (2.135)$$

while the t-part is

$$f_t = (\lambda U + U_x) + (\varepsilon U \bar{U} + \lambda^2) f + (\lambda \bar{U} - \bar{U}_x) f^2 \qquad (2.136)$$

with the identity

$$f_{xt} - f_{tx} = E + \varepsilon \bar{E} f^2. \qquad (2.137)$$

Equations (2.131) and (2.133) imply

$$u - U = 2(f_x - f^2 \bar{f}_x)/(1 - |f|^4). \qquad (2.138)$$

In all the above examples (KdV, MKdV, sG, NLS), the DT is defined with one (for KdV) or two (for the others) entire functions ψ. This distinction is the only relevant feature needed to obtain in an algorithmic way the Lax pair by methods linked to the singularity structure of these equations.

2.6.4   Higher-Order KdV-Type Equations

Among the fifth-order nonlinear evolution equations

$$u_t + (u_{xxxx} + (8\alpha - 2\beta)uu_{xx} - 2(\alpha + \beta)u_x^2 - (20/3)\alpha\beta u^3)_x = 0 \quad (2.139)$$

only three cases are integrable:

$$\frac{\beta}{\alpha} = -1 \; : \; u_t + \left(u_{xxxx} + 10\alpha uu_{xx} + 20\alpha^2 \frac{u^3}{3} \right)_x = 0, \qquad (2.140)$$

$$\frac{\beta}{\alpha} = -6 \; : \; u_t + \left(u_{xxxx} + 20\alpha uu_{xx} + 10\alpha u_x^2 + 40\alpha^2 u^3 \right)_x = 0, \qquad (2.141)$$

$$\frac{\beta}{\alpha} = -16 \; : \; u_t + \left(u_{xxxx} + 40\alpha uu_{xx} + 30\alpha u_x^2 + 320\alpha^2 \frac{u^3}{3} \right)_x = 0, \quad (2.142)$$

respectively named Sawada–Kotera (SK) or Caudrey–Dodd–Gibbon [20, 119], Lax's 5th-order KdV (KdV5) [82] and Kaup–Kupershmidt (KK) [68]. Their respective Lax representations (2.74) are [48, 50, 82]

(SK) $\qquad \alpha = 3, \quad L = \partial_x^3 + 6u\partial_x,$

$\qquad\qquad A = 9\partial_x^5 + 90u\partial_x^3 + 90u_x\partial_x^2 + (60u_{xx} + 180u^2)\partial_x; \qquad (2.143)$

(KdV5) $\qquad \alpha = \dfrac{1}{2}, \quad L = \partial_x^2 + u,$

$\qquad\qquad A = 16\partial_x^5 + 40u\partial_x^3 + 60u_x\partial_x^2 + (50u_{xx} + 30u^2)\partial_x$

$\qquad\qquad\quad + 15u_{xxx} + 30uu_x;$

(KK) $\alpha = \dfrac{3}{4}, \quad L = \partial_x^3 + 6u\partial_x + 3u_x,$

$A = 3(3\partial_x^5 + 30u\partial_x^3 + 45u_x\partial_x^2$

$\qquad + (35u_{xx} + 60u^2)\partial_x + 10u_{xxx} + 30uu_x).$ (2.144)

We discard the equation (2.141), for it has the same second-order scattering problem as the KdV equation, and we restrict to the two equations (2.140)–(2.142) possessing two different third-order scattering problems $L\psi = \lambda\psi$.
   The SK equation possesses the Darboux transformation [118]

$$u = U + \partial_x^2 \log\psi,   \qquad (2.145)$$

while for the KK equation this transformation is [86]

$$u = U + (1/2)\partial_x^2 \log\varphi, \quad \varphi = \psi\psi_{xx} - (1/2)\psi_x^2 + 3U\psi^2.   \qquad (2.146)$$

In the notation $w_x = u, W_x = U$, the x-part of the BT for SK [43, 118] is

$$(w - W)_{xx} + 3(v - W)(w + W)_x + (w - W)^3 = \lambda,   \qquad (2.147)$$

while for the KK equation it is

$$(w - W)_{xx} + 3(w - W)(w + W)_x - (3/4)(w - W)_x^2/(w - W)$$
$$+ (w - W)^3 = \lambda.   \qquad (2.148)$$

This last expression was obtained for the first time by Rogers and Carillo [116] in the particular case $\lambda = 0$.

3   Painlevé Analysis for PDEs

The WTC extension [134] of Painlevé analysis to partial differential equations consists of two parts:

1. generation of *necessary conditions* (Painlevé test) for the absence of movable critical singularities in the "general solution,"

2. explicit proof of *sufficiency* by finding the transformation that linearizes the PDE or yields an auto-BT or a BT to another PDE with the PP.

The methods relative to both parts are different.
   In the first part, for every noncharacteristic manifold $(\varphi(x,t) = 0, \varphi_x \neq 0)$, one tests the existence of all possible local representations of the "general solution" by a Laurent series in a neighborhood of $\varphi = 0$. This test may

- pass regardless of φ; the PDE may then have the PP;

- fail regardless of φ; this is typical of chaotic PDEs;

- pass with some constraints on φ; then there exists a particular Laurent series, and the PDE is called "partially integrable."

In the second part, the Weiss *truncation procedure* [130], using only the singular part of the Laurent series, may yield constructive results like

- the linearizing transformation or the BT, in case the PDE passes the Painlevé test for every φ,

- particular solutions, necessarily compatible with the singularity structure of the PDE, in case the Painlevé test is conditionally or not satisfied (see Section 4).

3.1 Necessary Conditions (Painlevé Test)

Contrary to the case of ODEs, the singularities in the complex domain of (x,t) are not isolated. Given a PDE $E(u,x,t)=0$ of order N polynomial in u and its partial derivatives (maybe after a preliminary change of variables), we consider the associated equation $\varphi(x,t)=0$ of the movable SM and an expansion of u and E as a Laurent series in χ in a neighborhood of $\varphi=0$. We distinguish between φ and the expansion variable χ and only require χ to vanish as φ:

$$u(x,t) = \sum_{j=0}^{+\infty} u_j(x,t)\chi^{j+p}, \quad E(u,x,t) = \sum_{j=0}^{+\infty} E_j(x,t)\chi^{j+q}, \qquad (3.1)$$

where (p,q) are two negative integers with $q \le p-1$, and (u_j, E_j) the Laurent series coefficients. The result of the Painlevé test (necessary conditions) is independent of the explicit expression for χ but some particular choices are better than others during the second part (sufficient conditions) when one looks for the Lax pair or tries to linearize the equation.

The main choices (gauges) for the expansion variable χ are

WTC gauge [134] $\chi = \varphi$, hence coefficients (u_j, E_j) rational in the derivatives $D\varphi$ of φ.

Dimensionless WTC gauge $\chi = \varphi/\varphi_x$, hence coefficients (u_j, E_j) rational in the derivatives $D\varphi$ of φ of homogeneity degree zero.

Kruskal gauge [67] $\chi = x - f(t)$, f arbitrary, hence the coefficients (u_j, E_j) independent of x and rational in the derivatives of f. This is the simplest choice for the test, but it cannot be used to obtain the Lax pair or particular solutions.

Conte gauge [28] $\chi = \varphi/(\varphi_x - \varphi_{xx}\varphi/(2\varphi_x)) \sim_{\varphi \to 0} \varphi/\varphi_x$, hence coefficients (u_j, E_j) rational in the derivatives of φ invariant under the group of homographic transformations

$$\varphi \mapsto (a\varphi + b)/(c\varphi + d), \quad (a,b,c,d) \text{ arbitrary complex constants.}$$

In this last case, the Riccati system satisfied by χ is

$$\chi_x = 1 + (S/2)\chi^2, \tag{3.2}$$

$$\chi_t = -C + C_x\chi - (1/2)(CS + C_{xx})\chi^2, \tag{3.3}$$

$$2((\chi_t^{-1})_x - (\chi_x^{-1})_t) = S_t + C_{xxx} + 2C_xS + CS_x = 0, \tag{3.4}$$

with

$$S = \{\varphi; x\} = (\varphi_{xx}/\varphi_x)_x - \frac{1}{2}(\varphi_{xx}/\varphi_x)^2, \quad C = -\varphi_t/\varphi_x. \tag{3.5}$$

The transformation $\chi = \psi/\psi_x$ linearizes this Riccati system into

$$\psi_{xx} + (S/2)\psi = 0 \tag{3.6}$$

$$\psi_t + C\psi_x - (C_x/2 + g(t))\psi = 0, \quad g \text{ an arbitrary function.} \tag{3.7}$$

This choice of gauge is equivalent to the expansion of (u, E) as

$$u = \sum_{j=0}^{+\infty} u_j(\psi/\psi_x)^{j+p}, \quad E = \sum_{j=0}^{+\infty} E_j(\psi/\psi_x)^{j+q}, \tag{3.8}$$

where the function ψ satisfies a second-order linear ODE in the x variable. To obtain the couples (u_0, p) one substitutes in the polynomial PDE

$$u \rightarrow u_0\chi^p, \quad \chi_x \rightarrow 1, \quad \chi_t \rightarrow -C, \quad Du \rightarrow u_0D(\chi^p). \tag{3.9}$$

One then determines the balance between the different terms of this polynomial expression. Each different solution (u_0, p) defines a *family*. For every $j \geq 1$ the recurrence relation determining u_j is

$$\forall j \geq 1 : P(j)u_j = Q_j(\{u_k, Du_k, \ k \in [0, j-1]\}), \tag{3.10}$$

where P is a polynomial of degree at most N.

The main requirements of the Painlevé test are

- the zeros of P (*Fuchs indices*, also named *Painlevé resonances*) are distinct integers;

- for every index i and every φ, the compatibility condition $Q_i = 0$ holds.

3.2   Methods for Proving Sufficiency

One distinguishes two main methods:

1. *the singular part transformation*, which may provide the explicit transformation linearizing the nonlinear PDE. If this is not the case, the transformation may yield an equation in a form more convenient than the original one to search for explicit solutions;

2. *the truncation procedure* of Weiss and its extensions for obtaining the BT and thus proving that the nonlinear PDE possesses the PP.

3.2.1   Singular Part Transformation

The method consists in transforming the PDE for u into an equation for φ by the nonlinear transformation

$$u = \mathcal{D} \log \varphi, \tag{3.11}$$

where $\mathcal{D}$ is the singular part operator associated with one of the families defined in the Painlevé test.

Example 3.1 (linearization). Burgers equation

$$u_t + u_{xx} + (u^2)_x = 0, \quad u = \varphi_x \varphi^{-1}, \tag{3.12}$$

$$u = \mathcal{D} \log \varphi = \partial_x \log \varphi, \varphi_t + \varphi_{xx} + K(t)\varphi = 0, \tag{3.13}$$

$K(t)$ an arbitrary function.

Example 3.2 (linearization). Liouville equation

$$v_{xt} - e^v = 0, \tag{3.14}$$

$$e^v = u, \quad uu_{xt} - u_x u_t - u^3 = 0, \quad u = 2(\varphi_x \varphi_t \varphi^{-2} - \varphi_{xt} \varphi^{-1}), \tag{3.15}$$

$$u = \mathcal{D} \log \varphi = -2\partial_{xt}^2 (\log \varphi), \quad \varphi_{xt} = 0. \tag{3.16}$$

Example 3.3 (linearization). Eckhaus equation [19, 25, 33]

$$iu_t + u_{xx} + q_r(|u|^4 + 2a(|u|^2)_x)u = 0, \quad a^2 = 1/q_r, \; q_r \in \mathcal{R}. \tag{3.17}$$

In the variables (w, θ) defined by $\theta = \arg u, w_x = |u|^2$, the equation (3.17) is equivalent to the system

$$\theta_x = -\frac{1}{2}\frac{w_t}{w_x}, \tag{3.18}$$

$$theta_t = \frac{1}{4}\left(2\frac{w_{xxx}}{w_x} - \frac{w_{xx}^2}{w_x^2}\right) - \frac{1}{4}\frac{w_t^2}{w_x^2} + q_r(w_x^2 + 2aw_{xx}), \tag{3.19}$$

whose compatibility condition is

$$\theta_{xt} - \theta_{tx} \equiv (w_{tt}w_x^2 + w_{xx}w_t^2)/2 + (w_{xxxx}w_x^2 + w_{xx}^3)/2 - w_x w_{xx}w_{xxx}$$
$$- w_t w_x w_{xt} + 2q_r(w_x^4 w_{xx} + aw_x^3 w_{xxx}) = 0. \tag{3.20}$$

Under the transformation $w = (a/2)\log \varphi$ defined by the singular part operator $\mathcal{D}$ of the equation for w, these three equations become

$$\theta_x = -\varphi_t/(2\varphi_x), \tag{3.21}$$

$$\theta_t = \varphi_{xxx}/(2\varphi_x) - \varphi_{xx}^2/(4\varphi_x^2) - \varphi_t^2/(4\varphi_x^2), \tag{3.22}$$

$$\theta_{xt} - \theta_{tx} \equiv (\varphi_{tt}\varphi_x^2 + \varphi_{xx}\varphi_t^2)/2 + (\varphi_{xxxx}\varphi_x^2 + \varphi_{xx}^3)/2$$
$$- \varphi_x \varphi_{xx}\varphi_{xxx} - \varphi_t \varphi_x \varphi_{xt} = 0. \tag{3.23}$$

The three equations (3.21), (3.22), (3.23) are deduced from the three previous ones (3.18), (3.19), (3.20) by the following simple operation: change w to φ and assign q_r to zero. Thus the transformation has linearized the Eckhaus equation (3.17) into the Schrödinger equation

$$i\nu_t + \nu_{xx} = 0, \quad \varphi_x = |\nu|^2, \quad \arg\nu = \arg u. \tag{3.24}$$

Because of the conservation of the phase, one finally has

$$u = (\sqrt{a/2})\nu / \sqrt{\int^x |\nu|^2 dx}, \quad |u|^2 = (a/2)\partial_x \log\varphi. \tag{3.25}$$

Example 3.4 (bilinearization). **Korteweg–de Vries** equation

$$u_t + u_{xxx} + 3(u^2)_x = 0, \quad u = -2\varphi_x^2\varphi^{-2} + 2\varphi_{xx}\varphi^{-1}, \tag{3.26}$$

$$u = \mathcal{D}\log\varphi = 2\partial_x^2\log\varphi, (D_x D_t + D_x^4)(\varphi \cdot \varphi) = 0. \tag{3.27}$$

The transformed equation, quadratic in φ (see the numerous papers of Hirota [58,59] for the definition of the bilinear operators D_x, D_t) is convenient to look for N-soliton solutions, auto-BTs, Miura transformations.

3.2.2   Weiss Method and Its Limitations

If a nonlinear PDE passing the Painlevé test is not linearizable, the idea of Weiss [130, 134] is that the principal part of this *local* Laurent series contains all the information for proving that the PDE possesses the Painlevé property through the knowledge of its BT (i.e., its DT and Lax pair). This method consists in truncating the Laurent series for u and $E(u)$ to their nonpositive powers in χ,

$$u_T = \sum_{j=0}^{-p} u_j \chi^{j+p}, \quad E_T = \sum_{j=0}^{-q} E_j \chi^{j+q}, \tag{3.28}$$

and identifying to zero the coefficients E_j of the χ-polynomial $\chi^{-q}E_T(\chi)$. Equations $E_j = 0$ for $j = 0, \ldots, -p$ determine the $p+1$ coefficients u_j as equal to those of the infinite expansion. After replacement of u_j by these values, the remaining equations are

$$E_j(D\varphi, u_i) = 0, \ j \in \{-p+1, \ldots, -q\}, \ j \neq \text{compatible indices},$$
$$i = \text{indices} \in \{0, \ldots, -p\}. \tag{3.29}$$

In the Conte gauge, the coefficients u_j, E_j depend on the derivatives of φ through the homographic invariants (S, C) and their derivatives. As the variable $\chi^{-1} = \psi_x/\psi$ satisfies a Riccati equation, one can connect the monomial $(\psi_x/\psi)^n$ with the derivatives $(\log \psi)_{jx}, j \leq n \in \mathcal{N}^+$, and show that

$$\sum_{j=0}^{-p-1} u_j(S,C)(\psi/\psi_x)^{j+p} \equiv \sum_{j=1}^{-p} \tilde{u}_j(S,C)(\log \psi)_{jx} + f(S,C). \qquad (3.30)$$

Then the relation

$$u_T - \tilde{u} = \sum_{j=1}^{-p} \tilde{u}_j(S,C)(\log \psi)_{jx} = \mathcal{D} \log \psi, \qquad (3.31)$$

where $\mathcal{D}$ is the singular part operator and $\tilde{u} = u_{-p}(S,C) + f(S,C)$, defines a Darboux transformation if $E(\tilde{u}) = 0$. For this reason, we call equations (3.29) *Painlevé–Darboux* equations. The elimination of the arbitrary functions u_i among this set must produce only one "independent" equation

$$F(S,C) = 0, \qquad (3.32)$$

called the *singular manifold equation*, modulo the ever present link between S and C given by equation (3.4).

The next step consists in finding a parametric representation for equation (3.32) under the form (S,C) depending on a function U and an arbitrary constant λ such that the cross-derivative condition (3.4) is identical to the original equation $E(U) = 0$ for U. If this is indeed the case and if U can be identified with $\tilde{u}$, the truncation will provide the DT (as a consequence of equation (3.31)), the Lax pair (as consequences of the linear system (3.6)–(3.7), and the parameterization of S and C) and thus the BT.

This method succeeds only for a few equations, like KdV [134], KdV5, AKNS [97], all belonging to the same hierarchy. Let us describe it for the KdV equation (1.1). This equation, which passes the Painlevé test, admits the single family $u \sim -2\chi^{-2}$ with Fuchs indices $-1, 4, 6$,

$$u = -2\chi^{-2} + (C - 4S)/6 - \frac{1}{6}(C - S)_x \chi + O(\chi^2). \qquad (3.33)$$

The algorithmic results of the Painlevé analysis for KdV are given in Table 8.2. They yield the SM equation

$$C - S + 6\lambda = 0, \quad \lambda = \text{ arbitrary constant.} \qquad (3.34)$$

Its parametric representation

$$S = 2(U + \lambda), \quad C = 2(U - 2\lambda) \qquad (3.35)$$

provides the second-order linear system (3.6)–(3.7)

$$\psi_{xx} + (U + \lambda)\psi = 0, \qquad (3.36)$$
$$\psi_t + 2(U - 2\lambda)\psi_x - U_x\psi = 0 \qquad (3.37)$$

satisfying the cross-derivative condition $\psi_{xxt} - \psi_{txx} \equiv 2\,\mathrm{KdV}(U)\psi = 0$. The map between two solutions of KdV coming out of the truncation is

$$u_T = 2(\log \psi)_{xx} + (C + 2S)/6 = 2(\log \psi)_{xx} + U. \qquad (3.38)$$

Thus, the Weiss truncation yields both the Lax pair (2.90)–(2.91) and the DT (2.93) of the KdV equation. The auto-BT (2.85)–(2.86) is obtained by substitution of the DT (3.38), i.e., $\psi_x/\psi = (w - W)/2$, into the couple (3.36) and (3.37) (notation $u_T = u = w_x, U = W_x$).

It happens that for other equations possessing either one family of movable singularities or several families with nonopposite residues like Boussinesq, Sawada–Kotera, Hirota–Satsuma [62] equations, the parameterization of (S, C) yields a condition (3.4) for U different from the original equation, which defines a transformation between u_T and U, called a *Miura transformation*, obtained by the elimination of (χ, S, C) between the four equations: $u_T =$ the truncation, the two equations of the parametric representation $(S, C) = f(U)$, and any one of the two (nonindependent) equations (3.2), (3.3). Then in order to obtain the auto-BT, one requires that the function ψ in equation (3.31) satisfy a linear third-order system whose coefficients are to be determined as functions of λ and another solution U of the analyzed PDE linked to u_T through the Darboux transformation.

3.2.3   Method for Third-Order Lax Pair

Let us denote by (a, b, c, d, e) the five unknown coefficients defining a third-order linear system for ψ,

$$\psi_{xxx} = a\psi_x + b\psi, \qquad (3.39)$$
$$\psi_t = c\psi_{xx} + d\psi_x + e\psi, \qquad (3.40)$$

whose compatibility condition is

$$(\psi_t)_{xxx} - (\psi_{xxx})_t \equiv X_0\psi + X_1\psi_x + X_2\psi_{xx} = 0, \qquad (3.41)$$
$$X_0 \equiv -b_t - ae_x + e_{xxx} + b_{xx}c,$$
$$+ 3bc_{xx} + 3b_xc_x + 3bd_x + b_xd = 0, \qquad (3.42)$$
$$X_1 \equiv -a_t + 3e_{xx} + 2b_xc + a_{xx}c + d_{xxx} + 3ac_{xx} + 2ad_x,$$
$$+ 3a_xc_x + 3bc_x + a_xd = 0, \qquad (3.43)$$
$$X_2 \equiv (2ac + c_{xx} + 3d_x + 3e)_x = 0. \qquad (3.44)$$

In the two independent components $Z_1 = \psi_x/\psi$, $Z_2 = \psi_{xx}/\psi$, the linear system (3.39)–(3.40) is equivalent to the projective Riccati system [7]

$$Z_{1,x} = (-Z_1)Z_1 + Z_2, \qquad (3.45)$$
$$Z_{2,x} = (-Z_1)Z_2 + aZ_1 + b, \qquad (3.46)$$
$$Z_{1,t} = (-dZ_1 - cZ_2)Z_1 + (ac + d_x)Z_1 + (c_x + d)Z_2 + e_x + bc, \qquad (3.47)$$

$$Z_{2,t} = (-dZ_1 - cZ_2)Z_2 + (2ac_x + a_xc + bc + d_{xx} + ad + 2e_x)Z_1,$$
$$+ (c_{xx} + 2d_x + ac)Z_2 + 2bc_x + b_xc + bd + e_{xx}. \tag{3.48}$$

The determining equations for the coefficients (a, b, c, d, e) of the Lax pair are generated by the expansion of $E_T = E(u_T)$ on the basis (Z_1, Z_2),

$$E_T = \sum_{l,m} C_{l,m} Z_1^l Z_2^m, \tag{3.49}$$

$$C_{l,m} \equiv C_{l,m}(a, b, c, d, e, U) = 0. \tag{3.50}$$

In case the solution of the determining equations does not lead to the expected solution, for a reason like the absence of a spectral parameter, the assumption to be changed is the order of the underlying scattering problem.

Let us give more details on the procedure [98, 99] for finding the BT of the Boussinesq and Sawada–Kotera equations.

Example 3.5 (First example: Boussinesq equation). Let us consider the Boussinesq (Bq) equation [124, 141]

$$E(u) \equiv u_{tt} + \varepsilon^2 \left((u + \alpha)^2 + (\beta^2/3)u_{xx}\right)_{xx} = 0, \tag{3.51}$$

with $(\alpha, \beta, \varepsilon)$ constant. The algorithmic results of the Painlevé analysis are $p = -2$, $q = -6$, indices -1, 4, 5, 6 compatible,

$$u_T = -2\beta^2\chi^{-2} - 2\beta^2 S/3 - \varepsilon^{-2}C^2/2, \tag{3.52}$$

χ defined by (3.2)–(3.3). The set of Painlevé–Darboux equations reduces to the single equation

$$E_3 \equiv (1/3)\beta^2\varepsilon^2 S_x - C_t + CC_x = 0, \tag{3.53}$$

which is the SM equation for the Bq equation [133] in the invariant formalism. This is a conservation law, which can be parameterized as

$$C = (\beta\varepsilon)^2 z_x, \quad S = 3z_t - \frac{3}{2}(\beta\varepsilon)^2 z_x^2. \tag{3.54}$$

The compatibility condition of the system (3.6)–(3.7) reads

$$3z_{tt} + (\beta\varepsilon)^2 z_{xxxx} + 6(\beta\varepsilon)^2 z_t z_{xx} - 6(\beta\varepsilon)^4 z_x^2 z_{xx} = 0, \tag{3.55}$$

which is *not* the Bq equation but another PDE called the modified Bq equation [49, 63]. The elimination of S between (3.52) and (3.2) yields the Miura transformation between the Bq and the modified equation,

$$u_T = -\frac{2}{3}\beta^2\chi^{-2} - \frac{1}{2}\varepsilon^{-2}C^2 + \frac{4}{3}\beta^2(\chi^{-1})_x, \tag{3.56}$$

while the assumption for a DT like (3.31) leads to

$$\tilde{u} = -\frac{2}{3}\beta^2 \chi^{-2} - \frac{1}{2}\varepsilon^{-2}C^2 - \frac{2}{3}\beta^2(\chi^{-1})_x, \qquad (3.57)$$

which does not coincide with (3.56). We then conclude that a second-order linear system is not convenient to represent the Lax pair of the Bq equation.

So, let us assume an underlying scattering problem of the third order for ψ and the existence of a DT given by the singular part operator

$$v_T = 2\beta^2 \log \psi + V, \quad \mathrm{Bq}(v_{T,xx}) = 0, \quad \mathrm{Bq}(V_{xx}) = 0. \qquad (3.58)$$

Defining the "second potential Bq" equation

$$F(v) \equiv v_{tt} + \varepsilon^2 \left((v_{xx} + \alpha)^2 + (\beta^2/3)v_{xxxx} \right) = 0, \qquad (3.59)$$

where $F(v_T)$ is a second-degree polynomial in (Z_1, Z_2):

$$F(v_T) \equiv C_{02}Z_2^2 + C_{11}Z_1 Z_2 + C_{20}Z_1^2 + C_{01}Z_2 + C_{10}Z_1 + C_{00} = 0, \qquad (3.60)$$

which we identify to zero. This provides

$$C_{02} \equiv 2((\beta\varepsilon)^2 - c^2) = 0 \implies c^2 = (\beta\varepsilon)^2 \qquad (3.61)$$

$$C_{11} \equiv -4cd = 0 \implies d = 0 \qquad (3.62)$$

$$C_{20} \equiv V_{xx} + \alpha + 2\beta^2 a/3 = 0 \implies a = -3(V_{xx} + \alpha)/(2\beta^2) \qquad (3.63)$$

$$C_{01} \equiv 2(\beta\varepsilon^{-1}ac + 2(V_{xx} + \alpha) + \beta^2 a/3) = 0 \implies c = \beta\varepsilon \qquad (3.64)$$

$$C_{10} \equiv 8(\beta\varepsilon)^2 a_x/3 + 4e_x c = 0 \implies e_x = \beta^{-1}\varepsilon V_{xxx}, \qquad (3.65)$$

$$C_{00} \equiv 2(\beta^{-2}V_{xxt} + \tfrac{4}{3}\beta\varepsilon b_x + e_{xx}) = 0$$

$$\implies b = g(t) - \frac{3}{4})(\beta^{-2}V_{xxx} + \beta^{-3}\varepsilon^{-1}V_{xt}). \qquad (3.66)$$

Finally, the compatibility condition $X_0 = 0$ implies that $g(t)$ is an arbitrary constant, denoted by λ. The coefficients a, b, c, d, e are

$$a = -\frac{3}{2}\beta^{-2}(U + \alpha), \quad c = \beta\varepsilon, \quad d = 0,$$

$$b = \lambda - \frac{3}{4}\beta^{-2}U_x - \frac{3}{4}\beta^{-3}\varepsilon^{-1}V_{xt}, \quad e = \beta^{-2}c(U + \alpha), \qquad (3.67)$$

i.e., the associated third-order Lax pair [95, 140, 141] of the derivative of (3.59) (notation $U = V_{xx}$).

Since $d = c_x = 0$, the BT obtained by eliminating Z_2 between (3.45)–(3.47) is

$$Z_{1,xx} + 3Z_1 Z_{1,x} + Z_1^3 - aZ_1 - b = 0, \qquad (3.68)$$

$$Z_{1,t} + c(Z_1 Z_{1,x} + Z_1^3 - aZ_1 - \beta^{-2}U_x - b) = 0, \qquad (3.69)$$

$$(Z_{1,xx})_t - (Z_{1,t})_{xx} = -\frac{3}{4}\beta^{-3}\varepsilon^{-1}(F(V))_x, \qquad (3.70)$$

or equivalently, with $U = W_x = V_{xx}$ and $Z_1 = (w - W)/(2\beta^2)$,

$$(w - W)_{xx} + 3\beta^{-2}(w - W)((w + W)_x + 2\alpha) + \beta^{-4}(w - W)^3$$
$$+ 3\beta^{-1}\varepsilon^{-1}(w + W)_t - 8\beta^2\lambda = 0, \tag{3.71}$$
$$(w + W)_{xx} + \beta^{-2}(w - W)(w - W)_x - \beta^{-1}\varepsilon^{-1}(w - W)_t = 0, \tag{3.72}$$

an extension to $\lambda \neq 0$ of the bilinear BT of Hirota and Satsuma [63, 64].

Example 3.6 (Second example: Sawada–Kotera equation).
In the same way, we can easily find the coefficients of the third-order Lax pair by processing the fifth-order potential equation

$$\mathrm{pSK}(v) \equiv v_t + v_{5x} + 30v_x v_{3x} + 60v_x^3 + F(t) = 0, \quad F(t) \text{ arbitrary.} \tag{3.73}$$

The algorithmic results of the Painlevé analysis are the following: Equation (3.73) possesses two families, each with five compatible indices. For the "principal" family

$$p = -1, \quad q = -6, \quad u_0 = 1, \quad \text{indices} \ -1, 1, 2, 3, 10 \text{ compatible}, \tag{3.74}$$

the truncation is

$$v_T = \chi^{-1} + v_1. \tag{3.75}$$

The assumption $\chi = \psi/\psi_x$ with ψ a solution of the second-order Lax pair (3.6)–(3.7) generates the Painlevé–Bäcklund equations [29]

$$E_4 \equiv C - 4S^2 + 9S_{xx} + 60Sv_{1,x} - 180v_{1,x}^2 - 30v_{1,xxx} = 0, \tag{3.76}$$
$$E_5 \equiv -C_x - 2SS_x + S_{xxx} + 30S_x v_{1,x} = 0, \tag{3.77}$$
$$E_6 \equiv \mathrm{pSK}(v_1) + (SE_4 - E_{5,x})/2 + (5/2)S_x(6v_{1,xx} - S_x) = 0. \tag{3.78}$$

Demanding that v_1 be another solution of pSK implies $v_{1,xx} = S_x/6$ and, after computation, provides a nongeneric solution. Note, however, that a *particular* solution of the truncation is [131]

$$S_{xx} + 4S^2 - C = 0, \quad v_{1,x} = S/3, \quad \mathrm{KK}(v_1) = 0, \tag{3.79}$$

which defines a Miura transformation between the SK and KK equations.
As in the preceding example, the hypothesis of the DT,

$$v = (\log \psi)_x + V, \tag{3.80}$$

with V another solution of pSK and ψ a solution of the third-order linear system (3.39)–(3.40), makes pSK(v)-pSK(V) a second-degree polynomial in (Z_1, Z_2) like (3.60). The six determining equations $C_{lm} = 0$, added to the three compatibility conditions (3.42)–(3.44), have a unique solution depending on an arbitrary constant λ,

$$a = -6V_x, \quad b = \lambda, \quad c = 9\lambda - 18V_{xx}, \tag{3.81}$$

$$d = -36V_x^2 + 6V_{3x}, e_x \qquad\qquad = 36\lambda V_{xx}, \tag{3.82}$$

a result that coincides with the Lax pair (2.143).

The x-part of the BT (2.147) is obtained by eliminating Z_2 between (3.45) and (3.46), then substituting $Z_1 = v - V$ as results from (3.80).

3.2.4 Third Example: Kaup–Kupershmidt Equation

In the potential equation, the hypothesis of the differential operator $\mathcal{D} = \partial_x$ for the DT, associated to the linear system (3.39)–(3.40), yields neither the Lax pair (2.144) nor the BT (2.148). This problem has been solved [101] by remarking that in his classification of second-order first-degree nonlinear ODEs possessing the Painlevé property, Gambier [53] mentions that the following

(G.5) : $\qquad Y_{1,xx} + 3Y_1 Y_{1,x} + Y_1^3 + 6UY_1 - \lambda = 0,$ $\qquad$ (3.83)

(G.25) : $\qquad Y_2 Y_{2,xx} - \dfrac{3}{4} Y_{2,x}^2 + \dfrac{3}{2} Y_2^2 Y_{2,x} + \dfrac{1}{4} Y_2^4 + 6UY_2^2 - 2\lambda Y_2 = 0,$

$$\tag{3.84}$$

are the only ones to be linearizable into third-order equations, namely,

(G.5) : $\qquad Y_1 = \psi_x/\psi, \quad \psi_{xxx} + 6U\psi_x - \lambda\psi = 0,$ $\qquad$ (3.85)

(G.25) : $\qquad Y_2^{-1} = \lambda^{-1}[(\psi_x/\psi)_x + (1/2)(\psi_x/\psi)^2 + 3U],$

$$\psi_{xxx} + 6U\psi_x + (3U_x - \lambda)\psi = 0, \tag{3.86}$$

corresponding to the scattering problem of, respectively, the SK equation for U and the KK equation for U. It can then be shown that the DTs

$$Y_1 = w - W, \qquad \text{with } \mathrm{SK}(w_x) = \mathrm{SK}(W_x) = 0, \tag{3.87}$$

$$Y_2 = 2(w - W), \quad \text{with } \mathrm{KK}(w_x) = \mathrm{KK}(W_x) = 0, \tag{3.88}$$

leading to the BTs (2.147) and (2.148), can be found by singularity analysis.

3.2.5 Fourth Example: Tzitzéica Equation

The x-part of the auto-Bäcklund transformation written in the variable

$$Y_2 = -\frac{1}{2}\int \left(e^u - e^U\right), \quad \text{with } \mathrm{Tzi}(u) = \mathrm{Tzi}(U) = 0, \tag{3.89}$$

is given by the nonlinear differential equation

$$Y_2 Y_{2,xx} - 2Y_{2,x}^2 - (U_x Y_2 - 3e^U)Y_{2,x} + \lambda e^{-U} Y_2^3 - e^{2U} = 0, \tag{3.90}$$

which belongs to the equivalence class of the fifth Gambier equation. This
can be linearized by the transformation

$$Y_2^{-1} = e^{-U} \partial_x \log(e^U \psi), \tag{3.91}$$

into the third-order linear equation

$$\psi_{xxx} + (2U_{xx} - U_x^2)\psi_x + ((2U_{xx} - U_x^2)_x/2 - \lambda)\psi = 0. \tag{3.92}$$

3.2.6 Two–Singular Manifold Method

For equations with two families of movable poles with opposite residues,
the truncation procedure that considers only one family of singularities
does not yield the auto-BT. An extension of the Weiss method consists in
considering two distinct functions ψ_1, ψ_2, assuming now a DT of the form

$$u_T - U = \mathcal{D}\log\psi_1 - \mathcal{D}\log\psi_2, \quad E(u_T) = E(U) = 0, \tag{3.93}$$

and, assuming that $Y = \psi_1/\psi_2$ satisfies the most general Riccati system

$$Y_x = R_0 + R_1 Y + R_2 Y^2, \tag{3.94}$$

$$Y_t = S_0 + S_1 Y + S_2 Y^2, \tag{3.95}$$

$$Y_{xt} - Y_{tx} \equiv X_0 + X_1 Y + X_2 Y^2, \tag{3.96}$$

$$X_0 \equiv R_{0,t} - S_{0,x} + R_1 S_0 - R_0 S_1, \tag{3.97}$$

$$X_1 \equiv R_{1,t} - S_{1,x} + 2(R_2 S_0 - R_0 S_2), \tag{3.98}$$

$$X_2 \equiv R_{2,t} - S_{2,x} + R_2 S_1 - R_1 S_2, \tag{3.99}$$

eliminating derivatives of Y and identifying E_T to the null polynomial in
Y.

The determining equations so generated for the six unknowns (R_i, S_i)
must have a solution such that each R_i is a linear function of U and an
arbitrary constant λ and such that at least one of the three cross-derivative
conditions $X_i = 0$ is identical to the original equation for U. In such a case,
one has found the DT and the Lax pair, i.e., the BT. In the case of two
simple poles with constant opposite residues $\pm u_0$ and opposite singular
part operators $\pm u_0 \partial_x$, the truncation

$$u_T = u_0 (\log Y)_x + U \tag{3.100}$$

becomes, by elimination of Y_x from (3.94),

$$u_T = u_0 (R_0 Y^{-1} + R_1 + R_2 Y) + U. \tag{3.101}$$

This represents an extension of the Weiss truncation to the positive powers of Y. A similar extension was previously made [108] in the variable
$\chi = \psi/\psi_x$ for obtaining particular solutions of nonlinear PDEs. The "two–
singular manifold" method is successful for finding the auto-BT of MKdV

TABLE 8.2. Algorithmic results of the Painlevé analysis. The integers (p, q) are defined in (3.28); for sG, the polynomial PDE is (3.113). The next column lists the indices, except -1. The column "PD equations" lists the subscripts of the not identically zero Painlevé–Darboux equations; in the sG and NLS cases, they depend on the arbitrary coefficients introduced at the index 2 (sG) and 0 (NLS).

Name	p	q	Indices	PD eq.	Singular manifold equation
KdV	-2	-5	$4, 6$	$3, 5$	$S - C = 6\lambda$
MKdV	-1	-4	$3, 4$	2	$S - C = 0$
sG	-2	-6	2	$3, 4, 5, 6$	$S + C^{-1}C_{xx}$ $-C^{-2}C_x^2/2$ $+2\lambda = 0$
NLS	$(-1, -1)$	$(-3, -3)$	$0, 3, 4$	$2, 2, 3$	$C_t + 3CC_x - S_x$ $+8\lambda C_x = 0$

and sine–Gordon equations [100] but only partially for the NLS equation.

Before detailing this result, let us first reproduce in Table 8.2 the algorithmic results of the Painlevé analysis for the four equations belonging to the AKNS scheme; these include the SM equation associated with these well-known NLPDEs which pass the Painlevé test.

Let us now use the information contained in the SME.

Modified Korteweg–de Vries. Equation (2.94) has two families $u \sim \pm a\chi^{-1}$, denoted by $u \sim a\chi^{-1}$, since a is defined by its square. The truncated expansion of a family is

$$u_T = a\chi^{-1}. \tag{3.102}$$

The SME $S - C = 0$ is parameterized as

$$S = 2v, \quad C = 2v, \quad \text{KdV}(v) = 0, \tag{3.103}$$

and the precise relation between u and v (Miura transformation) is obtained by eliminating χ between (3.102) and (3.2):

$$(u_T/a)_x + (u_T/a)^2 = -v. \tag{3.104}$$

In fact, there are two such Miura transformations, one for each sign of a, i.e., one for each family.

Let us first obtain the Darboux transformation for MKdV from that of KdV and show that it involves two SMs. The Darboux transformation for KdV has been obtained in Section 3.2.2, equation (3.38). The two Miura transformations (3.104) and the parameterization (3.103) imply

$$-\frac{S_1}{2} = \left(\frac{u_T}{a}\right)^2 + \left(\frac{u_T}{a}\right)_x = 2(\log\psi_1)_{xx} + \left(\frac{U}{a}\right)^2 + \left(\frac{U}{a}\right)_x, \qquad (3.105)$$

$$-\frac{S_2}{2} = \left(\frac{u_T}{a}\right)^2 - \left(\frac{u_T}{a}\right)_x = 2(\log\psi_2)_{xx} + \left(\frac{U}{a}\right)^2 - \left(\frac{U}{a}\right)_x, \qquad (3.106)$$

and the elimination of the nonlinear terms leads to

$$u_{T,x} = a(\log(\psi_1/\psi_2))_{xx} + U_x, \qquad (3.107)$$

which after one integration yields the Darboux transformation for MKdV

$$u_T = a(\log(\psi_1/\psi_2))_x + U. \qquad (3.108)$$

With this DT, the Lax pair is obtained as explained in the introduction of this section. Setting $Y = \psi_1/\psi_2$ and taking account of (3.94)–(3.95), every derivative of Y can be replaced by a polynomial in Y. Consequently, the Darboux transformation (3.108) becomes identical to

$$u_T = a(R_0 Y^{-1} + R_1 + R_2 Y) + U, \qquad (3.109)$$

and one must identify to zero the polynomial in Y

$$E(u_T) \equiv E_T = \sum_{j=0}^{8} E_j Y^{j-4}. \qquad (3.110)$$

Among the nine Painlevé–Darboux equations $E_j = 0$, only four ($j = 1$, 2, 6, 7) are not identically zero. Their resolution, as detailed in [100], yields the following parametric representation of the six unknowns R_i, S_i in which R_i is linear in U and the spectral parameter λ

$$R_0 = \lambda, \quad R_2 = -\lambda, \quad R_1 = -2U/a, \qquad (3.111)$$

$$S_0 = -4\lambda^2 + 2(U/a)^2 - 2U_x/a, \quad S_1 = 8\lambda^2 U/a - 4(U/a)^3 + 2U_{xx}/a,$$

$$S_2 = -4\lambda^2 + 2(U/a)^2 + 2U_x/a. \qquad (3.112)$$

This solution associated with the Riccati equations (3.94)–(3.95) reproduces the equations (2.99)–(2.100) for the pseudopotential of MKdV.

Sine–Gordon. The sine–Gordon equation (2.118), invariant by parity on u, is first transformed into a polynomial equation for $v = e^{iu}$, invariant under $v \mapsto 1/v$:

$$\mathrm{PsG}(v) \equiv 2vv_{xt} - 2v_x v_t - v^3 + v = 0, \quad v = e^{iu}. \qquad (3.113)$$

This PDE has two families of movable singularities $v = v_1 \sim -4C_1\chi^{-2}$ and $v = v_2^{-1} \sim -4C_2\chi^{-2}$. The truncation equations have the following general solution [28, 132]. For the first family $((v, S, C, \psi)$ are subscripted with 1),

$$S_1 = -v_{1,xx}/v_1 + v_{1,x}^2/(2v_1^2) - 2\lambda = -iu_{xx} + u_x^2/2 - 2\lambda, \qquad (3.114)$$

$$C_1 = -v_1/(4\lambda) = -e^{iu}/\lambda, \qquad (3.115)$$

$$v_1 = -4(\log \psi_1)_{xt} + V_1, \quad \mathrm{PsG}(V_1) = 0. \qquad (3.116)$$

For the second family $e^{-iu} = v_2 \sim -4C_2\chi^{-2}$,

$$S_2 = -v_{2,xx}/v_2 + v_{2,x}^2/(2v_2^2) - 2\lambda = iu_{xx} + u_x^2/2 - 2\lambda, \qquad (3.117)$$

$$C_2 = -v_2/(4\lambda) = -e^{-iu}/\lambda \qquad (3.118)$$

$$v_2 = -4(\log \psi_2)_{xt} + V_2, \quad \mathrm{PsG}(V_2) = 0. \qquad (3.119)$$

If one considers only one of these two equivalent SMs, the Schwarzian S_i, $i = 1$ or 2, does depend on an arbitrary constant λ, but it has two drawbacks: It is not invariant under parity on u, and it is not linear in the physical field u as requested for the Lax pair (3.6)–(3.7) to be a "good" one.

Since $v_1 - v_2 = 2i\sin u$, the difference of (3.116) and (3.119) reads

$$\sin u = 2i(\log(\psi_1/\psi_2))_{xt} + \sin U, \quad \mathrm{sG}(U) = 0, \qquad (3.120)$$

i.e., from the definition of the equation

$$u_{xt} = 2i(\log(\psi_1/\psi_2))_{xt} + U_{xt}. \qquad (3.121)$$

Integrating twice, we finally obtain the Darboux transformation of sG,

$$u = 2i\log(\psi_1/\psi_2) + U, \qquad (3.122)$$

defined in terms of *both* families. For the solution of the polynomial PDE $\mathrm{PsG}(v) = 0$ associated to the sG equation by $v = e^{iu}$, the DT is

$$v = VY^{-2}, \quad Y = \psi_1/\psi_2, \quad \mathrm{PsG}(V) = 0, \qquad (3.123)$$

and one must identify

$$E(v) = V^2 \sum_{i=0}^{4} E_i Y^{i-6} \qquad (3.124)$$

to the null polynomial in Y. Among the five Painlevé–Darboux equations, the equation E_2 is functionally dependent on (E_0, E_1), a consequence of the compatibility of the index 2. Their resolution yields the Riccati pseudopotential (2.122)–(2.123),

$$Y_x = \lambda(1 - Y^2) + iU_xY, \qquad (3.125)$$

$$Y_t = ((1 - Y^2)\cos U + i(1 + Y^2)\sin U)/(4\lambda), \qquad (3.126)$$

$$(Y_{xt} - Y_{tx})/Y = \mathrm{sG}(U), \qquad (3.127)$$

where the x-part is now linear in the spectral parameter λ and the field U associated with the DT (3.122).

3.2.7 Weiss Method Plus Homography

Pickering [109] remarks three drawbacks in the previous method:

i. No explicit relationship is given between the variable $Y \equiv \psi_1/\psi_2$ and the variable χ of the invariant Painlevé analysis, while the one between χ and φ is well-defined by the homographic transformation [28]

$$\chi = \frac{\varphi}{\varphi_x - \varphi_{xx}\varphi/(2\varphi_x)}. \tag{3.128}$$

ii. The result of the previous truncation for MKdV does not reveal any relationship between the MKdV and KdV equations as it would be,

iii. Knowledge of the DT is required in advance.

He notices that for finding the BT of MKdV and sine–Gordon equations it is sufficient to consider a Riccati system constructed from the nonlinearization of the second-order scalar linear system

$$\eta_{xx} = 2A\eta_x + B\eta, \tag{3.129}$$

$$\eta_t = -C\eta_x + \left(\int^x D\,dx'\right)\eta \tag{3.130}$$

by the transformation $Z^{-1} = \eta_x/\eta$. The corresponding nonlinear system

$$Z_x = 1 - 2A - BZ^2, \tag{3.131}$$

$$Z_t = -C + (C_x + 2AC)Z - (D - BC)Z^2 \tag{3.132}$$

depends on four functions A, B, C, D in place of six as with the system (3.94)–(3.95) considered in the 2-SM method. Its compatibility condition is

$$Z_{xt} - Z_{tx} \equiv X_1 Z + X_2 Z^2 = 0, \tag{3.133}$$

$$X_1 \equiv 2\left(D - (A_t + (AC)_x + C_{xx}/2)\right) = 0, \tag{3.134}$$

$$X_2 \equiv D_x - B_t - 2BC_x - B_x C - 2AD = 0. \tag{3.135}$$

The solution η of the linear ODE (3.129) is related to the solution ψ of (3.6) by the gauge transformation

$$\eta = \left(e^{\int^x A\,dx'}\right)\psi \tag{3.136}$$

with

$$S = -2(B + A^2 - A_x). \tag{3.137}$$

Then, computing the x-derivative of $\log \eta$, one gets the transformation

$$Z^{-1} = \chi^{-1} + A, \tag{3.138}$$

which means that the new expansion variable Z is related to χ by a homo-graphic transformation (as suggested in [98], formula (17)) such that in a neighborhood of $\chi = 0$, one has $Z \sim \chi$. Then the system (3.131)–(3.132) combined with the truncation in Z

$$u_T = a\partial_x \log Z + U \qquad (3.139)$$

(U a function of (x,t) and a constant) such that $u_T \sim a\chi^{-1}$ as $\chi \to 0$ extends the Weiss truncation to positive powers of Z. Solving the Painle-vé–Darboux equations associated with $E(u_T) = 0$, Pickering obtains for the MKdV equation (2.94) the following results:

$$A = U/a, \quad B = \lambda^2, \quad C = 2(U_x/a - (U/a)^2 + 2\lambda^2), \qquad (3.140)$$
$$D = 4\lambda^2 U_x/a, \quad \lambda = \lambda(t) \text{ an arbitrary integration function,} \qquad (3.141)$$

and the compatibility conditions (3.134)–(3.135) yield

$$X_1 \equiv -(2/a)(U_t + U_{xxx} - 2a^{-2}(U^3)_x) = 0, \qquad (3.142)$$
$$X_2 \equiv -(\lambda^2)_t = 0. \qquad (3.143)$$

From (3.137) and (3.140) one gets

$$S = 2(U_x/a - (U/a)^2 - \lambda^2), \qquad (3.144)$$
$$S - C + 6\lambda^2 = 0, \qquad (3.145)$$

the latter equation being the SM equation of the KdV equation.

Let us remark that the expressions of S and C in the function of the solution U of the MKdV equation and the constant parameter λ coincide with the relation (2.111) obtained previously. The expression (2.108) for Y with $\alpha - U/a$ implies the identification $Z = \lambda^{-1}Y$.

For the sine–Gordon equation (2.118), considering the truncation

$$u_T = 2i \log Z + U \qquad (3.146)$$

such that $u_T \sim 2i \log \chi$ as $\chi \to 0$, Pickering obtains

$$A = -(i/2)U_x, \quad B = \lambda^2, \quad D = -(i/2)\sin U, \quad \lambda \text{ arbitrary constant,} \qquad (3.147)$$
$$C = -\lambda^{-2}e^{iU}/4, \quad S = -iU_{xx} + (1/2)U_x^2 - 2\lambda^2, \quad sG(U) = 0, \qquad (3.148)$$
$$S + C_{xx}/C - (1/2)(C_x/C)^2 + 2\lambda^2 = 0. \qquad (3.149)$$

Again the expressions for S, C in the function of U and λ and equation (3.149) coincide with the relations (2.128) and (2.129). The identification $Z = \lambda^{-1}Y$ is also easy to obtain taking account that for sG, $\alpha = -iu_x/2$.

TABLE 8.3. Transformations of the dependent variable(s) conserving the equation(s), for the AKNS group PDEs (complex conjugation, phase shift, parity).

PDE	Transformation(s)
AKNS system	$(u, v, i) \mapsto (v, u, -i); \; \forall k : \; (u, v) \mapsto (ku, v/k)$
Sine-Gordon	$u \mapsto -u$
MKdV	$u \mapsto -u$
KdV	none

3.2.8   Weiss Method Plus Involutions

The AKNS system [2, 138]

$$E^{(1)} \equiv iu_t + p_r u_{xx} + q_r u^2 v = 0,$$
$$E^{(2)} \equiv -iv_t + p_r v_{xx} + q_r uv^2 = 0 \tag{3.150}$$

has the BT [22, 72, 81, 85]

$$a^2 = -2p_r/q_r, \quad R^2 = (u+U)(v+V)/a^2 - (\lambda - \mu)^2),$$
$$(u+U)_x = -(u-U)R - i(\lambda + \mu)(u+U),$$
$$(v+V)_x = -(v-V)R + i(\lambda + \mu)(v+V),$$
$$+ip_r^{-1}(u+U)_t = (u-U)_x R + (u+U)M + i(\lambda + \mu)(u+U)_x,$$
$$-ip_r^{-1}(v+V)_t = (v-V)_x R + (v+V)M - i(\lambda + \mu)(v+V)_x,$$
$$M = (uv + UV)/a^2, \tag{3.151}$$

with λ, μ arbitrary complex constants. Galilean invariance $(x, t, u, v) \mapsto (x - 2p_r ct, t, e^{i(cx - p_r c^2 t)} u, e^{-i(cx - p_r c^2 t)} v)$ allows one to choose $c = \lambda + \mu = 0$ [72].

The above BT cannot be found either by the one-SM method [133] or by the two-SM method [100], nor by the one-SM method plus homography [109]. The challenge of the Painlevé approach to find this BT by singularity analysis *only* is solved in [34] as follows.

As the one-SM method provides only a partial result $T(\chi, u, \lambda)$ for the truncation, one then considers all transformations on u conserving the equation $E(u) = 0$ in order to uncover a second solution U; see Table 8.3.

For the AKNS system (3.150), the one-family truncation

$$u = u_0 \chi^{-1} + u_1, \quad v = v_0 \chi^{-1} + v_1, \tag{3.152}$$

which has the general solution [100, 133] (λ arbitrary complex constant)

$$u = a(\chi^{-1} - f_x/(2f) - i\lambda)f, \tag{3.153}$$
$$v = a(\chi^{-1} + f_x/(2f) + i\lambda)/f, \tag{3.154}$$

$$f_x/f = -2i\lambda - (u/a)f^{-1} + (v/a)f, \tag{3.155}$$

$$ip_r^{-1}f_t/f = 2uv/a^2 + 4\lambda^2 + (u_x - 2i\lambda u)/(af),$$
$$+ (v_x + 2i\lambda v)f/a, \tag{3.156}$$

$$(f_{xt} - f_{tx})/f = (f^{-1}E^{(1)} + fE^{(2)})/a, \tag{3.157}$$

fails to introduce a second solution (U, V); see details in [100, Appendix C]. This is done by applying the two point transformations of Table 3.2 to the above truncation T_1 (3.153)–(3.156):

$$\left.\begin{array}{llllllll} T_1: & \chi_1 & u & v & i & f & \lambda & \text{(identity)} \\ T_2: & \chi_2 & v & u & -i & g & \mu & \text{(conjugation)} \\ T_3: & \chi_3 & kU & k^{-1}V & i & f & \lambda' & \text{(phase shift)} \\ T_4: & \chi_4 & k^{-1}V & kU & -i & g & \mu' & \text{(both)} \end{array}\right\} \tag{3.158}$$

These transformations act on (u, v, f, λ) as in Chen [22]. This is equivalent to successively processing the four families of the AKNS system by the one-SM method. In order that (u, v) and $(kU, V/k)$ be distinct, one must have $\lambda' = \mu$, $\mu' = \lambda$.

The four sets (3.153)–(3.154) define a system of eight equations in the eight unknowns $(\chi_1^{-1}, \chi_2^{-1}, \chi_3^{-1}, \chi_4^{-1}, u, v, kU, V/k)$. This system is linear with determinant $fg - 1/(fg)$, and it provides the DT straightforwardly (with the nonrestrictive choice $k = -1$):

$$u - U = 2a[\partial_x \log(g - 1/f) - i(\lambda + \mu)]/(g + 1/f), \tag{3.159}$$

$$v - V = 2a[\partial_x \log(f - 1/g) + i(\lambda + \mu)]/(f + 1/g), \tag{3.160}$$

$$u + U = 2ia(\lambda - \mu)/(g - 1/f), \tag{3.161}$$

$$v + V = 2ia(\lambda - \mu)/(f - 1/g) \tag{3.162}$$

(to stick to our definition, the DT comprises two equations, either (3.159)–(3.160) or (3.161)–(3.162)). The nonconstant factor of the logarithmic derivatives is similar to that of (P3), (P5), (P6) (see Section 7.1, Conte's chapter), while $\lambda + \mu$ is chosen as a real constant.

The Lax pair in its Riccati form comprises the four equations resulting from the action of T_3 and T_4 on (3.155)–(3.156). The BT comprises the four equations resulting from the elimination of the pseudopotentials (f, g) between the six equations defining the DT and the Lax pair, and these are precisely (3.151). This elimination is quite easy, since equations (3.161)–(3.162) are algebraic in (f, g):

$$f = ia(\lambda - \mu + R)/(v + V), \quad g = ia(\lambda - \mu + R)/(u + U). \tag{3.163}$$

Remark. The system [22] of two equations for (f, g), obtained by eliminating (u, v, U, V) between (3.159)–(3.162) and the PDE, is invariant under

TABLE 8.4. Reductions of the Darboux transformation of the AKNS system.

PDE	v, g, μ	χ^{-1}	$(u-U)/a$	$(u+U)/a$
NLS	$\bar{u}, f, \lambda$		(3.159)	$4(\mathrm{IM}\,\lambda)/(1/f - f)$
sG MKdV	$\varepsilon u, \varepsilon f, -\lambda$ $e^2 = \varepsilon$	$\dfrac{\lambda}{Y} - \dfrac{eU}{4a},$ $Y = \dfrac{ef - 1}{ef + 1}$	$(4/e)\dfrac{Y_x}{Y}$ $= \dfrac{2f_x}{\varepsilon f^2 - 1}$	$\dfrac{i\lambda}{e}\left(\dfrac{1}{Y} - Y\right)$ $= 4i\lambda\dfrac{f}{\varepsilon f^2 - 1}$
KdV	$1, \varepsilon/f, -\lambda$	$f - i\lambda$	$-2f_x$ $= -2(\chi^{-1})_x$	$2(f^2 - 2i\lambda f)$ $= 2(\chi^{-2} + \lambda^2)$

$(\lambda, \mu) \mapsto (\mu, \lambda)$. The elimination of g between this system provides the Broer-Kaup equation for $w = -i \log f$, a result also obtainable by the Weiss truncation [100]

$$p_r^{-1} w_{tt} + 4 w_x w_{xt} + 2 w_t w_{xx} + p_r(6 w_x^2 w_{xx} + w_{xxxx}) = 0. \qquad (3.164)$$

3.2.9 Reductions of the DT of AKNS System

The x-part of the AKNS spectral problem admits the three reductions $v = \bar{u}$, $v = \pm u$, $v = 1$, and the DT, obtained only from the x-part, must admit them. This is indeed the case: Equations (3.159)–(3.162) admit the two reductions $(v, V, g, \mu) = (\bar{u}, \overline{U}, \bar{f}, \bar{\lambda})$, $(\varepsilon u, \varepsilon U, \varepsilon f, -\lambda)$, $\varepsilon^2 = 1$, and one must add the case $g = \varepsilon/f$ when the determinant vanishes. Table 8.4 summarizes these reductions and the homographic link between f and the χ of the invariant analysis.

3.3 Cosgrove Classification for Semilinear PDEs of Second Order

In two papers [36, 37], Cosgrove classifies two cases of Painlevé-type semilinear PDEs of second order. The necessary conditions for the Painlevé property that he establishes combine the criteria of Painlevé and Gambier for ODEs and the WTC criteria for PDEs.

For **hyperbolic PDEs** in two independent variables of the type

$$u_{xt} = F(x, t, u, u_x, u_t) \qquad (3.165)$$

the equivalence classes are defined by the H-transformation

$$\tilde{u} = \frac{\alpha(x,t)u + \beta(x,t)}{\gamma(x,t)u + \delta(x,t)}, \qquad (3.166)$$

where

$$u = u(\tilde{x}, \tilde{t}), \quad \tilde{x} = X(x), \quad \tilde{t} = T(t), \quad \alpha\delta - \beta\gamma \neq 0, \qquad (3.167)$$

and the necessary conditions are

1. the dependence in u_x, u_t of F must be of the form

$$u_{xt} = A(x, t, u)u_x u_t + B(x, t, u)u_x + C(x, t, u)u_t + D(x, t, u); \quad (3.168)$$

2. as a function of u, the term A is the sum of at most three simple poles, at locations set to $u = 0$, 1, $H(x, t)$ (H an arbitrary function of x, t), while B, C, D cannot grow faster than, respectively, u, u, u^3;

3. the equation must pass the WTCK Painlevé test, i.e., all Fuchs indices are distinct integers and all positive indices are compatible, in order to guarantee the existence of local Laurent series in the Kruskal variable $\varphi(x, t) = x \pm f_1(t)$ or $\varphi(x, t) = t \pm f_2(x)$.

At the end, he obtains 22 canonical equations, reducible to

$$u_{xt} = \sin u \ \text{(sine--Gordon)} \quad \text{or} \quad u_{xt} = ae^{2u} + be^{-u} \ \text{(Tzitzéica)} \quad (3.169)$$

or linearizable by the *singular part* transformation.

For **parabolic PDEs** the equivalence class is a little bit larger than in the previous case in the sense that the new independent variables in the transformation (3.166) may be related to (x, t) as

$$\tilde{x} = X(x, t), \quad \tilde{t} = T(t). \quad (3.170)$$

The successive necessary conditions for having the Painlevé property yield the following results:

1. In two independent variables, the sole equation is

$$u_t + u_{xx} + 2uu_x = F(x, t), \quad F(x, t) \ \text{an arbitrary function}, \quad (3.171)$$

i.e., the Forsyth–Burgers equation is linearizable into the heat equation.

2. In more than two independent variables, only trivial soliton equations are obtained, i.e., nonlinear PDEs related to linear ones.

3.4 PDEs with Variable Coefficients

The Painlevé test can be applied to nonlinear PDEs with variable coefficients in order to determine the conditions under which the equation might be integrable. The sufficient part of the analysis entails the determination of the DT and the Lax pair, as well as the transformation that could relate the equation to its autonomous integrable counterpart. Many authors have considered the generalized KdV and NLS equations with variable coefficients due to their interest in many physical systems. The results of Brugarino [15] for the generalized variable coefficient KdV equation (VCKdV)

$$u_t + a(t)u + (b(x, t)u)_x + c(t)uu_x + d(t)u_{xxx} + e(x, t) = 0 \quad (3.172)$$

are

i. The equation passes the Painlevé test under the condition

$$b_t + (a - Lc)b + bb_x + db_{xxx} = 2ah + hL(d/c^2) + h' + ce$$
$$+ x \left(2a^2 + aL(d^3/c^4) + a' + L(d/c)L(d/c^2) + (L(d/c))'\right) \quad (3.173)$$

with $L = (d/dt)\log$ and $h(t)$ arbitrary.

ii. The solution of the Weiss truncation yields the DT and the Lax pair.

iii. With the transformation from (u, x, t) to (Θ, ξ, t),

$$u = ((a + L(d/c))\xi + g - b + \Theta)/c, \quad \xi = x - \int^t g(T)\, dT, \quad (3.174)$$

and the condition (3.173), he gets the equation

$$\Theta_t + \left(2a + L(d/c^2)\right)\Theta + \xi\left(a + L(d/c)\right)\Theta_\xi + \Theta\Theta_\xi$$
$$+ d\Theta_{\xi\xi\xi} = 0, \quad (3.175)$$

equivalent to [16, 60, 65] the KdV equation with constant coefficients. In the case $a = b = e \equiv 0$, the condition (3.173) becomes simply

$$d = c\left(K_1 \int^t c(T)\, dT + K_2\right), \quad K_1, K_2 \text{ arbitrary constants}, \quad (3.176)$$

i.e., the one given by Joshi [66] when performing the Painlevé test on this particular VCKdV equation. The same relation was obtained by Winternitz and Gazeau [136] using the symmetry group.

iv. Several equations of physical interest that satisfy (3.173) are presented.

Gagnon and Winternitz [52] have analyzed from the point of view of symmetries a variable coefficient nonlinear Schrödinger (VCNLS) equation

$$iu_t + f(x, t)u_{xx} + g(x, t)u|u|^2 + h(x, t)u = 0 \quad (3.177)$$

involving the three complex functions f, g, and h of the variables x, t. The symmetry group is shown to be five-dimensional iff the equation (3.177) is equivalent to NLS itself or to CGL3, and at most four-dimensional in all other cases. In this framework they give the allowed transformations of (3.177), in the case $f = 1 + if_2$, to the equation with constant coefficients

$$i\tilde{u}_{\tilde{t}} + (1 + if_2)\tilde{u}_{\tilde{x}\tilde{x}} + (\tilde{g}_1 + i\tilde{g}_2)\tilde{u}|\tilde{u}|^2 + (\tilde{h}_1 + i\tilde{h}_2)\tilde{u} = 0, \quad (3.178)$$

which are, in the special case $f_2 = 0$ (I, J, K, L arbitrary functions of t),

$$g = (\tilde{g}_1 + i\tilde{g}_2)\dot{T}I^{-2}, \tag{3.179}$$

$$\text{Re}\,h = (\dot{K} + 4K^2)x^2 + (\dot{L} + 4KL)x + \dot{J} + L^2 + \tilde{h}_1\dot{T}, \tag{3.180}$$

$$\text{IM}\,h = -\dot{I}I^{-1} - 2K + \tilde{h}_2\dot{T}, \tag{3.181}$$

$$\tilde{t} = T, \quad \dot{T} = T_0 e^{-8\int K dt}, \quad \tilde{x} = \sqrt{\dot{T}}x + \xi, \quad \dot{\xi} = -2\sqrt{\dot{T}}L, \tag{3.182}$$

$$u = \tilde{u}(\tilde{x}, \tilde{t})Ie^{i(Kx^2 + Lx + J)}. \tag{3.183}$$

One example given by the authors leading to the NLS equation ($\tilde{g}_2 = \tilde{h}_1 = \tilde{h}_2 = 0$) corresponds to the following choice of the arbitrary functions involved in the transformation:

$$\dot{I}I^{-1} = -2K, \quad J = L = 0, \quad \dot{K} + 4K^2 = K_0/4, \quad K_0 \text{ constant}, \tag{3.184}$$

and yields the VCNLS equation

$$iu_t + u_{xx} + \tilde{g}_1 T_0 \,\text{sech}(\sqrt{K_0}t)u|u|^2 + (K_0/4)x^2 u = 0. \tag{3.185}$$

This equation is related to NLS by the change of variables

$$\tilde{x} = \sqrt{T_0}x\,\text{sech}(\sqrt{K_0}t) + x_0, \quad \tilde{t} = T_0 \tanh(\sqrt{K_0}t)/\sqrt{K_0}, \tag{3.186}$$

$$u(x, t) = \tilde{u}(\tilde{x}, \tilde{t})\,\text{sech}^{1/2}(\sqrt{K_0}t)e^{(1/4)i\sqrt{K_0}x^2 \tanh(\sqrt{K_0}t)}. \tag{3.187}$$

Considering the other choice; $\dot{T}I^{-2} = 1$, one obtains

$$iu_t + u_{xx} + \tilde{g}_1 u|u|^2$$
$$+ \left((4K^2 + \dot{K})x^2 + (\dot{L} + 4KL)x + L^2 + \dot{J} + 2iK\right)u = 0, \tag{3.188}$$

equivalent to NLS by the transformation

$$\tilde{t} = T_0 \int^t dt' e^{-8\int^{t'} K ds}, \quad \tilde{x} = x\sqrt{T_0}e^{-4\int^t K dt'} + \xi,$$

$$\xi = \xi_0 - 2\sqrt{T_0} \int^t L(t')e^{-4\int^{t'} K(s)ds} dt', \tag{3.189}$$

$$u(x, t) = \tilde{u}(\tilde{x}, \tilde{t})e^{-4\int^t K dt'}e^{i(Kx^2 + Lx + J)}.$$

For $K(t) = -\beta(t)/2$, equation (3.188) and the transformation (3.189) coincide with the results of Clarkson [27] in the analysis of the PDE

$$iu_t + u_{xx} - 2u|u|^2 = a(x, t)u + b(x, t), \tag{3.190}$$

which passes the Painlevé test iff there exist functions $(\beta(t), \alpha_1(t), \alpha_0(t))$ such that

$$a = i\beta + x^2[\beta'/2 - \beta^2] + x\alpha_1 + \alpha_0, \quad b = 0. \tag{3.191}$$

As noticed by Clarkson, this result proves that the equation

$$iu_t + u_{xx} - 2u|u|^2 = \beta x^2 u, \quad \beta \text{ constant}, \quad \beta \neq 0, \qquad (3.192)$$

which does not satisfy the condition (3.191), is not integrable, while

$$iu_t + u_{xx} - 2u|u|^2 = (i\kappa - \kappa^2 x^2)u, \quad \kappa \text{ real}, \qquad (3.193)$$

can be transformed into NLS and hence is integrable.

4 Partially Integrable and Nonintegrable Equations

Let us consider a polynomial PDE (depending on a parameter μ) that does not pass the Painlevé test but possesses a singularity structure compatible with single-valued solutions. The reasons for the failure of the Painlevé test are the following:

i. The Fuchs indices are *all* integers (possibly for particular values of μ), but some of them do not satisfy the compatibility condition. Therefore, the existence of a Laurent series idependent on conditions on the SM.

ii. Whatever μ might be, some indices are irrational and thus unable to generate compatibility conditions. Therefore, the number of arbitrary functions that can be introduced in the Laurent series is lower than the order of the equation; this generally characterizes equations with chaotic behavior.

In the first case, we classify the equation as being partially integrable, while in the second case we classify it as being nonintegrable. We provide methods for finding particular closed-form solutions and illustrate them on some examples. In each case one looks for solutions related by a rational transformation to the general solutions of first-order nonlinear ODEs like Riccati, Weierstrass, or Jacobi possessing the PP, or solutions related by a nonlinear transformation like the logarithmic derivative to a second- or third-order linear system with constant coefficients.

4.1 Partially Integrable Equations

4.1.1 KPP Equation

The Kolmogorov, Petrovskii, and Piskunov equation (KPP) [47, 70, 102],

$$E \equiv u_t - u_{xx} + \frac{2}{d^2}(u - e_1)(u - e_2)(u - e_3) = 0, \quad e_j \text{ distinct}, \qquad (4.1)$$

possesses two opposite families of singularities:

$$u = d\chi^{-1} + s_1/3 - dC/6 + O(\chi), \quad \text{indices:} \ -1, 4, \quad , \mathcal{D} = d\partial_x, \qquad (4.2)$$

$$s_1 = e_1 + e_2 + e_3, \quad s_2 = e_2 e_3 + e_3 e_1 + e_1 e_2, \quad s_3 = e_1 e_2 e_3, \qquad (4.3)$$

with the condition on the SM coming from the index 4:

$$Q_4 \equiv C[-3(C_t + CC_x) + \prod_{k=1}^{3}(C + s_1 - 3e_k)] = 0. \qquad (4.4)$$

Denoting by (j, l, m) any permutation of $(1, 2, 3)$ and $k_i = (3e_i - s_1)/(3d)$, the one-family truncation $u = d\chi^{-1} + s_1/3 - dC/6$ yields the moving front solution

$$u = \frac{e_l + e_m}{2} + d\frac{k}{2} \tanh \frac{k}{2}(x - ct - x_0), \qquad (4.5)$$

$$k^2 = (k_l - k_m)^2, \quad c = -3(k_l + k_m)/2. \qquad (4.6)$$

The two-family truncation as for MKdV or sG yields the stationary pulse

$$u = e_1 + \frac{e_2 - e_3}{\sqrt{2}} \operatorname{sech} i\frac{e_2 - e_3}{d\sqrt{2}}(x - x_0), \quad 2e_1 - e_2 - e_3 = 0. \qquad (4.7)$$

A third very interesting solution representing the collision of two fronts [69] with different velocities can easily be found [32] with the assumption

$$u = \frac{s_1}{3} + u_0 \partial_x \log \psi, \qquad (4.8)$$

where ψ is the general solution of a linear system with constant coefficients,

$$\psi = \sum_{n=1}^{3} C_n \exp\left[k_n(x + b_2 t) + k_n^2 b_1 t\right], \quad C_n \text{ arbitrary}, \quad C_1 C_2 C_3 \neq 0,$$

$$\psi_{xxx} - a_1 \psi_x - a_2 \psi = 0, \quad \psi_t - b_1 \psi_{xx} - b_2 \psi_x = 0$$

with the following values of the coefficients and the constant k_n:

$$a_1 = (s_1^2 - 3s_2)/(3d^2), \quad a_2 = (2s_1^3 - 9s_1 s_2 + 27s_3)/(27d^3),$$

$$b_1 = -3, \quad b_2 = 0, \quad k_n = (3e_n - s_1)/(3d). \qquad (4.9)$$

4.2   Nonintegrable Equations

4.2.1   Kuramoto–Sivashinsky Equation

The Kuramoto–Sivashinsky (KS) equation

$$E \equiv u_t + uu_x + u_{xx} + u_{xxxx} = 0 \qquad (4.10)$$

describes, for instance, the fluctuation of the position of a flame front or the motion of a fluid going down a vertical wall or a spatially uniform oscillating chemical reaction in a homogeneous medium. For a review, see [88].

This equation possesses only one family of singularities: [30]

$$u = 120\chi^{-3} + 60(S + 1/19)\chi^{-1} + (C - 15S_x) + O(\chi), \qquad (4.11)$$

$$\text{indices: } -1, 6, (13 \pm i\sqrt{71})/2, \quad \mathcal{D} = 60\partial_x{}^3 + (60/19)\partial_x. \qquad (4.12)$$

Due to the existence of the two complex irrational Fuchs indices, the Laurent series depends only on two arbitrary functions (u_6 and the arbitrary function in the expansion variable χ) regardless of the SM. The WTC truncation

$$u = 120\chi^{-3} + 60(S + 1/19)\chi^{-1} + (C - 15S_x) \qquad (4.13)$$

generates three determining equations $E_4 = 0$, $E_5 = 0$, $E_7 = 0$, whose general solution is $C = $ arbitrary c, $S = -11/38, 1/38$. This corresponds to the well-known traveling-wave solution of Kuramoto and Tsuzuki [78] existing only for two values $k^2 = -1/19$ or $11/19$:

$$u = c + \left(\frac{30}{19}k - 15k^3\right)\tanh\frac{k}{2}(\xi - x_0) + 15k^3\tanh^3\frac{k}{2}(\xi - x_0), \quad (4.14)$$

where $\xi = x - ct$ and c, x_0 are arbitrary constants.

This solution can also be retrieved with the assumption that $u = c + \mathcal{D}\log\psi$ with ψ the general solution of a linear system with constant coefficients,

$$\psi_{xx} - (k^2/4)\psi = 0, quad \psi_t - c\psi_x = 0. \qquad (4.15)$$

Let us remark that the reduction $u(x,t) \mapsto c + U(x - ct)$ of the PDE (4.10) yields the nonintegrable ODE

$$U''' + U' + U^2/2 + K = 0, \quad K \text{ arbitrary}, \qquad (4.16)$$

for which we have found, in the case $K = -450k^2/19^2$, a Riccati subequation linearizable into the system (4.15). The challenge not yet solved is to find for *every* K a closed-form particular solution depending on one arbitrary constant.

Let us also mention the interesting work of Porubov [111, 112], who has found for a large class of nonlinear PDEs like

$$\eta_t + a_1\eta_x + a_2\eta\eta_x + a_3(\eta\eta_x)_x + a_4\eta_{xx} + a_5\eta_{xxx} + a_6\eta_{xxxx} = 0 \qquad (4.17)$$

and for some particular values of the constant parameters $\{a_i\}$, traveling-wave solutions in terms of Weierstrass elliptic functions or their logarithmic derivatives.

For the KS equation with an additional dispersive term

$$E \equiv u_t + uu_x + u_{xx} + bu_{xxx} + u_{xxxx} = 0, \qquad (4.18)$$

TABLE 8.5. Degeneracies of the known solutions of CGL3.

CGL3	NLS $(p_i = q_i = \gamma = 0)$	KPP $(p_r = q_r = 0)$
propagating hole [9, 32]	dark soliton (2.28)	
shock or front [32, 104]		front (4.5)
pulse or solitary wave [32, 107]		stationary pulse (4.7)
collision of two shocks $(p_r = 0)$ (4.26) [32, 103, 104]		collision of two fronts (4.8)

which belongs to the previous class (4.17), Kudryashov [74] has found a particular solution depending on three arbitrary constants in terms of the Weierstrass elliptic function and its derivative in the single case $b^2 = 16$. An easy way to find it is to assume the existence of a solution of the form

$$u = a_0 + a_2 \wp(x - ct - x_0, g_2, g_3) + a_3 \wp'(x - ct - x_0, g_2, g_3), \qquad (4.19)$$

compatible with the singularity structure of its Laurent series expansion. Enforcing in the lhs E the conditions $\wp'^2 = 4\wp^3 - g_2\wp - g_3$ and $\wp'' = 6\wp^2 - (g_2/2)$, one identifies to zero a polynomial of two variables $(\wp, \wp')$ of degree one in $\wp'$. This similarly generates six equations in the five unknowns a_j, g_2, g_3 and yields a nondegenerate elliptic solution only if $b^2 = 16$ and

$$a_0 = c - 4/b, \quad a_2 = -15b, \quad a_3 = -60, \quad g_2 = 1/12, \quad (g_3, c, x_0) \text{ arbitrary.}$$

4.2.2   Complex Ginzburg–Landau Equation CGL3

The one-dimensional cubic complex Ginzburg–Landau (CGL3) equation

$$E \equiv iu_t + pu_{xx} + q|u|^2 u - i\gamma u = 0, \ pq \neq 0, \ (u, p, q) \in \mathcal{C}, \quad \gamma \in \mathcal{R}, \quad (4.20)$$

with p, q, γ constants, describes pattern formation and coherent structures in many different domains: Taylor–Couette flows between coaxial rotating cylinders, wave propagation in optical fibers, and chemical reactions. For a review see [38]. This PDE is physically strongly connected to the KS equation, and it also possesses two irrational complex conjugate indices. As is the case for the AKNS system, the CGL3 equation possesses four families of singularities

$$u = A_0 \chi^{-1+i\alpha}(1 + A_1\chi + O(\chi^2)), \qquad (4.21)$$

$$\bar{u} = B_0 \chi^{-1-i\alpha}(1 + B_1\chi + O(\chi^2)), \qquad (4.22)$$

$$A_0 B_0 = 3|p^2|\alpha/D_i, \quad \alpha = D_r \pm \sqrt{D_r^2 + 8D_i^2/9}, \ D_i = p_r q_i - p_i q_r, \qquad (4.23)$$

$$D_r = (p_r q_r + p_i q_i), \quad \text{Fuchs indices:} \ -1, 0, \frac{7}{2} \pm \sqrt{1 - 24\alpha^2}/2. \qquad (4.24)$$

The important information we get from the singularity analysis is that neither $(u, \bar{u})$, nor $(|u|, \arg u)$ nor $(\operatorname{Re} u, \operatorname{IM} u)$ has a simple singularity structure. In this framework, better variables are (Z, θ) defined as

$$u = Ze^{i\theta}, \quad \theta = \alpha \log \psi + \theta_0, \quad \psi/\psi_x = \chi, \tag{4.25}$$

where θ_0 is an arbitrary function representing the index 0. Then Z and $\operatorname{grad}\theta$ behave like simple poles, so that the usual methods are applicable.

With ψ the general solution of a second-order or third-order system and with a one-family or two-family truncation for (Z, θ) one obtains [32] *all* the known closed-form solutions of this equation. Among them, a very interesting solution representing a "collision of two shocks" [103, 104]

$$u = A_0 \frac{k}{2} \frac{\sinh kx/2}{\cosh kx/2 + e^{-3\gamma(t-t_0)/2}} e^{i[\alpha \log(1+e^{3\gamma(t-t_0)/2} \cosh kx/2)]},$$
$$k^2 = -2\gamma/p_i, \quad p_r = 0, \tag{4.26}$$

is easily found by assuming that ψ satisfies a third-order linear system. For $q_r = 0$ (hence $\alpha = 0$) this solution degenerates to the "collision of two fronts" solution of KPP previously considered. Table 8.5 summarizes the known solutions of CGL3 and their degeneracies to the NLS and KPP equations. An open problem is to find the solution of CGL3 that degenerates for $p_i = q_i = \gamma = 0$ to the bright soliton (2.27) of NLS.

5 References

[1] M.J. Ablowitz and P.A. Clarkson, *Solitons, nonlinear evolution equations and inverse scattering* (Cambridge University Press, Cambridge, 1991).

[2] M.J. Ablowitz, D.J. Kaup, A.C. Newell, and H. Segur, *The inverse scattering transform-Fourier analysis for nonlinear problems*, Stud. Appl. Math. **53** (1974), 249–315.

[3] M.J. Ablowitz, A. Ramani, and H. Segur, *Nonlinear evolution equations and ordinary differential equations of Painlevé type*, Lett. Nuov. Cimento **23** (1978), 333–338.

[4] M.J. Ablowitz, A. Ramani, and H. Segur, *A connection between nonlinear evolution equations and ordinary differential equations of P-type. I*, J. Math. Phys. **21** (1980), 715–721; II, **21** (1980), 1006–1015.

[5] M.J. Ablowitz and H. Segur, *Exact linearization of a Painlevé transcendent*, Phys. Rev. Lett. **38** (1977), 1103–1106.

[6] M.J. Ablowitz and H. Segur, *Solitons and the Inverse Scattering Transform* (SIAM, Philadelphia, 1981).

[7] R. Anderson, J. Harnad, and P. Winternitz, *Systems of ordinary differential equations with nonlinear superposition principles*, Physica D **4** (1982), 164–182.

[8] A.V. Bäcklund, *Om ytor med konstant negativ krökning*, Lunds Universitets Arsskrift Avd. 2 **19** (1883), with an abstract in French.

[9] N. Bekki and K. Nozaki, *Formations of spatial patterns and holes in the generalized Ginzburg–Landau equation*, Phys. Lett. A **110** (1985), 133–135.

[10] T.B. Benjamin, J.L. Bona, and J.J. Mahoney, *Model equations for long waves in nonlinear dispersive systems*, Phil. Trans. Roy. Soc. A **272** (1972), 47–78.

[11] M. Boiti and F. Pempinelli, *Nonlinear Schrödinger equation, Bäcklund transformations and Painlevé transcendents*, Il Nuovo Cimento B **59** (1980), 40–58.

[12] M. Boiti and Tu G.-Z., *Bäcklund transformations via gauge transformations*, Il Nuovo Cimento B **71** (1982), 253–264.

[13] M. Boiti, C. Laddomada, F. Pempinelli, and Tu G.Z., *Bäcklund transformations related to the Kaup–Newell spectral problem*, Physica D **9**, 425–432.

[14] J. Boussinesq, *Théorie des ondes qui se propagent le long d'un canal rectangulaire horizontal, en communiquant au liquide contenu dans ce canal des vitesses sensiblement pareilles de la surface du fond*, J. de math. pures et appliquées **17** (1872), 55–108.

[15] T. Brugarino, *Painlevé property, auto-Bäcklund transformation, Lax pairs, and reduction to the standard form for the Korteweg–de Vries equation with nonuniformities*, J. Math. Phys. **30** (1989), 1013–1015.

[16] T. Brugarino and P. Pantano, *The integration of Burgers and Korteweg–de Vries equations with nonuniformities*, Phys. Lett. A **80** (1980), 223–224.

[17] F.J. Bureau, *Differential equations with fixed critical points*, Annali di Matematica pura ed applicata **LXVI** (1964), 1–116 [abbreviated as M.II].

[18] F.J. Bureau, *Équations différentielles du second ordre en Y et du second degré en Ÿ dont l'intégrale générale est à points critiques fixes*, Annali di Matematica pura ed applicata **XCI** (1972), 163–281 [abbreviated as M.III].

[19] F. Calogero and S. de Lillo, *The Eckhaus PDE* $i\psi_t + \psi_{xx} + 2(|\psi|^2)_x\psi + |\psi|^4\psi = 0$, Inverse problems **3** (1987), 633–681; **4** (1988), 571.

[20] P.J. Caudrey, R.K. Dodd, and J.D. Gibbon, *A new hierarchy of Korteweg-de Vries equations*, Proc. Roy. Soc. London A **351** (1976), 407–422.

[21] J. Chazy, *Sur les équations différentielles du troisième ordre et d'ordre supérieur dont l'intégrale générale a ses points critiques fixes*, Thèse, Paris (1910); Acta Math. **34** (1911), 317–385. Table des matières commentée avec index, R. Conte (1991), 6 pages.

[22] H.H. Chen, *General derivation of Bäcklund transformations from inverse scattering problems*, Phys. Rev. Lett. **33** (1974), 925–928.

[23] M.J. Clairin, *Sur quelques équations aux dérivées partielles du second ordre*, Ann. Toulouse **5** (1903), 437–458.

[24] P.A. Clarkson, J.B. McLeod, P.J. Olver, and A. Ramani, *Integrability of Klein-Gordon equations*, Siam J. Math. Anal. **17** (1986), 798–802.

[25] P.A. Clarkson and C.M. Cosgrove, *The Painlevé property and a generalised derivative nonlinear Schrödinger equation*, J. Phys. A **20** (1987), 2003–2024.

[26] P.A. Clarkson, A.S. Fokas, and M.J. Ablowitz, *Hodograph transformations on linearizable partial differential equations*, SIAM J. Appl. Math. **49** (1989), 1188–1209.

[27] P. Clarkson, *Painlevé analysis of the damped, driven nonlinear Schrödinger equation*, Proc. Roy. Soc. Edin. **109A** (1988), 109–126.

[28] R. Conte, *Invariant Painlevé analysis of partial differential equations*, Phys. Lett. A **140** (1989), 383–390.

[29] R. Conte, *Painlevé singular manifold equation and integrability*, Inverse methods in action, ed. P.C. Sabatier, Springer series "Inverse problems and theoretical imaging" (Springer, Berlin, 1990), pages 497–504.

[30] R. Conte and M. Musette, *Painlevé analysis and Bäcklund transformation in the Kuramoto-Sivashinsky equation*, J. Phys. A **22** (1989), 169–177.

[31] R. Conte and M. Musette, *Link between solitary waves and projective Riccati equations*, J. Phys. A **25** (1992), 5609–5623.

[32] R. Conte and M. Musette, *Linearity inside nonlinearity: exact solutions to the complex Ginzburg-Landau equation*, Physica D **69** (1993), 1–17.

[33] R. Conte and M. Musette, *Exact solutions to the partially integrable Eckhaus equation*, Teor. i Mat. Fiz. **99** (1994), 226–233; Theor. and Math. Phys. **99** (1994), 543–548.

[34] R. Conte and M. Musette, *Beyond the two-singular manifold method*, Nonlinear physics: theory and experiment, eds. E. Alfinito, M. Boiti, L. Martina and F. Pempinelli (World Scientific, Singapore, 1996), pages 67–74.

[35] R. Conte, M. Musette, and A.M. Grundland, *Bäcklund transformation of partial differential equations from the Painlevé–Gambier classification, II. Tzitzéica equation*, J. Math. Phys. **40** (1999), 2092–2106.

[36] C.M. Cosgrove, *Painlevé classification of all semilinear partial differential equations of the second-order I. Hyperbolic equations in two independent variables*, Stud. Appl. Math. **89** (1993), 1–61.

[37] C.M. Cosgrove, *Painlevé classification of all semilinear partial differential equations of the second-order II. Parabolic and higher dimensional equations*, Stud. Appl. Math. **89** (1993), 95–151.

[38] M.C. Cross and P.C. Hohenberg, *Pattern formation outside of equilibrium*, Rev. Mod. Phys. **65** (1993), 851–1112.

[39] M.M. Crum, *Associated Sturm–Liouville systems*, Quart. J. Math. Oxford **6** (1955), 121–127.

[40] G. Darboux, *Sur une proposition relative aux équations linéaires*, C. R. Acad. Sc. Paris **94** (1882), 1456–1459.

[41] G. Darboux, *Leçons sur la théorie générale des surfaces et les applications géométriques du calcul infinitésimal*, vol. III (Gauthier-Villars, Paris, 1894). Reprinted, *Théorie générale des surfaces* (Chelsea, New York, 1972). Reprinted (Gabay, Paris, 1993).

[42] R.K. Dodd and R.K. Bullough, *Polynomial conserved densities for the sine-Gordon equations*, Proc. Roy. Soc. London A **352** (1977), 481–503.

[43] R.K. Dodd and J.D. Gibbon, *The prolongation structure of a higher-order Korteweg-de Vries equation*, Proc. Roy. Soc. London A **358** (1977), 287–296

[44] P.G. Drazin and R.S. Johnson, *Solitons: An Introduction* (Cambridge University Press, Cambridge, 1989).

[45] F.B. Estabrook and H.D. Wahlquist, J. Math. Phys. **17** (1976), 1293–1297.

[46] A.S. Fokas and Q.-M. Liu, *Nonlinear interaction of traveling waves of nonintegrable equations*, Phys. Rev. Lett. **72** (1994), 3293–3296.

[47] R. FitzHugh, *Impulses and physiological states in theoretical models of nerve membrane*, Biophys. J. **1** (1961), 445–466.

[48] A.P. Fordy and J. Gibbons, *Factorization of operators I. Miura transformations*, J. Math. Phys. **21** (1980), 2508–2510.

[49] A.P. Fordy and J. Gibbons, *Factorization of operators. II*, J. Math. Phys. **22** (1981), 1170–1175.

[50] A.P. Fordy, *The Hénon–Heiles system revisited*, Physica D **52** (1991), 204–210.

[51] A.R. Forsyth 1906, *Theory of differential equations.* Part IV (vol. VI) Partial differential equations (Cambridge University Press, Cambridge, 1906). Reprinted (Dover, New York, 1959).

[52] L. Gagnon and P. Winternitz, *Symmetry classes of variable coefficient nonlinear Schrödinger equations*, J. Phys. A **26** (1993), 7061–7076.

[53] B. Gambier, Sur les équations différentielles du second ordre et du premier degré dont l'intégrale générale est à points critiques fixes, Thèse, Paris (1909); Acta Math. **33** (1910), 1–55.

[54] C.S. Gardner, J.M. Greene, M.D. Kruskal, and R.M. Miura, *Method for solving the Korteweg–de Vries equation*, Phys. Rev. Lett. **19** (1967), 1095–1097.

[55] E. Goursat, *Cours d'analyse mathématique* (Gauthier-Villars, Paris, 1924). English translation, *A course in Mathematical Analysis* (Dover, New York, 1956).

[56] E. Goursat, *Le problème de Bäcklund*, Mémorial des sciences mathématiques, fascicule 6 (Gauthier-Villars, Paris, 1925). (This volume contains an extensive bibliography of early papers on Bäcklund transformations.)

[57] S.E. Harris, *Conservation laws for a nonlinear wave equation*, Nonlinearity **9** (1996), 187–208.

[58] R. Hirota, *Exact solution of the Korteweg–de Vries equation for multiple collisions of solitons*, Phys. Rev. Lett. **27** (1971), 1192–1194.

[59] R. Hirota, *A new form of Bäcklund transformation and its relation to the inverse scattering*, Prog. Theor. Phys. **52** (1974), 1498-1512.

[60] R. Hirota, *Exact solutions to the equation describing "cylindrical solitons"*, Phys. Lett. A **71** (1979), 393–394.

[61] R. Hirota, *Direct methods in soliton theory*, Solitons, eds. R.K. Bullough and P.J. Caudrey, Springer Topics in Current Physics (Springer, Berlin, 1980), pages 157–176.

[62] R. Hirota and J. Satsuma, *N-soliton solutions of model equations for shallow water waves*, J. Phys. Soc. Japan Letters **40** (1976), 611–612.

[63] R. Hirota and J. Satsuma, *Nonlinear evolution equations generated from the Bäcklund transformation for the Boussinesq equation*, Prog. Theor. Phys. **57** (1977), 797–807.

[64] R. Hirota and J. Satsuma, *A simple structure of superposition formula of the Bäcklund transformation*, J. Phys. Soc. Japan **45** (1978), 1741–1750.

[65] L. Hlavatý, *The Painlevé analysis of damped KdV equation*, J. Phys. Soc. Japan **55** (1986), 1405–1406.

[66] N. Joshi, *Painlevé property of general variable-coefficient versions of the Korteweg-de Vries and non-linear Schrödinger equations*, Phys. Lett. A **125** (1987), 456–460.

[67] M. Jimbo, M.D. Kruskal, and T.Miwa, *Painlevé test for the self-dual Yang-Mills equation*, Phys. Lett. A **92** (1982), 59–60.

[68] D.J. Kaup, *On the inverse scattering problem for cubic eigenvalue problems of the class $\psi_{xxx} + 6Q\psi_x + 6R\psi = \lambda\psi$*, Stud. Appl. Math. **62** (1980), 189–216.

[69] T. Kawahara and M. Tanaka, *Interactions of traveling fronts: an exact solution of a nonlinear diffusion equation*, Phys. Lett. A **97** (1983), 311–314.

[70] A.N. Kolmogorov, I.G. Petrovskii, and N.S. Piskunov, *The study of a diffusion equation, related to the increase of the quantity of matter, and its application to one biological problem*, Bulletin de l'Université d'État de Moscou, série internationale, section A Math. Méc. **1** (1937), 1–26.

[71] M.D. Kruskal, R.M. Miura, C.S. Gardner, and N.J. Zabusky, *Korteweg-de Vries equation and generalisations. V. Uniqueness and nonexistence of polynomial conservation laws*, J. Math. Phys. **11** (1970), 952–960.

[72] K. Konno and M. Wadati, *Simple derivation of Bäcklund transformation from Riccati form of inverse method*, Prog. Theor. Phys. **53** (1975), 1652–1656.

[73] D.J. Korteweg and G. de Vries, *On the change of form of long waves advancing in a rectangular canal, and on a new type of long stationary waves*, Phil. Mag. **39** (1895), 422–443.

[74] N.A. Kudryashov, *On types of nonlinear nonintegrable equations with exact solutions*, Phys. Lett. A **155** (1991), 269–275.

[75] A. Kundu, *Landau–Lifshitz and higher-order nonlinear systems gauge generated from nonlinear Schrödinger type equations*, J. Math. Phys. **25** (1984), 3433–3438.

[76] A. Kundu, *Exact solutions to higher-order nonlinear equations through gauge transformation*, Physica D **25** (1987), 399–406.

[77] A. Kundu, *Comments on the Eckhaus PDE* $i\psi_t + \psi_{xx} + 2(|\psi|^2)_x\psi + |\psi|^4\psi = 0$, Inverse Problems **4** (1988), 1143–1144.

[78] Y. Kuramoto and T. Tsuzuki, *Persistent propagation of concentration waves in dissipative media far from thermal equilibrium*, Prog. Theor. Phys. **55** (1976), 356–369.

[79] G.L. Lamb Jr., *Propagation of ultrashort optical pulses*, Phys. Lett. A **25** (1967), 181–182.

[80] G.L. Lamb Jr., *Analytical descriptions of ultrashort optical pulse propagation in a resonant medium*, Rev. Mod. Phys. **43** (1971), 99–124.

[81] G.L. Lamb Jr., *Bäcklund transformations for certain nonlinear evolution equations*, J. Math. Phys. **15** (1974), 2157–2165.

[82] P.D. Lax, *Integrals of nonlinear equations of evolution and solitary waves*, Comm. Pure Appl. Math. **21** (1968), 467–490.

[83] D. Levi, O. Ragnisco, and A. Sym, *Bäcklund transformations vs. the dressing method*, Lett. Nuov. Cimento **33** (1982), 401–406.

[84] D. Levi and O. Ragnisco, *Bäcklund transformations for chiral field equations*, Phys. Lett. A **87**, 381–384.

[85] D. Levi, O. Ragnisco, and A. Sym, *Dressing method vs. classical Darboux transformation*, Il Nuovo Cimento B **83** (1984), 34–41.

[86] D. Levi and O. Ragnisco, *Nonisospectral deformations and Darboux transformations for third-order spectral problem*, Inverse Problems **4** (1988), 815–828.

[87] Q.M. Liu and A.S. Fokas, *Exact interaction of solitary waves for certain nonintegrable equations*, J. Math. Phys. **37** (1996), 324–345.

[88] P. Manneville, *The Kuramoto-Sivashinsky equation: a progress report*, Propagation in systems far from equilibrium, eds. J. Weisfreid, H.R. Brand, P. Manneville, G. Albinet, and N. Boccara (Springer, Berlin, 1988), pages 265–280.

[89] J.B. McLeod and P.J. Olver, *The connection between partial differential equations soluble by inverse scattering and ordinary differential equations of Painlevé type*, SIAM J. Math. Anal. **14** (1983), 488–506.

[90] Y. Matsuno, *Bilinear Transformation Method* (Academic Press, London, 1984).

[91] V.B. Matveev and M.A. Salle, *Darboux Transformations and Solitons*, Springer Series in Nonlinear Dynamics (Springer-Verlag, 1991).

[92] A.V. Mikhailov, *Integrability of a two-dimensional generalization of the Toda chain*, Pis'ma Zh. Eksp. Teor. Fiz. **30** (1979), 443–448; Soviet Physics JETP Letters **30** (1979), 414–418.

[93] A.V. Mikhailov, *The reduction problem and the inverse scattering method*, Physica D **3** (1981), 73–117.

[94] R.M. Miura, C.S. Gardner, and M.D. Kruskal, *Korteweg-de Vries equation and generalisations. II. Existence of conservation laws and constants of motion*, J. Math. Phys. **9** (1968), 1204–1209.

[95] H.C. Morris, *Prolongation structures and a generalized inverse scattering problem*, J. Math. Phys. **17** (1976), 1867–1869.

[96] Th.-F. Moutard, *Note sur les équations différentielles linéaires du second ordre*, C. R. Acad. Sc. Paris **80** (1875), 729–733, Journal de l'École Polytechnique **45** (1878), 1–11.

[97] M. Musette, *Insertion of the Darboux transformation in the invariant Painlevé analysis of nonlinear partial differential equations*, Painlevé Transcendents, Their Asymptotics and Physical Applications, eds. D. Levi and P. Winternitz (Plenum Publishing Corp., New York, 1992), pages 197–209.

[98] M. Musette and R. Conte, *Algorithmic method for deriving Lax pairs from the invariant Painlevé analysis of nonlinear partial differential equations*, J. Math. Phys. **32** (1991), 1450–1457.

[99] M. Musette and R. Conte, *Solitary waves and Lax pairs from polynomial expansions of nonlinear differential equations*, Nonlinear evolution equations and dynamical systems, 161–170, eds. M.Boiti, L. Martina, and F. Pempinelli (World Scientific, Singapore, 1992).

[100] M. Musette and R. Conte, *The two-singular manifold method, I. Modified KdV and sine-Gordon equations*, J. Phys. A **27** (1994), 3895–3913.

[101] M. Musette and R. Conte, *Bäcklund transformation of partial differential equations from the Painlevé-Gambier classification, I. Kaup-Kupershmidt equation*, J. Math. Phys. **39** (1998), 5617–5630.

[102] A.C. Newell and J.A. Whitehead, *Finite bandwidth, finite amplitude convection*, J. Fluid Mech. **38** (1969), 279–303.

[103] K. Nozaki and N. Bekki, *Pattern selection and spatiotemporal transition to chaos in the Ginzburg-Landau equation*, Phys. Rev. Lett. **51** (1983), 2171–2174.

[104] K. Nozaki and N. Bekki, *Exact solutions of the generalized Ginzburg-Landau equation*, J. Phys. Soc. Japan **53** (1984), 1581–1582.

[105] P.O. Olver, *Euler operators and conservation laws of the BBM equation*, Math. Proc. Camb. Phil. Soc. **85** (1979), 143–160.

[106] P. Painlevé, *Sur les équations différentielles du second ordre à points critiques fixes*, C. R. Acad. Sc. Paris **143** (1906), 1111–1117.

[107] N.R. Pereira and L. Stenflo, *Nonlinear Schrödinger equation including growth and damping*, Phys. Fluids **20** (1977), 1733–1743.

[108] A. Pickering, *A new truncation in Painlevé analysis*, J. Phys. A **26** (1993), 4395–4405.

[109] A. Pickering, *The singular manifold method revisited*, J. Math. Phys. **37** (1996), 1894–1927.

[110] D.W. McLaughlin and A.C. Scott, *A restricted Bäcklund transformation*, J. Math. Phys. **14** (1973), 1817–1828.

[111] A.V. Porubov, *Exact travelling wave solutions of nonlinear evolution equation of surface waves in a convecting fluid*, J. Phys. A **26** (1993), L797–L800.

[112] A.V. Porubov, *Periodical solution to the nonlinear dissipative equation for surface waves in a convecting liquid layer*, Phys. Lett. A **211** (1996), 391–394.

[113] G.R.W. Quispel, F.W. Nijhoff, and H.W. Capel, *Linearization of the Boussinesq equation and the modified equation*, Phys. Lett. A **91** (1982), 143–145.

[114] Lord Rayleigh, *On waves*, Phil. Mag. **1** (1876), 257–279.

[115] C. Rogers and W.F. Shadwick, *Bäcklund Transformations and Their Applications* (Academic Press, New York, 1982).

[116] C. Rogers and S. Carillo, *On reciprocal properties of the Caudrey-Dodd-Gibbon and Kaup-Kupershmidt hierarchies*, Physica Scripta **36** (1987), 865–869.

[117] J.S. Russell, *Report on waves*, Rep. 14th Meet. Brit. Assoc. Adv. Sci., York, 311–390 (London, John Murray, 1844).

[118] J. Satsuma and D.J. Kaup, *A Bäcklund transformation for a higher-order Korteweg-de Vries equation*, J. Phys. Soc. Japan **43** (1977), 692–697.

[119] K. Sawada and T. Kotera, *A method for finding N-soliton solutions of the K.d.V. equation and K.d.V.-like equation*, Prog. Theor. Phys. **51** (1974), 1355–1367.

[120] A.C. Scott, F.Y. Chu and D.W. McLaughlin, *The Soliton: A New Concept in Applied Science*, Proc. IEEE **61** (1973), 1443–1483.

[121] M. Tajiri, *Similarity reductions of the one and two-dimensional nonlinear Schrödinger equations*, J. Phys. Soc. Japan **52** (1983), 1908–1917.

[122] G. Tzitzéica, *Sur une nouvelle classe de surfaces*, Rendiconti del Circolo Matematico di Palermo **25** (1908), 180–187; **28** (1909), 210–216.

[123] G. Tzitzéica, *Sur une nouvelle classe de surfaces*, C. R. Acad. Sc. Paris **150** (1910), 955–956, **150** (1910), 1227–1229.

[124] F. Ursell, *The long-wave paradox in the theory of gravity waves*, Proc. Camb. Philos. Soc. **49** (1953), 685–694.

[125] M. Wadati, *Bäcklund transformation for solutions of the Modified Korteweg-de Vries equation*, J. Phys. Soc. Japan **36** (1979), 1498.

[126] M. Wadati, K. Konno, and Y.-H. Ichikawa, *A generalization of the inverse scattering method*, J. Phys. Soc. Japan **46** (1979), 1965–1966.

[127] M. Wadati, H. Sanuki, and K. Konno, *Relationships among inverse method, Bäcklund transformation and an infinite number of conservation laws*, Prog. Theor. Phys. **53** (1975), 419–436.

[128] H.D. Wahlquist and F.B. Estabrook, *Bäcklund transformation for solutions of the Korteweg-de Vries equation*, Phys. Rev. Lett. **31** (1973), 1386–1390.

[129] H.D. Wahlquist and F.B. Estabrook, *Prolongation structures of nonlinear evolution equations*, J. Math. Phys. **16** (1975), 1–7.

[130] J. Weiss, *The Painlevé property for partial differential equations. II: Bäcklund transformation, Lax pairs, and the Schwarzian derivative*, J. Math. Phys. **24** (1983), 1405–1413.

[131] J. Weiss, *On classes of integrable systems and the Painlevé property*, J. Math. Phys. **25** (1984), 13–24.

[132] J. Weiss, *The sine–Gordon equations: Complete and partial integrability*, J. Math. Phys. **25** (1984), 2226–2235.

[133] J. Weiss, *The Painlevé property and Bäcklund transformations for the sequence of Boussinesq equations*, J. Math. Phys. **26** (1985), 258–269.

[134] J. Weiss, M. Tabor, and G. Carnevale, *The Painlevé property for partial differential equations*, J. Math. Phys. **24** (1983), 522–526.

[135] G.B. Whitham, *Nonlinear dispersive waves*, Proc. Roy. Soc. London A, **283** (1965), 238–261.

[136] P. Winternitz and J.-P. Gazeau, *Allowed transformations and symmetry classes of variable coefficient Korteweg–de Vries equations*, Phys. Lett. A **167** (1992), 246–250.

[137] N. Zabusky and M.D. Kruskal, *Interaction of solitons in a collisionless plasma and the recurrence of initial states*, Phys. Rev. Lett. **15** (1965), 240–243.

[138] V.E. Zakharov and A.B. Shabat, *Exact theory of two-dimensional self-focusing and one-dimensional self-modulation of waves in nonlinear media*, Zh. Eksp. Teor. Fiz. **61** (1971), 118–134; Soviet physics JETP **34** (1972), 62–69.

[139] V.E. Zakharov and A.B. Shabat, *Interaction between solitons in a stable medium*, Zh. Eksp. Teor. Fiz. **64** (1973), 1627–1639; Soviet Physics JETP **37** (1973), 823–828.

[140] V.E. Zakharov, *On stochastization of one-dimensional chains of nonlinear oscillators*, Zh. Eksp. Teor. Fiz. **65** (1973), 219–225; Soviet Physics JETP **38** (1974), 108–110.

[141] V.E. Zakharov and A.B. Shabat, *A scheme for integrating the nonlinear equations of mathematical physics by the method of the inverse scattering problem. I*, Funktsional'nyi Analiz i Ego Prilozheniya **8** (1974), 43–53; Func. Anal. Appl. **8** (1974), 226–235.

9

On Painlevé and Darboux–Halphen-Type Equations

Mark J. Ablowitz, Sarby Chakravarty, and Rod Halburd

ABSTRACT It is now well known that a deep connection exists between soliton equations and ODEs of Painlevé type. As a consequence there has been a significant reemergence of interest in the study of such ODEs and related issues. In this paper we demonstrate that a novel class of nonlinear ODEs, Darboux–Halphen (DH) type systems, can be obtained as reductions of the self-dual Yang–Mills (SDYM) equations. We show how to find by reduction from SDYM the associated linear pair for DH. This linear system is found to be monodromy evolving, which is different from the linear systems associated with the Painlevé equations, which are isomonodromy. The solution of the DH system can be obtained in terms of Schwarzian equations, which are themselves linearizable. The DH system has solutions that are related to Painlevé equations, but the solutions can have complicated analytic singularities such as natural boundaries and dense branching.

1 Introduction

This paper emanates from the lectures one of us (MJA) gave at a meeting in Cargèse, Corsica, organized during the summer of 1996 focusing on Painlevé equations and related issues. The meeting was held partly in commemoration of P. Painlevé's work one century ago. Painlevé and his school were well known at the end of the nineteenth century and early twentieth century. But by the 1970s many of the results, although published in leading journals, were not known or appreciated by most mathematicians and physicists. The reader may wish to review Painlevé's collected works [17] for an in-depth discussion of his contributions and point of view.

However, this situation changed dramatically with the recognition that equations of "Painlevé type" were intimately connected with a class of integrable systems, i.e., soliton and related equations. Namely, it was demonstrated that ordinary differential equations (ODEs) obtained as reductions of the well-known soliton equations yielded ODEs with the *Painlevé Property*. Namely, the solutions of the resulting ODEs were free of movable branch points. Moreover, similarity reductions of the best-known soliton equations often resulted in one of the classical second-order Painlevé equations. Background information and a description of the research involving Painlevé equations and their relationship to integrable soliton systems can be found in [2,4]. Ever since the discovery that such integrable systems and Painlevé equations were deeply connected there has been a major research effort, which has involved many aspects of Painlevé equations. The scope of the work and results are far too numerous to discuss here.

In this paper we will:

i. Describe, by example, some of the salient features of the connection of integrable systems and Painlevé equations. Notably, we discuss the connection of the inverse scattering transform (IST) to Painlevé equations and how one can obtain the linearization and solutions from this connection.

ii. Discuss a generalized Darboux–Halphen (DH) system, which is a fifth-order ODE and which reduces to the classical third-order DH system first studied by Darboux in 1879 [11]. This equation was discussed in a recent letter [5]. This system, which is a reduction of the self-dual Yang–Mills equations (SDYM), has a compatible linear monodromy evolving system. The reason for this can be traced to the fact that the reduction from SDYM involves an infinite-dimensional algebra: $\mathfrak{sdiff}(3)$.

iii. Show how, for the first time, one can use the infinite-dimensional algebra in SDYM to deduce the compatible linear monodromy system for DH.

iv. Quote the main results of the monodromy analysis and demonstrate that although the generalized DH system is linearizable and hence integrable, nevertheless, generically speaking, the solution is densely branched. Only when a constant of the motion takes on a denumerably infinite set of values does the solution become single-valued. In this case the solution is expressible in terms of automorphic functions. Indeed, we note that in the special case of the classical third-order DH system the solution is always single-valued. However, in this case the solution contains natural boundaries in the complex plane.

2   Painlevé Equations and IST

In the Cargèse lectures the importance of similarity reductions of PDEs was reviewed. Asymptotic analysis of Fourier integrals shows that the long-time behavior of linear evolution equations with constant coefficients yields the fact that self-similar (i.e., similarity) solutions are leading-order asymptotic states in certain regions of space (e.g., [4]). In the well-known soliton systems a similar situation arises. For example, associated with the modified Korteweg–de Vries (mKdV) equation,

$$u_t - 6u^2 u_x + u_{xxx} = 0, \tag{2.1}$$

is the self-similar reduction

$$u(x,t) = w(x/(3t)^{1/3})/(3t)^{1/3}, \tag{2.2}$$

which upon substitution into equation (2.1) yields, after an integration, the second (P_{II}) of the six classical Painlevé equations,

$$w'' - zw - 2w^3 = \alpha, \tag{2.3}$$

where α is an arbitrary constant.

Ablowitz and Segur (see, e.g., [4]) showed that equation (2.3) with $\alpha = 0$ governs the dominant long-time asymptotic state of the Cauchy problem associated with mKdV equation (2.1), in the region $|x/(3t)^{1/3}| = O(1)$. Not only can one associate the second Painlevé equation with mKdV, but the association yields an integral equation governing a one-parameter family of solutions to P_{II}, which is relevant to the Cauchy problem of mKdV.

Namely, the inverse scattering transform (IST) shows that the mKdV equation can be solved (i.e., linearized) to the following integral equation:

$$K(x,y;t) - F(x+y;t)$$
$$- \int_{-\infty}^{x} \int_{-\infty}^{x} K(x,z;t) F(z+s;t) F(s+y;t) \, dz \, ds = 0,$$

where

$$F(x,t) = \frac{1}{2\pi} \int_{-\infty}^{\infty} r(k) \exp(ikx + 8ik^3)t \, dk,$$

and the solution to mKdV is given by

$$u(x,t) = 2K(x,x;t).$$

The long-time asymptotic analysis of mKdV shows that to leading order in the region $|x/(3t)^{1/3}| = O(1)$, the solution of mKdV (2.1) satisfies equation (2.3) with $\alpha = 0$, and the relevant solution of the P_{II} equation is linearized via

$$K_{\#}(x,y) - r_0 Ai(x+y)$$
$$- (r_0)^2 \int_{-\infty}^{x} \int_{-\infty}^{x} K_{\#}(x,z) Ai(z+s) Ai(s+y) dz \, ds = 0,$$

where Ai(x) is the well-known Airy function, whose integral representation is given by

$$\mathrm{Ai}(x) = \frac{1}{2\pi} \int_{-\infty}^{\infty} \exp(ikx + ik^3/3)dk,$$

and the solution of P_{II} is given by

$$w(x) = 2K_{\#}(x,x).$$

This solution satisfies the following "connection" formulae (cf. [2,4]):
As $x \to +\infty$,

$$w(x) \sim r_0 \mathrm{Ai}(x) \sim \frac{r_0}{2\sqrt{\pi}x^{1/4}} e^{-\frac{2}{3}x^{3/2}},$$

and as $x \to -\infty$,

$$w(x) \sim \frac{d_0}{(-x)^{1/4}} \sin\theta,$$

where

$$d_0 = -(1/\pi)\log(1 - (r_0)^2),$$
$$\theta = (2/3)(-x)^{3/2} - (3/4)d_0^2 \log(-x) + \theta_0,$$
$$\theta_0 = \pi/4 - (3/2)d_0^2 \log 2 - \arg[\Gamma(1 - id_0^2/2)],$$

where $\Gamma(x)$ is the usual gamma function.

Not only does similarity reduction yield special solutions, but it is also the mechanism for one to be able to obtain the isomonodromy problems, which govern the general solution of the Painlevé equation. Indeed, Flaschka and Newell [12] showed that the similarity reduction of the compatible linear system for mKdV also yields a compatible linear system monodromy problem governing P_{II}. Namely, equation (2.3) is the compatibility condition $(\Psi_{t\zeta} = \Psi_{\zeta t})$ for the following system:

$$\Psi_\zeta = \left(\begin{pmatrix} -i(4\zeta^2 + z + 2w^2) & 4\zeta w + 2iw_z + \alpha\zeta^{-1} \\ 4\zeta w - 2iw_z + \alpha\zeta^{-1} & i(4\zeta^2 + z + 2w^2) \end{pmatrix} \right) \Psi,$$
$$\Psi_z = \left(\begin{pmatrix} -i\zeta & w \\ w & i\zeta \end{pmatrix} \right) \Psi.$$

In principle, the compatible isomonodromy system yields the complete solution of the mKdV equation.

We shall not dwell on this aspect of the problem, since we prefer to discuss a novel reduction of the self-dual Yang–Mills (SDYM) integrable system: the generalized Darboux–Halphen (gDH) system. SDYM reductions alone occupy a major aspect of the integrable systems literature. We only point out here that lower-dimensional reductions of the four-dimensional SDYM system yield virtually all the well-known soliton equations (see, e.g., [2]) and all the classical Painlevé equations with their monodromy problems (see, e.g., [15]). Just as the reduction (2.2) of the mKdV linear system gives rise to the isomonodromy problem for P_{II}, the SDYM linear system under this novel reduction yields a new class of *monodromy evolving* problems underlying the gDH systems. The evolution of this monodromy system is such that the temporal evolution of the scattering data can be tracked exactly, and this allows us to find the exact solution of the gDH system.

The results associated with DH-type systems also have an important connection in terms of Painlevé equations. Namely, we will see that there is a parameter in their associated linear systems. When this parameter vanishes, the compatible linear system of DH reduces to an isomonodromy system associated with one of the classical Painlevé equations. This isomonodromic system is essential in solving the DH problem.

In this paper we present the solution of the following system of ODEs, which we refer to as the *generalized DH system*:

$$
\begin{aligned}
\dot{\omega}_1 &= \omega_2\omega_3 - \omega_1(\omega_2 + \omega_3) + \phi^2, \\
\dot{\omega}_2 &= \omega_3\omega_1 - \omega_2(\omega_3 + \omega_1) + \theta^2, \\
\dot{\omega}_3 &= \omega_1\omega_2 - \omega_3(\omega_1 + \omega_2) - \theta\phi, \\
\dot{\phi} &= \omega_1(\theta - \phi) - \omega_3(\theta + \phi), \\
\dot{\theta} &= \omega_2(\phi - \theta) - \omega_3(\theta + \phi),
\end{aligned}
\tag{2.4}
$$

where the dots denote differentiation with respect to t. They correspond to an off-diagonal Bianchi IX metric with self-dual Weyl curvature (see [5]). In Chakravarty, Ablowitz, and Takhtajan [8], it was shown that this system arises as a reduction of the SDYM equations. From the reduction process we obtain a linear problem for the DH system from the linear problem for SDYM. As outlined earlier in Chakravarty and Ablowitz [5], this linear problem, which has nonconstant monodromy, allows us to solve the initial value problem for the system (2.4) in terms of solutions of Schwarzian equations, which arise in the theory of conformal mappings.

We remark that this system with $\theta = \phi = 0$, i.e.,

$$
\begin{aligned}
\dot{\omega}_1 &= \omega_2\omega_3 - \omega_1(\omega_2 + \omega_3), \\
\dot{\omega}_2 &= \omega_3\omega_1 - \omega_2(\omega_3 + \omega_1), \\
\dot{\omega}_3 &= \omega_1\omega_2 - \omega_3(\omega_1 + \omega_2),
\end{aligned}
\tag{2.5}
$$

which we call the *classical DH system*, was originally analyzed by Darboux [11] in the context of certain triply orthogonal surfaces. Shortly thereafter, the solution was found by Halphen [13]. The system is equivalent to the Einstein field equations for a (complex) self-dual Bianchi IX spacetime with diagonal metric. We also note that the symmetric combination of variables $y := -2(\omega_1 + \omega_2 + \omega_3)$ satisfies the Chazy equation (Chazy [9, 10]),

$$\frac{\mathrm{d}^3 y}{\mathrm{d}t^3} = 2y\frac{\mathrm{d}^2 y}{\mathrm{d}t^2} - 3\left(\frac{\mathrm{d}y}{\mathrm{d}t}\right)^2, \tag{2.6}$$

whose general solution possesses a natural boundary.

3   Darboux–Halphen Systems and Their Linear Problems as Reductions of SDYM

As mentioned in Section 1, the self-dual Yang–Mills equations (SDYM) play a central role in the theory of integrable systems. In Chakravarty, Ablowitz, and Clarkson [6, 7] it was shown that the classical Darboux–Halphen system (2.5) is a reduction of SDYM with the gauge algebra $\mathfrak{sdiff}(\mathrm{SU}(2))$. In this section we discuss the SDYM equations and their reduction to the generalized Darboux–Halphen system (2.4). We then go on to show that a linear problem for the Darboux–Halphen system can be obtained from that for SDYM. To our knowledge, all previously studied linear problems associated with integrable systems of ODEs have been *isomonodromy problems*. The linear problem obtained in this section, however, has nonconstant, i.e., evolving monodromy, data. It is important to stress that we do not impose the evolving monodromy condition; rather it is a logical consequence of the SDYM reduction.

For a given Lie algebra $\mathfrak{g}$ corresponding to the Lie group G, let the *gauge potential 1-form* $\mathbf{A}$ be given by

$$\mathbf{A} = \sum_\mu A_\mu \, \mathrm{d}x_\mu,$$

where the x_μ are coordinates on $\mathbb{R}^4$, $A_\mu : \mathbb{R}^4 \to \mathfrak{g}$, $\mu = 0, \ldots, 3$. The *curvature 2-form* $\mathbf{F}$ is given by

$$\mathbf{F} = \frac{1}{2}\sum_{\mu,\nu} F_{\mu\nu}\mathrm{d}x_\mu \wedge \mathrm{d}x_\nu,$$

where

$$F_{\mu\nu} = \partial_\mu A_\nu - \partial_\nu A_\mu - [A_\mu, A_\nu]$$

and we define $\partial_\mu \equiv \partial/\partial x_\mu$.

With respect to the standard coordinates on $\mathbb{R}^4$, the *SDYM equations associated with the gauge algebra* $\mathfrak{g}$ are

$$F_{01} = F_{23}, \qquad F_{02} = F_{31}, \qquad F_{03} = F_{12}.$$

Alternatively, in terms of the so-called null coordinates

$$\alpha = x_0 + ix_1, \quad \bar{\alpha} = x_0 - ix_1, \quad \beta = x_3 + ix_2, \quad \bar{\beta} = x_3 - ix_2,$$

SDYM becomes

$$F_{\alpha\beta} = F_{\bar{\alpha}\bar{\beta}} = F_{\alpha\bar{\alpha}} + F_{\beta\bar{\beta}} = 0, \tag{3.1}$$

where

$$A_\alpha = \frac{1}{2}(A_0 - iA_1), \qquad A_\beta = \frac{1}{2}(A_3 - iA_2),$$
$$A_{\bar{\alpha}} = \frac{1}{2}(A_0 + iA_1), \qquad A_{\bar{\beta}} = \frac{1}{2}(A_3 + iA_2).$$

It is especially easy to see that equations (3.1) are the compatibility of the system

$$(\partial_\alpha - \lambda\partial_{\bar{\beta}})\Psi = (A_\alpha - \lambda A_{\bar{\beta}})\Psi, \tag{3.2}$$
$$(\partial_\beta + \lambda\partial_{\bar{\alpha}})\Psi = (A_\beta + \lambda A_{\bar{\alpha}})\Psi, \tag{3.3}$$

since

$$0 = [\partial_\alpha - \lambda\partial_{\bar{\beta}}, \partial_\beta + \lambda\partial_{\bar{\alpha}}]\Psi = \{F_{\alpha\beta} + \lambda(F_{\alpha\bar{\alpha}} + F_{\beta\bar{\beta}}) + \lambda^2 F_{\bar{\alpha}\bar{\beta}}\}\Psi.$$

The SDYM equations are invariant under the *gauge transformation*

$$A_\mu \mapsto h^{-1}A_\mu h - h^{-1}\partial_\mu h,$$

for any G-valued function h on $\mathbb{R}^4$, since this induces the map $\mathbf{F} \mapsto h^{-1}\mathbf{F}h$. This gauge invariance extends to the linear problem (3.2)–(3.3) by the transformation $\Psi \mapsto h^{-1}\Psi$.

Consider the one-dimensional reduction in which the A_μ's are functions of $x_0 =: t$ only. Choosing a gauge in which $A_0 = 0$, the SDYM equations reduce to

$$\partial_t A_1 + [A_2, A_3] = 0,$$
$$\partial_t A_2 + [A_3, A_1] = 0, \tag{3.4}$$
$$\partial_t A_3 + [A_1, A_2] = 0.$$

This is known as the *Nahm system* (see Nahm [16]).

If we choose the Lie algebra $\mathfrak{g}$ to be SU(2) and take

$$A_i(t) = -\omega_i(t)X_i,$$

where the $\{X_i\}_{i=1,2,3}$ is a basis for $\mathfrak{su}(2)$ satisfying $[X_i, X_j] = \sum_k \varepsilon_{ijk} X_k$, then we obtain the system

$$\dot{\omega}_1 = \omega_2\omega_3, \qquad \dot{\omega}_2 = \omega_3\omega_1, \qquad \dot{\omega}_3 = \omega_1\omega_2, \tag{3.5}$$

where the dots denote differentiation with respect to t. This system, which dates back to Lagrange, admits a general solution in terms of elliptic functions. Equation (3.5) can be viewed (very loosely speaking) as an "unperturbed" version of the system (2.5).

Actually, since we are interested in the "off-diagonal" system (2.4), we first consider the "unperturbed" off-diagonal system associated with it, which is obtained by assuming

$$
\begin{aligned}
-A_1(t) &\equiv T_1(t) = \omega_1(t)X_1 + \theta(t)X_2, \\
-A_2(t) &\equiv T_2(t) = \phi(t)X_1 + \omega_2(t)X_2, \\
-A_3(t) &\equiv T_3(t) = \omega_3(t)X_3,
\end{aligned}
\tag{3.6}
$$

whereupon (3.4) implies

$$
\begin{aligned}
\dot{\omega}_1 &= \omega_2\omega_3, \\
\dot{\omega}_2 &= \omega_3\omega_1, \\
\dot{\omega}_3 &= \omega_1\omega_2 - \theta\phi, \\
\dot{\theta} &= -\phi\omega_3, \\
\dot{\phi} &= -\theta\omega_3.
\end{aligned}
\tag{3.7}
$$

As with equations (3.5), equations (3.7) are solvable in terms of elliptic functions. Indeed, we note that

$$
\theta^2 - \phi^2 = C^2, \qquad \omega_1^2 - \omega_2^2 = E^2,
$$

where C and E are constants. The parametrizations

$$
\theta = C\cosh\psi(t), \quad \phi = C\sinh\psi(t), \quad \omega_1 = E\cosh\mu(t), \quad \omega_2 = E\sinh\mu(t),
$$

imply

$$
\psi + \mu = k \quad \text{and} \quad \ddot{\psi} = \alpha\sinh(2\psi - \beta),
$$

where k is a constant, $\alpha^2 = (C^4 + E^4 + 2C^2E^2\cosh 2k)/4$, and $\tanh\beta = E^2\sinh 2k/(C^2 + E^2\cosh 2k)$. This equation can be solved via elliptic functions. In our study of equation (2.4) we will deform (3.6) in equation (3.4) appropriately.

In fact, the Darboux–Halphen system arises from the Nahm system (equation 3.4) with a different choice of gauge algebra [6,8]. We also again mention that unlike equations (3.5) or (3.7), the solution of equation (2.4) or (2.5) is expressible in terms of Schwarzian functions; hence the structure of the solution is more complicated.

We begin by briefly reviewing the double covering of SO(3) by SU(2). Let

$$
\mathbf{x} = \left(\begin{pmatrix} x_3 & x_1 - ix_2 \\ x_1 + ix_2 & -x_3 \end{pmatrix} \right) = \sum_{i=1}^{3} x_i\sigma_i
$$

be any trace-free Hermitian matrix, where $x_i \in \mathbb{R}$ and

$$\sigma_1 = \left(\begin{pmatrix} 0 & 1 \\ 1 & 0 \end{pmatrix} \right), \qquad \sigma_2 = \left(\begin{pmatrix} 0 & -i \\ i & 0 \end{pmatrix} \right), \qquad \sigma_3 = \left(\begin{pmatrix} 1 & 0 \\ 0 & -1 \end{pmatrix} \right) \quad (3.8)$$

are the Pauli matrices. For any $g \in \mathrm{SU}(2)$ (i.e., g is a unit determinant 2×2 matrix over $\mathbb{C}$ satisfying $gg^\dagger = I$),

$$\mathbf{y} = \left(\begin{pmatrix} y_3 & y_1 - iy_2 \\ y_1 + iy_2 & -y_3 \end{pmatrix} \right) := g\mathbf{x}g^{-1} \qquad (3.9)$$

is also a trace-free Hermitian matrix. Furthermore, $y_1^2 + y_2^2 + y_3^2 = -\det \mathbf{y} = -\det \mathbf{x} = x_1^2 + x_2^2 + x_3^2$. In other words, the mapping $\mathbf{O} : (x_1, x_2, x_3) \mapsto (y_1, y_2, y_3)$ is in $\mathrm{SO}(3)$. Using the standard representation of $\mathbf{O}$ as a 3×3 matrix, the components, O_{ij}, are given by

$$g^{-1}\sigma_i g = \sum_j O_{ij}(g)\sigma_j \qquad (3.10)$$

for all $g \in \mathrm{SU}(2)$. The matrices g and O_{ij} can be parametrized by the Euler angles θ, ϕ, ψ (see, for example, Vilenkin [19]):

$$g = \left(\begin{pmatrix} e^{-i(\phi+\psi)/2} \cos(\theta/2) & -ie^{i(\psi-\phi)/2} \sin(\theta/2) \\ -ie^{i(\phi-\psi)/2} \sin(\theta/2) & e^{i(\phi+\psi)/2} \cos(\theta/2) \end{pmatrix} \right)$$

and

$$\mathbf{O} = \left(\begin{pmatrix} \cos\phi\cos\psi - \sin\phi\sin\psi\cos\theta & -\cos\phi\sin\psi - \sin\phi\cos\psi\cos\theta & \sin\phi\sin\theta \\ \sin\phi\cos\psi + \cos\phi\sin\psi\cos\theta & -\sin\phi\sin\psi + \cos\phi\cos\psi\cos\theta & -\cos\phi\sin\theta \\ \sin\psi\sin\theta & \cos\psi\sin\theta & \cos\theta \end{pmatrix} \right),$$

where $0 < \theta < \pi$, $0 \le \phi < 2\pi$, and $-2\pi \le \psi < 2\pi$. Note that for this range of the Euler angles $\mathrm{SU}(2)$ (parametrized by g) is covered once and $\mathrm{SO}(3)$ (parametrized by $\mathbf{O}$) is covered twice.

Choose the gauge algebra to be $\mathfrak{sdiff}(\mathrm{SU}(2))$ with now

$$A_i(t) = -\tilde{T}_i, \quad \text{where} \quad \tilde{T}_i := \sum_j O_{ij} T_j, \qquad (3.11)$$

and the T_i are given by (3.6), where $\{X_i\}$ are the standard left-invariant vector fields generating $\mathrm{SU}(2)$, which satisfy

$$X_j(g) = \frac{1}{2i} g\sigma_j. \qquad (3.12)$$

In this notation the X_j are given explicitly as vector fields in the Euler angles (θ, ϕ, ψ) as

$$X_1 = \cos\psi \frac{\partial}{\partial\theta} + \sin\psi \cosec\theta \frac{\partial}{\partial\phi} - \cot\theta \sin\psi \frac{\partial}{\partial\psi},$$

$$X_2 = -\sin\psi \frac{\partial}{\partial\theta} + \cos\psi \cosec\theta \frac{\partial}{\partial\phi} - \cot\theta \cos\psi \frac{\partial}{\partial\psi},$$

$$X_3 = \frac{\partial}{\partial\psi}.$$

Using the properties (3.12) (or equivalently $X_i(O_{jk}) = \sum_l \varepsilon_{ikl}O_{jl}$) it can be shown that under the present reduction, the SDYM equations become the gDH system (2.4) (see also [6,8]). We note that if O_{ij} is replaced by δ_{ij} in equation (3.11), we recover the system (3.7). It is in this sense that equations (2.4) are a perturbed form of equations (3.7).

Next we will find a linear problem associated with equation (2.4) from the linear problem for SDYM. In (3.2)–(3.3) we take

$$-2A_\alpha = i\tilde{T}_3 = 2A_{\bar{\alpha}}, \quad 2iA_\beta = \tilde{T}_1 - i\tilde{T}_2 \equiv \tilde{T}_-, \quad -2iA_{\bar{\beta}} = \tilde{T}_1 + i\tilde{T}_2 \equiv \tilde{T}_+.$$

Noting $\partial_\alpha \Psi = \partial_{\bar{\alpha}}\Psi = \partial_t \Psi / 2$, $\partial_\beta \Psi = \partial_{\bar{\beta}} \Psi = 0$ and defining $\tilde{T}_0 \equiv \partial / \partial t$, we then obtain

$$\left(\tilde{T}_0 + i\tilde{T}_3 + i\lambda\tilde{T}_+\right)\Psi = 0, \tag{3.13}$$

$$\left(\lambda\tilde{T}_0 + i\tilde{T}_- - i\lambda\tilde{T}_3\right)\Psi = 0. \tag{3.14}$$

We will find it convenient to parametrize λ by writing it as a "projective coordinate" $\lambda = \pi^1/\pi^0$. Assuming Ψ to be a function of t and λ only, we obtain the following linear pair from (3.13)–(3.14):

$$l_1\Psi = l_2\Psi = 0,$$

where

$$((l_1 \quad l_2)) = \left(\left(\pi^0(\tilde{T}_0 + i\tilde{T}_3) + i\pi^1\tilde{T}_+ \quad i\pi^0\tilde{T}_- + \pi^1(\tilde{T}_0 - i\tilde{T}_3)\right)\right)$$

$$= ((\pi^0 \quad \pi^1)) \left(\begin{pmatrix} \tilde{T}_0 + i\tilde{T}_3 & i\tilde{T}_- \\ i\tilde{T}_+ & \tilde{T}_0 - i\tilde{T}_3 \end{pmatrix}\right) = ((\pi^0 \quad \pi^1))\,\tilde{\mathbf{T}},$$

$\tilde{T}_0 = \partial_t$, and $\tilde{T}_\pm = \tilde{T}_1 \pm i\tilde{T}_2$.

This linear problem is unfortunately too complicated to work with concretely because $\mathbf{T} \in \mathfrak{sdiff}(SU(2))$. In order to simplify it, we first note from equation (3.9) that equation (3.11) can be written as an SU(2) action:

$$\tilde{\mathbf{T}} = g\mathbf{T}g^{-1}, \tag{3.15}$$

where

$$\mathbf{T} = \left(\begin{pmatrix} T_0 + iT_3 & iT_- \\ iT_+ & T_0 - iT_3 \end{pmatrix} \right),$$

$T_0 = \partial_t$, $T_\pm := T_1 \pm iT_2$. Equation (3.15) says that $\widetilde{\mathbf{T}}(\phi) = g\mathbf{T}(\phi)g^{-1}$ for all SU(2)-valued functions ϕ.

Next we take two independent linear combinations of l_1 and l_2, only one of which, M, contains a derivative with respect to t:

$$L := \left((l_1 \quad l_2) \right) \left(\begin{pmatrix} -\pi^1 \\ \pi^0 \end{pmatrix} \right) = \left((\pi^0 \quad \pi^1) \right) g \mathbf{T} g^{-1} \left(\begin{pmatrix} -\pi^1 \\ \pi^0 \end{pmatrix} \right), \qquad (3.16)$$

$$M := \left((l_1 \quad l_2) \right) \left(\begin{pmatrix} -v^1 \\ v^0 \end{pmatrix} \right) = \left((\pi^0 \quad \pi^1) \right) g \mathbf{T} g^{-1} \left(\begin{pmatrix} -v^1 \\ v^0 \end{pmatrix} \right), \qquad (3.17)$$

where $\pi^0 v^1 - \pi^1 v^0 = 1$ and we have used equation (3.15).

Define $\tilde{\pi}^A$, $\tilde{v}^A$, which we interpret as spinors (see, e.g., [18]), as

$$\left(\begin{pmatrix} \tilde{\pi}^0 & \tilde{\pi}^1 \\ \tilde{v}^0 & \tilde{v}^1 \end{pmatrix} \right) = \left(\begin{pmatrix} \pi^0 & \pi^1 \\ v^0 & v^1 \end{pmatrix} \right) g,$$

and clearly $\tilde{\pi}^0 \tilde{v}^1 - \tilde{\pi}^1 \tilde{v}^0 = 1$. In this way the g-action is absorbed into the spinors $\tilde{\pi}^A$, $\tilde{v}^A$. From (3.16)–(3.17) we see that these operators have an elegant representation in terms of the spinors π^A, v^A, $\tilde{\pi}^A$, and $\tilde{v}^A$:

$$L = \sum_{A,B} \pi^A \widetilde{\mathbf{T}}_A{}^B \pi_B = \sum_{A,B} \tilde{\pi}^A \mathbf{T}_A{}^B \tilde{\pi}_B,$$

$$M = \sum_{A,B} \pi^A \widetilde{\mathbf{T}}_A{}^B v_B = \sum_{A,B} \tilde{\pi}^A \mathbf{T}_A{}^B \tilde{v}_B,$$

and where the dual spinors π_A, v_A, etc., are given by

$$\pi_B = \sum_A \pi^A \varepsilon_{AB}, \qquad \varepsilon_{AB} = \left(\begin{pmatrix} 0 & 1 \\ -1 & 0 \end{pmatrix} \right).$$

We also note that by multiplication, equations (3.16)–(3.17) become

$$-L = i \left\{ (\tilde{\pi}^1)^2 T_+ + 2\tilde{\pi}^0 \tilde{\pi}^1 T_3 - (\tilde{\pi}^0)^2 T_- \right\}, \qquad (3.18)$$

$$-M = \partial_t + i \left\{ \tilde{\pi}^1 \tilde{v}^1 T_+ + (\tilde{\pi}^0 \tilde{v}^1 + \tilde{\pi}^1 \tilde{v}^0) T_3 - \tilde{\pi}^0 \tilde{v}^0 T_- \right\}. \qquad (3.19)$$

Notice that in the operators L and M we have moved the explicit g-dependence from the operators $\widetilde{\mathbf{T}}^A{}_B$ to the spinors $\tilde{\pi}^A$, $\tilde{v}^A$.

This will help us simplify the linear problem.

We need to take into account the action of $\mathbf{T}$ on the spinors $\tilde{\pi}^A$ and $\tilde{v}^A$, which depend on g. Without this action of $\mathbf{T}$ on $\tilde{\pi}^A$, $\tilde{v}^A$, i.e., replacing $\tilde{\pi}^A \mapsto \pi^A$, $\tilde{v}^A \mapsto v^A$, we actually reduce to the unperturbed system (3.7).

Specifically, using equation (3.12), we obtain a convenient *linear* representation of the $X_j =: (1/2i)\widehat{X}_j$ on the space of $\tilde{\pi}^A$ and $\tilde{v}^A$:

$$\widehat{X}_j = \sum_{A,B} \left(\tilde{\pi}^A \sigma_{jA}{}^B \frac{\partial}{\partial \tilde{\pi}^B} + \tilde{v}^A \sigma_{jA}{}^B \frac{\partial}{\partial \tilde{v}^B} \right), \qquad (3.20)$$

where $\sigma_{jA}{}^B$ is the (A, B)–entry of the Pauli matrix σ_j (cf. 3.8). Explicitly,

$$\widehat{X}_1 = \tilde{\pi}^1 \frac{\partial}{\partial \tilde{\pi}^0} + \tilde{\pi}^0 \frac{\partial}{\partial \tilde{\pi}^1} + \tilde{v}^1 \frac{\partial}{\partial \tilde{v}^0} + \tilde{v}^0 \frac{\partial}{\partial \tilde{v}^1},$$

$$\widehat{X}_2 = i \left(\tilde{\pi}^1 \frac{\partial}{\partial \tilde{\pi}^0} - \tilde{\pi}^0 \frac{\partial}{\partial \tilde{\pi}^1} + \tilde{v}^1 \frac{\partial}{\partial \tilde{v}^0} - \tilde{v}^0 \frac{\partial}{\partial \tilde{v}^1} \right),$$

$$\widehat{X}_3 = \tilde{\pi}^0 \frac{\partial}{\partial \tilde{\pi}^0} - \tilde{\pi}^1 \frac{\partial}{\partial \tilde{\pi}^1} + \tilde{v}^0 \frac{\partial}{\partial \tilde{v}^0} - \tilde{v}^1 \frac{\partial}{\partial \tilde{v}^1}.$$

Using equations (3.18)–(3.19), the operators L and M can be written as

$$L = -\left(\tilde{\pi}^1\right)^2 \mathcal{L}, \qquad M = -(\partial_t + \mathcal{M}) - \tilde{\pi}^1 \tilde{v}^1 \mathcal{L},$$

where

$$\mathcal{L} = i\left\{ T_+ - 2i\tilde{\lambda}T_3 + \tilde{\lambda}^2 T_- \right\}, \qquad \mathcal{M} = \tilde{\lambda}T_- - iT_3, \quad \text{and} \quad \tilde{\lambda} = i\tilde{\pi}^0/\tilde{\pi}^1.$$

The operators L, M still have the group action contained, but the representations are greatly simplified. Observe that the compatibility of the system $L\Psi = M\Psi = 0$ is, after recombination and some calculation,

$$[L, M] = (\tilde{\pi}^1)^2 \left(-\mathcal{L}_t + [\mathcal{L}, \mathcal{M}] + A\mathcal{L} \right) = 0, \qquad (3.21)$$

where $A = \tilde{\pi}^1 \mathcal{L}\tilde{v}^1 - \tilde{v}^1 \mathcal{L}\tilde{\pi}^1 - (2/\tilde{\pi}^1)\mathcal{M}\tilde{\pi}^1$, which depends *only* on $\tilde{\lambda}$ and t. It is important to note that since the T_i (through the $\widehat{X}_j$) act on the $\tilde{\pi}^A$ and $\tilde{v}^A$, we cannot replace the equation $L\Psi = 0$ with $\mathcal{L}\Psi = 0$ without changing the compatibility conditions. Thus equation (3.21) reduces the compatibility of L, M to a compatibility equation involving $\mathcal{L}$, $\mathcal{M}$, which depend on the "unperturbed" operators T_i and $\tilde{\lambda}$.

Using the representation (3.20), we see that

$$A = -\frac{1}{2} \left\{ [(\omega_1 + \omega_2 + 2\omega_3) - i(\theta - \phi)] - \tilde{\lambda}^2 [(\omega_1 - \omega_2) - i(\theta + \phi)] \right\},$$

and

$$\widehat{X}_1(\tilde{\lambda}) = i(\tilde{\lambda}^2 + 1), \qquad \widehat{X}_2(\tilde{\lambda}) = \tilde{\lambda}^2 - 1, \qquad \widehat{X}_3(\tilde{\lambda}) = 2\tilde{\lambda}.$$

We replace the operators $\widehat{X}_i$ by the operators $\widehat{Y}_i := \sigma_i + \widehat{X}_i$ in order to calculate the compatibility condition (3.21). In this extended algebra the $\widehat{Y}_i$ form a suitable basis that can be multiplied by $\tilde{\lambda}$-dependent functions (the original $\widehat{X}_i$ acting on $\Psi(\tilde{\lambda}, t)$ are not linearly independent when multiplied by such functions). Explicitly, the $\widehat{Y}_i$ have the form

$$\widehat{Y}_1 = \sigma_1 + i(\tilde{\lambda}^2 + 1)I\partial_{\tilde{\lambda}},$$
$$\widehat{Y}_2 = \sigma_2 + (\tilde{\lambda}^2 - 1)I\partial_{\tilde{\lambda}},$$
$$\widehat{Y}_3 = \sigma_3 + 2\tilde{\lambda}I\partial_{\tilde{\lambda}},$$

where the σ_i are the Pauli matrices (3.8). Thus equation (3.21) can now be written as the compatibility condition for a system in t and $\tilde{\lambda}$ only:

$$\mathcal{L}\Psi = \frac{i}{2}\mu\Psi, \tag{3.22}$$

$$\partial_t \Psi = -(\mathcal{M} + \nu I)\Psi, \tag{3.23}$$

subject to the condition

$$\mathcal{L}_d(\nu) = A\mu, \tag{3.24}$$

where

$$\mathcal{L} = \frac{i}{2}\left\{P(\tilde{\lambda})\frac{\partial}{\partial\tilde{\lambda}} + l\right\} = \frac{i}{2}(\mathcal{L}_d + l), \tag{3.25}$$

$$\mathcal{M} = \frac{1}{2}\left\{(\alpha_-\lambda^3 + \beta_+\lambda)\frac{\partial}{\partial\tilde{\lambda}} + l_1\right\}, \tag{3.26}$$

$$\alpha_\pm = (\omega_1 - \omega_2) \pm i(\theta + \phi), \quad \beta_\pm = \omega \pm i(\theta - \phi), \tag{3.27}$$

$$\omega = \omega_1 + \omega_2 - 2\omega_3, \quad P(\tilde{\lambda}) = \alpha_+ + \tilde{\lambda}^2(\beta_+ + \beta_-) + \tilde{\lambda}^4\alpha_-, \tag{3.28}$$

and the matrices l_1 and l are given by

$$l_1 = \tilde{\lambda}X + Z, \qquad l = \tilde{\lambda}^2 X + 2\tilde{\lambda}Z + Y,$$
$$X = -(\phi + i\omega_1)\sigma_1 - (\omega_2 + i\theta)\sigma_2,$$
$$Y - (\phi - i\omega_1)\sigma_1 + (\omega_2 - i\theta)\sigma_2,$$
$$Z = -\omega_3\sigma_3.$$

4 The Monodromy Evolving System and the Solution of the Generalized DH System

In this section we recapitulate the main result, and we shall summarize the results of the monodromy analysis. Full details of the monodromy analysis will be published separately. From (3.22)–(3.24) the linear monodromy evolving system is given by

$$\Psi_\lambda = (1/P)(\mu I - l)\Psi, \tag{4.1}$$

$$\Psi_t = -(\nu I + \frac{1}{2}l_1 + \frac{1}{2}f_1\partial_\lambda)\Psi, \tag{4.2}$$

where

$$f_1 = \alpha_- \lambda^3 + \beta_+ \lambda,$$

and we have dropped the tildes on the λ's. From (3.24) the parameter ν satisfies

$$\frac{\partial \nu}{\partial \lambda} = \frac{A}{P}\mu. \tag{4.3}$$

Detailed analysis of the linear system (4.1)–(4.2) shows that (4.3) governs the evolution of the monodromy associated with $\Psi(\lambda)$ [5].

We remark upon the important point that when $\mu = 0$, the system (4.1)–(4.2) is isomonodromic. Indeed, equation (4.1) has four singular points corresponding to the zeros of P :

$$\lambda_1 = \sqrt{(-r+v)/\delta}, \qquad\qquad \lambda_2 = -\sqrt{(-r-v)/\delta},$$
$$\lambda_3 = -\sqrt{(-r+v)/\delta}, \qquad\qquad \lambda_4 = \sqrt{(-r-v)/\delta}, \tag{4.4}$$

where we have defined $r = \omega/\sqrt{\alpha_+ \alpha_-}$, $\delta^2 = \alpha_-/\alpha_+$, and $v^2 = r^2 - 1$. The fact that equation (4.1) has four distinct singular points indicates that it is related to the isomonodromy problem for the sixth Painlevé equation, P_{VI} [14]. The canonical form of this isomonodromy problem has singularities at 0, 1, ∞, and s: the independent variable of P_{VI}. To find s we map λ to a new spectral parameter, $\hat{\lambda}$, given by

$$\hat{\lambda}(\lambda) = \frac{(\lambda_2 - \lambda_3)(\lambda - \lambda_1)}{(\lambda_2 - \lambda_1)(\lambda - \lambda_3)}.$$

So $\hat{\lambda}(\lambda_1) = 0$, $\hat{\lambda}(\lambda_2) = 1$, $\hat{\lambda}(\lambda_3) = \infty$, and we define the isomonodromy variable s by

$$s := \hat{\lambda}(\lambda_4) = \frac{r+1}{r-1}. \tag{4.5}$$

This variable plays a central role in the solution of the, gDH system. Although we will not go through the analysis in this paper, we note that the Lax pair (4.1)–(4.2) can be used to express ω_i, θ, and ϕ in terms of s and its derivatives. The field equations (i.e., the system (2.4)) are then used to show that s must satisfy a third-order Schwarzian equation. Namely, the analysis of (4.1)–(4.2) establishes the following:

$$\frac{(\phi - \theta)^2}{\omega^2 - \alpha_+ \alpha_-} = C_0^2, \tag{4.6}$$

where C_0 is a constant. Also, it can be shown that

$$\omega = \frac{1}{2}\left(\frac{s+1}{s-1}\right)\frac{\dot{s}}{s}, \tag{4.7}$$

$$\omega_3 = -\frac{1}{2}\left(\frac{\ddot{s}}{\dot{s}} - \frac{\dot{s}}{s}\right), \tag{4.8}$$

$$\theta - \phi = \frac{C_0\dot{s}}{\sqrt{s}(s-1)}, \tag{4.9}$$

$$\alpha_\pm = \kappa_\pm \frac{\dot{s}}{s} e^{\pm iu(t)}, \tag{4.10}$$

where $\kappa_\pm$ are constants satisfying $\kappa_+\kappa_- = \frac{1}{4}$ and

$$u(t) = C_0 \ln\left(\frac{\sqrt{s}-1}{\sqrt{s}+1}\right).$$

Equations (4.7)–(4.10) show that $(\omega_1+\omega_2)$, ω_3, $(\phi-\theta)^2$, and $(\omega_1-\omega_2)^2 + (\phi+\theta)^2$ are rational functions of s, $\dot{s}$, and $\ddot{s}$. We see that we can solve for $\Omega = (\omega_1,\omega_2,\omega_3,\theta,\phi)$ in terms of s and its first and second derivatives. A direct calculation shows that Ω is a solution of the gDH system if and only if the following Schwarzian equation is satisfied:

$$\{s,t\} + \frac{\dot{s}^2}{2}V(s) = 0, \tag{4.11}$$

where

$$\{s,t\} \equiv \frac{d}{dt}\left(\frac{\ddot{s}}{\dot{s}}\right) - \frac{1}{2}\left(\frac{\ddot{s}}{\dot{s}}\right)^2$$

is the Schwarzian derivative and

$$V(s) = \frac{1}{s^2} - \frac{1+C_0^2}{s(s-1)} + \frac{1+C_0^2}{(s-1)^2}.$$

This solution can be verified by direct substitution. It can be shown (see, e.g., Ablowitz and Fokas [3]) that although this equation is linearizable, in general the solutions are densely branched! The general solution is single-valued only when $C_0 = 0$ or $C_0 = i/n$ for some integer n.

5 Discussion

In this paper we have demonstrated how reductions of the SDYM equations with an infinite-dimensional gauge algebra lead to DH-type systems. These equations are linearized by monodromy evolving systems. Thus the gDH system (2.4) is a reduction of the SDYM equations and is solvable via an associated linear problem. The gDH system has been concretely linearized and therefore must be considered to be integrable in terms of real variables. It does not, however, share one of the other properties normally associated with integrable systems obtained by reduction from soliton equations. In particular, solutions of the gDH system do not, in general, have "nice" singularity structure in the complex plane. In the well-known special case of the classical DH system ($\theta = \phi = 0$) (2.5) admits solutions with a movable natural boundary—a circle on the complex sphere across which the solution cannot be analytically continued and whose center and radius depend on initial conditions. Although these solutions possess movable singularities other than poles, nevertheless they are single-valued in their domain of existence; hence the classical system can still be considered to possess the Painlevé property. The general solution of the system (2.4), however, is densely branched in the complex plane and is definitely not of Painlevé type. This example shows that integrability as solvability via an associated linear problem does not imply integrability in the complex plane (see also [1]). Nevertheless, the gDH system is deeply connected to the equations of Painlevé type, as is demonstrated here.

6 REFERENCES

[1] M.J. Ablowitz, S. Chakravarty, and B.M. Herbst, *Integrability, computability and applications*, Acta Appl. Math. **39** (1995), 5–37.

[2] M.J. Ablowitz and P.A. Clarkson, *Solitons, Nonlinear Evolution Equations and Inverse Scattering*, LMS Lecture Notes **149** (Cambridge Univ. Press, Cambridge, 1991).

[3] M.J. Ablowitz and A.S. Fokas, *Complex Variables: Introduction and Applications* (Cambridge University Press, Cambridge, 1997).

[4] M.J. Ablowitz and H. Segur, *Solitons and the Inverse Scattering Transform*, SIAM, Philadelphia, 1981.

[5] S. Chakravarty and M.J. Ablowitz, *Integrability, monodromy, evolving deformations, and self-dual Bianchi IX systems*, Phys. Rev. Lett. **76** (1996), 857–860.

[6] S. Chakravarty, M.J. Ablowitz, and P.A. Clarkson, *Reductions of self-dual Yang-Mills fields and classical systems*, Phys. Rev. Lett. **65** (1990), 1085–1087.

[7] _____ , *One dimensional reductions of self-dual Yang–Mills fields and classical systems*, Recent Advances in General Relativity: Essays in Honor of Ted Newman (Pittsburg, 1990), eds. A.I. Janis and J.R. Porter (Birkhäuser, Boston, Basel, 1991).

[8] S. Chakravarty, M.J. Ablowitz, and L.A. Takhtajan, *Self-dual Yang-Mills equation and new special functions in integrable systems*, Nonlinear Evolution Equations and Dynamical Systems, eds. M. Boiti, L. Martina, and F. Pempinelli (World Scientific, Singapore, 1992).

[9] J. Chazy, *Sur les équations différentielles dont l'intégrale générale possède une coupure essentielle mobile*, C.R. Acad. Sc. Paris **150** (1910), 456–458.

[10] _____, *Sur les équations différentielles du troisième ordre et d'ordre supérieur dont l'intégrale générale a ses points critiques fixes*, Acta Math. **34** (1911), 317–385.

[11] G. Darboux, *Sur la théorie des coordonnées curvilignes et les systèmes orthogonaux*, Ann. Ec. Normale Supér. **7** (1878), 101–150.

[12] H. Flaschka and A.C. Newell, *Monodromy- and spectrum-preserving deformations. I*, Commun. Math. Phys. **76** (1980), 65–116.

[13] G. Halphen, *Sur un système d'équations différentielles*, C. R. Acad. Sci. Paris **92** (1881), 1101–1103.

[14] M. Jimbo and T. Miwa, *Monodromy preserving deformation of linear ordinary differential equations with rational coefficients II*, Phys. D **2** (1981), 407–448.

[15] L.J. Mason and N.M.J. Woodhouse, *Integrability, Self-Duality, and Twistor Theory*, LMS Monograph, New Series **15** (Oxford University Press, Oxford, 1996).

[16] W. Nahm, *The algebraic geometry of multimonopoles*, Group Theoretical Methods in Physics, eds. M. Serdaroglu and E. Inonu, Lecture Notes in Physics **180** (Springer, Berlin, 1982), pages 456–466.

[17] P. Painlevé, *Oeuvres de Paul Painlevé. I, II, III* (Éditions du Centre national de la recherche scientifique, Paris, 1973–77).

[18] R. Penrose and W. Rindler, *Spinors and Space-Time*. vol. 1 (Cambridge University Press, Cambridge, 1987).

[19] N.J. Vilenkin, *Special Functions and the Theory of Group Representations* (USSR, 1965); (AMS, Providence RI, 1968).

10

Symmetry Reduction and Exact Solutions of Nonlinear Partial Differential Equations

Peter A. Clarkson
Pavel Winternitz

1 Introduction

The purpose of these lectures is to show how the method of symmetry reduction can be used to obtain certain classes of exact analytic solutions of systems of partial differential equations. We use the words "symmetry reduction" in a rather broad sense. Namely, the reduction of a partial differential equation (PDE) or a system of such equations to a "reduced" system, involving fewer independent variables.

The reason why these lectures were included in a school dedicated to Painlevé analysis is that this singularity analysis can often be used to help solve the reduced equations, once they are obtained. In other words, symmetry reduction, combined with Painlevé analysis, very often provides physically interesting solutions, for both integrable and nonintegrable equations.

Lie group theory plays an important role in symmetry reduction. Indeed, the traditional method for reducing the number of independent variables in an equation is to require that a solution be invariant under some subgroup of the symmetry group of the system.

The first part of this chapter will be devoted to such group invariant solutions (Sections 2–5). Section 6 will be dedicated to other reduction methods that can also be viewed in a group-theoretical context.

In Section 2 we present the classical algorithm for calculating the symmetry group of a system of differential equations. It goes back to S. Lie and has been reviewed in many recent books and lectures series [15, 66, 91, 131, 138, 159, 167, 179, 180]. Section 3 is devoted to specific examples of calculating symmetry groups, i.e., groups of local Lie point transformations, transforming solutions of the considered equation among each other. In Section 4 we first outline the method of symmetry reduction, then apply it to examples. An essential part of the method of symmetry reduction is the classification of subalgebras of the "symmetry algebra," i.e., of the Lie algebra of the symmetry group. This is described in Section 5. A direct method of performing symmetry reduction is discussed in Section 6, where it is also interpreted in terms of "conditional symmetries." These correspond to transformations that do not leave the entire solution set invariant, but only a subset, subject to an additional condition, the surface condition.

2   Algorithm for Calculating the Symmetry Group of a Differential System

Let us consider a system of differential equations

$$
\begin{aligned}
&E^i(x, u, u^{(1)}, u^{(2)}, \ldots, u^{(n)}) = 0 \\
&x \in \mathbb{R}^p, u \in \mathbb{R}^q, \quad i = 1, \ldots, m,
\end{aligned}
\tag{2.1}
$$

where $u^{(k)}$ denotes all partial derivatives of order k of all components u_α of u. The system is quite general: The numbers p, q, m and n are all arbitrary finite positive integers.

We now wish to find all local point transformations of the form

$$
\tilde{x} = \Lambda_g(x, u), \quad \tilde{u} = \Omega_g(x, u)
\tag{2.2}
$$

such that a solution $u = f(x)$ of the system (2.1) is transformed into a solution $\tilde{u} = \tilde{f}(\tilde{x})$. The transformations are called "point" ones because the new variables $(\tilde{x}, \tilde{u})$ depend only on the old ones (x, u), i.e., on a point in the space

$$
M \subset X \times U, \quad X \sim \mathbb{R}^p, \quad U \sim \mathbb{R}^q,
\tag{2.3}
$$

of independent and dependent variables.

More general transformations, in which $\tilde{x}$ and $\tilde{u}$ depend also on derivatives u_x, u_{xx}, etc., or integrals $D^{-1}u$, ... , will not be considered here, though they are also of considerable interest.

The subscript g in (2.2) denotes a finite or infinite number of group parameters, and the transformations form a local Lie group. The word "local" in this context means that the transformations need only be defined and invertible for g close to the identity element of the group and for (x, u) close to the origin in $X \times U$ space.

A method for determining the functions Λ and Ω in (2.2) is due to Sophus Lie and is described in many books [15, 66, 91, 131, 138, 159, 167, 179, 180].

In principle, one could use (2.2) directly to calculate derivatives like $\tilde{u}_{\tilde{x}}$, etc. Substituting back into (2.1), one would get differential equations for the functions Λ and Ω. This approach is not fruitful; the equations determining Λ and Ω are at least as difficult to solve as the original system. Lie's outstanding contribution was that he showed that nearly all the relevant information can be obtained using an infinitesimal approach. Instead of (2.2) we consider transformations

$$\tilde{x}_i = x_i + \varepsilon \xi_i(x, u), \quad \tilde{u}_\alpha = u_\alpha + \varepsilon \phi_\alpha(x, u) \tag{2.4}$$

and obtain equations for ξ_i and ϕ_α, ignoring all terms of order ε^p, $p \geq 2$. This provides us with a system of *linear* equations (due to the restriction to ε^0 and ε^1 terms). Solving these determining equations, we obtain the Lie algebra L of the symmetry group G, realized by vector fields

$$\hat{X} = \sum_{i=1}^p \xi_i(x, u) \partial_{x_i} + \sum_{\alpha=1}^q \phi_\alpha(x, u) \partial_{u_\alpha}. \tag{2.5}$$

Each vector field $\hat{X}$ generates a one-parameter subgroup of the symmetry group, obtained by integrating the vector field

$$\begin{aligned}
\frac{d\tilde{x}_i}{d\varepsilon} &= \xi_i(\tilde{x}, \tilde{u}) \quad \tilde{x}_i|_{\varepsilon=0} = x_i, \\
\frac{d\tilde{u}_\alpha}{d\varepsilon} &= \phi(\tilde{x}, \tilde{u}) \quad \tilde{u}_\alpha|_{\varepsilon=0} = u_\alpha.
\end{aligned} \tag{2.6}$$

The general group transformation is obtained by composing the individual one-parameter transformations.

We shall present an algorithm for calculating the "symmetry algebra," i.e., the vector fields $\hat{X}$ of (2.5) that generate the symmetry group G. It will be described in detail; all proofs can be found in the literature, in particular in P.J. Olver's book [131].

The basic tool is prolongation theory. The symmetry group G acts on the manifold M:

$$G: \{x, u\} \in M \mapsto \{\tilde{x}, \tilde{u}\} \in M. \tag{2.7}$$

Thus, it takes functions into functions:

$$u = f(x) \mapsto \tilde{u} = \tilde{f}(\tilde{x}) = g \cdot f(x). \tag{2.8}$$

The nth prolongation of G also takes derivatives of orders up to n into derivatives:

$$\mathrm{pr}^{(n)} G: \{x, f(x), f^{(1)}(x), \ldots, f^{(n)}(x)\}$$
$$\mapsto \{\tilde{x}, \tilde{f}(\tilde{x}), \tilde{f}^{(1)}(\tilde{x}), \ldots, \tilde{f}^{(n)}(\tilde{x})\}. \tag{2.9}$$

We assume that all functions $f(x)$ are sufficiently smooth (locally) for all considered derivatives to exist. The expression $\mathrm{pr}^{(n)} G$ contains no new information. If we know how variables and functions transform, we can calculate how derivatives transform.

The vector field $\widehat{X}$ acts on functions of the variables x and u. Its nth prolongation will act on functions of x, u and all derivatives u_x, ... , u_{nx}. The form of the prolongation is determined by the requirement that if we integrate $\mathrm{pr}^{(n)}\,\widehat{X}$, we obtain the prolongation of the group action $\mathrm{pr}^{(n)}\,G$.

This determines the form of the prolongation of the vector field $\widehat{X}$ to be

$$\mathrm{pr}^{(n)}\,\widehat{X} = \widehat{X} + \sum_{\alpha=1}^{q} \sum_{k=1}^{n} \sum_{J} \phi_\alpha^J \frac{\partial}{\partial u_J^\alpha}. \tag{2.10}$$

Here J is a set of indices:

$$J \equiv J(k) = (j_1, \ldots, j_b), \quad 1 \le j_k \le p, \quad k = j_1 + j_2 + \cdots + j_k. \tag{2.11}$$

The coefficients ϕ_α^J are expressed in terms of ξ, ϕ and their derivatives up to order k. They are hence functions of $\{x, u, u_x, \ldots, u_{kx}\}$. Explicit formulas for ϕ_α^J are given by Olver [131], as are recursion relations. We find the recursion relations more useful, and we shall reproduce those.

For the first prolongation we have

$$\mathrm{pr}^{(1)}\,\widehat{X} = \widehat{X} + \sum_{\alpha=1}^{q} \sum_{i=1}^{p} \phi_\alpha^i(x, u, u_x) \frac{\partial}{\partial u_{x_i}^\alpha}, \tag{2.12}$$

$$\phi_\alpha^i = D_{x_i}\phi_\alpha - \sum_{j=1}^{p}(D_{x_i}\xi^j)u_{\alpha,x_j}, \tag{2.13}$$

where D_{x_i} is the total derivative operator:

$$D_{x_i} = \frac{\partial}{\partial x_i} + \sum_{\alpha=1}^{q} \frac{\partial u_\alpha}{\partial x_i} \frac{\partial}{\partial u_\alpha} + \sum_{\alpha=1}^{q} \sum_{j=1}^{p} \frac{\partial u_{\alpha,x_j}}{\partial x_i} \frac{\partial}{\partial u_{\alpha,x_j}} + \cdots. \tag{2.14}$$

If the nth prolongation is known, the $(n+1)$th is given by

$$\mathrm{pr}^{(n+1)}\,\widehat{X} = \mathrm{pr}^{(n)}\,\widehat{X} + \sum_{\alpha=1}^{q} \sum_{i_1\,\mathrm{nth}\ldots,i_{n+1}=1}^{p} \phi_\alpha^{i_1\ldots i_{n+1}} \frac{\partial}{\partial u_{\alpha,x_{i_1},\ldots,x_{i_{n+1}}}} \tag{2.15}$$

$$\phi_\alpha^{i_1\ldots i_n i_{n+1}} = D_{x_{i_{n+1}}}\phi_\alpha^{x_1,\ldots,i_n} - \sum_{j=1}^{p}\left(D_{x_{i_{n+1}}}\xi^j\right)u_{\alpha,x_{i_1}\ldots x_{i_n}x_j}. \tag{2.16}$$

The vector fields $\widehat{X}$ form a Lie algebra. Their prolongations realize an isomorphic Lie algebra, since we have

$$\mathrm{pr}^{(n)}(a\widehat{X}_1 + b\widehat{X}_2) = a\,\mathrm{pr}^{(n)}\,\widehat{X}_1 + b\,\mathrm{pr}^{(n)}\,\widehat{X}_2, \tag{2.17}$$

$$[\mathrm{pr}^{(n)}\,X_1, \mathrm{pr}^{(n)}\,X_2] = \mathrm{pr}^{(n)}[X_1, X_2]. \tag{2.18}$$

The algorithm for determining the symmetry algebra of the system (2.1) can now be stated quite simply. The nth prolongation of the vector field must annihilate the equations on their solution set:

$$\mathrm{pr}^{(n)}\, \widehat{X} \cdot E^i|_{E^k=0} = 0, \quad i = 1,\ldots,m, \quad k = 1,\ldots,m. \qquad (2.19)$$

The equations (2.1) are viewed as a set of algebraic equations on the nth jet space with coordinates x, u, u_x, u_{xx}, $\ldots$, u_{nx}. The functions to be determined from (2.19) are the coefficients ξ_i and ϕ_α. They depend only on x and u. Equation (2.19) will also involve the derivatives u_x, $\ldots$, u_{nx} explicitly. Hence, the coefficients multiplying each linearly independent expression in the derivatives must vanish identically. This provides us with a system of linear differential equations of order n (or less), the so-called determining equations for the functions ξ_i and ϕ_α. These equations are linear, even if the system (2.1) is nonlinear. Their linearity is due to the infinitesimal approach; all higher powers of the parameter ε in (2.4) are dropped (or not even introduced in the prolongation approach).

Let us sum up the algorithm for finding the symmetry algebra L of (2.1).

1. Calculate the nth prolongation (2.10) of the vector field (2.5). This does not depend on the equation (2.1), only on the number of independent and dependent variables and on the order n of the system.

2. Solve system (2.1) for m of the highest derivatives in the system. We call them $v_1, \ldots, v_m$. They must be so chosen that they can be unambiguously obtained from (2.1), that no derivative of v_i figures in the system, no v_i is a derivative of another v_j, the set is linearly independent, and each v_i involves at least one derivative (of some u^α).

3. Implement (2.19) and substitute for all v_i the expressions calculated in step 2.

4. Identify all linearly independent expressions in the remaining derivatives of u_α and set the coefficients of these expressions equal to zero. This gives us the determining equations.

5. Solve the determining equations and obtain $\xi_i(x,u)$ and $\phi_\alpha(x,u)$.

Numerous computer programs realizing the above algorithm exist. All of them realize steps 1–4 completely, step 5 to varying degrees. We mention a REDUCE program [162], a MACSYMA one [28, 29], MATHEMATICA program [13], and review articles on this topic [83, 84], containing many further references.

The system of determining equations is nearly always overdetermined and should be transformed to some canonical form, including all compatibility conditions. Finite algorithms exist that provide us with the dimension and other basic properties of the solution set [154–158].

The following possibilities occur:

1. The only solution is the trivial one: $\xi_i = 0$, $\phi_\alpha = 0$ for all i and α. No nontrivial symmetry group exists and the method is not applicable.

2. The general solution of the determining equations depends on N significant integration constants. The dimension of the symmetry algebra (and the symmetry group) is equal to $N < \infty$.

3. The general solution depends on arbitrary functions of some variables. The symmetry group is then infinite dimensional. That occurs for all linear PDEs (as a manifestation of the linear superposition principle), but also for integrable nonlinear equations in 3 dimensions [178] (e.g., the Kadomtsev–Petviashili equation [50, 51]) and all equations in the KP series [137].

3   Examples of Symmetry Groups

3.1   The Forced Korteweg–de Vries Equation

As a first example, let us consider the equation

$$E \equiv u_t + uu_x + u_{xxx} - f(x,t) = 0, \tag{3.1}$$

where $f(x,t)$ is some known, but unspecified, function. For $f = 0$ this is the famous Korteweg–de Vries equation, describing the unidirectional propagation of water waves in shallow water of constant depth and density. The KdV is the prototype integrable nonlinear evolution equation, solvable by the inverse spectral transform, having soliton solutions [1].

The forcing term $f(x,t)$ allows for a varying depth, in principle both space and time dependent. We shall call (3.1) the forced KdV equation (FKdV).

The vector fields (2.5) realizing the symmetry algebra will in this case have the form

$$\widehat{X} = \xi\partial_x + \tau\partial_t + \phi\partial_u, \tag{3.2}$$

where ξ, τ, and ϕ are functions of x, t, and u.

To obtain the symmetry algebra of (3.1) we calculate the third prolongation $\mathrm{pr}^{(3)}\,\widehat{X}$ of $\widehat{X}$ using (2.10). The equation

$$\mathrm{pr}^{(3)}\,\widehat{X} \cdot E|_{E=0} = 0 \tag{3.3}$$

in this case is

$$\phi^t + \phi u_x + u\phi^x + \phi^{xxx} - \xi\frac{\partial f}{\partial x} - \tau\frac{\partial f}{\partial t}\bigg|_{u_{xxx}=-u_t-uu_x+f} = 0 \tag{3.4}$$

with

$$\phi^t = D_t\phi - (D_t\tau)u_t - (D_t\xi)u_x$$
$$= \phi_t - \xi_t u_x + (\phi_u - \tau_t)u_t - \xi_u u_x u_t - \tau_u u_t^2, \tag{3.5}$$
$$\phi^x = \phi_x + (\phi_u - \xi_x)u_x - \tau_x u_t - \xi_u u_x^2 - \tau_u u_x u_t, \tag{3.6}$$
$$\phi^{xx} = D_x\phi^x - (D_x\xi)u_{xx} - (D_x\tau)u_{xt}, \tag{3.7}$$
$$\phi^{xxx} = D_x\phi^{xx} - (D_x\xi)u_{xxx} - (D_x\tau)u_{xxt}$$
$$= \phi_{xxx} + (3\phi_{xxu} - \xi_{xxx})u_x - \tau_{xxx}u_t$$
$$+ 3(\phi_{xuu} - \xi_{xxu})u_x^2 - 3\tau_{xxu}u_x u_t$$
$$+ (\phi_{uuu} - 3\xi_{xuu})u_x^3 - 3\tau_{xuu}u_x^2 u_t - \xi_{uuu}u_x^4$$
$$- \tau_{uuu}u_x^3 u_t + 3(\phi_{xu} - \xi_{xx})u_{xx} - 3\tau_{xx}u_{tx}$$
$$+ 3(\phi_{uu} - 3\xi_{xu})u_x u_{xx} - 6\tau_{xu}u_x u_{xt}$$
$$- 3\tau_{ux}u_t u_{xx} - 6\xi_{uu}u_x^2 u_{xx}$$
$$- 3\tau_{uu}u_x u_t u_{xx} - 3\tau_{uu}u_x^2 u_{xt} - 3\xi_u u_{xx}^2$$
$$- 3\tau_u u_{tx}u_{xx} + (\phi_u - 3\xi_x)u_{xxx} - 3\tau_x u_{txx}$$
$$- \tau_u u_t u_{xxx} - 4\xi_u u_x u_{xxx} - 3\tau_u u_x u_{xxt}. \tag{3.8}$$

We substitute (3.5), (3.6), (3.7), and (3.8) into (3.4) and eliminate u_{xxx} using (3.1). The coefficients of each different term of the type $u_t^a u_x^b u_{xx}^c \times u_{tx}^d u_{xxt}^e$ must vanish, and this provides the determining equations for the coefficients ξ, τ, and ϕ in (3.2). The coefficients of $u_x u_{xxt}$, u_{xxt}, u_{xx}^2, and $u_x u_{xx}$ yield, respectively,

$$\tau_u = 0, \quad \tau_x = 0, \quad \xi_u = 0, \quad \phi_{uu} = 0, \tag{3.9}$$

and we obtain

$$\xi = \xi(x,t), \quad \tau = \tau(t), \quad \phi = A(x,t)u + B(x,t). \tag{3.10}$$

Thus, the dependence on u is completely specified. We substitute this intermediate result into (3.4) and require that the coefficients of u^2, u_{xx}, u_x, $u u_x$, u_t, and u vanish. We obtain

$$A_x = 0, \quad \xi_{xx} = 0, \quad -\xi_t + B = 0, \quad A_t + B_x = 0,$$
$$A + 2\xi_x = 0, \quad -\frac{d\tau}{dt} + 3\xi_x = 0. \tag{3.11}$$

Thus, for any function $f(x,t)$ we find that the coefficients τ, ξ, and ϕ must have the form

$$\tau = c_1 + 3c_2 t, \quad \xi = c_2 x + \gamma(t), \quad \phi = -2c_2 u + \frac{d\gamma}{dt}, \tag{3.12}$$

where c_1 and c_2 are constants and $\gamma(t)$ is a yet undetermined function of time t. The terms in (3.4) that do not depend on u or its derivatives give the remaining determining equation, namely

$$-\frac{d^2\gamma}{dt^2} + 5c_2 f + [c_2 x + \gamma(t)]f_x + [c_1 + 3c_2 t]f_t = 0. \qquad (3.13)$$

Equation (3.13) depends crucially on the forcing term, i.e., the function $f(x,t)$.

To proceed further we can take two different approaches.

1. Specify the function $f(x,t)$ and solve (3.13) for c_1, c_2, and $\gamma(t)$.

2. Assume that a nontrivial symmetry algebra exists. Then find the function $f(x,t)$ that admits it.

As an example of the first approach, consider the KdV equation itself, i.e., $f(x,t) = 0$. Equation (3.13) then reduces to $d^2\gamma/dt^2 = 0$, i.e., $\gamma = c_3 t + c_4$. Thus the symmetry algebra of the KdV equation is four-dimensional with

$$\tau = c_1 + 3c_2 t, \quad \xi = c_2 x + c_3 t + c_4, \quad \phi = -2c_2 u + c_3. \qquad (3.14)$$

A basis for this Lie algebra is given by the vector fields

$$P_0 = \partial_t, \quad D = x\partial_x + 3t\partial_t - 2u\partial_u, \quad B = t\partial_x + \partial_u, \quad P_1 = \partial_x. \qquad (3.15)$$

The transformations corresponding to P_0, P_1, B, and D are time and space translations, Galilei transformations and dilations, respectively. To obtain the symmetry group transformations we proceed as in (2.6). We denote the group parameters corresponding to time and space translations, Galilei transformations and dilations by t_0, x_0, v, and λ respectively, to obtain

$$\tilde{x} = e^\lambda[x + x_0 + v(t + t_0)], \quad \tilde{t} = e^{3\lambda}(t + t_0), \quad \tilde{u} = e^{-2\lambda}(u + v). \qquad (3.16)$$

The statement then is; If $u(x,t)$ is a solution of the KdV equation, then so is

$$\tilde{u}(\tilde{x}, \tilde{t}) = e^{-2\lambda}[u(x,t) + v],$$
$$x = e^{-\lambda}\tilde{x} - x_0 - ve^{-3\lambda}\tilde{t}, \quad t = e^{-3\lambda}\tilde{t} - t_0. \qquad (3.17)$$

Thus, each solution yields a family of solutions, depending on up to 4 parameters. The above results for the KdV equation are, of course, all well known [131].

Let us now consider the second approach, leaving $f(x,t)$ arbitrary. Let us classify the forced KdV equations under a group of "allowed transformations," taking (3.1) into another equation of the same type, possibly with a different function $\tilde{f}(\tilde{x}, \tilde{t})$.

We shall consider only transformations of the form

$$u(x,t) = W\big(\tilde{u}(\tilde{x},\tilde{t}),x,t\big), \quad \tilde{x} = \tilde{x}(x,t), \quad \tilde{t} = \tilde{t}(x,t), \tag{3.18}$$

i.e., fiber-preserving transformations (for which the new independent variables do not depend on the old dependent one).

Substituting into (3.1) and requiring that $\tilde{u}(\tilde{x},\tilde{t})$ should satisfy

$$\tilde{u}_{\tilde{t}} + \tilde{u}\tilde{u}_{\tilde{x}} + \tilde{u}_{\tilde{x}\tilde{x}\tilde{x}} - \tilde{f}(\tilde{x},\tilde{t}) = 0, \tag{3.19}$$

we obtain that the allowed transformations (3.18) have the form

$$u(x,t) = A\tilde{u}(\tilde{x},\tilde{t}) - \frac{1}{\alpha}\frac{d\beta}{dt}, \quad \tilde{x} = \alpha x + \beta(t),$$
$$\tilde{t} = c_1 t + c_0, \quad A\alpha c_1 \neq 0, \tag{3.20}$$

where β is an arbitrary function of t and A, α, c_1, and c_0 are constants. The rescalings and translations corresponding to A, α, c_1, and c_0 can be considered separately, so we set $A = \alpha = c_1 = 1$, $c_0 = 0$ and use only the transformation

$$u(x,t) = \tilde{u}(\tilde{x},\tilde{t}) - \frac{d\beta}{d\tilde{t}}, \quad \tilde{t} = t, \quad \tilde{x} = x + \beta(t). \tag{3.21}$$

Equation (3.1) is transformed into

$$\tilde{u}_{\tilde{t}} + \tilde{u}\tilde{u}_{\tilde{x}} + \tilde{u}_{\tilde{x}\tilde{x}\tilde{x}} - \tilde{f}(\tilde{x},\tilde{t}) = 0, \tag{3.22}$$

$$\tilde{f}(\tilde{x},\tilde{t}) = f\big(\tilde{x} - \beta(\tilde{t}),\tilde{t}\big) + \frac{d^2\beta}{d\tilde{t}^2}. \tag{3.23}$$

A vector field

$$\widehat{X} = (c_1 + 3c_2 t)\partial_t + \big(c_2 x + \gamma(t)\big)\partial_x + \left(-2c_2 u + \frac{d\gamma}{dt}\right)\partial_u \tag{3.24}$$

belonging to the symmetry algebra of (3.1) will be transformed by an allowed transformation (3.21) into

$$\widehat{X} = (c_1 + 3c_2\tilde{t})\partial_{\tilde{t}} + \left\{c_2\big(\tilde{x} - \beta(\tilde{t})\big) + \gamma(\tilde{t}) + \frac{d\beta}{d\tilde{t}}(c_1 + 3c_2\tilde{t})\right\}\partial_{\tilde{x}}$$
$$+ \left\{-2c_2\left[\tilde{u} - \frac{d\beta}{d\tilde{t}}\right] + \frac{d\gamma}{d\tilde{t}} + [c_1 + 3c_2\tilde{t}]\frac{d^2\beta}{d\tilde{t}^2}\right\}\partial_{\tilde{u}}. \tag{3.25}$$

The free function $\beta(t)$ can be used to simplify the studied equation and/or the vector field (3.24), and hence (3.13).

Let us look at specific examples.

1. $f(x,t) = f(t)$.

   Choosing $d^2\beta/dt^2 = -f(t)$ in (3.23) we obtain $\tilde{f} = f + d^2\beta/dt^2 = 0$. Thus, a forcing term depending only on time can be transformed away. The symmetry algebra of (3.1) is in this case isomorphic to that of the free KdV. The 4 generators (3.15) are transformed into the basis elements of the symmetry algebra of the FKdV equation (3.1) with $f = f(t)$, namely

   $$P_0 = \partial_t + \frac{d\beta}{dt}\partial_x + \frac{d^2\beta}{dt^2}\partial_{\tilde{u}}, \quad P_1 = \partial_x, \quad B = t\partial_x + \partial_u,$$

   $$D = 3t\partial_t + \left(x - \beta + 3t\frac{d\beta}{dt}\right)\partial_x + \left(-2u + 2\frac{d\beta}{dt} + 3t\frac{d^2\beta}{dt^2}\right)\partial_u, \quad (3.26)$$

   $$\frac{d^2\beta}{dt^2} = -f(t).$$

2. Let us assume that $f(x,t)$ is such that the equation allows a symmetry operator (3.25) with $c_2 \neq 0$. We can then set $c_2 = 1$, translate time to set $c_1 = 0$, and use $\beta(t)$ to annul $\gamma(t)$. We obtain a dilation D as in (3.15). Equation (3.13) with $c_2 = 1$, $c_1 = \gamma = 0$ can be solved for $f(x,t)$, and we obtain the most general forcing for which the FKdV is scaling invariant, namely

   $$f(x,t) = t^{-5/3}F(\xi), \quad \xi = xt^{-1/3}. \quad (3.27)$$

3. Now let us assume that $c_2 = 0$, $c_1 = 1$. The function $\gamma(t)$ can be removed by an appropriate choice of $\beta(t)$, and we obtain the (trivial) result that for $f = f(x)$ the FKdV equation is invariant under time translations generated by $P_0 = \partial_t$. More generally, the FKdV equation with

   $$f(x,t) = f(x - \beta(t)) + \frac{d^2\beta}{dt^2} \quad (3.28)$$

   is invariant under transformations generated by

   $$\hat{X} = \partial_t + \frac{d\beta}{dt}\partial_x + \frac{d^2\beta}{dt^2}\partial_u. \quad (3.29)$$

4. Finally, let $c_1 = c_2 = 0$, in (3.25). Equation (3.13) then implies that the forcing term and corresponding symmetry operator are

   $$f(x,t) = \frac{x}{\gamma(t)}\frac{d^2\gamma}{dt^2} + \alpha(t),$$
   $$\hat{X} = \gamma(t)\partial_x + \frac{d\gamma}{dt}\partial_u, \quad (3.30)$$

   where $\gamma(t)$ and $\alpha(t)$ are arbitrary functions of time t.

   For a similar analysis of a variable coefficient KdV equation

   $$u_t + f(x,t)uu_x + g(x,t)u_{xxx} = 0$$

see [181].

3.2 The Modified Kadomtsev–Petviashvili Equation

Let us now consider an equation with 1 dependent variable u and 3 independent ones, x, y, and t, namely the modified Kadomtsev–Petviashvili equation (MKP):

$$(u_{xxx} - 2u_x^3 - 4u_t)_x - 6u_{xx}u_y + 3u_{yy} = 0. \qquad (3.31)$$

This is one of the equations known to be integrable by the inverse spectral transform [94].

The vector field (2.5) in this case will have the form

$$\widehat{X} = \xi\partial_x + \eta\partial_y + \tau\partial_t + \phi\partial_u, \qquad (3.32)$$

where ξ, η, τ, and ϕ depend on x, y, t, and u.

We have calculated the symmetry group using a MACSYMA package [29]. The package calculates $\mathrm{pr}^{(4)}\, X$, applies it to the MKP equation as in (2.19), eliminates the variable u_{xt} using (3.31), and prints out the following system of 20 determining equations (with $x = x_1$, $y = x_2$, $t = x_3$, $\eta_1 = \xi$, $\eta_2 = \eta$, $\eta_3 = \tau$, $\phi_1 = \phi$):

(E1) $\dfrac{\partial}{\partial u_1}(\eta_3) = 0,$

(E2) $\dfrac{\partial}{\partial u_1}(\eta_2) = 0,$

(E3) $\dfrac{\partial}{\partial x_1}(\eta_3) = 0,$

(E4) $\dfrac{\partial}{\partial u_1}(\eta_1) = 0,$

(E5) $\dfrac{\partial}{\partial x_1}(\eta_2) = 0,$

(E6) $\dfrac{\partial}{\partial x_2}(\eta_3) = 0,$

(E7) $\dfrac{\partial^2}{\partial u_1^2}(\phi_1) = 0,$

(E8) $\dfrac{\partial^2}{\partial u_1 \partial x_1}(\phi_1) = 0,$

(E9) $\dfrac{\partial^2}{\partial x_1^2}(\phi_1) = 0,$

(E10) $\dfrac{\partial^2}{\partial x_1^2}(\eta_1) = 0,$

(E11) $\quad 2\left(\dfrac{\partial}{\partial x_1}(\phi_1)\right) - \dfrac{\partial}{\partial x_2}(\eta_1) = 0,$

(E12) $\quad \dfrac{\partial^2}{\partial x_2^2}(\eta_2) - 2\left(\dfrac{\partial^2}{\partial u_1 \partial x_2}(\phi_1)\right) = 0,$

(E13) $\quad 2\left(\dfrac{\partial}{\partial x_3}(\eta_2)\right) - 3\left(\dfrac{\partial}{\partial x_2}(\eta_1)\right) = 0,$

(E14) $\quad 3\left(\dfrac{\partial}{\partial x_2}(\phi_1)\right) - 2\left(\dfrac{\partial}{\partial x_3}(\eta_1)\right) = 0,$

(E15) $\quad \dfrac{\partial}{\partial x_3}(\eta_3) - 3\left(\dfrac{\partial}{\partial x_1}(\eta_1)\right) = 0,$

(E16) $\quad 3\left(\dfrac{\partial^2}{\partial x_2^2}(\phi_1)\right) - 4\left(\dfrac{\partial^2}{\partial x_1 \partial x_3}(\phi_1)\right) = 0,$

(E17) $\quad \dfrac{\partial}{\partial x_3}(\eta_3) - 2\left(\dfrac{\partial}{\partial x_2}(\eta_2)\right) + \dfrac{\partial}{\partial x_1}(\eta_1) = 0,$

(E18) $\quad \dfrac{\partial}{\partial x_3}(\eta_3) + 2\left(\dfrac{\partial}{\partial u_1}(\phi_1)\right) - 3\left(\dfrac{\partial}{\partial x_1}(\eta_1)\right) = 0,$

(E19) $\quad 4\left(\dfrac{\partial^2}{\partial u_1 \partial x_3}(\phi_1)\right) + 3\left(\dfrac{\partial^2}{\partial x_2^2}(\eta_1)\right) - 4\left(\dfrac{\partial^2}{\partial x_1 \partial x_3}(\eta_1)\right) = 0,$

(E20) $\quad \dfrac{\partial}{\partial x_3}(\eta_3) - \dfrac{\partial}{\partial x_2}(\eta_2) + \dfrac{\partial}{\partial u_1}(\phi_1) - \dfrac{\partial}{\partial x_1}(\eta_1) = 0.$

This system is easy to solve in order to obtain the coefficients of the vector field (3.32). The results can be summed up as follows. The symmetry algebra is infinite-dimensional, since the solution of the system (E1), ..., (E20) depends on 4 arbitrary functions of time t. These functions appear as integration "constants" when we solve the determining equations. The vector fields (3.32), realizing the symmetry algebra of the MKP equation, can be written as

$$X = T(f) + X(h) + Y(g) + U(k), \tag{3.33}$$

$$T(f) = f\partial_t + \left(\frac{1}{3}x\frac{\mathrm{d}f}{\mathrm{d}t} + \frac{2}{9}y^2\frac{\mathrm{d}^2f}{\mathrm{d}t^2}\right)\partial_x + \frac{2}{3}y\frac{\mathrm{d}f}{\mathrm{d}t}\partial_y$$

$$+ \left(\frac{2}{9}xy\frac{\mathrm{d}^2f}{\mathrm{d}t^2} + \frac{4}{81}y^3\frac{\mathrm{d}^3f}{\mathrm{d}t^3}\right)\partial_u, \tag{3.34}$$

$$X(h) = h\partial_x + \frac{2}{3}y\frac{\mathrm{d}h}{\mathrm{d}t}\partial_u, \tag{3.35}$$

$$Y(g) = g\partial_y + \frac{2}{3}y\frac{\mathrm{d}g}{\mathrm{d}t}\partial_x + \frac{1}{3}\left(x\frac{\mathrm{d}g}{\mathrm{d}t} + \frac{2}{3}y^2\frac{\mathrm{d}^2g}{\mathrm{d}t^2}\right)\partial_u, \tag{3.36}$$

$$U(k) = k\partial_u, \tag{3.37}$$

where f, h, g, and k are arbitrary functions of t. A basis of the symmetry algebra is obtained, e.g., by expanding these functions in power series (which may involve both positive and negative powers).

The nonzero commutation relations for this algebra are

$$[T(f_1), T(f_2)] = T\left(f_1 \frac{df_2}{dt} - \frac{df_1}{dt} f_2\right), \tag{3.38}$$

$$[X(h), Y(g)] = \frac{1}{3} U\left(h \frac{dg}{dt} - \frac{dh}{dt} g\right),$$

$$[Y(g_1), Y(g_2)] = \frac{2}{3} X\left(g_1 \frac{dg_2}{dt} - \frac{dg_1}{dt} g_2\right), \tag{3.39}$$

$$[T(f), X(h)] = X\left(f \frac{dh}{dt} - \frac{1}{3} \frac{df}{dt} h\right),$$

$$[T(f), Y(g)] = Y\left(f \frac{dg}{dt} - \frac{2}{3} \frac{df}{dt} g\right), \tag{3.40}$$

$$[T(f), U(k)] = U\left(f \frac{dk}{dt}\right). \tag{3.41}$$

These commutation relations characterize a centerless Kac–Moody–Virasoro algebra [74, 178]. The Virasoro part is generated by $T(f)$. The Kac–Moody one by $X(h)$, $Y(g)$, and $U(k)$. The Kac–Moody algebra is an ideal in the entire structure. The existence of a Kac–Moody–Virasoro Lie point symmetry algebra is typical for integrable nonlinear evolution equations in $2 + 1$ dimensions [30, 50, 51, 104, 109, 122, 137, 161, 178].

Finally, let us mention that some obvious symmetries are obtained by restricting the functions f, g, h, k to be linear. Thus, the translations are

$$P_0 = T(1) = \partial_t, \quad P_1 = X(1) = \partial_x, \quad P_2 = Y(1) = \partial_y.$$

Similarly, we have

$$D = T(t) = t\partial_t + \frac{1}{3} x\partial_x + \frac{2}{3} y\partial_y,$$

$$B_1 = X(t) = t\partial_x + \frac{2}{3} y\partial_u,$$

$$B_2 = Y(t) = t\partial_y + \frac{2}{3} y\partial_x + \frac{1}{3} x\partial_u,$$

where, D, B_1, and B_2 correspond to dilations and Galilei transformations in the x and y directions, respectively.

The symmetry $U(k)$ simply tells us that if $u(x, y, t)$ is a solution of the MKP equation, then so is $\tilde{u} = u + k(t)$.

4 Symmetry Reduction, Group Invariant Solutions, Partially Invariant Solutions

4.1 Outline of the Method

The most important application of the symmetry group G of Lie point transformations, leaving a system of PDEs invariant, is to perform symmetry reduction. In the case of PDEs that means a reduction of the number of independent variables in the equation. In particular, it may be possible and desirable to reduce to an ODE or even to an algebraic equation.

The basic idea of symmetry reduction is to take some subgroup $G_0 \subseteq G$ of the symmetry group G and look for solutions that are invariant under G_0 (rather than transformed into other solutions). Requiring invariance is equivalent to imposing additional first-order linear equations on solutions. These can be solved and the result substituted into the original equations. This provides the reduced systems to be solved. The procedure involves an obvious loss of generality. We obtain particular solutions for which boundary conditions can be imposed on surfaces invariant under the chosen group G_0, rather than on arbitrary surfaces.

The procedure for performing symmetry reduction can be outlined as an algorithm, consisting of the following steps.

1. Find the symmetry group G of local point transformations (2.2), leaving the considered system (2.1) invariant, and obtain the corresponding Lie algebra L of vector fields (2.5). The method for doing this was presented in Section 2 above. As mentioned above, this step has been computerized [13, 28, 84, 155, 156, 162, 163].

2. Identify the symmetry algebra L as an abstract Lie algebra. In step 1 one obtains a finite or infinite set of linearly independent vector fields. From these one can calculate the structure constants C_{ik}^l of the Lie algebra. The structure constants are basis-dependent. The aim is to extract basis-independent information from them. The idea is to transform to a "canonical basis," in which all the basis-independent properties are obvious. In particular, if the Lie algebra is decomposable, it should be decomposed into the direct sum of indecomposable Lie algebras,

$$L = L_1 \oplus L_2 \oplus \cdots \oplus L_n. \tag{4.1}$$

For each indecomposable component L_1 one should obtain its Levi decomposition [93]

$$L = S \niplus R, \quad [S, S] = S, \quad [S, R] \subseteq R, \quad [R, R] \subseteq R, \tag{4.2}$$

where S is semisimple and R is the radical of L (maximal solvable ideal). For each solvable Lie algebra R it is useful to identify its nilradical (maximal nilpotent ideal) [93].

For low-dimensional Lie algebras it is possible to go further and to completely identify their isomorphism class.

Algorithms exist for identifying a Lie algebra from its structure constants [153], and they have been at least partly computerized [149, 150, 152]. They are valid for finite-dimensional Lie algebras, and need to be generalized to infinite-dimensional ones.

3. Classify the subalgebras of L into conjugacy classes under the action of the Lie group G. This is a classification under inner automorphisms of L, if G is restricted to being the connected component $\exp L$ of G. In some cases it is convenient to enlarge the classification group G to include discrete transformations, leaving the studied equations invariant. This problem will be discussed below in Section 5.

The reason this classification is needed is that each subgroup G_0, corresponding to a different conjugacy class of subalgebras L_0, will give a different type of invariant solution.

4. Consider a subalgebra $L_0 \subseteq L$, representing a class of subalgebras found in step 3, and the corresponding subgroup $G_0 \subseteq G$. The group G_0 acts on the space $M \sim X \times U$ of independent and dependent variables. Find the invariants of this action, i.e., the functionally independent solutions

$$\tilde{I}_j(x, u), \quad j = 1, \ldots, N, \tag{4.3}$$

of the set of first-order linear PDEs

$$X_i F(x, u) = 0, \quad i = 1, \ldots, n_0, \tag{4.4}$$

where X_i are vector fields of the form (2.5) that form a basis of the Lie algebra L_0. The number of invariants N is equal to the codimension of the generic orbits of G_0 in M:

$$N = p + q - d,$$

where d is the dimension of these orbits.

The following cases can arise:

A. Among the invariants $\tilde{I}_j(x, u)$ it is possible to choose q functions $I_j(x, u)$ that provide an invertible mapping to the dependent variables. The Jacobian determinant then satisfies

$$J \equiv \left(\frac{\partial(I_1, \ldots, I_q)}{\partial(u_1, \ldots, u_q)} \right), \quad \det J \neq 0. \tag{4.5}$$

The remaining $k = N - q$ invariants can be chosen to depend only on the independent variables, and we denote them by

$$\xi_1(x), \ldots, \xi_k(x), \quad k < p. \tag{4.6}$$

Now let us restrict to the solution set of (2.1). We consider u_j as functions of x, and this can be imposed by setting

$$I_i = F_i(\xi_1, \ldots, \xi_k). \tag{4.7}$$

Using condition (4.5) we solve (4.7) for the dependent variables and obtain

$$u_i(x) = U_i(x, F_i(\xi)). \tag{4.8}$$

Upon substitution into (2.1) we obtain a set of equations involving only the functions F_i, $i = 1, \ldots, q$, the variables ξ_a ($a = 1, \ldots, k$), and derivatives of F_i with respect to ξ_a. Since the original equation is G-invariant and (4.4) provides a complete set of G_0 invariants, the noninvariant quantities x in (4.8) must drop out.

Since we have $k < p$, we have reduced the number of independent variables. If the reduced equations are solved for $F_i(\xi)$, substitution into (4.8) provides solutions of the original system.

B. Equation (4.5) is satisfied, but the complementary variables ξ_i of (4.6) also depend upon u. We proceed as above. However, substitution of $F_i(\xi)$ with $\xi = \xi(x, u)$ yields implicit solutions, rather than explicit ones, i.e., (4.8) is a functional equation for u. In some cases this can be solved, and we obtain further explicit solutions.

C. The condition (4.5) on the Jacobian is not satisfied. Let the rank of the Jacobian J be

$$1 \leq r(J) = q' < q. \tag{4.9}$$

We choose q' of the invariants $\tilde{I}_j(x, u)$ such that we can invert q' relations of the type (4.7) for q' dependent variables, say $u_1, \ldots, u_{q'}$. We then have

$$u_i(x) = U_i(x, F_i(\xi)), \quad i = 1, \ldots, q'. \tag{4.10}$$

The remaining variables $u_{q'+1}, \ldots, u_q$ depend on all the original ones $x_1, \ldots, x_p$. Substituting into system (2.1) we obtain a system of PDEs in which some of the dependent variables $F_i(\xi)$ depend on fewer variables than the remaining ones. This imposes consistency conditions on the reduced equations. Solutions of the reduced equations then provide "partially invariant solutions" of the original equations. This concept was introduced by Ovsiannikov [138]. The definition used here is different from the one given by Ovsiannikov, but the concepts are related. Partially invariant solutions very often turn out to coincide with invariant ones, i.e., the compatibility conditions force solutions to be invariant under some subgroup of the symmetry group G. Genuinely partially invariant solutions do exist in some cases [121, 123], and we shall give examples of them.

5. Solve the reduced equations. This step is, of course, less algorithmic than the previous ones. The reduced equation may be integrable, even if the original one was not. Thus, it may be transformable into a linear equation, or solvable by inverse spectral transform techniques [1]. If necessary, group theory can be applied once more to the reduced equation. If it is a PDE, we can further reduce the number of independent variables. For ODEs one can reduce the order of the equation. An approach that is often very fruitful is Painlevé analysis: an analysis of the singularity structure of the solutions of the reduced PDE, or ODE [2,3,108,131,151,175,176]. See also R. Conte [47], M. Musette [126], and Grammaticos et al. [77] in this volume.

6. The last step is nonalgorithmic, namely to use the obtained solutions: analyze their stability, their asymptotic behavior, calculate observable quantities, etc.

4.2 Example of the KdV Equation

Let us consider the KdV equation

$$u_t + uu_x + u_{xxx} = 0 \tag{4.11}$$

and apply the method described in Section 4.1.

4.2.1 Step 1

The symmetry algebra was already found and is given in (3.15).

4.2.2 Step 2

The commutation relations for this Lie algebra are

$$[P_0, D] = 3P_0, \quad [B, D] = -2B, \quad [P_1, D] = P_1,$$
$$[P_0, B] = P_1, \quad [P_0, P_1] = 0, \quad [B, P_1] = 0. \tag{4.12}$$

We have an indecomposable Lie algebra that is solvable. Its nilradical (maximal nilpotent ideal) is $\{B, P_0, P_1\}$, itself isomorphic to the Heisenberg algebra. We have $L \sim D \ni \{B, P_0, P_1\}$.

4.2.3 Step 3

Methods for classifying subalgebras of Lie algebras will be presented below. Here we only need one-dimensional subalgebras of L, and the classification is quite simple. A one-dimensional subalgebra will have the form

$$X = aD + bB + cP_0 + dP_1. \tag{4.13}$$

We shall classify subalgebras under the action of the group of inner automorphisms. We shall use the Baker–Campbell–Hausdorff formula [185]

$$e^{\alpha Y} X e^{-\alpha Y} = X + \alpha[Y, X] + \frac{\alpha^2}{2!}[Y, [Y, X]] + \cdots. \qquad (4.14)$$

For $a \neq 0$ in (4.13) we set $a = 1$, choose $Y = \alpha_1 B + \alpha_2 P_0 + \alpha_3 P_1$ in (4.14), and find that we can annul b, c, and d by an appropriate choice of α_1, α_2, α_3.

Next, let us consider $a = 0$, $b = 1$. Choosing $Y = \lambda P_0$ in (4.14), we can annul d in (4.13). Setting $Y = \mu D$ we can scale c to $c = \pm 1$ for $c \neq 0$. For $a = 0$, $b = 0$, $c = 1$ we can transform $P_0 + dP \mapsto P_0$.

Finally, we obtain a list of representatives of one-dimensional subalgebras of L, namely

$$\begin{aligned} L_{1,1} &\sim \{D\}, \quad L_{1,2}^{(a)} \sim \{B + aP_0\} \quad a = 0, 1, -1, \\ L_{1,3} &= \{P_0\}, \quad L_{1,4} = \{P_1\}. \end{aligned} \qquad (4.15)$$

4.2.4 Step 4

Let us now perform the symmetry reduction, i.e., find solutions invariant under the individual subgroups.

Translationally Invariant Solutions

For convenience we combine $L_{1,3}$ and $L_{1,4}$. The invariants of the corresponding group action are obtained by solving the equation

$$(P_0 + cP_1)F(x, t, u) = 0. \qquad (4.16)$$

We obtain two elementary invariants

$$I_1 = \xi = x - ct, \quad I_2 = u. \qquad (4.17)$$

Hence, on the solution set we have

$$u = F(\xi), \quad \xi = x - ct. \qquad (4.18)$$

Substituting (4.18) into the KdV equation, we obtain an ordinary differential equation

$$-c\frac{dF}{dt} + F\frac{dF}{dt} + \frac{d^3F}{dt^3} = 0, \qquad (4.19)$$

where the prime is a derivative with respect to ξ. Integrating (4.19) twice, we obtain

$$\left(\frac{dF}{dt}\right)^2 = -\frac{1}{3}F^3 + cF^2 + kF + l, \qquad (4.20)$$

where k and l are integration constants. This can be rewritten as

$$\left(\frac{dF}{dt}\right)^2 = -\frac{1}{3}(F - F_1)(F - F_2)(F - F_3) \tag{4.21}$$

with the roots F_1, F_2, and F_3 satisfying

$$F_1 + F_2 + F_3 = 3c, \quad F_1 F_2 + F_2 F_3 + F_3 F_1 = -3k,$$
$$F_1 F_2 F_3 = 3l. \tag{4.22}$$

For F_1, F_2, and F_3 real and distinct we put $F_1 < F_2 < F_3$, and (4.21) has a real nonsingular solution

$$u(x, t) = (F_3 - F_2)\,\mathrm{cn}^2[\alpha(x - ct) + \delta, p] + F_2, \tag{4.23}$$

$$\alpha = \frac{1}{2}\sqrt{\frac{F_3 - F_2}{3}}, \quad p = \sqrt{\frac{F_3 - F_2}{F_3 - F_1}}, \quad \delta = \text{const},$$

where $\mathrm{cn}(u, p)$ is a Jacobi elliptic function.

For $F_1 = F_2$ we obtain the limiting case

$$F = (F_3 - F_2)\,\mathrm{sech}^2\left[\frac{1}{2}\sqrt{\frac{1}{3}(F_3 - F_1)}(x - ct) + \delta\right] + F_2. \tag{4.24}$$

For $F_1 = F_2 = 0$, $F_3 = 3c$, i.e., $k = l = 0$ in (4.20), we obtain

$$F = 3c\,\mathrm{sech}^2\left[\frac{1}{2}\sqrt{c}(x - ct) + \delta\right], \quad c > 0. \tag{4.25}$$

Thus, symmetry reduction has provided us with some very important solutions of the KdV equation. The "cnoidal waves" (4.23) are periodic waves on a constant background. Equation (4.24) describes a soliton of amplitude $(F_3 - F_2)$ on a constant background F_2, and (4.25) gives the usual soliton satisfying $u(x, t) \to 0$ for $t \to \infty$ for any fixed value of x. We mention the (well-known) fact that the amplitude $3c$ and the velocity c of the soliton are proportional (i.e., the bigger the wave, the faster it goes).

Space Independent Solutions

The algebra $P_1 = \partial_x$ leads to the invariants t and u. Putting $u = F(t)$ into the KdV we obtain the trivial solution $u = u_0 = \text{const}$.

Galilei-Invariant Solutions

Let us consider the algebra $L_{1,2}(a)$ for $a = 0$. The equation $BF(x, t, u) = 0$ in this case implies that the invariants are $I_1 = \xi = t$, $I_2 = u - x/t$. The reduction formula hence is

$$u = \frac{x}{t} + F(t), \quad \frac{dF}{dt} + \frac{1}{t}F = 0. \tag{4.26}$$

The Galilei-invariant solutions of the KdV equation hence are

$$u = \frac{x + x_0}{t}, \quad x_0 = \text{const.} \tag{4.27}$$

Invariance under $L_{1,2}(a)$, $a \neq 0$, implies

$$I_1 = \xi = x - \frac{1}{2a}t^2, \quad I_2 = u - \frac{t}{a}. \tag{4.28}$$

We hence put $I_2 = F(\xi)$, substitute into the KdV, and obtain an ordinary differential equation. Integrating once we obtain

$$\frac{d^2F}{dt^2} = -\frac{1}{2}F^2 + \frac{1}{a}\xi + b. \tag{4.29}$$

Rescaling F and ξ after shifting $\xi \mapsto \xi - ab$, we obtain the equation for the first Painlevé transcendent P_I [92, 108],

$$\frac{d^2w}{dz^2} = 6w^2 + z. \tag{4.30}$$

Scaling Invariant Solutions

Finally, let us consider the subalgebra $L_{1,1}$. To find the invariants we put

$$DF = (x\partial_x + 3t\partial_t - 2u\partial_u)F = 0 \tag{4.31}$$

and obtain

$$I_1 = \xi = xt^{1/3}, \quad I_2 = ut^{2/3}. \tag{4.32}$$

The reduction formula and reduced equation hence are

$$u(x,t) = t^{-2/3}F(\xi), \quad \xi = \frac{x}{t^{1/3}} \tag{4.33}$$

$$\frac{d^3F}{dt^3} + F\frac{dF}{dt} - \frac{1}{3}\xi\frac{dF}{dt} - \frac{2}{3}F = 0. \tag{4.34}$$

This equation can be reduced [131] to the equation for the second Painlevé transcendent P_{II} [92, 108],

$$\frac{d^2w}{dz^2} = 2w^3 + zw + \alpha, \tag{4.35}$$

where α is an arbitrary constant.

4.3 Example of a Nonlinear Schrödinger Equation in $3 + 1$ Dimensions

Let us now consider a nonlinear partial differential equation that is not integrable and hence not expected to have the Painlevé property. The equation is the nonlinear Schrödinger equation in $3 + 1$ dimensions with cubic and quintic nonlinear terms

$$i\psi_t + \psi_{xx} + \psi_{yy} + \psi_{zz} = a_1|\psi|^2\psi + a_2|\psi|^4\psi$$
$$a_1, a_2 \in \mathbb{R}, \ \psi(\vec{r}, t) \in \mathbb{C}, \ (a_1, a_2) \neq (0, 0). \quad (4.36)$$

The symmetry group of this equation is known [67–70] and was obtained using the algorithm described in Section 2.

If the coefficients a_1 and a_2 both satisfy $a_1 \neq 0$, $a_2 \neq 0$, then the symmetry group is the extended Galilei group. Its Lie algebra is given by the vector fields

$$P_1 = \partial_x, \quad P_2 = \partial_y, \quad P_3 = \partial_z, \quad P_0 = \partial_t,$$
$$J_1 = z\partial_y - y\partial_z, \quad J_2 = x\partial_z - z\partial_x, \quad J_3 = y\partial_x - x\partial_y,$$
$$K_1 = t\partial_x - \frac{1}{2}x(u_2\partial_{u_1} - u_1\partial_{u_2}), \quad K_2 = t\partial_y - \frac{1}{2}y(u_2\partial_{u_1} - u_1\partial_{u_2}),$$
$$K_3 = t\partial_z - \frac{1}{2}z(u_2\partial_{u_1} - u_1\partial_{u_2}),$$
$$M = u_2\partial_{u_1} - u_1\partial_{u_2},$$
$$(4.37)$$

where u_1 and u_2 are the real and imaginary parts of ψ.

If $a_1 = 0$, or $a_2 = 0$, we have an additional symmetry, namely dilations:

$$D = 2t\partial_t + (x\partial_x + y\partial_y + z\partial_z) - \alpha(u_1\partial_{u_1} + u_2\partial_{u_2}) + 2a_0t(u_2\partial_{u_1} - u_1\partial_{u_2}),$$
$$\alpha = \begin{cases} \frac{1}{2} & \text{for } u_1 = 0, a_2 \neq 0 \\ 1 & \text{for } a_1 \neq 0, a_2 = 0. \end{cases} \quad (4.38)$$

We see that P_1, P_2, P_3, and P_0 generate translations, J_i generate rotations, K_i Galilei transformations in the x, y, and z directions, respectively, whereas M corresponds to adding a constant to the phase of ψ.

For a subalgebra classification we refer to the original articles [67–70]. Here let us just consider the situation where boundary conditions are imposed on a sphere. The boundary conditions will break the symmetry of the equation, and the remaining symmetry of the system will be a subgroup G_0 of the symmetry group G.

Let the boundary conditions have the form

$$\psi(\vec{r}, t)|_{t=t_0, r=R} = m_0 e^{i\phi_0}, \quad \frac{\partial\psi}{\partial r}(\vec{r}, t)\bigg|_{t=t_0, r=R} = n_0 e^{i\theta_0}. \quad (4.39)$$

This means that for the time $t = t_0$, ψ must depend on the radius $r = |\vec{r}|$ alone.

The boundary is invariant under rotations. This leads us to subgroups involving $\{J_1, J_2, J_3\}$ and possibly other elements.

For $a_1 \neq 0$, $a_2 \neq 0$, there is only one class of such algebras, namely

$$\{J_1, J_2, J_3, P_0 + aM\}, \quad a = \text{const.} \tag{4.40}$$

For $a_1 a_2 = 0$, $a_1^2 + a_2^2 \neq 0$, there are two more, namely

$$\{J_1, J_2, J_3, D + bM\}, \quad \{J_1, J_2, J_3, P_0, D + bM\}. \tag{4.41}$$

Using just rotational invariance we find that the invariants implied by $\{J_1, J_2, J_3\}$ are $\{r, t, u_1, u_2\}$. Putting $\psi = \psi(r, t)$ we obtain the "spherical" nonlinear Schrödinger equation

$$i\psi_t + \psi_{rr} + \frac{2}{r}\psi_r = a_1\psi|\psi|^2 + a_2\psi|\psi|^4. \tag{4.42}$$

To reduce to an ordinary differential equation we need a further symmetry. Let us consider the algebra (4.40), i.e., add the operator $P_0 + aM$. The function $\psi(r, t)$ satisfying (4.42) and invariant under these additional transformations satisfies

$$\psi(r, t) = M(r)e^{i\chi(r)}e^{-iat}. \tag{4.43}$$

We substitute (4.43) into (4.42) and separate the real and imaginary parts of the reduced equation. The obtained equations can be decoupled and partially solved. We obtain

$$\chi = S_0 \int \frac{dr}{r^2 M^2} + \chi_0, \tag{4.44}$$

$$\frac{d^2 M}{dt^2} - \frac{S_0^2}{r^4 M^3} + \frac{2}{r}\frac{dM}{dt} = -aM + a_1 M^3 + a_2 M^5, \tag{4.45}$$

where S_0 and X_0 are real constants.

To solve equation (4.45) in general is difficult; in particular, this ODE has no further symmetries to help us reduce its order.

Instead, we try to transform it into an equation with the Painlevé property. Let us consider the case $a_2 \neq 0$, and subject the equation to the Painlevé test [2, 3]. The test indicates that a transformation is needed of the type $M(r) = H(r)^{1/2}$ to be able to balance the leading powers after the expansion:

$$H = H_{-1}(r - r_0)^{-1} + H_0 + H_1(r - r_0)^1 + \cdots. \tag{4.46}$$

The Painlevé test is passed only if we have $a = a_1 = 0$.

Putting

$$M(r) = \sqrt{\frac{1}{r}W(\eta)}, \quad \eta = \left(\frac{4}{3}a_2\right)^{1/2}\ln r, \tag{4.47}$$

we obtain

$$\frac{d^2W}{dt^2} = \frac{1}{2W}\left(\frac{dW}{dt}\right)^2 + \frac{3}{2}W^3 + \frac{3}{8a_2}W + \frac{3S_0^2}{2a_2}\frac{1}{W}. \tag{4.48}$$

This equation is listed as PXXX in Ince's book [92]. It is one of the equations that has a rational first integral C and can thus be reduced to the equation for elliptic functions:

$$\left(\frac{dW}{dt}\right)^2 = W^4 + \frac{3}{4a_2}W^2 + CW - \frac{3S_0^2}{a_2}$$
$$= (W - W_1)(W - W_2)(W - W_3)(W - W_4). \tag{4.49}$$

We thus obtain a large number of solutions in terms of elliptic functions [67–70], or in the case of multiple roots, elementary functions.

For instance, for $a_2 < 0$, $S_0 = C = 0$ we obtain

$$\psi = \left(-\frac{3}{a_2}\right)^{1/4}\left(\frac{r_0}{r^2 + r_0^2}\right)^{1/2}e^{i\chi_0}. \tag{4.50}$$

4.4 Example of Partially Invariant Solutions of a Nonlinear Schrödinger Equation

Partially invariant solutions, as discussed in Section 4.1, can occur only for systems of two or more equations. As an example, let us consider one complex equation, amounting to a system of two real ones. The equation is a rather general nonlinear Schrödinger equation in $1 + 1$ dimensions

$$iu_t + u_{xx} = (F + iK)u + (G + iL)u_x, \tag{4.51}$$

where $u(x, t)$ is a complex function of two real variables x, t, and F, G, K and L are real functions of $|u|$ and $|u|_x$.

Partially invariant solutions of this equation were studied in [121]. The "direct method" of dimensional reduction has been applied to a special case of equation (4.51), namely the one with

$$F = -(a_1 + b_1)|u|_x^2 - c|u|^4 - d|u|^2, \quad G = -a_1|u|^2$$
$$L = -a_2|u|^2, \quad K = -(a_2 + b_2)|u|_x^2, \tag{4.52}$$

with a_1, a_2, b_1, b_2, c, and d real constants [32].

Let us first rewrite (4.51) as a pair of real equations. We put

$$u(x,t) = \rho(x,t)e^{i\omega(x,t)}, \quad 0 \le \rho < \infty, \quad 0 \le \omega \le 2\pi,$$

and transform (4.51) to

$$\rho_{xx} - \rho\omega_x^2 - \rho\omega_x = F\rho + G\rho_x + L\rho\omega_x,$$
$$\rho_t + 2\rho_x\omega_x + \rho\omega_{xx} = K\rho + G\rho\omega_x + L\rho_x. \tag{4.53}$$

Applying the symmetry group algorithm, we find that (4.53) (and (4.51)) is invariant for arbitrary F, G, K, and L under the abelian group generated by

$$P_0 = \partial_t, \quad P_1 = \partial_x, \quad \text{and} \quad W = \partial_\omega. \tag{4.54}$$

In special cases the symmetry algebra is larger. Thus, for $G = L = 0$ it includes Galilei transformations

$$B = t\partial_x + \frac{1}{2}x\partial_\omega, \tag{4.55}$$

in other cases, e.g., dilations.

Let us consider the general case of (4.53) and take a two-dimensional subalgebra of the symmetry algebra, namely $\{P_1, W\}$. The invariants in the space $\{x, t, \rho, \omega\}$ are just $I_1 = t$, $I_2 = \rho$. Thus, the phase ω is not involved. Hence, we put

$$\rho = \rho(t), \quad \omega = \omega(x, t). \tag{4.56}$$

Substituting into (4.53) we obtain

$$\omega_t = -F + L\omega_x - \omega_x^2, \tag{4.57}$$

$$\omega_{xx} - G\omega_x = K - \frac{\rho_t}{\rho}. \tag{4.58}$$

Let us consider two cases separately.

4.4.1 $G \ne 0$

The general solution of (4.58) can be written as

$$\omega(x,t) = \beta(t)e^{xG} + \frac{1}{G}\left(\frac{\rho_t}{\rho} - K\right)x + \alpha(t) \tag{4.59}$$

(in view of (4.56) we now have $G_x = F_x = K_x = L_x = 0$). Next, we substitute (4.59) into (4.57) and compare the coefficients of e^{2xG}, e^{xG}, xe^G, x, and x^0. We obtain

$$\beta(t) = 0, \quad \frac{1}{G}\left(\frac{\rho_t}{\rho} - K\right) = a = \text{const.} \tag{4.60}$$

The solution is

$$\omega(x,t) = ax + \alpha(t) \tag{4.61}$$

and is hence invariant under a different subgroup of the symmetry group, namely the one generated by $P_1 + aW$.

4.4.2 $G = 0$

The solution of (4.58) in this case is

$$\omega = \frac{1}{2}\left(K - \frac{\rho_t}{\rho}\right)x^2 + \alpha(t)x + \beta(t). \tag{4.62}$$

Using (4.57) and comparing coefficients of x^2, x, and x^0 we obtain

$$\rho_t = \left[K(\rho,0) - \frac{1}{2t}\right]\rho, \tag{4.63}$$

$$\alpha(t) = \frac{\alpha_0}{t} + \frac{1}{2t}\int L(\rho,0)dt, \tag{4.64}$$

$$\beta(t) = \int \left[-F(\rho,0) + L(\rho,0)\alpha - \alpha^2\right]dt + \beta_0. \tag{4.65}$$

We thus have a solution of (4.53) with

$$
\begin{aligned}
\rho &= \rho(t,c_1), \\
\omega &= \frac{x^2}{4t} + \left[\alpha_0 + \frac{1}{2}\int L(\rho,0)dt\right]\frac{x}{t} + \beta(t),
\end{aligned}
\tag{4.66}
$$

where $\rho(t,c_1)$ is a solution of (4.63) and $\alpha(t)$, $\beta(t)$ are given above.

The solution (4.66) is obtained for any function $L(|u|, |u|_x)$. If we have $L = 0$ (in addition to $G = 0$), the studied equation (4.51) is Galilei-invariant, and solution (4.66) is invariant under the group generated by $B \mid aW$. However, for $L \neq 0$, (4.66) still provides a solution, not invariant under any subgroup of the symmetry group. Hence, for $L \neq 0$ we have a genuine partially invariant solution. One way of putting it is that (4.66) provides a Galilei invariant solution of an equation that is not Galilei invariant.

For other partially invariant solutions, see [78–80, 121, 123].

5   Classification of the Subalgebras of a Finite-Dimensional Lie Algebra

5.1   Formulation of the Problem

We have seen in Section 4 that a classification of the subalgebras of the symmetry algebra L of a system of differential equations is an essential part of the systematic application of symmetry reduction. Such a classification has many other applications. For instance, in group representation theory one can use chains of subgroups to label basis functions of representations. When separating variables in a linear partial differential equation, chains of subgroups can be used to introduce certain types of coordinates [124]. A systematic study of symmetry breaking in physics involves a study of the subgroups of the symmetry group of the considered system [12, 19].

The classification problem can be formulated as follows.

Let L be a finite-dimensional Lie algebra and G a group of automorphisms of L. Symbolically, we write

$$GLG^{-1} \sim L. \tag{5.1}$$

Two subalgebras $L_i \subset L$ and $L_i' \subset L$ are mutually conjugate if we have

$$GL_iG^{-1} = L_i'. \tag{5.2}$$

Our aim is to produce a representative list of G-conjugacy classes of subalgebras of L, i.e., a list $S(L)$ such that every subalgebra of L is conjugate to precisely one algebra in the list.

Two different approaches have been applied to this classification problem. The first is a structural one. Restrictions are imposed on the type of the Lie algebra L considered (e.g., L is simple) and type of subalgebra L_i considered. Then it may not be necessary to specify the dimension of L. Thus, Dynkin classified all semisimple subalgebras of the complex semisimple Lie algebras [54] and also all of their maximal subalgebras [55]. Similar work for the real semisimple Lie algebras was done by Cornwell [48] and Komrakov [98,99]. Similarly, a large body of work exists on maximal abelian subalgebras of the classical Lie algebras [89, 90, 96, 100, 130, 144, 168–170].

The other approach is enumerative. The algebra L is completely specified (e.g., $\mathrm{sl}(3, \mathbb{C})$, or $\mathrm{o}(3, 2)$), and we wish to classify all subalgebras $L_i \subset L$, irrespectively of their character.

In this chapter we concentrate on the second approach. The classification group G will be the group of inner automorphisms of L, $G \sim \langle \exp L \rangle$, though often it is convenient to use a larger group $\widetilde{G}$, e.g., $\widetilde{G} = G_D \rtimes G$ where G_D is a discrete group, also leaving the considered equation invariant.

Three different cases occur, and three different algorithms exist for treating them:

1. The Lie algebra L is simple (i.e., contains no nontrivial ideals).

2. The Lie algebra L is the direct sum of two algebras:

$$L = L_1 \oplus L_2, \quad [L_i, L_i] \subseteq L_i, \; i = 1, 2, \quad [L_1, L_2] = 0. \tag{5.3}$$

3. The Lie algebra L is a semidirect sum of two algebras

$$L \sim F \rtimes N, \quad [F, F] \subseteq F, \quad [N, N] \subseteq N, \quad [F, N] \subseteq N,$$
$$F \neq \varnothing, \quad N \neq \varnothing. \tag{5.4}$$

5.2   Subalgebras of a Simple Lie Algebra

Let L be a simple Lie algebra. We start the subalgebra classification by finding all *maximal* subalgebras $L_M \subset L$. A subalgebra is maximal if

$$L_M \subseteq \tilde{L} \subseteq L, \quad [\tilde{L}, \tilde{L}] \subseteq \tilde{L} \tag{5.5}$$

implies

$$\tilde{L} = L_M \quad \text{or} \quad \tilde{L} = L. \tag{5.6}$$

To find and classify all maximal subalgebras of L we first choose a finite dimensional faithful linear representation $E(L)$ of L and the corresponding representation $E(G)$ of the classifying group G. $E(L)$ and $E(G)$ act on the linear space V, and we have $\dim V = N_0 < \infty$. (It is advantageous to choose N_0 as small as possible.)

Any subalgebra $L_0 \subset L$, in particular any maximal subalgebra $L_M \subset L$, will be embedded in $E(L)$ either reducibly or irreducibly.

If $E(L_M)$ is embedded in $E(L)$ *reducibly*, then it must leave some proper subspace $V_0 \subset V$ invariant,

$$E(L_M)V_0 \subseteq V_0. \tag{5.7}$$

The procedure then is to classify subspaces into conjugacy classes under $E(G)$ and to choose a representative subspace V^R for each class:

$$E(G)V_0 = V_0^R, \quad V_0^R \subseteq V. \tag{5.8}$$

Once the spaces V_0^R are chosen, we obtain $E(L_m^R)$ (and hence L_m^R) as the maximal subalgebra leaving V_0^R invariant

$$E(L_M^R)V_0^R \subseteq V_0^R. \tag{5.9}$$

Determining V_0^R and $E(L_M^R)$ is thus reduced to linear algebra. Moreover, the subspaces V_0 are completely characterized by their dimension alone if no G-invariant form on V exists (e.g., for $\mathrm{sl}(n, \mathbb{R})$ or $\mathrm{sl}(n, \mathbb{C})$). If an invariant form exists, then V_0 is characterized by its signature, the number of positive length, negative length, and isotropic vectors in any orthogonal basis for V_0 (e.g., for the pseudo-orthogonal Lie algebra $\mathrm{o}(p, q)$).

If $E(L_M)$ is embedded irreducibly in $E(L)$, then it must be simple, semisimple, or reductive, i.e., a direct sum of simple Lie algebras and an abelian one. The semisimple subalgebras of all finite-dimensional semisimple Lie algebras are known, both in the complex [54,55] and the real [48,98,99] cases. The maximal reductive subalgebras L_M of L are found from the maximal semisimple ones $\tilde{L}_M$ as their centralizers,

$$L_M = \mathrm{cent}(\tilde{L}_M, L) = \{x \in L \mid [x, \tilde{L}_M] = 0\}. \tag{5.10}$$

Thus, all irreducibly embedded maximal subalgebras are either simple or direct sums. The reducibly embedded ones can be simple or direct sums; they can be solvable or have a nontrivial Levi decomposition (4.2) with $S \neq 0$, $R \neq 0$.

If the maximal subalgebra L_M is simple, we apply the same method to it. If it is a direct sum, we apply the method described in Section 5.3 below. If it has a nontrivial ideal, we apply the method referred to in Section 5.4.

As an example, let us consider the pseudo-orthogonal Lie algebra o(4, 2). This is, by definition, the Lie algebra of real matrices $X \in \mathbb{R}^{6 \times 6}$, satisfying

$$XK + KX^T = 0, \quad K = K^T \in \mathbb{R}^{6 \times 6}, \tag{5.11}$$

where K is a matrix of signature $(4, 2)$, i.e.,

$$GKG^T = I_{4,2}, \quad I_{4,2} = \operatorname{diag}(1, 1, 1, 1, -1, -1), \tag{5.12}$$

for some $G \in \mathrm{SL}(6, \mathbb{R})$.

Let us first find all reducibly embedded subalgebras. In this case $E(L) = E\big(o(4, 2)\big)$ is the representation (5.11). Invariant subspaces of V_6 of dimension 1, 2, and 3 must be considered. Those of dimension 4 and 5 will not give new results, since if a subspace of V_6 is invariant, so is its orthogonal complement. The spaces that lead to maximal subalgebras of o(4, 2) have signatures $(+)$, $(-)$, (0), $(++)$, $(--)$, (00), $(+++)$, and $(++-)$. A subalgebra leaving a space of signature $(+0)$, $(+-)$, (-0), $(++0)$, $(+-0)$, and $(+00)$ invariant will also leave the space (0) or (00) invariant and will not be maximal. In this manner we obtain 8 distinct maximal subalgebras, listed in Table 10.1, together with the signature of the invariant subspace.

Now let us look at the irreducibly embedded subalgebras in this representation. The only simple Lie algebra that has a 6-dimensional real irreducible representation leaving a metric of signature $(+, +, +, +, -, -)$, invariant is su(2, 1). It is hence a subalgebra of o(4, 2). It is, however, not a maximal subalgebra of o(4, 2), since it has a larger centralizer, u(2, 1) $\subset$ o(4, 2). Thus, the reductive algebra u(2, 1) is the only maximal subalgebra of o(4, 2), irreducibly embedded in o(4, 2). The corresponding representation is realized by the following matrices Y, satisfying (5.11) with $K = I_{4,2}$:

$$Y \sim \begin{pmatrix} 0 & a & c & d & e & f \\ -a & 0 & -d & c & -f & e \\ -c & d & 0 & b & g & h \\ -d & -c & -b & 0 & -h & g \\ e & -f & g & -h & 0 & -a-b \\ f & e & h & g & a+b & 0 \end{pmatrix} \oplus \begin{pmatrix} 0 & x & & & & \\ -x & 0 & & & & \\ & & 0 & x & & \\ & & -x & 0 & & \\ & & & & 0 & x \\ & & & & -x & 0 \end{pmatrix}. \tag{5.13}$$

Thus we obtain precisely 9 maximal subalgebras of o(4, 2). They are all listed in Table 10.1.

TABLE 10.1. Maximal subalgebras of o(4, 2).

No.	Signature of invariant subspace	L_M	$\dim L_M$	Type of subalgebra
1	$(+)$	$O(3,2)$	10	simple
2	$(-)$	$O(4,1)$	10	simple
3	$(++)$	$O(4) \oplus O(2)$	7	reductive
4	$(--)$	$O(2,2) \oplus O(2)$	7	reductive
5	$(+++)$	$O(3) \oplus O(1,2)$	6	semisimple
6	$(++-)$	$O(2,1) \oplus O(2,1)$	6	semisimple
7	(0)	$\mathrm{sim}(3,1)$	11	maximal parabolic
8	(00)	$\mathrm{opt}(3,1)$	10	maximal parabolic
9	none	$u(2,1)$	10	reductive

A maximal parabolic subalgebra of a simple Lie algebra is one that contains a maximal solvable subalgebra. Here we have two distinct ones, namely $\mathrm{sim}(3,1)$, the Lie algebra of the similitude group (the Poincaré group extended by dilations) of $(3+1)$-dimensional Minkowski space $M(3,1)$ and $\mathrm{opt}(3,1)$, the Lie algebra of the optical group of $M(3,1)$. We refer to the articles [25, 140–144] for a further discussion.

5.3 Subalgebras of Direct Sums. The Goursat Method

Goursat proposed a method for classifying subgroups of direct products of discrete groups [76]. The method has been adapted to the classification of subalgebras of direct sums of Lie algebras [140].

The algorithm consists of several steps. Let us consider a Lie algebra L satisfying

$$L = A \oplus B, \quad \dim A = n_a, \ \dim B = n_b, \ 1 \le n_a < \infty, \ 1 \le n_b < \infty, \tag{5.14}$$

where A and B are not necessarily indecomposable (the method can be applied iteratively to sums of several indecomposable Lie algebras). We wish to classify all subalgebras of L into conjugacy classes under the action of the direct product group

$$G = G_A \otimes G_B, \quad G_A = \exp A, \quad G_B = \exp B. \tag{5.15}$$

We distinguish two types of subalgebras of the direct sum (5.14).

Nontwisted subalgebras These are direct sums or conjugate to direct sums of subalgebras $A_0 \subseteq A$, $B_0 \subseteq B$ and can be represented by direct sums:

$$L_0 = A_0 \oplus B_0, \quad A_0 \subseteq A, \ B_0 \subseteq B. \tag{5.16}$$

Twisted subalgebras Subalgebras that are not conjugate to algebras of
the form (5.16). In any basis there will be elements with nonzero
projections onto both A and B.

To provide a representative list of all G-conjugacy classes of subalgebras
of L, we proceed in 4 steps.

5.3.1   Step 1

Find representatives of all G_A and G_B conjugacy classes of subalgebras of
A and B, respectively. Denote them by

$$A_{j,a} \subseteq A, \quad j = 1, \ldots, n_a, \ a = 1, 2, \ldots,$$
$$B_{k,b} \subseteq B, \quad k = 1, \ldots, n_b, \ b = 1, 2, \ldots, \tag{5.17}$$

where the first subscript is equal to the dimension of the subalgebra, while
the second one distinguishes nonconjugate algebras of the same dimension.
We put

$$A_{0,1} = B_{0,1} = \{\varnothing\}, \quad A_{n_a,1} = A, \quad B_{n_b,1} = B, \tag{5.18}$$

i.e., the trivial subalgebras are included.

For each representative subalgebra find its normalizer group in G_A or
G_B, respectively:

$$\mathrm{Nor}(A_{j,a}, G_A) = \{g \in G_A \mid gA_{j,a}g^{-1} \subseteq A_{j,a}\},$$
$$\mathrm{Nor}(B_{k,b}, G_B) = \{g \in G_B \mid gB_{k,b}g^{-1} \subseteq B_{k,b}\}. \tag{5.19}$$

5.3.2   Step 2

Form a representative list of all nontwisted subalgebras of L:

$$S_1 = \{A_{j,a} \oplus B_{k,b}\}, \quad j = 0, \ldots, n_a, \ k = 0, \ldots, n_b,$$
$$a = 1, 2, \ldots, \ b = 1, 2, \ldots. \tag{5.20}$$

5.3.3   Step 3

Form a representative list S_2 of all twisted subalgebras of L. Two representative subalgebras $A_{j,a} \subseteq A$ and $B_{k,b} \subseteq B$ can be twisted together (the
"Goursat twist") if a homomorphism (that may be an isomorphism) exists
from one to the other, say

$$\tau(A_{j,a}) = B_{k,b}, \quad j \geq k \geq 1. \tag{5.21}$$

If a homomorphism τ exists, then choose a basis for $A_{j,a} = \{a_1, \ldots, a_j\}$
and construct the most general mapping

$$\tau : a_i \mapsto \tau(a_i) \in B_{k,b}. \tag{5.22}$$

A twisted subalgebra is obtained by taking

$$\tilde{L}_{j,a} = \{a_i + \tau(a_i)\}, \quad i = 1, \ldots, j. \tag{5.23}$$

The mapping (5.22) will in general contain free parameters that make their appearance in (5.23). In order to classify the subalgebras $\tilde{L}_{j,a}$ we apply the normalizer group

$$\text{Nor}(A_{j,a}, G_A) \otimes \text{Nor}(B_{k,b}, G_B). \tag{5.24}$$

This will transform the subalgebras $\tilde{L}_{j,a}$ for different values of the subscript a among each other and change the values of the free parameters. We annul as many of the parameters as possible and standardize as many as possible of the remaining ones. If all parameters can be annulled, the algebra is not twisted and is conjugate to one in the list S_1.

5.3.4   Step 4

Form a final representative list of all subalgebras of L by merging the lists S_1 and S_2. The final list can always be ordered by dimension and isomorphism class. Moreover, the representative subalgebras can be so chosen that the final list is "normalized." This means that the normalizer algebra $\text{nor}(L_{j,a}, L)$ of each algebra in the list is also in the list, where

$$\text{nor}(L_{j,a}, L) = \{x \in L \mid [x, L_{j,a}] \subseteq L_{j,a}\}. \tag{5.25}$$

Let us consider a simple example, namely the direct sum of two real Lie algebras

$$L = o(2, 1) \oplus o(3). \tag{5.26}$$

We have

$$\begin{aligned}
A \sim o(2, 1) &\sim \{K_1, K_2, J_3\}, \quad B \sim o(3) \sim \{L_1, L_2, L_3\}, \\
[K_1, K_2] &= -J_3, \quad [J_3, K_1] = K_2, \quad [J_3, K_2] = -K_1, \\
[L_1, L_2] &= L_3, \quad [L_3, L_1] = L_2, \quad [L_3, L_2] = -L_1.
\end{aligned} \tag{5.27}$$

Step 1. The representative subalgebras of A and B and their normalizer groups are shown in Tables 10.2 and 10.3

Step 2. The list of nontwisted subalgebras S_1 of L consists of all pairs of the form (5.20) with $A_{j,a}$ from Table 10.2 and $B_{k,b}$ from Table 10.3.

Step 3. Construct the homomorphisms $A_{j,a} \to B_{k,b}$

$$A_{1,1} : \tau(J_3) = \sum_{i=1}^{3} \lambda_i L_i,$$

TABLE 10.2. $A \sim SO(2,1)$

$A_{j,a}$	$\mathrm{Nor}\{A_{j,a}, SO(2,1)\}$
$A_{0,1} = \{0\}$	$\exp A_{3,1} \sim SO(2,1)$
$A_{1,1} = \{J_3\}$	$\exp A_{1,1} \sim SO(2)$
$A_{1,2} = \{K_1\}$	$[\exp A_{1,2} \cup e^{\pi J_3}] \sim SO(1,1)$
$A_{1,3} = \{J_3 + K_2\}$	$\exp A_{2,1}$
$A_{2,1} = \{K_1, J_3 + K_2\}$	$\exp A_{2,1}$
$A_{3,1} = \{K_1, K_2, J_3\}$	$\exp A_{3,1} \sim SO(2,1)$

TABLE 10.3. $B \sim SO(3)$

$B_{j,a}$	$\mathrm{Nor}\{B_{j,a}, SO(3)\}$
$B_{0,1} = \{0\}$	$\exp B_{3,1} \sim SO(3)$
$B_{1,1} = \{L_3\}$	$\exp B_{1,1} \sim SO(2)$
$B_{3,1} = \{L_1, L_2, L_3\}$	$\exp B_{3,1} \sim SO(3)$

$$A_{1,2} : \tau(K_1) = \sum_{i=1}^{3} \lambda_i L_i,$$

$$A_{1,3} : \tau(J_3 + K_2) = \sum_{i=1}^{3} \lambda_i L_i,$$

$$A_{2,1} : \tau(K_1) = \sum_{i=1}^{3} \lambda_i L_i,$$

$$\tau(J_3 + K_2) = 0,$$

$$A_{3,1} : \text{none (over } \mathbb{R}).$$

We obtain the following list of twisted subalgebras of L:

$$\tilde{L}_{1,1} = J_3 + aL_3, \qquad\qquad a > 0, \qquad (5.28)$$

$$\tilde{L}_{1,2} = K_1 + aL_3, \qquad\qquad a > 0, \qquad (5.29)$$

$$\tilde{L}_{1,3} = J_3 + K_2 + \varepsilon L_3, \qquad\qquad \varepsilon = \pm 1. \qquad (5.30)$$

Step 4. Merge the two lists.

For other examples, see [180] and [140].

5.4   Subalgebras of Semidirect Sums

The method for classifying subalgebras of direct sums was elaborated and applied in a series of articles [25, 140–144] and reviewed, with examples, in earlier lecture series [179, 180]. We simply refer to those articles here.

6   Direct Reductions and Conditional Symmetries

6.1   Introductory Remarks

There have been several generalizations of the classical Lie group method for symmetry reductions. Ovsiannikov [138] developed the method of partially invariant solutions; recently, Ondich [136] has shown that this method can be considered as a special case of the method of differential constraints introduced by Yanenko [182] and Olver and Rosenau [133, 134]. Bluman and Cole [14], in their study of symmetry reductions of the linear heat equation, proposed the so-called nonclassical method of group-invariant solutions; this technique is also known as the "method of conditional symmetries" (cf. [107]) and the "method of partial symmetries of the first type" (cf. [174]).

Suppose $(x, t) \in \mathbb{R}^2$ are the independent variables, $u \in \mathbb{R}$ the dependent variable, and $\boldsymbol{u}^{(\ell)}(x, t)$ denotes the set of all the partial derivatives of order ℓ of u, and consider the general Nth-order partial differential equation

$$E = E\big(x, t, u, \boldsymbol{u}^{(1)}(x, t), \ldots, \boldsymbol{u}^{(N)}(x, t)\big) = 0. \qquad (6.1)$$

In the nonclassical method, the original partial differential equation (6.1) is augmented with the invariant surface condition

$$\psi \equiv \xi(x, t, u)u_x + \tau(x, t, u)u_t - \phi(x, t, u) = 0, \qquad (6.2)$$

which is associated with the vector field

$$\mathbf{v} = \xi(x, t, u)\partial_x + \tau(x, t, u)\partial_t + \phi(x, t, u)\partial_u \qquad (6.3)$$

(see (2.5)). By requiring that the set of simultaneous solutions of (6.1) and (6.2) be invariant under the transformation

$$
\begin{aligned}
\tilde{x} &= x + \varepsilon\xi(x, t, u) + O(\varepsilon^2), \\
\tilde{t} &= t + \varepsilon\tau(x, t, u) + O(\varepsilon^2), \\
\tilde{u} &= u + \varepsilon\phi(x, t, u) + O(\varepsilon^2),
\end{aligned}
\qquad (6.4)
$$

where ε is the group parameter, one obtains an overdetermined *nonlinear* system of equations, as opposed to a linear system in the classical case, for the infinitesimals ξ, τ, and ϕ, which appear in both the transformations (6.4) and the supplementary condition (6.2). The number of determining equations arising in the nonclassical method is smaller than for the classical method, since there are fewer linearly independent expressions in the derivatives. Since all solutions of the classical determining equations necessarily satisfy the nonclassical determining equations, the solution set may be larger in the nonclassical case. For some equations, such as the Korteweg–de Vries equation (3.1), which is a soliton equation solvable by the inverse scattering method [73], the infinitesimals arising from the classical and nonclassical methods coincide. It should be emphasized that the vector fields associated with the nonclassical method do not form a vector space, still less a Lie algebra, since the invariant surface condition (6.2) depends upon the particular reduction. For example, the sum of two nonclassical symmetry operators is not, in general, a symmetry operator at all; similarly, the commutator of two nonclassical symmetry operators or the sum of a classical symmetry operator and a nonclassical symmetry operator is not, in general, a symmetry operator.

Subsequently, these methods were further generalized by Olver and Rosenau [133,134] to include "weak symmetries" and, even more generally, "side conditions" or "differential constraints." However, their framework appears to be too general to be practical, and they concluded that "the unifying theme behind finding special solutions of partial differential equations is not, as is commonly supposed, group theory, but rather the more analytic subject of overdetermined systems of partial differential equations."

Motivated by the fact that symmetry reductions of the Boussinesq equation were known that are not obtainable using the classical Lie group method [128, 133, 134, 148, 160], Clarkson and Kruskal [36] developed a direct, algorithmic method for finding symmetry reductions (in the sequel referred to as the *direct method*—also known as the "method of Ansatzes" [64], which they used to obtain previously unknown reductions of the Boussinesq equation (see Section 6.2 for details).

The nonclassical method lay dormant for several years, essentially until the papers by Olver and Rosenau [133, 134]. However, following the development of the direct method, there has been renewed interest in the nonclassical method, and recently both these methods have been used to generate many new symmetry reductions and exact solutions for several physically significant partial differential equations, which represents significant and important progress (cf. [33, 34, 37, 63, 65] and the references therein).

Recent generalizations of the direct method include those due to Burdé [22,23], Galaktionov [71], and Hood [88]. Generalizations of the nonclassical method are discussed by Bluman and Shtelen [16], Burdé [24], and Olver and Vorob'ev [135].

6.2 Example of the Boussinesq Equation

In this section we discuss symmetry reductions of the Boussinesq equation

$$E \equiv u_{tt} + uu_{xx} + u_x^2 + u_{xxxx} = 0, \qquad (6.5)$$

which is also a soliton equation solvable by inverse scattering (cf. [52]). The Boussinesq equation arises in several physical applications: propagation of long waves in shallow water [17, 18, 173], one-dimensional nonlinear lattice waves [171,183], vibrations in a nonlinear string [184], and ion sound waves in a plasma [164].

6.2.1   Classical Lie Method

To apply the classical Lie method to the Boussinesq equation (6.5) we require that the set $\mathcal{S}_E := \{u : E = 0\}$ of solutions of (6.5) be invariant under the transformation (6.4). As described above, this yields the *determining equations*, a *linear, homogeneous* system of PDEs for ξ, τ, and ϕ, and is accomplished by requiring that (2.19) hold.

Hence we obtain the following twelve determining equations for the infinitesimals:

$$\xi_t = 0, \quad \xi_u = 0, \quad \tau_x = 0, \quad \tau_u = 0,$$
$$\phi_{uu} = 0, \quad 2\xi_x - \tau_t = 0, \quad \phi_u + 2\xi_x = 0,$$
$$2\phi_{xu} - 3\xi_{xx} = 0, \quad 2\phi_{tu} - \tau_{tt} = 0,$$
$$\phi_{tt} + \phi_{xxxx} + u\phi_{xx} = 0, \tag{6.6}$$
$$\phi + 2u\xi_x - 4\xi_{xxx} + 6\phi_{xxu} = 0,$$
$$2\phi_x - u\xi_{xx} - \xi_{xxxx} + 2u\phi_{xu} + 4\phi_{xxxu} = 0.$$

These equations were generated using the MACSYMA package `symmgrp.max` [28] and have the general solution

$$\xi = \alpha x + \beta, \quad \tau = 2\alpha t + \gamma, \quad \phi = -2\alpha u, \tag{6.7}$$

where α, β, and γ are arbitrary constants [128, 160] and the associated vector field is

$$\begin{aligned} \mathbf{v} &= (\alpha x + \beta)\partial_x + (2\alpha t + \gamma)\partial_t - 2\alpha u\partial_u \\ &= \alpha(x\partial_x + 2t\partial_t - 2u\partial_u) + \beta\partial_x + \gamma\partial_t. \end{aligned}$$

The generators of this are $\mathbf{v}_1 = \partial_x$ (corresponding to space translational invariance), $\mathbf{v}_2 = \partial_t$ (time translational invariance), and $\mathbf{v}_3 = x\partial_x + 2t\partial_t - 2u\partial_u$ (scaling or dilational invariance). Consequently, there are two canonical (classical) symmetry reductions.

Case I. $\alpha = 0$

In this case we set $\gamma = 1$, without loss of generality (for $\gamma \neq 0$), and obtain the traveling wave reduction

$$u(x,t) = w(z), \quad z = x - \beta t, \tag{6.8}$$

where $w(z)$ satisfies

$$\frac{d^4 w}{dz^4} + w\frac{d^2 w}{dz^2} + \left(\frac{dw}{dz}\right)^2 + \beta^2 \frac{d^2 w}{dz^2} = 0. \tag{6.9}$$

Integrating this with respect to z twice yields

$$\frac{d^2w}{dz^2} + \beta^2 w + \frac{1}{2}w^2 = Az + B,$$

with A and B arbitrary constants, which is solvable in terms of the first Painlevé equation P_I (see 4.30) (if $A \neq 0$) and elliptic or elementary functions (if $A = 0$).

Case II. $\alpha = 1$

In this case we set $\beta = \gamma = 0$, without loss of generality, and obtain the scaling reduction

$$u(x,t) = t^{-1}w(z), \quad z = x/t^{1/2}, \tag{6.10}$$

where $w(z)$ satisfies

$$\frac{d^4w}{dz^4} + w\frac{d^2w}{dz^2} + \left(\frac{dw}{dz}\right)^2 + \frac{1}{4}z^2\frac{dw}{dz} + \frac{7}{4}z\frac{dw}{dz} + 2w = 0. \tag{6.11}$$

This equation is solvable in terms of the fourth Painlevé equation (P_{IV})

$$\frac{d^2y}{dx^2} = \frac{1}{2y}\left(\frac{dy}{dx}\right)^2 + \frac{3}{2}y^3 + 4xy^2 + 2(x^2 + a)y + \frac{b}{y}, \tag{6.12}$$

where a and b are arbitrary constants [36].

However, as noted by several authors [128, 133, 134, 148, 160], the Boussinesq equation (6.5) also possesses the accelerating wave solution

$$u(x,t) = w(z) - 4\mu^2 t^2, \quad z = x + \mu t^2, \tag{6.13}$$

where μ is an arbitrary constant and $w(z)$ satisfies

$$\frac{d^3w}{dz^3} + w\frac{dw}{dz} + 2\mu w = 8\mu^2 z + A,$$

with A an arbitrary constant, which is solvable in terms of the second Painlevé equation (P_{II}) (see (4.35)). Associated infinitesimals for this reduction are $\xi = 2\mu t$, $\tau = -1$, $\phi = 8\mu^2 t$, which are clearly *not* a special case of (6.7), with associated vector field $\mathbf{v} = 2\mu t\partial_x - \partial_t + 8\mu^2 t\partial_u$.

6.2.2   Direct Method

Clarkson and Kruskal [36] developed the direct method in an attempt to understand the symmetry reduction (6.13) and derive it systematically (the previous derivations had been by seemingly ad hoc techniques). The basic idea of the direct method is to seek a solution of a partial differential equation such as (6.1) in the form

$$u(x,t) = F(x,t,w(z(x,t))), \tag{6.14}$$

and require that $w(z)$ satisfy an ordinary differential equation. This imposes conditions upon $F(x, t, w)$, $z(x, t)$, and their derivatives in the form of an overdetermined system of equations, whose solution yields the desired reductions. The novel characteristic about the direct method, in comparison to the others mentioned above, is that it involves no use of group theory. We remark that the direct method has certain resemblances to the so-called method of free parameter analysis (cf. [81]); though in this latter method the boundary conditions are crucially used in the determination of the reduction, whereas they are not used in the direct method.

For the Boussinesq equation (6.5), Clarkson and Kruskal [36] show that it is sufficient to seek a solution in the linear form

$$u(x, t) = \beta(x, t)w(z(x, t)) + \alpha(x, t), \tag{6.15}$$

rather than the more general form (6.14). There are two cases to consider: $z_x \not\equiv 0$ and $z_x \equiv 0$.

Case I. $z_x \not\equiv 0$

In this case substituting (6.15) into the Boussinesq equation (6.5) yields

$$\beta z_x^4 \frac{d^4 w}{dz^4} + (6\beta z_x^2 z_{xx} + 4\beta_x z_x^3) \frac{d^3 w}{dz^3}$$

$$+ [\alpha\beta z_x^2 + \beta z_t^2 + \beta(3z_{xx}^2 + 4z_x z_{xxx}) + 12\beta_x z_x z_{xx} + 6\beta_{xx} z_x^2] \frac{d^2 w}{dz^2}$$

$$+ [\beta z_{xxxx} + 4\beta_x z_{xxx} + 6\beta_{xx} z_{xx} + 4\beta_{xxx} z_x + 2\alpha_x \beta z_x + 2\alpha\beta_x z_x$$

$$+ \alpha\beta z_{xx} + 2\beta_t z_t + \beta z_{tt}] \frac{dw}{dz}$$

$$+ [\beta_{xxxx} + 2\alpha_x \beta_x + \alpha\beta_{xx} + \alpha_{xx}\beta + \beta_{tt}] w$$

$$+ \beta^2 z_x^2 w \frac{d^2 w}{dz^2} + \beta(4\beta_x z_x + \beta z_{xx}) w \frac{dw}{dz} + (\beta_x{}^2 + \beta\beta_{xx}) w^2$$

$$+ \beta^2 z_x^2 \left(\frac{dw}{dz}\right)^2 + \alpha_{tt} + \alpha\alpha_{xx} + \alpha_x^2 + \alpha_{xxxx} = 0.$$

For this to be an ordinary differential equation for $w(z)$, the coefficients must be of the form $\beta z_x^4 \Gamma(z)$ (using the coefficient of $d^4 w/dz^4$ as the normalizing coefficient). This requirement generates the following overdetermined system of equations for $\alpha(x, t)$, $\beta(x, t)$, and $z(x, t)$:

$$\beta z_x^4 \Gamma_1(z) = \beta^2 z_x^2,$$

$$\beta z_x^4 \Gamma_2(z) = \beta(4\beta_x z_x + 6\beta z_{xx}),$$

$$\beta z_x^4 \Gamma_3(z) = \beta_x{}^2 + \beta\beta_{xx},$$

$$\beta z_x^4 \Gamma_4(z) = 6\beta z_x^2 z_{xx} + 4\beta_x z_x^3,$$

$$\beta z_x^4 \Gamma_5(z) = \alpha\beta z_x^2 + \beta z_t^2 + \beta(3z_{xx}^2 + 4z_x z_{xxx})$$

$$+ 12\beta_x z_x z_{xx} + 6\beta_{xx} z_x^2, \tag{6.16}$$

$$\beta z_x^4 \Gamma_6(z) = \beta z_{xxxx} + 4\beta_x z_{xxx} + 6\beta_{xx} z_{xx} + 4\beta_{xxx} z_x$$
$$+ 2\alpha_x \beta z_x + 2\alpha \beta_x z_x + \alpha \beta z_{xx} + 2\beta_t z_t + \beta z_{tt},$$
$$\beta z_x^4 \Gamma_7(z) = \beta_{xxxx} + 2\alpha_x \beta_x + \alpha \beta_{xx} + \alpha_{xx} \beta + \beta_{tt},$$
$$\beta z_x^4 \Gamma_8(z) = \alpha_{tt} + \alpha \alpha_{xx} + \alpha_x^2 + \alpha_{xxxx},$$

where $\Gamma_1(z), \ldots, \Gamma_8(z)$ are functions to be determined. Solving (6.16) yields the following generic symmetry reduction of the Boussinesq equation (6.5) given by

$$u(x,t) = \theta^2(t)w(z) - \frac{1}{\theta^2(t)} \left(x \frac{d\theta}{dt} + \frac{d\phi}{dt} \right)^2, \qquad (6.17)$$

$$z(x,t) = x\theta(t) + \phi(t), \qquad (6.18)$$

where $\theta(t)$ and $\phi(t)$ are any solutions of

$$\frac{d^2\theta}{dt^2} = A\theta^5, \qquad \frac{d^2\phi}{dt^2} = (A\phi + B)\theta^4, \qquad (6.19)$$

A and B are arbitrary constants, and $w(z)$ satisfies

$$\frac{d^4w}{dz^4} + w\frac{d^2w}{dz^2} + \left(\frac{dw}{dz}\right)^2 + (Az + B)\frac{dw}{dz} + 2Aw = 2(Az + B)^2. \quad (6.20)$$

Depending upon the choice of the constants, this equation is solvable in terms of the first, second, and fourth Painlevé equations. In general, (6.20) is equivalent to P_{IV} (6.12), since if we make the transformation

$$w(z) = -3\sqrt{A}\left(\frac{dy}{dx} + y^2 + 2xy + \frac{2}{3}x^2\right) + 2\sqrt{A}(\alpha - 1),$$

$$x = \left(\frac{1}{4}A\right)^{1/4}\left(z + \frac{B}{A}\right),$$

where $y(x)$ is a solution of P_{IV}, then $w(z)$ satisfies (6.20). In fact, there is a one-to-one relationship between solutions of (6.20), with $A \neq 0$ and those of P_{IV} (6.12) [36]. When $A = 0$ and $B \neq 0$, (6.20) is equivalent to P_{II} (4.35), since if

$$w(z) = -6B^{2/3}\left(\frac{dy}{dx} + y^2\right) + Bz, \qquad x = -B^{1/3}z, \qquad (6.21)$$

where $y(x)$ is a solution of P_{II}, then $w(z)$ satisfies (6.20). In fact, there is a one-to-one relationship between solutions of (6.20), with $A = 0, B \neq 0$, and those of P_{II} (4.35) [61]. Finally, if $A = 0$ and $B = 0$, then (6.20) is equivalent to either P_I (4.30) or the Weierstrass elliptic function equation

$$\left(\frac{dP}{dx}\right)^2 = 4P^3 - g_2P - g_3, \qquad (6.22)$$

with g_2 and g_3 constants.

Solving (6.19) yields six canonical types of symmetry reductions:

$$u(x,t) = w_1(z), \quad z = x + \mu_1 t, \tag{6.23}$$

$$u(x,t) = t^2 w_2(z) - \frac{x^2}{t^2}, \quad z = xt, \tag{6.24}$$

$$u(x,t) = w_3(z) - 4\mu_3^2 t^2, \quad z = x + \mu_3 t^2, \tag{6.25}$$

$$u(x,t) = t^2 w_4(z) - \frac{(x + 6\mu_4 t^5)^2}{t^2}, \quad z = xt + \mu_4 t^6, \tag{6.26}$$

$$u(x,t) = t^{-1} w_5(z) - \frac{(x - 3\mu_5 t^2)^2}{4t^2}, \quad z = xt^{-1/2} + \mu_5 t^{3/2}, \tag{6.27}$$

$$u(x,t) = \frac{1}{\wp(t)} \left\{ w(z) - \left[\frac{1}{2} z \frac{d\wp}{dt} + \mu_6 \wp^{3/2}(t) \right]^2 \right\}, \tag{6.28}$$

$$z = \wp^{-1/2}(t)[x + \mu_6 \zeta(t)], \tag{6.29}$$

where μ_1, μ_3, $\ldots$, μ_6 are arbitrary constants, $\wp(t) = \wp(t + t_0; 0, g_3)$ is the Weierstrass elliptic function, $\zeta(t) = \zeta(t + t_0; 0, g_3)$ the Weierstrass zeta function; $w_1(z)$ and $w_2(z)$ satisfy an equation equivalent to P_I (4.30), $w_3(z)$ and $w_4(z)$ satisfy an equation equivalent to P_{II} (4.35), and $w_5(z)$ and $w_6(z)$ satisfy an equation equivalent to P_{IV} (6.12). By using known rational and special solutions of P_{II} and P_{IV} (cf. [10, 11] and the references therein), these symmetry reductions generate new rational and special solutions of the Boussinesq equation (6.5) expressible in terms of elementary, Airy, and Weber–Hermite functions (see [31] for further details).

It can be shown that for the symmetry reductions of the Boussinesq equation (6.5) that are *not* obtainable using the classical Lie group method, the associated group transformation does not map (6.5) into itself, whereas the symmetry reductions obtained by the classical Lie group method do. For example, consider the symmetry reduction (6.24). A one-parameter group associated with this reduction is

$$x^* = e^\varepsilon x, \quad t^* = e^{-\varepsilon} t, \quad u^* = e^{-2\varepsilon} u + e^{-2\varepsilon}(1 - e^{6\varepsilon}) \frac{x^2}{t^2}, \tag{6.30}$$

where ε is the group parameter. This maps solutions of (6.5) into solutions of

$$u_{tt} + uu_{xx} + u_x^2 + u_{xxxx} = (e^{-6\varepsilon} - 1)t^{-2} \Phi, \tag{6.31}$$

where

$$\Phi := x^2 u_{xx} + 4x u_x + 2u - t^2 u_{tt} + 6\frac{x^2}{t^2}. \tag{6.32}$$

If u is given by the symmetry reduction (6.24), then it is easily seen that $\Phi \equiv 0$, i.e., the group (6.30), maps (6.5) into (6.31), with (6.32) identically zero, and so (6.31) is identical to (6.5) when u is given by (6.24). To illustrate why the perturbation Φ necessarily vanishes identically, consider the associated infinitesimals for the symmetry reduction (6.24) given by

$$\xi = x, \quad \tau = -t, \quad \phi = -2u - 6\frac{x^2}{t^2}. \tag{6.33}$$

The symmetry reduction necessarily satisfies the invariant surface condition $\xi u_x + \tau u_t = \phi$, i.e., if $u(x,t)$ is given by (6.30), then

$$\psi := xu_x - tu_t + 2u + 6\frac{x^2}{t^2} = 0. \tag{6.34}$$

It is easily shown that

$$\Phi = x\psi_x + t\psi_t + \psi, \tag{6.35}$$

and so if $u(x,t)$ satisfies the invariant surface condition $\psi = 0$, then it also satisfies $\Phi = 0$.

Case II. $z_x \equiv 0$

This case is discussed by Clarkson [31] and Lou [111], when it suffices to seek reductions in the form

$$u(x,t) = \alpha(x,t) + \beta(x,t)w(t) \tag{6.36}$$

and require that the result be an ordinary differential equation for $w(t)$. There are two canonical types of reduction:

Subcase a.

$$u(x,t) = x^2 w_1(t) - \frac{12}{x^2}, \tag{6.37}$$

where $w_1(t)$ satisfies the Weierstrass elliptic function equation

$$\frac{d^2 w_1}{dt^2} + 6w_1^2 = 0, \tag{6.38}$$

which has general solution $w_1(t) = -\wp(t+t_0; 0, g_3)$, with t_0 and g_3 arbitrary constants [177].

Subcase b.

$$u(x,t) = w_2(t) + x\psi(t) + x^2\phi(t), \tag{6.39}$$

where $\phi(t)$ satisfies the Weierstrass elliptic function equation (6.38), i.e., $\phi(t) = -\wp(t + t_0; 0, g_3)$ with t_0 and g_3 arbitrary constants, $\psi(t)$ satisfies the Lamé equation [177]

$$\frac{d^2\psi}{dt^2} - 6\wp(t + t_0; 0, g_3)\psi = 0,$$

which has general solution

$$\psi(t) = A\wp(t + t_0; 0, g_3) + B\wp(t + t_0; 0, g_3) \int^t \frac{\mathrm{d}s}{\wp^2(s + t_0; 0, g_3)}, \qquad (6.40)$$

where A and B are arbitrary constants, and $w_2(t)$ satisfies the inhomogeneous Lamé equation

$$\frac{\mathrm{d}^2 w_2}{\mathrm{d}t^2} - 2\wp(t + t_0; 0, g_3)w_2 = -\psi^2(t), \qquad (6.41)$$

with $\psi(t)$ as in (6.40). The general solution of the homogeneous Lamé equation

$$\frac{\mathrm{d}^2 \eta}{\mathrm{d}t^2} - 2\wp(t + t_0; 0, g_3)\eta = 0 \qquad (6.42)$$

is given by

$$\eta(t) = c_1 \exp\{-\tau\zeta(a)\}\frac{\sigma(\tau + a)}{\sigma(\tau)} + c_2 \exp\{\tau\zeta(a)\}\frac{\sigma(\tau - a)}{\sigma(\tau)},$$

where $\tau = t + t_0$, and c_1, c_2 are arbitrary constants, in which $\zeta(z)$, $\sigma(z)$ are the Weierstrass zeta and sigma functions defined by

$$\zeta(z) := -\int^z \wp(s)\,\mathrm{d}s, \quad \sigma(z) := \exp\left\{\int^z \zeta(s)\,\mathrm{d}s\right\},$$

together with the conditions

$$\lim_{z \to 0}\left(\zeta(z) - \frac{1}{z}\right) = 0, \quad \lim_{z \to 0}\left(\frac{\sigma(z)}{z}\right) = 1,$$

respectively (cf. [5,56]), and a is any solution of the transcendental equation

$$\wp(a) = 0,$$

i.e., a is a zero of the Weierstrass elliptic function (cf. [92]). Hence the general solution of (6.41) is given by

$$w(t) = c_1 w_1(\tau) + c_2 w_2(\tau)$$
$$-\frac{1}{\kappa}\int^\tau \{w_1(\tau)w_2(s) - w_2(\tau)w_1(s)\}\phi^2(s)\,\mathrm{d}s, \qquad (6.43)$$

where $\tau = t + t_0$ and

$$\kappa = \frac{\partial w_1}{\partial t}w_2 - \frac{\partial w_2}{\partial t}w_1 = \wp'(a)\sigma^2(a),$$

and t_0, c_1, and c_2 are arbitrary constants. We remark that in order to verify that (6.43) is a solution of (6.41) one uses the following addition theorems for Weierstrass elliptic, zeta, and sigma functions:

$$\wp(s+t) = -\wp(s) - \wp(t) + \frac{1}{4}\left[\frac{\wp'(s) - \wp'(t)}{\wp(s) - \wp(t)}\right]^2,$$

$$\zeta(s \pm t) = \zeta(s) \pm \zeta(t) + \frac{1}{2}\left[\frac{\wp'(s) \mp \wp'(t)}{\wp(s) - \wp(t)}\right]^2,$$

$$\sigma(s+t)\sigma(s-t) = \sigma^2(s)\sigma^2(t)\left[\wp(s) - \wp(t)\right]$$

(cf. [5, 56]).

6.2.3 Nonclassical Method

In the concluding discussion of [36], Clarkson and Kruskal expressed the "hope that a group-theoretical explanation of the [direct] method will be possible in due course." Levi and Winternitz [107] quickly gave such an explanation of Clarkson and Kruskal's results by showing that all their new reductions of the Boussinesq equation could also be obtained using the nonclassical method of Bluman and Cole [14], as we show in this subsection. Recently, Olver [132] has given a proof of the precise relationship between the direct methods, which we discuss below (see also [9, 146]).

In the nonclassical method it is required that the infinitesimal transformation (6.4) leave invariant the set of simultaneous solutions of the Boussinesq equation (6.5) and the invariant surface condition (6.2), where ξ, τ, and ϕ are the same as in the transformation (6.4). That is, we require that the subset of $\mathcal{S}_E$ given by

$$\mathcal{S}_{E,\psi} = \{u(x,t) : E(u) = 0, \psi(u) = 0\} \tag{6.44}$$

be invariant under the transformation (6.4). Thus "nonclassical symmetries," or "conditional symmetries," of a system of partial differential equations Σ are transformations that leave only the subset $\mathcal{S}_{E,\psi}$ of the solution set $\mathcal{S}_E$ of the system invariant. Other solutions of Σ that are not in the subset $\mathcal{S}_{E,\psi}$ are *not* necessarily transformed to the set $\mathcal{S}_E$.

The usual method of applying the nonclassical method (e.g., as described in [107]), to the Boussinesq equation (6.5) involves applying the prolongation $\mathrm{pr}^{(4)}\,\mathbf{v}$ to the system of equations given by (6.5) and the invariant surface condition (6.2) and requiring that the resulting expressions vanish for $u \in \mathcal{S}_{E,\psi}$, i.e.,

$$\mathrm{pr}^{(4)}\,\mathbf{v}(E)|_{E=0,\psi=0} = 0, \quad \mathrm{pr}^{(1)}\,\mathbf{v}(\psi)|_{E=0,\psi=0} = 0. \tag{6.45}$$

It is easily shown that

$$\mathrm{pr}^{(1)}\,\mathbf{v}(\psi) = -(\xi_u u_x + \tau_u u_t - \phi_u)\psi,$$

which vanishes identically, when $\psi = 0$, without imposing any conditions upon ξ, τ, and ϕ. However, as shown by Clarkson and Mansfield [39], this procedure for applying the nonclassical method can create difficulties, in particular in the implementation of symbolic manipulation programs. These difficulties often arise for equations such as (6.5), which require the use of differential consequences of the invariant surface condition (6.2). In [39] Clarkson and Mansfield proposed an algorithm for calculating the determining equations associated with the nonclassical method that avoids many of the difficulties commonly encountered.

There are two cases to consider: (i) $\tau \neq 0$; and (ii) $\tau = 0$ and $\xi \neq 0$.

Case I. $\tau \neq 0$

In this case we set $\tau = 1$, without loss of generality, and then from the invariant surface condition (6.2) obtain

$$u_t = \phi - \xi u_x,$$
$$u_{xt} = \phi_x + \phi_u u_x - (\xi_x u_x + \xi_u u_x^2 + \xi u_{xx}),$$
$$u_{tt} = \phi_t + \phi_u u_t - \xi_t u_x - \xi_u u_x u_t - \xi u_{xt}$$
$$= \phi_t + \phi_u(\phi - \xi u_x) - \xi_t u_x - \xi_u u_x(\phi - \xi u_x)$$
$$- \xi[\phi_x + \phi_u u_x - (\xi_x u_x + \xi_u u_x^2 + \xi u_{xx})].$$

Hence eliminating u_{tt} in (6.5) yields

$$\phi_t + \phi_u(\phi - \xi u_x) - \xi_t u_x - \xi_u u_x(\phi - \xi u_x) - \xi[\phi_x + \phi_u u_x - (\xi_x u_x + \xi_u u_x^2 + \xi u_{xx})]$$
$$+ uu_{xx} + u_x^2 + u_{xxxx} = 0, \quad (6.46)$$

which essentially is an ordinary differential equation for $u(x)$ with t a parameter, since only x-derivatives of u arise. Now we apply the classical Lie algorithm to this equation, i.e., we require that it be invariant under the transformation (6.4) with $\tau = 1$, and then use (6.46) to eliminate u_{xxxx}. This yields the following seven determining equations:

$$\xi_u = 0, \quad \phi_{uu} = 0,$$
$$2\phi_{xu} - 3\xi_{xx} = 0, \quad \phi_u + 2\xi_x = 0,$$
$$6\phi_{xxu} + \phi + 2\xi\xi_t - 4\xi_{xxx} + 4\xi^2\xi_x + 2u\xi_x = 0,$$
$$\phi_{tt} + 4\xi_x\phi_t + \phi_{xxxx} + u\phi_{xx} - 2\xi_t\phi_x - 4\xi\xi_x\phi_x \quad (6.47)$$
$$+ 2\phi\phi_{xu} + 4\xi_x\phi\phi_u = 0,$$
$$2u\phi_{xu} - u\xi_{xx} + 2\phi_x + 4\phi_{xxxu} - 8\xi\xi_x\phi_u - 2\xi_t\phi_u$$
$$- 2\xi\phi_{tu} - \xi_{xxxx} + 4\xi\xi_x^2 - 2\xi_t\xi_x - \xi_{tt} = 0.$$

These equations were calculated using the MACSYMA package symmgrp.max [28], and their solution is

$$\xi = xf(t) + g(t), \quad (6.48)$$

$$\phi = -\left\{2f(t)u + 2x^2 f(t)\left[\frac{df}{dt} + 2f^2(t)\right] + 2x\left[\frac{df}{dt}g(t) + f(t)\frac{dg}{dt} + 4f^2(t)g(t)\right]\right.$$
$$\left. + 2g(t)\left[\frac{dg}{dt} + 2f(t)g(t)\right]\right\}, \quad (6.49)$$

where

$$f(t) = \frac{1}{2p(t)}\frac{dp}{dt}, \quad g(t) = \frac{\kappa_1}{2p(t)}\frac{dp}{dt} + \kappa_0 \int^t \frac{p(s)}{[p'(s)]^2}\,ds,$$

and $p(t)$ satisfies

$$\left(\frac{dp}{dt}\right)^2 = \kappa_3 p^3 + \kappa_2, \quad (6.50)$$

and κ_3, κ_2, κ_1, and κ_0 are arbitrary constants. Equation (6.50) is solvable in terms of the Weierstrass elliptic function $\wp(x;0,g_3)$ if $\kappa_3\kappa_2 \neq 0$ and in terms of elementary functions otherwise.

Solving (6.50) for $p(t)$ yields the six canonical symmetry reductions (6.23)–(6.29) of the Boussinesq equation (6.5) that were derived by Clarkson and Kruskal [36] using the direct method, as discussed in the previous section.

Case II. $\tau = 0$ and $\xi \neq 0$

In this case we set $\xi = 1$, and then the invariant surface condition (6.2) simplifies to

$$u_x = \phi(x,t,u).$$

Hence using this to eliminate u_x, u_{xx}, and u_{xxxx} in the Boussinesq equation (6.5) yields

$$u_{tt} + u(\phi_x + \phi\phi_u) + \phi^2 + \phi_{xxx} + \phi_u\phi_{xx} + 3\phi\phi_x\phi_{uu} + 3\phi_x\phi_{xu} + \phi_x\phi_u^2$$
$$+ \phi^3\phi_{uuu} + 3\phi^2\phi_{xuu} + 4\phi^2\phi_u\phi_{uu} + 3\phi\phi_{xxu} + 5\phi\phi_u\phi_{xu} + \phi\phi_u^3 = 0,$$

which essentially is an ordinary differential equation for $u(t)$ with x a parameter. Now we apply the classical Lie algorithm to this equation, i.e., we require that it be invariant under the transformation (6.4) with $\tau = 0$ and $\xi = 1$, and then eliminate u_{tt}. This yields the following three determining equations:

$$\phi_{uu} = 0, \quad \phi_{tu} = 0,$$
$$\phi_{tt} + \phi_{xxxx} + 4\phi_{xx}\phi_{xu} + u\phi_{xx} + 6\phi_{xxu}\phi_x + 4\phi_x\phi_u\phi_{xu} + 3\phi\phi_x$$
$$+ 4\phi\phi_{xxxu} + 6\phi\phi_u\phi_{xxu} + 8\phi\phi_{xu}^2 + 4\phi\phi_u^2\phi_{xu} + 2u\phi\phi_{xu} + 2\phi^2\phi_u = 0.$$

These equations were calculated using the MACSYMA package symmgrp.max [28], and there are two solutions.

Subcase i $\phi = 2u/(x+x_0) + 48/(x+x_0)^3$, where x_0 is an arbitrary constant. In this case we obtain the reduction (6.37) given above.

Subcase ii $\phi = 2x\psi_2(t) + \psi_1(t)$, where $\psi_1(t)$ and $\psi_0(t)$ satisfy

$$\frac{\mathrm{d}^2\psi_2}{\mathrm{d}t^2} + 6\psi_2^2 = 0, \quad \frac{\mathrm{d}^2\psi_1}{\mathrm{d}t^2} + 6\psi_2\psi_1 = 0.$$

In this case we obtain the reduction (6.39) given above.

Hence, using the nonclassical method, we have obtained all the symmetry reductions of the Boussinesq equation (6.5) that were obtained above using the direct method.

6.3   *Differential Gröbner Bases*

An effective method for solving the determining equations for the infinitesimals in both the classical and nonclassical cases appears to be the use of differential Gröbner bases (DGB). Given a system of partial differential equations Σ with n independent variables $\{x_1, x_2, \ldots, x_n\}$ and m dependent variables $\{u_1, u_2, \ldots, u_m\}$, a DGB is defined to be a basis $\mathcal{B}$ of the differential ideal $I(\Sigma)$ generated by the system Σ such that every member of the ideal pseudo-reduces to zero with respect to $\mathcal{B}$ (see [38, 116, 117] for further details and definitions). Applying the method of DGB to a given system of partial differential equations is analogous to reducing a matrix to its echelon form. This method provides a systematic framework for finding integrability and compatibility conditions of an overdetermined system of partial differential equations. It avoids the problems of infinite loops in reduction processes, and yields, as far as is currently possible, a "triangulation" or "standard" form of the system from which the solution set can be derived [119, 154–156]. In a sense, a DGB provides the maximum amount of information possible using elementary differential and algebraic processes in a finite time.

The rationale behind using the MAPLE package diffgrob2 [114, 115] to apply the DGB method is to find the easiest integration problem equivalent to the given one. The algorithms implemented are based on Buchberger's algorithm for a Gröbner basis of a polynomial ideal (cf. [20]). Special cases of the parameters or arbitrary functions for which extra solutions exist are found, and the algorithms are guaranteed to terminate. The output can be used to obtain a representation of the general solution that is, in some sense, simplest. The drawback of the algorithms is their complexity, and intermediate expressions swell. However, these are also problems with integration heuristics, which can suffer from blowout in the number of functions of integration and the subsidiary conditions they satisfy. Of course, there are still systems that are effectively intractable by either differential algebra or integration heuristics, since even in the so-called triangulation output by diffgrob2, the equations are too hard to solve.

In pseudo-reduction, one is allowed to multiply the expression being reduced by differential, that is, nonconstant, coefficients of the highest-derivative terms of the reducing equations. The reason one must do this is that on nonlinear systems, the algorithms for calculating the differential analogue of a Gröbner basis will not terminate if only strict reduction is allowed. What this means is that such coefficients are assumed to be nonzero. To obtain solutions of the system that evaluate to zero one of these coefficients, one needs to include it with the system from the start of the calculation; such a solution is called a singular integral for the obvious reason.

As mentioned above, the major problems with the DGB method in practice are its poor complexity and expression swell. However, on systems where the process can be completed within reasonable limits, by which is meant that the length of the expressions obtained is small enough to be meaningful, the output is extremely useful. Comparing the determining equations for classical symmetries and a triangulation for that system demonstrates this point; compare the determining equations for the Boussinesq equation (6.6) and the equivalent system given in following example.

We illustrate the DGB method in the following example.

Example 6.1. Consider the system of determining equations obtained applying the classical Lie method to the Boussinesq equation (6.5), i.e., the system of determining equations (6.6). Applying the method of DGB to (6.6), the system simplifies to the following seven equations

$$\xi_{xx} = 0, \quad \xi_t = 0, \quad \xi_u = 0,$$
$$\tau_x = 0, \quad \tau_u = 0, \quad \tau_t - 2\xi_x = 0, \tag{6.51}$$
$$2u\xi_x + \phi = 0,$$

which are easily solved to yield (6.7). To show this, from (6.6) we have $2\xi_x - \tau_t = 0$, and $\tau_x = 0$, which implies that $\xi_{xx} = 0$, hence $\phi_{xu} = 0$, and so $\phi_x = 0$. Thus it is easily seen that the system (6.6) simplifies to

$$\xi_{xx} = 0, \quad \xi_t = 0, \quad \xi_u = 0,$$
$$\tau_x = 0, \quad \tau_u = 0, \quad 2\xi_x - \tau_t = 0,$$
$$\phi + 2u\xi_x = 0, \tag{6.52}$$
$$\phi_x = 0, \quad 2\phi_{tu} - \tau_{tt} = 0, \quad \phi_{tt} = 0.$$

On solving the first seven equations we find that the last three equations are identically satisfied. Therefore, we have shown that the method of DGB reduces the system (6.6) to (6.51).

The important point is that (6.51) is considerably simpler to solve than (6.6). Furthermore, the equations are *ordered*, so that one knows a priori the optimal sequence in which to solve them. This makes the procedure entirely algorithmic and obviates any ad hoc method for solving the determining equations that might be used. The system (6.51) was calculated using the MAPLE package diffgrob2 [114]. Reid in MAPLE [154–156] and Schwarz in REDUCE [163] have developed similar triangulation packages for linear overdetermined systems of partial differential equations.

For nonlinear systems, despite the complexity problem, DGB have been used effectively to solve the determining equations for nonclassical symmetries [38–43, 45, 115–118, 120], using various strategies that address the complexity problem and that minimize the number of singular integral cases to be considered, i.e.,that minimize the differential coefficients used in the pseudo-reduction processes.

Example 6.2. Applying the nonclassical method to the Boussinesq equation (6.5) yields the set of seven determining equations (6.47). Then applying the method of DGB to (6.47), the system simplifies to

$$\xi_{xx} = 0,$$
$$\xi_{tt} + 2\xi_x\xi_t - 4\xi\xi_x^2 = 0,$$
$$\xi_u = 0,$$
$$\phi + 2u\xi_x + 4\xi^2\xi_x + 2\xi\xi_t = 0,$$
(6.53)

which are easily solved to yield (6.48), (6.49).

As for the classical case, the output (6.53) is obtained literally by "pressing return." It should be emphasized that it is quite surprising that this is sufficient to obtain the DGB of a nonlinear system such as (6.47). More commonly one has to use diffgrob2 interactively, as is illustrated in the following example.

Example 6.3. Consider the Fitzhugh–Nagumo equation

$$u_t = u_{xx} + u(1-u)(u-a),$$
(6.54)

where a is an arbitrary parameter, which arises in population genetics [7,8] and models the transmission of nerve impulses [59,127]. Applying the nonclassical method to (6.54) yields the following four determining equations:

$$\xi_{uu} = 0,$$
(6.55)
$$\phi_{uu} - 2(\xi_{xu} - \xi\xi_u) = 0,$$
(6.56)
$$2\phi_{xu} - 2\xi_u\phi - 3u(u-a)(u-1)\xi_u - \xi_{xx} + 2\xi\xi_x + \xi_t = 0,$$
(6.57)
$$\phi_t - \phi_{xx} + 2\phi\xi_x + (2\xi_x - \phi_u)u(u-a)(u-1)$$
$$+ [3u^2 - 2(a+1)u + a]\phi = 0.$$
(6.58)

A preliminary triangulation of this system leads to the condition

$$\xi_u(2\xi_u^2 - 9) = 0,$$
(6.59)

TABLE 10.4. Nonclassical infinitesimals for the Fitzhugh–Nagumo equation when $\xi_u = 0$.

a	ξ	ϕ
arbitrary	c_0	0
$a = -1$	$-\dfrac{3}{\sqrt{2}}\dfrac{c_1 \exp(\sqrt{2}x) + c_2}{c_1 \exp(\sqrt{2}x) - c_2}$	$-\dfrac{6uc_1c_2 \exp(\sqrt{2}x)}{[c_1 \exp(\sqrt{2}x) - c_2]^2}$
$a = \dfrac{1}{2}$	$-\dfrac{3}{\sqrt{2}}\dfrac{c_1 \exp(\sqrt{2}x) + c_2}{c_1 \exp(\sqrt{2}x) - c_2}$	$-\dfrac{3uc_1c_2 \exp(\sqrt{2}x)}{[c_1 \exp(\sqrt{2}x) - c_2]^2}$
$a = 2$	$-\dfrac{3}{2\sqrt{2}}\dfrac{c_1 \exp(\sqrt{2}x/2) + c_2}{c_1 \exp(\sqrt{2}x/2) - c_2}$	$-\dfrac{3(2u - 1)c_1c_2 \exp(\sqrt{2}x/2)}{4[c_1 \exp(\sqrt{2}x/2) - c_2]^2}$

which is obtained by differentiating (6.57) three times with respect to u and using (6.55), (6.56). The objective in this strategy, referred to as the "direct search" strategy in [115], is to eliminate the x- and t-derivatives in favor of u-derivatives and eliminate derivatives of ϕ in favor of derivatives of ξ. The condition (6.59) shows that there are two cases to consider: (a) $\xi_u = \frac{3}{2}\sqrt{2}$ and (b) $\xi_u = 0$.

Case a. If $\xi_u = \frac{3}{2}\sqrt{2}$, then the triangulation yields the infinitesimals

$$\xi = \frac{1}{2}\sqrt{2}[3u + (a + 1)], \quad \phi = -\frac{3}{2}u(u - a)(u - 1).$$

Case b. If $\xi_u = 0$, then one obtains the triangulation

$$3\phi + (3u - a - 1)\xi_x = 0,$$
$$\xi_t - 3\xi_{xx} + 2\xi\xi_x = 0,$$
$$(1 - a + a^2)\xi_x + 3\xi_{xxx} - 3\xi\xi_{xx} = 0,$$
$$\xi_{xx}(3\xi_{xx} - 2\xi\xi_x) = 0,$$
$$\xi_x^3\xi^2(1 + a^2 - a)[3(a^2 - a + 1) - 2\xi^2 + 6\xi_x] = 0,$$
$$\xi_x(a - 2)(2a - 1)(a + 1) = 0.$$

$$(6.60)$$

We now solve these in reverse order to obtain the infinitesimals given in the Table 10.4, in which c_0, c_1, and c_2 are arbitrary constants.

A much older method of finding a basis for the ideal of a system from which formal solutions may be easily obtained, due to Janet, has been implemented for linear systems [163,172]. Also for linear systems (and linear differential–difference systems), the differential analogue of Buchberger's algorithm [20,21] for calculating an algebraic Gröbner basis has been implemented [139]. For orthonomic systems, those whose members are solvable for their leading-derivative term, the Reid–Wittkopf differential algebra package [157] will calculate the standard form of the system, calculate the number of arbitrary constants and functions which a formal solution depends on (see also [163]), and can then calculate the formal power series solution to any order [156]. This package handles equations with nontrivial coefficients of the leading derivative terms, provided that MAPLE can solve the expression (algebraically) for the leading term. One can then systematically go through the singular integrals using the `divpivs` command.

The triangulations (6.51), (6.53), and (6.60) were performed using the MAPLE package `diffgrob2` [114, 115]. This package was written specifically to handle fully nonlinear equations (in contrast to the packages of Reid [154–156] and Schwarz, [163], which are primarily for linear equations). All calculations are strictly "polynomial," that is, there is no division. Implemented there are the Kolchin–Ritt algorithm, the differential analogue of Buchberger's algorithm using pseudo-reduction instead of reduction, and extra algorithms needed to calculate a DGB (as far as possible using the current theory), for those cases where the Kolchin–Ritt algorithm is not sufficient [119]. Designed to be used interactively as well as algorithmically, the package `diffgrob2` has proved useful for solving some fully nonlinear systems (see, for example, [38–45, 115–118, 120]). As yet, however, algorithmic methods for finding the most efficient orderings, the best method of choosing the sequence of pairs to be cross-differentiated, for deciding when to integrate and read off coefficients of independent functions in one of the variables, for finding the best change of coordinates, and so on, are still the subject of much investigation.

6.4   Relationship Between Classical, Direct, and Nonclassical Methods

In Section 6.2 above it was shown that applying the nonclassical method yields all the new symmetry reductions of the Boussinesq equation (6.5) that were derived by Clarkson and Kruskal [36] using the direct method (see also [107]). In [46] we obtained a similar result for the Kadomtsev–Petviashvili (KP) equation

$$(u_t + uu_x + u_{xxx})_x + \sigma^2 u_{yy} = 0, \quad \sigma^2 = \pm 1, \tag{6.61}$$

which arises in many physical applications including weakly two-dimensional long waves in shallow water [4, 165], where the sign of σ^2 depends upon the relevant magnitudes of gravity and surface tension. We showed that the direct and nonclassical methods yield the same symmetry reductions for the KP equation (6.61) [46]. The results in [46, 107] suggested that maybe the direct and nonclassical methods were equivalent, i.e., they yield the same reductions. Indeed Clarkson and Kruskal [36] posed the question on the relationship between these two methods in the conclusion of their original paper on the direct method.

This question was investigated by Nucci and Clarkson [129], who applied both the direct and nonclassical methods to the Fitzhugh–Nagumo equation (6.54).

Example 6.4. Applying the classical Lie method to the Fitzhugh–Nagumo equation yields the traveling wave solution

$$u(x,t) = w(z), \quad z = \mu x - \lambda t. \tag{6.62}$$

Applying the direct method to the Fitzhugh–Nagumo equation yields, in addition to the traveling wave solution (6.62), the exact solutions of the Fitzhugh–Nagumo equation expressed in terms of Jacobi elliptic functions for $a = -1$, $a = \frac{1}{2}$, and $a = 2$. For example, for $a = -1$,

$$u(x,t) = z_x \operatorname{ds}\left(z; \frac{1}{2}\sqrt{2} \right),$$

$$z(x,t) = c_1 \exp\left[\frac{1}{2}(\sqrt{2}\,x + 3t) \right] + c_2 \exp\left[\frac{1}{2}(-\sqrt{2}\,x + 3t) \right] + c_3, \tag{6.63}$$

where $\operatorname{ds}(z;k)$ is the Jacobi elliptic function satisfying

$$(\eta')^2 = \eta^4 + (2k^2 - 1)\eta^2 + k^2(k^2 - 1).$$

Applying the nonclassical method to the Fitzhugh–Nagumo equation yields, in addition to the traveling wave solution (6.62) and the elliptic functions solution (6.63), the following exact solution of the Fitzhugh–Nagumo equation, for $a \neq 0$, $a \neq 1$:

$$u(x,t) = \frac{ac_1 \exp\left\{ \frac{1}{2}(\pm\sqrt{2}\,ax + a^2t) \right\} + c_2 \exp\left\{ \frac{1}{2}(\pm\sqrt{2}\,x + t) \right\}}{c_1 \exp\left\{ \frac{1}{2}(\pm\sqrt{2}\,ax + a^2t) \right\} + c_2 \exp\left\{ \frac{1}{2}(\pm\sqrt{2}\,x + t) \right\} + c_3 \exp(at)}, \tag{6.64}$$

where c_1, c_2, and c_3 are arbitrary constants. If $a = 0$ or $a = 1$, then similar solutions are obtained.

This demonstrated that the nonclassical method is more general than the direct method, at least as it was originally formulated. We note that the solution (6.64) was obtained by Kawahara and Tanaka [97] using Hirota's bilinear method [86]; Hereman [82] also found it using a truncated Painlevé expansion method, sometimes called the singular manifold method [175, 176].

Estévez [57] has shown that the exact solution (6.64) can be obtained using a generalization of the direct method. Substituting the standard direct method ansatz

$$u(x,t) = \beta(x,t)w(z) + \alpha(x,t),$$

where $\alpha(x,t)$, $\beta(x,t)$, and $z(x,t)$ are functions to be determined, into the Fitzhugh–Nagumo equation yields

$$\beta z_x^2 \frac{d^2 w}{dz^2} + (2\beta_x z_x + \beta z_{xx} - \beta z_t)\frac{dw}{dz} - \beta^3 w^3 + \beta^2(a + 1 - 3\alpha)w^2$$
$$+ \left[2(a+1)\alpha\beta - 3\beta\alpha^2 - a\beta + \beta_{xx} - \beta_t \right] w$$
$$+ \alpha_{xx} + \alpha_t - \alpha(\alpha - a)(\alpha - 1) = 0. \tag{6.65}$$

If we set $\beta = z_x$ and require that $z(x,t)$ and $\alpha(x,t)$ satisfy

$$3z_{xx} - z_t = \pm\sqrt{2}\,(a+1-3\alpha)\,z_x, \tag{6.66}$$

$$z_{xxx} - z_{xt} = \left[3\alpha^2 - 2(a+1)\alpha + a\right]z_x, \tag{6.67}$$

$$\alpha_{xx} - \alpha_t + \alpha(\alpha-1)(a-\alpha) = 0, \tag{6.68}$$

then (6.65) reduces to

$$z_x\left(\frac{\mathrm{d}^2w}{\mathrm{d}z^2} - w^3\right) + (a+1-3\alpha)\left(\pm\sqrt{2}\frac{\mathrm{d}w}{\mathrm{d}z} + w^2\right) = 0.$$

Therefore, if $w(z)$ satisfies the overdetermined system of equations

$$\frac{\mathrm{d}^2w}{\mathrm{d}z^2} - w^3 = 0, \quad \pm\sqrt{2}\frac{\mathrm{d}w}{\mathrm{d}z} + w^2 = 0, \tag{6.69}$$

which have common solution

$$w(z) = \pm\frac{\sqrt{2}}{z-z_0}, \tag{6.70}$$

with z_0 an arbitrary constant, then we obtain an exact solution of (6.54). Equation (6.68) has three constant solutions, $\alpha = 0$, $\alpha = 1$, and $\alpha = a$. Then, given α, solving (6.66), (6.67) for $z(x,t)$ and substituting in (6.70) yields (6.64) for all three values of α. Hence in order to obtain the exact solution (6.64), Estévez [57] reduced a single partial differential equation to a coupled system of ordinary differential equations. This procedure is similar to that advocated by Galaktionov [71, 72]. We remark also that (6.66)–(6.68) are the same equations as those that arise in the truncated Painlevé expansion (singular manifold) method [82].

More generally, the singular manifold method [175, 176] has proven to be an effective method for generating exact solutions of several partial differential equations. In particular, we mention the work of Cariello and Tabor [26, 27] and Estévez [57], who have shown that the so-called singularity method plays the role of a similarity variable and that the singular manifold method is capable of yielding solutions corresponding to both classical and nonclassical symmetries. The relationship between nonclassical symmetries and the singular manifold method is considered in depth by Estévez and Gordoa [58].

Since the original development of the direct method it has been known that there exist exact solutions of partial differential equations that are not obtainable using the direct method; for example, the two-soliton solution (and more generally the N-soliton solution) of the Boussinesq equation (6.5) given by

$$u(x,t) = 12\frac{\partial^2}{\partial x^2}\{\ln[1 + \exp(\eta_1) + \exp(\eta_2)$$
$$+ A_{12}\exp(\eta_1+\eta_2)]\} - 1, \tag{6.71}$$

where $\eta_j = \kappa_j(x + \mu_j t) + \delta_j$ for $j = 1, 2$, and

$$A_{12} = \frac{\mu_1\mu_2 + 2\kappa_1^2 - 3\kappa_1\kappa_2 + 2\kappa_2^2 - 1}{\mu_1\mu_2 + 2\kappa_1^2 + 3\kappa_1\kappa_2 + 2\kappa_2^2 - 1},$$

with $\mu_1 = (1 - \kappa_1^2)^{1/2}$, $\mu_2 = (1 - \kappa_2^2)^{1/2}$, and δ_1 and δ_2 arbitrary constants. It should be noted that the special case of the solution (6.71) with $A_{12} = 0$ can be obtained using the truncated Painlevé expansion (singular manifold) method. Hirota and Ito [87] refer to this special case of the general soliton solution as being the "resonant state" where either two solitons fuse together after colliding with each other or a single soliton splits into two solitons.

We remark that the solution (6.64) of the Fitzhugh–Nagumo equation has certain similarities to a "two-soliton" solution, and as mentioned above, it has also been obtained using Hirota's bilinear method [97]; since the Fitzhugh–Nagumo equation is thought to be nonintegrable, the solution (6.64) is not strictly a "two-soliton."

These results pose the following important open question: For which partial differential equations does the nonclassical method yield more symmetry reductions than the direct method? Furthermore, it remains an open question to determine a priori which partial differential equations possess symmetry reductions that are not obtainable using the classical Lie group approach.

The ansatz $u(x, t) = F(x, t, w(z))$ with $z = z(x, t)$ used in the direct method assumes that the symmetry variable z does not depend upon u. Consequently, it is implicitly assumed that the ratio of infinitesimals ξ/τ is independent of u. For the exact solution (6.64), this ratio of infinitesimals is dependent upon u. However, even if the ratio is dependent upon u, this does not guarantee that the associated symmetry solution is not obtainable using the direct method (see, for example, the example given in [34]).

Recently, Olver [132] (see also [9, 35, 146, 187]) has proved the precise relationship between the direct and nonclassical methods. Consider the second-order partial differential equation

$$E(x, t, u, u_x, u_t, u_{xx}, u_{xt}, u_{tt}) = 0 \qquad (6.72)$$

(the following results are easily extended to general scalar nth order partial differential equations). Equation (6.72) admits a *direct reduction* if there exist functions $z = z(x, t)$ and $u = U(x, t, w)$ such that the Clarkson–Kruskal ansatz

$$u(x, t) = U(x, t, w(z)) \qquad (6.73)$$

reduces (6.72) to a single ordinary differential equation for $w(z)$. (Note that U is not uniquely determined, since we can incorporate any functions of the similarity variable z into w.) Olver [132] proved the following two theorems.

Theorem 6.1. *There is a one-to-one correspondence between the ansätze of the direct method* (6.73) *with* $U_w \neq 0$ *and the quasi-linear first-order differential constraint*

$$\mathbf{v}(u) \equiv \xi(x,t)u_x + \tau(x,t)u_t = \phi(x,t,u). \tag{6.74}$$

Theorem 6.2. *The ansatz* (6.73) *will reduce the partial differential equation* (6.72) *to a single ordinary differential equation for* $w(z)$ *if and only if the overdetermined system of partial differential equations defined by* (6.72) *and* (6.74) *is compatible.*

Thus, there is a one-to-one correspondence between direct reductions of the partial differential equation (6.72) and compatible first-order quasi-linear differential constraints. Solutions of (6.74) are the functions that are invariant under the one-parameter group generated by the vector field

$$\mathbf{w} = \xi(x,t)\partial_x + \tau(x,t)\partial_t + \phi(x,t,u)\partial_u. \tag{6.75}$$

Hence $\mathbf{w}$ generates a group of "fiber-preserving transformations", since ξ and τ are independent of u.

In the direct method one requires that the ansatz (6.73) reduce the partial differential equation (6.72) to a single ordinary differential equation. In the nonclassical method, one requires that the differential constraint (6.74) that requires the solutions to be invariant under the group generated by $\mathbf{w}$ be compatible with the original partial differential equation (6.72) in the sense that the overdetermined system of partial differential equations defined by (6.72) and (6.74) has no integrability conditions. The general nonclassical method, which allows arbitrary point transformation symmetry groups, so that ξ and τ in (6.75) can also depend upon u, is similarly equivalent to the more general (though considerable harder to deal with) ansatz

$$u(x,t) = U\big(x,t,w(z)\big), \quad z = z(x,t,u).$$

Applying the direct method with the ansatz (6.73) does *not* always find all reductions that are obtained using the classical methods, as shown in the following example due to Ludlow [112].

Example 6.5. Consider the equation

$$u_x u_{xx} - (\alpha u u_x - \beta u_t)(1 - t u_x)^3 = 0, \tag{6.76}$$

with α and β arbitrary constants. Applying the classical method to this equation yields the infinitesimals,

$$\xi = \frac{1}{3}(\kappa_1 + \kappa_3)(x + 2tu) + \kappa_4 u - \kappa_2 t + \kappa_5,$$
$$\tau = \kappa_3 t + \kappa_4, \tag{6.77}$$
$$\phi = \kappa_1 u + \kappa_2 \frac{\beta}{\alpha},$$

where $(\alpha + \beta)(\kappa_1 + \kappa_3) = 0$ and $\kappa_1, \kappa_2, \ldots, \kappa_5$ are arbitrary constants. In the special case where $\kappa_1 = \kappa_2 = \kappa_3 = \kappa_5 = 0$ and $\kappa_4 = 1$, the invariant surface condition becomes

$$uu_x + u_t = 0,$$

which has solution

$$u(x,t) = w(z), \quad z = x - ut. \tag{6.78}$$

Substituting this back into (6.76) yields

$$\frac{dw}{dt}\left[\frac{d^2w}{dt^2} - (\alpha + \beta)w\right] = 0.$$

For convenience we set $\alpha = \beta = \frac{1}{2}$, and thus we obtain the implicit solution

$$u(x,t) = A\exp\{x - u(x,t)t\} + B\exp\{-x + u(x,t)t\}, \tag{6.79}$$

where A and B are arbitrary constants, which defines $u(x,t)$ by a transcendental function. Applying the direct method with the ansatz (6.73) will not obtain such a reduction.

In addition, it is not clear how the direct method, developed by Clarkson and Kruskal [36] for finding symmetry reductions of partial differential equations, may be applied to equations that contain arbitrary functions such as the nonlinear heat equation

$$u_t = u_{xx} + f(u), \tag{6.80}$$

where $f(u)$ is an arbitrary sufficiently differentiable function and subscripts denote partial derivatives. This equation arises in several important physical applications including microwave heating (where $f(u)$ is the rate of absorption of microwave energy; cf. [145,166]), in the theory of chemical reactions (where $f(u)$ is the temperature-dependent reaction rate; cf. [6,7,62]) and in mathematical biology (where $f(u)$ represents the reaction kinetics in a diffusion process; cf. [125]). Clarkson and Mansfield [38] used the nonclassical method in conjunction with the method of differential Gröbner bases (cf. [119]) to find the conditions on $f(u)$ in (6.80) under which symmetries other than the trivial spatial and temporal translational symmetries exist, and then solve the determining equations for the infinitesimals. Clarkson and Mansfield [38] give a complete catalogue of symmetry reductions for the nonlinear heat equation (6.80) and in particular a classification of exact solutions of (6.80) for $f(u) = (u - a)(u - b)(u - c)$ expressed in terms of the roots a, b, and c of the cubic.

The use of DGB has made the analysis of overdetermined systems of partial differential equations, such as those arising as the determining equations for classical and nonclassical symmetries, more tractable. While the `diffgrob2` [114] package needs to be used interactively at present, nevertheless it has proved effective in solving such overdetermined systems (cf. [38–45, 115–118, 120]).

It appears to be the case that for some partial differential equations one of the direct or nonclassical methods is simpler to apply, and vice versa for other equations. One difference between the two methods is that the direct method yields the symmetry reduction in one-step, whereas in the nonclassical method, one first solves for the infinitesimals and then, given the infinitesimals, one solves the invariant surface condition, which is a two-step procedure.

To conclude this section we make some remarks comparing the classical Lie, direct, and nonclassical methods.

Classical Lie Method: The positive aspects of this method are that the determining equations are linear and the associated vector fields have a Lie algebraic structure, which has many useful applications. In particular, this Lie algebra gives rise to group transformations, transforming solutions into solutions. Moreover, there exist several symbolic manipulation programs. However, as we have seen, the method does not find all reductions for all partial differential equations.

Clarkson–Kruskal Direct Method: This method is more general than the classical Lie method, has no associated group framework, and one can choose the dimension of the reduced equation (i.e., the number of independant variables obtained after the reduction). Furthermore, the direct method is a one-step procedure. However, the determining equations are nonlinear, the associated vector fields have no Lie algebraic structure, and there are only limited symbolic manipulation programs available.

Nonclassical Method, Conditional Symmetries: This method is even more general than the other two and can be viewed as a modification of the classical theory. As for the direct method, the determining equations are nonlinear, the associated vector fields have no Lie algebraic structure, and there are only limited symbolic manipulation programs, though in contrast to the direct method it is a two-step procedure.

7 Conclusions

As the title of this chapter indicates, its main purpose has been to review the different methods that can be used to reduce nonlinear partial differential equations to ordinary ones. The most important feature of these Lie algebraic tools and their generalizations is their universality. They can be applied to arbitrary systems of partial differential equations involving any number of independent and dependent variables and derivatives of any order. On the other hand, the reduction techniques for nonlinear equations involve a loss of generality. The obtained solutions are compatible with a restricted class of initial and boundary conditions. For linear partial differential equations symmetry reduction essentially boils down to the separation of variables. Since this can be combined with the linear superposition principle, very general classes of solutions can be expanded in terms of separated ones.

The two major sources of exact analytical solutions of nonlinear partial differential equations are group theory, in the general sense used in this chapter, and the inverse scattering techniques of soliton theory. In a nutshell, a comparison between the two can be stated as follows. The integrability techniques of soliton theory provide larger classes of solutions, e.g., n-soliton ones, but are applicable to a restricted class of equations, namely those that are, by definition, integrable. Symmetry methods provide more restricted solutions, but for a much larger class of equations. For integrable equations, symmetry methods often provide solutions other than the inverse scattering techniques, in particular dilationally invariant solutions.

A common feature of both methods is the usefulness of Painlevé analysis. Indeed, quite often nonintegrable partial differential equations provide reduced equations that do have the Painlevé property, either in general, or for certain values of some parameters in the original or reduced equations.

Not presented in these lecture notes are extensions of Lie-algebraic methods, including conditional symmetries, to discrete equations. Much progress has been made in the study of symmetries of difference and differential difference equations, some of it discussed in this volume. Here we limit ourselves to giving a few relevant references [53, 60, 75, 85, 95, 101–103, 105, 106, 110, 113, 147, 186]

Acknowledgments: It is a pleasure to thank Mark Ablowitz, Andrew Bassom, George Bluman, Pilar Estévez, Victor Galaktionov, Willy Hereman, Martin Krustal, Decio Levi, Elizabeth Mansfield, Clara Nucci, Peter Olver, and Greg Reid for their contributions, helpful comments, and illuminating discussions.

8   References

[1] M.J. Ablowitz and P.A. Clarkson, *Solitons, Nonlinear Evolution Equations and Inverse Scattering* (Cambridge Univ. Press, Cambridge, 1991).

[2] M.J. Ablowitz, A. Ramani, and H. Segur, *A connection between non-*

linear evolution equations and ordinary differential equations of P-type, J. Math. Phys. **21** (1980), 715–721.

[3] M.J. Ablowitz, A. Ramani, and H. Segur, *A connection between non-linear evolution equations and ordinary differential equations of P-type*, J. Math. Phys. **21** (1980), 1006–1015.

[4] M.J. Ablowitz and H. Segur, *On the evolution of packets of water waves*, J. Fluid Mech. **92** (1979), 691–715.

[5] M. Abramowitz and I.A. Stegun, *Handbook of Mathematical Functions* (Dover, New York, 1965).

[6] R. Aris, *The Mathematical Theory of Diffusion and Reaction in Permeable Catalysts*, vols. I and II (Oxford Univ. Press, Oxford, 1975).

[7] D.G. Aronson and H.F. Weinberger, *Nonlinear diffusion in population genetics, combustion and nerve propagation*, Partial Differential Equations and Related Topics, ed. J.A. Goldstein (Springer-Verlag, Berlin, 1975), Lect. Notes in Math., vol. 446 pages 5–49.

[8] D.G. Aronson and H.F. Weinberger, *Multidimensional nonlinear diffusion arising in population genetics*, Adv. Math. **30** (1978), 33–76.

[9] D. Arrigo, P. Broadbridge, and J.M. Hill, *Nonclassical symmetry solutions and the methods of Bluman–Cole and Clarkson–Kruskal*, J. Math. Phys. **34** (1993), 4692–4703.

[10] A.P. Bassom, P.A. Clarkson, and A.C. Hicks, *Bäcklund transformations and solution hierarchies for the fourth Painlevé equation*, Stud. Appl. Math. **95** (1995), 1–71.

[11] A.P. Bassom, P.A. Clarkson, and A.C. Hicks, *On the application of solution hierarchies of the fourth Painlevé equation to several physically significant nonlinear partial differential equations*, Advances in Differential Equations **1** (1996), 175–198.

[12] J. Beckers, J. Patera, M. Perroud, and P. Winternitz, *Subgroups of the Euclidean group and symmetry breaking in nonrelativistic quantum mechanics*, J. Math. Phys. **18** (1977), 72–83.

[13] D. Bérubé and M. de Montigny, *A* MATHEMATICA *program for the calculation of Lie point symmetries of systems of differential equations*, Preprint CRM-1822, Montréal, 1992.

[14] G.W. Bluman and J.D. Cole, *The general similarity of the heat equation*, J. Math. Mech. **18** (1969), 1025–1042.

[15] G.W. Bluman and S. Kumei, *Symmetries and Differential Equations* (Springer-Verlag, Berlin, 1989), Appl. Math. Sci., vol. 81.

[16] G.W. Bluman and V. Shtelen, *Developments in similarity methods related to pioneering work of Julian Cole*, Mathematics is for Solving Problems, eds. L.P. Cook, V. Roytburd, and M. Tulin (SIAM, Philadelphia, 1996), pages 105–117.

[17] J. Boussinesq, *Théorie de l'intumescence appelée onde solitaire ou de translation se propageant dans un canal rectangulaire*, Comptes Rendus **72** (1871), 755–759.

[18] J. Boussinesq, *Théorie des ondes et des remous qui se propagent le long d'un canal rectangulaire horizontal, en communiquant au liquide contenu dans ce canal des vitesses sensiblement pareilles de la surface au fond*, J. Math. Pures Appl. Ser. 2. **17** (1872), 55–.

[19] C. P. Boyer, R. T. Sharp, and P. Winternitz, *Symmetry breaking interactions for the time dependent Schrödinger equation*, J. Math. Phys. **17** (1976), 1439–1451.

[20] B. Buchberger, *A survey on the method of Gröbner bases for solving problems in connection with systems of multivariate polynomials*, Symbolic and Algebraic Computation by Computers, eds. N. Inada and T. Soma (World Scientific, Singapore, 1985), pages 69–83.

[21] B. Buchberger, *Applications of Gröbner bases in nonlinear computational geometry*, Mathematical Aspects of Scientific Software, ed. J. Rice (Springer Verlag, New York, 1988), pages 59–87.

[22] G.I. Burdé, *The construction of special explicit solutions of the boundary-layer equations. Steady flows*, Q. J. Mech. Appl. Math **47** (1994), 247–260.

[23] G.I. Burdé, *The construction of special explicit solutions of the boundary-layer equations. Unsteady flows*, Q. J. Mech. Appl. Math **48** (1995), 611–633.

[24] G.I. Burdé, *New similarity solutions of the steady-state boundary-layer equations*, J. Phys. A **1996** (29), 1665–1683.

[25] G. Burdet, J. Patera, M. Perrin, and P. Winternitz, *The optical group and its subgroups*, J. Math. Phys. **19** (1978), 1758–1780.

[26] F. Cariello and M. Tabor, *Painlevé expansions for nonintegrable evolution equations*, Physica D **39** (1989), 77–94.

[27] F. Cariello and M. Tabor, *Similarity reductions from extended Painlevé expansions for nonintegrable evolution equations*, Physica D **53** (1991), 59–70.

[28] B. Champagne, W. Hereman, and P. Winternitz, *The computer calculation of Lie point symmetries of large systems of differential equations*, Comp. Phys. Commun. **66** (1991), 319–340.

[29] B. Champagne and P. Winternitz, *A* MACSYMA *program for calculating the symmetry group of a system of differential equations*, Preprint CRM-1278, Montréal, 1985.

[30] B. Champagne and P. Winternitz, *On the infinite-dimensional symmetry group of the Davey–Stewartson equation*, J. Math. Phys. **29** (1988), 1–8.

[31] P.A. Clarkson, *New exact solutions for the Boussinesq equation*, Europ. J. Appl. Math. **1** (1990), 279–300.

[32] P.A. Clarkson, *Dimensional reductions and exact solutions of a generalized nonlinear Schrödinger equation*, Nonlinearity **5** (1992), 453–472.

[33] P.A. Clarkson, *Nonclassical symmetry reductions of nonlinear partial differential equations*, Math. Comp. Model. **18** (1993), 45–68.

[34] P.A. Clarkson, *Nonclassical symmetry reductions of the Boussinesq equation*, Chaos, Solitons and Fractals **5** (1995), 2261–2301.

[35] P.A. Clarkson and S. Hood, *Nonclassical symmetry reductions and exact solutions of the Zabalotskaya–Khokhlov equation*, Europ. J. Appl. Math. **3** (1992), 381–414.

[36] P.A. Clarkson and M.D. Kruskal, *New similarity solutions of the Boussinesq equation*, J. Math. Phys. **30** (1989), 2201–2213.

[37] P.A. Clarkson, D.K Ludlow and T.J. Priestley, *The classical, direct and nonclassical methods for symmetry reductions of nonlinear partial differential equations*, Meth. Appl. Anal. **4** (1997), 173–195.

[38] P.A. Clarkson and E.L. Mansfield, *Symmetry Reductions and Exact Solutions of a class of Nonlinear Heat Equations*, Physica D **70** (1994), 250–288.

[39] P.A. Clarkson and E.L. Mansfield, *Algorithms for the nonclassical method of symmetry reductions*, SIAM J. Appl. Math. **54** (1994), 1693–1719.

[40] P.A. Clarkson and E.L. Mansfield, *On a shallow water wave equation*, Nonlinearity **7** (1994), 975–1000.

[41] P.A. Clarkson and E.L. Mansfield, *Symmetry reductions and exact solutions of shallow water wave equations*, Acta Appl. Math. **39** (1995), 245–276.

[42] P.A. Clarkson, E.L. Mansfield, and A.E. Milne, *Symmetries and Exact Solutions of a 2 + 1-dimensional Sine-Gordon System*, Phil. Trans. R. Soc. Lond. A **354** (1996), 1807–1835.

[43] P.A. Clarkson, E.L. Mansfield, and T.J. Priestley, *Symmetries of a class of nonlinear third-order partial differential equations*, Math. Comp. Model. **25** (1997), 195–212.

[44] P.A. Clarkson and T.J. Priestley, *Symmetries of a class of nonlinear fourth-order partial differential equations*, J. Nonlinear Math. Phys. (6), 1999.6698

[45] P.A. Clarkson and T.J. Priestley, *Shallow water wave systems*, Studies in Applied Mathematics (101), 1999.389432

[46] P.A. Clarkson and P. Winternitz, *Nonclassical symmetry reductions for the Kadomtsev–Petviashvili equation*, Physica D **49** (1991), 257–272.

[47] R. Conte, Chapter 3, this volume.

[48] J.F. Cornwell, *Group Theory in Physics, vol. II* (Academic Press, New York, 1984).

[49] C.M. Cosgrove, *Corrections and annotations to E.L. Ince*, Ordinary differential equations, Chapter 14, on the classification of Painlevé differential equations, unpublished (1993).

[50] D. David, N. Kamran, D. Levi, and P. Winternitz, *Subalgebras of loop algebras and symmetries of the Kadomtsev–Petviashvili equation*, Phys. Rev. Lett. **55** (1985), 2111–2113.

[51] D. David, N. Kamran, D. Levi, and P. Winternitz, *Symmetry reduction for the Kadomtsev–Petviashvili equation using a loop algebra*, J. Math. Phys. **27** (1986), 1225–1237.

[52] P. Deift, C. Tomei, and E. Trubowitz, *Inverse scattering and the Boussinesq equation*, Commun. Pure Appl. Math. **35** (1982), 567–628.

[53] V.A. Dorodnitsyn, *Transformation groups in net spaces*, Itogi Nauki i Tekhniki **34** (1989), 149–191; J. Soviet Math. **55** (1991), 1490–1517.

[54] E.B. Dynkin, *Semisimple subalgebras of semisimple Lie algebras*, Mat. Sbornik **30** (1952), 349–462; Amer. Math. Soc. Transl. **6** (1957), 111–244.

[55] E.B. Dynkin Maximal subgroups of the classical groups, Trudy Moskov. Mat. Obšč **1** (1952), 39–166; Amer. Math. Soc. Transl. **6** (1957), 245–378.

[56] A. Erdélyi, W. Magnus, F. Oberhettinger, and F.G. Tricomi, *Higher Transcendental Functions.* II (McGraw-Hill, New York, 1953).

[57] P.G. Estévez, *Nonclassical symmetries and the singular manifold method for the Fitzhugh–Nagumo equation*, Phys. Lett. **171A** (1992), 259–261.

[58] P.G. Estévez and P.R. Gordoa, *Nonclassical symmetries and the singular manifold method: theory and six examples*, Stud. Appl. Math. **95** (1995), 73–113.

[59] R. Fitzhugh, *Impulses and physiological states in theoretical models of nerve membrane*, Biophysical J. **1** (1961), 445–466.

[60] R. Floreanini and L. Vinet, *Lie symmetries of finite-difference equations*, J. Math. Phys. **36** (1995), 7024–7042.

[61] A.S. Fokas and M.J. Ablowitz, *On a unified approach to transformations and elementary solutions of Painlevé equations*, J. Math. Phys. **23** (1983), 2033–2042.

[62] D.A. Frank-Kamenetskii, *Diffusion and Heat Exchange in Chemical Kinetics* (Princeton University Press, Princeton, 1955).

[63] W.I. Fushchich, *Conditional symmetry of the equations of Mathematical Physics*, Ukrain. Math. J. **43** (1991), 1456–1470.

[64] W.I. Fushchich, *Ansatz '95*, Nonlinear Math. Phys. **2** (1995), 216–235.

[65] W.I. Fushchich, W.M. Shtelen, and N.I. Serov, *Symmetry Analysis and Exact Solutions of the Equations of Mathematical Physics* (Kluwer, Dordrecht, 1993).

[66] G. Gaeta, *Nonlinear Symmetries and Nonlinear Equations* (Kluwer, Dordrecht, 1994).

[67] L. Gagnon and P. Winternitz, *Lie symmetries of a generalised nonlinear Schrödinger equation: I–The symmetry group and its subgroups*, J. Phys. A **21** (1988), 1493–1511.

[68] L. Gagnon and P. Winternitz, *Lie symmetries of a generalised nonlinear Schrödinger equation: II–Exact solutions*, J. Phys. A **22** (1989), 469–497.

[69] L. Gagnon and P. Winternitz, *Lie symmetries of a generalised nonlinear Schrödinger equation: II–Exact solutions*, J. Phys. A **22** (1989), 469–497.

[70] L. Gagnon and P. Winternitz, *Exact solutions of the spherical quintic nonlinear Schrödinger equation*, Phys. Lett. **134A** (1989), 276–281.

[71] V.A. Galaktionov, *On new exact blow-up solutions for nonlinear heat conduction equations with source and applications*, Diff. and Int. Eqns. **3** (1990), 863–874.

[72] V.A. Galaktionov, *Quasilinear heat equations with first-order sign-invariants and new explicit solutions*, Nonlinear Anal., Theory, Meth., and Appl. **23** (1994), 1595–1621.

[73] C.S. Gardner, J.M. Greene, M.D. Kruskal, and R.M. Miura, *Method for solving the KdV equation*, Phys. Rev. Lett. **19** (1967), 1095–1097.

[74] P. Goddard and D. Olive, *Kac Moody and Virasoro algebras in relation to quantum physics*, Int. J. Mod. Phys. A **1** (1986), 303–414.

[75] D. Gómez-Ullate, S. Lafortune, and P. Winternitz, *Symmetries of discrete dynamical systems involving two species*, J. Math. Phys., 40.199927822804

[76] E. Goursat, *Sur les substitutions orthogonales et les divisions régulières de l'espace*, Ann. Sci. Ec. Norm. Sup (3) **6** (1880), 9–102.

[77] B. Grammaticos, F. W. Nijhoff, and A. Ramani, Chapter 7, this volume.

[78] A. M. Grundland and L. Lalague, *Lie subgroups of the symmetry group of the equations describing a nonstationary and isotropic flow: invariant and partially invariant solutions*, Can. J. Phys. **72** (1994), 362–374.

[79] A. M. Grundland and L. Lalague, *Lie subgroups of symmetry groups of fluid dynamics and magnetohydrodynamics equations*, Can. J. Phys. **73** (1995), 463–477.

[80] A. M. Grundland and L. Lalague, *Invariant and partially invariant solutions of the equations describing a nonstationary and isentropic flow for an ideal and incompressible fluid in $(3+1)$ dimensions*, J. Phys. A. Math. gen **29** (1996), 1723–1739.

[81] A.G. Hansen, *Similarity Analyses of Boundary Value Problems in Engineering* (Prentice-Hall, Englewood Cliffs, 1964).

[82] W. Hereman, *Application of a* MACSYMA *program for the Painlevé test to the Fitzhugh–Nagumo equation*, Partially Integrable Evolution Equations in Physics, eds. R. Conte and N. Boccara (Kluwer, Dordrecht, 1990), NATO ASI Series C, vol. 310 pages 585–586.

[83] W. Hereman, *Review of symbolic software for the computation of Lie symmetries of differential equations*, Euromath. Bull **2** (1993), 45–79.

[84] W. Hereman, *Symbolic software for Lie symmetry analysis*, CRC Handbook of Lie Group Analysis of Differential Equations, ed. N. Ibragimov (CRC Press, Boca Raton, 1996), vol. 3 pages 367–522.

[85] R. Hernández Heredero, D. Levi, and P. Winternitz, *Point symmetries and generalized symmetries of nonlinear difference equations*, Preprint CRM-2568, Montréal, 1998.

[86] R. Hirota, *Direct methods in soliton theory*, Solitons, eds. R.K. Bullough and P.J. Caudrey (Springer-Verlag, Berlin, 1980), Topics in Current Physics, vol. 17 pages 157–176.

[87] R. Hirota and M. Ito, *Resonance of solitons in one dimension*, J. Phys. Soc. Japan **52** (1983), 744–748.

[88] S. Hood, *New exact solutions of Burgers's equation—an extension to the direct method of Clarkson and Kruskal*, J. Math. Phys. **36** (1995), 1971–1990.

[89] V. Hussin, P. Winternitz, and H. Zassenhaus, *Maximal abelian subalgebras of complex orthogonal Lie algebras*, Lin. Alg. Appl. **141** (1990), 183–220.

[90] V. Hussin, P. Winternitz, and H. Zassenhaus, *Maximal abelian subalgebras of pseudo-orthogonal Lie algebras*, Lin. Alg. Appl. **173** (1992), 125–163.

[91] N.H. Ibragimov, *Transformation Groups Applied to Mathematical Physics* (Reidel, Dordrecht, 1985).

[92] E.L. Ince, *Ordinary Differential Equations* (1926, Reprinted (Dover, New York, 1956). See errata in [49].), Longmans, Green and co., London and New York.

[93] N. Jacobson, *Lie algebras* (Dover, New York, 1979).

[94] M. Jimbo and T. Miwa, *Solitons and infinite dimensional Lie algebras*, Publ. RIMS, Kyoto **19** (1983)), 943–1001.

[95] N. Joshi and P.J. Vassiliou, *The existence of Lie symmetries for first-order analytic discrete dynamical systems*, J. Math. Anal. Appl. **195** (1995), 872–887.

[96] E.G. Kalnins and P. Winternitz, *Maximal abelian subalgebras of complex Euclidean Lie algebras*, Can. J. Phys. **72** (1994), 389–404.

[97] T. Kawahara and M. Tanaka, *Interactions of traveling fronts—an exact solution of a nonlinear diffusion equation*, Phys. Lett. **97A** (1983), 311–314.

[98] B.P. Komrakov, *Reductive subalgebras of semisimple real Lie algebras*, Doklady AN SSSR **308** (1989), 521–525; Soviet Math. Dokl. **40** (1990), 329–333.

[99] B.P. Komrakov *Maximal subalgebras of real Lie algebras and a problem of Sophus Lie*, Doklady AN SSSR **311** (1990), 538–532; Soviet Math. Dokl. **41** (1990), 269–273.

[100] B. Kostant, *On the conjugacy of real Cartan subalgebras*, Proc. Nat. Academy Sci. USA **41** (1955), 967–970.

[101] D. Levi and M.A. Rodríguez, *Symmetry group of partial differential equations and of differential-difference equations: the Toda lattice versus the Korteweg-de Vries equation*, J. Phys. A **25** (1992), L975–L979.

[102] D. Levi, L. Vinet, and P. Winternitz, *Lie group formalism for difference equations*, J. Phys. A **30** (1991), 633–649.

[103] D. Levi, L. Vinet and P. Winternitz, eds., *Symmetries and integrability of difference equations* (AMS, Providence, R.I., 1996), CRM proceedings and lecture notes, Vol 9.

[104] D. Levi and P. Winternitz, *The cylindrical Kadomtsev-Petviashvili equation: its Kac-Moody-Virasoro algebra and relation to KP equation*, Phys. Lett. **129A** (1988), 165–167.

[105] D. Levi and P. Winternitz, *Continuous symmetries of discrete equations*, Phys. Lett. **152A** (1991), 335–338.

[106] D. Levi and P. Winternitz, *Symmetries of discrete dynamical systems*, J. Math. Phys. **37** (1996), 5551–5576.

[107] D. Levi and P. Winternitz, *Nonclassical symmetry reduction: example of the Boussinesq equation*, J. Phys. A **22** (1989), 2915–2924.

[108] D. Levi and P. Winternitz, eds., *Painlevé transcendents, their asymptotics and physical applications* (Plenum, New York, 1992).

[109] D. Levi and P. Winternitz, *Symmetries and conditional symmetries of differential-difference equations*, J. Math. Phys. **34** (1993), 3713–3730.

[110] D. Levi and R. Yamilov, *Conditions for the existence of higher symmetries of evolutionnary equations on the lattice*, J. Math. Phys. **38** (1997), 6648–6674.

[111] S.-Y. Lou, *A note on the new similarity reductions of the Boussinesq equation*, Phys. Lett. **151A** (1990), 133–135.

[112] D.K. Ludlow, *Nonclassical similarity reductions of the Navier–Stokes equations*, Ph.D. thesis (Department of Mathematics, University of Exeter, UK), 1994.

[113] S. Maeda, *The similarity method for difference equations*, IMA J. Appl. Math. **38** (1987), 129–134.

[114] E.L. Mansfield, *diffgrob2: A symbolic algebra package for analysing systems of PDE using* MAPLE, 1993 ftp ftp.ukc.ac.uk, login: anonymous, password: your email address, directory: pub/Liz/Maple, files: diffgrob2_src.tar.Z, diffgrob2_man.tex.Z.

[115] E.L. Mansfield, *diffgrob2: A symbolic algebra package for analysing systems of PDE using* MAPLE, preprint, Department of Mathematics, University of Exeter, U.K., 1993.

[116] E.L. Mansfield, *The differential algebra package diffgrob2*, MAPLE Technical Newsletter **3** (1996), 33–37.

[117] E.L. Mansfield, *Applications of the differential algebra package diffgrob2 to classical symmetries of PDEs*, J. Symb. Comp. **23** (1997), 517–533.

[118] E.L. Mansfield, *Symmetries and exact solutions for a 2 + 1-dimensional shallow water wave equation*, Math. Comp. Simul. **43** (1997), 39–55.

[119] E.L. Mansfield and E. Fackerell, *Differential Gröbner Bases*, preprint no. 92/108, Macquarie Univerisity, Sydney, Australia, 1992.

[120] E.L. Mansfield, G.J. Reid and P.A. Clarkson, *Nonclassical reductions of a coupled nonlinear Schrödinger system*, Comp. Phys. Comm. (115), 1999.460488

[121] L. Martina, G. Soliani, and P. Winternitz, *Partially invariant solutions of a class of nonlinear Schrödinger equations*, J. Phys. A **25** (1992), 4425–4435.

[122] L. Martina and P. Winternitz, *Analysis and applications of the symmetry group of the multidimensional three-wave resonant interaction problem*, Ann. Phys., NY **196** (1989), 231–277.

[123] L. Martina and P. Winternitz, *Partially invariant solutions of nonlinear Klein–Gordon and Laplace equations*, J. Math. Phys. **33** (1992), 2718–2727.

[124] W. Miller Jr., J. Patera, and P. Winternitz, *Subgroups of Lie groups and separation of variables*, J. Math. Phys. **22** (1981), 251–260.

[125] J.D. Murray, *Mathematical Biology* (Springer-Verlag, New York, 1989).

[126] M. Musette, Chapter 8, this volume.

[127] J.S. Nagumo, S. Arimoto, and S. Yoshizawa, *An active pulse transmission line simulating nerve axon*, Proc. IRE **50** (1962), 2061–2070.

[128] T. Nishitani and M. Tajiri, *On similarity solutions of the Boussinesq equation*, Phys. Lett. **89A** (1982), 379–380.

[129] M.C. Nucci and P.A. Clarkson, *The nonclassical method is more general than the direct method for symmetry reductions: an example of the Fitzhugh-Nagumo equation*, Phys. Lett. **164A** (1992), 49–56.

[130] M.A. del Olmo, M.A. Rodríguez, P. Winternitz, and H. Zassenhaus, *Maximal abelian subalgebras of pseudounitary Lie algebras*, Lin. Alg. Appl. **135** (1990), 79–151.

[131] P.J. Olver, *Applications of Lie Groups to Differential Equations* (Springer-Verlag, New York, 1993), 2nd edition, Graduate Texts Math., vol. 107.

[132] P.J. Olver, *Direct reduction and differential constraints*, Proc. R. Soc. Lond. A **444** (1994), 509–523.

[133] P.J. Olver and P. Rosenau, *The construction of special solutions to partial differential equations*, Phys. Lett. **114A** (1986), 107–112.

[134] P.J. Olver and P. Rosenau, *Group-invariant solutions of differential equations*, SIAM J. Appl. Math. **47** (1987), 263–275.

[135] P.J. Olver and E.M. Vorob'ev, *Nonclassical and conditional symmetries*, CRC Handbook of Lie Groups Analysis of Differential Equations. III., Theoretical Developments an Computational Methods (ed. N.H. Ibragimov, CRC Press, Boca Raton), pages 1996–291.328

[136] J. Ondich, *A differential constraints approach to partial invariance*, Europ. J. Appl. Math. **6** (1995), 631–637.

[137] A.Yu. Orlov and P. Winternitz, *Algebra of pseudodifferential operators and symmetries of equations in the Kadomtsev-Petviashvili hierarchy*, J. Math. Phys. **38** (1997), 4644–4674.

[138] L.V. Ovsiannikov, *Group Analysis of Differential Equations* (Academic, New York, 1982).

[139] E.V. Pankrat'ev, *Computations in differential and difference modules*, Acta Appl. Math. **16** (1989), 167–189.

[140] J. Patera, R.T. Sharp, P. Winternitz, and H. Zassenhaus, *Continuous subgroups of the fundamental groups of physics. III. The de Sitter groups*, J. Math. Phys. **18** (1977)), 2259–2288.

[141] J. Patera, P. Winternitz, and H. Zassenhaus, *Continuous subgroups of the fundamental groups of physics. III. General method and the Poincaré group*, J. Math. Phys. **16** (1975), 1597–1614.

[142] J. Patera, P. Winternitz, and H. Zassenhaus, *Continuous subgroups of the fundamental groups of physics. III. The similitude group*, J. Math. Phys. **16** (1975), 1615–1624.

[143] J. Patera, P. Winternitz, and H. Zassenhaus, *Quantum numbers for particles in de Sitter space*, J. Math. Phys. **17** (1976), 717–728.

[144] J. Patera, P. Winternitz, and H. Zassenhaus, *Maximal abelian subalgebras of real and complex symplectic Lie algebras*, J. Math. Phys. **24** (1983), 1973–1985.

[145] A.H. Pincombe and N.F. Smyth, *Initial boundary value problems for the Korteweg-de Vries equation*, Proc. Roy. Soc. Lond. A **433** (1991), 479–498.

[146] E. Pucci, *Similarity reductions of partial differential equations*, J. Phys. A **25** (1992), 2631–2640.

[147] G. R. W. Quispel, H. W. Capel, and R. Sahadevan, *Continuous symmetries of differential-difference equations*, Phys. Lett. **170** (1992), 379–383.

[148] G.R.W. Quispel, F.W. Nijhoff, and H.W. Capel, *Linearization of the Boussinesq equation and the modified Boussinesq equation*, Phys. Lett. **91A** (1982), 143–145.

[149] D. Rand, *Pascal Programs for Identification of Lie Algebras, I: RADICAL, A Program to Calculate the Radical & Nilradical of Parameter-Free and Parameter-Dependent Lie Algebras*, Comp. Phys. Commun. **41** (1986), 105–125.

[150] D. Rand, *Pascal Programs for Identification of Lie Algebras, III: Levi Decomposition and Canonical Basis*, Comp. Phys. Commun. **46** (1987), 311–322.

[151] D. Rand and P. Winternitz, *ODEPAINLEVE A Macsyma Package for Painlevé Analysis of Ordinary Differential Equations*, Comp. Phys. Commun. **42** (1986), 359–383.

[152] D. Rand, P. Winternitz, and H. Zassenhaus, *Pascal Programs for Identification of Lie Algebras, II: SPLIT, A Program to Decompose Parameter-Free and Parameter-Dependent Lie Algebras into Direct Sums*, Comp. Phys. Commun. **46** (1987), 297–309.

[153] D. Rand, P. Winternitz, and H. Zassenhaus, *On the Identification of a Lie Algebra Given by its Structure Constants, I: Direct Decompositions, Levi Decompositions and Nilradicals*, Lin. Alg. Appl. **109** (1988), 197–246.

[154] G.J. Reid, *A triangularization algorithm which determines the Lie symmetry algebra of any system of PDEs*, J. Phys. A **23** (1990), L853–L859.

[155] G.J. Reid, *Algorithms for reducing a system of PDEs to standard form, determining the dimension of its solution space and calculating its Taylor series solution*, Europ. J. Appl. Math. **2** (1991), 293–318.

[156] G.J. Reid, *Finding abstract Lie symmetry algebras of differential equations without integrating determining equations*, Europ. J. Appl. Math. **2** (1991), 319–340.

[157] G.J. Reid and A. Wittkopf, *Finding abstract Lie symmetry algebras of differential equations without integrating determining equations*, A Differential Algebra Package for MAPLE, 1993 ftp 137.82.36.21 login: anonymous, password: your email address, directory: pub/standardform.

[158] G.J. Reid, A.D. Wittkopf, and A. Boulton, *Reduction of systems of nonlinear partial differential equations to simplified involutive forms*, Europ. J. Appl. Math. **7** (1996), 635–666.

[159] C. Rogers and W.F. Ames, *Nonlinear Boundary Value Problems in Science and Engineering* (Academic Press, New York, 1989).

[160] P. Rosenau and J.L. Schwarzmeier, *On similarity solutions of Boussinesq type equations*, Phys. Lett. **115A** (1986), 75–77.

[161] J. Rubin and P. Winternitz, *Point symmetries and conditionally integrable nonlinear evolution equations*, J. Math. Phys. **31** (1990), 2085–2090.

[162] F. Schwarz, *A REDUCE package for determining Lie symmetries of ordinary and partial differential equations*, Comp. Phys. Commun. **27** (1982), 179–186.

[163] F. Schwarz, *An algorithm for determining the size of symmetry groups*, Computing **49** (1992), 95–115.

[164] A.C. Scott, *The application of Bäcklund transforms to physical problems*, Bäcklund Transformations, R.M. Miura, ed. (Lect. Notes in Math., vol. 515, Springer, Berlin, 1975), pages 80–105.

[165] H. Segur and A. Finkel, *An analytical model of periodic waves in shallow water*, Stud. Appl. Math. **73** (1985), 183–220.

[166] N.F. Smyth, *The effect of conductivity on hotspots*, J. Aust. Math Soc., Ser. B **33** (1992), 403–413.

[167] H. Stephani, *Differential Equations, Their Solutions Using Symmetries* (Cambridge Univ. Press, Cambridge, 1989).

[168] D.A. Suprunenko and R.I. Tyshkevich, *Commutative Matrices* (Academic Press, New York, 1968).

[169] Z. Thomova and P. Winternitz, *Maximal Abelian Subgroups of the Isometry and Conformal Groups of Euclidean and Minkowski Spaces*, J. Phys. A **31** (1998), 1831–1858.

[170] Z. Thomova and P. Winternitz, *Maximal Abelian subalgebras of pseudo-Euclidean Lie algebras*, Lin. Alg. Annl. (1999), (to appear).

[171] M. Toda, *Studies of a nonlinear lattice*, Phys. Rep. **8** (1975), 1–125.

[172] V.L. Topunov, *Reducing systems of linear–differential systems to passive form*, Acta Appl. Math. **16** (1989), 191–206.

[173] F. Ursell, *The long-wave paradox in the theory of gravity waves*, Proc. Camb. Phil. Soc. **49** (1953), 685–694.

[174] E.M. Vorob'ev, *Symmetries of compatibility conditions for systems of differential equations*, Acta Appl. Math. **24** (1991), 1–24.

[175] J. Weiss, *The Painlevé property for partial differential equations. II. Bäcklund transformation, Lax pairs, and the Schwarzian derivative*, J. Math. Phys. **24** (1983), 1405–1413.

[176] J. Weiss, M. Tabor, and G. Carnevale, *The Painlevé property for partial differential equations*, J. Math. Phys. **24** (1983), 522–526.

[177] E.E. Whittaker and G.M. Watson, *Modern Analysis* (4th edition, Cambridge Univ. Press, Cambridge, 1927).

[178] P. Winternitz, *Kac–Moody–Virasoro symmetries of integrable nonlinear partial differential equations*, Symmetries and Nonlinear Phenomena, eds. D. Levi and P. Winternitz (World Scientific, Singapore, 1988), pages 358–375.

[179] P. Winternitz, *Group theory and exact solutions of partially integrable differential systems*, Partially Integrable Evolution Equations in Physics, eds. R. Conte and N. Boccara (Kluwer, Dordrecht, 1990), pages 515–567.

[180] P. Winternitz, *Lie groups and solutions of nonlinear partial differential equations*, Integrable Systems, Quantum Groups and Quantum Field Theories, eds. L.A. Ibort and M.A. Rodríguez (Kluwer, Dordrecht, 1993), pages 429–495.

[181] P. Winternitz and J.-P. Gazeau, *Allowed transformations and symmetry classes of variable coefficient Korteweg-de Vries equations*, Phys. Lett. **167A** (1992), 246–250.

[182] N.N. Yanenko, *Theory of consistency and methods of integrating systems of nonlinear partial differential equations*, Proceedings of the Fourth All-Union Mathematics Congress, Leningrad (1964, 247), pages 259–

[183] N.J. Zabusky, *A synergetic approach to problems of nonlinear dispersive wave propagation and interaction*, Nonlinear Partial Differential Equations, ed. W.F. Ames (Academic Press, New York, 1967), pages 233–258.

[184] V.E. Zakharov, *On stochastization of one-dimensional chains of nonlinear oscillations*, Sov. Phys. JETP **38** (1974), 108–110.

[185] H. Zassenhaus, *Lie groups, Lie algebras and representation theory* (Les Presses de l'Université de Montréal, Montréal, 1981).

[186] Zhuhan Jiang, *Lie symmetries and their local determinacy for a class of differential-difference equations*, Phys. Lett. **240A** (1998), 137–143.

[187] S. Zidowitz, *Conditional symmetries and the direct reduction of partial differential equations*, Modern Group Analysis: Advanced Analytical and Computational Methods in Mathematical Physics, eds. N.H. Ibragimov, M. Torrisi, and A. Valenti (Kluwer, Dordrecht, 1993), pages 387–393.

11

Painlevé Equations in Terms of Entire Functions

Jarmo Hietarinta

ABSTRACT In this chapter we discuss how the Painlevé equations can
be written in terms of entire functions, and then in the Hirota bilinear
(or multilinear) form. Hirota's method, which has been so useful in soliton
theory, is reviewed, and connections from soliton equations to Painlevé
equations through similarity reductions are discussed from this point of
view. In the main part we discuss how the singularity structure of the
solutions and formal integration of the Painlevé equations can be used to
find a representation in terms of entire functions. Sometimes the final result
is a pair of Hirota bilinear equations, but for P_{VI} we need also a quadrilinear
expression. The use of discrete versions of the Painlevé equations is also
discussed briefly. It turns out that with discrete equations one gets better
information on the singularities, which can then be represented in terms of
functions with a simple zero.

1 Introduction

Hirota's bilinear method has turned out to be very efficient in constructing
multisoliton solutions to integrable evolution equations. But since Painlevé
equations do not have soliton solutions, why should we care about writing
them in the Hirota bilinear form? In this chapter we will show that this
method is relevant even for integrable ODEs.

The fundamental idea behind Hirota's direct method is the following:

*Change into new variables in which the solutions have the simplest
form.*

This transformation will change at least the dependent variable, and may
sometimes be rather complicated. For solitons the nicest possible form is the
one for which the soliton solution is given as a polynomial of exponentials
with exponents linear in the independent variables (see Section 2).

Painlevé equations do not usually have solutions that can be written as
polynomials of exponentials, and although there are other special solutions
(rational or solutions made of special functions) to which Hirota's method
is relevant, there are other more general features that lead to the same

solutions. Indeed, the idea of solutions being "as nice as possible" can be extended to ODEs: We can demand that the solutions be expressed in terms of *entire functions*. This is not a new idea; it was studied by Painlevé himself in [1,2]. In his paper of 1902, he writes [3, p. 14]::

> In fact the integrals $y(x)$ of the previous equations P_I, P_{II}, P_{III} in the present notation) are meromorphic functions in the entire plane. It is clearly obvious that they are representable by the quotient of two entire functions. However, it is important to notice that one can choose these functions in a such way that they satisfy a simple differential equation of third order.

It should not be surprising that Painlevé's explicit results for P_I, P_{II}, and P_{III} in [3] are in the bilinear form. More recently, solutions to the Painlevé equations in terms of entire functions were considered by Lukashevich [7], and in this school K. Okamoto will give still another method of bilinearizing the Painlevé equations.

The outline of this chapter is the following. In Section 2 we will introduce Hirota's direct (bilinear) method by discussing the soliton solutions of the Korteweg–de Vries equation. In this case "niceness" is obvious, because the explicit soliton solutions have the simplest possible form. In the subsequent sections we will write the Painlevé equations in terms of entire functions, using three methods. First, in Section 3 we use the fact that many Painlevé equations can be obtained by similarity reductions from soliton equations with already known bilinear forms. In Section 4, which is the main part, we discuss how quadratic and quartic forms can be derived by studying the singularity structure of the solution and then writing the equations in terms of entire functions. (For a previous study along these lines, see [4].) In Section 5 we discuss briefly how discrete Painlevé equations (cf. Chapter 7) can be used as starting points, because somehow the discrete formulation is more sensitive to the singularity structure.

Finally, in this introduction let us list the equations under discussion:

$$P_I: \quad y'' = 6y^2 + z, \tag{1.1}$$

$$P_{II}a: \quad y'' = 2y^3 + xy + \alpha, \tag{1.2}$$

$$P_{III}: \quad y'' = \frac{1}{y}\,y'^2 - \frac{1}{z}\,y' + y^3 + \frac{1}{z}\,(\alpha y^2 + \beta) - \frac{1}{y}, \tag{1.3}$$

$$eP_{III}: \quad u'' = \frac{1}{u}\,u'^2 + e^{2x}u^3 + e^x(\alpha u^2 + \beta) - e^{2x}\frac{1}{u}, \tag{1.4}$$

$$P_{IV}: \quad y'' = \frac{1}{2y}\,y'^2 + \frac{3}{2}\,y^3 + 4zy^2 + 2(z^2 - \alpha)y + \beta\,\frac{1}{y}, \tag{1.5}$$

$$P_V: \quad y'' = \left(\frac{1}{2y} + \frac{1}{y-1}\right) y'^2 - \frac{1}{z}\,y'$$
$$+ \alpha\frac{y(y-1)^2}{z^2} + \beta\frac{(y-1)^2}{z^2 y} + \gamma\frac{y}{z} + \delta\frac{y(y+1)}{y-1}, \tag{1.6}$$

$$eP_V: \quad u'' = \frac{1}{2}\left(\frac{1}{u} + \frac{1}{u-1}\right)u'^2$$

$$- \left(\alpha\frac{u}{u-1} + \beta\frac{u-1}{u} + \gamma e^x u(u-1) + \delta e^{2x} u(u-1)(2u-1)\right), \quad (1.7)$$

$$P_{VI}: \quad y'' = \frac{1}{2}\left(\frac{1}{y} + \frac{1}{y-1} + \frac{1}{y-z}\right)y'^2 - \left(\frac{1}{z} + \frac{1}{z-1} + \frac{1}{y-z}\right)y'$$

$$+ \frac{y(y-1)(y-z)}{z^2(z-1)^2}\left(\alpha + \beta\frac{z}{y^2} + \gamma\frac{z-1}{(y-1)^2} + \delta\frac{z(z-1)}{(y-z)^2}\right), \quad (1.8)$$

$$eP_{VI}: \quad u'' = \frac{1}{2}\left(\frac{1}{u} + \frac{1}{u-1} + \frac{1}{u-e^x}\right)u'^2 - e^x\left(\frac{1}{e^x-1} + \frac{1}{u-e^x}\right)$$

$$+ \frac{u(u-1)(u-e^x)}{(e^x-1)^2}\left[\alpha + \frac{e^x\beta}{u^2} + \frac{(e^x-1)\gamma}{(u-1)^2} + \frac{e^x(e^x-1)\gamma}{(u-e^x)^2}\right]. \quad (1.9)$$

The exponential versions are obtained by $y(z) = u(x)$ for P_{III} and P_{VI}, and $y(z) = u(x)/(u(x)-1)$ for P_V, where $z = e^x$, and the primes of u stand for differentiation with respect to x.

2 Hirota's Bilinear Method for Soliton Equations

Here we will briefly discuss Hirota's method [5] for constructing multisoliton solutions to integrable equations (for a review, see e.g. [6,8]).

2.1 Definitions

The first step in the construction is to transform the equation into the Hirota form. As an example let us consider the Korteweg – de Vries (KdV) equation

$$u_{xxx} + 6uu_x + u_t = 0. \quad (2.1)$$

Let us introduce the dependent-variable transformation

$$u = 2\partial_x^2 \log F \quad (2.2)$$

(we will see below that the new function F is regular and simple for soliton solutions), and then one can write (2.1), in the following quadratic form (after one integration):

$$F_{xxxx}F - 4F_{xxx}F_x + 3F_{xx}^2 + F_{xt}F - F_xF_t = 0. \quad (2.3)$$

This does not look simpler than (2.1), but one can write it in a condensed form using the Hirota D operator:

$$(D_x^4 + D_xD_t)F \cdot F = 0, \quad (2.4)$$

where D is a kind of antisymmetric derivative,

$$D_x^n f \cdot g = (\partial_{x_1} - \partial_{x_2})^n f(x_1) g(x_2) \big|_{x_2 = x_1 = x}. \tag{2.5}$$

The minus sign, which differentiates D from Leibniz's rule, is crucial. We have

$$D_x^2 u \cdot u = u u'' - u'^2, \quad D_x^4 u \cdot u = u u'''' - 4 u' u''' + 3 u''^2,$$

etc. In the context of ODEs it is worth recalling that Borel and Chazy arrived to these expressions by invariance theory [9, 10], and observed that they yield equations whose solutions are entire. For later soliton computations note that $P(D)e^{px} \cdot e^{qx} = P(p - q)e^{(p+q)x}$.

2.2 Multisoliton Solutions

The KdV equation is the prototypical representative of the class

$$P(D_x, D_y, \dots)F \cdot F = 0, \quad P(0) = 0, \tag{2.6}$$

for which the multisoliton solutions are indeed simple in terms of F, as opposed to u. The general method of construction of multisoliton solutions is by considering the formal expansion

$$F = 1 + \varepsilon f_1 + \varepsilon^2 f_2 + \varepsilon^3 f_3 + \cdots, \tag{2.7}$$

where ε is the expansion parameter, and truncating this at some order. The vacuum, or zero-soliton, solution (0SS) is given by $F = 1$. For the one-soliton solution (1SS) only one term is needed: It is easy to see that due to the antisymmetry in (2.5), $F_1 = 1 + e^\eta$ ($\eta = p \cdot x$) is a solution of (2.6) if the parameters p satisfy a dispersion relation $P(p) = 0$. Here F_1 is the one-soliton solution (1SS), and substitution in (2.2) yields the standard result for u.

The two-soliton solution (2SS) for (2.6) is obtained from the truncation $F_2 = 1 + f_1 + f_2$, where $f_1 = e^{\eta_1} + e^{\eta_2}$. In order to fix f_2 we note that when we stay in the comoving frame of one soliton while the other one goes to $\pm\infty$ (that is, when the other η approaches $\pm\infty$) we should get the 1SS again. This means that we should try

$$F_2 = 1 + e^{\eta_1} + e^{\eta_2} + A_{12} e^{\eta_1 + \eta_2}, \tag{2.8}$$

and substituting this into (2.6) and using the dispersion relation for the parameters p_i we find that the equation is satisfied if the "phase factor" is given by

$$A_{12} = -\frac{P(p_1 - p_2)}{P(p_1 + p_2)}. \tag{2.9}$$

An important point to observe is that the above works for *any* polynomial P. In fact, there are still other classes of equations for which the generic form has 2SS, but it should be noted that the existence of 2SS does not imply integrability.

Although 2SS can be constructed for the whole class, 3SS works only for certain equations, namely for the integrable ones. It turns out that the ansatz for a possible 3SS is fixed by the 2SS: The only one compatible with the 2SS (2.8) is

$$F = 1 + e^{\eta_1} + e^{\eta_2} + e^{\eta_3} + A_{12}e^{\eta_1+\eta_2} + A_{13}e^{\eta_1+\eta_3} + A_{23}e^{\eta_2+\eta_3}$$
$$+ A_{12}A_{13}A_{23}e^{\eta_1+\eta_2+\eta_3}, \quad (2.10)$$

where $\eta_i = p_i \cdot x + \eta_i^0$, and the parameters p_i satisfy the dispersion relation $P(p_i) = 0$ and A_{ij} are given by (2.9). This form is dictated by the requirement that as one of the solitons goes to infinity (i.e., the corresponding η approaches $\pm\infty$) the other two should form a 2SS (2.8).

When the ansatz (2.10) is substituted into (2.6), one obtains the condition

$$\sum_{\sigma_i=\pm 1} P(\sigma_1\vec{p}_1 + \sigma_2\vec{p}_2 + \sigma_3\vec{p}_3)$$
$$\times P(\sigma_1\vec{p}_1 - \sigma_2\vec{p}_2)P(\sigma_2\vec{p}_2 - \sigma_3\vec{p}_3)P(\sigma_3\vec{p}_3 - \sigma_1\vec{p}_1) = 0 \quad (2.11)$$

on the manifold defined by the dispersion relations $P(p_i) = 0$. Since the ansatz was completely fixed, there are no free coefficients, and since our principle is that we should be able to combine *any* three solitons into a 3SS, we cannot impose any new conditions on the parameters $\vec{p}_i$ either. Thus the condition is on the *equation*.

The three-soliton condition (2.11) can be used to search for integrable equations within the class (2.6) [11]. Note that in this kind of search there are no initial assumptions about the number of independent variables and no preferred time. [This is in contrast with searches assuming a structure like $u_t = F(u, u_x, u_{xx}, \dots)$.] The (nontrivial) results of this search are as follows [11]:

$$(D_x^4 - 4D_xD_t + 3D_y^2)F \cdot F = 0, \quad (2.12)$$
$$(D_x^3D_t + aD_x^2 + D_tD_y)F \cdot F = 0, \quad (2.13)$$
$$(D_x^4 - D_xD_t^3 + aD_x^2 + bD_xD_t + cD_t^2)F \cdot F = 0, \quad (2.14)$$
$$(D_x^6 + 5D_x^3D_t - 5D_t^2 + D_xDy)F \cdot F = 0. \quad (2.15)$$

Three of these were known before: (2.12) is the Kadomtsev–Petviashvili (KP) equation, (2.13) is the Hirota–Satsuma–Ito equation, and (2.15) the Sawada–Kotera–Ramani equation. Equation (2.14) is the only new equation, and it is obvious that this equation could not have been found by any ansatz assuming simple t-dependence. All of these equations have also 4SS and pass the Painlevé test [12].

Similar analysis of 2SSs and 3SSs have been performed on other types of bilinear equations (mKdV [13], sG [14], nlS [15], and BO [15]).

2.3 Gauge Invariance and Generalization to Multilinearity

We have so far discussed Hirota's method only from the soliton point of view, but it has been found useful in other approaches as well. In particular, the τ-functions ($= F$ above) have been essential in the Kyoto school approach to integrable PDEs [16].

One recent important observation is that Hirota forms are intimately related to gauge invariance. It is easy to show that if F, G, ... solve some bilinear equations, so do $e^{ax}F$, $e^{ax}G$, ... But the reverse is true as well [17]: If some quadratic expression is gauge invariant, then all derivatives must appear as Hirota derivatives. The proof is simple. Consider the quadratic homogeneous combination $A_n(f,g) := \sum_{i=0}^{n} c_i \left(\partial_x^i f\right)\left(\partial_x^{n-i} g\right)$. From the gauge invariance $A_n(e^{ax}f, e^{ax}g) = e^{2ax}A_n(f,g)$ we can solve for the constants c_i and find that $c_i = (-1)^i \binom{N}{i}c_0$, so that A_n can indeed be written in terms of bilinear derivatives D: $A_n(f,g) = c_0 D_x^n f \cdot g$.

This gauge principle can be applied to higher multilinear expressions [17]. For the cubic case one finds that for gauge-invariant expressions the derivatives appear through

$$T = \partial_1 + j\partial_2 + j^2\partial_3, \quad T^* = \partial_1 + j^2\partial_2 + j\partial_3, \qquad (2.16)$$

where the subscript indicates on which factor the derivative operates, and $j = e^{2i\pi/3}$. The multilinear generalization is

$$M_n^m = \sum_{k=0}^{n-1} e^{2\pi ikm/n}\partial_{k+1}, \quad \text{where } 0 < m < n.$$

One can now search for integrable equations from the class

$$P(T, T^*)F \cdot F \cdot F = 0, \qquad (2.17)$$

and new equations have been found in [18], for example a generalization of the KP equation

$$(T_x^4 T_y^* + 8\,T_x^3 T_y T_x^* + 27\,T_y^3 - 36\,T_x^2 T_t)F \cdot F \cdot F = 0, \qquad (2.18)$$

or in the nonlinearized form obtained with $F = e^g$,

$$g_{xxxxy} + 8g_{xxy}g_{xx} + 4g_{xy}g_{xxx} + 3g_{yyy} - 4g_{xxt} = 0. \qquad (2.19)$$

3 Bilinear Forms and Similarity Reduction

Similarity reductions of PDEs to ODEs of Painlevé type are very important theoretically; in fact, the ARS conjecture [19] states that if an integrable PDE is reduced to an ODE, the ODE should be of Painlevé type. We will now follow this reduction path, but with a different purpose: Since the bilinear formalism has been so useful and is well known for soliton equations, we will use such equations as starting points and then apply similarity reductions in order to derive bilinear forms for ODEs. We will not consider here all possible similarity reductions to Painlevé equations, but just some typical cases with direct reduction. [In many cases the connection to Painlevé equations goes through rather complicated (differential) transformations, which probably does not help in the present objective of getting bilinear forms.] For further references about similarity reductions, see [20, Sections 6.5.15 and 7.2].

For bilinear variables the similarity reduction is always assumed to be of the form $F(x, t) = \phi(z)e^{a(x,t)}, \dots$, where the exponents are to be determined so that the bilinear equation is in terms of z only.

3.1 P_I

To get a similarity reduction to P_I let us consider the KdV equation (2.1). If one substitutes into it the ansatz [21]

$$u(x, t) = 2t - 2y(z), \quad z = x - 6t^2, \tag{3.1}$$

then one gets for $y(z)$ the equation

$$y''' = 12yy' + 1, \tag{3.2}$$

which integrates to P_I (1.1).

As was shown before, the bilinearization of KdV proceeds through the dependent-variable transformation (2.2), so that we now have the relation

$$t - y(x - 6t^2) = \partial_x^2 \log F. \tag{3.3}$$

This suggest that for P_I we should introduce a new dependent variable ϕ by

$$y = -(\log \phi)'', \tag{3.4}$$

and from (3.3), (3.4) we find that the similarity reduction for the bilinear dependent variable corresponding to (3.1) should be

$$F = \phi(x - 6t^2)e^{tx^2/2 + a(t)x + b(t)}. \tag{3.5}$$

(Note the free functions a and b, on which we have no information at the moment.) When this is substituted into (2.4) we obtain something that is a function of z alone, if we choose $a(t) = -4t^3$ (b drops out). The result is then

$$(D_z^4 + 2z)\phi \cdot \phi = 0, \tag{3.6}$$

which is the standard bilinear form for P_{I}. The notable feature in the above process is the necessity of the gauge factor, in this case $e^{tx(x-8t^2)/2}$.

We could also start from the Boussinesq equation

$$u_{xxxx} + 3(u^2)_{xx} + u_{xx} - u_{tt} = 0, \tag{3.7}$$

and using similarity reduction $u = -2y(x - t)$ [22] we immediately obtain $y'''' = 6(y^2)''$, which can be integrated twice to yield (1.1) with suitable integration constants. Equation (3.7) can be bilinearized as KdV with (2.2), which yields (after two x integrations)

$$(D_x^4 + D_x^2 - D_t^2)F \cdot F = 0. \tag{3.8}$$

The similarity reduction for F should now be of the form

$$F = \phi(x - t)e^{xa(t)+b(t)}, \tag{3.9}$$

and indeed the bilinear form (3.6) follows, with $z = x - t$, if we use the gauge $e^{-t^2x/2+t^3/6}$.

3.2 P_{II}

The second Painlevé equation can be obtained by a similarity reduction from the mKdV equation

$$u_{xxx} - 6u^2u_x + u_t = 0, \tag{3.10}$$

by the reduction ansatz [22]

$$u = (3t)^{-1/3}y(z), \quad z = x/(3t)^{1/3}. \tag{3.11}$$

This yields for y the equation

$$y''' = 6y^2y' + zy' + y, \tag{3.12}$$

which can be integrated to (1.2).

There are two ways to bilinearize mKdV. In the conventional approach we have to use the potential form, as for KdV. Thus let us introduce v by $u = \partial_x v$, substitute this into (3.10), and integrate the result with respect to x. This yields

$$v_{xxx} - 2(v_x)^3 + v_t = 0. \tag{3.13}$$

(The integration constant can be absorbed into v, since it is defined up to an additional function of t.) The bilinearizing dependent variable transformation is

$$v = \log \frac{G}{F},\tag{3.14}$$

and substitution into (3.13) yields

$$-FG[(D_x^3 + D_t)F \cdot G] + 3[(D_x^2)F \cdot G][D_x F \cdot G] = 0.\tag{3.15}$$

At this point we have one equation for two functions, so in principle we can introduce extra conditions for them. Recall that F and G are defined only up to a common multiplicative factor, so this is the origin of the freedom we now have. For soliton solutions it turns out that the best way to fix this factor is to demand $D_x^2 F \cdot G = 0$. Then we get the bilinear form

$$(D_x^3 + D_t)F \cdot G = 0,$$
$$D_x^2 F \cdot G = 0.\tag{3.16}$$

The 1SS for this class of equations is given by $F = 1 + e^\eta$, $G = 1 - e^\eta$ with dispersion relation given by the odd polynomial.

Maybe a general comment on equation splitting is in order here. It should be noted that for some other kind of solutions the above might not be the best way to split (3.15). The general method is to put $D_x^2 F \cdot G = \lambda FG$ where λ is an arbitrary function, which yields the pair

$$(D_x^3 + D_t - 3\lambda)F \cdot G = 0,$$
$$(D_x^2 - \lambda)F \cdot G = 0.$$

If we now make a gauge change

$$F \mapsto e^\theta F, \quad G \mapsto e^\theta G,$$

the above equation changes to

$$(D_x^3 + D_t - 3(\lambda - 2\theta))F \cdot G = 0,$$
$$(D_x^2 - (\lambda - 2\theta))F \cdot G = 0.$$

For a given type of solution (rational, soliton) one needs a specific form of $(\lambda - 2\theta)$; for soliton solutions this term should vanish.

The other bilinearization of (3.10) is obtained by substituting $u = g/f$ directly into it, and the result can then be split into an NLS type bilinear equation

$$(D_x^3 + D_t)f \cdot g = 0,$$
$$D_x^2 f \cdot f + 2g^2 = 0.\tag{3.17}$$

The 1SS of this system is given by $f = 1 - e^{2\eta}$, $g = -2pe^{\eta}$.

The dependent variables of these two forms (3.16, 3.17) are related by

$$g = D_x G \cdot F, \quad f = GF. \tag{3.18}$$

Let us now see how the above bilinear forms can be used to bilinearize P_{II}. In the first case with bilinearization through $u = \partial_x \log(G/F)$ the natural ansatz is $y = \frac{d}{dz} \log \frac{\psi}{\phi}$, because then

$$\partial_x \log \frac{G}{F} = u = \frac{1}{(3t)^{1/3}} y(z) = \frac{1}{(3t)^{1/3}} \frac{d}{dz} \log \frac{\psi}{\phi} = \partial_x \log \frac{\psi}{\phi}$$

(note the partial derivatives), so that we could just try

$$G(x,t) = \psi(z), \quad F(x,t) = \phi(z), \quad \text{with } z = x/(3t)^{1/3}. \tag{3.19}$$

Indeed this works, and we get from (3.16)

$$\begin{aligned} (D_z^3 - zD_z)\phi \cdot \psi &= 0, \\ D_z^2 \phi \cdot \psi &= 0. \end{aligned} \tag{3.20}$$

In the second case with $u = g/f$,

$$\frac{g}{f} = u = \frac{1}{(3t)^{1/3}} y(z) = \frac{1}{(3t)^{1/3}} \frac{\Psi(z)}{\Phi(z)}$$

suggests that we should try

$$g = a(x,t)\Psi(z), \quad f = a(x,t)(3t)^{1/3}\Phi(z), \tag{3.21}$$

and then from (3.17) we get a bilinear form depending only on z, if we just choose $a = 1$:

$$\begin{aligned} (D_z^3 - zD_z + 1)\Phi \cdot \Psi &= 0, \\ D_z^2 \Phi \cdot \Phi + 2\Psi^2 &= 0. \end{aligned} \tag{3.22}$$

It is easy to check that the substitution $\Psi = \phi\psi$, $\Phi = D_z\phi \cdot \psi$ reduces (3.22) to (3.20).

3.3 P_{III}

Special cases of P_{III} can be obtained by similarity reductions [22, 23] from the sineGordon (sG) equation

$$u_{xt} = \sin u. \tag{3.23}$$

The first similarity ansatz is

$$u(x,t) = -i \log y(z), \quad z = xt, \tag{3.24}$$

and substitution to (3.23) leads to the the special case

$$y'' = \frac{1}{y} y'^2 - \frac{1}{z} y' + \frac{1}{2z}(y^2 - 1). \qquad (3.25)$$

The sG equation (3.23) can be bilinearized using

$$u = -2i \log \frac{f + ig}{f - ig}, \qquad (3.26)$$

yielding

$$(D_x D_t - 1)g \cdot f = 0,$$
$$D_x D_t(f \cdot f - g \cdot g) = 0. \qquad (3.27)$$

The similarity reductions for f, g should be $g(x, t) = \phi(z)$, $f = \psi(z)$, and they yield [23]

$$y = \left(\frac{\psi + i\phi}{\psi - i\phi} \right)^2, \qquad (3.28)$$

with

$$(zD_z^2 + \partial_z - 1)\phi \cdot \psi = 0,$$
$$(zD_z^2 + \partial_z)(\phi \cdot \phi - \psi \cdot \psi) = 0. \qquad (3.29)$$

This, however, is not satisfactory, because it contains ordinary derivatives. The trick to eliminate them is to change the dependent variables by $\phi(z) = \bar{\phi}(\xi)$, $\psi(z) = \bar{\psi}(\xi)$, $z = e^\xi$, because then we get

$$(D_\xi^2 - \xi)\bar{\phi} \cdot \bar{\psi} = 0,$$
$$D_\xi^2(\bar{\phi} \cdot \bar{\phi} - \bar{\psi} \cdot \bar{\psi}) = 0. \qquad (3.30)$$

Another similarity ansatz for (3.23) is

$$u(x, t) = -2i \log w(\zeta), \quad \zeta = 2\sqrt{xt}, \qquad (3.31)$$

leading to

$$w'' = \frac{1}{w} w'^2 - \frac{1}{z} w' + \tfrac{1}{4}(w^3 - 1/w). \qquad (3.32)$$

This is related to the above as follows: If $\bar{\phi}(\xi)$, $\bar{\psi}(\xi)$ solve (3.30), then

$$w = \frac{\bar{\psi}(2 \log(\zeta/2)) + i\bar{\phi}(2 \log(\zeta/2))}{\bar{\psi}(2 \log(\zeta/2)) - i\bar{\phi}(2 \log(\zeta/2))}. \qquad (3.33)$$

Thus we have the same basic bilinear equation (3.30) corresponding to two different nonlinear ones.

4   Solutions in Terms of Entire Functions

As was mentioned before, Painlevé considered already quite early the question of representing the solutions in terms of entire functions [1–3]. But how
could we find such entire functions? (Painlevé does not give any constructive method, but just the solutions.) One direct way is by studying the
singularities of the solutions and then doing some manipulations on them
so that their entireness becomes manifest [4].

Here we would like to present an additional aspect to the introduction of
the entire functions: By choosing these functions properly one can actually
integrate the equation once.

Suppose we have an equation of the form

$$y'' = \alpha y'^2 + \beta y' + \gamma, \tag{4.1}$$

where α, β, γ are functions of z and y, and primes stand for derivatives
with respect to z. We want to integrate it to the form

$$I := Ay'^2 + By' + C - \int_c^z D\,d\zeta, \tag{4.2}$$

i.e., to find functions A, B, C, D of z and y such that

$$\frac{dI}{dz} \equiv y''(2Ay' + B) + A_y y'^3 + (A_z + B_y)y'^2 \tag{4.3}$$

$$+ (B_z + C_y)y' + C_z - D \tag{4.4}$$

$$= (2Ay' + B)(y'' - \alpha y'^2 - \beta y' - \gamma) = 0. \tag{4.5}$$

(Here the subscripts stand for partial derivatives.) This immediately yields
the set of equations

$$A_y = -\,2\alpha A,$$
$$A_z + B_y = -\,2\beta A - \alpha B,$$
$$B_z + C_y = -\,2\gamma A - \beta B,$$
$$C_z - D = -\,\gamma B. \tag{4.6}$$

In the following, it often turns out that

$$f := e^{\iint D\,dz\,dz} \tag{4.7}$$

is an entire function, and that the other entire function can be obtained
from $g := yf$. With this definition of f we get two equations from the
above:

$$\frac{Q_1}{f^2} \equiv (\log f)'' - D(z, g/f) = 0 \tag{4.8}$$

and

$$\frac{R}{\varrho} \equiv (\log f)' - \int_c^z D\,dz \tag{4.9}$$

$$= (\log f)' - \left[A\left(z, \frac{g}{f}\right)\left(\frac{g}{f}\right)'^2 + B\left(z, \frac{g}{f}\right)\left(\frac{g}{f}\right)' + C\left(z, \frac{g}{f}\right) \right] - c_1 \tag{4.10}$$

$$= 0, \tag{4.11}$$

where c_1 is a constant.

Below we will show that for the Painlevé equations, R defined above is quartic (with ϱ a simple quartic polynomial of f and g (no derivatives)) and Q_1 quadratic in f, g and their derivatives. From these two equations further equations can be derived, including those in Hirota bilinear form.

4.1 P_I

For P_I the situation is special, and one entire function is enough. Using the above method we find that with $\alpha = \beta = 0$ and $\gamma = 6y^2 + z$ one solution to (4.6) is given by

$$A = \tfrac{1}{2}, \quad B = 0, \quad C = -(2y^3 + zy), \quad D = -y. \tag{4.12}$$

Painlevé mentions that $f := e^{\int\int D\,dz}$ is entire when $\int D\,dz = \tfrac{1}{2}y'^2 - 2y^3 - yz + c_1$, in accordance with (4.12). This is easy to prove: Near any singularity the solution y of (1.1) behaves as [4]

$$y = \frac{1}{(z - z_0)^2} + O((z - z_0)^2), \tag{4.13}$$

so that at that point f [as defined by (4.7) with $D = -y$] behaves smoothly:

$$f = (z - z_0) \cdot [\text{const} + O(z - z_0)]. \tag{4.14}$$

Then from (4.8) (using y in place of g/f) we get [3]

$$y = -(\log f)'', \tag{4.15}$$

and when this is substituted into P_I we get for f an equation in Hirota's bilinear form

$$(D_z^4 + 2z)f \cdot f = 0. \tag{4.16}$$

In this chapter we also want keep track of the integration constants. Equation (4.16) is of fourth order, so it has two additional constants of integration. They are related to the gauge invariance under $f \mapsto e^{\alpha + z\beta} f$, which is a common property of equations in Hirota form (in [17] it was argued that the gauge invariance is the *defining* property of Hirota form).

In the following the final results for other Painlevé equations cannot be written as one quadratic equation but rather as a pair, so let us do it here also. For this purpose we take another solution of (4.6) with A, C, D multiplied by 2 of what was given in (4.12). This yields $f = \exp(-2 \int dz \int dz\, y) = (z - z_0)^2 \cdot [\text{const} + O(z - z_0)]$, which is needed to guarantee that $g := yf$ is also entire. Then from (4.8)

$$Q_1 \equiv f''f - f'^2 + 2fg \equiv \tfrac{1}{2} D_z f \cdot f + 2fg = 0, \qquad (4.17)$$

and from (4.11) ($\varrho = -f^4$)

$$R \equiv (f'g - g'f)^2 - f^3 f' - 4g^3 f - 2zgf^3 - c_1 f^4 = 0. \qquad (4.18)$$

These equations are equivalent to (1.1) in the following sense:

$$2(f'g - fg')P_I = Q_1 + f^2 \left(R/f^4 \right)'. \qquad (4.19)$$

The pair (4.17), (4.18) is of third order, and since two constants of integration are accounted for (c_1 and the overall scale of f, g), only one more constant of integration remains, and in this sense this pair represents the once integrated P_I.

Further equations can be derived as follows: Considering $(R/(f^2 g^2))' = 0$ and using (4.17) to simplify the result, we get another quadratic equation,

$$Q_2 \equiv gg'' - g'^2 + ff' + zfg + c_1 f^2 = 0, \qquad (4.20)$$

which appears in [7]. However, it is not gauge invariant and therefore not expressible in Hirota form. Furthermore, the pair $Q_1 = 0$, $Q_2 = 0$ is not equivalent to (1.1) (one would need $R = 0$ as well). If one instead considers the combination $Q_3 := (g^2 Q_1 + f^2 Q_2 + R)/(fg)$, one obtains

$$Q_3 \equiv f''g - 2f'g' + fg'' - zf^2 - 2g^2 \equiv D_z^2 f \cdot g - zf^2 - 2g^2 = 0, \quad (4.21)$$

and (1.1) is equivalent to $Q_1 = 0$, $Q_3 = 0$. Furthermore, this pair is in the Hirota form, and the two integration constants are related to the gauge invariance $(f, g) \mapsto (e^{\alpha + x\beta} f, e^{\alpha + x\beta} g)$.

Thus in terms of the entire functions f and g, P_I can be expressed by one fourth-order equation in Hirota form (4.16) or by the third-order pair (4.17), (4.18), or by the pair of second-order equations in Hirota form (4.17), (4.21).

4.2 P_{II}

For P_{II} the expansion around a singularity is [4]

$$y = \pm \frac{1}{z - z_0} \mp \frac{z_0}{6}(z - z_0) + \cdots , \qquad (4.22)$$

and if we just consider y^2 we get entire functions from

$$f := e^{-\iint y^2\, dz\, dz} = z - z_0 + \cdots, \quad g := yf = \pm 1 + \cdots. \qquad (4.23)$$

The integration yields the solution

$$A = 1, \quad B = 0, \quad C = -(y^4 + zy^2 + 2\alpha y), \quad D = -y^2, \qquad (4.24)$$

agreeing with the above. (Painlevé considers $\int D\, dz = y'^2 - y^4 - zy^2 - 2\alpha y$ in [1,3]). Then from (4.8) we get the equation

$$Q_1 \equiv ff'' - f'^2 + g^2 \equiv \tfrac{1}{2}D_z^2 f \cdot f + gg = 0, \qquad (4.25)$$

and from (4.11) $(\varrho = -f^4)$

$$R \equiv (f'g - g'f)^2 - f^3 f' - g^4 - zg^2 f^2 - 2\alpha gf^3 - c_1 f^4 = 0 \qquad (4.26)$$

(both given by Painlevé in [3]). Equation (4.19) holds also for P_{II}.

As before, another quadratic equation is obtained from $(R/(f^2 g^2))' = 0$ [7]:

$$Q_2 \equiv gg'' - g'^2 + ff' + \alpha gf + c_1 f^2 = 0. \qquad (4.27)$$

(Note that here z is absent.) Instead of this, one could consider the gauge-invariant (and c_1-independent) expression $Q_3 := (g^2 Q_1 + f^2 Q_2 + R)/(fg)$, i.e.,

$$Q_3 \equiv f''g - 2f'g' + fg'' - \alpha f^2 - zfg \equiv (D_z^2 - z)f \cdot g - \alpha f^2 = 0. \qquad (4.28)$$

The Hirota bilinear pair (4.25), (4.28) is the same as given in [4]. The counting of integration constants is as before.

Still another form is obtained if we take $f = FG$, $g = D_z F \cdot G$ corresponding to $y = F'/F - G'/G$ [3], which leads to the bilinear form

$$D_z^2 F \cdot G = 0,$$
$$(D_z^3 - zD_z - \alpha)\, F \cdot G = 0. \qquad (4.29)$$

This is of fifth order, and there are 3 obvious integration constants related to the invariance under $F \mapsto ae^{cx}F$, $G \mapsto be^{cx}G$.

4.3 P_{III}

For P_{III} the expansion around a movable singularity is

$$y = \pm \frac{1}{z - z_0} - \frac{\alpha \pm 1}{2z_0} + \cdots, \qquad (4.30)$$

and if one consider the combination $z(y^2 + (\alpha/z)y) = z_0/(z - z_0)^2 + O(1)$, one finds that

$$f := e^{-\int dz/z \int z(y^2 + (\alpha/z)y)dz}, \quad g := yf \tag{4.31}$$

are entire [4].

The term dz/z above suggests that it might be better to work with the exponential version (1.4) (as was done by Painlevé [3]). Then one solution to the integration problem is

$$\bar{A} = 1/u^2, \quad \bar{B} = 0,$$
$$\bar{C} = 2e^x(\beta/u - \alpha u) - e^{2x}(1/u^2 + u^2),$$
$$\bar{D} = 2e^x(\beta/u - \alpha u) - 2e^{2x}(1/u^2 + u^2). \tag{4.32}$$

This corresponds to Painlevé's 2ζ in [3, p. 15]. However, this does not directly lead to entire functions, and Painlevé adds some ad hoc operations, which in fact amount to using another solution:

$$A = 1/(4u^2), \quad B = -1/(2u),$$
$$C = \frac{1}{2}(e^x(\beta/u - \alpha u) - \frac{1}{2}e^{2x}(1/u^2 + u^2)),$$
$$D = -(e^x\alpha u + e^{2x}u^2). \tag{4.33}$$

This leads directly to the desired result: $f := e^{\iint D d^2x}$ and $g := uf$ are entire, and we get

$$Q_1 \equiv f''f - f'^2 + \alpha e^x fg + e^{2x}g^2 \tag{4.34}$$
$$\equiv \tfrac{1}{2}D_x^2 f \cdot f + \alpha e^x fg + e^{2x}g^2 = 0 \tag{4.35}$$

and $(\varrho = -4f^2g^2)$

$$R \equiv (f'g - g'f)^2 - 2fg(f'g + g'f) + f^2g^2 - 2e^x(\alpha fg^3 - \beta f^3 g) \tag{4.36}$$
$$- e^{2x}(g^4 + f^4) - 4c_1 f^2 g^2. \tag{4.37}$$

As usual, by considering $(R/(f^2g^2))' = 0$ we get another equation, which now happens to be in the Hirota form:

$$Q_2 \equiv gg'' - g'^2 - \beta e^x fg - \delta e^{2x} f^2 \tag{4.38}$$
$$\equiv \tfrac{1}{2}D_x^2 g \cdot g - \beta e^x fg + e^{2x}f^2 = 0. \tag{4.39}$$

Thus P_{III} is equivalent either to the pair (4.35), (4.37) or (4.35), (4.39). In the first case the system is of third order and there are two integration constants, the overall scale and c_1; in the second case the system is of fourth order with two-parameter gauge freedom.

Note that P_{III} is invariant under $u \mapsto 1/u$ accompanied by the parameter changes $\alpha \mapsto -\beta$, $\beta \mapsto -\alpha$. This corresponds to $f \leftrightarrow g$, and we see from the above that it is indeed a symmetry of the bilinear equations. Thus it might be said that the zeros of u are as important singularities as its poles, and to handle all of them at the same time one could define functions F, G, K, M by [3]

$$f = FG, \quad g = KM, \quad e^x u = \frac{G'}{G} - \frac{F'}{F}, \quad \frac{e^x}{u} = \frac{M'}{M} - \frac{K'}{K}. \quad (4.40)$$

In that case we get four equations for four entire functions, two from the above definitions,

$$D_x G \cdot F = e^x KM, \quad D_x M \cdot K = e^x FG, \quad (4.41)$$

and two from Q_1, Q_2,

$$D_x^2 F \cdot G = -\alpha e^x KM, \quad D_x^2 M \cdot K = \beta e^x FG. \quad (4.42)$$

The system is now of sixth order with four scale-related integration constants: $F \mapsto abe^{cx} F$, $G \mapsto a/be^{cx} G$, $K \mapsto kme^{cx} K$, $M \mapsto k/me^{cx} M$. The result (4.41), (4.42) is more symmetric, but it involves twice as many dependent variables. Whether it is more useful in practical applications depends on the problem.

4.4 P_{IV}

For P_{IV} the expansion reads

$$y = \frac{\pm 1}{z - z_0} - z_0 + \cdots, \quad (4.43)$$

and one finds that

$$f := e^{-\iint \mathrm{d}z(y^2 + 2zy)}, \quad g := yf, \quad (4.44)$$

define entire functions [4].

The integration method again works with the simplest choice. If we use the solution

$$A = \frac{1}{4y}, \quad B = 0, \quad C = -\tfrac{1}{4}(y^3 + 4zy^2 + 4(z^2 - \alpha)y + 2\beta/y),$$
$$D = -(y^2 + 2zy), \quad (4.45)$$

agreeing with (4.44), we get the equations

$$Q_1 \equiv ff'' - f'^2 + g^2 + 2zfg \equiv \tfrac{1}{2}D_z^2 f \cdot f + g^2 + 2zfg \doteq 0 \quad (4.46)$$

and ($\varrho = -4f^3 g$)

$$R \equiv (f'g - g'f)^2 - 4f^2 gf' - g^4 - 4zfg^3 - 4(z^2 - \alpha)f^2 g^2$$
$$\qquad\qquad + 2\beta f^4 - 2c_1 f^3 g = 0, \quad (4.47)$$

and the other quadratic equation is

$$Q_2 \equiv g''g - g'^2 + 2gf' - 2\beta f^2 + c_1 fg = 0. \qquad (4.48)$$

The gauge-invariant combination of the above turns out to be trilinear:

$$T \equiv T_z T_z^* f \cdot g \cdot g - 2(\beta f^3 + 2(z^2 - \alpha)fg^2 + zg^3) = 0 \qquad (4.49)$$

$(T_z T_z^* f \cdot g \cdot g = gD_z\, g \cdot f + \frac{1}{2}fD_z\, g \cdot g)$. One can now show that P_{IV} can be expressed as a linear combination of either Q_1 and $(R/(f^3 g))'$ or Q_1 and T, with the usual accounting of integration constants.

At this point we would like to return to the question of gauge trasformations, briefly mentioned before. The point is that the function

$$f := e^{-\int dz \int dz(y^2 + 2zy) + p(z)}$$

is entire for any fixed polynomial p of z. For example, note that P_{IV} has the polynomial solution $y = -\frac{2}{3}z$, and then from $(\log f)'' = -y^2 - 2zy$ we would get $f \propto e^{(8/27)z^4}$. It would clearly be desirable to have polynomial f, g as well. This could be obtained by a proper choice of p; see [26, p. 68], for details. The same problem exists for P_{III}; see [26, p. 90].

4.5 P_V

In this case again the nicest results are obtained for a specific form of the equation. Computations with the standard form reveal that one should instead consider the equation (1.7) obtained from the standard one by $y(z) = u(x)/(u(x) - 1)$, $z = e^x$ [4]. The expansion for u is given by

$$u = \pm \frac{i/\sqrt{2\delta}}{z - z_0} - \frac{\pm i\sqrt{2\delta} + \gamma - 2\delta z_0}{4\delta z_0} + \cdots, \qquad (4.50)$$

and using the method of [4] leads one to the entire functions

$$f := e^{\int dx \int dx(\gamma e^x u + 2\delta e^{2x} u(u-1))}, \quad g := uf. \qquad (4.51)$$

The integration has the corresponding solution

$$A = \frac{1}{2u(u-1)}, \quad B = 0,$$

$$C = -\alpha \frac{1}{u-1} - \beta \frac{1}{u} + \gamma e^x u + \delta e^{2x} u(u-1),$$

$$D = \gamma e^x u + 2\delta e^{2x} u(u-1), \qquad (4.52)$$

and from (4.8) we get the first equation

$$Q_1 \equiv \frac{1}{2}D_x^2 f \cdot f - \gamma e^x fg - 2\delta e^{2x} g(f - g) = 0, \qquad (4.53)$$

and from (4.11) ($\varrho = -2f^2g(g-f)$)

$$R \equiv (f'g - fg')^2 - 2fg(g-f)f' - c_1f^2g(g-f) - 2f^3[\alpha g + \beta(g-f)]$$
$$+ 2(g-f)g^2[\gamma e^x f - \delta e^{2x}(g-f)] = 0. \quad (4.54)$$

From the derivative $(R/(f^2g^2))' = 0$ one obtains the equation

$$Q_2 \equiv g''g - g'^2 - gf' - 2\beta f^2 + (\alpha + \beta - c_1/2)fg = 0, \quad (4.55)$$

and the gauge-invariant linear combination of the above is again trilinear:

$$T \equiv T_xT_x^*(f - \tfrac{2}{3}g) \cdot g \cdot g - 2\alpha fg^2 - 2\beta f(f-g)^2 - \gamma e^x g^2(2f-g)$$
$$- 2\delta e^{2x}g^2(f-g) = 0 \quad (4.56)$$

$(T_xT_x^*(f - \tfrac{2}{3}g) \cdot g \cdot g = gD_x^2f \cdot g + (f/2 - g)D_x^2g \cdot g)$. Furthermore, one finds that P_V is expressible as a linear combination of Q_1 and $(R/(f^2g(f-g)))'$ or of Q_1 and T.

4.6 P_{VI}

For P_{VI} the situation is more complicated. In fact, the method used in [4] does not work: There are no polynomials of u alone from which entire functions can be built. On the other hand, Painlevé in [24] proposes an expression that is supposed to yield an entire function, but this expression also involves u'. Presumably, one can search for such expressions using the expansion around the singularity, but we will here take a different route.

It turns out that Lukashevich [7] obtained some quadratic and quartic expressions for P_{VI} as well, but we could not verify the precise forms given in [7]. Using these results as a guide we searched for two quadratic and a quartic expression with similar properties as before, using (1.9). This resulted in

$$Q_1 := (e^x - 1)^2(f''f - f'^2) + (e^x - 1)fg' + 2\alpha g(g-f)$$
$$- (\alpha + c_1)(e^x - 1)fg, \quad (4.57)$$
$$Q_2 := e^{-x}(e^x - 1)^2(g''g - g'^2) + (e^x - 1)f'g + \beta e^x f(g-f)$$
$$- (\beta - \delta - \gamma + c_1)(e^x - 1)fg, \quad (4.58)$$

and

$$R := (e^x - 1)^2(f'g - fg')^2 - 2(e^x - 1)fg(f-g)(e^xf' - g') \quad (4.59)$$
$$- 2\alpha g^2(f-g)(e^xf - g) + 2\beta e^x f^2(f-g)(e^xf - g) \quad (4.60)$$
$$- 2\gamma(e^x - 1)f^2g(e^xf - g) - 2\delta e^x(e^x - 1)f^2g(f-g) \quad (4.61)$$
$$+ 2c_1(e^x - 1)fg(f-g)(e^xf - g). \quad (4.62)$$

The relationships between these expressions and the $P_{\rm VI}$ equation are as follows:

$$2(e^x - 1)^2 f^3 g(f - g)(e^x f - g) P_{\rm VI}$$
$$= 2(-g^2 Q_1 + e^x f^2 Q_2)(f - g)(e^x f - g) + (e^x f^2 - g^2) R \quad (4.63)$$

and

$$M\, P_{\rm VI} = 2e^x (Q_1 - Q_2) + (e^x f^2 - g^2)(e^x - 1)^2 (R/U)', \quad\quad (4.64)$$

where

$$U := (e^x - 1) f g (f - g)(e^x f - g),$$
$$M := -2(e^x - 1)^3 f^2 \left[(e^x - 1)(f'g - fg')(e^x f^2 - g^2) - e^x f g (f - g)^2\right] / U$$

(note a spurious solution: M vanishes if u solves $(e^x - 1) u'(u^2 - e^x) = e^x u (u - 1)^2$).

Finally, we have a relation between Q_1, Q_2, and R as

$$B Q_1 + C Q_2 + (e^x - 1) f g V (R/V)' = 0, \quad\quad (4.65)$$

where

$$B := -2g^2 \left[(e^x - 1)(f'g - fg') - e^x f(f - g)\right],$$
$$C := 2f^2 \left[e^x (e^x - 1)(f'g - fg') - e^x f(f - g)\right],$$
$$V := (e^x - 1)^2 f^2 g^2,$$

As far as gauge-invariant expressions are concerned, one finds that

$$X Q_1 + Y Q_2 + Z R$$

is gauge-independent whenever

$$X + Y = 2Z(f - g)(e^x f - g),$$

and then the c_1 terms vanish as well. One possibility is to take

$$Q_1 - Q_2 \equiv \tfrac{1}{2}(e^x - 1)^2 (D_x^2\, f \cdot f - e^{-x} D_x^2\, g \cdot g) + (e^x - 1) D_x\, g \cdot f$$
$$+ 2(\alpha g - \beta e^x f)(g - f) - (\alpha - \beta + \gamma + \delta) f g (e^x - 1) = 0, \quad (4.66)$$

which is bilinear. For the other expression we could not get any simplification, so it is quadrilinear; for example,

$$(e^x - 1)^2 (f - g)(e^x f - g) D_x^2 f \cdot f + (e^x - 1)^2 (D_x f \cdot g)^2 \quad\quad (4.67)$$
$$- 2e^x (e^x - 1) f (f - g)(D_z f \cdot g) - 2\alpha g (f - g)(e^x f - g)((e^x - 1) f - g) \quad\quad (4.68)$$
$$+ 2\beta e^x f^2 (f - g)(e^x f - g) - 2\gamma (e^x - 1) f^2 g (e^x f - g) \quad\quad (4.69)$$
$$- 2\delta e^x (e^x - 1) f^2 g (f - g) = 0. \quad\quad (4.70)$$

Let us now return to the integration procedure and try to understand why the straightforward procedure failed. Substituting $g = uf$ into (4.62) shows that

$$\frac{R}{-2(e^x - 1)fg(f - g)(e^x f - g)} = (\log f)' - [Au'^2 + Bu' + C] - c_1,$$

(4.71)

where

$$A = \frac{e^x - 1}{2u(u - 1)(u - e^x)}, \quad B = \frac{-1}{u - e^x},$$

$$C = -\frac{\alpha u}{e^x - 1} + \frac{e^x \beta}{(e^x - 1)u} + \frac{\gamma}{u - 1} + \frac{e^x \delta}{u - e^x}.$$

(4.72)

This expression is, in fact, in [24], (equation (3), $m = Au'^2 + Bu' + C$ from above), and Painlevé states that $e^{\int m}$ has no singularities, apart from the fixed ones (for z they are at 0, 1, and ∞, for x ($= \log z$) at $-\infty$, 0, ∞).

The integration procedure therefore works as before up to this point, and the problem is in D of (4.2); it is no longer a function of u only. Indeed, if one writes again

$$I := Au'^2 + Bu' + C - \Delta$$

(4.73)

and identifies Δ with $(\log f)'$, then from Q_1 we get

$$(e^x - 1)\Delta' + u(\Delta - \alpha - c_1) + \frac{2\alpha u(u - 1)}{e^x - 1} = 0.$$

(4.74)

(Similar expressions were considered in [25].)

For P_{VI} the situation has then turned out to be quite different from the others, as might have been expected. Nevertheless, even in this case the final result can be written in multilinear form, in which case we need one bilinear (4.66) and one quadrilinear (4.70) expression.

5   Discrete Painlevé

At the moment the most interesting developments in the field of integrable systems seem to take place in the area of integrable *difference* equations. Most properties of continuous integrable systems can be extended to the discrete case, e.g., Lax pairs, existence of solitons (for partial difference equations), and the Painlevé test. [For an overview see Chapter 7.]

In order to define discrete Painlevé equations, one should have a definition of the discrete Painlevé property. Grammaticos, Ramani, and Papageorgiou [27] have proposed that *singularity confinement* is the proper discrete analogue of the Painlevé property. Singularity confinement means that if a mapping leads to a singularity, then after a finite number of steps one should get ride of this singularity, and this should take place without essential loss of information. Singularity confinement has subsequently been used to generate discrete forms of Painlevé equations, in fact several families of them. (Before calling some difference equation a discrete version of a differential equation, one must verify at least that its continuum limit is the original continuous equation, but the continuous and discrete equations should share some other properties as well.)

The bilinear approach has a natural analogue in the discrete case. It is best stated using the gauge principle: If the expression is homogeneous in the dependent variables $F, G, \ldots$ and invariant under $F(n) \mapsto F(n)e^{pn}$, $G(n) \mapsto G(n)e^{pn}, \ldots$ then we say that it is in Hirota form.

As an example let us consider dP_I [30]. One version is

$$\overline{w} + w + \underline{w} = \frac{z}{w} + a, \tag{5.1}$$

where $\overline{w} = w(n+1)$, $w = w(n), \underline{w} = w(n-1)$, and $z = \alpha n + \beta$. If in this equation one hits a singularity, it is by first arriving somehow to $w(k) = 0$, and a closer study indicates that the sequence of special w values are $\{0, \infty, \infty, 0\}$, after which regular values are again obtained. This pattern of w values is obtained from

$$w(n) = \frac{F(n+2)F(n-1)}{F(n+1)F(n)} \tag{5.2}$$

if F has a simple zero $F(k-1) = 0$. The expression (5.2) is homogeneous and gauge invariant, and if one substitutes it to the discrete derivative of (5.1), one arrives at

$$F(n+3)F(n-1)F(n-2) - F(n+2)F(n+1)F(n-3)$$
$$= z(n)F(n+1)^2 F(n-2) - z(n-1)F(n+2)F(n-1)^2. \tag{5.3}$$

This is the trilinear version of dP_I.

If one now applies the continuous limit to the above, the previous results are obtained [30]: If $a = 6$, $w = 1 + \varepsilon^2 y$, $z = -3 + \varepsilon^4 \zeta$, $\zeta = n\varepsilon$, one finds that $y = 2(\log F)_{\zeta\zeta}$ and (5.3) becomes the z derivative of (4.16) divided by f^2.

The most important aspect of the above is the way the complicated singularity pattern of w is obtained from a simple zero of F. This is the discrete analogue of expressing the original solution in terms of entire functions. In the discrete case the process is much clearer, and this is an indication that discrete systems are more fundamental.

If the system has several singularity patterns, we need more functions to handle them. In general, the number of singularity patterns is the same as the number of entire functions. In [30] this idea was followed to its logical conclusion: For dP_{VI} the authors obtained a set of bilinear equations involving 8 functions. We will not repeat all of their results here, just some illustrative examples.

One version of dP_{II} is

$$\overline{w} + \underline{w} = \frac{zw + a}{1 - w^2}. \tag{5.4}$$

It has two singularity patterns, $\{-1, \infty, 1\}$ and $\{1, \infty, -1\}$. The entrance to the first pattern and exit from the second is described by

$$w(n) = -1 + \frac{F(n+1)G(n-1)}{F(n)G(n)}, \tag{5.5}$$

while for the remaining part we would get

$$w(n) = -1 - \frac{F(n-1)G(n+1)}{F(n)G(n)}. \tag{5.6}$$

Equating these two expressions we get the first equation

$$F(n+1)G(n-1) + F(n-1)G(n+1) - 2F(n)G(n) = 0, \tag{5.7}$$

whose continuous limit yields the first equation of (4.29). The second equation is obtained from (5.4). In this case we have two singularity patterns and functions, and the structure of the patterns determines one equation between the functions.

For the higher discrete Painlevé equations the singularity patterns sometimes determine everything. More precisely, the singularity patterns suggest different ways of writing the dependent variable in terms of functions with simple zeros, and comparing these expressions one gets enough equations. As an example, let us consider dP_{III},

$$\overline{w}\underline{w} = \frac{cd(w - az)(w - bz)}{(w - c)(w - d)}, \tag{5.8}$$

where $z = \lambda^n$ (corresponding to the change of variables $z = e^x$ in the continuous form); the singularity patterns are $\{c, \infty, d\}$, $\{d, \infty, c\}$, $\{az, 0, bz\}$, and $\{bz, 0, az\}$. This suggest the representations [30]

$$w = c\left(1 - \frac{F(n+1)G(n-1)}{F(n)G(n)}\right) = d\left(1 - \frac{F(n-1)G(n+1)}{F(n)G(n)}\right)$$

$$= \frac{HK}{FG}, \tag{5.9}$$

$$\frac{1}{w} = \frac{1}{az}\left(1 - \frac{H(n+1)K(n-1)}{H(n)K(n)}\right) = \frac{1}{bz}\left(1 - \frac{H(n-1)K(n+1)}{H(n)K(n)}\right)$$

$$= \frac{FG}{HK}. \tag{5.10}$$

Equating these expressions and taking suitable sums and differences yields four bilinear equations in a nice symmetric form:

$$F(n+1)G(n-1) + F(n-1)G(n+1) - 2F(n)G(n)$$
$$= -\left(\frac{1}{c} + \frac{1}{d}\right) H(n)K(n),$$

$$F(n+1)G(n-1) - F(n-1)G(n+1) = \left(\frac{1}{d} - \frac{1}{c}\right) H(n)K(n),$$

$$H(n+1)K(n-1) + H(n-1)K(n+1) - 2H(n)K(n)$$
$$= -z(a+b)F(n)G(n),$$

$$H(n+1)K(n-1) - H(n-1)K(n+1) = z(b-a)F(n)G(n).$$

The continuous limit is obtained with $z = e^{2\varepsilon n} = e^{2x}$, $a = \varepsilon - a_0\varepsilon^2$, $b = -\varepsilon - b_0\varepsilon^2$, $c = 1/\varepsilon + c_0$, $d = -1/\varepsilon + d_0$, and yields

$$D_x F \cdot G = -HK,$$
$$D_x H \cdot K = -e^{2x}FG,$$
$$D_x^2 F \cdot G = (c_0 + d_0)HK,$$
$$D_x^2 H \cdot K = (a_0 + b_0)e^{2x}FG.$$

One can now verify that if one uses the substitution

$$u = -e^{-x} \frac{\mathrm{d}}{\mathrm{d}x} \log \frac{F}{G}$$

in (1.4), the result vanishes due to the above equations.

The important point in the above construction is that the singularity patterns determined *all* of the bilinear equations. The Painlevé equation is then just a way to represent the singularity patterns through one function and its equation. This situation continues with some of the higher discrete Painlevé equations.

The approach of taking the discrete singularity patterns seriously and using them to derive bilinear forms [30] is systematic and powerful. In this way the Hirota bilinearization of P_{VI} was first obtained. The drawback is in the proliferation of dependent variables: For P_{VI}, eight functions are needed, one for each singularity type (the singularities are at $y = 0$, ∞, 1, and z, with the next term in the expansion having $\pm$ sign). In the bilinear+quadrilinear form (4.66), (4.70) we manage with two functions, and this is enough also for Okamoto's method (truly bilinear, but using also ordinary derivatives, and therefore not gauge-invariant). Which form is best will then depend on the practical problem on hand, and it is useful to keep all alternatives in mind.

Acknowledgments: I would like to thank M. Kruskal for discussions and B. Grammaticos for comments on the manuscript. This work was supported in part by the Academy of Finland, project 31445.

6   References

[1] P. Painlevé, C.R. Acad. Sci. Paris **126**, 1697 (1898).

[2] P. Painlevé, C.R. Acad. Sci. Paris **127**, 945 (1898).

[3] P. Painlevé, Acta Mathematica **25**, 1 (1902).

[4] J. Hietarinta and M. Kruskal, *Hirota forms for the six Painlevé equations from singularity analysis*, Painlevé transcendent: their asymptotics and physical applications, eds. D. Levi and P. Winternitz, NATO ASI Series, B 278 (Plenum, 1992), page 175.

[5] R. Hirota, Phys. Rev. Lett. **27**, 1192 (1971).

[6] J. Hietarinta, Partially Integrable Evolution Equations in Physics, eds. R. Conte and N. Boccara (Kluwer Academic, 1990), page 459.

[7] N.A. Lukashevich, *The theory of Painlevé's equations*, Diff. Urav. **6** (1970), 425–430; Diff. Eq. **6** (1970), 329–333.

[8] J. Hietarinta, *Introduction to the Hirota Bilinear Method*, Integrability of Nonlinear Systems eds. Y. Kosmann-Schwarzbach, B. Grammaticos and K.M. Tamizhmani, Lecture Notes in Physics **495** (Springer, Berlin, 1997), pages 95–103.

[9] J. Chazy, Acta Math. **34**, 317 (1911).

[10] M.E. Borel, C.R. Acad. Sci. Paris **138**, 337 (1904).

[11] J. Hietarinta, J. Math. Phys. **28**, 1732 (1987).

[12] B. Grammaticos, A. Ramani, and J. Hietarinta, J. Math. Phys. **31**, 2572 (1990).

[13] J. Hietarinta, J. Math. Phys. **28**, 2094 (1987).

[14] J. Hietarinta, J. Math. Phys. **28**, 2586 (1987).

[15] J. Hietarinta, J. Math. Phys. **29**, 628 (1988).

[16] M. Jimbo and T. Miwa, Publ. RIMS, Kyoto Univ. **19**, 943 (1983).

[17] B. Grammaticos, A. Ramani, and J. Hietarinta, Phys. Lett. A **190** 65 (1994).

[18] J. Hietarinta, B. Grammaticos, and A. Ramani, NEEDS '94, eds. V. Makhankov et. al. (World Scientific, 1995), page 54.

[19] M. Ablowitz, A. Ramani, and H. Segur, J. Math. Phys. **21**, 715, 1006 (1980).

[20] M. Ablowitz and P. Clarkson: *Solitons, Nonlinear Evolution Equations and Inverse Scattering* (Cambridge U P, 1991).

[21] A. Fokas and M. Ablowitz, J. Math. Phys. **23**, 2033 (1982).

[22] M. Ablowitz and H. Segur, Phys. Rev. Lett. **38**, 1103 (1977).

[23] S. Oishi, J. Phys. Soc. Japan, **49**, 1647 (1980).

[24] P. Painlevé, C.R. Acad. Sci. Paris **143**, 1111 (1906).

[25] R. Garnier, C.R. Acad. Sci. Paris **162**, 939 (1916).

[26] V.I. Gromak and N.A. Lukashevich: *The Analytical Solutions of the Painlevé Equations* (Universitetskoye Publishers, Minsk, 1990).

[27] B. Grammaticos, A. Ramani, and V. Papageorgiou, Phys. Rev. Lett. **67**, 1825 (1991).

[28] A. Ramani, B. Grammaticos, and J. Hietarinta, Phys. Rev. Lett. **67**, 1829 (1991).

[29] A. Ramani, B. Grammaticos, and J. Satsuma, J. Phys. A: Math. Gen **28**, 4655 (1995).

[30] Y. Ohta, A. Ramani, B. Grammaticos, and K.M. Tamizhmani, Phys. Lett. A. **216**, 255 (1996).

12

Bäcklund Transformations of Painlevé Equations and Their Applications

Valerii I. Gromak

ABSTRACT The Painlevé equations (P_1)–(P_6) were first derived around the turn of the century in an investigation by Painlevé and his colleagues. Nonlinear ordinary differential equations have the property that the singularities other than poles of any of the solutions are independent of the particular solution and so are independent only of the equation; this property is known as the Painlevé property. Although first discovered from strictly mathematical considerations, the Painlevé equations have appeared in various physical applications. There has been considerable interest in Painlevé equations over the last few years due the fact that they arise as reductions of solutions of soliton equations solvable by the inverse scattering method. The Painlevé equations may also be thought of as nonlinear analogues of the classical special functions. There is currently much interest in studying the Painlevé equations using the isomonodromy deformation method. In this approach, the Painlevé equation is written as an integrability condition of a linear system.

There have been many recent studies of integrable mappings and discrete systems, including discrete analogues of the Painlevé equations. Therefore, any success in the study of Painlevé equations provides new results in the associated problems.

In this chapter we shall discuss the recurrence relations and their application to the construction of various kinds of solutions of the Painlevé equations. In the context of the Painlevé equations, recurrence relations are usually referred to as Bäcklund transformations. Informally, a Bäcklund transformation is defined as a system of equations relating one solution of a given equation either to another solution of the same equation, possibly with different values of the parameters, or to a solution of another equation. In this approach one can derive various properties of the Painlevé equations, including exact solutions (rational, algebraic, classical), the fundamental domain of the parameter space, special integrals, and transcendence. This is the purpose of the present chapter.

It is organized as follows. In the introduction we discuss the Painlevé property and Bäcklund transformations. In Section 2 we review several important general properties of the second Painlevé equation (P_2), such as the character of the singularities, the representation of the solutions as a ratio of entire functions, τ-functions, the number of poles of the solutions.

Then from the Hamiltonian system we obtain differential and integral forms of Bäcklund transformations of (P_2). Using this result we obtain algebraic nonlinear superposition of solutions (Bianchi formula or a discrete analogue of (P_2)) and the fundamental domain of the parameter space. Since one of the principal aims of our investigation is to obtain solution hierarchies for the Painlevé equations by making multiple applications of Bäcklund transformations, in Sections 3 and 4 we obtain the rational solutions and classical solutions that can be expressed in terms of Airy functions. Then in Section 5 we discuss the transcendence of (P_2).

In the sixth section we consider the higher analogue of (P_2) that arises as an exact reduction of the higher analogue of the Korteweg–de Vries equation. We show that the equations of the higher analogue have the same properties of solutions as (P_2).

In Sections 7–9 we give a brief discussion of elementary properties of (P_4) and construction of a Bäcklund transformation of (P_4). Then we describe the effect on the parameters of (P_4) of repeated application of Bäcklund transformation. Then we apply these results to finding the parameter criterion of rational solutions and solutions that can be expressed by means of solutions of the Weber–Hermite equation. This family includes as special cases solutions expressible in terms of the error function. Then we obtain the Weyl chamber in parameter space and discuss the transcendence of (P_4).

In Section 10 we review several important properties of the third Painlevé equation (P_3), and include here a number of elementary scaling transformations, the first integrals, and the special Riccati equation. With the exception of the cases where (P_3) is fully integrable ($\alpha = \gamma = 0$ or $\beta = \delta = 0$), we are able to restrict our attention to two specialized cases, i.e., $\gamma\delta \neq 0$ (with no loss of generality $\gamma = 1$, $\delta = -1$) and $\gamma = 0$, $\alpha\delta \neq 0$ ($\alpha = 1$, $\delta = -1$), which simplifies our ensuing discussion.

In Sections 10–13 we obtain the Bäcklund transformations for the two considered cases of (P_3). This information characterizes the parameter sets for which exact solutions exist. We conduct a systematic study of these solutions and illustrate how they can be categorized into three hierarchies; one of solutions rational in z, a second of algebraic solutions that are rational in $z^{1/3}$, and finally a family whose members can be expressed in terms of suitable Bessel functions.

In Sections 12, 14 we show the reducibility of equation (P_3) to (P_5) and back for some values of the parameters. Then we obtain the fundamental domain and discuss the transcendence of (P_3). Finally, in Sections 14, 15 we give a brief consideration of integrability of (P_5) and (P_6), Bäcklund transformations, and solution hierarchies of these equations.

1   Introduction

1.1   Painlevé Property

Singular points and their classifications. Cauchy's theorem. Fixed

and movable singularities, examples. Painlevé property. First-order
P-type equations. Painlevé and Fuchs theorems. Second-order P-type
equations. Painlevé list. Painlevé equations (P_1)–(P_6).

In this chapter, we shall consider nonlinear ODEs (and PDEs) in the complex plane

$$w' = f(z, w), \qquad (1.1)$$

$f : D \subset \mathbb{C}^{n+1} \to \mathbb{C}^n$, i.e., equations (or systems of equations) that are regarded as analytic functions of complex variables. Therefore, a solution of equation (1.1) is an analytic function, which is determined by its singular points.

It is necessary to say some words about the classification of the singular points of an analytic function $w = w(z)$.

Definition 1.1. A point at which $w(z)$ fails to be analytic is called a singular point or singularity of $w(z)$.

Singular points can belong to the following classes ("a" and "b" are opposites).

1a. Isolated singular points.

1b. Nonisolated singular points.

2a. Single-valued points, for which the function does not change its value as z goes around a given initial point z_0.

2b. Multivalued or branch (or critical) points.

3a. The points for which the function has a limit, whether finite or infinite, as $z \to z_0$.

3b. The points for which the function $w = w(z)$ has no limit as $z \to z_0$, i.e., z_0 is an essential singular point.

A singular point may have more than one of the above properties. For example, it may be isolated and single-valued (it is then either a pole or a removable singularity or an essential singular point). It may be multivalued, for example, finite-valued and isolated (e.g., algebraic critical point, algebraic pole). Algebraic branch points, poles, and algebraic poles are called algebraic singularities.

Theorem 1.1. *Consider the system*

$$\frac{dw_j}{dz} = f_j(z, w_1, \ldots, w_n), \quad w_j(z_0) = w_j^0, \quad j = 1, \ldots, n,$$

and let f be analytic in the domain $|z - z_0| \leq a$, $|w_j - w_j^0| \leq b$ and M be the upper bound of the set f in this domain. Then the system admits a *unique solution $w(z)$, which is analytic within the circle $|z - z_0| \leq \rho$ and which reduces to w_0 when $z \to z_0$, $\rho = a(1 - e^{-b/M(n+1)})$. In the linear case ρ is equal to a.*

If (z_0, w_0) is a singularity of f, then the solution $w(z)$ can have a singularity at z_0.

Singularities of solutions of the above system are divided into two classes: fixed singularities and movable singularities.

Example 1.1.

$$w' = 1/z, \quad w = \ln z + C.$$

In this example $z = 0$ and $z = \infty$ are singularities for every solution, and the coefficients of the equation are not analytic at $z = 0, \infty$. Those singularities are fixed ones.

Example 1.2.

$$w' = -w^2, \quad w = (z + C)^{-1}.$$

In this case the solution $w(z)$ defined by the initial data w_0, z_0 has a pole at the point $-z_0 + 1/w_0$. In this example $z = -C$ is a simple pole for the general solution. It is a movable singularity.

Example 1.3.

$$w' = e^{-w}, \quad w = \ln(z + C).$$

In this example $z = -C$ is a transcendental movable singularity.

Singularities of ODEs are of two types: fixed and movable.
Consider the nth-order equation in the complex domain

$$x^{(n)} + \sum_{j=1}^{n-j} p_j(t) x^{(n-j)} = 0.$$

If the coefficients $p_j(t)$ are all analytic in a neighborhood of a point t_0, $|t - t_0| < \rho$, then t_0 is a regular point of the ODE, and there exists a unique solution that is expressed as a Taylor series in $t - t_0$ and that is convergent at least within some neighborhood of t_0, $|t - t_0| < \rho$. Therefore, the singularities of solutions of linear equations can be located only at singularities of the coefficients. These singularities are said to be fixed, since their location does not vary when we go from one particular solution to another. For example, solutions of the hypergeometric equation

$$t(t - 1)\frac{d^2 x}{dt^2} + (\gamma - (\alpha + \beta + 1)t)\frac{dx}{dt} - \alpha\beta x = 0$$

can have singularities only at the points 0, 1, ∞. It is a general property of linear ODEs that all singularities of solutions are fixed. Nonlinear ODEs lose this property. Consider a simple ODE,

$$x' + x^2 = 0.$$

Its general solution is $x(t) = 1/(t - t_0)$, where t_0 is an integration constant, which also denotes the location of the singularity. This singularity, a pole, is said to be movable, since its location depends on the constant of integration t_0.

Definition 1.2. The movable singularities of the solution are the singularities whose location depends on the constant of integration. Fixed singularities occur at points where the coefficients of the equation are singular.

Nonlinear ODEs can have both movable and fixed singularities. For example, the equation $w' = w^2/z$ has a general solution $w = -1/\ln Cz$. In this case $z = 0$ is a fixed singular point, but $z = 1/C$ is a movable singularity. A solution of an ODE can have various kinds of singularities. For example, the equations

$$x' + x^3 = 0, \quad x = 1/\sqrt{2(t - t_0)}, \tag{1.2}$$

$$xx'' - x'+ = 0, \quad x = (t - t_0)\ln(t - t_0) + \alpha(t - t_o), \tag{1.3}$$

$$axx'' + (1 - a)x'^2 = 0, \quad x = \alpha(t - t_0)^a, \tag{1.4}$$

$$(xx'' - x'^2)^2 + 4xx'^3 = 0, \quad x = \alpha e^{(t-t_0)^{-1}}, \tag{1.5}$$

$$(1 + x^2)x'' + (1 - 2x)x'^2 = 0, \quad x = \tan\ln\alpha(t - t_0), \tag{1.6}$$

where α and t_0 are the constants of integration, have different types of singularities. In the first example t_0 is a movable algebraic critical pole. In the second example t_0 is a movable logarithmic branch point. In the third one t_0 is a transcendental singular point unless a is not rational. In the fourth example t_0 is a movable isolated essential singularity. In the fifth one t_0 is a nonisolated essential singularity. In the last example $x(t)$ has no limit as $t \to t_0$ along special paths.

Definition 1.3. The ODE is said to possess the *Painlevé property* when all movable singularities are single-valued, i.e., when solutions are free from movable critical points but can have fixed multivalued singularities.

ODEs that possess the Painlevé property are said to be of Painlevé type (*P*-type).

Painlevé proved that for the first-order equation of the form

$$P(w', w, z) = 0,$$

where P is polynomial in w', w and analytic in t, the only movable singularities of the solutions are poles and/or algebraic branch points.

L. Fuchs showed that an equation of P-type of the form

$$w' = \frac{P(z, w)}{Q(z, w)},$$

where P and Q are polynomials in x with coefficients analytic in z, is necessarily the Riccati equation. Fuchs solved this problem in the general case for first-order ODEs.

Picard posed the problem of determining the equations of the form

$$w'' = R(w', w, z),$$

where R is a rational function of w', w and an analytic function of z, that have no solutions with movable critical points. The problem was solved by Painlevé and his colleagues. The general method used to find equations whose solutions are free from such singularities is known as the α-method (or method of small parameter). Painlevé showed that there are only 50 canonical equations that have the property of having no movable critical points. He and his colleagues further showed that forty-four equations were integrable in terms of known functions (such as elliptic functions and functions that are solutions of linear equations) or were reducible to one of six new nonlinear differential equations:

$$w'' = 6w^2 + z, \tag{P_1}$$

$$w'' = 2w^3 + zw + \alpha, \tag{P_2}$$

$$w'' = \frac{w'^2}{w} - \frac{w'}{z} + \frac{1}{z}(\alpha w^2 + \beta) + \gamma w^3 + \frac{\delta}{w}, \tag{P_3}$$

$$w'' = \frac{w'^2}{2w} + \frac{3}{2}w^3 + 4zw^2 + 2(z^2 - \alpha)w + \frac{\beta}{w}, \tag{P_4}$$

$$w'' = \frac{3w - 1}{2w(w - 1)}w'^2 - \frac{1}{z}w' + \frac{1}{z^2}(w - 1)^2\left(\alpha w + \frac{\beta}{w}\right) + \frac{\gamma}{z}$$
$$+ \frac{\delta w(w + 1)}{w - 1}, \tag{P_5}$$

$$w'' = \frac{1}{2}\left(\frac{1}{w} + \frac{1}{w - 1} + \frac{1}{w - z}\right)w'^2 - \left(\frac{1}{z} + \frac{1}{z - 1} + \frac{1}{w - z}\right)w'$$
$$+ \frac{w(w - 1)(w - z)}{z^2(z - 1)^2}\left(\alpha + \frac{\beta z}{w^2} + \frac{\gamma(z - 1)}{(w - 1)^2} + \frac{\delta z(z - 1)}{(w - z)^2}\right) \tag{P_6}$$

with α, β, γ, δ arbitrary constants.

These equations are known as the Painlevé equations. The first three equations (P_1), (P_2), (P_3) were discovered by Painlevé; the fourth and fifth, (P_5) and (P_6), were added by Gambier; and the general form of the sixth, (P_6), was given by R. Fuchs. The solutions of (P_1), (P_2), (P_4) are meromorphic functions of z. If the substitution $z = e^t$ is made in (P_3) and

(P_5), then the solutions become meromorphic functions of t. However, for (P_6) the points 0, 1, ∞ are fixed critical points, and hence the general solution of (P_6) is not meromorphic throughout the complex plane. In fact, equation (P_6) contains the other five, since these may be obtained from (P_6) by passing to the limit $(P_6) \rightarrow (P_5) \rightarrow \cdots \rightarrow (P_1)$. For example, if in (P_2) $w \rightarrow \varepsilon w + \varepsilon^{-5}$, $z \rightarrow \varepsilon^2 z - 6\varepsilon^{10}$, $\alpha \rightarrow 4\varepsilon^{-5}$, and $\varepsilon \rightarrow 0$, then $(P_2) \rightarrow (P_1)$. Since Painlevé proved that solutions of P_1 define new transcendental functions, it then follows that the solutions of (P_2)–(P_6) also do this, except for special values of the constants α, β, γ, δ. But what are these parametric values? In this chapter we investigate this problem and the properties of solutions for (P_2)–(P_6). The main approach to the investigation is the method of Bäcklund transformations.

1.2   Bäcklund Transformation

Definition of B-transformations and examples: Cole Hopf, Miura transformations; Bessel, Riccati, KdV, and sine–Gordon equations.

We start with some examples of Bäcklund transformations.

1. The solutions of the linear and Riccati equations

$$x'' + p_1(t)x' + p_2(t)x = 0, \quad y' + y^2 + p_1(t)y + p_2(t) = 0$$

are related by the formulae

$$x = \exp \int y \, dt, \quad y = x'/x.$$

2. The Cole Hopf transformation $u = (\ln v)_x$ reduces the Burgers's equation (it is the simplest and rather interesting nonlinear evolution, a one-dimensional reduction of the Navier Stokes equation)

$$u_t = u_{xx} + 2uu_x$$

to the linear diffusion equation $v_t = v_{xx}$.

3. The transformation between the Liouville equation $u_{xy} = \exp u$ and the wave equation $\bar{u}_{xy} = 0$ has the following form:

$$u_x + \bar{u}_x = \sqrt{2} \exp(u - \bar{u})/2,$$
$$u_y - \bar{u}_y = \sqrt{2} \exp(u + \bar{u})/2.$$

These transformations are examples of Bäcklund transformations between two different equations. In case the equations are the same, these transformations are called auto-Bäcklund transformations.

4. It is known that there exist recurrence relations between solutions of the Bessel equation with different values of the parameters,

$$z^2 w'' + zw' + (z^2 - \nu^2)w = 0,$$

$$J_{\nu+1} = \frac{2\nu}{z} J_\nu - J_{\nu-1}, \quad J_{\nu+1} = J_{\nu-1} - 2J_\nu'.$$

5. Having a solution ϕ of the sine Gordon equation $\phi_{xy} = \sin\phi$, it is possible to construct a new solution ψ by means of the following equations:

$$\psi_x = \phi_x + 2a\sin\frac{\phi+\psi}{2},$$

$$\psi_t = -\phi_t + \frac{2}{a}\sin\frac{\phi+\psi}{2}.$$

6. The KdV equation

$$u_t - 6uu_x + u_{xxx} = 0$$

can be reduced to the equation

$$w_t - 3w_x^2 + w_{xxx} = 0$$

by the transformation $u = w_x$ and one integration. Then

$$w_x + \bar{w}_x = -m + (w - \bar{w})^2/2,$$
$$w_t + \bar{w}_t = (w - \bar{w})(w_{xx} - \bar{w}_{xx}) + 2(u^2 + u\bar{u} + \bar{u}^2),$$

where w and $\bar{w}$ are the solutions and $u, \bar{u}$ are the corresponding solutions of KdV equation.

Definition 1.4. A relation

$$R(u, v, u_x, v_x, \dots) = 0,$$

or a system of such relations, or the explicit formulae

$$u = R_1(v, v_x, v_t, \dots), \quad v = R_2(u, u_x, u_t, \dots),$$

is called a Bäcklund transformation of equations

$$E(u) = 0, \quad D(v) = 0,$$

if v is a solution of $D(v) = 0$ whenever u is a solution of $E(u) = 0$, or vice versa.

2 The Second Painlevé Equation

2.1 General Properties of the Solutions of the Equation (P_2)

Character of singularities. Representation of the solutions as a ratio of entire functions. τ-functions. 1-solutions. Hamiltonian form of an equivalent system. Hamiltonian of (P_2). The number of solution poles.

The second Painlevé equation has the form

$$w'' = 2w^3 + zw + \alpha. \qquad (P_2)$$

It has one parameter and the simplest form among the Painlevé equations. Painlevé proved that any solution of (P_2) is a meromorphic function of z. Therefore, the singularities of any solution are poles. In a neighborhood of each pole z_0, a solution has the following expansion

$$w = \frac{\varepsilon}{\tau} - \frac{\varepsilon}{6} z_0 \tau - \frac{\alpha + \varepsilon}{4} \tau^2 + h\tau^3 - \frac{3\alpha + \varepsilon}{72} z_0 \tau^4 + \cdots, \quad \varepsilon^2 = 1, \qquad (2.1)$$

where $z - z_0 = \tau$, and h is an arbitrary constant.

The theorem on meromorphic functions implies that each solution of (P_2) can be expressed as the ratio of two entire functions, i.e.,

$$w(z) = \frac{v(z)}{u(z)}. \qquad (2.2)$$

From (2.1) there follow the relations

$$w^2 = \frac{1}{(z - z_0)^2} - z_0/3 + O(z - z_0) \qquad (2.3)$$

and

$$\int_{z_1}^{z} w^2 \mathrm{d}z = -\frac{1}{z - z_0} + O(z - z_0).$$

Therefore, the function

$$u(z) = \exp\left(- \int_{z_1}^{z} \mathrm{d}z \int_{z_1}^{z} w^2 \mathrm{d}z\right), \qquad (2.4)$$

where the path of integration does not pass through singularities of $w(z)$, is for an arbitrary solution an entire function, and it has a simple zero at z_0. Therefore, the function $v(z)$ is entire, too. From (2.4) and equation (P_2), it follows that the equations for $u(z)$ and $v(z)$ have the form

$$uu'' = u'^2 - v^2,$$

$$v''u^2 + vu'^2 - 2u'v'u = v^3 + zvu^2 + \alpha u^3. \qquad (2.5)$$

It can be proved that all the solutions of system (2.5) are entire and that if (u, v), $u \neq 0$, is an arbitrary solution of (2.5), then (2.2) determines a solution of (P_2).

Proposition 2.1. *Any nonrational solution of (P_2) has an infinite number of simple poles.*

This result follows from the Wittich theorem, stating that an algebraic differential equation $P(z, w, w', \ldots, w^{(n)}) = 0$ with one dominant term has no entire transcendental solutions.

Equation (P_2) has a one-parameter family of solutions (1-solutions), which is defined by the general solution of some Riccati equation

$$w' = a(z)w^2 + b(z)w + c(z). \tag{2.6}$$

Indeed, from (2.6) we have

$$w'' = a'w^2 + b'w + c' + (2aw + b)(aw^2 + bw + c).$$

Therefore, the coefficients a, b, c are determined by the system

$$a^2 = 1, \quad a' + 3ab = 0, \quad b' + 2ac + b^2 = z, \quad c' + bc = \alpha.$$

Then $a^2 = 1$, $b = 0$, $c = z/2a$, $\alpha = a/2$.

Proposition 2.2. *All the solutions of the equation*

$$w' = \varepsilon w^2 + \varepsilon z/2, \quad \alpha = \varepsilon/2, \quad \varepsilon^2 = 1, \tag{2.7}$$

are solutions of (P_2).

If $w = -\varepsilon u'/u$ in (2.7), then $u'' + (z/2)u = 0$. The last equation is an Airy equation. Therefore, 1-solutions of (P_2) for $\alpha = \varepsilon/2$ can be expressed by means of the Airy function.

Let us consider the system

$$
\begin{aligned}
w' &= \varepsilon w^2 + \varepsilon z/2 + v \\
v' &= \alpha - \varepsilon/2 - 2\varepsilon wv, \quad \varepsilon^2 = 1.
\end{aligned}
\tag{2.8}
$$

Elimination of v from (2.8) yields (P_2). On the other hand, (2.8) yields

$$w = \frac{\varepsilon}{2v}(\alpha - \varepsilon/2 - v'), \quad v \neq 0, \tag{2.9}$$

and

$$v = w' - \varepsilon w^2 - \varepsilon z/2, \tag{2.10}$$

where

$$v'' = \frac{v'^2}{2v} - zv - 2\varepsilon v^2 - \frac{(\alpha - \varepsilon/2)^2}{2v}. \tag{2.11}$$

The first equation of the system (2.8) is a Riccati equation, and it may be considered as the Miura transformation of (P_2). In this case equation (2.5) is the result of the Miura transformation. Also, the system (2.8) is a linearization of the equation (P_2). System (2.8) is a Hamiltonian system with Hamiltonian

$$H(z, w, v) = \frac{v^2}{2} + \varepsilon(w^2 + z/2)v - \left(\alpha - \frac{\varepsilon}{2}\right)w.$$

Now we can set $h(z) = H(z, w(z), v(z))$ and $\tau(z) = \exp \int h(z)dz$. In this case $\tau(z)$ is an entire function, and $h(z)$ satisfies the equation

$$h''^2 + 4h'^3 + 2h'(zh' - h) - (\alpha + \tfrac{1}{2})^2/4 = 0$$

and

$$w = \frac{2h'' + \alpha + \tfrac{1}{2}}{4h'}, \quad v = -2h', \quad \varepsilon = -1.$$

Note that if in the system (2.8) v vanishes, then we have 1-solutions (2.7) for $\alpha = \varepsilon/2$.

As for equation (2.5), it has the P-property. This equation is the 34th in the Painlevé list, and its general solution is

$$v = -\varepsilon(y' + y^2 + z/2), \tag{2.12}$$

where y is the general solution of equation

$$y'' = 2y^3 + zy + \alpha - \varepsilon/2 - \frac{1}{2}. \tag{2.13}$$

We can verify this fact by differentiating and substituting (2.12) into equation (2.11). Equation (2.13) is the second Painlevé equation with parameter $\alpha - (\varepsilon + 1)/2$.

2.2 Bäcklund Transformations of (P_2)

T, T^{-1}, S-transformations. Integral form of Bäcklund transformation. Group of Bäcklund transformations. Nonlinear superposition of solutions (Bianchi's formula). Discrete form of (P_2). Fundamental domain of parameter space (Weyl chamber).

Now we can construct the Bäcklund transformation of (P_2). First of all, from (2.8) and (2.12) we have

$$\varepsilon w' - w^2 + y' + y^2 = 0. \tag{2.14}$$

This equation relates solutions of (P_2) with two different values of the parameters.

Differentiating equation (2.14), we obtain

$$y = -\varepsilon w + \frac{(\alpha - 1/2)(\varepsilon + 1)}{2(\varepsilon w' - w^2 - z/2)}.$$

Of course, we can find w by means of y. In this case we have

$$w = -\varepsilon y - \frac{\alpha(\varepsilon + 1) - 1}{2\varepsilon(y' + y^2 + z/2)}.$$

Proposition 2.3 (Bäcklund transformations). *If* $w_\alpha = w(z, \alpha)$ *is a solution of* (P_2) *for a given value of the parameter* α, *then the transformations*

$$S : w(z, \alpha) \to w(z, -\alpha) = -w(z, \alpha), \tag{2.15}$$

$$T : w_{\alpha-1} \to w_\alpha = -w_{\alpha-1} - \frac{\alpha - \frac{1}{2}}{w'_{\alpha-1} + w^2_{\alpha-1} + z/2}, \tag{2.16}$$

$$T^{-1} : w_\alpha \to w_{\alpha-1} = -w_\alpha + \frac{\alpha - \frac{1}{2}}{w'_\alpha - w^2_\alpha - z/2} \tag{2.17}$$

determine other solutions of (P_2).

In (2.15)–(2.16) $TT^{-1} = T^{-1}T = \mathrm{id}$, $S^2 = \mathrm{id}$. We can regard S, T, T^{-1} as the transformations on the set $F = \{\alpha, w\}$, where $\alpha \in \mathbb{C}$ and w is a solution of (P_2). In this case S and T generate a transformation group G_2 on F. In addition, we define the transformations

$$\tau : \mathbb{C} \to \mathbb{C}, \quad \alpha \mapsto -\alpha,$$

$$\tau_+ : \mathbb{C} \to \mathbb{C}, \quad \alpha \mapsto \alpha + 1$$

associated with the transformations S, T. This is a group $\widetilde{G_2}$.

Proposition 2.4. $\widetilde{G_2} \simeq G_2 \simeq W(A_1)$, *where* $W(A_1)$ *is the affine Weyl group associated with the Lie algebra of type* A_1.

Let $\alpha = \varepsilon/2, v \neq 0$ in (2.1). This condition is equivalent to

$$R(z, \varepsilon/2) \equiv w'(z, \varepsilon/2) - \varepsilon w^2(z, \varepsilon/2) - \varepsilon/2 \neq 0,$$

and the equation for v is

$$v'' = \frac{v'^2}{2v} - zv - 2\varepsilon v^2.$$

Let v be $v = \lambda y^2$. Then

$$y'' = -\varepsilon \lambda y^3 - zy/2.$$

Let z be $z = -2^{1/3}\tau, \lambda = -\varepsilon 2^{1/3}$. Then

$$y'' = 2y^3 + \tau y.$$

Thus we have obtained the following result.

Proposition 2.5. *The solutions* $w(z, \varepsilon/2)$, $w(z, 0)$ *of equation* (P_2) *are linked by the formulae*

$$\lambda w^2(\tau, 0) = w'(z, \varepsilon/2) - \varepsilon w^2(z, \varepsilon/2) - \varepsilon z/2,$$

$$w(z, \varepsilon/2) = -\frac{\varepsilon}{2v} v'_z = 2^{-1/3} \varepsilon \frac{w'(\tau, 0)}{w(\tau, 0)}. \tag{2.18}$$

The following result is a consequence of Propositions 2.3, 2.5.

Proposition 2.6. *In order to find the general solution of* (P_2) *for arbitrary values of the parameter* α*, it is sufficient to consider the values of* α *satisfying*

$$0 \leq \operatorname{Re} \alpha \leq \frac{1}{2}. \tag{2.19}$$

This is the fundamental domain of the parameter space (or Weyl chamber). Note that if w is an unknown function in (2.14), then $w = -\varepsilon y$ is a particular solution. Therefore, we can solve the Riccati equation (2.14). If $w(z_0), w'(z_0)$ are the initial data of the solution w, then

$$w_{\alpha\varepsilon-1} = -\varepsilon w_\alpha + \frac{(\alpha\varepsilon - \frac{1}{2})J(w_\alpha)}{w'_0\varepsilon - w_0^2 - z_0/2 + (\alpha - \frac{1}{2})\int_{z_0}^z J(w_\alpha)dz}, \tag{2.20}$$

where

$$J(w_\alpha) = \exp\left(2\int_{z_0}^z \varepsilon w_\alpha dz\right).$$

This is the integral form of the Bäcklund transformation.

There are some results for the system (2.16), (2.17).

1. Let $\alpha \to \alpha + 1$ in (2.16). Then

$$w_{\alpha+1} = -w_\alpha - \frac{\alpha + \frac{1}{2}}{w'_\alpha + w_\alpha^2 + z/2}.$$

From (2.17) we obtain

$$w'_\alpha = w_\alpha^2 + z/2 + \frac{\alpha - \frac{1}{2}}{w_{\alpha-1} + w_\alpha}.$$

Then we have

$$w_{\alpha+1} = -w_\alpha - \frac{\alpha + \frac{1}{2}}{2w_\alpha^2 + z + (\alpha - \frac{1}{2})/(w_\alpha + w_{\alpha-1})}. \tag{2.21}$$

This result can be regarded as a Bianchi formula or a discrete analogue of (P_2). It is a consequence of the following proposition.

Proposition 2.7. *Assume that the functions w_α, $w_{\alpha-1}$ satisfy the system (2.16), (2.17). Then they and $w_{\alpha+1}$ from (2.21) are solutions of equation (P_2) for the parameters α, $\alpha - 1$, $\alpha + 1$.*

The transformations (2.16), (2.17) allow us to construct solutions for different values of the parameters. But if $w(z, \alpha)$ is a solution of (P_2), then

$$S_j : w(z, \alpha) = w_j(z, \alpha) = \mu w(\mu z, \alpha), \quad \mu = \exp(2\pi i j/3).$$

This is an example of an auto-Bäcklund transformation.

Proposition 2.8. *Let $w(z, \alpha)$ be a solution of (P_2). Then*

$$\tilde{w}(z, \alpha) = T^\alpha S T^{-\alpha} w(z, \alpha)$$

is a solution of (P_2), too.

It is easy to show that $\tilde{w}(z, \alpha) = w(z, \alpha)$ if and only if $w(z, \alpha)$ is a rational solution.

3   Rational Solutions of (P_2) (0-Solutions)

> *General form of 0-solutions. System for polynomials $P(z)$ and $Q(z)$. Yablonskii–Vorob'ev polynomials. Necessary and sufficient conditions.* ∎
> *List of 0-solutions.*

Proposition 3.1. *An arbitrary 0-solution has the form*

$$w(z) = P(z)/Q(z), \tag{3.1}$$

where $P(z), Q(z)$ are polynomials with $\deg P = n - 1$, $\deg Q = n$.

Consequence. An arbitrary 0-solution has an expansion

$$w(z) = \sum_{j=1}^{n} \frac{\varepsilon_j}{z - z_j}, \tag{3.2}$$

where $\mathrm{res}_{z=z_j}\, w(z) = \varepsilon_j, \varepsilon_j^2 = 1$.

Proposition 3.2. *The solution w from (3.1) is a 0-solution of (P_2) if and only if $P(z), Q(z)$ are solutions of the system*

$$QQ'' = Q'^2 - P^2, \tag{3.3}$$

$$P''Q^2 + PQ'^2 = P^3 + 2P'Q'Q + zPQ^2 + \alpha Q^3.$$

Proof. To prove this proposition we can use the system (2.5) in the entire functions $u(z), v(z)$. Indeed, if (3.1) is a solution of (P_2), then system (2.5) has a solution

$$u(z) = Q(z)\exp(g(z)), \quad v(z) = P(z)\exp(g(z)), \tag{3.4}$$

where $g(z)$ is an entire function. Substituting (3.4) into (2.6), we obtain $g''(z) = 0$, and $P(z), Q(z)$ satisfy the system (3.3). □

Let P, Q have the expansions

$$P = p_{n-1}z^{n-1} + p_{n-2}z^{n-2} + \cdots, \quad Q = q_n z^n + q_{n-1}z^{n-2} + \cdots, \quad q_n \neq 0.$$

Then from system (3.3) we have

$$nq_n^2 = p_{n-1}^2, p_{n-1} = -\alpha q_n \implies n = \alpha^2.$$

Therefore, we obtain the following results.

Proposition 3.3. (P_2) *has a unique rational solution if* $\alpha \in [0, \frac{1}{2})$. *It is* $w = 0$.

Proposition 3.4. (P_2) *has a unique rational solution if and only if* α *is an integer.*

Proof. This fact follows from the Bäcklund transformation (2.16), (2.17) and Proposition 3.3. Indeed, if we start from the rational solution $w = 0$ with $\alpha = 0$ and apply T, T^{-1}, we can construct the 0-solution for any integer α. Moreover, by the property $TT^{-1} = T^{-1}T = \text{id}$ and the uniqueness of $w = 0$ for $\alpha = 0$, (P_2) cannot have more than one 0-solution for every $\alpha \in \mathbb{Z}$. If we admit the existence of a 0-solution for $\alpha \neq n$, $n \in \mathbb{Z}$, this contradicts Proposition 3.3 due to the transformations T, T^{-1}. □

Remark. Yablonski and Vorob'ev showed that the 0-solution of (P_2) has the form

$$w(z) = \frac{P'_\alpha}{P_\alpha} - \frac{Q'_\alpha}{Q_\alpha}$$

and $P_{\alpha+1} = Q_\alpha, Q_{\alpha+1} = (zQ_\alpha^2 + 4Q_\alpha'^2 - 4Q_\alpha Q_\alpha'')/P_\alpha, P_0 = Q_0 = 1$. If $\alpha = 1$, then $P_1 = 1$, $Q_1 = z$. If $\alpha = 2$, then $P_2 = z$, $Q_2 = z^3 + 4$.

The polynomials P_α, Q_α are called Yablonski–Vorob'ev polynomials.

Every rational solution has only simple poles with residue $+1$ or -1. Now we can find how many poles there are with residue $+1$ and how many poles there are with residue -1. Let l_+ and l_- be the numbers of poles with residue $+1$ and -1, respectively; then $l_+ + l_- = n = \alpha^2$. Now we can find the residue of $z = \infty$. Let z be $z = 1/t$. Then $w'_z = -t^2 w'_t$,

TABLE 12.1. A list of 0-solutions of (P_2).

α	$w(z, \alpha)$
0	0
± 1	$\mp 1/z$
± 2	$\mp \dfrac{2(z^3 - 2)}{z(z^3 + 4)}$
± 3	$\mp \dfrac{3z^2(z^6 + 8z^3 + 160)}{(z^3 + 4)(z^6 + 20z^3 - 80)}$
± 4	$\mp \dfrac{4(z^{15} + 50z^{12} + 1000z^9 - 22400z^6 - 112000z^3 - 224000)}{z(z^6 + 20z^3 - 80)(z^9 + 60z^6 + 11200)}$

$w''_{z^2} = 2t^3 w'_t + t^4 w''_{t^2}$. Therefore, equation ($P_2$) in a neighborhood of infinity has the form

$$t^4 w'' + 2t^3 w' = 2w^3 + \frac{w}{t} + \alpha.$$

Let the 0-solution have an expansion $w = a_1 t + a_2 t^2 + \cdots$ (around $z = \infty$). Then $a_1 = -\alpha, a_2 = 0, a_3 = 0$, and $\mathrm{res}_{z=\infty} w(z) = \alpha$. Therefore,

$$l_+ - l_- = -\alpha,$$

because the sum of the residues of the poles is $l_+ - l_-$. Hence,

$$l_+ + l_- = \alpha^2, \quad l_+ - l_- = -\alpha$$

and

$$l_+ = \frac{\alpha(\alpha - 1)}{2}, \quad l_- = \frac{\alpha(\alpha + 1)}{2}.$$

4   One-Parameter Families of Classical Solutions (1-Solutions)

Family of solutions expressible by Airy's functions. List of 1-solutions. ■

In the previous section we showed that (P_2) has 1-solutions generated by the general solution of the Riccati equation

$$w' = \varepsilon w^2 + \varepsilon z/2, \quad \alpha = \varepsilon/2, \quad \varepsilon^2 = 1 \tag{4.1}$$

and that are expressed by means of the Airy function

$$w = -\varepsilon u'/u, \quad u'' + \frac{z}{2}u = 0.$$

Now we use the Bäcklund transformations for the construction of 1-solutions for other values of the parameter α. We start with a solution (4.1) for $\alpha = \frac{1}{2}$. Then from (2.17) with $\alpha = \frac{3}{2}$ we obtain

$$w_{1/2} = -y + (y' - y^2 - z/2)^{-1}, \qquad (4.2)$$

where $y = w(z, \frac{3}{2})$. Substituting (4.2) into (4.1) with $\varepsilon = 1$, we obtain

$$y'^3 - \left(y^2 + \frac{z}{2}\right)y'^2 - \left(y^4 + zy^2 + 4y + \frac{z^2}{4}\right)y' + y^6 + \frac{3}{2}zy^4 + 4y^3 + \frac{3}{4}z^2y^2$$

$$+ 2yz + 2 + \frac{z^3}{8} = 0. \quad (4.3)$$

All solutions of this equation are simultaneously solutions of (P_2) with $\alpha = \frac{3}{2}$. From (2.6) with $\alpha = \frac{3}{2}$ we have

$$y = w(z, \tfrac{3}{2}) = w_{1/2} - (2w_{1/2}^2 + z)^{-1}, \quad w_{1/2} = -\frac{u'}{u}, u'' + \frac{z}{2}u = 0. \quad (4.4)$$

By analogy to the above we can construct 1-solutions for $\alpha = \frac{5}{2}$ and for all $\alpha = n + \frac{1}{2}$, $n \in \mathbb{N}$.

Finally, using the property $w(z, -\alpha) = -w(z, \alpha)$ we obtain the following result.

Proposition 4.1. (P_2) *admits a one-parameter family of particular solutions represented by rational functions of the solution of an Airy equation and its derivative for* $\alpha = n + \frac{1}{2}$, $n \in \mathbb{Z}$. *These 1-solutions satisfy the Fuchsian equation*

$$w'^n + \sum_{j=1}^{n} P_j(z, w)w'^{n-j} = 0, \quad n \in \mathbb{N}, \qquad (4.5)$$

where

$$P_j(z, w) \equiv \sum_{k=0}^{2j} p_{jk}(z)w^{2j-k}, \quad j = 1, 2, \ldots, n. \qquad (4.6)$$

5   Algebraic Nonintegrability of (P_2)

Arbitrary constants of the general solution of (P_2). *Transcendence of* (P_2).

Now we shall prove that the condition $\alpha = n + \frac{1}{2}$, $n \in \mathbb{Z}$, is necessary, i.e., if we have the equation

$$P(w', w, z) = 0, \qquad (5.1)$$

TABLE 12.2. Airy solutions of (P_2), where $U = u'/u$ and u is any solution of the Airy equation $u'' + (z/2)u = 0$.

α	$w(z, \alpha)$
$\pm\frac{1}{2}$	$\mp U$
$\pm\frac{3}{2}$	$\pm\dfrac{2U^3 + zU - 1}{2U^2 + z}$
$\pm\frac{5}{2}$	$\pm\dfrac{4zU^4 + 6U^3 + 4z^2U^2 + 3zU + z^3 - 1}{(2U^2 + z)(4U^3 + 2zU - 1)}$

where P is a polynomial in w', w that determines 1-solutions of (P_2), then (5.1) coincides with some equation from (4.5). Simultaneously we shall prove that equation (P_2) has no first integral of the form (5.1). To prove these facts it is sufficient to show that equation (5.1) does not determine 1-solutions in the fundamental domain $0 \le \mathrm{Re}(\alpha) \le \frac{1}{2}$.

If equation (5.1) generates 1-solutions of (P_2), then (5.1) is a first integral of (P_2), and (5.1) is an equation of P-type. In this case it has the form of equation (4.5). In a neighborhood of a pole we have

$$w' = a_2 w^2 + a_1 w + a_0 + \sum_{j=1}^{\infty} a_{-j} w^{-j}. \qquad (5.2)$$

Substituting (5.2) in (P_2), we find that

$$a_2 = \varepsilon, \quad \varepsilon^2 = 1, \quad a_1 = 0, \quad a_0 = \varepsilon z/2, \quad a_{-1} = \alpha \varepsilon - \tfrac{1}{2}, \quad a_{-2} = h,$$

where h is an arbitrary constant and $a_j = 0$, $j \le 3$, are determined by the recurrence relations

$$a_{-n-1} = \frac{\varepsilon}{n-1}\left(\sum_{j=1}^{n-1} j a_{-j} a_{1+j-n} - a'_{-n}\right), \quad n \ge 2. \qquad (5.3)$$

It follows from (5.3) that $a_{-j} = 0$, $j \ge 3$, if $a_{-1} = 0$, $h = 0$. Each root w of (5.1) can be expressed in the form (5.2). Hence, if $G(w, z, \alpha, \varepsilon, h)$ denotes the right-hand side of (5.2), then

$$P(w', w, z) = \prod_{i=1}^{l_+} \prod_{j=1}^{l_-} (w' - S(w, z, \alpha, 1, h_i))(w' - S(w, z, \alpha, -1, h_j)), \qquad (5.4)$$

where $l_+, l_- \in \mathbb{Z}_0$ are the numbers of roots of (5.1) with $\varepsilon = 1$ and $\varepsilon = -1$ $(l_+ + l_- = n, n \ge 1)$, respectively, and h_ν is a value of the arbitrary coefficient a_2 for the νth root. Expanding (5.4), we obtain

$$P(w', w, z) = w'^n - \left(S_2 w^2 + S_1 w + S_0 + \sum_{j=1}^{\infty} \frac{S_{-j}}{w^j}\right) w'^{n-1} + \cdots, \qquad (5.5)$$

where $S_2 = l_+ - l_-$, $S_1 = 0$, $S_0 = (z/2)(l_+ - l_-)$,

$$S_{-1} = \alpha(l_+ - l_-) - (n/2), \quad S_{-j} = \sum_{k=0}^{n} \overset{(k)}{_{-j}}, \quad j \geq 2.$$

The left-hand side of (5.5) is a polynomial in w', w. Hence, $S_{-j} = 0$, $j \geq 1$. In particular, from condition $S_{-1} = 0$ we have

$$\alpha(l_+ - l_-) = n/2, \quad n = l_+ + l_-, \quad n \in \mathbb{N}. \tag{5.6}$$

From (5.6) it follows immediately that (P_2) with $\alpha = 0$ has no 1-solutions given by (5.1) and $l_+ - l_- \neq 0$. Consider the relation (5.6) in the fundamental domain $0 \leq \operatorname{Re}(\alpha) \leq \frac{1}{2}$. Indeed, if either (P_2) has 1-solutions defined by (5.1) or (5.1) is a first integral, then there exist $l_+, l_- \in \mathbb{Z}_0$ such that α defined by (5.6) belongs to the fundamental domain, i.e.,

$$0 \leq \frac{l_+ + l_-}{l_+ - l_-} < 1. \tag{5.7}$$

Expanding (5.7), we see that this is not possible.

Proposition 5.1. *Equation (P_2) has no algebraic first integral and no 1-solutions different from those obtained from the Riccati equation by application of the T, T^{-1} transformations.*

Thus all solutions of (P_2) can be divided into three classes:

(a) 0-solutions for $\alpha \in \mathbb{Z}$, generated by $w(z, 0) = 0$.

(b) 1-solutions for $\alpha - \frac{1}{2} \in \mathbb{Z}$, generated by the Riccati equation.

(c) solutions that do not coincide with the solutions of classes (a) and (b). They define the second Painlevé transcendental functions.

6   Higher Analogue of (P_2)

Relation between KdV and (P_2) equations. $(_mP_2)$ equations. Properties of solutions of $(_mP_2)$.

Consider the higher analogue of the Korteweg de Vries equation

$$(2m - 1)u_t = X_m u,$$

$$X_m u = (2u + 2DuD^{-1} - D^2)X_{m-1}u = D\frac{\partial H_m}{\partial u}, \tag{6.1}$$

where $X_1 = Du = u_x = D\partial H_1/\partial u$.

If $m = 2$, then

$$3u_t = X_2 u = (2u + 2DuD^{-1} - D^2)u_x = 6uu_x - u_{xxx}. \qquad (6.2)$$

For $m = 3$ we find, that

$$5u_t = 30u^2 u_x - 20u_x u_{2x} - 10uu_{3x} + u_{5x}.$$

The variable changes

$$z = xt^{-1/(2m-1)}, \quad u(x,t) = t^{-2/(2m-1)}\left(\frac{dy(z)}{dz} + y^2(z)\right)$$

reduce (6.2) to the ODE

$$D^{-1}S_y^{m-1}(y') + zy + \alpha = 0, \qquad (_m P_2)$$

where $D = d/dz$, $S_y = 4y^2 + 4y'D^{-1}y - D^2$, and α is an arbitrary parameter. If $m = 2$, then $y'' = 2y^3 + zy + \alpha$. Therefore, if $m = 2$, we have equation (P_2).

Theorem 6.1 (M. Ablowitz). *If $w = w(z)$ is a solution of (P_2), then the function*

$$u(x,t) = t^{-2/3}(w'(z) + w^2(z)), \quad z = xt^{-1/3} \qquad (6.3)$$

is a solution of the Korteweg de Vries equation

$$3u_t = 6uu_x - u_{xxx}. \qquad (\text{KdV})$$

By applying relations (6.3) we can construct special classes of solutions of the KdV equation. Moreover, 0-solutions of (P_2) generate solutions of KdV rational in x and t. We can use 1-solutions of (P_2) for the construction of special classes of solutions of KdV expressed by means of Airy functions. The solution properties of equation $(_m P_2)$ are the same as those for (P_2). First of all, we note that equation $(_m P_2)$ can be rewritten in the form of the equivalent system (H. Airault)

$$q = y' + y^2, \qquad (6.4)$$

$$y = \left(z + 2\frac{\partial H_{m-1}}{\partial q}\right) + \alpha = X_{m-1}q. \qquad (6.5)$$

Using this result we can obtain the Bäcklund transformation for $(_m P_2)$.

Theorem 6.2. *Let $y(z)$ be a solution to $(_m P_2)$ such that*

$$z + 2\frac{\partial H_{m-1}}{\partial r} \neq 0, \quad r = y^2 - y'.$$

Then the transformations

$$T: y \to \tilde{y} = -y - (2\alpha - 1)\left(z + 2\frac{\partial H_{m-1}}{\partial r}\right)^{-1},$$

$$T^{-1}: \tilde{y} \to y = -\tilde{y} - (2\alpha + 1)\left(z + 2\frac{\partial H_{m-1}}{\partial \tilde{q}}\right)^{-1}, \qquad (6.6)$$

where $\tilde{q} = \tilde{y}^2 + \tilde{y}'$, determine solutions of $(_mP_2)$ for $\tilde{\alpha} = \alpha - 1$.

The order of equation $(_mP_2)$ is equal to $2m - 2$. It follows from (6.6) that $(_mP_2)$ has a $(2m - 3)$-parameter family of solutions when $\alpha = n + \frac{1}{2}$, $n \in \mathbb{Z}$. This family is generated by the general solution of the equation

$$2\frac{\partial H_{m-1}}{\partial r} + z = 0. \qquad (6.7)$$

Rational solutions exist if and only if α is an integer. The poles of 0-solutions are simple with residues $-1, +1, -2, +2, \ldots, -(m-1), +(m-1)$.

Finally, we remark that the fundamental domain of equation $(_mP_2)$ is $0 \le \mathrm{Re}(\alpha) < \frac{1}{2}$.

Theorem 6.3. *Let $y(z, \alpha)$ be a nonrational solution of (0_mP_2). Then the function*

$$\tilde{y}(z, \alpha) = T^{-\alpha}ST^{\alpha}y(z, \alpha)$$

is also a solution of $(_mP_2)$ for $\tilde{\alpha} = \alpha - 1$ and $\tilde{y} \neq y$.

Now we briefly consider the equation $(_3P_2)$

$$y^{(4)} = 10y^2y'' + 10yy'^2 - 6y^5 - zy - \alpha. \qquad (_3P_2)$$

The system equivalent to $(_3P_2)$ has the form

$$q = y' + y^2, \quad y(6q^2 + z - 2q'') = 6qq' - q''' - \alpha,$$

and the Bäcklund transformation of $(_3P_2)$ is the following:

$$T: y \to \tilde{y} = -y - (2\alpha - 1)R^{-1}(r),$$

$$T^{-1}: \tilde{y} \to y = -\tilde{y} - (2\alpha + 1)R^{-1}(\tilde{q}),$$

where $R(r) = z + 6r^2 - 2r''$, $r = y^2 - y'$, $\tilde{q} = y' + \tilde{y}^2$. The poles of the solution are simple and have residues $-1, +1$ or $-2, +2$. 0-solutions exist and are unique if and only if α is an integer.

It follows from the equivalent system that $(_3P_2)$ has a 3-parameter family of solutions (3-solutions), defined by the system

$$y' = -\varepsilon y^2 + w, \quad w'' = 3\varepsilon w^2 + \varepsilon z/2, \quad \varepsilon^2 = 1, \quad \alpha = -\varepsilon/2.$$

TABLE 12.3. List of some rational solutions of $(_3P_2)$.

α	$y(z,\alpha)$
0	0
± 1	$\mp 1/z$
± 2	$\mp 2/z$
± 3	$\pm(2 - 5z^5)/(z(z^5 + 144))$

The second equation of this system is the first Painlevé equation. Let $y = \varepsilon u'/u$; then

$$u'' + \varepsilon wu = 0, \quad w'' = 3\varepsilon w^2 + \varepsilon z/2.$$

This result can be compared with the Lamé equation

$$y'' + (a\wp(z) + b)y = 0,$$

where $\wp$ is the Weierstrass elliptic function.

7   The Fourth Painlevé Equation

Character of singularities. System for entire functions $u(z)$, $v(z)$. Riccati equation. Simple symmetry. Bäcklund transformations. General formula for parameters of B-transformations. Weyl chamber.

The fourth Painlevé equation has the form

$$w'' = \frac{w'^2}{2w} + \frac{3}{2}w^3 + 4zw^2 + 2(z^2 - \alpha)w + \frac{\beta}{w}, \qquad (P_4)$$

where α, β are parameters and $w(z)$ is a meromorphic function.

Although the Painlevé equations were first discovered from strictly mathematical considerations, they have appeared in many physical applications. In particular, it has been demonstrated that (P_4) has relevance within the fields of fluid mechanics, nonlinear optics and quantum gravity. In addition, (P_4) also appears as a symmetry reduction from a large class of significant PDEs including the Boussinesq and modified Boussinesq equations, the dispersive long wave equation, KP equation, cubic nonlinear Schrödinger equation, among others. Now we consider some general properties of the solutions.

1. Every solution $w(z)$ has for its only singularities simple poles. If a solution has a pole at z_0, then $w(z)$ must have an expansion

$$w(z) = \frac{\varepsilon}{z - z_0} - z_0 - \frac{\varepsilon}{3}(z_0^2 + 2\alpha - 4\varepsilon)(z - z_0) + h(z - z_0)^2$$
$$+ O((z - z_0)^3), \quad (7.1)$$

where ε denotes 1 or -1 and h is an arbitrary constant.

2. From (7.1) it follows that

$$w(z) + z = \varepsilon/(z - z_0) + (z - z_0)\phi_1(z, z_0)$$

and

$$(w(z) + z)^2 = (z - z_0)^{-2} + \phi_2(z - z_0),$$

where $\phi_j(z, z_0), j = 1, 2$, is a holomorphic function around z_0. Hence,

$$u(z) = \exp\left(-\int_{z_1}^z dz \int_{z_1}^z (w^2(z) + 2zw(z))dz\right)$$

is an entire function and $v(z) = w(z)u(z)$ is also an entire function.
The system for the entire functions $u(z)$ and $v(z)$ is the following:

$$uu'' - u'^2 + v^2 + 2zuv = 0,$$

$$2uv(uv'' - u'v') = u^2v'^2 - u'^2v^2 + v^4 + 4zuv^3 + 4(z^2 - \alpha)u^2v^2 + 2\beta u^4.$$

3. If we substitute the Riccati form

$$w' = a(z)w^2 + b(z)w + c(z)$$

in (P_4), we obtain the following result.

All solutions of the Riccati equation

$$w' - \varepsilon_1 w^2 + 2\varepsilon_1 zw - 2(1 + \alpha\varepsilon_1), \quad \beta = -2(1 + \alpha\varepsilon_1)^2, \quad \varepsilon_1^2 = 1$$

are at the same time solutions of (P_4).

Application of the linearizing transformation $w = -\varepsilon_1 u'/u$ to this equation yields the Weber–Hermite equation

$$u'' - 2\varepsilon_1 zu' - 2(\alpha + \varepsilon_1)u = 0.$$

Further, if we perform the transformation

$$u(z) = \eta(\xi)\exp(\varepsilon_1\eta^2/4), \quad \xi = \sqrt{2}z,$$

then we obtain the equation for the parabolic cylinder function

$$\frac{d^2\eta}{d\xi^2} = \left(\frac{\xi^2}{4} + \alpha + \frac{\varepsilon_1}{2}\right)\eta, \quad \varepsilon_1^2 = 1.$$

This equation has the general solution

$$\eta(\xi) = C_1 D_\nu(\xi) + C_2 D_\nu(-\xi),$$

where $\nu = -\alpha - (1 + \varepsilon_1)/2$, constants C_1 and C_2 are arbitrary, and $D_\nu(\eta)$ is the parabolic cylinder function solution of

$$\frac{d^2 D_\nu}{d\xi^2} = \left(\frac{\xi^2}{4} + \alpha + \frac{\varepsilon_1}{2}\right) D_\nu$$

satisfying

$$D_\nu(\xi) \sim \xi^\nu \exp(-\xi^2/4), \quad \xi \to +\infty,$$

and

$$D_\nu(\xi) \sim -\frac{\sqrt{2\pi}}{\Gamma(-\nu)} \exp(i\pi\nu)\xi^{-\nu-1} \exp(\xi^2/4), \quad \xi \to -\infty,$$

provided that ν is not a positive integer. When ν is a positive integer, then

$$D_\nu(\xi) = \mathrm{He}_n(\xi) \exp(-\xi^2/4),$$

where $\mathrm{He}_n(\xi)$ is the Hermite polynomial of degree n given by

$$\mathrm{He}_n(\xi) = (-1)^n \exp(\xi^2/2)\frac{d^n}{d\xi^n}(\exp(-\xi^2/2)).$$

Using this result we can obtain the solutions

$$w(z, -\nu, -2(\nu + 1)^2) = -\frac{\sqrt{2}(C_1 D_{\nu+1}(z\sqrt{2}) - C_2 D_{\nu+1}(-z\sqrt{2}))}{C_1 D_\nu(z\sqrt{2}) + C_2 D_\nu(-z\sqrt{2})},$$

$$w(z, -\nu - 1, -2\nu^2) = -2z + \frac{\sqrt{2}(C_1 D_{\nu+1}(z\sqrt{2}) - C_2 D_{\nu+1}(-z\sqrt{2}))}{C_1 D_\nu(z\sqrt{2}) + C_2 D_\nu(-z\sqrt{2})},$$

provided that ν is not a positive integer. We remark that in order to obtain these solutions we have used the recurrence relation

$$\frac{dD_\nu}{d\xi} = \frac{1}{2}\xi D_\nu(\xi) - D_{\nu+1}(\xi).$$

In particular, (P_4) has a solution

$$w(z, 1, 0) = \frac{2A \exp(-z^2)}{\sqrt{\pi}(1 - A\operatorname{erfc}(z))}, \quad \operatorname{erfc}(z) = \frac{2}{\sqrt{\pi}} \int_z^\infty \exp(-t^2)dt.$$

4. The equation (P_4) has the symmetry

$$S: w(z, \alpha, \beta) \to \lambda w(\lambda z, \alpha\lambda^2, \beta), \quad \lambda^4 = 1.$$

5. Bäcklund transformation of (P_4).

Consider the system

$$w' = q + 2\varepsilon zw + \varepsilon w^2 + 2\varepsilon wu,$$
$$u' = p - 2\varepsilon zu - \varepsilon u^2 - 2\varepsilon wu, \tag{7.2}$$

where $q^2 = -2\beta$, $p = -1 - \varepsilon\alpha - q/2$, $\varepsilon^2 = 1$.

This system is a system of two coupled Riccati equations. If u is eliminated between these equations, it is easy to show that w satisfies (P_4). However, if w is eliminated, then we see that u is also a solution of (P_4) but with parameter

$$\tilde{\alpha} = \frac{1}{4}(2\varepsilon - 2\alpha + 3\varepsilon\sqrt{-2\beta}), \quad \tilde{\beta} = -\frac{1}{2}\left(1 + \alpha\varepsilon + \frac{1}{2}\sqrt{-2\beta}\right)^2. \tag{7.3}$$

Hence the following proposition.

Proposition 7.1. *Let $w = w(z)$ be a solution of (P_4). Then*

$$T : w(z, \alpha, \beta) \to \tilde{w}(z, \tilde{\alpha}, \tilde{\beta}) = R(w)/(2\varepsilon w), \tag{7.4}$$

where $R(w) = w' - \varepsilon w^2 - 2\varepsilon zw - q$, $\varepsilon^2 = 1$, $q^2 = -2\beta$.

The inverse transformation follows from system (7.2):

$$T^{-1} : \tilde{w}(z, \tilde{\alpha}, \tilde{\beta}) \to w(z, \alpha, \beta) = -\tilde{R}(\tilde{w})/(2\varepsilon\tilde{w}), \tag{7.5}$$

where $\tilde{R}(\tilde{w}) = \tilde{w}' - \tilde{q} + 2\varepsilon z\tilde{w} + \varepsilon\tilde{w}^2$, $\tilde{q}^2 = -2\tilde{\beta}$, and

$$\alpha = (-2\varepsilon - 2\tilde{\alpha} - 3\varepsilon\tilde{q})/4, \quad \beta = -(\tilde{\alpha}\varepsilon - 1 - \tilde{q}/2)^2/2.$$

It is easy to show that $TT^{-1} = T^{-1}T = \text{id}$.

Our aim is to construct solution hierarchies. It is then important to consider the result of the application of more than two Bäcklund transformations. This allows us to deduce the general form of α and β after applying an arbitrary number of Bäcklund transformations.

Proposition 7.2. *Successive applications of transformations T and T^{-1} to a solution $w(z, \alpha, \beta)$ of (P_4) lead to a solution $w(z, \tilde{\alpha}, \tilde{\beta})$, where either*

$$\tilde{\alpha} = \alpha + n_1, \quad \tilde{\beta} = -(2n_2 + \sqrt{-2\beta})^2/2 \tag{7.6}$$

or

$$\tilde{\alpha} = n_1 + (\varepsilon - \alpha)/2 + 3\varepsilon\sqrt{-2\beta}/4,$$
$$\tilde{\beta} = -(2n_2 + 1 + \varepsilon\alpha + \sqrt{-2\beta}/2)^2/2 \tag{7.7}$$

and $n_1 + n_2 = 2n_3$, $n_j \in \mathbb{Z}$, $j \in \{1, 2, 3\}$, $\varepsilon^2 = 1$. This result is proved by induction.

From the above statement, there results the fundamental domain of parameter space.

Corollary 7.1. *In order to construct the general solution with arbitrary values of the parameters, it is enough to construct it for every α, β in the domain*

$$G : 0 \leq \operatorname{Re}\alpha < 1, \quad \operatorname{Re}\sqrt{-2\beta} > 0, \quad \operatorname{Re}(\sqrt{-2\beta} + 2\alpha) \leq 2.$$

G is the Weyl chamber.

8 Classical Solutions of (P_4)

Riccati equation and 1-solutions. Special equation of the second degree in w'. The values of parameters of 1-solutions. Complementary error function solutions. Transcendence of (P_4).

Now we describe various specific one-parameter families of solutions of (P_4) and start by examining solutions of the Riccati equation

$$R_1(v) = v' - \mu v^2 - 2\mu z v - q\delta = 0, \tag{8.1}$$

which determines 1-solutions of (P_4) with parameters

$$\alpha\mu + 1 + q\sigma/2 = 0, \quad \mu^2 = \sigma^2 = 1, \quad q^2 = -2\beta. \tag{8.2}$$

Now we apply the Bäcklund transformation

$$T : v \to \tilde{w} = R(v)/(2\varepsilon v) \tag{8.3}$$

to the solutions defined by (8.1). Of course, we choose ε and the branch of $\sqrt{-2\beta}$ in such a way that $R(v) \neq R_1(v)$.

From the inverse transformation we have

$$v = \tilde{R}(\tilde{w})/(2\varepsilon\tilde{w}). \tag{8.4}$$

Now we can find the equation for the function $\tilde{w}(z)$. For this purpose we substitute (8.4) into (8.1). For example, if $\varepsilon = -\mu$, $\nu = -\sigma$, then we have

$$\tilde{w}'^2 + 4\tilde{w} - \tilde{w}^4 - 4z\tilde{w}^3 - 4(z^2 - \alpha)\tilde{w}^2 + 4 = 0, \tag{8.5}$$

where $\tilde{\beta} = -2$. Equation (8.5) determines 1-solutions of (P_4). These solutions are linked to solutions of (8.1) by the formula

$$\tilde{w} = 2(\alpha - \mu)/v - 2z - v, \quad \tilde{\alpha} = -2\alpha - 2\mu. \tag{8.6}$$

This was the first step in applying the Bäcklund transformation T to solutions of (8.1).

Now we can obtain the values of the parameters α, β for which the equation (P_4) has 1-solutions defined by the Riccati equation (8.1).

TABLE 12.4. Complementary error functions solutions of the form $Q(z, \Psi)/R(z, \Psi)$, $\Psi \equiv \psi'/\psi$, $\psi = 1 - A\,\mathrm{erfc}(z)$.

α	β	$w(z, \alpha, \beta)$
-2	-2	$(\Psi^2 + 2z\Psi - 2)/(\Psi + 2z)$
2	-2	$2/(\Psi + 2z)$
3	-8	$2(\Psi + 2z)/(z\Psi + 2z^2 + 1)$

Proposition 8.1. *Equation* (P_4) *with values of parameters*

$$\beta = -2(\alpha\varepsilon + 2n - 1)^2, \quad n \in \mathbb{N}, \tag{8.7a}$$

or

$$\beta = -2n^2, \quad n \in \mathbb{N}, \tag{8.7b}$$

has 1-solutions expressed by means of Weber–Hermite functions. The formulae (8.7) are obtained by substituting the 1-solution parameters of the Riccati equation (8.1) into the formulae (7.6), (7.7) of successive applications of the Bäcklund transformations.

Let us now consider one particular case. Let $\beta = 0$ and $\alpha = 1$. Then $v' = -v^2 - 2zv$ and

$$v = \frac{\exp(z^2)}{C - \mathrm{erfc}(z)}, \quad \mathrm{erfc}(z) = \int_z^\infty \exp(-t^2)\mathrm{d}t.$$

Therefore, (P_4) has the one-parameter family of solutions

$$w(z, 1, 0) = \frac{2A\exp(-z^2)}{\sqrt{\pi}(1 - A\,\mathrm{erfc}(z))}, \quad w(z, -1, 0) = \frac{-2iA\exp(-z^2)}{\sqrt{\pi}(1 - A\,\mathrm{erfc}(iz))},$$

where A is an arbitrary constant.

Setting $\alpha = 1$ and $\beta = 0$ in (7.6), (7.7), we see that solutions of this set will have parameter values given by

$$\alpha = m_1 + 1, \quad \beta = -2m_2^2, \quad m_1 + m_2 = 2m_3, \quad m_j \in \mathbb{Z}.$$

For details see also [5].

Let us now prove the necessity of the condition (8.7) for (P_4) to have 1-solutions generated by Riccati equations. For the proof we use the method proposed for (P_2). To prove this fact it is sufficient to show that in the fundamental domain G the equation (P_4) has 1-solutions defined by the equation

$$P(w', w, z) = 0 \tag{8.8}$$

only under the condition $\sqrt{-2\beta} + 2\alpha - 2 = 0$.

As for (P_2), in a neighborhood of poles of w we have

$$w' = a_2 w^2 + a_1 w + a_0 + \frac{a_{-1}}{w} + \frac{a_{-2}}{w^2} + O(w^{-3}), \tag{8.9}$$

where $a_2 = \varepsilon$, $a_1 = 2\varepsilon z$, $a_0 = -2 - 2\alpha\varepsilon$, $a_{-1} = h$ is arbitrary,

$$a_{-2} = \varepsilon a'_{-1} - 2za_{-1} - \varepsilon(a_0^2/2 + \beta),$$

$$a_{-3} = \varepsilon a'_{-2}/2 - 2za_{-2} - \varepsilon a_0 a_{-1},$$

$$a_{-j-1} = \frac{\varepsilon}{j} a'_{-j} - 2za_{-j} - \varepsilon a_0 a_{-j+1} - \varepsilon a_{-1} a_{-j+2} - \cdots$$

$$- \varepsilon a_{-r+1} a_{-r}, \ j = 2r,$$

$$a_{-j-1} = l\frac{\varepsilon}{j} a'_{-j} - 2za_{-j} - \varepsilon a_0 a_{-j+1} - \varepsilon a_{-1} a_{-j+2} - \cdots - \varepsilon a_{-r+1} a_{-r-1}$$

$$- \varepsilon a_{-r}^2/2, \ j = 2r + 1.$$

It follows from these recurrence relations that $a_{-j} = 0$, $j \geq 2$, if $a_{-1} = h = 0$ and

$$2(1 + \alpha\varepsilon)^2 + \beta = 0. \tag{8.10}$$

Hence, if $G(z, w, \varepsilon, h)$ denotes the right-hand side of (8.9), then

$$P(w', w, z) = \prod_{i=1}^{l_+} \prod_{i=1}^{l_-} (w' - G(z, w, 1, h_i))(w' - G(z, w, -1, h_j)), \tag{8.11}$$

where l_+ and l_- are the number of roots w' of (8.8) with $\varepsilon = 1$ and $\varepsilon = -1$, respectively, and h_ν is the arbitrary coefficient a_{-1} for the νth root. Expanding (8.11), we obtain

$$P(w', w, z) = w'^n - \left(S_2 w^2 + S_1 w + S_0 + \frac{S_{-1}}{w} + \sum_{j=2}^{\infty} \frac{S_j}{w^j} \right) w'^{n-j}$$

$$+ \cdots = 0, \tag{8.12}$$

where $S_2 = l_+ - l_-$, $S_1 = 2z(l_+ - l_-)$, $S_0 = -2n + 2\alpha(l_- - l_+)$. Hence the coefficients of the polynomial

$$P_1(z, w) = p_{10}(z)w^2 + p_{11}(z)w + p_{12}(z) \tag{8.13}$$

are

$$p_{10} = l_- - l_+, \quad p_{11} = 2z(l_- - l_+), \quad p_{12} = 2n - 2\alpha(l_- - l_+),$$

and $S_{-j} = 0$, since the left-hand side of (8.12) is a polynomial in w.

If we make the substitution $w = 1/u$ in equations (P_4) and (8.8), we have

$$u'' = \frac{3u'^2}{2u} - \frac{3}{2u} - 4z - 2(z^2 - \alpha)u - \beta u^3, \qquad (P_4')$$

$$\tilde{P}(u', u, z) = u'^n + (\tilde{p}_{10}u^2 + \tilde{p}_{11}u + \tilde{p}_{12})u'^{n-1} + \cdots = 0, \qquad (8.14)$$

where

$$\tilde{p}_{10} = -p_{12}, \quad \tilde{p}_{11} = -p_{11}, \quad \tilde{p}_{12} = -p_{10}. \qquad (8.15)$$

As for (P_4), we obtain the expansion (8.14) by means of roots of u putting

$$u' = b_2 u^2 + b_1 u + b_0 + \frac{b_{-1}}{u} + \cdots$$

in (P_4'), and find that $b_2 = \mu\sqrt{-2\beta}$, $\mu^2 = 1$, and b_1 is arbitrary.

Hence, if $\tilde{l}_+$ and $\tilde{l}_-$ denote the number of roots of (8.14) with $\mu = 1$ and $\mu = -1$, respectively, then, as in the case of (8.8), we have $(l_+ + l_- = n)$

$$\tilde{p}_{10} = (\tilde{l}_- - \tilde{l}_+)\sqrt{-2\beta}. \qquad (8.16)$$

Relations (8.13), (8.15), (8.16) yield the condition

$$(\tilde{l}_- - \tilde{l}_+)q = 2\alpha(l_- - l_+) - 2n. \qquad (8.17)$$

First of all, we see that $\tilde{l}_- - \tilde{l}_+$ and $l_- - l_+$ do not vanish simultaneously.

1. Suppose that $l_- - l_+ = 0$ and $\tilde{l}_- - \tilde{l}_+ \neq 0$. Then,

$$q = 2\frac{\tilde{l}_- + \tilde{l}_+}{\tilde{l}_+ - \tilde{l}_-}.$$

In G we have $0 < q \leq 2$. Hence, when there exist 1-solutions given by (8.8), one must have $l_- = l_+$, for which case $\tilde{l}_-, \tilde{l}_+ \in \mathbb{Z}_0$ and

$$0 < \frac{\tilde{l}_- + \tilde{l}_+}{\tilde{l}_+ - \tilde{l}_-} \leq 1.$$

Thus, $\tilde{l}_- = 0, \forall \tilde{l}_+ \in \mathbb{N}$. Then $q = 2$, and in this case G contains only one value (α, q), i.e., $(0, 2)$.

2. Suppose that $l_+ - l_- \neq 0$ and $\tilde{l}_+ - \tilde{l}_- = 0$. Then, as before, $\alpha = (l_+ + l_-)/(l_+ - l_-)$ and

$$0 \leq \frac{l_+ + l_-}{l_+ - l_-} \leq 1$$

in G; hence $l_- = 0, \forall l_+ \in \mathbb{N}$. Thus, $\alpha = 1$ and $\beta = 0$, but this is not possible in G.

3. Let $l_+ - l_- \neq 0$ and $\tilde{l}_+ - \tilde{l}_- \neq 0$. In this case it can easily be shown that the parameters for which 1-solutions given by (8.8) exist and these parameters satisfy the condition $\sqrt{-2\beta}+2\alpha-2 = 0$ if they are considered in G. Therefore, the following result holds.

Proposition 8.2. *Equation* (P_4) *has 1-solutions defined by* (8.8) *if and only if the parameters* α, β *satisfy* (8.7).

Proposition 8.3. *Equation* (P_4) *has no 1-solutions given by* (8.8) *and that are distinct from those formed by Riccati equation* (8.1) *under condition* (8.2).

This result can be proved by applying the following fact. If we have (8.1), then $q - 2\alpha = 2$ in G, and it follows from this that $S_{-j} = 0$, $j \geq 1$. Hence, the infinite series on the right-hand side of (8.9) does not contain the negative powers of w, and it determines the Riccati equation (8.8). This proves Proposition 8.3.

From the above considerations we have the following result.

Proposition 8.4. *The general solution of* (P_4) *is an essential function of two arbitrary constants of integration for generic values of the parameters* α, β.

9   Rational Solutions of (P_4)

> *General form of 0-solutions. The first and the second necessary conditions for 0-solutions. Criterion for 0-solutions. List of 0-solutions.*

Any solution of (P_4) is a meromorphic function. Hence, rational solutions of (P_4) exist if and only if the point $z = \infty$ is either a pole or a holomorphic point.

Let $zt = 1$. In variables (t, w), (P_4) has the form

$$w'' = \frac{w'^2}{2w} - \frac{2w'}{t} + \frac{3w^3}{2t^4} + \frac{4w^2}{t^5} + 2(1 - \alpha t^2)\frac{w}{t^6} + \frac{\beta}{t^4 w}. \tag{9.1}$$

Suppose that $t = 0$ is a pole of the solution $w(t)$. Then in a neighborhood of the pole the solution has the form

$$w(t) = \frac{a_{-1}}{t} + \sum_{j=0}^{\infty} a_j t^j, a_{-1} \neq 0, \tag{9.2}$$

where a_j can be found from the system

$$a_{-1}^2(3a_{-1}^2 + 8a_{-1} + 4) = 0,$$

$$a_{-1}a_1(3a_{-1}^2 + 6a_{-1} + 2) = \alpha a_{-1}^2,$$

$$4a_{-1}a_3(3a_{-1}^2 + 6a_{-1} + 2) = -a_{-1}^2 - 18a_{-1}^2 - 24a_{-1}a_1^2 - 4a_1^2 + 8\alpha a_{-1}a_1 - 2\beta,$$

$$\vdots$$

$$a_{-1}a_{2j+1}(3a_{-1}^2 + 8a_{-1} + 2) = P_{2j+1}(a_{-1}, a_1, \ldots, a_{2j-1}, \alpha, \beta),$$

where $a_{2j} = 0$.

Therefore, two cases are possible:

1. $a_{-1} = -2$, and in this case $a_1 = -\alpha$.

2. $a_{-1} = -\frac{2}{3}$ and $a_1 = \alpha$.

Suppose that $t = 0$ is a holomorphic point for the solution $w(z)$. Substituting $w = \sum_{j=0}^{\infty} b_j t^j$ in (9.1) and comparing the coefficients with equal degrees of t we obtain

$$\beta + 2b_1^2 = 0, \quad 2b_3 = \alpha b_1 - 2b_1^2, \quad b_{2j} = 0.$$

The poles of $w(z)$ are simple with residues 1 or -1, and the residue of $z = \infty$ is equal to α or $-\alpha$ in the first case and $\sqrt{-\beta/2}$ in the second case. This is why the following statement is true.

Proposition 9.1. *The rational solutions that have $z = \infty$ as a pole can exist only when $\alpha \in \mathbb{Z}$, and the 0-solutions that have $z = \infty$ as a holomorphic point can exist only when $\sqrt{-\beta/2} \in \mathbb{Z}$.*

This is the first necessary condition for the existence of 0-solutions. Of course, this condition is not sufficient.

From the above expansions of 0-solutions there follows the form of any 0-solution

$$w(z) = \lambda z + \xi(z)/\eta(z),$$

where $\lambda = 0$ or $\lambda = -2$, or $\lambda = -\frac{2}{3}$ and

$$\xi(z) = \sum_{j=0}^{m-1} \xi_j z^j, \quad \xi_{m-2} = \xi_{m-4} = \cdots = 0,$$

$$\eta(z) = \sum_{j=0}^{m} \eta_j z^j, \quad \eta_{m-1} = \eta_{m-3} = \cdots = 0.$$

If $\xi(z) = 0$, then the rational solutions of (P_4) are the following:

$$w = 0, \quad \beta = 0;$$
$$w = -2z, \quad \alpha = 0, \quad \beta = -2; \qquad (9.3)$$
$$w = -2z/3, \quad \alpha = 0, \quad \beta = -2/9.$$

We can now use the solutions (9.3) as initial solutions for the Bäcklund transformation. But we shall first find the number of solution poles, and this will be the second necessary condition for 0-solutions. Putting

$$v(z) = P(z)\exp(g(z)), \quad u(z) = Q(z)\exp(g(z)),$$

where $P(z)$, $Q(z)$ are polynomials of nth and mth degree, respectively, and $g(z)$ is an entire function, we obtain the system

$$QQ'' - Q'^2 + g''Q^2 + 2zPQ + P^2 = 0,$$
$$2PQ(P''Q - P'Q') + 2P^2Q^2 g'' \qquad (9.4)$$
$$= P'^2Q^2 - P^2Q'^2 + 4zP^3Q + P^4 + 4(z^2 - \alpha)P^2Q^2 + 2\beta Q^4.$$

Thus, the equation (P_4) has a 0-solution if and only if the system (9.4) has a polynomial solution (P, Q) under some choice of the polynomial $g(z)$. From (9.4) it follows that either (1) $n = m + 1$ if $\lambda = -2$ or $\lambda = -\frac{2}{3}$, or (2) $n = m - 1$ if $\lambda = 0$. In the first case $\deg g \leq 4$ if $n = m + 1$ and $\deg g \leq 2$ if $n = m - 1$.

Substituting $g(z) = \sum_{j=0}^{4} g_j z^j$ and

$$P(z) = p_0 z^{m+1} + p_2 z^{m-1} + \cdots, \quad Q(z) = z^m + q_2 z^{m-2} + \cdots$$

in the system (9.4), we find the following possibilities:

1. If $p_0 = 0$, i.e., $w = P_{m-1}/Q_m$, then $g_4 = g_3 = 0$, and

$$m = \left| \alpha\sqrt{-\beta/2} + \frac{\beta}{2} \right|. \qquad (9.5)$$

2. If $p = -2$, then $q_4 = g_3 = 0$, $g_2 \neq 0$, and

$$m = -\frac{\alpha^2}{2} - \frac{\beta}{4} - \frac{1}{2}. \qquad (9.6)$$

3. If $p_0 = -\frac{2}{3}$, then $g_4 \neq 0$ and

$$m = \frac{\alpha^2}{2} - 3\frac{\beta}{4} - \frac{1}{6}. \qquad (9.7)$$

Therefore, a second necessary condition for the existence of 0-solutions is $m \in \mathbb{N}$.

Now we can apply the Bäcklund transformations to the initial solutions (9.3). In cases (9.3a) and (9.3b) we have a set of 0-solutions with parameters

$$\alpha = n_1, \quad \beta = -2(1 + 2n_2 - n_1)^2, \quad n_1, n_2 \in \mathbb{Z}. \qquad (9.8)$$

TABLE 12.5. The first rational solutions of (P_4) of the form $w(z, \alpha, \beta)$ $= P_{m-1}(z)/Q_m(z)$.

α	β	$w(z, \alpha, \beta)$
± 2	-2	$\pm 1/z$
± 3	-8	$\pm 4z/(2z^2 \pm 1)$
± 4	-2	$\pm(2z^2 \pm 1)/(z(2z^2 \mp 1))$

In the case (9.3c) we have a set of 0-solutions with parameters

$$\alpha = n_1, \quad \beta = -\frac{2}{9}(6n_2 - 3n_1 + \varepsilon)^2, \quad \varepsilon^2 = 1, \quad n_1, n_2 \in \mathbb{Z}. \tag{9.9}$$

We note that these values of parameters can be obtained by substitution of the parameters of the initial solutions (9.3) into the formula for the general form of parameters of the Bäcklund transformation in (7.6), (7.7). If we consider the second necessary condition of 0-solutions for parameters (9.8) in the fundamental domain

$$G: \quad 0 \leq \operatorname{Re} \alpha < 1, \quad \operatorname{Re} \sqrt{-2\beta} > 0, \quad \operatorname{Re}(\sqrt{-2\beta} + 2\alpha) \leq 2,$$

it is easy to check that $m \in \mathbb{N}$ in G only for the initial solution (9.3). Hence, the following result is true.

Proposition 9.2. *Rational solutions of (P_4) exist if and only if parameter relations (9.8), (9.9) are true.*

The uniqueness of the constructed solutions follows from the uniqueness of the initial solutions (9.3). It must be noted that the initial solutions $w = 0$ with $\beta = 0$ and $w = -2z$ with $\alpha = 0, \beta = -2$ are solutions of the Riccati equation (8.1), which defines 1-solutions of (P_4). Therefore, these solutions generate the 0-solutions of the Weber–Hermite equation. They have the form

$$w = -\sqrt{\varepsilon}\,\mathrm{He}_n'(t)/\,\mathrm{He}_n(t), \alpha = -\varepsilon(1 + n), \beta = -2n^2, z = \sqrt{\varepsilon}t;$$
$$w = -2z + i\sqrt{\varepsilon}\,\mathrm{He}_n'(t)/\,\mathrm{He}_n(t), \alpha = n\varepsilon, \beta = -2(1 + n)^2, z = i\sqrt{\varepsilon}t, \varepsilon^2 = 1.$$

10   The Third Painlevé Equation

General properties. The first integrals. Special Riccati equation. T-transformations. General condition for rational and algebraic solutions.

TABLE 12.6. The first rational solutions of (P_4) of the form $w(z, \alpha, \beta)$ $= -2z + P(z)/Q(z)$.

α	β	$2z + w(z, \alpha, \beta)$
0	-2	0
0	-18	$-8z/((2z^2 + 1)(2z^2 - 1))$
± 1	-8	$\mp 1/z$
± 2	-18	$\mp 4z/(2z^2 \pm 1)$

TABLE 12.7. The first rational solutions of (P_4) of the form $w(z, \alpha, \beta) =$ $-2z/3 + P(z)/Q(z)$.

α	β	$2z/3 + w(z, \alpha, \beta)$
0	$-2/9$	0
± 1	$-32/9$	$\pm(2z^2 \pm 3)/z(2z^2 \pm 3)$
± 1	$-8/9$	$\pm 1/z$
± 2	$-2/9$	$\pm 4z/(2z^2 \pm 3)$

In this section we consider the third Painlevé equation

$$w'' = \frac{w'^2}{w} - \frac{w'}{z} + \frac{1}{z}(\alpha w^2 + \beta) + \gamma w^3 + \frac{\delta}{w}, \qquad (P_3)$$

where $\alpha, \beta, \gamma, \delta$ are arbitrary constants.

First of all we indicate two cases where (P_3) has first integrals.

Indeed, after some transformation and one integration, (P_3) can be written either in the form

$$(w'/w)^2 + G(w, x) = 2 \int [(\delta/w^2 - \gamma w^2)e^{2x} + (\beta/w - \alpha w)e^x]dx \quad (10.1)$$

or

$$w'/w = \int [(\delta/w^2 + \gamma w^2)e^{2x} + (\beta/w + \alpha w)e^x]dx, \qquad (10.2)$$

where $G(w, x) = (\delta/w^2 - \gamma w^2)e^{2x} + 2(\beta/w - \alpha w)e^x$, $z = e^x$. The sum and difference of (10.1) and twice equation (10.2) read

$$(w'/w)^2 - 2w'/w + G(w, x) = -4 \int (e^{2x}\gamma w^2 + e^x \alpha w)dx \qquad (10.3)$$

and

$$(w'/w)^2 + 2w'/w + G(w, x) = 4 \int [(e^{2x}\delta + e^x \beta w)/w^2]dx. \qquad (10.4)$$

One sees from (10.1) that if in equation (P_3) all parameters are zero, the general solution is $w = C_1 z^{C_2}$. If $\delta = \beta = 0$ in (10.4) or $\gamma = \alpha = 0$ in (10.3), we have two algebraic first integrals:

$$
\begin{aligned}
(w'/w)^2 + 2w'/w - 2\alpha w e^x - \gamma w^2 e^{2x} &= C_1, \\
(w'/w)^2 - 2w'/w + \delta e^{2x}/w^2 + 2\beta e^x/w &= C_2.
\end{aligned}
\tag{10.5}
$$

Of course, equations (10.5) can be integrated with elementary functions. Painlevé knew these first integrals. But what about the other algebraic first integrals? We shall prove the uniqueness of integrals (10.5).

Consider the Riccati equation

$$
w' = a(z)w^2 + b(z)w + c(z).
\tag{10.6}
$$

If all solutions of equation (10.6) are at the same time solutions of (P_3), then $\gamma \neq 0$ and

$$
a^2 = \gamma, \quad b = \frac{\alpha - a}{az}, \quad c^2 = -\delta, \quad a\beta + c(\alpha - 2a) = 0.
$$

In this case we have the equation

$$
w' = aw^2 + \frac{\alpha - a}{az}w + c,
\tag{10.7}
$$

which can be reduced by the linearizing transformation $w = -u'/(au)$, $z = \lambda\tau$, $\lambda^2 = (ac)^{-1}$, $c \neq 0$, to the equation for $u(\tau)$

$$
u'' - \frac{\alpha - a}{a\tau}u' + u = 0.
\tag{10.8}
$$

The general solution of (10.8) is

$$
u = \tau^\nu [C_1 J_\nu(\tau) + C_2 Y_\nu(\tau)],
\tag{10.9}
$$

where $\nu = \alpha/2a$ and $J_\nu(\tau)$, $Y_\nu(\tau)$ are Bessel functions. In particular cases the general solution of (10.8) can be an elementary function. One such case is $\alpha = a(1 + 2n)$, $n \in \mathbb{Z}_0$, for which

$$
u = \tau^{n+1/2}\left(C_1 \frac{d^n}{(\tau d\tau)^n}\frac{\cos\tau}{\tau} + C_2 \frac{d^n}{(\tau d\tau)^n}\frac{\sin\tau}{\tau} \right).
$$

Note that the equation (10.7) may have rational solutions. Below we shall find the condition for the existence of such solutions.

Now let us demonstrate one more property of (P_3). Let $\phi(z, \alpha, \beta, \gamma, \delta)$ be a solution of (P_3) and

$$
T: \quad w \to \sigma_1 w^{k_1}, \quad z \to \sigma_2 z^{k_2}, \quad \sigma_i \neq 0, \quad k_i \neq 0.
$$

Then the transformations

$$T_1(\sigma_1,\sigma_2) : \sigma_1 \phi(\sigma_2 z, \alpha\sigma_1^{-1}\sigma_2, \beta\sigma_1\sigma_2, \gamma\sigma_1^{-2}\sigma_2, \delta\sigma_1^2\sigma_2^2),$$
$$T_2 : \phi \to \phi^{-1}(z, -\beta, -\alpha, -\delta, -\gamma),$$
$$T_3 : \phi(z, \alpha, \beta, 0, 0) \to \phi^{1/2}(z^2, 0, 0, 2\alpha, 2\beta)$$

give solutions of (P_3).

Therefore, if we exclude the integrable cases (a) $\alpha = \beta = \gamma = \delta = 0$, (b) $\alpha = \gamma = 0$, (c) $\beta = \delta = 0$, there are only four cases to consider:

(1)	$\gamma\delta \neq 0$,	(2)	$\gamma = 0,\quad \alpha\delta \neq 0$,
(3)	$\delta = 0,\quad \beta\gamma \neq 0$,	(4)	$\delta = \gamma = 0,\quad \alpha\beta \neq 0$.

By means of transformations T_2 and T_3, cases (4) and (3) reduce to (1) and (2). Hence we are left with only two cases to consider: (1) $\gamma\delta \neq 0$ (without loss of generality $\gamma = 1$, $\delta = -1$); (2) $\gamma = 0$, $\alpha\delta \neq 0$ ($\gamma = 1$, $\delta = -1$).

Let us now investigate the point $z = \infty$. Putting $z = 1/t$ in (P_3) we obtain

$$w'' = \frac{w'^2}{w} - \frac{w'}{t} + \frac{1}{t^3}(\alpha w^2 + \beta) + \frac{\gamma}{t^4}w^3 + \frac{\delta}{t^4 w}. \tag{10.10}$$

Substituting the expansion

$$w(t) = t^{r/s}\sum_{j=0}^{\infty} a_j t^{j/s}, \quad a_0 \neq 0,$$

where r and $s > 0$ are integers, into (10.10), we obtain the following possible cases:

1. $r = 0$, $s = 1$, $\gamma\delta \neq 0$,

2. $r = -1$, $s = 3$, $\gamma = 0$, $\alpha\delta \neq 0$,

3. $r = 1$, $s = 3$, $\delta = 0$, $\beta\gamma \neq 0$,

provided that the integrable cases are excluded. Therefore, the following statement holds.

Proposition 10.1. *Under the above condition, equation (P_3) can have rational solutions for $\gamma\delta \neq 0$ and algebraic solutions that are rational solutions in $z^{1/3}$ for $\gamma = 0, \alpha\delta \neq 0$.*

11   Equation (P_3) for $\gamma = 0$, $\alpha\delta \neq 0$

Equivalent system. Bäcklund transformation. Fundamental domain. Algebraic solutions.

We consider equation (P_3) in the case $\gamma = 0$, $\alpha\delta \neq 0$ ($\alpha = 1$, $\delta = -1$), i.e.,

$$w'' = \frac{w'^2}{w} - \frac{w'}{z} + \frac{1}{z}(w^2 + \beta) - \frac{1}{w}. \tag{11.1}$$

We can rewrite (11.1) in the form of the equivalent system

$$\begin{aligned}
zw' &= \varepsilon z + (1 - \varepsilon\beta)w + zw^2v, \\
zv' &= 1 - (2 - \varepsilon\beta)v - zwv^2,
\end{aligned} \tag{11.2}$$

where $\varepsilon = 1$. From the first equation of (11.2) it follows that the equation

$$v'' = \frac{v'^2}{v} - \frac{v'}{z} - \varepsilon v^2 + \frac{2 - \varepsilon\beta}{z^2} - \frac{1}{z^2 v}$$

is of P-type. This equation determines the function v. Hence we can reduce it to the canonical form. Indeed, if we put $v = -w/z$, the function w satisfies equation (11.1) but with the parameter $\tilde\beta = \beta - 2\varepsilon$. Therefore, we have proved the following result.

Proposition 11.1. *Let $w = w(z, \beta)$ be a solution of (11.1). Then the function $\tilde w(z, \tilde\beta)$ defined by*

$$\tilde w(z, \tilde\beta) = w^{-2}[(\varepsilon - \beta)w + z - \varepsilon z w'] \tag{11.3}$$

is also a solution of (11.1) with the parameter

$$\tilde\beta = \beta - 2\varepsilon, \quad \varepsilon^2 = 1.$$

Corollary 11.1. *In order to construct the general solution of (11.1) for arbitrary values of the parameter β, it is sufficient to construct it for all β in the fundamental domain*

$$r \leq \operatorname{Re}\beta < r + 2, \quad \forall r \in R. \tag{11.4}$$

With the help of Proposition 11.1 let us construct the algebraic solutions of the equation (P_3).

Since the only movable singularities of equation (P_3) are poles, according to Proposition 10.1 algebraic solutions can exist only if either (a) $\gamma = 0, \alpha\delta \neq 0$, or (b) $\delta = 0, \beta\delta \neq 0$, if the integrable cases are excluded. But case (b) reduces to case (a) by a T_2-transformation. Hence, we can consider only the possibility $\gamma = 0, \alpha\delta \neq 0$ ($\alpha = 1, \delta = -1$). In this case the algebraic solutions are rational solutions in z. Let us start the construction by putting $z = \tau^3$. Then instead of (11.1) and (11.3) we have

$$w'' = \frac{w'^2}{w} - \frac{w'}{\tau} + 9\tau^2\left(w^2 + \beta - \frac{\tau^3}{w}\right), \tag{11.5}$$

$$\tilde w(z, \tilde\beta) = \left((\varepsilon - \beta)w + \tau^3 - \varepsilon\tau w'/3\right)/w^2. \tag{11.6}$$

Proposition 11.2. *For the existence of rational solutions of* (11.5), *it is necessary and sufficient that*

$$\beta = 2k, \quad k \in \mathbb{Z}. \tag{11.7}$$

Proof. It is easy to see that (11.5) has the solution

$$w = \lambda\tau, \quad \lambda^3 = 1, \quad \beta = 0. \tag{11.8}$$

Hence, using repeatedly the relation (11.6), we get

$$w_1 = \frac{2\varepsilon\lambda + 3\tau^2}{3\lambda^2\tau}, \quad \beta = -2\varepsilon,$$

$$w_2 = \frac{9\lambda\tau^5 + 24\varepsilon\lambda^2\tau^3 + 20\tau}{(2\varepsilon\lambda + 3\tau^2)^2}, \quad \beta = -4\varepsilon,$$

and similarly for all $\beta = 2k$, $k \in \mathbb{Z}$.

The necessity of condition (11.7) can be proved by using the system for the entire functions $u(z), v(z)$ of any solution of (P_3) when $z = e^x$. The same has been done for equations (P_2) and (P_4). In this case we have the following result. □

If $w = P(\tau)/Q(\tau)$ is a rational solution of (11.5), then $n = m - 1$, where m and n are the degrees of polynomials $P(\tau)$ and $Q(\tau)$, respectively, and

$$n = \beta^2/4. \tag{11.9}$$

Investigating the last relation in the fundamental domain (11.4) we obtain the sufficiency of the condition (11.7) and the following result.

Corollary 11.2. *Equation* (11.5) *with* $\beta = 2k$, $k \in \mathbb{Z}$, *has only three rational solutions. They are defined by* (11.9) *depending on the value* λ *from* (11.8).

According to Proposition 10.1 and the T_2-transformation, the rational solutions of (11.5) contain all the algebraic solutions of (P_3) if one excludes the integrable cases.

12   Equation (P_3) for $\gamma\delta \neq 0$

Relation between (P_3) and (P_5). Bäcklund transformation between (P_3) with $\gamma\delta \neq 0$ and (P_5) with $\delta = 0$, $\gamma \neq 0$. Weyl chamber.

Consider the equation (P_3) with $\gamma\delta \neq 0$ (without loss of generality $\gamma = 1$, $\delta = -1$),

$$w'' = \frac{w'^2}{w} - \frac{w'}{z} + \frac{1}{z}(\alpha w^2 + \beta) + w^3 - \frac{1}{w}. \tag{12.1}$$

We shall show that equation (12.1) can be reduced to the fifth Painlevé equation by some transformations.

Equation (12.1) is equivalent to the system

$$zw' = (\alpha\varepsilon - 1)w + \varepsilon zw^2 + zv,$$
$$zwv' = \beta w - z + (\alpha\varepsilon - 2)wv + zv^2. \tag{12.2}$$

Indeed, the elimination of v from the system (12.2) yields (12.1). Since it can be expressed rationally in terms of w' and w, the function v has no movable critical points. Let us find the equation for v.

Eliminating w from (12.2), we obtain

$$w'' = \frac{v}{v^2 - 1}v'^2 + \frac{v'}{z} + \Theta(z, v), \tag{12.3}$$

where

$$\Theta(z, v) = \varepsilon(1 - v^2) - \frac{1}{z}\beta(\alpha\varepsilon - 2) - \frac{2\beta(\alpha\varepsilon - 2)}{z^2(v^2 - 1)} - \frac{\beta^2 + (\alpha\varepsilon - 2)^2}{z^2}\frac{v}{v^2 - 1}.$$

The function $v/(v^2 - 1)$ has two poles $v = 1$ and $v = -1$. Putting

$$v = -\frac{u(z) + 1}{u(z) - 1}, \quad z = \sqrt{2\tau}, \quad u(z) \neq 1, \tag{12.4}$$

in (12.3), we get

$$u'' = \frac{3u - 1}{2u(u - 1)}u'^2 - \frac{u'}{\tau}$$
$$+ \left[(\beta - \alpha\varepsilon + 2)^2 - \frac{(\beta + \alpha\varepsilon - 2)^2}{u}\right]\frac{(u - 1)^2}{32\tau^2} - \frac{\varepsilon u}{\tau}, \tag{12.5}$$

where $\varepsilon^2 = 1$.

Equation (12.5) is the fifth Painlevé equation. From system (12.2) and the transformation (12.4) we thus obtain a link between the solutions of (12.1) and (12.5).

Theorem 12.1. *Let $w = w(z)$ be a solution of the third Painlevé equation (12.1) such that*

$$R(z) \equiv w' - \varepsilon w^2 - (\alpha\varepsilon - 1)\frac{w}{z} + 1 \neq 0. \tag{12.6}$$

Then the function

$$u(\tau) = 1 - 2R^{-1}(\sqrt{2\tau}) \tag{12.7}$$

is a solution of the fifth Painlevé equation with parameters

$$a = \frac{1}{32}(\beta - \alpha\varepsilon + 2)^2, \quad b = -\frac{1}{32}(\beta + \alpha\varepsilon - 2)^2, \quad c = -\varepsilon, \quad d = 0. \tag{12.8}$$

From the system (12.2) it follows that all solutions of the equation

$$w' = \varepsilon w^2 + (\alpha\varepsilon - 1)\frac{w}{z} - 1 \tag{12.9}$$

with

$$\alpha\varepsilon - \beta - 2 = 0 \tag{12.10}$$

and all solutions of the equation

$$w' = \varepsilon w^2 + (\alpha\varepsilon - 1)\frac{w}{z} + 1 \tag{12.11}$$

with

$$\alpha\varepsilon + \beta - 2 = 0 \tag{12.12}$$

define solutions of (12.1). Now let us find the solution w in terms of the solution u.

Theorem 12.2. *Let* $u = u(\tau)$ *be a solution of the fifth Painlevé equation for some parameter values* $a, b, c^2 = 1, d = 0$ *such that*

$$\Phi \equiv \tau\frac{du}{d\tau} - \sqrt{2a}u^2 + (\sqrt{2a} + \sqrt{-2b})u - \sqrt{-2b} \neq 0. \tag{12.13}$$

Then the function

$$w(z) = \begin{cases} \sqrt{2\tau}u/\Phi(u), \\ 2\tau = z^2 \end{cases}$$

is a solution of the third Painlevé equation (12.1) *with*

$$\alpha = 2c(\sqrt{2a} - \sqrt{-2b} - 1), \quad \beta = 2\sqrt{2a} + 2\sqrt{-2b},$$
$$\gamma = 1, \quad \delta = -1. \tag{12.14}$$

From theorems 12.1 and 12.2, one solution w defines two solutions of (P_5) and one solution u defines four solutions of (P_3), depending on the choice of the branches of $\sqrt{2a}$, $2\sqrt{-2b}$ and the choice of ε in (12.6), (12.7).

By virtue of this branching we can obtain a link between the solutions with different values of parameters.

Theorem 12.3. *Let* $w = w(z)$ *be a solution of* (12.1) *for some values of* α *and* β; *then the function*

$$\tilde{w}(z, \tilde{\alpha}, \tilde{\beta}) = \frac{2zR(R - 2)}{2z dR/dz + (\varepsilon_3 A - \varepsilon_2 B)R - 2\varepsilon_3 A} \tag{12.15}$$

is a solution of (12.1) *with*

$$\tilde{\alpha} = \varepsilon_1(\varepsilon_2 B - \varepsilon_3 A + 4)/2, \quad \tilde{\beta} = \varepsilon_2 B/2 + \varepsilon_3 A/2,$$
$$B = \beta + \alpha\varepsilon - 2, \quad A = \beta - \alpha\varepsilon + 2, \quad \varepsilon_j^2 = 1, \quad j \in \{1, 2, 3\}.$$

Theorem 12.4. *If $u(\tau)$ is a solution of the fifth Painlevé equation for the parameter values a, b, $c^2 = 1$ and $d = 0$, then the function*

$$\tilde{u}(\tau) = 1 - 2M^2/\Phi,$$

where

$$M \equiv \tau u' - \sqrt{2a}u^2 + (\sqrt{2a} + \sqrt{-2b})u - \sqrt{-2b},$$
$$\Phi \equiv 2\tau u M'_\tau + M^2 - [2\tau u' - 2(\sqrt{2a} - \sqrt{-2b} - 2)u]M + 2\tau cu^2 \not\equiv 0,$$

is a solution of the same equation for the parameters

$$\tilde{a} = (\sqrt{2a} - 1)^2/2, \quad \tilde{b} = -(\sqrt{-2b} + 1)^2/2, \quad \tilde{c} = c.$$

13   Rational and Classical Solutions of (P_3) for $\gamma\delta \neq 0$

Rational and 1-solutions; their criterion. Fundamental domain. List of 0- and 1-solutions.

Let $w = P(z)/Q(z)$ be a rational solution of equation (12.1), where $P(z)$, $Q(z)$ are irreducible polynomials of degrees n and m, respectively. Replacing w into (12.1), we get $n = m$. In the case $n = 0$ equation (12.1) has the solution

$$w = \lambda, \quad \alpha\lambda^2 + \beta = 0, \quad \lambda^4 = 1. \tag{13.1}$$

Solution (13.1) is the unique rational solution when $\alpha\lambda^2 + \beta = 0$, $\lambda^4 = 1$, and this follows from the expansion of the solution in a neighborhood of the point $z = \infty$. If we take the solution (13.1) as the the initial solution for the Bäcklund transformation, we have four types of rational solutions depending on the choice of λ. After the first step of the B-transformation we get

$$w_1 = \frac{z + a}{\lambda z + a\lambda - \varepsilon}, \quad \alpha_1 = \lambda^2(2a\lambda - 3\varepsilon), \quad \beta_1 = -2a\lambda - \varepsilon, \quad \lambda^4 = 1.$$

Note that $\alpha_1 + \beta_1 = -4\varepsilon$ in the case $\lambda^2 = 1$, and $\alpha_1 - \beta_1 = 4\varepsilon$ in the case $\lambda^2 = -1$. This property of the rational solutions parameters is also valid in the general case.

Theorem 13.1. *Equation (12.1) has rational solutions if and only if*

$$\alpha + \beta\varepsilon = 4n, \quad n \in \mathbb{Z}. \tag{13.2}$$

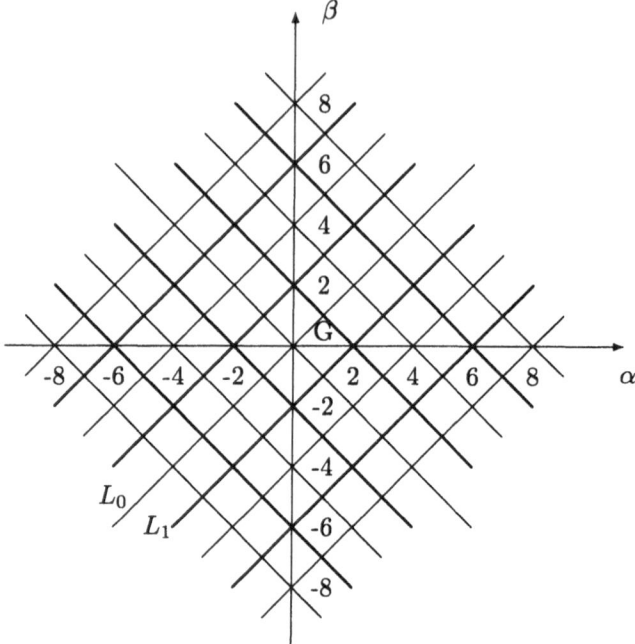

FIGURE 12.1. Parameter space (α, β) of the Painlevé third equation (12.1).

The sufficiency of this theorem can be proved by induction, and the necessity can be proved by examining the point $z = \infty$, as has already been done for equations (P_2) and (P_4). Moreover, we can find the number of poles of any rational solution

$$m = (\alpha + \beta\lambda^2)^2/16, \quad \lambda^4 = 1.$$

If we take solutions of the Riccati equation as initial solutions of the Bäcklund transformation, we get 1-solutions of equation (12.1) for parameters

$$\alpha + \beta\varepsilon = 4n + 2, \quad \varepsilon^2 = 1, \quad n \in \mathbb{Z}. \tag{13.3}$$

These solutions can be expressed with the Bessel function and its derivative. The fundamental domain of equation (12.1) is defined by the relation

$$G : \quad \operatorname{Re}\beta \geq 0, \quad \operatorname{Re}(\alpha - \beta) \geq 0, \quad \operatorname{Re}(\alpha + \beta) < 2. \tag{13.4}$$

For real values of α, β, the parameter space of equation (12.1) is pictured in Figure 12.1.

Here G is, the fundamental domain, L_0 are the lines of existence of rational solutions, and L_1 are the lines of existence of one-parameter solutions.

14   The Fifth Painlevé Equation

The first integral. Riccati equation. Simple symmetries. Bäcklund transformation in the case $\delta \neq 0$. Rational and one-parameter families of solutions. Some supplementary results on the link between (P_3) and (P_5).

The fifth Painlevé equation has the following form:

$$ w'' = \frac{3w-1}{2w(w-1)}w'^2 - \frac{w'}{z} + \frac{(w-1)^2}{z^2}\left(\alpha w + \frac{\beta}{w}\right) + \frac{\gamma}{z}w + \frac{\delta w(w+1)}{w-1}, \qquad (P_5) $$

where α, β, γ, δ are constants.

In the general case (P_5) is not integrable with known classical transcendental functions. But if $\gamma = \delta = 0$, then (P_5) has the first integral

$$ w'^2 = (w-1)^2(2\alpha w^2 + Cw - 2\beta), \quad z = e^t. \qquad (14.1) $$

Equation (14.1) belongs to the class of first-order equations whose general solution has single-valued movable singularities.

Let us note that (P_5) is invariant under the transformations

$$ \begin{aligned} T_1 &: \ w(z, \alpha, \beta, \gamma, \delta) \mapsto w(-z, \alpha, \ \beta, -\gamma, \delta), \\ T_2 &: \ w(z, \alpha, \beta, \gamma, \delta) \mapsto w^{-1}(z, -\beta, -\alpha, -\gamma, \delta). \end{aligned} \qquad (14.2) $$

Equation (P_5) has 1-solutions defined by the Riccati equation

$$ zw' = \sqrt{2\alpha}\,w^2 + b(z)w + \sqrt{-2\beta}, \qquad (14.3) $$

where $b(z) = \sqrt{-2\delta}\,z - \sqrt{2\alpha} - \sqrt{-2\beta}$ and the following condition is true:

$$ \gamma = \sqrt{-2\delta}(1 + \sqrt{-2\beta} - \sqrt{2\alpha}). \qquad (14.4) $$

This equation can be reduced to Whittaker's equation

$$ 4x^2 y'' = (x^2 - 4kx + 4m^2 - 1)y. $$

As shown before, in the case $\delta = 0, \gamma \neq 0$, equation (P_5) is reducible to equation (P_3) with $\gamma\delta \neq 0$.

Let us now consider (P_5) with $\delta \neq 0$. In this case it can be rewritten as the equivalent system

$$ \begin{aligned} \tau \frac{du}{d\tau} &= -a - (a+c)u - uv - u^2 v, \\ \tau \frac{dv}{d\tau} &= \delta\tau^2 - \gamma\tau + (a+c)v + 2\delta\tau^2 u + v^2/2 + uv^2, \end{aligned} \qquad (14.5) $$

where $a^2 = -2\beta$, $c^2 = 2\alpha$, $u = w/(1-w)$, $\tau = z$, and $w(z)$ is a solution of (P_5).

Since $v(z)$ is expressed rationally in terms of w, w', the equation that defines $v(\tau)$ has P-type

$$v'' = \frac{v}{v^2 + 2\delta\tau^2}v'^2 - \frac{v'}{\tau}\frac{v^2 - 2\delta\tau^2}{v^2 + 2\delta\tau^2} + \frac{1}{\tau}\Theta(v, \tau), \qquad (14.6)$$

where $\Theta(v, \tau)$ is some rational function of v, τ. The function $v^2 + 2\delta$ has two poles $v = k\tau, k^2 = -2\delta$, since $\delta \neq 0$. Putting

$$y(\tau) = \frac{v + k\tau}{v - k\tau}, \quad v \neq k\tau, \qquad (14.7)$$

we obtain that $y(\tau)$ is a solution of (P_5) with parameters

$$\tilde{\alpha} = -\frac{(\gamma + k(1 - a - c))^2}{16\delta}, \quad \tilde{\beta} = \frac{(\gamma - k(1 - a - c))^2}{16\delta},$$

$$\tilde{\gamma} = k(a - c), \quad \tilde{\delta} = \delta. \quad (14.8)$$

According to the system (14.5) and the transformation (14.7) we have the following result.

Theorem 14.1. *Let $w = w(z)$ be a solution of (P_5) for some fixed values of parameters α, β, γ, δ, such that*

$$\Phi_1(w) \equiv zw' - cw^2 + (c - a + kz)w + a \neq 0. \qquad (14.9)$$

Then the function

$$w_1 = 1 - 2kz/\Phi_1 \qquad (14.10)$$

is a solution of (P_5) with parameters (14.8).

Theorem 14.1 defines the Bäcklund transformation of (P_5) for $\delta \neq 0$. When applied to the construction of 1-solutions and 0-solutions, it gives the following result.

Theorem 14.2. *Equation (P_5) has 1-solutions that can be expressed in terms of Whittaker's function and its derivatives with parameters constrained either by*

$$\gamma = 2n - 1 + \sqrt{2\alpha} + \sqrt{-2\beta}, \quad n \in \mathbb{Z},$$

or by

$$(2\alpha - n^2)(2\beta + n^2) = 0, \quad n \in \mathbb{N}.$$

Equation (P_5) can have rational solutions. They have the structure

$$w(z) = \lambda z + \mu + \frac{P_{n-1}(z)}{Q_n(z)},$$

where λ, μ are constants, $P_{n-1}(z)$, $Q_n(z)$ polynomials. If $P_{n-1} = 0$, then (P_5) has the following solutions:

$$w = -1, \quad \gamma = 0, \quad \alpha + \beta = 0;$$
$$w = k^2 z/\gamma + 1, \quad \gamma \neq 0, \ \beta = -\tfrac{1}{2}, \quad 4\alpha\delta + \gamma^2 = 0, \quad k^2 = -2\delta; \quad (14.11)$$
$$w = kz + \sqrt{-2\beta}, \ \alpha = \tfrac{1}{2}, \ \sqrt{-2\beta} \neq 1, \ \gamma = 2k - k\sqrt{-2\beta}.$$

Applying the Bäcklund transformation to the solutions (14.11), we obtain the following result:

Theorem 14.3. *For the existence of rational solutions of equation (P_5) of the form*

$$w = \frac{P_{n+1}(z)}{Q_n(z)}, \quad w = \frac{P_n(z)}{Q_{n+1}(z)},$$

it is necessary that either

$$2\alpha = -l^2, \quad \gamma + \sqrt{-2\beta} = m, \quad \delta = -\tfrac{1}{2}, \quad l + m = 2r + 1$$

or

$$2\beta = -l^2, \quad \gamma + \sqrt{2\alpha} = m, \quad \delta = -\tfrac{1}{2}, \quad l + m = 2r + 1,$$

where $l, m, r \in \mathbb{Z}$.

If $|\gamma| + |\beta| \neq 0, |\gamma| + |\alpha| \neq 0$, then these conditions are sufficient. For the existence of rational solutions of the form

$$w = \frac{P_n(z)}{Q_n(z)}$$

it is necessary and sufficient that either

$$\sqrt{2\alpha} + \sqrt{-2\beta} = l, \quad \gamma = m, \quad \delta = -\tfrac{1}{2}, \quad l + m = 2r$$

or

$$8\alpha = (2l + 1)^2, \quad 8\beta = -(2m + 1)^2, \quad l, m, r \in \mathbb{Z}.$$

Let us now obtain some results on the link between equations (P_3) and (P_5) with different values of the parameters.

Theorem 14.4. *Let $u(\tau)$ be a solution of (P_5) with $a = b = 0$, and let*

$$u = [(w + 1)/(w - 1)]^2.$$

Then the function $w(\tau)$ satisfies (P_3) with $\alpha = -\beta = -c/4$, $\gamma = -\delta = -d/8$.

Theorem 14.5. *Let $u(z)$ be a solution of (P_5) for $a = b = d = 0$. Then $\tilde{u}(\tau)$ is a solution of (P_5) for $\tilde{a} = \tilde{b} = \tilde{c} = 0, \tilde{d} = 2c$, with*

$$u = \frac{(\tilde{u}+1)^2}{4\tilde{u}}, \quad \tau = \tilde{\tau}^2/2.$$

This result defines a relation between solutions of (P_5) with $d = 0$ and $d \neq 0$. It can be extended.

Theorem 14.6. *Let $u(z) \neq 0$ be a solution of (P_5) with parameters*

$$a = p^2/8, \quad b = -p^2/8, \quad c = 0, \quad d \neq 0.$$

Then the function

$$\tilde{u}(\tau) = \frac{2f^2(\sqrt{\tau})}{d + 2f^2(\sqrt{\tau})}, \quad f(z) = \frac{u'(z)}{2u(z)} - p\left(u(z) - \frac{1}{u(z)}\right)\bigg/(4z)$$

is a solution of (P_5) with parameters

$$\tilde{a} = (1+p)^2/2, \quad \tilde{b} = 0, \quad \tilde{c} = d/4, \quad \tilde{d} = 0.$$

15   The Sixth Painlevé Equation

Group of discrete transformations. Integrable case. Equivalent system. Bäcklund transformation. Extension of integrability.

The sixth Painlevé equation has the form

$$
\begin{aligned}
w'' = {} & \frac{1}{2}\left(\frac{1}{w} + \frac{1}{w-1} + \frac{1}{w-z}\right)w'^2 - \left(\frac{1}{z} + \frac{1}{z-1} + \frac{1}{w-z}\right)w' \\
& + \frac{w(w-1)(w-z)}{z^2(z-1)^2}\left(\alpha + \frac{\beta z}{w^2} + \frac{\gamma(z-1)}{(w-1)^2} + \frac{\delta z(z-1)}{(w-z)^2}\right), \quad (P_6)
\end{aligned}
$$

where α, β, γ, δ are parameters.

Picard (1889) proved that if $\alpha = \beta = \gamma = 0$, $\delta = \frac{1}{2}$, the general solution has the form

$$w = \Lambda(C_1\omega_1 + C_2\omega_2, z),$$

where $\Lambda(u, z)$ is a function defined by the elliptic integral

$$u = \int_0^\Lambda \frac{dw}{\sqrt{w(w-1)(w-z)}}$$

with periods $2\omega_1$, $2\omega_2$ functions of z. Equation (P_6) admits a group of discrete transformations, generated by the T_j-transformations

$$t_j : w \mapsto w_j, \quad j \in \{1, 2, 3\},$$

where

$$w_1(z, -\beta, -\alpha, \gamma, \delta) = w^{-1}(1/z);$$
$$w_2(z, -\beta, -\gamma, \alpha, \delta) = 1 - w^{-1}\big(1/(1-z)\big);$$
$$w_3(z, -\beta, -\alpha, -\delta + \tfrac{1}{2}, -\gamma + \tfrac{1}{2}) = z/w(z).$$

The order of this group is 24.

Equation (P_6) has the equivalent system

$$
\begin{aligned}
z(z-1)w' &= \lambda z + \big[(r-\lambda)z - (1+\lambda+q)\big]w + C_2 w^2 \\
&\quad + C_1 w(w-1)(w-z)v, \\
z(z-1)v' &= \frac{2\alpha - C_2^2}{2C_1} - \big[(r-\lambda)z - (1+\lambda+q)\big]v \\
&\quad - 2C_2 wv - C_1(3w^2 - 2zv - 2w + z)v^2/2,
\end{aligned}
\tag{15.1}
$$

where

$$\lambda^2 = -2\beta, \quad r^2 = 2\gamma, \quad q^2 = 1 - 2\delta, \quad C_2 = 1 + \lambda - r + q.$$

From this system it follows that (P_6) has 1-solutions defined by the Riccati equation

$$z(z-1)w' = \lambda z + \big[(r-\lambda)z - (1+\lambda+q)\big]w + C_2 W^2, \quad 2\alpha = C_2^2.$$

The last equation is integrated by the hypergeometric function. Note that the system (15.1) is Hamiltonian. It can be used for the construction of a Bäcklund transformation. There are several forms of B-transformations. Using the idea of A. Fokas and Y. Yortsos [6], who were the first to obtain the B-transformation for (P_6), we restrict to the following form

Theorem 15.1 ([1]). *Let $w(z)$ be a solution of (P_6) such that*

$$\Phi(w) \equiv 2f' + I/z + \{kf(z+1)\}/z(z-1)\} \not\equiv 0,$$

where

$$f = z\frac{w'}{w} + \frac{\lambda - k - 1}{2(z-1)}w + \frac{\lambda + k + 1}{2w(z-1)} - \frac{\lambda(z+1)}{2(z-1)} - \left(\frac{1}{2} + \frac{\mu}{4}\right),$$
$$I = f^2 + f\mu/2 + \nu, \quad k = \beta_1 - \alpha_1 - 1 \neq 0, \quad \lambda = \alpha_1 + \beta_1, \quad \alpha_1^2 = 2\alpha,$$
$$\beta_1^2 = -2\beta, \quad \mu = -4(\gamma + \beta - 1/2)/k, \quad \nu = 2\delta - 1 + (\mu/4 + k/2)^2.$$

Then the function $\tilde{w}(z)$ defined by

$$\tilde{w} = w + \big(2(z+1)f' - 4wf'\big)/\Phi(w)$$

is a solution of (P_6) for parameters

$$\tilde{\alpha} = (\sqrt{2\alpha} + 1)^2/2, \quad \tilde{\beta} = -(\sqrt{-2\beta} - 1)^2/2, \quad \tilde{\gamma} = \gamma, \quad \tilde{\delta} = \delta.$$

This result can now be applied to construct 0- and 1-solutions and to obtain new cases of integrability of (P_6) by means of elliptic functions. For instance, from the B-transformation and the integrability of (P_6) for $\alpha = \beta = \gamma = 0$, $\delta = \frac{1}{2}$, we have the following result.

Theorem 15.2. *Equation (P_6) is integrated by elliptic functions for the following parameters:*

$$\alpha = (m_1 + 1)^2/2, \quad \beta = -(2m_2 - m_1)^2/2, \quad \gamma = (2n_2 - n_1)^2/2,$$
$$\delta = (1 - n_1^2)/2, \; n_j, m_j \in \mathbb{Z}.$$

16   References

[1] V.I. Gromak and N.A. Lukashevich, *Analytical properties of solutions of the Painlevé equations* (Minsk, Universitetskoe, 1990).

[2] K. Okamoto, *Studies of the Painlevé equations.* Math. Ann. **275** (1986), 221–255.

[3] H. Airault, *Rational solutions of Painlevé equations.* Studies in Appl. Math. **61** (1979), 31–53.

[4] M.D. Kruskal and P.A. Clarkson, *The Painlevé–Kowalevski and poly–Painlevé test for integrability.* Studies in Appl. Math. **86** (1992), 87–165.

[5] A.P. Bassom, P.A. Clarkson, and A.C. Hicks, *Bäcklund transformations and solution hierarchies for the fourth Painlevé equation,* Studies in Appl. Math. **95** (1995), 1–71.

[6] A.S. Fokas and Y.C. Yortsos, *The transformation properties of the sixth Painlevé equation and one-parameter families of solutions,* Lett. Nuovo Cimento **30** (1981), 539–544.

13

The Hamiltonians Associated to the Painlevé Equations

Kazuo Okamoto

1 Introduction

The present article is based on a series of lectures by the author on the occasion of the summer school held at Cargèse in June 1996. We will study the Hamiltonian structure associated with the Painlevé equations and show certain new results on the generalization of the second Painlevé equation to the case of two independent variables.

In what follows, we refer to each of the six Painlevé equations as P_J ($J =$ I, II, ... , VI). The *Hamiltonian structure* associated with P_J is the quartet: $\mathcal{H}_J = (q, p, H, t)$, where H is the Hamiltonian function associated with P_J. We denote by H_J the Hamiltonian of P_J; H_J is a polynomial in (q, p, t). Then the equation P_J can be written in the form

$$Dq = \frac{\partial H}{\partial p}, \quad Dp = -\frac{\partial H}{\partial q}, \quad H = H_J.$$

Here we put, for $J =$ I, II, IV,

$$D = \frac{\mathrm{d}}{\mathrm{d}t},$$

for $J =$ III, V,

$$D = t\frac{\mathrm{d}}{\mathrm{d}t},$$

and, for $J =$ VI,

$$D = t(t-1)\frac{\mathrm{d}}{\mathrm{d}t}.$$

The Hamiltonian system associated with P_J is called the *Painlevé system*.

We will determine the explicit form of the Hamiltonian function H_J by using the following three methods.

- the theory of the *holonomic deformation* of linear ordinary differential equations of the second order;

- the *step-by-step degeneration* starting from the sixth Painlevé equation, P_{VI};

- the *singularity analysis* around a pole of a solution of P_J.

The main subject of the present article is the Hamiltonian structure associated with the second Painlevé equation. We will deal with the space of initial conditions of the second Painlevé equations. By using the Hamiltonian structure $\mathcal{H}_J$, we can construct the space of initial conditions of each P_J, which gives a geometrical interpretation of the Painlevé equations. A solution of each Painlevé equation is called a *Painlevé transcendental function*, which is a *new* function except for classical or rational solutions. We will discuss the concept of *classical functions* established recently by H. Umemura and the *irreducibility* of Painlevé transcendental functions for the first and the second Painlevé equations.

This note consisits of two parts; Sections 2 to 4 are devoted to an expository survey of known results on the Hamiltonian structure of each P_J. Moreover, in Sections 5 and 6, by using these known facts, we will study the τ-functions related to the second Painlevé system and its generalization to the case of two variables. The results given in these two sections were partly announced in the lectures.

In Section 2 we will study the Hamiltonian structure for the second Painlevé equation P_{II}, by using the method of Painlevé analysis. The space of initial conditions of P_{II} is the main subject of Section 3. We will construct this space by means of blowing-ups Hirzebruch surfaces. Section 4 concerns the irreducibility of Painlevé transcendental functions.

In Section 5 we will investigate in detail the τ-functions of P_{II}; in particular, we will obtain algebraic relations among several adjacent τ-functions. We will deal with a generalization of the second Painlevé system to the case of two variables in Section 6. We will introduce the concept of the A_2 system and give explicit forms of birational canonical transformations of the A_2 system. The results obtained in Section 6 are an extension of those of Section 5 to the case of the A_2 system.

2 Hamiltonians and Painlevé Analysis

2.1 The Painlevé Equations

We give a table of the Painlevé equations

$$P_I \qquad \frac{\mathrm{d}^2 q}{\mathrm{d}t^2} = 6q^2 + t;$$

$$P_{II} \qquad \frac{\mathrm{d}^2 q}{\mathrm{d}t^2} = 2q^3 + tq + \alpha;$$

TABLE 13.1.

P_J	1	2	3	4	5	6
D	$\dfrac{\mathrm{d}}{\mathrm{d}t}$	$\dfrac{\mathrm{d}}{\mathrm{d}t}$	$t\dfrac{\mathrm{d}}{\mathrm{d}t}$	$\dfrac{\mathrm{d}}{\mathrm{d}t}$	$t\dfrac{\mathrm{d}}{\mathrm{d}t}$	$t(t-1)\dfrac{\mathrm{d}}{\mathrm{d}t}$
number of parameters	0	1	2	2	3	4

$$P'_{III} \qquad \frac{\mathrm{d}^2 q}{\mathrm{d}t^2} = \frac{1}{q}\left(\frac{\mathrm{d}q}{\mathrm{d}t}\right)^2 - \frac{1}{t}\frac{\mathrm{d}q}{\mathrm{d}t} + \frac{q^2}{4t^2}(\gamma q + \alpha) + \frac{\beta}{4t} + \frac{\delta}{4q};$$

$$P_{IV} \qquad \frac{\mathrm{d}^2 q}{\mathrm{d}t^2} = \frac{1}{2q}\left(\frac{\mathrm{d}q}{\mathrm{d}t}\right)^2 + \frac{3}{2}q^3 + 4tq^2 + 2\left(t^2 - \alpha\right)q + \frac{\beta}{q};$$

$$P_V \qquad \frac{\mathrm{d}^2 q}{\mathrm{d}t^2} = \left(\frac{1}{2q} + \frac{1}{q-1}\right)\left(\frac{\mathrm{d}q}{\mathrm{d}t}\right)^2 - \frac{1}{t}\frac{\mathrm{d}q}{\mathrm{d}t} + \frac{(q-1)^2}{t^2}\left(\alpha q + \frac{\beta}{q}\right)$$
$$+ \gamma\frac{q}{t} + \delta\frac{q(q+1)}{q-1};$$

$$P_{VI} \qquad \frac{\mathrm{d}^2 q}{\mathrm{d}t^2} = \frac{1}{2}\left(\frac{1}{q} + \frac{1}{q-1} + \frac{1}{q-t}\right)\left(\frac{\mathrm{d}q}{\mathrm{d}t}\right)^2 - \left(\frac{1}{t} + \frac{1}{t-1} + \frac{1}{q-t}\right)\frac{\mathrm{d}q}{\mathrm{d}t}$$
$$+ \frac{q(q-1)(q-t)}{t^2(t-1)^2}\left\{\alpha + \beta\frac{t}{q^2} + \gamma\frac{t-1}{(q-1)^2} + \delta\frac{t(t-1)}{(q-t)^2}\right\}.$$

Here we assume that $\gamma\delta \neq 0$ for P'_{III} and that $\delta \neq 0$ for P_V; hence we put in what follows $\gamma = 4$, $\delta = -4$ for P'_{III}, and $\delta = -\frac{1}{2}$ for P_V.

By replacing t by t^2 and q by tq in P'_{III}, we obtain the equation

$$P_{III} \qquad \frac{\mathrm{d}^2 q}{\mathrm{d}t^2} = \frac{1}{q}\left(\frac{\mathrm{d}q}{\mathrm{d}t}\right)^2 - \frac{1}{t}\frac{\mathrm{d}q}{\mathrm{d}t} + \frac{1}{t}\left(\alpha q^2 + \beta\right) + \gamma q^3 + \frac{\delta}{q},$$

which is usually called the *third* Painlevé equation. We give in Table 13.1 the differential D associated with each of the Painlevé equations and the number of parameters contained in P_J.

2.2 Painlevé Analysis and P_I

Given a polynomial function $g = g(t)$, consider the differential equation of second order

$$\frac{\mathrm{d}^2 q}{\mathrm{d}t^2} = 6q^2 + g(t). \tag{2.1}$$

By use of the method of Painlevé analysis, we can prove that if the equation (2.1) is free from movable critical points, then $g(t)$ is linear in t, that is, $g(t) = at + b$, a and b being constants.

We show firstly this fact by considering the Hamiltonian function

$$H(t; q, p) = \frac{1}{2}p^2 - 2q^3 - g(t)q \tag{2.2}$$

of (2.1), which is equivalent to the Hamiltonian system

$$\frac{dq}{dt} = \frac{\partial H}{\partial p}, \quad \frac{dp}{dt} = -\frac{\partial H}{\partial q} \tag{2.3}$$

with the Hamiltonian $H = H(t; q, p)$ given by (2.2). Since

$$\frac{dH}{dt} = \frac{dq}{dt}\frac{\partial H}{\partial q} + \frac{dp}{dt}\frac{\partial H}{\partial p} + \frac{\partial H}{\partial t} = \frac{\partial H}{\partial t}$$

along a solution of (2.3), we have

$$\frac{dH}{dt} = -g'(t)q, \quad \frac{d^2H}{dt^2} = -g''(t)q - g'(t)p, \tag{2.4}$$

where $' = d/dt$. Under the assumption that $g'(t) \neq 0$, we can verify easily the following proposition.

Proposition 2.1. *The Hamiltonian function satisfies the nonlinear differential equation*

$$\left[\frac{d^2H}{dt^2} - \frac{g''}{g'}\frac{dH}{dt}\right]^2 + \frac{4}{g'}\left(\frac{dH}{dt}\right)^3 - 2g'\left(g'H - g\frac{dH}{dt}\right) = 0. \tag{2.5}$$

A general solution of (2.1) is obtained from that of (2.5) by (2.4), while the latter equation has the singular solution

$$H = \lambda g(t) + \mu,$$

where λ and μ are constants such that $4\lambda^3 - 2\mu = 0$.

Now we suppose that in a neighborhood of a point $t_0 \in \mathbb{C}$, the polynomial $g'(t)$ is of the form

$$g'(t) \approx a(t - t_0).$$

Then it follows from (2.5) that

$$\frac{dH}{dt} \approx -\frac{a}{t - t_0};$$

hence

$$H \approx -a\log(t - t_0).$$

Therefore, if (2.1) is free from movable critical points, then $g'(t) \neq 0$, that is,

$$g(t) = at + b.$$

And in this case, by a suitable change of variables, (2.1) is reduced to the equation

$$P_I \qquad\qquad \frac{d^2 q}{dt^2} = 6q^2 + t,$$

or to the equation of elliptic functions

$$\frac{d^2 q}{dt^2} = 6q^2 - \frac{1}{2} g_2.$$

2.3 The Hamiltonian and the τ-Function of P_I

The Hamiltonian function of P_I is

$$H(t; q, p) = \frac{1}{2} p^2 - 2q^3 - tq;$$

the function $H = H(t; q, p)$ satisfies the equation

$$\left(\frac{d^2 H}{dt^2} \right)^2 + 4 \left(\frac{dH}{dt} \right)^3 - 2 \left(H - t\frac{dH}{dt} \right) = 0. \qquad (2.6)$$

This equation has a singular solution of the form

$$H = \lambda t + \mu, \qquad 4\lambda^3 - 2\mu = 0,$$

and the general solution of (2.6) gives a solution of P_I by

$$\frac{dH}{dt} = -q.$$

H is a meromorphic funcrtion on $\mathbb{C}$. On the other hand, if we define the τ-function, $\tau = \tau(t)$, by

$$H = \frac{d}{dt} \log \tau, \qquad (2.7)$$

up to a multiplicative constant, then the function τ is entire on $\mathbb{C}$. We have again the well-known formula

$$q = -\frac{d^2}{dt^2} \log \tau$$

for the general solution of P_I.

2.4   The Hamiltonian of P_{II}

We next study the second Painlevé equation

$$P_{II} \qquad\qquad \frac{d^2q}{dt^2} = 2q^3 + tq + \alpha,$$

whose Hamiltonian $H(t; q, p)$ considered in the following is

$$H_{II} \qquad\qquad \frac{1}{2}p^2 - \left(q^2 + \frac{t}{2}\right)p - \kappa q.$$

Now, it is apparent that

$$K(t; q, p_1) = \frac{1}{2}p_1^2 - \frac{1}{2}q^4 - \frac{t}{2}q^2 - \alpha q \qquad (2.8)$$

is also a Hamiltonian for P_{II}. We will explain shortly why the Hamiltonian H_{II} is adapted to our studies of the Painlevé equations, instead of (2.8).

By using the method of Panlevé analysis, it is easy to show that we have local expansions of solutions of P_{II}, around a movable pole t_0, of the form

$$q = \frac{1}{z} - \frac{t_0}{6}z - \frac{\alpha + 1}{4}z^2 + hz^3 + \cdots, \qquad (2.9)$$

$$q = -\frac{1}{z} + \frac{t_0}{6}z - \frac{\alpha - 1}{4}z^2 + hz^3 + \cdots, \qquad (2.10)$$

where $z = t - t_0$ and h is an arbitrary constant. The Hamiltonian function $K = K(t; q, p_1)$ has a local expansion

$$K = \frac{1}{2}z + \cdots$$

around $t = t_0$, since by (2.8)

$$\frac{dK}{dt} = -\frac{1}{2}q^2.$$

Then the function $\exp \int^t K dt$ is not holomorphic on $\mathbb{C}$, and we cannot distinguish the two kinds of poles, (2.9) and (2.10), by considering the Hamiltonian (2.8).

On the other hand, when considering the Hamiltonian H_{II}, we have

$$\frac{dq}{dt} = p - q^2 - \frac{t}{2}, \quad \frac{dp}{dt} = 2qp + \alpha + \frac{1}{2}, \quad \frac{dH}{dt} = -\frac{1}{2}p,$$

where $H = H(t; q, p)$ is a Hamiltonian function associated with H_{II}. Now, we deduce from (2.9) that

$$\frac{dq}{dt} + q^2 + \frac{t}{2} = -\left(\alpha + \frac{1}{2}\right)z + \left(\frac{t_0^2}{36} + 5h\right) + \cdots.$$

It follows that the function $p = dq/dt + q^2 + t/2$ is holomorphic around a movable pole $t = t_0$ of q, of the type (2.9), and that so is the function H. Moreover, we can show that H has a simple pole with residue 1 at t_0 for an expansion of the form (2.10).

2.5   The τ-Functions of P_{II}

By considering H_{II}, we can thus distinguish the two types of solutions, (2.9) and (2.10), and the τ-function, τ, defined by (2.7), is entire on $\mathbb{C}$. The Hamiltonian H_{II} regularizes all the poles of the type (2.9), but those of the type (2.10) remain singularities of the Hamiltonian function. By considering a solution of the form (2.10) in the place of (2.9), we obtain the other Hamiltonian

$$\overline{H} = \frac{1}{2}\bar{p}^2 + \left(q^2 + \frac{t}{2}\right)\bar{p} - \left(\alpha - \frac{1}{2}\right)q \tag{2.11}$$

of the second Painlevé equation, P_{II}.

The two Hamiltonians, H_{II} and (2.11), are mutually connected by the change of variables

$$p = \bar{p} + 2q^2 + t, \quad H = \overline{H} - q, \tag{2.12}$$

which defines a canonical transformation. Let us define the other τ-function, $\bar{\tau}$, by

$$\overline{H} = \frac{\mathrm{d}}{\mathrm{d}t}\log\bar{\tau}.$$

Then $\bar{\tau}$ is an entire function, and we deduce from (2.12) the following expression of a solution of P_{II}:

$$q = \frac{\mathrm{d}}{\mathrm{d}t}\log\frac{\bar{\tau}}{\tau}.$$

By considering the canonical transformation

$$p = p_1 + \frac{1}{2}q^2 + \frac{t^2}{8}, \quad K = H + \frac{1}{2}q + \frac{t^2}{8},$$

we obtain the Hamiltonian $K(t; q, p_1)$ of the form (2.8) and then

$$K = \frac{\mathrm{d}}{\mathrm{d}t}\log\sqrt{\tau\bar{\tau}}.$$

This expression shows again that $\exp\int^t K\,\mathrm{d}t$ is not holomorphic on $\mathbb{C}$.

2.6   The Hamiltonian of the Painlevé Systems

The *Painlevé system* associated with P_J is by definition the Hamiltonian system

$$Dq = \frac{\partial H}{\partial p}, \quad Dp = -\frac{\partial H}{\partial q}, \quad H = H_J.$$

The Hamiltonians H_J considered in what follows are given by

$$H_I \quad \frac{1}{2}p^2 - 2q^3 - tq, \qquad D = \frac{d}{dt};$$

$$H_{II} \quad \frac{1}{2}p^2 - \left(q^2 + \frac{t}{2}\right)p - \kappa q, \qquad D = \frac{d}{dt}; \quad \kappa = \alpha + \frac{1}{2};$$

$$H'_{III} \quad q^2 p^2 - \left(q^2 + \theta_0 q - t\right)p - \kappa q, \qquad D = t\frac{d}{dt};$$

$$\alpha = -4\theta_\infty, \quad \beta = 4\left(\theta_0 + 1\right), \quad \gamma = 4, \quad \delta = -4, \quad \kappa = \frac{1}{2}\left(\theta_0 + \theta_\infty\right);$$

$$H_{IV} \quad 2q^2 p^2 - \left(q^2 + 2tq + 2\kappa_0\right)p + \theta_\infty q, \qquad D = \frac{d}{dt};$$

$$\alpha = -\kappa_0 + 2\theta_\infty + 1, \quad \beta = -2\kappa_0^2;$$

$$H_V \quad q(q-1)^2 p^2 - \left\{\kappa_0(q-1)^2 + \theta q(q-1) - tq\right\}p + \kappa(q-1), \qquad D = t\frac{d}{dt};$$

$$\alpha = \frac{1}{2}\kappa_\infty^2, \quad \beta = -\frac{1}{2}\kappa_0^2, \quad \gamma = -\theta - 1, \quad \delta = -\frac{1}{2},$$

$$\kappa = \frac{1}{4}\left(\kappa_0 + \theta\right)^2 - \frac{1}{4}\kappa_\infty^2;$$

$$H_{VI} \quad q(q-1)(q-t)p^2 - \left\{\kappa_0(q-1)(q-t) + \kappa_1 q(q-t) + (\theta-1)q(q-1)\right\}p$$

$$+ \kappa(q-t), \qquad D = t(t-1)\frac{d}{dt};$$

$$\alpha = \frac{1}{2}\kappa_\infty^2, \quad \beta = -\frac{1}{2}\kappa_0^2, \quad \gamma = \frac{1}{2}\kappa_1^2, \quad \delta = -\frac{1}{2}\left(1 - \theta^2\right),$$

$$\kappa = \frac{1}{4}\left(\kappa_0 + \kappa_1 + \theta - 1\right)^2 - \frac{1}{4}\kappa_\infty^2.$$

2.7 Remarks on the Hamiltonians

By the definition (2.7), the τ-functions of the Painlevé systems are holomorphic at each point except for the fixed singularities and have only simple zeros. This fact has been established in [22], by the use of the well-known fact that Painlevé transcendental functions are free from movable critical points. In [26] we have studied the determination of the Hamiltonian, H_{II}, by using the local expansions, (2.9) and (2.10), of solutions of P_{II}.

The method explained in this section can be applied to determine the Hamiltonians of the other Painlevé equations. For example, to obtain the Hamiltonian, H_{VI}, of the sixth Painlevé equation, P_{VI}, we have to consider the following four types of expansions:

$$q(t) = \frac{\pm\kappa_0}{t_0 - 1}z + hz^2 + \cdots, \tag{2.13}$$

$$q(t) = 1 \pm \frac{\kappa_1}{t_0}z + hz^2 + \cdots, \tag{2.14}$$

$$q(t) = t_0 + (1 \pm \theta)z + hz^2 + \cdots, \tag{2.15}$$

$$\frac{1}{q(t)} = \frac{\pm\kappa_\infty}{t_0(t_0 - 1)}z + hz^2 + \cdots, \tag{2.16}$$

where $z = t - t_0$ and h is an arbitrary constant. We can deduce from one determination of the first expansion the system of differential equations

$$\begin{cases} \dfrac{dq}{dt} = \dfrac{-\kappa_0}{t-1} + qp_0, \\[2mm] \dfrac{dp_0}{dt} = A_0(t,q)p_0^2 + B_0(t,q)p_0 + C_0(t,q), \end{cases} \tag{2.17}$$

such that

$$A_0(t,q) = \frac{q^2 - t}{2(q-1)(q-t)},$$

$$B_0(t,q) = \frac{-\kappa_0}{t-1}\left(\frac{1}{q-1} + \frac{1}{q-t}\right) - \frac{1}{t} - \frac{1}{t-1} - \frac{1}{q-t},$$

$$C_0(t,q) = \frac{1}{2}\left(\frac{\kappa_0}{t-1}\right)^2\left(-\frac{1}{t} + \frac{1}{q-1} + \frac{1}{t(q-t)}\right) + \frac{\kappa_0}{t-1}\frac{1}{t(q-t)}$$
$$+ \frac{1}{2}\left[\frac{\kappa_\infty^2}{t(t-1)} + \frac{\kappa_1^2}{t(q-1)^2} + \frac{1-\theta^2}{(q-t)^2}\right].$$

Note that the rational functions of the left-hand sides of (2.17) are holomorphic at $q = 0$, while $q = 0$ is a singularity of the differential equation P_{VI}.

We can obtain, for $\xi = 0, 1, t, \infty$, the four systems

$$\begin{cases} \dfrac{dq}{dt} = b_\xi(t,q) + c_\xi(t,q)p_\xi, \\[2mm] \dfrac{dp_\xi}{dt} = A_\xi(t,q)p_\xi^2 + B_\xi(t,q)p_\xi + C_\xi(t,q), \end{cases}$$

where b_ξ, c_ξ, A_ξ, B_ξ, and C_ξ are rational functions in (t,q), holomorphic at $q = \xi$. Using these systems, we can arrive in a natural way at the Hamiltonian system (2.3) with the Hamiltonian H_{VI}; for details, see [20] and [4].

On the other hand, by means of the step-by-step degeneration starting from the sixth Painlevé equation P_{VI}, we have the Hamiltonian H_{II}. This can be characterized also by the theory of holonomic deformation of a linear ordinary differential equation of the second order; see [23]. We can adapt the latter method for obtaining the Hamiltonian systems of several variables; see Section 6.

3   The Space of Initial Conditions

3.1   Riccati Equations

Consider first the *Riccati equation*

$$\frac{dq}{dt} = a(t)q^2 + b(t)q + c(t), \tag{3.1}$$

$a(t)$, $b(t)$, and $c(t)$ being entire functions on $\mathbb{C}$. Given any $t_0 \in \mathbb{C}$, a solution $q(t; q_0)$ of (3.1) is determined uniquely by the initial condition

$$q(t_0) = q_0. \tag{3.2}$$

Let us denote by $S(t_0)$ the *space of initial conditions* of (3.1); we have from (3.2) that $S(t_0) \supset \mathbb{C}$. Since a solution of (3.1) is in general a meromorphic function on $\mathbb{C}$, we obtain $S(t_0)$ by adding to $\mathbb{C}$ a point corresponding to the initial condition

$$q(t_0) = \infty. \tag{3.3}$$

If we define a new variable Q by

$$q = \frac{1}{Q}, \tag{3.4}$$

then (3.1) is converted to a Riccati equation of the other form

$$-\frac{dQ}{dt} = a(t) + b(t)Q + c(t)Q^2,$$

and the initial condition (3.3) is written as

$$Q(t_0) = 0.$$

The space $S(t_0)$ is a patch of two copies of $\mathbb{C}$, the q-space, and the Q-space, through the relation (3.4).

Therefore, the space of initial conditions of the Riccati equation (3.1) is

$$S(t_0) = \mathbb{C} \cup \mathbb{C} = \boldsymbol{P}^1.$$

3.2   Poles of P_{II}

The general solution of the second Painlevé equation,

$$P_{II} \qquad\qquad \frac{d^2q}{dt^2} = 2q^3 + tq + \alpha,$$

is meromorphic on $\mathbb{C}$ and so is the general solution of the Hamiltonian system

$$\frac{dq}{dt} = \frac{\partial H}{\partial p}, \quad \frac{dp}{dt} = -\frac{\partial H}{\partial q}, \tag{3.5}$$

with the Hamiltonian $H = H(t; q, p)$ given by

$$H_{\text{II}} \qquad \frac{1}{2}p^2 - \left(q^2 + \frac{t}{2}\right)p - \kappa q.$$

In what follows, we consider the space $S(t_0)$ of initial conditions of the differential equation P_{II}, or equivalently that of the Hamiltonian system (3.5).

It is clear that $S(t_0) \supset \mathbb{C}^2$; in fact, for any $(q_0, p_0) \in \mathbb{C}^2$, the initial condition

$$q(t_0) = q_0, \quad p(t_0) = p_0,$$

determines a unique meromorphic solution $q(t; q_0, p_0)$ of (3.5). In order to construct the space $S(t_0)$, we have to study an initial condition of the type (3.3), that is, to investigate a solution having a pole at $t = t_0$. In this case, a solution of P_{II} has a pole of the type (2.9) or of the type (2.10).

By considering a pole of the type (2.9), we can show the following proposition.

Proposition 3.1. *The Hamiltonian system* (3.5) *is equivalent to the system*

$$\begin{cases} \dfrac{dQ}{dt} = 1 + \dfrac{t}{2}Q^2 + \kappa Q^3 + Q^4 P, \\ \dfrac{dP}{dt} = -2Q^3 P^2 - (3\kappa Q^2 + tQ)P - \kappa^2 Q^2 - \kappa\dfrac{t}{2}, \end{cases} \tag{3.6}$$

with $\kappa = \alpha + \frac{1}{2}$, *where we put in* (3.5)

$$q = \frac{1}{Q}, \quad p = -\kappa Q - Q^2 P. \tag{3.7}$$

Equation (3.6) is the Hamiltonian system with the Hamiltonian

$$H = \frac{1}{2}Q^4 P^2 + \left[\kappa Q^3 + \frac{t}{2} + 1\right]P + \kappa^2 Q^2 + \kappa\frac{t}{2}Q,$$

since the transformation (3.7) is a canonical one. The initial condition of (3.6)

$$Q(t_0) = 0, \quad P(t_0) = P_0,$$

defines a meromorphic solution of the form (2.9) while, in the case where $q(t)$ has an expansion of the type (2.10), $P(t)$ has again a pole at $t = t_0$.

3.3 Hirzebruch Surfaces

We now consider the patch of two copies of $\mathbb{C}^2$,

$$\Sigma_{(\kappa)} = \mathbb{C}^2 \cup \mathbb{C}^2,$$

with the equations

$$qQ = 1, \quad p = -\kappa Q - Q^2 P.$$

By the above arguments, $\Sigma_{(\kappa)}$ is a subset of the space $S(t_0)$ of initial conditions of $\mathrm{P_{II}}$, and $S(t_0) \setminus \Sigma_{(\kappa)}$ is the set of initial conditions such that

$$Q(t_0) = 0, \quad P(t_0) = \infty. \tag{3.8}$$

By adding to $\Sigma_{(\kappa)}$ all the points corresponding to a solution of the type (2.10), we will arrive at the space $S(t_0)$.

$\Sigma_{(\kappa)}$ can be regarded as an affine fiber bundle over $\boldsymbol{P}^1$ with the fiber $\mathbb{C}$. By compactifying the fiber, we obtain from $\Sigma_{(\kappa)}$ the manifold $\overline{\Sigma}_{(\kappa)}$, which is now a $\boldsymbol{P}^1$-bundle over $\boldsymbol{P}^1$. In general, $\boldsymbol{P}^1$-bundles over $\boldsymbol{P}^1$ are called *Hirzebruch surfaces*; $\overline{\Sigma}_{(\kappa)}$ is an example of a Hirzebruch surface.

$\overline{\Sigma}_{(\kappa)}$ is a complex analytic manifold of complex dimension two, consisting of the four coordinate coverings

$$\overline{\Sigma}_{(\kappa)} = \cup_{i=1}^4 V_i, \quad V_i = \mathbb{C}^2.$$

Let us denote by (q_i, p_i) the coordinates of V_i such that

$$(q_1, p_1) = (q, p);$$

we consider the Hamiltonian system (3.5) with the Hamiltonian $\mathrm{H_{II}}$. V_1 and V_2 are patched by the equation

$$q_1 q_2 = 1, \quad p_1 = -\kappa q_2 - q_2^2 p_2,$$

that is, the system of equation (3.6) is defined in V_2 and we have

$$\Sigma_{(\kappa)} = V_1 \cup V_2.$$

The relation between V_1 and V_3 is given by

$$q_1 = q_3, \quad p_1 p_3 = 1,$$

and we patch V_2 and V_4 by the equation

$$q_2 = q_4, \quad p_2 p_4 = 1.$$

3.4 Flows Defined by $\mathrm{P_{II}}$

We now investigate the flows defined by $\mathrm{P_{II}}$ in the product space $\overline{\Sigma}_{(\kappa)} \times \mathbb{C}$. We have seen above that for any point $x_0 \in \Sigma_{(\kappa)}$ and $t_0 \in \mathbb{C}$, there exists a unique regular solution passing through x_0 at $t = t_0$.

In the case $\kappa = 0$, $\Sigma_{(\kappa)}$ contains a compact submanifold Z, defined by the equation

$$p_1 = p_2 = 0.$$

Z is isomorphic to $\boldsymbol{P}^1$, and a solution starting at $t = t_0$ from $x_0 \in Z$ stays in Z for any $t \in \mathbb{C}$. In fact, if $\kappa = 0$, then (3.5) is written as follows:

$$\frac{dq}{dt} = p - q^2 - \frac{t}{2}, \quad \frac{dp}{dt} = 2pq.$$

This equation has a particular solution of the form

$$p = 0, \quad \frac{dq}{dt} = -q^2 - \frac{t}{2}, \tag{3.9}$$

which is a Riccati equation. As we have seen above, the space of initial conditions of a Riccati equation is $\boldsymbol{P}^1$, and Z is nothing but the submanifold corresponding to the well-known classical solutions (3.9) of $\mathrm{P_{II}}$.

Returning to studies on the generic case of $\mathrm{P_{II}}$, consider a one-parameter family of solutions of the form

$$\begin{cases} q(t) = -\dfrac{1}{z} + \dfrac{t_0}{6} z - \dfrac{\alpha - 1}{4} z^2 + h z^3 + \cdots, \\[2mm] p(t) = \dfrac{2}{z^2} \left[1 + \dfrac{t_0}{6} z + \cdots \right], \end{cases} \tag{3.10}$$

with $z = t - t_0$, h being an arbitrary constant. Such solutions do not pass through any point of $\Sigma_{(\kappa)}$ at t_0 but they accumulate at the point of V_4

$$(q_4, p_4) = (0, 0);$$

see (3.8). This point is a singularity of the foliation defined by the Hamiltonian system (3.5). We have to separate the one-parameter family (3.10) of solutions by using the method of blowing-up of singularities.

3.5 Singularities of Differential Equations

Let a, b, f, and g be holomorphic functions in $(x, y; t) \in X \times U$ and consider flows defined by the equation

$$a\frac{dx}{dt} = f, \quad b\frac{dy}{dt} = g. \tag{3.11}$$

Define two one-forms ω_i $(i = 1, 2)$ by

$$\omega_1 = adx - fdt, \quad \omega_2 = bdy - gdt;$$

we have

$$\omega_1 \wedge \omega_2 = abdx \wedge dy - agdx \wedge dt + bfdy \wedge dt.$$

A point of the subset

$$\{ab = 0\} \cap \{ag = 0\} \cap \{bf = 0\}$$

is called a *singularity* of the system (3.11) or of the system

$$\omega_1 = \omega_2 = 0.$$

There is no singular point of (3.11), provided that $a \neq 0$ and $b \neq 0$ for any $(x, y; t) \in X \times U$.

On the other hand, if $f(0, 0; t) = 0$ for any $t \in U$, then the system

$$y\frac{dx}{dt} = f, \quad \frac{dy}{dt} = g \tag{3.12}$$

has a set of singular points

$$\{(0, 0)\} \times U \subset X \times U. \tag{3.13}$$

Moreover, for any $t_0 \in U$, (3.12) admits a particular solution of the form

$$y = 0, \quad t = t_0,$$

which is called a *vertical solution*. The system of differential equations

$$y\frac{dx}{dt} = f, \quad x\frac{dy}{dt} = g$$

has two vertical solutions given by

$$x = 0, \quad y = 0,$$

and a set of singular points of the form (3.13).

Let W_i $(i = 1, 2)$ be copies of $\mathbb{C}^2$ with coordinates (x_i, y_i) $(i = 1, 2)$ and $W = W_1 \cup W_2$ be the patch of W_1 and W_2 such that

$$x_1 y_2 = 1. \tag{3.14}$$

Consider the map

$$Q_p : W \longrightarrow \mathbb{C}^2,$$

given by

$$x = x_1 y_1, \quad y = y_1; \quad x = x_2, \quad y = x_2 y_2, \tag{3.15}$$

where (x, y) is the coordinate system of $\mathbb{C}^2$ and p is the origin of $\mathbb{C}^2$: $(x, y) = (0, 0)$.

Then $Q_p^{-1}(p)$ is a submanifold, X, of W defined by the equation

$$y_1 = x_2 = 0,$$

which is isomorphic to $\boldsymbol{P}^1$ by means of (3.14). The restriction of Q_p

$$W \setminus X \longrightarrow \mathbb{C}^2 \setminus \{p\}$$

is a biholomorphic map. This process of obtaining W from $\mathbb{C}^2$ is called the *blowing-up* at the origin, p, and Q_p is called a *quadratic map*. By surgery, we have from $\mathbb{C}^2$ the space $W = (\mathbb{C}^2 \setminus \{p\}) \cup \boldsymbol{P}^1$. By using blowing-ups, we study a resolution of singularities of a system of differential equations.

3.6   Resolution of Singularities

As an example, we consider the system

$$\begin{cases} y\dfrac{dx}{dt} = 2ax + by + 2cx^2 + exy + f_3 y^3, \\[2mm] \dfrac{dy}{dt} = a + 2cx + ey + f_2 y^2, \end{cases} \tag{3.16}$$

where $a \neq 0$, b, c, e, f_2, and f_3 are constants. The system (3.16) has the vertical solution

$$y = 0, \quad t = t_0, \tag{3.17}$$

for any $t_0 \in \mathbb{C}$ and the set of singular points

$$\{(0,0)\} \times \mathbb{C} \subset \mathbb{C}^2 \times \mathbb{C}. \tag{3.18}$$

For any t_0, we make a blowing-up at the origin, $p : (x, y) = (0, 0)$, of (3.16). We deduce from (3.15) the following two systems of differential equations

$$\begin{cases} y_1 \dfrac{dx_1}{dt} = ax_1 + b + (f_3 - f_2 x_1) y_1^2, \\[2mm] \dfrac{dy_1}{dt} = a + 2cx_1 y_1 + ey_1 + f_2 y_1^2, \end{cases} \tag{3.19}$$

defined in $W_1 \times \mathbb{C}$, and

$$\begin{cases} y_2 \dfrac{dx_2}{dt} = 2a + by_2 + 2cx_2 + ex_2 y_2 + f_3 x_2^2 y_2^3, \\[2mm] x_2 \dfrac{dy_2}{dt} = -a - by_2 + (f_2 - f_3 y_2) x_2^2, \end{cases} \tag{3.20}$$

in $W_2 \times \mathbb{C}$. These systems admit the two vertical solutions

$$y_1 = x_2 = 0, \quad t = t_0; \quad y_2 = 0, \quad t = t_0,$$

where the latter is a proper transform of (3.17) and the former is a new one obtained by the blowing-up.

The singular point of (3.20),

$$(x_2, y_2) = (0, 0), \quad t = t_0,$$

is the intersection of the two vertical solutions, and there is no other solution passing through this singularity. On the other hand, (3.19) has the other singular point

$$(x_1, y_1) = \left(-\frac{b}{a}, 0\right), \quad t = t_0,$$

for which we have to perform another blowing-up.

Putting in (3.19)

$$y = y_1, \quad x = x_1 + \frac{b}{a},$$

we obtain a system of the form

$$
\begin{cases}
y\dfrac{dx}{dt} = ax + (f_4 - f_2 x)y^2, \\[2mm]
\dfrac{dy}{dt} = a + 2cxy + e_1 y + f_2 y^2,
\end{cases}
$$

f_4, e_1 being constants. The quadratic map (3.15) at $(x, y) = (0, 0)$ gives us the two systems

$$
\begin{cases}
\dfrac{dx_1}{dt} = f_4 - e_1 x_1 + f_2 y_1^2 + 2cx_1^2 y_1, \\[2mm]
\dfrac{dy_1}{dt} = a + e_1 y_1 + f_2 y_1^2 + 2cx_1 y_1^2, \\[2mm]
y_2\dfrac{dx_2}{dt} = a + (f_4 - f_2 x_2)x_2 y_2^2, \\[2mm]
\dfrac{dy_2}{dt} = e_1 y_2 + 2f_2 x_2 y_2^2 - f_4 y_2^2 + 2cx_2 y_2.
\end{cases}
\qquad (3.21)
$$

By this surgery, we do not obtain a new vertical solution but a line, ℓ, on which (3.16) is converted to the system (3.21), which has neither a singular point nor a vertical solution.

This fact shows that all solutions of (3.16) passing through the singular points (3.18) separate out on the line ℓ. We have thus completed the process of resolution of singularities of (3.16).

3.7   The Space of Initial Conditions

Let us return to the case of the Painlevé equation P_{II}, and construct the space $S(t_0)$ of initial conditions. The system of differential equations associated with P_{II} is written in $V_4 \times \mathbb{C}$ as

$$p_4 \frac{dq_4}{dt} = q_4^4 + \cdots, \quad \frac{dp_4}{dt} = 2q_4^3 + \cdots.$$

In order to separate all solutions of the form (3.10) merged at

$$(q_4, p_4) = (0, 0),$$

we have to perform a blowing-up eight times. By such a process of successive blowing-ups, we finally obtain a manifold $\overline{\Sigma}(t_0)$, which contains the set $\mathcal{D}(t_0)$ of eight vertical solutions. And since none of the vertical solutions represents a solution of P_{II}, the space of initial conditions of P_{II} is given by

$$S(t_0) = \overline{\Sigma}(t_0) \setminus \mathcal{D}(t_0). \tag{3.22}$$

The space $S(t_0)$ can be characterized by the product of eight blowing-ups and by the configuration of vertical solutions; it is known that the set $\mathcal{D}(t_0)$ is described by the *extended Dynkin diagram* of the type $\boldsymbol{E}_7$; see [26].

As for the other Painlevé equations, the space $S(t_0)$ of initial conditions can be constructed in a similar manner. We start from the Hirzebruch surface

$$\overline{\Sigma}_{(\varepsilon)} = \cup_{i=1}^4 V_i, \quad V_i = \mathbb{C}^2,$$

such that V_1 and V_2 are patched by the equation

$$q_1 q_2 = 1, \quad p_1 = -\varepsilon q_2 - q_2^2 p_2,$$

ε being a constant. By eight successive blowing-ups for each P_J, we obtain the space $\overline{\Sigma}(t_0)$ and the set $\mathcal{D}(t_0)$ of vertical solutions contained in $\overline{\Sigma}(t_0)$. Then the space $S(t_0)$ is given by (3.22). We do not enter into the details of the construction of $S(t_0)$; see [20].

For P_I, we have

$$\varepsilon = 0,$$

and in $\overline{\Sigma}_{(\varepsilon)}$, two vertical solutions and a singular point, a_1. By making eight successive blowing-ups at singularities starting from a_1, we obtain $\overline{\Sigma}(t_0)$. As explained above, we have

$$\varepsilon = \kappa = \alpha + \frac{1}{2}.$$

for P_{II} and to construct $\overline{\Sigma}(t_0)$, we make eight successive blowing-ups for the singular point a_1 in $\overline{\Sigma}_{(\varepsilon)}$.

For P_{III} there exist a vertical solution and two singular points, a_1 and a_2, in $\overline{\Sigma}_{(\varepsilon)}$ with

$$\varepsilon = \frac{1}{2}\left(\theta_0 + \theta_\infty\right).$$

We have $\overline{\Sigma}(t_0)$ by virtue of four successive blowing-ups for each point of a_1, a_2. We have also for P_{IV} a vertical solution and two singular points, a_1 and a_2, in $\overline{\Sigma}_{(\varepsilon)}$ such that

$$\varepsilon = \frac{1}{2}\left(\kappa_0 + \theta_\infty\right).$$

We make two successive blowing-ups for one of the singular points, a_1, but six successive ones for the other one, a_2, and then obtain $\overline{\Sigma}(t_0)$.

For P_V, we have

$$\varepsilon = \frac{1}{2}\left(\kappa_0 + \kappa_\infty + \theta\right),$$

and in $\overline{\Sigma}_{(\varepsilon)}$ a vertical solution and three singular points, a_1, a_2, and a_3. By making two successive blowing-ups for each of the two singular points, a_1 and a_2, and four successive ones for a_3, we arrive at $\overline{\Sigma}(t_0)$.

The sixth Painlevé equation P_{VI} has a nice symmetry. In $\overline{\Sigma}_{(\varepsilon)}$ with

$$\varepsilon = \frac{1}{2}\left(\kappa_0 + \kappa_1 + \theta + 1 + \kappa_\infty\right),$$

we have four singular points, a_i $(i = 1, 2, 3, 4)$ on a vertical solution. $\overline{\Sigma}(t_0)$ is obtained by two successive blowing-ups for each a_i.

3.8 Remarks on the Space of Initial Conditions

The space $S(t_0)$ of initial conditions of P_J was studied for the first time in [20]. When this paper was written, the Hamiltonian structure for P_J was not yet recognized as a main subject of study in the Painlevé equations. Although a Hamiltonian system equivalent to P_J is obtained in [20] by means of a geometrical viewpoint of the foliation defined by the Painlevé systems, geometric structures of the space $S(t_0)$ have not been investigated sufficiently. We notice the recent works of K. Takano et al. [14, 30], where the symplectic structure of the space $S(t_0)$ is studied systematically; see also [15].

An intersection of two vertical solutions in $\overline{\Sigma}(t_0)$ is a singular point. There exists no solution of P_J passing through this point. Local behaviors of solutions of P_J around such a singularity and around vertical solutions are studied in [21].

TABLE 13.2.

P_J	1	2	3	4	5	6
diagram of singularities	E_8	E_7	D_6	E_6	D_5	D_4
m	9	8	7	7	6	5

The configuration of the set $\mathcal{D}(t_0)$ of vertical solutions for each Painlevé equation is associated with the extended Dynkin diagram; see [26]. We give in Table 13.2 the diagrams associated with $\mathcal{D}(t_0)$, and the number m of vertical solutions included in $\overline{\Sigma}(t_0)$.

4 The Irreducibility of P_{II}

4.1 Hamiltonian Vector Fields

The purpose of this section is to study the irreducibility of the second Painlevé equation P_{II}. We begin this section by preparing some notation on Hamiltonian vector fields.

For the Hamiltonian system

$$\frac{dq}{dt} = \frac{\partial H}{\partial p}, \quad \frac{dp}{dt} = -\frac{\partial H}{\partial q}, \tag{4.1}$$

we denote by $\mathcal{X}$ the *Hamiltonian vector field* for the Hamiltonian system (4.1):

$$\mathcal{X} = \frac{\partial H}{\partial p}\frac{\partial}{\partial q} - \frac{\partial H}{\partial q}\frac{\partial}{\partial p} + \frac{\partial}{\partial t}.$$

For P_{II}, the Hamiltonian H and the Hamiltonian vector field $\mathcal{X}$ are given by

$$H = \frac{1}{2}p^2 - \left(q^2 + \frac{t}{2}\right)p - \kappa q, \quad \kappa = \alpha + \frac{1}{2},$$

$$\mathcal{X} = \left(p - q^2 - \frac{t}{2}\right)\frac{\partial}{\partial q} + (2qp + \kappa)\frac{\partial}{\partial p} + \frac{\partial}{\partial t}. \tag{4.2}$$

Let K be a field and D be a linear transformation such that for $f, g \in K$,

$$D(fg) = f(Dg) + (Df)g.$$

Such a field K is called a differential field with differential D. The field $\mathbb{C}(t)$ of rational functions is a differential field with the differential

$$D = \frac{d}{dt}.$$

In what follows, we consider a differential field K, containing the field $\mathbb{C}(t)$ of rational functions as a differential subfield. When studing the Hamiltonian vector field $\mathcal{X}$ for the other Painlevé system, we replace $D = \mathrm{d}/\mathrm{d}t$ by a suitable differential operator according to Table 13.1.

The Hamiltonian vector field $\mathcal{X}$ can be naturally regarded as a differential operator acting on the polynomial ring $K[q,p]$ under the identification $D = \partial/\partial t$.

We say that a nonzero polynomial $F = F(q,p)$ in $K[q,p]$ defines an *invariant divisor* for the Hamiltonian system (4.1) if there exists some $G \in K[q,p]$ such that

$$\mathcal{X}F = GF.$$

For example, consider the case where $\kappa = 0$ in (4.2). Then we have

$$\mathcal{X}p = 2qp;$$

in this case, $F = p$ is an invariant divisor for (4.1) associated with P_{II}.

4.2   Classical Functions

A function $f \in K$ is said to be *classical* if it is obtained by finitely many applications of the following operators to an element of $\mathbb{C}(t)$.

(1) $f(t) \longmapsto \dfrac{\mathrm{d}}{\mathrm{d}t} f(t)$;

(2) $\begin{matrix} f(t) \\ g(t) \end{matrix} \longmapsto \begin{matrix} f(t) \pm g(t), \\ f(t)g(t), \\ f(t)/g(t) \ (g(t) \not\equiv 0); \end{matrix}$

(3) $\begin{matrix} f_1(t) \\ \vdots \\ f_n(t) \end{matrix} \longmapsto g(t)$ such that $g^n + f_1 g^{n-1} + \cdots + f_n = 0$;

(4) $f(t) \longmapsto F(t)$ such that $\dfrac{\mathrm{d}}{\mathrm{d}t} F(t) = f(t)$;

(5) $\begin{matrix} f_1(t) \\ \vdots \\ f_n(t) \end{matrix} \longmapsto g(t)$ such that $\dfrac{\mathrm{d}^n g}{\mathrm{d}t^n} + f_1 \dfrac{\mathrm{d}^{n-1} g}{\mathrm{d}t^{n-1}} + \cdots + f_n g = 0$;

Let $A = \mathbb{C}^n/\Gamma$ be an abelian variety with a lattice $\Gamma \subset \mathbb{C}^n$ and $\pi : \mathbb{C}^n \to A$ the projection. Then

(6) $\begin{matrix} f_1(t) \\ \vdots \\ f_n(t) \end{matrix} \longmapsto g = \varphi \circ \pi \circ F$ where φ is meromorphic on A and $F = (f_1, \ldots f_n)$.

We can state the above given conditions 1, 2, 4, 5, and 6 as one property by using the notion of algebraic group and integration of an invariant vector field of the algebraic group; cf. [31]. A function f is called irreducible over $\mathbb{C}(t)$ if it is not classical.

Since P. Painlevé insisted that the first Painlevé equation P_I defines a new transcendental function, the irreducibility of P_I has been one of the main problems of study on the Painlevé equations; it was even a subject of dispute, see [29, III, 77–109]. The irreducibility of P_I is a good example and an indispensable prototype of differential Galois theory of nonlinear differential equations, in the sense of Lie Picard Vessiot; on this viewpoint of the irreducibiblity, see [32] and the articles quoted in this paper.

The problem of the irreducibility of P_I has finally been settled in the affirmative by K. Nishioka [17] and then by H. Umemura [31]. The idea of classical functions was already mentioned in *Leçons de Stockholm* of P. Painlevé [29, I]. In order to give rigorous definitions of classical functions and irreducibility, and to confirm the irreducibility of P_I, we absolutely need the latest mathematical developments, in particular modern notions of algebraic geometry.

4.3   The Irreducibility of P_I

By the general scheme of H. Umemura [31], it is known that the problem of irreducibility of a Hamiltonian system (4.1) with polynomial Hamiltonian can be reduced to the following two practical problems:

(a) Determine the semigroup of invariant divisors of (4.1), for all differential extensions K of $\mathbb{C}(t)$.

(b) Find all the rational solutions to (4.1).

To make clear the idea of irreducibility, we first study the problems a and b for the first Painlevé equation

$$P_I \qquad\qquad \frac{d^2 q}{dt^2} = 6q^2 + t.$$

It is not difficult to show that there exists no rational solution of P_I. We restrict ourselves to consideration of the problem b, that is, to the verification of the nonexistence of an invariant divisor for the Hamiltonian vector field

$$\mathcal{X} = p\frac{\partial}{\partial q} + (6q^2 + t)\frac{\partial}{\partial p} + \frac{\partial}{\partial t}. \qquad (4.3)$$

By introducing weights $w(\cdot)$ for the canonical variables such that

$$w(q) = 2, \quad w(p) = 3, \quad w(t) = 0, \qquad (4.4)$$

we have a weight decomposition of cX of the form

$$\mathcal{X} = \mathcal{X}_1 + \mathcal{X}_0 + \mathcal{X}_{-3},$$

where

$$\mathcal{X}_1 = p\frac{\partial}{\partial q} + 6q^2\frac{\partial}{\partial p}, \quad \mathcal{X}_0 = \frac{\partial}{\partial t}, \quad \mathcal{X}_{-3} = t\frac{\partial}{\partial p}.$$

Suppose that $F \in K[q,p]$ is an invariant divisor for the Hamiltonian vector field (4.3):

$$\mathcal{X}F = GF, \quad G \in K[q,p].$$

Then we have the weight decomposition

$$F = F_m + F_{m-1} + \cdots$$

of F, where m is the highest weight of F with respect to the weight (4.4). Since the highest weight of $\mathcal{X}_1 F$ is $m+1$, G has a weight decomposition of the form

$$G = G_1 + G_0.$$

On the other hand, there is no polynomial in $K[q,p]$ such that

$$w(G_1) = 1, \quad G_1 \neq 0,$$

by (4.4); hence $G_1 = 0$, and we have the equation

$$\mathcal{X}_1 F_m = 0. \tag{4.5}$$

We deduce from (4.5) that

$$F_m = f \cdot \left(p^2 - 4q^3\right)^l, \quad f \in K^\times, \quad m = 6l.$$

By replacing, if necessary, F by $f \cdot F$, we can assume without loss of generality that $f = 1$. Then we have

$$F_m = \left(p^2 - 4q^3\right)^l, \quad m = 6l,$$

and the equation for F_{m-1}

$$\mathcal{X}_1 F_{m-1} + \mathcal{X}_0 F_m = G_0 F_m. \tag{4.6}$$

By taking into consideration the uniqueness of a solution of (4.6) and $\mathcal{X}_0 F_m = 0$, we can write F_{m-1} in the form

$$F_{m-1} = f \cdot qp \left(p^2 - 4q^3\right)^{l-1}, \quad f \in K.$$

Since $\mathcal{X}_1(qp) = p^2 + 6q^3$, we obtain $f = 0$ and hence

$$F_{m-1} = 0, \quad G_0 = 0.$$

It follows that

$$F_{m-2} = F_{m-3} = 0,$$

and that

$$\mathcal{X}_1 F_{m-4} + \mathcal{X}_{-3} F_m = 0. \tag{4.7}$$

By putting in (4.7)

$$F_{m-4} = A \cdot \left(p^2 - 4q^3\right)^{l-1},$$

we have

$$\mathcal{X}_1 A + 2ltp = 0, \quad w(A) = 2,$$

and then

$$F_{m-4} = -2ltq \left(p^2 - 4q^3\right)^{l-1}. \tag{4.8}$$

The next term F_{m-5} satisfies the equation

$$\mathcal{X}_1 F_{m-5} + \mathcal{X}_0 F_{m-4} = 0, \tag{4.9}$$

from which we see that F_{m-5} is of the form

$$F_{m-5} = B \cdot \left(p^2 - 4q^3\right)^{l-1}, \quad w(B) = 1.$$

On the one hand, since there is no nontrivial polynomial such that $w(B) = 1$, we obtain $F_{m-5} = 0$. On the other hand, we deduce from (4.8)

$$\mathcal{X}_0 F_{m-4} \neq 0,$$

which contradicts equation (4.9). We have thus arrived at a contradiction and thereby completed the proof of the inexistence of an invariant divisor for the vector field (4.3).

4.4   The Invariant Divisors of P_{II}

As to the second Painlevé equation, the answer to the problem (b) of determining all rational solutions is already known by [16]. We are thus led to investigate the invariant divisors for the Hamiltonian systems H_{II}. We now state the theorem.

Theorem 4.1. *Let K be any differential extension of $\mathbb{C}(t)$. Let $F = F(q, p)$ be a nonzero polynomial in $K[q, p]$ and suppose that*

$$\mathcal{X}F = GF \quad \text{for some } G \in K[q, p].$$

(1) If the complex parameter α satisfies the condition

$$\kappa \notin \{a \in \mathbb{Q}; \ a(a - 1) \geq 0\},$$

then $F = f$ for some $f \in K^{\times}$.

(2) If $\kappa = 0$, then $F = fp^n$ for some $f \in K^{\times}$ and $n = 0, 1, \ldots$.

We present in what follows an outline of a proof of the first assertion stated in Theorem 4.1. Consider the weight decomposition of the polynomial ring $K[q, p]$ by defining the weights of t, q, p, to be 0, 1, 2, respectively. The Hamiltonian vector field (4.2) is decomposed by the weights as follows:

$$\mathcal{X} = \mathcal{X}_1 + \mathcal{X}_0 + \mathcal{X}_{-1} + \mathcal{X}_{-2},$$

where

$$\mathcal{X}_0 = (p - q^2)\frac{\partial}{\partial q} + 2qp\frac{\partial}{\partial p}, \quad \mathcal{X}_0 = \frac{\partial}{\partial t}, \quad \mathcal{X}_{-1} = -\frac{1}{2}t\frac{\partial}{\partial q}, \quad \mathcal{X}_{-2} = \kappa\frac{\partial}{\partial p}.$$

Let F be an invariant divisor for (4.2),

$$\mathcal{X}F = GF, \quad G \in K[q, p]. \tag{4.10}$$

Since $\mathcal{X}$ has weight 1, the weight of G is at most 1. If F is of weight m, then we have the weight decomposition

$$F = F_m + F_{m-1} + \cdots, \quad G = G_1 + G_0.$$

Since G_1 is of weight 1, it can be written in the form

$$G_1 = a \cdot q, \quad a \in K.$$

We rewrite (4.10) as

$$(\mathcal{X}_1 - G_1) F_k = -(\mathcal{X}_0 - G_0) F_{k+1} - \mathcal{X}_{-1}F_{k+2} - \mathcal{X}_{-2}F_{k+3}, \tag{4.11}$$

for $k \geq -1$, where we agree to put $F_k = 0$ for $k < 0$ and for $k > m$. By solving the equation (4.11) for $k = m$ and then $k = m - 1$, we can establish the following proposition.

Proposition 4.1. *If F and G satisfy (4.10) with (4.2), then m is even and there exists an integer l such that*

$$G_1 = 2(-n + 2l)q, \quad F_m = (2q^2 - p)^{n-l}p^l \quad (0 \leq l \leq n), \ m = 2n. \tag{4.12}$$

We can pursue our study of the determination of F_k by considering step by step equation (4.12). In fact, we obtain

$$G_0 = 0, \quad F_{m-1} = 0, \tag{4.13}$$

and then

$$F_{m-2} = (n - l)t \left(2q^2 - p\right)^{n-l-1} p^l. \tag{4.14}$$

Moreover, by means of (4.11), (4.12), (4.13), and (4.14), F_{m-3} satisfies the equation

$$
\begin{aligned}
(\mathcal{X}_1 - G_1)F_{m-3} \\
&= -\mathcal{X}_0 F_{m-2} - \mathcal{X}_{-2} F_m \\
&= -\kappa k \left(2q^2 - p\right)^{n-l} p^{l-1} + (\kappa - 1)(n - l) \left(2q^2 - p\right)^{n-l-1} p^l. \tag{4.15}
\end{aligned}
$$

On the other hand, we have the identity

$$
\begin{aligned}
(\mathcal{X}_1 - G_1) \left\{ 2q \left(2q^2 - p\right)^{n-l-1} p^{l-1} \right\} \\
&= -\left(2q^2 - p\right)^{n-l} p^{l-1} + \left(2q^2 - p\right)^{n-l-1} p^l. \tag{4.16}
\end{aligned}
$$

By comparing the identity (4.16) with equation (4.15), we obtain the following necessary condition for the existence of an invariant divisor:

$$\kappa l = (\kappa - 1)(n - l). \tag{4.17}$$

We see easily that a value of the parameter κ satisfying (4.17) lies in the set

$$\{a \in \mathbb{Q}; \ a(a - 1) \geq 0\},$$

as desired. We do not enter into details; see [18].

4.5 Classical Solutions of P_{II}

The polynomial ring $K[q, p]$ can be regarded as a differential ring by means of the Hamiltonian system (4.1). For an arbitrarily fixed value α of the parameter, we consider the specialization

$$K[q, p] \quad \longrightarrow \quad K[q_0, p_0],$$

(q_0, p_0) being a particular solution of (4.1) in some differential extension of K. Let $L_H(\alpha; q_0, p_0)$ be the quotient field of $K[q_0, p_0]$; $L_H(\alpha; q_0, p_0)$ is a differential extension of K with respect to $D = d/dt$, by virtue of (4.1). Theorem 4.1 implies that if $\kappa \notin \{a \in \mathbb{Q}; a(a - 1) \geq 0\}$, then

$$\text{trans.} \deg_K L_H(\kappa; q_0, p_0) = 2 \text{ or } 0.$$

On the other hand, we define (q_1, p_1) and (q_{-1}, p_{-1}) as follows:

$$\begin{cases} q_1 = -q + \dfrac{\kappa - 1}{p - 2q^2 - t}, \\ p_1 = -p + 2q^2 + t, \end{cases} \qquad \begin{cases} q_{-1} = -q + \dfrac{\kappa}{p}, \\ p_{-1} = -p + 2q_{-1}^2 + t. \end{cases} \qquad (4.18)$$

Then the following result is known by [25].

Proposition 4.2. $L_H(\kappa - 1; q_1, p_1) \cong L_H(\kappa; q_0, p_0) \cong L_H(\kappa + 1; q_{-1}, p_{-1})$.

From Theorem 4.1 we deduce the following theorem.

Theorem 4.2. *For any particular solution* (q_0, p_0), *we have*

$$\text{trans.} \deg_K L_H(\kappa; q_0, p_0) = 2 \ or \ 0$$

if $\kappa \notin \{n; n \in \mathbb{Z}\}$.

We consider classical solutions of P_{II}. When $\kappa = 0$, the Hamiltonian system (4.1) admits a family of particular solutions such that

$$\frac{dq_0}{dt} = -q_0^2 - \frac{t}{2}, \quad p_0 \equiv 0. \qquad (4.19)$$

In this case, we have

$$\text{trans.} \deg_{\mathbb{C}(t)} L_H(0; q_0, p_0) = 1.$$

Therefore, by Proposition 4.2, we have the family of differential fields such that

$$\text{trans.} \deg_{\mathbb{C}(t)} L_H(0; q_n, p_n) = 1, \quad n \in \mathbb{Z}.$$

The second statement of Theorem 4.1 implies that if $\kappa = 0$, then

$$\text{trans.} \deg_{\mathbb{C}(t)} L_H(0; q_0, p_0) = 2,$$

unless the Riccati equation (4.19) is satisfied.

4.6 Rational Solutions of P_{II}

When considering rational solutions of P_{II}, we adopt the parameter α instead of κ; note that

$$\kappa = \alpha + \frac{1}{2}.$$

When $\alpha = 0$, it is easy to see that P_{II} has the particular solution $q_0 = 0$. Hence,

$$(q_0, p_0) = \left(0, \frac{t}{2} \right) \qquad (4.20)$$

solves (4.1), when $\alpha = 0$. Moreover, if $\alpha \in \mathbb{Z}$, (4.1) has the only rational solution that is reduced to (4.20) by the transformation (4.18) given above.

Summarizing all the above given considerations, we obtain the following results on the irreducibility of P_{II}:

(1) When $\alpha \in \mathbb{Z}$, the system (4.1) has only one rational solution.

(2) When $\kappa \in \mathbb{Z}$, (4.1) has a one-parameter family of solutions expressed as rational functions of solutions of the Riccati equation (4.19).

(3) Any solution other than those described in 1 and 2 generates a differential field with transcendence degree 2 over $\mathbb{C}(t)$.

For the irreducibility of P_{II}, see also [34].

4.7 Remarks on Classical Solutions

The Painlevé equations except for the first one contain parameters. Let us denote by v the set of parameters contained in the Hamiltonian H_J and by V_J the space of all parameters. When considering the Hamiltonian structure at an arbitrarily fixed value v of parameters, we write it as $\mathcal{H}_J(v)$.

For $J = II, \ldots , VI$, we will give a group G_J of affine transformations of V_J such that $\mathcal{H}_J(g(v))$ is birationally canonical to $\mathcal{H}_J(v)$ for each $g \in G_J$. In particular, we have the birational canonical transformations associated to the translations of the space V_J contained in G_J. For example, we have $V_{II} = \mathbb{C}$ for the second Painlevé system, and (4.18) is the birational canonical transformation associated to the translation

$$\kappa \longmapsto \kappa \pm 1.$$

Such a birational canonical transformation plays an important role in the study of the Hamiltonian structure, as well as in the investigation of the irreducibility of a solution of the Painlevé equations; see [24].

The group G_J has a description in terms of the affine Weyl groups. A fundamental region in V_J with respect to G_J is a Weyl chamber, whose boundaries are called walls of the chamber. If a value of the parameter v of V_J is contained in a wall, then P_J has particular solutions represented by classical transcendental functions. Table 13.3 concerns the group G_J and the classical transcendental functions associated with P_J.

As mentioned above, the irreducibility was established for the first time by [17] and [31]. We can show the *transcendence* of all other Painlevé equations P_J by using the birational canonical transformation. Here we say that the transcendence of P_J is established if the following problems are completely settled; see Theorem 4.1:

TABLE 13.3.

P_J	1	2	3	4	5	6
affine Weyl group	—	A_1	B_2	A_2	A_3	D_4
particular solitions	—	Airy	Bessel	Hermite–Weber	Confluent hypergeo-metric	Gauss's hypergeo-metric

(1) to show the inexistence of an invariant divisor for a generic value of the parameter v;

(2) to give the set of all exceptional values of v such that the Hamiltonian vector field $\mathcal{X}_J$ associated with P_{rJ} admits an invariant divisor;

(3) to determine invariant divisors of $\mathcal{X}_J$.

The transcendence of P_{II} and that of P_{IV} are shown by [18]; see also [34]. As to the transcendence of P_{III}, we refer to [35]. The transcendence of P_V and that of P_{VI} are shown by [36] and by [37], respectively.

P. Painlevé showed the inexistence of a rational solution of P_I, and all rational solutions of P_{II} and those of P_{IV} are determined by [16]; see also [34]. Having proved the transcendence and determined all rational solutions, we thus establish the irreducibility for the Painlevé equations, P_I, P_{II}, and P_{IV}. By studying the algebraic structure of rational solutions of P_{II} and P_{IV}, we arrive in a natural way at the bilinear relations of the τ-functions related to the Hamiltonians H_{II} and H_{IV}, respectively; see [1].

The other Painlevé equations, P_{III}, P_V, and P_{VI}, admit as classical solution an *algebraic* function of t, as well as a rational solution. To establish the irreducibility of each P_J of these equations, we have to determine all algebraic solutions, but this problem remains unsettled. On the other hand, we know that P_J possesses a fruitful structure of these algebraic solutions; cf. [3, 19, 33].

5   The τ-Functions of the Second Painlevé System

5.1   The Second Painlevé System

The second Painlevé system is, by definition, the Hamiltonian system

$$\frac{dq}{dt} = \frac{\partial H}{\partial p}, \quad \frac{dp}{dt} = -\frac{\partial H}{\partial q}, \tag{5.1}$$

with the Hamiltonian

$$H_{II}H = \frac{1}{2}p^2 - \left(q^2 + \frac{t}{2}\right)p - \kappa q,$$

κ being a complex parameter. The second Painlevé system is obtained by the step-by-step degeneration from the sixth Painlevé system; see [4], [23].

Consider the second Painlevé system with an arbitrarily fixed value κ of the parameter. We will write the canonical variables as $q(\kappa)$, $p(\kappa)$ and the Hamiltonian as $H(\kappa)$, in order to describe the transformation. We have the following contiguity relations of the canonical variables:

$$\begin{cases} q(\kappa - 1) = -q(\kappa) + \dfrac{\kappa - 1}{p(\kappa) - 2q(\kappa)^2 - t}, \\ p(\kappa - 1) = -p(\kappa) + 2q(\kappa)^2 + t, \end{cases} \tag{5.2}$$

and the change of the Hamiltonian functions

$$H(\kappa - 1) = H(\kappa) + q(\kappa); \tag{5.3}$$

see [25]. It is easy to see that (5.2) and (5.3) define the birational canonical transformation

$$\mathcal{H}(\kappa) \longrightarrow \mathcal{H}(\kappa - 1)$$

of the second Painlevé system, where

$$\mathcal{H}(\kappa) = (q(\kappa), p(\kappa), H(\kappa), t).$$

The τ-function of the second Painlevé system is defined by

$$\frac{d}{dt}\log\tau(\kappa) = H(\kappa),$$

up to multiplicative constants. It is known by [25] that the τ-functions satisfy the Toda equation

$$\frac{d^2}{dt^2}\log\tau(\kappa) = \frac{\tau(\kappa - 1)\tau(\kappa + 1)}{\tau(\kappa)^2}. \tag{5.4}$$

Consider the sequence

$$\ldots, \tau(\kappa + 1), \tau(\kappa), \tau(\kappa - 1), \ldots \tag{5.5}$$

of the τ-functions related to the second Painlevé system. The aim of this section is to show that (5.5) are not algebraically independent.

In fact, there exists a polynomial

$$\Phi \in \mathbb{Z}[x_0, x_1, x_2, x_3, x_4, x_5; K],$$

homogeneous with respect to the first six elements, such that

$$\Phi\left(\tau(\kappa+2), \tau(\kappa+1), \tau(\kappa), \tau(\kappa-1), \tau(\kappa-2), \tau(\kappa-3); \kappa\right) = 0.$$

Here we denote by $\mathbb{Z}$ the ring of rational integers. Such a relation between τ-functions, related to the second Painlevé system, was established in [5] by use of the Schlesinger transformations of linear systems of ordinary differential equations.

5.2 Algebraic Relations

The following three contiguity relations are established in [25] for the second Painlevé system

$$q(\kappa-1) + q(\kappa) = -\frac{\kappa-1}{p(\kappa-1)}, \tag{5.6}$$

$$\begin{cases} p(\kappa-1) + p(\kappa) = 2q(\kappa)^2 + t, \\ p(\kappa-2) + p(\kappa-1) = 2q(\kappa-1)^2 + t. \end{cases} \tag{5.7}$$

We deduce from (5.6) and (5.7) the following equality:

$$p(\kappa-2) - p(\kappa) = -2\left(q(\kappa-1) - q(\kappa)\right)\frac{\kappa-1}{p(\kappa-1)}.$$

It follows from (5.6) that

$$q(\kappa) = -\frac{\kappa-1}{2p(\kappa-1)} + \frac{1}{4(\kappa-1)}p(\kappa-1)\left(p(\kappa-2) - p(\kappa)\right), \tag{5.8}$$

$$q(\kappa-1) = -\frac{\kappa-1}{2p(\kappa-1)} - \frac{1}{4(\kappa-1)}p(\kappa-1)\left(p(\kappa-2) - p(\kappa)\right). \tag{5.9}$$

By replacing $\kappa-1$ by κ in (5.9) we have

$$q(\kappa) = -\frac{\kappa}{2p(\kappa)} - \frac{1}{4\kappa}p(\kappa)\left(p(\kappa-1) - p(\kappa+1)\right). \tag{5.10}$$

And then comparing (5.10) with (5.8), we arrive at the following Theorem.

Theorem 5.1. *The four functions*

$$p(\kappa-2), \quad p(\kappa-1), \quad p(\kappa), \quad p(\kappa+1) \tag{5.11}$$

satisfy an algebraic equation of the form

$$2\kappa^2(\kappa-1)p(\kappa-1) + (\kappa-1)p(\kappa)^2 p(\kappa-1)\left[p(\kappa-1) - p(\kappa+1)\right]$$
$$= 2\kappa(\kappa-1)^2 p(\kappa) + \kappa p(\kappa)p(\kappa-1)^2\left[p(\kappa-2) - p(\kappa)\right]. \tag{5.12}$$

For the Hamiltonian H_{II}, we have

$$\frac{dH(\kappa)}{dt} = \left(\frac{\partial}{\partial t}\right) H(\kappa) = -\frac{1}{2}p(\kappa),$$

by the Hamiltonian system (5.1). Then the following expression results from the Toda equation (5.4):

$$p(\kappa) = -2\frac{\tau(\kappa-1)\tau(\kappa+1)}{\tau(\kappa)^2}. \tag{5.13}$$

Therefore, (5.12) defines by (5.13) an algebraic relation of the form

$$\Phi\left(\tau(\kappa+2), \tau(\kappa+1), \tau(\kappa), \tau(\kappa-1), \tau(\kappa-2), \tau(\kappa-3); \kappa\right) = 0 \tag{5.14}$$

among the six adjacent τ-functions

$$\tau(\kappa+2), \quad \tau(\kappa+1), \quad \tau(\kappa), \quad \tau(\kappa-1), \quad \tau(\kappa-2), \quad \tau(\kappa-3). \tag{5.15}$$

Here $\Phi \in \mathbb{Z}[x_0, x_1, x_2, x_3, x_4, x_5; K]$; the homogeneity of the polynomial Φ with respect to $x_0, \ldots, x_5$ is clear. When κ is not an integer, we can rewrite (5.12) in the following symmetric form

$$\frac{2\kappa}{p(\kappa)} + \frac{1}{\kappa}p(\kappa)\left[p(\kappa-1) - p(\kappa+1)\right]$$
$$= \frac{2(\kappa-1)}{p(\kappa-1)} - \frac{1}{\kappa-1}p(\kappa-1)\left[p(\kappa-2) - p(\kappa)\right].$$

By means of (5.10) and (5.13), the canonical variables $p(\kappa)$ and $q(\kappa)$ are rational functions of the six τ-functions (5.15). We show that the Hamiltonian

$$H(\kappa) = \frac{1}{2}p(\kappa)^2 - \left(q(\kappa)^2 + \frac{t}{2}\right)p(\kappa) + \kappa q(\kappa)$$

and the independent variable t can be described as rational functions of $p(\kappa-1), p(\kappa), p(\kappa+1)$. By the use of (5.7), it is easy to see that

$$H(\kappa) = -\frac{1}{2}p(\kappa)p(\kappa-1) + \kappa q(\kappa).$$

It follows from (5.10) that

$$H(\kappa) = \frac{\kappa^2}{2p(\kappa)} - \frac{1}{4}p(\kappa)\left(p(\kappa+1) + p(\kappa-1)\right).$$

On the other hand, by putting (5.10) into (5.7), we have

$$2t = V(\kappa), \tag{5.16}$$

where

$$V(\kappa) = 2\left(p(\kappa - 1) + p(\kappa)\right)$$

$$- \left[\frac{\kappa}{p(\kappa)} + \frac{p(\kappa)}{2\kappa}\left(p(\kappa - 1) - p(\kappa + 1)\right)\right]^2. \quad (5.17)$$

It is easy to show that (5.12) induces the equality $V(\kappa) = V(\kappa - 1)$. It follows that all the canonical variables of $\mathcal{H}(\kappa)$ can be written as rational functions of (5.15).

By taking into consideration the ambiguity of multiplicative constants for the τ-functions, we regard (5.15) as a point in the complex projective space $\boldsymbol{P}^5$ of dimension 5. Then the algebraic relation (5.14) defines a hypersurface $\boldsymbol{X}$ of $\boldsymbol{P}^5$, and the canonical variables of $\mathcal{H}(\kappa)$ are rational functions on $\boldsymbol{X}$. In particular, $V(\kappa)$ given by (5.17) is rational on $\boldsymbol{X}$, and for any $t = t_0$ fixed, (5.16) also defines a hypersurface $\boldsymbol{Y}(t_0)$ in $\boldsymbol{P}^5$. By virtue of the results obtained in what precedes, we have thus a rational map from $\boldsymbol{X} \cap \boldsymbol{Y}(t_0)$ to $\overline{\Sigma}(t_0)$, where $\overline{\Sigma}(t_0)$ is the compact manifold considered in Section 3.

Because of (5.13), the four functions given by (5.11) play an important role in the study of the second Painlevé system. We give in the rest of this subsection miscellaneous differential equations satisfied by these functions.

Since we have the equation

$$\frac{dp(\kappa)}{dt} = -\frac{\partial H(\kappa)}{\partial q(\kappa)} = 2q(\kappa)p(\kappa) + \kappa,$$

$p(\kappa)$ satisfies the differential equation

$$\frac{dp(\kappa)}{dt} = -\frac{1}{\kappa}p(\kappa)^2\left(p(\kappa - 2) - p(\kappa)\right).$$

On the other hand, by using (5.12), we compute the derivative of $p(\kappa - 1)$ as follows:

$$\frac{dp(\kappa - 1)}{dt} = -\frac{1}{\kappa - 1}p(\kappa - 1)^2\left(p(\kappa - 2) - p(\kappa)\right)$$

$$= 2\kappa\frac{p(\kappa - 1)}{p(\kappa)} - 2(\kappa - 1)$$

$$+ \frac{1}{\kappa}p(\kappa)p(\kappa - 1)\left(p(\kappa - 1) - p(\kappa + 1)\right).$$

It follows that $p(\kappa)$, $p(\kappa - 1)$ satisfy the differential equation

$$\frac{d}{dt}\left(p(\kappa)p(\kappa - 1)\right) = 2\kappa p(\kappa - 1) - 2(\kappa - 1)p(\kappa).$$

It is interesting to investigate relations of τ-functions related to the Painlevé equation P_J. Algebraic relations among τ-functions were studied in [5] for P_{III} by means of the Schlesinger transformations of linear systems of ordinary differential equations. For P_{IV} we can obtain analogous results to what has been obtained above for P_{II}; see [27].

5.3 The Bilinear Forms

We express the second Painlevé equation P_{II} as bilinear forms in terms of the Hirota derivatives

$$\mathcal{D}g \cdot f = (Dg)f - g(Df), \quad \mathcal{D}^2 g \cdot f = (D^2 g)f - 2(Dg)(Df) + g(D^2 f),$$
$$\mathcal{D}^3 g \cdot f = (D^3 g)f - 3(D^2 g)(Df) + 3(Dg)(D^2 f) - g(D^3 f), \quad \cdots.$$

For P_{II}, we treat only the case

$$D = \frac{d}{dt}, \tag{5.18}$$

and when considering the other Painlevé equations, we agree to replace (5.18) by the differentials according to Table 13.1.

In general, let H, H_1 be Hamiltonians such that

$$X = H_1 - H \tag{5.19}$$

is a function of the canonical variables. The τ-functions f and g related to these Hamiltonians are defined by

$$H_1 = D \log g, \quad H = D \log f, \tag{5.20}$$

respectively; we have from (5.19)

$$X = D \log \frac{g}{f}. \tag{5.21}$$

It is easy to verify the following fundamental formulae of differentiation:

$$DH_1 + DH = \frac{\mathcal{D}^2 g \cdot f}{g \cdot f} - \left(\frac{\mathcal{D}g \cdot f}{g \cdot f} \right)^2,$$

$$D^2 H_1 - D^2 H = \frac{\mathcal{D}^3 g \cdot f}{g \cdot f} - 3 \frac{\mathcal{D}^2 g \cdot f}{g \cdot f} \frac{\mathcal{D}g \cdot f}{g \cdot f} + 2 \left(\frac{\mathcal{D}g \cdot f}{g \cdot f} \right)^3.$$

If the left-hand sides of the fundamental formulae can be represented in terms of X, H_1, and H, then we deduce from (5.20) and (5.21) certain relations among f, g, and their derivatives.

In the case of P_{II}, we put

$$H = H(\kappa), \quad H_1 = H(\kappa - 1),$$

that is, $f = \tau(\kappa)$, $g = \tau(\kappa - 1)$. And by (5.3) we have $X = q(\kappa)$. By the use of the differential equations of P_{II}

$$Dq = p - q^2 - \frac{t}{2}, \quad Dp = 2qp + \alpha + \frac{1}{2},$$

we compute

$$DH = -\frac{1}{2}p, \quad DH_1 = \frac{1}{2}p - q^2 - \frac{t}{2}. \tag{5.22}$$

Here we write $q = q(\kappa)$ and so on, for the sake of simplicity of presentation. It follows from (5.22) that

$$DH_1 + DH = -q^2 - \frac{t}{2},$$

and we obtain the first bilinear relation

$$\mathcal{D}^2 g \cdot f + \frac{t}{2} g \cdot f = 0, \tag{5.23}$$

by means of (5.21) and the fundamental formulae. Moreover, since $H_1 - H = q$, we have

$$D^2 H_1 - D^2 H = 2q^3 + tq + \alpha, \quad \kappa = \alpha + \frac{1}{2},$$

from which we deduce

$$\mathcal{D}^3 g \cdot f - 3\frac{\mathcal{D}^2 g \cdot f}{g \cdot f} \mathcal{D}g \cdot f = t\mathcal{D}g \cdot f + \alpha g \cdot f.$$

Therefore, the first bilinear relation implies the second one,

$$\mathcal{D}^3 g \cdot f + \frac{t}{2}\mathcal{D}g \cdot f = \alpha g \cdot f. \tag{5.24}$$

Finally, we arrive at the bilinear forms (5.23)–(5.24) associated with P_{II}, corresponding to the translation

$$\kappa \longmapsto \kappa - 1.$$

It is easy to verify that (5.23)–(5.24) are equivalent to P_{II}.

If we put $f = 1$ in the bilinear forms, then the Hirota derivative $\mathcal{D}$ is reduced to the ordinary derivative D. In the case of P_{II}, (5.23)–(5.24) are converted into

$$\begin{cases} D^2 g + \dfrac{t}{2}g = 0, \\[2mm] D^3 g + \dfrac{t}{2}Dg = \alpha g. \end{cases}$$

By differentiating the first equation with respect to t, we deduce from the second equation the constraint

$$\alpha = -\frac{1}{2}.$$

The function $q = D \log g$ satisfies the Riccati equation

$$Dq = q^2 + \frac{t}{2},$$

which gives classical solutions of P_{II}.

Return to the bilinear forms (5.23)–(5.24) and replace f and g by F and G, respectively, such that

$$f = \exp\left(-\frac{1}{24}t^3\right)F, \quad g = \exp\left(-\frac{1}{24}t^3\right)G.$$

Then we obtain the other form of bilinear relations

$$\begin{cases} \mathcal{D}^2 G \cdot F = 0, \\ \mathcal{D}^3 G \cdot F = t\mathcal{D}G \cdot F + \alpha G \cdot F, \end{cases} \tag{5.25}$$

which is known as the usual bilinear form of P_{II}; cf. [1, 2].

The general process explained above for P_{II} works effectively even for the other Painlevé equations P_J. In fact, by suitably choosing the Hamiltonians H and H_1, we will obtain the bilinear forms associated with P_J, except for P_I; recall that P_I contains no parameter.

Consider, for example, the sixth Painlevé equation

$$P_{VI} \quad \frac{d^2 q}{dt^2} = \frac{1}{2}\left(\frac{1}{q} + \frac{1}{q-1} + \frac{1}{q-t}\right)\left(\frac{dq}{dt}\right)^2 - \left(\frac{1}{t} + \frac{1}{t-1} + \frac{1}{q-t}\right)\frac{dq}{dt}$$
$$+ \frac{q(q-1)(q-t)}{t^2(t-1)^2}\left\{\alpha + \beta\frac{t}{q^2} + \gamma\frac{t-1}{(q-1)^2} + \delta\frac{t(t-1)}{(q-t)^2}\right\}.$$

For studying the Hamiltonian structure of P_{VI}, it is convenent to use the parameters b_k $(k = 1, 2, 3, 4)$ such that

$$b_1 = \frac{1}{2}(\kappa_0 + \kappa_1), \qquad\qquad b_2 = \frac{1}{2}(\kappa_0 - \kappa_1),$$
$$b_3 = \frac{1}{2}(\theta - 1 + \kappa_\infty), \qquad\qquad b_4 = \frac{1}{2}(\theta - 1 - \kappa_\infty).$$

Then the bilinear forms fo P_{VI} associated with the translation

$$\begin{pmatrix} b_1 \\ b_2 \\ b_3 \\ b_4 \end{pmatrix} \longmapsto \begin{pmatrix} b_1 \\ b_2 \\ b_3 \\ b_4 \end{pmatrix} + \begin{pmatrix} 0 \\ 0 \\ 0 \\ 1 \end{pmatrix}$$

can be written as follows:

$$\mathcal{D}^2 g \cdot f + A(t)\mathcal{D}g \cdot f - (2t - 1)g \cdot Df + t(t - 1)\lambda g \cdot f = 0,$$

$$\mathcal{D}^3 g \cdot f + A(t)\mathcal{D}^2 g \cdot f - (2t-1)\mathcal{D}g \cdot Df + t(t-1)\lambda \mathcal{D}g \cdot f$$
$$= t(t-1)\left[2b_4 g \cdot Df + 2(b_4+1)\mathcal{D}g \cdot f - 2\mu \mathcal{D}g \cdot f - B(t)g \cdot f\right].$$

Here the Hirota derivative concerns the differential

$$D = t(t-1)\frac{d}{dt},$$

and we use the following notation:

$$A(t) = (b_1 + b)(2t-1) - b_2,$$
$$B(t) = (b_1 + b_3)\mu(2t-1),$$
$$\lambda = (b_1 + b_3)(b_1 + b_3 + 2b_4 + 1),$$
$$\mu = (b_1 + b_4)b + (b_1 + b_4 + 1)(b+1),$$
$$b = b_3 + b_4.$$

Note that if we put $f = 1$, then the first bilinear relation is reduced to the hypergeometric differential equation of Gauss with respect to the unknown function g. Studies of the bilinear forms of the Painlevé equations will be published in the near future by the author of the present note.

5.4 Remarks on the Holonomic Deformation

Consider the linear ordinary differential equation

$$\frac{d^2y}{dx^2} + p_1(x,t)\frac{dy}{dx} + p_2(x,t)y = 0, \qquad (5.26)$$

with the coefficients

$$\begin{cases} p_1(x,t) = -2x^2 - t - \dfrac{1}{x-q}, \\[2mm] p_2(x,t) = -2\kappa x - 2H + \dfrac{p}{x-q}. \end{cases} \qquad (5.27)$$

The linear differential equation (5.26)–(5.27) admits an irregular singular point, $x = \infty$, of Poincaré rank 2, and a regular singular point, $x = q$. Assuming that $x = q$ is an apparent singularity, we can determine the parameter H as a function of q, p, and t, by means of the Frobenius method. In fact, we see that $H = H(t; q, p)$ is exactly the same polynomial as the Hamiltonian of the second Painlevé system

$$H = \frac{1}{2}p^2 - \left(q^2 + \frac{t}{2}\right)p - \kappa q. \qquad (H_{II})$$

Moreover, it is known by [23] that the holonomic deformation of (5.26)–(5.27) is governed by the Hamiltonian system

$$\frac{dq}{dt} = \frac{\partial H}{\partial p}, \quad \frac{dp}{dt} = -\frac{\partial H}{\partial q}, \qquad (5.28)$$

with the Hamiltonian H_{II}; this fact gives a characterization of H_{II}.

Let $(q, p) = (q(t), p(t))$ be a solution of (5.28). Then there exists a fundamental system of solutions of (5.26)–(5.27) such that its monodromy data are independent of t. In particular, we consider a particular solution of the form

$$\begin{cases} q = \dfrac{1}{z} - \dfrac{t_0}{6}z - \dfrac{\alpha+1}{4}z^2 + hz^3 + \cdots, \\ p = -\kappa z + \cdots, \\ H = c + \cdots, \end{cases}$$

where $z = t - t_0$ and c is a constant depending on an arbitrary constant h. If we take the limit $t \to t_0$, under the assumption that x stays in a compact subset of the complex x-plane $\mathbb{C}$, then (5.26)–(5.27) tends to the following linear ordinary differential equation:

$$\frac{\mathrm{d}^2 y}{\mathrm{d}x^2} + (2x^2 + t_0)\frac{\mathrm{d}y}{\mathrm{d}x} - 2\kappa y = 2cy. \tag{5.29}$$

By the use of the Hamiltonian $H(t; q, p)$ of the second Painlevé system, (5.29) can be written in the form

$$H\left(t_0; x, \frac{\mathrm{d}}{\mathrm{d}x}\right)y = cy.$$

We have such a relation between (5.26)–(5.27) and the quantization of H_{II}, but have little knowledge about what this fact means.

6   The Painlevé System of Two Variables

6.1   Hamiltonian Systems of Two Variables

One of the extensions of the sixth Painlevé system, called the *Garnier system*, was studied in [23]. The Garnier system is characterized by the holonomic deformation of linear ordinary differential equations of the Fuchsian type. We have, in fact, a Hamiltonian system of several variables whose Hamiltonians are rational functions of the canonical variables. It is known by [10] that the Garnier system possesses a polynomial Hamiltonian structure; by a canonical transformation, we can convert the system to a Hamiltonian system with Hamiltonian functions that are polynomial with respect to the unknown variables of the equations. In [9], the step-by-step degeneration of the Garnier system was investigated in the case of two variables, and a family of Hamiltonian systems was proposed as an extension of the Painlevé systems to the case of two variables.

Among them, the following completely integrable system is considered in [28] to be an extension of the second one:

$$\frac{\partial q_k}{\partial t_j} = \frac{\partial H_j}{\partial p_k}, \quad \frac{\partial p_k}{\partial t_j} = -\frac{\partial H_j}{\partial q_k} \quad (j, k = 1, 2). \tag{6.1}$$

Here the Hamiltonian functions are given by

$$\begin{cases} H_1 = \left(q_2^2 - q_1 - t_1\right) p_1^2 + 2q_2 p_1 p_2 + p_2^2 \\ \quad + 2\left(q_1^2 - t_1^2 + t_2 q_2\right) p_1 + 2\left(q_1 q_2 + t_1 q_2 + t_2\right) p_2 + 2\kappa q_1, \\ H_2 = q_2 p_1^2 + 2p_1 p_2 + 2\left(q_1 q_2 + t_1 q_2 + t_2\right) p_1 \\ \quad + 2\left(q_2^2 - q_1 + t_1\right) p_2 + 2\kappa q_2. \end{cases} \tag{6.2}$$

Now we will make a brief survey of results on this system and introduce a new form of the Hamiltonians that is convenient for studying the canonical transformations and the τ-functions. Consider the linear equation

$$\frac{d^2 y}{dx^2} + p_1(x, t)\frac{dy}{dx} + p_2(x, t)y = 0, \tag{6.3}$$

such that

$$\begin{cases} p_1(x, t) = -P(x, t) - \dfrac{1}{x - \lambda_1} - \dfrac{1}{x - \lambda_2}, \\ p_2(x, t) = -2\kappa x^2 - 2K_1 x - 2K_2 + \dfrac{\mu_1}{x - \lambda_1} + \dfrac{\mu_2}{x - \lambda_2}, \end{cases} \tag{6.4}$$

where $t = (t_1, t_2)$ and

$$P(x, t) = 2x^3 + 2t_2 x + t_1. \tag{6.5}$$

Under the assumption that $x = \lambda_1$ and $x = \lambda_2$ are apparent singularities of (6.3)–(6.4), we have

$$\begin{cases} K_1 = \dfrac{1}{2}\sum_{k=1}^{2} \dfrac{1}{\Lambda'(\lambda_k)} \left[\mu_k^2 - P(\lambda_k)\mu_k - 2\kappa\lambda_k^2\right], \\ K_2 = \sum_{k=1}^{2} \dfrac{Q(\lambda_k)}{\Lambda'(\lambda_k)} \left[\mu_k^2 - \left(P(\lambda_k) + \dfrac{1}{Q(\lambda_k)}\right)\mu_k - 2\kappa\lambda_k^2\right], \end{cases} \tag{6.6}$$

with $\Lambda(x) = (x - \lambda_1)(x - \lambda_2)$ and $Q(x) = x - \lambda_1 - \lambda_2$, $\Lambda'(x)$ being the derivative of $\Lambda(x)$ with respect to x. The holonomic deformation of (6.3)–(6.4) has been studied in [9]; in this case we obtain the completely integrable Hamiltonian system

$$\frac{\partial \lambda_k}{\partial t_j} = \frac{\partial K_j}{\partial \mu_k}, \quad \frac{\partial \mu_k}{\partial t_j} = -\frac{\partial K_j}{\partial \lambda_k} \quad (j, k = 1, 2), \tag{6.7}$$

with the Hamiltonians (6.6). Moreover, we can associate with (6.7) the polynomial Hamiltonian structure with the Hamiltonians

$$
\begin{cases}
H_1 = -\dfrac{1}{2} q_1 p_2^2 + p_1 p_2 \\[2mm]
\qquad - \left(q_1^2 + q_2 + \dfrac{1}{2} t_2 \right) p_1 - \left(q_1 q_2 - \dfrac{1}{2} t_2 q_1 + \dfrac{1}{2} t_1 \right) p_2 - \kappa q_1, \\[3mm]
H_2 = \dfrac{1}{2} p_1^2 + \dfrac{1}{2} \left(q_1^2 + q_2 - \dfrac{1}{2} t_2 \right) p_2^2 - q_1 p_1 p_2 \\[2mm]
\qquad - \left(q_1 q_2 - \dfrac{1}{2} t_2 q_1 + \dfrac{1}{2} t_1 \right) p_1 - \left(q_2^2 - \dfrac{1}{2} t_1 q_1 - \dfrac{1}{4} t_2^2 \right) p_2 - \kappa q_2.
\end{cases}
\tag{6.8}
$$

This Hamiltonian structure is equivalent to that with the Hamiltonians (6.2); in fact, by replacing in (6.2) (q_1, q_2) by $(-q_2, -q_1)$, (p_1, p_2) by $(-p_2, -p_1)$, $(2t_1, 2t_2)$ by (t_2, t_1), and (H_1, H_2) by $(2H_2, 2H_1)$, we obtain (6.8). It is easy to see that this replacement defines a canonical transformation. The Hamiltonian system (6.1)–(6.8) can be characterized on the one hand by the step-by-step degeneration from the two-dimensional Garnier system, on the other hand by the holonomic deformation of the linear equation (6.3)–(6.4).

Since we can regard (6.5) as the polynomial representing the versal deformation of the simple singularity of A_2-type, we call the Hamiltonian system (6.1)–(6.8) the A_2-*system*. According to this terminology, the second Painlevé system can be called the A_1-*system*. We add another reason why the system (6.1)–(6.8) should be called the A_2-system. When $\kappa = 0$, the second Painlevé system, that is, the A_1-system, has particular solutions of the form

$$
p = 0, \quad q = \frac{\mathrm{d}}{\mathrm{d}t} \log u;
$$

u solves the holonomic system of partial differential equations and is written as follows:

$$
u = \int_\gamma \exp \left[\int_0^\xi -P_1(\eta, t)\, \mathrm{d}\eta \right] \mathrm{d}\xi.
$$

Here γ denotes a path in the complex plane $\mathbb{C}$, and

$$
P_1(\eta, t) = 2\eta^2 + t;
$$

see [25]. Moreover, for the A_2-system (6.1)–(6.8) with $\kappa = 0$, we have the particular solutions

$$
p_1 = p_2 = 0, \quad q_i = \frac{\partial}{\partial t_i} \log u \quad (i = 1, 2),
$$

where u is given by the integral representation

$$u = \exp\left(-\frac{1}{4}t_2^2\right) \int_\gamma \exp\left[\int_0^\xi -P(\eta, t)\,d\eta\right]\,d\xi.$$

This result was established for the first time by [28] with the Hamiltonian (6.2), and we rewrite it by using (6.8) instead of (6.2).

The generalization of the A_2-system to the case of several variables is called the A_g-system, which is considered in [11–13]. Moreover, Kawamuko is studying the fourth Painlevé system with several variables; see [6–8].

The aim of the present section is to study the birational canonical transformations of the A_2-system. We will show below that the contiguity relation for the A_2-system is given as follows:

$$\begin{cases} p_1(\kappa) - 2q_1(\kappa)^3 - 4q_1(\kappa)q_2(\kappa) - t_1 = q_1(\kappa - 1)p_2(\kappa - 1), \\ p_2(\kappa) - 2q_1(\kappa)^2 - 2q_2(\kappa) - t_2 = -p_2(\kappa - 1), \\ q_1(\kappa) = -\dfrac{p_1(\kappa - 1)}{p_2(\kappa - 1)}, \\ q_2(\kappa) = -q_2(\kappa - 1) + \dfrac{1}{2}p_2(\kappa - 1) - \dfrac{q_1(\kappa - 1)p_1(\kappa - 1) + \kappa - 1}{p_2(\kappa - 1)}. \end{cases}$$

Here the canonical variables of the A_2-system at an arbitrarily fixed value κ of the parameter are written as $q_1(\kappa)$, $q_2(\kappa)$, and so on.

6.2   Canonical Transformations and the A_2-System

We now recall briefly results on a Hamiltonian system of several variables; see, for example, [4]. Consider in general the system

$$\frac{\partial q_k}{\partial t_j} = \frac{\partial H_j}{\partial p_k}, \quad \frac{\partial p_k}{\partial t_j} = -\frac{\partial H_j}{\partial q_k} \quad (j, k = 1, \ldots, N). \tag{6.9}$$

This system concerns the canonical variables $q = (q_1, \ldots, q_N)$, $p = (p_1, \ldots, p_N)$, the multi-times $t = (t_1, \ldots, t_N)$, and the multi-Hamiltonians $H = (H_1, \ldots, H_N)$. The fundamental form of (6.9) is the two-form

$$\Omega = \Omega(q, p, H, t) = \sum_{k=1}^{N} dp_k \wedge dq_k - \sum_{j=1}^{N} dH_j \wedge dt_j. \tag{6.10}$$

This can be written as

$$\Omega = \sum_{i<j} \Omega_{ij}\,dt_i \wedge dt_j,$$

$$\Omega_{ij} = \sum_{k=1}^{N}\left[\frac{\partial H_j}{\partial q_k}\frac{\partial H_i}{\partial p_k} - \frac{\partial H_i}{\partial q_k}\frac{\partial H_j}{\partial p_k}\right] + \left(\frac{\partial}{\partial t_i}\right)H_j - \left(\frac{\partial}{\partial t_j}\right)H_i, \tag{6.11}$$

where $(\partial/\partial t_i)$ denotes differentiation with respect to t_j such that q, p are viewed to be independent of t. It is known that the system (6.9) is completely integrable if and only if the Ω_{ij}'s do not depend on q, p. Moreover, a transformation of the form

$$(q, p, H, t) \quad \longmapsto \quad (q', p', H', t')$$

is canonical if and only if

$$\Omega(q, p, H, t) = \Omega(q', p', H', t'),$$

where $q' = (q'_1, \ldots, q'_N)$, and so on.

Return to consideration of the A_2-system (6.1)–(6.8) and let the 2-form $\Omega = \Omega(q, p, H, t)$ be the fundamental form (6.10) for the A_2-system. We prove the following proposition.

Proposition 6.1. Ω *vanishes along a solution of the Hamiltonian system* (6.1).

In fact, by means of (6.8), we have

$$\left(\frac{\partial}{\partial t_1} \right) H_2 = \left(\frac{\partial}{\partial t_2} \right) H_1. \tag{6.12}$$

Moreover, we can verify by computation

$$\sum_{k=1,2} \left[\frac{\partial H_j}{\partial q_k} \frac{\partial H_i}{\partial p_k} - \frac{\partial H_i}{\partial q_k} \frac{\partial H_j}{\partial p_k} \right] = 0 \quad (i, j = 1, 2). \tag{6.13}$$

The proposition follows from (6.11) immediately.

On the other hand, since (6.11) can be written in the form

$$\Omega_{12} = \frac{1}{2} \left[\frac{\partial H_2}{\partial t_1} - \frac{\partial H_1}{\partial t_2} + \left(\frac{\partial}{\partial t_1} \right) H_2 - \left(\frac{\partial}{\partial t_2} \right) H_1 \right],$$

we have

$$\frac{\partial H_2}{\partial t_1} = \frac{\partial H_1}{\partial t_2}. \tag{6.14}$$

Hence we arrive at the following result.

Proposition 6.2. *The 1-form* $\omega \equiv H_1 dt_1 + H_2 dt_2$ *is closed.*

By virtue of this proposition, we define, up to multiplicative constants, the τ-function related to the A_2-system as follows:

$$\omega = d \log \tau. \tag{6.15}$$

As was mentioned above, when $\kappa = 0$, the A_2-system possesses particular solutions such that $p_1 = p_2 = 0$. In this case, we have $\omega = 0$, and then we can put $\tau = 1$.

6.3 The Hamiltonians of the A_2-System

To obtain a contiguity relation for the A_2-system, we follow the case of the studies on the second Painlevé system; in [25], we have established the relation (5.2) by introducing the variable

$$\bar{p} = p - 2q^2 - t.$$

For the A_2-system, we consider the auxiliary variables defined as follows:

$$\begin{cases} \bar{p}_1 = p_1 - 2q_1^3 - 4q_1 q_2 - t_1, \\ \bar{p}_2 = p_2 - 2q_1^2 - 2q_2 - t_2. \end{cases} \tag{6.16}$$

Proposition 6.3. *The change (6.16) of variables extends to the canonical transformation*

$$(q, p, H, t) \quad \longmapsto \quad (q, \bar{p}, \overline{H}, t),$$

where $\bar{p} = (\bar{p}_1, \bar{p}_2)$, $\overline{H} = (\overline{H}_1, \overline{H}_2)$.

In fact, put

$$\overline{H}_j = H_j + q_j \quad (j = 1, 2). \tag{6.17}$$

Let $\Omega\left(q, \bar{p}, \overline{H}, t\right)$ be a 2-form of the form (6.10) with $\bar{p}$, $\overline{H}$. Then it is easy to see by computation the equality

$$\Omega(q, p, H, t) = \Omega\left(q, \bar{p}, \overline{H}, t\right),$$

which proves the proposition.

Observe that the relation (6.16) can be written in the form

$$p_1 - \bar{p}_1 - t_1 = 2 \det \begin{pmatrix} q_1 & -q_2 & 0 \\ 1 & q_1 & -q_2 \\ 0 & 1 & q_1 \end{pmatrix},$$

$$p_2 - \bar{p}_2 - t_2 = 2 \det \begin{pmatrix} q_1 & -q_2 \\ 1 & q_1 \end{pmatrix}.$$

We can deduce from (6.16)–(6.17) the explicit forms of $\overline{H}_j$'s:

$$\begin{cases} \overline{H}_1 = -\dfrac{1}{2} q_1 \bar{p}_2^2 + \bar{p}_1 \bar{p}_2 \\ \qquad + \left(q_1^2 + q_2 + \dfrac{1}{2} t_2 \right) \bar{p}_1 + \left(q_1 q_2 - \dfrac{1}{2} t_2 q_1 + \dfrac{1}{2} t_1 \right) \bar{p}_2 - (\kappa - 1) q_1, \\ \overline{H}_2 = \dfrac{1}{2} \bar{p}_1^2 + \dfrac{1}{2} \left(q_1^2 + q_2 - \dfrac{1}{2} t_1 \right) \bar{p}_2^2 - q_1 \bar{p}_1 \bar{p}_2 \\ \qquad + \left(q_1 q_2 - \dfrac{1}{2} t_2 q_1 + \dfrac{1}{2} t_1 \right) \bar{p}_1 + \left(q_2^2 - \dfrac{1}{2} t_1 q_1 - \dfrac{1}{4} t_2^2 \right) \bar{p}_2 - (\kappa - 1) q_2. \end{cases} \tag{6.18}$$

6.4 Derivatives of the Hamiltonians

We compute derivatives of the Hamiltonian H_1. Since by (6.13)

$$
\frac{\partial H_i}{\partial t_j} = \sum_{k=1,2}\left[\frac{\partial H_i}{\partial q_k}\frac{\partial H_j}{\partial p_k} - \frac{\partial H_j}{\partial q_k}\frac{\partial H_i}{\partial p_k}\right] + \left(\frac{\partial}{\partial t_j}\right)H_i
$$

$$
= \left(\frac{\partial}{\partial t_j}\right)H_i \quad (i,j=1,2),
$$

(6.19)

we deduce from (6.12), (6.14), and (6.19) the following equations:

$$
\frac{\partial H_1}{\partial t_1} = -\frac{1}{2}p_2, \quad \frac{\partial H_1}{\partial t_2} = \frac{\partial H_2}{\partial t_1} = -\frac{1}{2}p_1 + \frac{1}{2}q_1 p_2.
$$

(6.20)

Then by means of (6.1), (6.8) we can compute higher derivatives of H_1 as follows

$$
\frac{\partial^2 H_1}{\partial t_1^2} = -\frac{1}{2}p_1 - \frac{1}{2}q_1 p_2,
$$

(6.21)

$$
\frac{\partial^2 H_1}{\partial t_1 \partial t_2} = \frac{1}{4}p_2^2 - \frac{1}{2}q_1 p_1 - q_2 p_2 - \frac{1}{2}\kappa,
$$

(6.22)

$$
\frac{\partial^2 H_1}{\partial t_2^2} = -\frac{1}{4}q_1 p_2^2 + \frac{1}{2}\left(q_1^2 - q_2 + \frac{1}{2}t_2\right)p_1 + \frac{1}{2}\left(q_1 q_2 + \frac{1}{2}t_2 q_1\right)p_2 + \frac{1}{2}\kappa q_1,
$$

and so on. Moreover, we obtain from (6.20), (6.21), (6.22) the following expressions:

$$
\begin{cases}
p_2 = -2\dfrac{\partial H_1}{\partial t_1}, \quad p_1 = -\dfrac{\partial H_1}{\partial t_2} - \dfrac{\partial^2 H_1}{\partial t_1^2}, \\[3mm]
q_1 = -\dfrac{\dfrac{\partial H_1}{\partial t_2} - \dfrac{\partial^2 H_1}{\partial t_1^2}}{2\dfrac{\partial H_1}{\partial t_1}}, \\[6mm]
q_2 = \dfrac{2\dfrac{\partial^2 H_1}{\partial t_1 \partial t_2} - 2\left(\dfrac{\partial H_1}{\partial t_1}\right)^2 + \kappa}{4\dfrac{\partial H_1}{\partial t_1}} + \dfrac{\left(\dfrac{\partial H_1}{\partial t_2}\right)^2 - \left(\dfrac{\partial^2 H_1}{\partial t_1^2}\right)^2}{8\left(\dfrac{\partial H_1}{\partial t_1}\right)^2}.
\end{cases}
$$

(6.23)

Therefore, we have established the following result.

Proposition 6.4. *The Hamiltonian functions H_j ($j = 1, 2$) and their derivatives are rational functions of*

$$
\frac{\partial H_1}{\partial t_1}, \quad \frac{\partial H_1}{\partial t_2}, \quad \frac{\partial^2 H_1}{\partial t_1^2}, \quad \frac{\partial^2 H_1}{\partial t_1 \partial t_2}.
$$

For example, we have

$$\frac{\partial^2 H_1}{\partial t_2^2} + q_1 \frac{\partial^2 H_1}{\partial t_1 \partial t_2} - \left(q_2 - \frac{1}{2}t_2\right)\frac{\partial^2 H_1}{\partial t_1^2} = 0,$$

$$\frac{\partial H_2}{\partial t_2} + \frac{\partial^2 H_1}{\partial t_1 \partial t_2} + \frac{1}{2}\kappa - 2\left(q_2 - \frac{1}{2}t_2\right)\frac{\partial H_1}{\partial t_1} = 0,$$

and so on.

Since for the Hamiltonian $\overline{H}_j$ $(j = 1, 2)$

$$\left(\frac{\partial}{\partial t_1}\right)\overline{H}_2 = \left(\frac{\partial}{\partial t_2}\right)\overline{H}_1,$$

we can pursue computations similar to those of the preceding subsection. In fact, we have instead of (6.20), (6.21), (6.22) the following equalities:

$$\begin{cases} \dfrac{\partial \overline{H}_1}{\partial t_1} = \dfrac{1}{2}\bar{p}_2, \quad \dfrac{\partial \overline{H}_1}{\partial t_2} = \dfrac{1}{2}\bar{p}_1 - \dfrac{1}{2}q_1\bar{p}_2, \\[2mm] \dfrac{\partial^2 \overline{H}_1}{\partial t_1^2} = -\dfrac{1}{2}\bar{p}_1 - \dfrac{1}{2}q_1\bar{p}_2, \\[2mm] \dfrac{\partial^2 \overline{H}_1}{\partial t_1 \partial t_2} = -\dfrac{1}{4}\bar{p}_2^2 - \dfrac{1}{2}q_1\bar{p}_1 - q_2\bar{p}_2 + \dfrac{\kappa - 1}{2}. \end{cases} \qquad (6.24)$$

It follows that

$$\begin{cases} \bar{p}_2 = 2\dfrac{\partial \overline{H}_1}{\partial t_1}, \quad \bar{p}_1 = \dfrac{\partial \overline{H}_1}{\partial t_2} - \dfrac{\partial^2 \overline{H}_1}{\partial t_1^2}, \\[3mm] q_1 = -\dfrac{\dfrac{\partial \overline{H}_1}{\partial t_2} + \dfrac{\partial^2 \overline{H}_1}{\partial t_1^2}}{2\dfrac{\partial \overline{H}_1}{\partial t_1}}, \\[5mm] q_2 = -\dfrac{2\dfrac{\partial^2 \overline{H}_1}{\partial t_1 \partial t_2} + 2\left(\dfrac{\partial \overline{H}_1}{\partial t_1}\right)^2 - (\kappa - 1)}{4\dfrac{\partial \overline{H}_1}{\partial t_1}} + \dfrac{\left(\dfrac{\partial \overline{H}_1}{\partial t_2}\right)^2 - \left(\dfrac{\partial^2 \overline{H}_1}{\partial t_1^2}\right)^2}{8\left(\dfrac{\partial \overline{H}_1}{\partial t_1}\right)^2}. \end{cases}$$

$$(6.25)$$

The Hamiltonians and their derivatives are rational functions of

$$\frac{\partial \overline{H}_1}{\partial t_1}, \quad \frac{\partial \overline{H}_1}{\partial t_2}, \quad \frac{\partial^2 \overline{H}_1}{\partial t_1^2}, \quad \frac{\partial^2 \overline{H}_1}{\partial t_1 \partial t_2}.$$

For example, we have

$$\frac{\partial^2 \overline{H}_1}{\partial t_2^2} + q_1 \frac{\partial^2 \overline{H}_1}{\partial t_1 \partial t_2} - \left(q_2 - \frac{1}{2}t_2\right)\frac{\partial^2 \overline{H}_1}{\partial t_1^2} = 0,$$

$$\frac{\partial \overline{H}_2}{\partial t_2} - \frac{\partial^2 \overline{H}_1}{\partial t_1 \partial t_2} + \frac{1}{2}(\kappa - 1) - 2\left(q_2 - \frac{1}{2}t_2\right)\frac{\partial \overline{H}_1}{\partial t_1} = 0,$$

and so on.

6.5 Contiguity Relations

Given a function H_1, we regard (6.23) as defining new functions (q, p) and write this correspondence as

$$(q, p) = \Phi(H_1).$$

Now, by using the Hamiltonian $\overline{H}_1$, we introduce new variables $Q = (Q_1, Q_2)$, $P = (P_1, P_2)$ by

$$(Q, P) = \Phi(\overline{H}_1).$$

By virtue of (6.24), (Q, P) can be written as functions of $(q, \bar{p})$. In fact, we verify the following equalities:

$$\begin{cases} P_2 = -\bar{p}_2, \quad P_1 = q_1 \bar{p}_2, \quad Q_1 = -\dfrac{\bar{p}_1}{\bar{p}_2}, \\[3mm] Q_2 = -q_2 - \dfrac{1}{2}\bar{p}_2 - \dfrac{q_1 \bar{p}_1 - \kappa + 1}{\bar{p}_2}. \end{cases} \tag{6.26}$$

These relations define a birational correspondence; in fact, we obtain from (6.26)

$$\begin{cases} \bar{p}_2 = -P_2, \quad \bar{p}_1 = Q_1 P_2, \quad q_1 = -\dfrac{P_1}{P_2}, \\[3mm] q_2 = -Q_2 + \dfrac{1}{2}P_2 - \dfrac{Q_1 P_1 + \kappa - 1}{P_2}. \end{cases} \tag{6.27}$$

Now we substitute (6.27) into (6.18) and then look at the expressions of the Hamiltonians $\overline{H}_1$ and $\overline{H}_2$ as functions of (Q, P). It is not difficult, even by hand, to verify that $\overline{H}_1$ and $\overline{H}_2$ are exactly the same polynomials as H_1 and H_2, respectively, except for the replacement of the parameter

$$\kappa \quad \longleftrightarrow \quad \kappa - 1.$$

Since we deduce from (6.26) the equality

$$\sum_{k=1,2} \mathrm{d}\bar{p}_k \wedge dq_k = \sum_{k=1,2} \mathrm{d}P_k \wedge dQ_k,$$

$\overline{H}_j$ $(j = 1, 2)$ are nothing but the Hamiltonians of the A_2-system with the value $\kappa - 1$ of the parameter. Hence we arrive at the following Theorem.

Theorem 6.1. *There exists a birational canonical transformation*

$$\mathcal{H}(\kappa) \longrightarrow \mathcal{H}(\kappa - 1)$$

of the A_2-system, where $\mathcal{H}(\kappa) = (q(\kappa), p(\kappa), H(\kappa), t)$.

Here we denote the canonical variables by $q(\kappa) = (q_1(\kappa), q_2(\kappa))$, and so on, in order to emphasize their dependence on the parameter. The explicit form of the birational canonical transformation is given by (6.16), (6.17), and (6.26), since $\overline{H}_j = H_j(\kappa - 1)$ and

$$P_1 = p_1(\kappa - 1), \quad P_2 = p_2(\kappa - 1), \quad Q_1 = q_1(\kappa - 1), \quad Q_2 = q_2(\kappa - 1). \tag{6.28}$$

We write again the contiguity relation for the A_2-system:

$$\begin{cases} p_1(\kappa - 1) = -q_1(\kappa)\bar{p}_2(\kappa), \quad p_2(\kappa - 1) = -\bar{p}_2(\kappa), \\ q_1(\kappa - 1) = -\dfrac{\bar{p}_1(\kappa)}{\bar{p}_2(\kappa)}, \\ q_2(\kappa - 1) = -q_2(\kappa) - \dfrac{1}{2}\bar{p}_2(\kappa) - \dfrac{q_1(\kappa)\bar{p}_1(\kappa) - \kappa + 1}{\bar{p}_2(\kappa)}. \end{cases} \tag{6.29}$$

Here $\bar{p}_1(\kappa) = p_1(\kappa) - 2q_1(\kappa)^3 - 4q_1(\kappa)q_2(\kappa) - t_1$, $\bar{p}_2(\kappa) = p_2(\kappa) - 2q_1(\kappa)^2 - 2q_2(\kappa) - t_2$. The inverse of (6.29) is given by

$$\begin{cases} \bar{p}_1(\kappa + 1) = -q_1(\kappa)p_2(\kappa), \quad \bar{p}_2(\kappa + 1) = -p_2(\kappa), \\ q_1(\kappa + 1) = -\dfrac{p_1(\kappa)}{p_2(\kappa)}, \\ q_2(\kappa + 1) = -q_2(\kappa) + \dfrac{1}{2}p_2(\kappa) - \dfrac{q_1(\kappa)p_1(\kappa) + \kappa}{p_2(\kappa)}. \end{cases} \tag{6.30}$$

6.6   The Toda Equation

We now study the τ-functions of the A_2-system (6.1) with the Hamiltonian (6.8). The definition of the τ-functions, $\tau(\kappa)$, for the A_2-system (6.1)–(6.8) is given by (6.15).

We consider the A_2-system

$$\mathcal{H}(\kappa) = (q(\kappa), p(\kappa) \ H(\kappa), t)$$

at an arbitrarily fixed value κ of the parameter. We have from (6.15)

$$H_1(\kappa) = \frac{\partial}{\partial t_1} \log \tau(\kappa), \quad H_2(\kappa) = \frac{\partial}{\partial t_2} \log \tau(\kappa).$$

Let $\overline{H} = (\overline{H}_1, \overline{H}_2)$ be the Hamiltonians given by (6.16)–(6.17). We have shown that $\overline{H} = H(\kappa - 1)$, and then we have the following result.

Proposition 6.5. For $q_j(\kappa) = q_j(t_1, t_2; \kappa)$ $(j = 1, 2)$,

$$q_j(\kappa) = \frac{\partial}{\partial t_j} \log \frac{\tau(\kappa - 1)}{\tau(\kappa)}. \tag{6.31}$$

In the rest of this subsection we will prove the following theorem.

Theorem 6.2. *The τ-functions satisfy the Toda equation*

$$\frac{\partial^2}{\partial t_1^2} \log \tau(\kappa) = \frac{\tau(\kappa - 1)\tau(\kappa + 1)}{\tau(\kappa)^2}. \tag{6.32}$$

In fact, by taking into consideration (6.28), we obtain from (6.23)

$$q_1(\kappa - 1) = -\frac{\dfrac{\partial \overline{H}_1}{\partial t_2} - \dfrac{\partial^2 \overline{H}_1}{\partial t_1^2}}{2\dfrac{\partial \overline{H}_1}{\partial t_1}},$$

and furthermore, by means of (6.25),

$$q_1(\kappa) = -\frac{\dfrac{\partial \overline{H}_1}{\partial t_2} + \dfrac{\partial^2 \overline{H}_1}{\partial t_1^2}}{2\dfrac{\partial \overline{H}_1}{\partial t_1}}.$$

Hence we have

$$q_1(\kappa - 1) - q_1(\kappa) = \frac{\dfrac{\partial^2 \overline{H}_1}{\partial t_1^2}}{\dfrac{\partial \overline{H}_1}{\partial t_1}} = \frac{\partial}{\partial t_1} \log \frac{\partial \overline{H}_1}{\partial t_1}. \tag{6.33}$$

On the other hand, by using again (6.23), (6.25), and (6.29), we can verify the following equation:

$$q_2(\kappa - 1) - q_2(\kappa) = \frac{\partial}{\partial t_2} \log \frac{\partial \overline{H}_1}{\partial t_1}. \tag{6.34}$$

It follows from (6.31), (6.33), (6.34) that

$$\frac{\partial}{\partial t_1} \log \left(\frac{\partial \overline{H}_1}{\partial t_1} \frac{\tau(\kappa - 1)^2}{\tau(\kappa)\tau(\kappa - 2)} \right) = \frac{\partial}{\partial t_2} \log \left(\frac{\partial \overline{H}_1}{\partial t_1} \frac{\tau(\kappa - 1)^2}{\tau(\kappa)\tau(\kappa - 2)} \right) = 0.$$

Therefore, we have a nonzero constant $c(\kappa - 1)$ such that

$$\frac{\partial \overline{H}_1}{\partial t_1} = c(\kappa - 1)\frac{\tau(\kappa)\tau(\kappa - 2)}{\tau(\kappa - 1)^2}. \tag{6.35}$$

The definition of the τ-functions implies

$$\frac{\partial \overline{H}_1}{\partial t_1} = \frac{\partial^2}{\partial t_1^2} \log \tau(\kappa - 1);$$

hence, by taking into consideration (6.35) and then by replacing $\kappa - 1$ by κ, we obtain the following equation:

$$\frac{\partial^2}{\partial t_1^2} \log \tau(\kappa) = c(\kappa) \frac{\tau(\kappa - 1)\tau(\kappa + 1)}{\tau(\kappa)^2},$$

$c(\kappa)$ being a nonzero constant. By choosing suitably multiplicative constants of the τ-functions, we can put $c(\kappa) = 1$ and then have the Toda equation (6.32).

6.7   The Sequence of τ-Functions

Consider the sequence

$$\ldots, \ \tau(\kappa + 1), \ \tau(\kappa), \ \tau(\kappa - 1), \ \ldots \tag{6.36}$$

of the τ-functions related to the A_2-system. We will see in what follows that (6.36) are not algebraically independent. In fact, there exists a polynomial

$$\Psi \in \mathbb{Z}[x_0, x_1, \ldots, x_6; T_1, T_2; K]$$

of ten variables such that

$$\Psi\left(\tau(\kappa + 3), \tau(\kappa + 2), \ldots, \tau(\kappa - 3); t_1, t_2; \kappa\right) = 0, \tag{6.37}$$

$\mathbb{Z}$ being the ring of rational integers. This fact means that the sequence (6.36) is subject to the algebraic relation (6.37), as well as the Toda equation (6.32). Moreover, the polynomial Ψ has homogeneity of degrees with respect to the first seven variables x_j. We can regard (6.37) as defining a hypersurface in the complex projective space $\boldsymbol{P}^6$ of dimension six, with the parameter (t_1, t_2, κ). Let us denote by $\Psi(\kappa)$ the left-hand side of (6.37). We can eliminate t_1 and t_2 from

$$\Psi(\kappa + 1) = 0, \quad \Psi(\kappa) = 0, \quad \Psi(\kappa - 1) = 0,$$

and then obtain an algebraic relation with respect to the adjacent nine τ-functions

$$\tau(\kappa + 4), \quad \tau(\kappa + 3), \quad \ldots, \quad \tau(\kappa - 4).$$

The algebraic variety defined by this relation may not be irreducible. We expect that an irreducible component of this variety should characterize the τ-functions of the A_2-system among solutions of the Toda equation (6.32). This kind of problem remains unsettled for the A_2-system.

In the rest of this subsection we establish algebraic relations among the seven adjacent τ-functions

$$\tau(\kappa + 3), \quad \tau(\kappa + 2), \quad \ldots, \quad \tau(\kappa - 2), \quad \tau(\kappa - 3). \tag{6.38}$$

Since $p_2 = -2\partial H_1/\partial t_1$, we obtain from (6.32)

$$p_2(\kappa) = -2\,\frac{\tau(\kappa-1)\tau(\kappa+1)}{\tau(\kappa)^2}. \tag{6.39}$$

Therefore, in order to study algebraic relations among τ-functions, it is sufficient to consider algebraic relations of $p_2(\kappa)$, $p_2(\kappa-1)$, ... , and so on. We deduce from the contiguity relations the following equalities:

$$\begin{cases} p_2(\kappa) + p_2(\kappa-1) - t_2 = 2q_1(\kappa)^2 + 2q_2(\kappa), \\ p_2(\kappa+1) + p_2(\kappa) - t_2 = 2q_1(\kappa+1)^2 + 2q_2(\kappa+1), \end{cases}$$

and

$$q_2(\kappa+1) + q_2(\kappa) = \frac{1}{2}p_2(\kappa) - \frac{q_1(\kappa)p_1(\kappa)}{p_2(\kappa)} - \frac{\kappa}{p_2(\kappa)}.$$

Then it is not difficult to show the equality

$$A(\kappa) = p_1(\kappa)^2 - q_1(\kappa)p_1(\kappa)p_2(\kappa) + q_1(\kappa)^2 p_2(\kappa)^2, \tag{6.40}$$

where we put

$$A(\kappa) \equiv \frac{1}{2}p_2(\kappa)^2 \left[p_2(\kappa+1) + p_2(\kappa) + p_2(\kappa-1) - 2t_2\right].$$

By the use of the canonical transformation (6.30), we can rewrite (6.40) as follows:

$$\frac{A(\kappa)}{p_2(\kappa)^2} = q_1(\kappa+1)^2 + q_1(\kappa+1)q_1(\kappa) + q_1(\kappa)^2. \tag{6.41}$$

By using carefully the contiguity relations, we can verify the following equality:

$$\begin{aligned} A(\kappa) - A(\kappa-1) &= q_1(\kappa)\left(p_2(\kappa) - p_2(\kappa-1)\right) \\ &\quad - \{p_2(\kappa)\left(q_1(\kappa+1) + q_1(\kappa)\right) + q_1(\kappa)p_2(\kappa-1) - F(\kappa)\} \\ &\quad \times \{3q_1(\kappa)\left(p_2(\kappa) + p_2(\kappa-1)\right) - 4F(\kappa)\}, \end{aligned} \tag{6.42}$$

where we put

$$F(\kappa) \equiv q_1(\kappa)^3 + t_2 q_1(\kappa) - \frac{1}{2}t_1.$$

Now we will prove our next theorem.

Theorem 6.3. *There exists an algebraic relation*

$$\Gamma\left(p_2(\kappa+2), p_2(\kappa+1), p_2(\kappa), p_2(\kappa-1), p_2(\kappa-2); t_1, t_2; \kappa\right) = 0$$

among the five adjacent functions

$$p_2(\kappa+2), \quad p_2(\kappa+1), \quad p_2(\kappa), \quad p_2(\kappa-1), \quad p_2(\kappa-2),$$

where $\Gamma \in \mathbb{Z}\,[x_0, x_1, x_2, x_3, x_4; T_1, T_2; K]$.

In fact, let R_0 be the polynomial ring $\mathbb{Q}[t; \kappa]$, $\mathbb{Q}$ being the field of rational numbers. By eliminating $q_1(\kappa + 1)$ from (6.41) and (6.42), we obtain a polynomial $G_1 \in R_0[x_0, x_1, x_2, x_3]$ such that

$$G_1(q_1(\kappa), A(\kappa), A(\kappa - 1), p_2(\kappa)) = 0.$$

On the other hand, we replace in (6.41)–(6.42) κ by $\kappa - 1$ and now eliminate $q_1(\kappa - 1)$ from these relations. This elimination yields the other algebraic relation

$$G_2(q_1(\kappa), A(\kappa - 1), A(\kappa - 2), p_2(\kappa - 1)) = 0.$$

By taking into consideration the equality

$$F(\kappa - 1) - F(\kappa) = (q_1(\kappa - 1) - q_1(\kappa)) \left(\frac{A(\kappa - 1)}{p_2(\kappa - 1)} + t_2 \right),$$

we can give the exact forms of the polynomials G_1, G_2 without difficulty; however it is not interesting to write it down explicitly. Now we regard G_1 and G_2 as polynomials of $q_1(\kappa)$ and consider their resultant $R(G_1, G_2)$. Then we have the algebraic relation

$$R(G_1, G_2) = 0,$$

where $R(G_1, G_2) \in R_0[p_2(\kappa + 1), p_2(\kappa), p_2(\kappa - 1), p_2(\kappa - 2), p_2(\kappa - 3)]$. By replacing κ by $\kappa + 1$, we complete the proof of Theorem 6.3.

By means of (6.39), the algebraic relation obtained in Theorem 6.3 concerns the seven τ-functions (6.38). If we write it down in terms of the τ-functions, this relation defines an algebraic variety in the projective space $\boldsymbol{P}^6$. In fact, through the ambiguity of multiplicative constants, we can regard (6.38) as homogeneous coordinates of $\boldsymbol{P}^6$.

7   REFERENCES

[1] S. Fukutani, K. Okamoto, and H. Umemura, *Special polynomials associated with the rational solutions and the Hirota bilinear relations of the 2nd and the 4th Painlevé equations*, preprint (1998).

[2] J. Hietarinta and M. Kruskal, *Hirota forms for the six Painlevé equations from singularity analysis*, Painlevé transcendents, their asymptotics and physical applications, eds. D. Levi and P. Winternitz (Plenum press, New York, 1992), pages 175–185.

[3] N. Hitchin, *Poncelet polygons and the Painlevé equations*, Geometry and Analysis, ed. S. Ramanan (Oxford University Press, Bombay, 1995).

[4] K. Iwasaki, H. Kimura, S. Shimomura, and M. Yoshida, From Gauss to Painlevé: a modern theory of special functions (Vieweg Verlag, Braunschweig, 1991).

[5] M. Jimbo and T. Miwa, *Monodromy preserving deformation of linear ordinary differential equations with rational coefficients, II*, Physica, **2D** (1981), 407–448.

[6] H. Kawamuko, *On the holonomic deformation of linear differential equations*, Proc. Japan Acad. Ser. A Math. Sci. **73** (1997), 152–154.

[7] H. Kawamuko, *On the polynomial Hamiltonian structure associated with the $L(1, g + 2; g)$ type*, Proc. Japan Acad. Ser. A Math. Sci. **73** (1997), 155–157.

[8] H. Kawamuko, Studies on the fourth Painlevé equation in several variables (Thesis, Univ. Tokyo, 1997).

[9] H. Kimura, *The degeneration of the two-dimensional Garnier system and the polynomial Hamiltonian structure*, Ann. Mat. Pura Appl., **155** (1989), 25–74.

[10] H. Kimura and K. Okamoto, *On the polynomial Hamiltonian structure of the Garnier system*, J. Math. Pures et Appl., **63** (1984), 129–146.

[11] D. Liu, *On the holonomic deformation of linear differential equations of A_3 type*, J. Math. Sci. Univ. Tokyo. **5** (1998), 435–458.

[12] D. Liu, *On the holonomic deformation of linear differential equations of A_4 type*, Kyushu J. Math., **51** (1997), 393–412.

[13] D. Liu, Holonomic deformation of linear differential equations of A_g type and polynomial Hamiltonian structure (Thesis, Univ. Tokyo, 1997).

[14] T. Matano, A. Matumiya, and T. Takano, *On some Hamiltonian structures of Painlevé systems, II*, to appear in J. Math. Japan.

[15] A. Matumiya, *On some Hamiltonian structures of Painlevé systems, III*, Kumamoto J. Math. **10** (1997), 45–73.

[16] Y. Murata, *Rational solutions of the second and the fourth equations of Painlevé* , Funkc. Ekvacioj. Ser. Int. **28** (1985), 1–32.

[17] K. Nishioka, *A note on the transcendence of Painlevé first transcendent*, Nagoya Math. J., **109** (1988), 63–67.

[18] M. Noumi and K. Okamoto, *Irreducibility of the second and the fourth Painlevé equations*, Funkc. Ekvacioj. Ser. Int. **40** (1997), 139–163.

[19] M. Noumi and Y. Yamada, *Symmetries in the fourth Painlevé equation and Okamoto polynomials*, to appear in Nagoya Math. J., **153** (1999). q-alg/9708018.

[20] K. Okamoto, *Sur les feuilletages associés aux équations du second ordre à points critiques fixes de P. Painlevé* , Japan J. Math., **5** (1979), 1–79.

[21] K. Okamoto, *Sur la convergence de solutions formelles d'équations différentielles avec singularités de première classe*, J. Fac. Sci. Univ. Tokyo, Sect. IA, **26** (1979), 183–198.

[22] K. Okamoto, *On the τ-functions of the Painlevé equations*, Physica **2D** (1981), 525–535.

[23] K. Okamoto, *Isomonodromic deformation and the Painlevé equations, and the Garnier system*, J. Fac. Sci. Univ. Tokyo Sect.IA Math., **33** (1986), 575–618.

[24] K. Okamoto, *Studies on the Painlevé equations, I*, Ann. Mat. Pura Appl., (4), **146** (1987), 337–381; *II*, Japan. J. Math., **13** (1987), 47–76; *IV*, Funkc. Ekvacioj. Ser. Int. **30** (1987), 305–332.

[25] K. Okamoto, *Studies on the Painlevé equations, III, Second and fourth Painlevé equations, P_{II} and P_{IV}*, Math. Ann. **275** (1986), 221–255.

[26] K. Okamoto, *The Painlevé Equations and the Dynkin Diagrams*, 299–313, Painlevé transcendents, their asymptotics and physical applications, eds. D. Levi and P. Winternitz (Plenum press, New York, 1992).

[27] K. Okamoto, *Algebraic relations among six adjacent τ-functions related to the fourth Painlevé system*, Kyushu J. Math., **50** (1996), 221–255.

[28] K. Okamoto and H. Kimura, *On particular solutions of the Garnier systems and the hypergeometric functions of several variables*, Quarterly J. Math., **37** (1986), 61–80.

[29] P. Painlevé, *Œuvres, I, II, III*, (Éditions du CNRS, Paris, 1973, 1974, 1976).

[30] T. Shioda and T. Takano, *On some Hamiltonian structures of Painlevé systems, I*, Funkc. Ekvacioj. Ser. Int. **40** (1997), 271–291.

[31] H. Umemura, *On the irreducibility of the first differential equation of Painlevé* , 771–789, Algebraic Geometry and Commutative Algebra, in honor of Masayoshi NAGATA (Kinokuniya, Tokyo, 1987).

[32] H. Umemura, *On the second proof of the irreducibility of the first differential equation of Painlevé* , Nagoya Math. J., **117** (1990), 125–171.

[33] H. Umemura, *Special polynomials associated with the Painlevé equations, I*, Theory of nonlinear special functions: the Painlevé transcendents, eds. L. Vinet and P. Winternitz (Springer, New York, 2000).

[34] H. Umemura and H. Watanabe, *Solutions of the second and fourth Painlevé equations, I*, Nagoya Math. J. **148** (1997), 151–198.

[35] H. Umemura and H. Watanabe, *Solutions of the third Painlevé equation, I*, Nagoya Math. J. **151** (1998), 1–24.

[36] H. Watanabe, *Solutions of the fifth Painlevé equation, I*, Hokkaido Math. J., **24** (1995), 231–267.

[37] H. Watanabe, *Birational canonical transformations and classical solutions of the sixth Painlevé equation*, to appear in Ann. Scuola Norm. Sup. Pisa, Cl. Sci.

14

"Completeness" of the Painlevé Test—General Considerations—Open Problems

Martin D. Kruskal[1]

1 Cultures in Mathematics

Though it may sound strange to laypersons and professionals alike, mathematics exhibits a variety of cultures, some of them known by a common name even though they are quite distinct. For example, in some circles if you mention "dynamics," it is taken for granted that you are referring to Hamiltonian systems, very likely even time-independent ones. To some other mathematicians, "dynamics" is taken to deal with evolutionary systems, those for which there is a "time variable" and which are well posed in the sense of Hadamard, i.e., whose solutions exist and are uniquely determined by an initial condition (and possibly boundary conditions, of course) and depend continuously on the given data—what are often called "marching problems"; the characteristic mathematical structure of the evolution in time is the semigroup.

There are many similar examples, but usually the confusion is merely semantic, arising from terminological laziness, or, more charitably, a natural tendency to abbreviate carried a bit too far. But here I want to call attention to a similar confusion that is pernicious because it is not widely understood, and therefore often causes real confusion, which cannot be resolved just by refining some terminology.

The subject I refer to is asymptotics, or from a broader perspective what I called "asymptotology" in an exposition [10] many years ago. To many classically trained mathematicians asymptotics is encountered mostly in the context of what is often called complex analysis, or complex function theory. But this is already an example of the laziness or brevity mentioned above, since the functions are generally what, for want of a better word, I have taken to calling "mostly analytic," meaning analytic everywhere in some

---

[1]Lecture notes prepared by the author and Christian Scheen.

domain of interest except for some lower-dimensional loci of singularity (a more general concept than meromorphicity).

To many applied mathematicians and physicists, asymptotics occurs more often in the context of real analysis, in which a function may be considered "smooth" if it has a derivative or two. For short, let me call these two varieties of asymptotics "complex" and "real" asymptotics (at the risk of indulging in similar laziness or undue brevity); this is certainly somewhat simplistic, and each can appear in the other's more characteristic type of situation, but we do need names to distinguish them.

To hammer home the difference between these two points of view in a simple context, consider the linear second-order ordinary differential equation

$$y_{xx} + \frac{a(x)}{x}y_x + \frac{b(x)}{x^2}y = 0 \qquad (1.1)$$

(here and throughout differentiation is denoted by subscripts), and assume that $a(x)$ and $b(x)$ are analytic at the point $x = 0$. Then this point is a regular singular point of the differential equation; Frobenius theory applies, and it does not much matter whether we are working in complex or real asymptotics.

But now modify the equation by adding an extra term to obtain

$$y_{xx} + \frac{a(x)}{x}y_x + \frac{b(x) + x^{20/3}}{x^2}y = 0. \qquad (1.2)$$

From the "classical" (complex) point of view we have altered the situation dramatically, since now $x = 0$ is an irregular singular point, if indeed we can speak of it that way at all, considering the multivaluedness that we have introduced. But from the point of view of real asymptotics the extra term has only a very minor effect (because the exponent $20/3$ is sufficiently large), and the methods of deriving asymptotic representations of the solutions (done properly) are unchanged. So $x = 0$ should still be considered a regular singular point—although this attitude will hardly be found in the literature, so little appreciated is the distinction I am making.

Incidentally, the branchedness is essentially irrelevant to the point I am making. I could have added to $b(x)$ the single-valued term $e^{-x^{-2}}$, which for $x \to 0$ is almost invisible in usual real asymptotics (small to all orders), but in complex asymptotics is catastrophically large for $x \to 0$ in some directions.

Analyticity in sectors of some angular width near a singular point (cf. Watson's lemma) is characteristic of complex asymptotics.

The theory of "integrability" of nonlinear ordinary differential equations introduced by Painlevé and continued by his colleagues and followers, what may be called the "Painlevé school," before and after the beginning of the twentieth century, is almost entirely within the complex tradition. The characteristic issue is, which equations have "analytic" (holomorphic) solutions—and which of them are not expressible in terms of already known functions and are therefore "new functions" available for future use?

It has been told many times how the Painlevé school found some fifty such second-order equations, six of which provided new functions, called the Painlevé transcendents. The basic idea of the method used was to examine the local behavior of solutions in a neighborhood of singular points of the equation to see whether the solutions are single-valued (unbranched). For reasons not usually explained or even mentioned, at least not clearly (but with some justification I won not discuss here), multivaluedness is tolerated around *fixed* singularities, which do not change position as the solution is varied, for example by varying some imposed initial conditions.

What has come to be called the Painlevé property, then, is, roughly, the absence of *movable* branch points in all solutions (except possibly for what are called singular solutions, the word "singular" here not referring to singular points, but rather to the solution being special, that is, not in the full-dimensional manifold of the "general" solution). For a more complete and accurate definition see [12].

What is called the Painlevé test of an equation, which must be in, or first transformed into, analytic polynomial form, is perhaps best thought of as the search for appropriately generalized asymptotic expansions of solutions at singular points of the equation, with the object of determining whether any branching appears (usually in the form of logarithmic or noninteger-power terms—thus the need for generalizing Poincaré's classical definition of asymptotic expansion). It is conceptually simpler to seek formal asymptotic expansions satisfying the equation, even if they contain terms you do not want to accept ultimately, than to restrict yourself to single-valued terms from the outset. This personal view is by no means uniquely my own, but I do not know of anyone else who has so explicitly formulated and emphasized it.

Now, I would like to propose an "eighth asymptotological principle" (an addendum to the seven principles enunciated, described, and illustrated in [10]), which says that a formal asymptotic solution of a problem is the asymptotic expansion of a true solution. This is a very subtle and tricky matter, and has to be stated and applied with great care to avoid apparent counterexamples—and even then there will probably be rare exceptions. Nevertheless, properly understood, this principle has been found very widely valid and useful. I will call it the principle of faithfulness, or of faithful representation, and hope to discuss it in detail on another occasion.

Its present relevance is that we can rely on the formal expansions to tell the truth about branching. If we find branching formally, there is almost surely branching in the true solutions. And if we find no branching, and our formal solution contains the requisite number of parameters for the order of the equation (else we may merely be missing some solutions with more complicated structure than contemplated), then there is almost surely no branching in the true solutions. (One of the subtleties to watch out for is the possibility, which can really occur, that a single parameter-free formal expansion, to infinitely many orders in powers of the distance from the singular point, may represent a whole family of true solutions that have a parameter hidden "beyond all orders" in exponentially small terms, which are not taken account of in classical asymptotics à la Poincaré. And if there are exponentially small terms, they may exhibit branching not visible to conventional asymptotics.)

The crucial point of all this is that the Painlevé property implies "integrability," namely single-valuedness of solutions in the large (except for branching at fixed singularities). This was proved for the six Painlevé equations by Painlevé himself [1]–[4] and his school, and recapitulated by Ince, with some flaws patched up by Cosgrove [5]–[6]. The method is purely complex-function-theoretic, and is widely considered difficult to follow and rather obscurely motivated.

An alternative, more transparent proof, based on a "direct method," has been given by Joshi and Kruskal [7]–[9]. It employs techniques of "real asymptotics," such as (generalized) Picard iteration, in addition to classical complex techniques. Basically, this method is threefold. The first step is to generate the formal asymptotic infinite series for the solution around a singular point, and recognize that (for the Painlevé equations, at least) they are simple integer-power series (thus exhibiting no branching). The second step is to prove convergence by contraction mapping methods, based on solving from the dominant terms, which produces a formal asymptotically appropriate recursion formula; so long as familiar safeguards are observed, such as including the highest-derivative term among the terms being solved from even if it is not actually among the formally dominant terms, the recursion process will be strongly contractive and easily shown to converge. The third step is to show that the sum of the formally generated series, the limit of the recursion process, does indeed satisfy the differential equation.

2   The Painlevé Test

To give an idea of how the Painlevé test works in a very simple example, let us take the first of Painlevé's celebrated six equations,

$$u_{tt} = 6u^2 + t, \tag{2.1}$$

and generalize it by introducing an arbitrary (mostly analytic) function $f(t)$,

$$u_{tt} = 6u^2 + f(t). \tag{2.2}$$

The only fixed singularities here are those of f, if any, and we stay away from them. To study the asymptotic behavior of the general solution in the vicinity of a movable singularity, at t_0 say, we assume a leading or dominant term of the general "algebraic" form

$$u(t) \sim a(t - t_0)^p, \qquad a \neq 0, \quad a, p \in \mathbb{C}. \tag{2.3}$$

We assume, what is very often the case (but not always), that the asymptotic representation can be differentiated formally, and substitute this into the equation. Writing

$$\tau := t - t_0 \tag{2.4}$$

for simplicity, we obtain

$$ap(p-1)\tau^{p-2} \sim 6a^2\tau^{2p} + f(t_0). \tag{2.5}$$

The principle of maximal balance (see [10]) requires that two (at least) terms be of equal maximally large order as $\tau \to 0$ (although a single term can be of maximal order formally if it is a derivative, which acts like a difference in this context and so counts as two or more terms). The orders of the three terms of (2.2) are determined by their powers of τ,

$$p - 2 \quad : \quad 2p \quad : \quad 0. \tag{2.6}$$

If we try to balance the last two terms, we get $p = 0$, and they both behave like τ^0, which is not a *maximal* balance, since the first term then formally dominates because of its smaller (more negative) power -2. (This consideration applies in spite of the fact that the coefficient of that term then vanishes.)

If we balance the first and last terms, we get $p = 2$, and again both terms behave like τ^0, which is indeed a maximal balance, since the middle term is then smaller, having power 4. However, it is easily seen that this balance leads to a Taylor series at t_0, which constitutes a *nonsingular* solution such as we know a priori must occur everywhere, since every solution is analytic at almost every point, and our procedure must produce the corresponding expansions in addition to the singular ones we are primarily interested in.

Finally, if we balance the first two terms, we get $p = -2$, which is a maximal balance, since they both then behave like τ^{-4}, larger than the last term of order τ^0. We have thus discovered a double-pole leading term.

Having balanced the powers maximally, we now have to balance the coefficients of these two terms,

$$ap(p-1) = 6a^2, \qquad \text{or} \quad 6a = 6a^2. \tag{2.7}$$

So $a = 1$ (since $a \neq 0$), and the leading behavior of $u(t)$ is given by

$$u \sim \tau^{-2} =: u_0. \tag{2.8}$$

To continue the series we can use a very unpretentious but effective (if slow) procedure I inelegantly call the "squeezing-out method," which seeks to find (squeeze out) one new asymptotic correction term at a time, sequentially. Thus we find the next term, which we will generically call $\hat{u}$ at every step, by setting

$$u =: \tau^{-2} + \hat{u}, \qquad \hat{u} \ll \tau^{-2}, \tag{2.9}$$

substituting this into the equation, and seeking the dominant behavior of $\hat{u}$. After an obvious cancellation we obtain, keeping only possibly dominant terms,

$$\hat{u}_{tt} \sim 12\tau^{-2}\,\hat{u} + f(t_0), \tag{2.10}$$

where we can assume $f(t_0) \neq 0$ because t_0 is a generic point and f is supposed not to vanish identically.

It is apparent that maximal balance requires $\hat{u}$ to have order τ^2 and all three terms to be "finite" (of order τ^0). Indeed, no other (even nonmaximal) balance works, since if $\hat{u}$ is taken smaller, there is nothing to balance $f(t_0)$, and if $\hat{u}$ is taken larger, $f(t_0)$ is negligible and $\hat{u}$ satisfies an Euler equation whose two independent solutions are τ^{-3} and τ^4, the former of which violates the assumed upper bound on the size of $\hat{u}$, and the larger violates the just derived lower bound on its size. We immediately see that

$$\hat{u} \sim -\left(\frac{1}{10}\right)\tau^2. \tag{2.11}$$

We continue to squeeze out higher terms the same way. Thus we take

$$u =: \tau^{-2} - \left(\frac{1}{10}\right)\tau^2 + \hat{u}, \qquad \hat{u} \ll \tau^2, \tag{2.12}$$

where $\hat{u}$ is hereby redefined, and find that $\hat{u}$ satisfies

$$\hat{u}_{tt} \sim 12\tau^{-2}\hat{u} + f_t(t_0)\tau, \tag{2.13}$$

so that

$$\hat{u} \sim -\left(\frac{1}{6}\right)f_t(t_0)\tau^3. \tag{2.14}$$

This begins to seem monotonous. But let us go one step further. We set

$$u =: \tau^{-2} - \left(\frac{1}{10}\right)\tau^2 - \left(\frac{1}{6}\right)f_t(t_0)\tau^3 + \hat{u}, \tag{2.15}$$

and obtain for the again redefined $\hat{u}$ the equation

$$\hat{u}_{tt} \sim 12\tau^{-2}\hat{u} + \left(\frac{1}{2}\right)f_{tt}(t_0)\tau^2. \tag{2.16}$$

If we assume that $\hat{u} \sim c\tau^4$ in analogy with the previous steps, we cannot find a constant c that works; in fact, c drops out of the equation! We have hit what is known as a resonance. We can continue only if $f_{tt}(t_0) = 0$. If it is, we have clear sailing from this step on—there are no more resonances (for this equation)—and we generate a formal Laurent series expansion with leading term for the solution around t_0 (which can, in fact, be shown to converge).

However, this happy situation holds for all possible singular points only if f_{tt} vanishes identically, namely if f is a linear function of t. But in that case the equation we started from is nothing but the first Painlevé equation, up to a trivial transformation (translation of the independent variable, and scalings of both variables).

So what happens if f is not linear, if $f_{tt}(t_0) \neq 0$? We can still solve for the leading part of $\hat{u}$, since it satisfies a nonhomogeneous linear differential equation whose associated homogeneous equation has the obvious solution τ^4. Following a form of the method of "variation of constants" we write

$$\hat{u}(t) =: \tau^4 v(t), \tag{2.17}$$

substitute, find that undifferentiated v cancels out, and obtain a first-order differential equation for the unknown v_t. This is then easily solved for v_t, which when integrated gives

$$v(t) \sim \left(\frac{1}{14}\right) f_{tt}(t_0) \ln \tau + c\tau^{-7} + k. \tag{2.18}$$

Here the first constant of integration c has to be 0, or v will be inconsistently large, while the second one, k, gives the usual free term expected at the resonance.

The point, however, is that multivaluedness has popped into view in the form of a logarithm! The equation has failed the Painlevé test.

Instead of this slow rigmarole, in practice it is more usual simply to assume an integer-power (Laurent) series with unknown leading power and coefficients,

$$u(t) = \sum_{j=p}^{\infty} a_j \tau^j, \qquad a_p \neq 0, \quad \tau := t - t_0, \tag{2.19}$$

and then seek to determine the unknowns by substituting into the equation. In the present case,

$$\sum_{j=p}^{\infty} a_j j(j-1)\tau^{j-2} \sim 6\left(\sum_{j=p}^{\infty} a_j \tau^j\right)^2 + \sum_{j=0}^{\infty} \frac{\tau^j}{j!} f_0^{(j)}. \tag{2.20}$$

Keeping only the possibly dominant terms gives

$$a_p p(p-1)\tau^{p-2} \sim 6a_p^2 \tau^{2p} + f_0^{(0)}. \tag{2.21}$$

After each possible leading power is determined, a recursion relation for the successive coefficients is obtained. A resonance is where the sought new coefficient cancels out of the relation. In that case either (1) the relation happens to be satisfied (vacuously, because there is no inhomogeneous term), reducing to $0 = 0$, so the new coefficient can be chosen arbitrarily, after which the search for coefficients can be continued at least to the next step; or else (2) the relation is violated (because there is an inhomogeneous term), so that no Laurent series can satisfy the differential equation. In case (2), in general, there is still a formal series solution starting as already found, but it is of more general form, and contains branching from a logarithmic term.

It is also possible to have noninteger power resonances, real or complex. These are terms that formally satisfy to dominant order the homogeneous linearized perturbation equation around the leading-power term. They also indicate branching, but algebraic rather than logarithmic.

To illustrate in the present case the method of finding general resonances, rather than just those at integers, which are all that can be found by assuming a Laurent series in advance, we write

$$u(t) \sim \tau^{-2} + \cdots + b\tau^r + \cdots, \qquad b, r \in \mathbb{C}, \qquad (2.22)$$

where the first term on the right is, of course, the leading term of the expansion at the singularity found previously. Substituting into the equation gives, from the b terms, $r(r-1) = 12$, so that the resonances are $r = -3, 4$. The resonance $r = 4$ has already been discussed. The resonance $r = -3$ is the so-called "universal resonance," appearing in general at $p - 1$ (here -3, since $p = -2$), and it corresponds to the free parameter implicit in the expansion variable $\tau \equiv t - t_0$. Indeed, if we formally perturb t_0, replacing it by $t_0 + \delta$ and expanding in small δ, we obtain

$$(t - t_0 - \delta)^{-2} \sim (t - t_0)^{-2} - 2\delta(t - t_0)^{-3} + \cdots. \qquad (2.23)$$

At first thought it may seem inconsistent to treat the δ term as a small perturbation of the leading term, since it has a more negative power of τ, which is approaching zero, but (see [12]) it can be justified by confining τ to a small annulus centered at t_0 (rather than to a disk, possibly punctured), fixing the size of the inner radius (say at half the outer one), and then choosing δ to be small enough. This is the heuristic key to understanding the so-called negative resonances (i.e., relatively negative: resonance powers r with real part less than the leading term's power p).

One of the main points I have been trying to make is that there is an alternative to the usual philosophy. The latter consists in assuming a Laurent series with leading term, and then worrying about whether as many free parameters appear as expected. This is a tricky business, since, though it is usually expected to get the full number of parameters (given by the order of the equation) in the general expansion from at least one leading power, there are simple examples showing that leading terms can sometimes be "deficient," generating fewer than the full number of parameters, and even that all of them can be deficient.

The alternative philosophy is to start from each leading term and generate whatever *generalized* asymptotic series develops, allowing whatever branching comes along. This is sometimes more laborious, but can also be more enlightening. If any branching occurs, one can immediately stop the procedure and declare that the test has been failed. But one can also continue, and quickly get a feel for how serious the failure is. If logarithms appear and pile up in powers from the nonlinearity of the equation, the failure appears serious. But it can happen, for instance, that a single logarithmic term appears, with no concomitant power, which suggests that the exponential of the dependent variable may satisfy an equation with the Painlevé property. If it does, the original equation may be considered as effectively, or virtually, satisfying the test, even though it does not do so in the strict sense.

The essential feature of generalized asymptotic series (in τ, here) is to allow any powers of τ (not just integer powers), real or complex, and also logarithmic and iterated logarithmic terms, the most general such term being

$$a\tau^{q_0}(\ln\tau)^{q_1}(\ln\ln\tau)^{q_2}\cdots(\ln\ldots\ln\tau)^{q_n}, \qquad (2.24)$$

where the q_j are any complex numbers. These terms are to be size-ordered primarily according to the real part of q_0; when that is indecisive, then according to the real part of q_1; when that is still indecisive, then according to the real part of q_2; etc.

Even further generalization is at least in principle appropriate, such as allowing suitable exponentially large or small terms (see [12]).

3   The PolyPainlevé Test

If an equation fails the Painlevé test, the question arises whether the branching can be removed by some simple transformation like introducing as a new variable the exponential of the originally given dependent variable. A crucial issue is how bad the branching is. If the multivaluedness of the solution that has been found is expressed as a discrete finite or infinite set of possible values for the dependent variable at a given value of the independent variable, such as typically occurs near a singularity of algebraic or logarithmic type, respectively, then the branching is relatively mild. But the branching referred to must not be just for the function confined to a vicinity of the singularity. Two (or more) singularities, each with simple branching, can interact to give "bad" branching globally (nonlocally), viz. in a region containing both of them. (This description is necessarily rather free-form, since going around one branched singularity leads to another branch of the Riemann surface of the solution where one may no longer find the other singularity, so to speak.)

The polyPainlevé test [13]–[14] was devised to investigate whether the global branchedness is discrete (that is, whether the values of the dependent variable at a given argument form a discrete set or a dense set, the two commonly found cases). To illustrate it we will give an idea how to work out about as simple a nontrivial example as possible. Consider the first-order differential equation

$$u_t = u^3 + t. \tag{3.1}$$

We immediately see that this equation does not possess the Painlevé property. Indeed, substituting an assumed leading behavior

$$u(t) \sim a(t - t_0)^p, \qquad a \neq 0, \tag{3.2}$$

gives a maximal balance

$$ap(t - t_0)^{p-1} \sim a^3(t - t_0)^{3p}, \tag{3.3}$$

$$u(t) \sim \pm i \left(\frac{\sqrt{2}}{2} \right) (t - t_0)^{-1/2}. \tag{3.4}$$

Locally there is two-branching; what about globally?

We wish to perform a perturbation expansion that will encompass two or more singular points simultaneously. (This explains the "poly" in the name of the test—the Painlevé test itself studies solutions in the vicinity of a single point.) A simple change of variables will put the equation into suitable form. Replacing

$$t \Rightarrow \lambda t + \mu, \qquad u \Rightarrow \nu u, \tag{3.5}$$

we get

$$u_t = \lambda \nu^2 u^3 + \lambda^2 \nu^{-1} t + \lambda \mu \nu^{-1}, \tag{3.6}$$

where we have divided through by ν/λ to normalize the coefficient of u_t to 1. We now normalize the coefficient of u^3 by setting

$$\lambda = \nu^{-2} \tag{3.7}$$

in order to have the same local isolated singular behavior as before. We then let $\nu \to \infty$ to knock the explicitly t-dependent term out from among the leading-order terms, and take

$$\mu = \nu^3 \tag{3.8}$$

to balance the constant term with the other leading-order terms.

Setting $\varepsilon := \nu^{-5}$ for convenience we obtain an equation equivalent to the one we started with but now suitable for expansion in the small parameter ε:

$$u_t = u^3 + 1 + \varepsilon t \tag{3.9}$$

We should then seek a formal solution in the form of a power series in ε,

$$u(t) = \sum_{n=0}^{\infty} \varepsilon^n u^{(n)}(t). \tag{3.10}$$

However, it is actually much simpler to carry out the formalism if we interchange the variables, treating u as the independent variable and t as the dependent one, so that the equation becomes

$$t_u = \left(u^3 + 1 + \varepsilon t\right)^{-1} = \sum_{n=0}^{\infty} (-t)^n \left(u^3 + 1\right)^{-n-1} \varepsilon^n. \tag{3.11}$$

To zeroth order this becomes

$$t_u^{(0)} = \left(u^3 + 1\right)^{-1}, \tag{3.12}$$

which is solvable by a simple quadrature, easily effected explicitly by partial fractions and producing a sum of three logarithmic terms whose coefficients add to zero (as is made obvious by considering $u \to \infty$), so that the terms can be combined into two logarithms. Although this is branched, the branching is mild, nondense, since the "uncertainty" in the answer is a linear combination with integer coefficients of only two complex numbers, namely a lattice in the complex plane that consists of infinitely many points stretching to infinity in all directions but clustering nowhere in the finite plane. This is a prime instance of acceptable mild, nondense branching.

In other words, the equation with the ε term omitted, whether for $t(u)$ or $u(t)$, should be considered effectively integrable.

For the full equation above we seek a formal solution $t(u)$ as a power series in ε,

$$t(u) = \sum_{n=0}^{\infty} \varepsilon^n t^{(n)}(u). \tag{3.13}$$

We find that to the next (ε) order the branching is still nondense. But at the second order the branching becomes dense, and the equation fails the polyPainlevé test. For the details I refer you to the references.

In contrast to this, the polyPainlevé test shows that the very similar-seeming equation $u_t = u^3 + ut$ does not have any dense accumulation of branch points (see Kruskal and Clarkson [14]). And indeed, multiplying through by u and taking $v := u^2$ gives $\frac{1}{2}v_t = v^2 + vt$, which is a Riccati equation and can be linearized and is therefore effectively integrable.

4   Asymptotic Expansions

Very simple equations can be difficult to solve asymptotically even to lowest order. Consider, for example, the differential equation

$$uu_{tt} = f(t). \tag{4.1}$$

If we proceed in the usual way in the vicinity of a generic point t_0 with the assumption $u \sim a\tau^p$, where $\tau := t - t_0$, we get for the leading terms

$$a^2 p(p-1)\tau^{2p-2} \sim f_0 := f(t_0). \tag{4.2}$$

Balancing the exponents gives $2p - 2 = 0$ or $p = 1$, but for $p = 1$ the left side drops out, and we have an inconsistency.

We could resolve this difficulty by allowing logarithmic factors, as mentioned in a previous section, but let me show instead how in this case the method of "solving from the dominant terms" produces those factors systematically. The dominant terms of the original equation are just as shown, except that $f(t)$ becomes the constant f_0. Without any a priori assumptions on the behavior of u, we obtain

$$uu_{tt} \sim f_0. \tag{4.3}$$

This has the integrating factor u_t/u, which we use for the full equation, integrating to obtain

$$\left(\frac{1}{2}\right) u_t^2 = f_0 \ln ku + \int_{t_0}^{t} (f - f_0)\frac{u_t}{u} dt, \tag{4.4}$$

where k is a constant of integration. Leaving out the small terms coming from the integral to avoid writing messy formulas, though they must be kept and "iterated on" if one wants to generate higher-order terms in the solution, we note that the dominant part of the equation is "autonomous," that is, free of explicit t-dependence, so we solve for dt and integrate:

$$\int_0^u \frac{d\bar{u}}{\sqrt{2f_0 \ln k\bar{u}}} = t - t_0 + \cdots, \tag{4.5}$$

where the new constant of integration has been determined by the requirement that $u = 0$ at $t = t_0$ in accord with the formal result $p = 1$ found earlier.

The simplest way to find the asymptotic behavior of the integral as $u \to 0$ is to integrate by parts; doing so in the most obvious and natural way works and yields

$$\frac{u}{\sqrt{2f_0 \ln ku}} + \mathcal{O}\left(\frac{u}{(2f_0 \ln ku)^{3/2}}\right) \sim t - t_0 + \cdots, \tag{4.6}$$

where the higher-order terms can be worked out as desired. Solving for the "sensitive" u in the numerator, treating the insensitive one inside the logarithm as if it were known (a constant), leads to

$$u \sim \sqrt{2f_0 \ln ku}\,(t - t_0) + \cdots. \tag{4.7}$$

Finally, we get an explicit leading-order asymptotic formula for u by substituting this recursively into itself:

$$u \sim (t - t_0)\sqrt{2f_0\left(\ln k(t - t_0)\right)}. \tag{4.8}$$

This far from immediately obvious asymptotic solution of the equation we started from shows the usefulness of such methods.

By finding branching already in the leading term we see that the equation lacks the Painlevé property. Of course, standard Painlevé theory would easily show that anyway, but it would not show how the failure takes place, how the multivaluedness appears—which is rather tricky in this instance.

Now, the reason for my emphasis on determining formal asymptotic solutions is the "eighth principle" mentioned earlier: the belief that (virtually always) a formal solution is, with certain caveats, in fact the asymptotic representation of a true solution. This is a concept generally unknown to analysts, as far as I can tell, and understandably so, since it is not a theorem, it is not true in full generality, and there are subtleties in distinguishing the cases that should work from those that do not. Nevertheless, I consider it an important principle.

For instance, I am confident that the result obtained above is correct, that there *is* a solution of the given equation with the rather strange root-logarithmic behavior found, and in fact a family of such solutions parametrized by k as shown.

As a simple and enlightening example of this principle of faithfulness, consider the differential equation

$$u_x - u = -x^{-1} + x^{-2}P(u), \tag{4.9}$$

x and u both real, with boundary condition that $u(x) \to 0$ as $x \to \infty$. We seek the behavior of u in the limit $x \to \infty$.

At first let us take the very special case $P(u) \equiv 0$. An appropriate formal recursive iteration scheme is obtained by writing the equation in the form $u = x^{-1} + u_x$, and as usual solving at each step for the new, improved formal approximation of u on the left side by plugging the previous result into the right side, successively obtaining the expressions

$$
\begin{aligned}
u(x) &= x^{-1} + u_x \\
&= x^{-1} + \left(x^{-1} + u_x\right)_x = x^{-1} - x^{-2} + u_{xx} \\
&= x^{-1} - x^{-2} + \left(x^{-1} + u_x\right)_{xx} \sim x^{-1} - 1!x^{-2} + 2!x^{-3} - 3!x^{-4} + \cdots \\
&\sim \sum_{j=0}^{\infty} \frac{(-1)^j j!}{x^{j+1}}.
\end{aligned}
\tag{4.10}
$$

Note, incidentally, how strikingly the characteristic divergence of many formal infinite asymptotic expansions appears, by virtue of the factorial in the numerator of the summand.

However, the main issue is whether this asymptotic series is relevant (faithful) to the actual solution of the equation under study. For this toy problem we can solve the differential equation exactly using the obvious integrating factor:

$$u(x) = e^x \left(k + \int_x^{+\infty} \frac{e^{-t}}{t} dt \right), \tag{4.11}$$

where the constant of integration k must be zero to satisfy the boundary condition. And we can then verify that the formal asymptotic series really does represent the solution asymptotically.

Less trivially, now, let us take $P(u) \equiv u^3$, say. Treating the new non-linear but manifestly small term (small because of the imposed boundary condition as well as the factor x^{-2}) as if it were known, we "solve from the dominant terms" (integrate) just as before and obtain now the recursion relation

$$u^{[n+1]}(x) = e^x \int_x^{+\infty} \frac{e^{-t}}{t} \left(1 - t^{-1} [u^{[n]}(t)]^3 \right) dt, \quad n = 0, 1, 2, \ldots . \tag{4.12}$$

We can start from any reasonable initial function, say $u^{[0]} = 0$. When a suitable approximation $u^{[n]}(x)$ of $u(x)$ is plugged into the right side of this recursion relation, its left side then constitutes a new (and formally better) approximation $u^{[n+1]}(x)$, and so on to higher and higher n. This process is somewhat analogous to Picard recursion, but only at the formal level.

The differential equation does have a solution satisfying the boundary condition, and that solution does have the formal asymptotic series as its asymptotic expansion. This can be shown, for example, by the "shooting method," a standard technique of "real" asymptotics. However, the generalization of this method to complex-plane analysis encounters the (already mentioned) generic problem of the Stokes phenomenon: In general, there cannot exist any *globally* valid formal asymptotic series *with a single formal expression*.

If we now consider the opposite limit $x \to -\infty$ (x real and negative), with the analogous boundary condition that $u(x) \to 0$ as $x \to -\infty$, we are led to the same formal series as before, except that now the term ke^x remains, because the constant of integration is not determined by the boundary condition. We obtain the recursion relation

$$u^{[n+1]}(x) = -e^{-x} \int_{x_0}^x \frac{e^{+t}}{t} \left(1 - t^{-1} u^{[n]}(t)^3 \right) dt, \tag{4.13}$$

where the constant of integration appears equivalently as x_0, one of the limits of integration. A *family* of representations is obtained, one for each value of x_0. The boundary condition does not single out a particular member of the family, since it is satisfied for all of them.

The asymptotic method presented here is "beyond all orders": If the series does converge, the small term containing the parameter lies beyond all orders. In cases where it does not, the information in that exponentially small term is invisible to classical Poincaré asymptotics, but its significance can be uncovered by examining the difference between two solutions.

5 REFERENCES

[1] P. Painlevé, *Mémoire sur les équations différentielles dont l'intégrale générale est uniforme*, Bull. Soc. Math. France **28** (1900), 201–261.

[2] P. Painlevé, *Sur les équations différentielles du second ordre et d'ordre supérieur dont l'intégrale générale est uniforme*, Acta Math. **25** (1902), 1–85.

[3] P. Painlevé, *Sur les équations différentielles du second ordre à points critiques fixes*, C. R. Acad. Sc. Paris **143** (1906), 1111–1117.

[4] B. Gambier, *Sur les équations différentielles du second ordre et du premier degré dont l'intégrale générale est à points critiques fixes*, Thèse (1909) Paris, Acta Math. **33** (1910), 1–55.

[5] E.L. Ince, *Ordinary Differential Equations* (Longmans, Green and co., London and New York, 1926). Reprinted (Dover, New York, 1956).

[6] C.M. Cosgrove, unpublished.

[7] N. Joshi and M.D. Kruskal, *A Simple Proof that Painlevé Equations Have no Movable Essential Singularities*, Preprint CMA–R06–90, Center for Mathematical Analysis, Australian National University (1990).

[8] N. Joshi and M.D. Kruskal, *A direct proof that solutions of the first Painlevé equation have no movable singularities except poles*, Nonlinear evolution equations and dynamical systems, eds. M. Boiti, L. Martina, and F. Pempinelli (Word Scientific, Singapore, 1992), pages 310–317.

[9] N. Joshi and M.D. Kruskal, *A Direct Proof that Solutions of the Six Painlevé Equations Have no Movable Singularities Except Poles*, Stud. Appl. Math. **93** (1994), 187–207.

[10] M.D. Kruskal, *Asymptotology*, Mathematical Models in Physical Sciences, ed. S. Drobot, (Prentice-Hall, 1963).

[11] C.M. Bender and S. A. Orszag, *Advanced Mathematical Methods for Scientists and Engineers* (McGraw-Hill, New-York, 1978).

[12] M.D. Kruskal *Flexibility in Applying the Painlevé Test*, in Painlevé Transcendents, eds. by D. Levi and P. Winternitz (Plenum Press, New-York, 1992), pages 187–195.

[13] M.D. Kruskal, A. Ramani, and B. Grammaticos, *Singularity Analysis and its Relation to Complete, Partial and Non-Integrability*, Partially Integrable Evolution Equations in Physics, eds. R. Conte and N. Boccara (Kluwer, Dordrecht, 1990), pages 321–372.

[14] M.D. Kruskal and P.A. Clarkson, *The Painlevé–Kowalevskaya and PolyPainlevé Tests for Integrability*, Stud. Appl. Math. **86** (1992), 87–165.

[15] S.L. Ziglin, *Branching of solutions and nonexistence of first integrals in Hamiltonian mechanics, I*, Funktsional'nyi Analiz i Ego Prilozheniva **16** (1982), 30–41; Funct. Anal. Appl. **16** (1983) 181–189.

[16] S.L. Ziglin, *Branching of solutions and noexistence of first integrals in Hamiltonian mechanics, II*, Funktsional'nyi Analiz i Ego Prilozheniva **17** (1982), 8–23; Funct. Anal. Appl. **17** (1983), 6–17.

[17] P.Painlevé, *Leçons sur la théorie analytique des équations différentielles*, Leçons de Stockholm (1895), (Hermann, Paris, 1897).

[18] J. Malmquist, *Sur les fonctions à un nombre fini de branches définies par les équations différentielles du premier ordre*, Acta Math. **36** (1913), 297–343.

Index